Topics in Applied Physics

Volume 138

Topics in Applied Physics is a well-established series of review books, each of which presents a comprehensive survey of a selected topic within the domain of applied physics. Since 1973 it has served a broad readership across academia and industry, providing both newcomers and seasoned scholars easy but comprehensive access to the state of the art of a number of diverse research topics.

Edited and written by leading international scientists, each volume contains high-quality review contributions, extending from an introduction to the subject right up to the frontiers of contemporary research.

Topics in Applied Physics strives to provide its readership with a diverse and interdisciplinary collection of some of the most current topics across the full spectrum of applied physics research, including but not limited to:

- Quantum computation and information
- Photonics, optoelectronics and device physics
- Nanoscale science and technology
- Ultrafast physics
- Microscopy and advanced imaging
- Biomaterials and biophysics
- Liquids and soft matter
- Materials for energy
- Geophysics
- Computational physics and numerical methods
- Interdisciplinary physics and engineering

We welcome any suggestions for topics coming from the community of applied physicists, no matter what the field, and encourage prospective book editors to approach us with ideas. Potential authors who wish to submit a book proposal should contact Zach Evenson, Publishing Editor:

zachary.evenson@springer.com

Topics in Applied Physics is indexed by SCOPUS and books in this series are submitted for indexing to Web of Science.

2018 Impact Factor: 0.746

More information about this series at http://www.springer.com/series/560

Eugene Kamenetskii
Editor

Chirality, Magnetism and Magnetoelectricity

Separate Phenomena and Joint Effects in Metamaterial Structures

Springer

Editor
Eugene Kamenetskii
Department of Electrical and Computer Engineering
Ben-Gurion University of the Negev
Beer-Sheva, Israel

ISSN 0303-4216 ISSN 1437-0859 (electronic)
Topics in Applied Physics
ISBN 978-3-030-62843-7 ISBN 978-3-030-62844-4 (eBook)
https://doi.org/10.1007/978-3-030-62844-4

This Springer imprint is published by the registered company Springer Nature Switzerland AG
The registered company address is: Gewerbestrasse 11, 6330 Cham, Switzerland

Preface

In this book, the reader will find a wide range of problems related to properties of symmetry in the study of electromagnetic fields, materials, and the field-matter interactions. The book contains 19 chapters covering various aspects of these effects. The chapters are written by international experts who have contributed to the advancement of science and engineering of the chirality, magnetism, and magnetoelectricity in optical and microwave systems. What problems and challenges are analyzed and discussed in the book? The fields of the studies are new and certain statements may be disputable. The reader can see that current research leaves some questions open to further discussion. Some chapters discuss similar issues from different physical points of view. In this book, we aim to provide the reader with an overview of the interdisciplinary research. Since many studies in these areas do not fit into a well-established classification, we do not separate chapters into topical sections.

The most interesting physical phenomena arise from effects associated with combinations of different types of symmetry and symmetry breakings. It is known that chirality is a geometric property. An object is called chiral when it exists in two enantiomeric forms, which can be superimposed only by a parity operation (mirror image). Chirality properties can be observed both in material structures and in novel types of engineered fields. It is also known that ferromagnetism breaks time-reversal symmetry. The spin momentum of magnons is based on the time-reversal symmetry breaking of the magnetic order. However, the magnetic dipole precession has neither left-handed nor right-handed quality, that is to say, no chirality. Where from magnetic symmetry can inherit chirality? One of the important issues is that magnetic symmetry can derive chirality from the crystal structure. In magnetism, the curvilinear geometry manifests Dzyaloshinskii–Moriya-like interaction. For magnons, the Dzyaloshinskii–Moriya interaction accounts for spin–orbit interaction and causes a nontrivial topology.

The hybridization of magnetism and topology is a significant problem. It is related to various topological structures of magnetic moments (spins), known as spin textures. This is related also to the question of topological phases. Whereas the topological phases were initially proposed for electron waves, then it became evident for other types of waves. The topological phase is of particular importance for the topologically nontrivial edge modes. Topological spin waves have been predicted to occur in mesoscopic crystals with artificial patterned ferromagnetic structures and also in atomic scale magnetic insulators with the kagome-lattice structures. Such structures can topologically protect unidirectional surface spin waves inside spin-wave band gaps. Chiral edge states arise due to broken time-reversal symmetry. These are the "one-way waveguides" which allow energy to flow in one direction only. In this book, the question of unidirectional propagation of energy of spin waves is discussed in Chaps. 1 and 14. In Chap. 1, the authors consider unidirectional spin waves generated by chiral spin pumping. In Chap. 14, the authors analyze unidirectional surfaces spin waves topologically protected by the band gap in the reciprocal space.

One of the questions that the reader may have in these studies is that in the case of unidirectional propagation of energy in structures with broken time-reversal symmetry, certain constraints should be imposed due to a joint analysis of the reciprocity and unitarity relations.

In chiral materials, magnetic moments (spins) can form various structures known as spin textures. Chapter 14 is devoted to general consideration of such topological spin textures. In Chap. 7, we have a good review on the mechanisms of manipulation of magnetic textures via spin-transfer and spin–orbit-torques. In Chap. 8, the author has presented the reader with recent theoretical studies on the microwave-induced physical phenomena and device functions of magnetic skyrmions. It is demonstrated that the spin-wave excitations give rise to translational motion of the skyrmions. New aspects of topological dynamics of spin texture-based metamaterials are considered in Chap. 15. In Chap. 16, the authors provide the reader with new studies on antiferromagnetic skyrmions and bimerons. These topological spin textures have attracted a lot of attentions because they have small size and low depinning current. For creation of new magnetic materials with specific engineered properties, a new field of research based on Floquet engineering is presented in Chap. 11. It is shown that such basic quantities of magnetic insulators as magnetization, spin chirality, and spin current can be controlled by the Floquet technique. In Chap. 3, the authors discuss plasmon resonances of spin-polarized magnetic nanoparticles. Interesting results of this study may raise the reader's question: can the coupling between linear (plasmon) and angular momenta (electron spin) cause the appearance of intrinsic chiral wavefunctions?

In the book, the chiral interaction between light and matter is considered in various aspects of research, concerning both non-magnetic and magnetic chiral matter. A lot of attention is paid to plasmonic chiral metasurfaces and superchiral optical near-fields for effective detection and differentiation of enantiomers. It is discussed that chiral sensitive techniques can be used to probe the fundamental symmetries. In Chap. 2, the authors demonstrate how surface plasmon resonance can be employed as a new research tool for chiral sensing. The interactions of chirality of light and plasmonic chiral metasurfaces are studied in Chap. 4. The author shows that the phase and amplitude of the circularly polarized light can be effectively modulated by optical chiral metasurfaces. In Chap. 5, the authors argue that the optical response of matter manifests itself not only in the form of optical signals, but also in the form of a mechanical force acting on the system of matter. Consequently, the plasmonic near field with a nano-scale radius of rotation is intended to control the motion of the center of mass of nano-objects. They show that all the basic elements of motion control of nano-objects can be realized using optical manipulation. Recent advancements in nano- and micro-fabrication technology have allowed the realization of artificially structured materials with strong electromagnetic chirality, far-exceeding natural chiral materials. Implementation of artificial chirality in micro-/nano-scale three-dimensional plasmonic structures is the subject of Chap. 10. Chiroptical light-matter interaction is largely boosted in the surroundings of complex-shaped metallic nanostructures. Theoretical generalization of the optical chirality to the case of arbitrary dispersive and lossy optical media is a subject of studies of Chap. 13. In Chap. 9, the authors are aimed to create the background for establishing a connection between the optics of chiral metamaterials and magnonics. This is a challenging task. Using formal similarity of the electromagnetic field and spin waves in an antiferromagnetic insulator, the authors discuss the possibility to generalize the notion of optical chirality in antiferromagnets with broken chiral symmetry. Concerning the studies of light-matter interaction, it is very important to note that when analyzing the interaction of chiral light with a helical wavefront (due to spin and orbital angular momenta) with chiral nanostructures, it is necessary to keep in mind the effects of nonlocality. The reader may find that simple models based on the dipole approximation are inapplicable when analyzing such nonlocal optical response effects. It is also worth noting that while in optics, a possible way to detect the effect of material chirality is through the transmission and reflection of polarized light, in microwaves, chirality parameters are obtained via far-field measurement of the scattering-matrix characteristics.

Any coupling between the magnetic and the electric properties of a material is denoted, in its most general form, as the magnetoelectric (ME) effect. The efficient energy conversion from the electric to the magnetic domain and vice versa still remains challenging. Chapter 12 is devoted to study of one of the most promising devices for this conversion: ME transducers, consisting of composite materials with piezoelectric and magnetostrictive layers. In Chap. 6, the authors have presented the reader with the role of chirality as the key factor in magnetoelectric phenomena observed in magnetic structures with various topological properties. In Chap. 17, the authors present the theoretical framework that allows to describe the electromagnetic response of magnetoelectric media by means of axion-like extended electrodynamics. In Chap. 19, it is shown that topological singularities originated from magnetic dipolar mode oscillations in ferrite disk particles have unique properties of the magnetoelectric near-fields. Quantized ME fields suggest a conceptually new microwave functionality for material characterization.

The studies shown in the book are enriched by Chap. 18. The analysis of the Purcell effect in PT-symmetric waveguides presented in this chapter is a valuable contribution in terms of basic aspects of symmetry in electromagnetic systems.

We may hope that the book will be a valuable aid to understand the current research of chirality, magnetism, and magnetoelectric phenomena in metamaterial structures for scientists, researchers, and graduate students working in the fields of electronic engineering, material science, and condense matter physics.

Beer Sheva, Israel Eugene Kamenetskii

Contents

Contributors

Christoph Adelmann Imec, Leuven, Belgium

Ali E. Aliev A. G. MacDiarmid NanoTech Institute, University of Texas at Dallas, Richardson, TX, USA

K. S. Antipin M.V. Lomonosov Moscow State University, Moscow, Russia

Trevon Badloe Department of Mechanical Engineering, Pohang University of Science and Technology (POSTECH), Pohang, Republic of Korea

Gerrit E. W. Bauer Institute for Materials Research & WPI-AIMR & CSRN, Tohoku University, Sendai, Japan;
Kavli Institute of Nanoscience, Delft University of Technology, Delft, The Netherlands

Lykourgos Bougas Institut für Physik, Johannes Gutenberg-Universität Mainz, Mainz, Germany

M. Cambiaso Universidad Andres Bello, Departamento de Ciencias Fisicas, Facultad de Ciencias Exactas, Santiago, Chile

Yunshan Cao School of Electronic Science and Engineering, University of Electronic Science and Technology of China, Chengdu, China

Florin Ciubotaru Imec, Leuven, Belgium

Vladimir P. Drachev Skolkovo Institute of Science and Technology, Moscow, Russia;
University of North Texas, Denton, TX, USA

Sotiris Droulias Institute of Electronic Structure and Laser, FORTH, Heraklion, Crete, Greece

Karin Everschor-Sitte Johannes Gutenberg University, Institute of Physics, Mainz, Germany

Motohiko Ezawa Department of Applied Physics, The University of Tokyo, Tokyo, Japan

T. T. Gareev M.V. Lomonosov Moscow State University, Moscow, Russia

Junho Gwak Department of Mechanical Engineering, Pohang University of Science and Technology (POSTECH), Pohang, Republic of Korea

Masayuki Hoshina Department of Physics and Electronics, Osaka Prefecture University, Sakai, Osaka, Japan

Kun Huang Department of Optics and Optical Engineering, University of Science and Technology of China, Hefei, China

Hajime Ishihara Department of Materials Engineering Science, Osaka University, Toyonaka, Osaka, Japan;
Department of Physics and Electronics, Osaka Prefecture University, Sakai, Osaka, Japan;
Quantum Information and Quantum Biology Division, Institute for Open and Transdisciplinary Research Initiatives, Osaka University, Toyonaka, Osaka, Japan

Heonyeong Jeong Department of Mechanical Engineering, Pohang University of Science and Technology (POSTECH), Pohang, Republic of Korea

Eugene Kamenetskii Microwave Magnetic Laboratoty, Ben Gurion University of the Negev, Beersheba, Israel

A. S. Kaminskiy M.V. Lomonosov Moscow State University, Moscow, Russia

Alina Karabchevsky School of Electrical and Computer Engineering, Ben-Gurion University of the Negev, Beer-Sheva, Israel

Inki Kim Department of Mechanical Engineering, Pohang University of Science and Technology (POSTECH), Pohang, Republic of Korea

Minkyung Kim Department of Mechanical Engineering, Pohang University of Science and Technology (POSTECH), Pohang, Republic of Korea

Yeseul Kim Department of Mechanical Engineering, Pohang University of Science and Technology (POSTECH), Pohang, Republic of Korea

D. P. Kulikova M.V. Lomonosov Moscow State University, Moscow, Russia

Sergey V. Levchenko Skolkovo Institute of Science and Technology, Moscow, Russia

Zhixiong Li School of Electronic Science and Engineering, University of Electronic Science and Technology of China, Chengdu, China

Xue Liang School of Science and Engineering, The Chinese University of Hong Kong, Shenzhen, Guangdong, China

Alejandro Martínez Nanophotonics Technology Center, Universitat Politècnica de València, Valencia, Spain

A. Martín-Ruiz Instituto de Ciencias Nucleares, Universidad Nacional Autónoma de México, Ciudad de México, Mexico

Jan Masell RIKEN Center for Emergent Matter Science, Wako, Saitama, Japan

Masahito Mochizuki Department of Applied Physics, Waseda University, Shinjuku-ku, Tokyo, Japan

Fyodor Morozko School of Electrical and Computer Engineering, Ben-Gurion University of the Negev, Beer-Sheva, Israel;
Department of Theoretical Physics and Astrophysics, Belarusian State University, Minsk, Belarus

Jungho Mun Department of Chemical Engineering, Pohang University of Science and Technology (POSTECH), Pohang, Republic of Korea

A. V. Nikolaev M.V. Lomonosov Moscow State University, Moscow, Russia;
Skolkovo Institute of Science and Technology, Moscow, Russia

E. P. Nikolaeva M.V. Lomonosov Moscow State University, Moscow, Russia

Andrey Novitsky Department of Theoretical Physics and Astrophysics, Belarusian State University, Minsk, Belarus

Maria Pogodaeva Skolkovo Institute of Science and Technology, Moscow, Russia

Igor Proskurin Department of Physics & Astronomy, University of Manitoba, Winnipeg, Canada;
Institute of Natural Sciences and Mathematics, Ural Federal University, Ekaterinburg, Russia

A. P. Pyatakov M.V. Lomonosov Moscow State University, Moscow, Russia

Junsuk Rho Department of Mechanical Engineering, Pohang University of Science and Technology (POSTECH), Pohang, Republic of Korea;
Department of Chemical Engineering, Pohang University of Science and Technology (POSTECH), Pohang, Republic of Korea

Masahiro Sato Department of Physics, Ibaraki University, Mito, Ibaraki, Japan

A. S. Sergeev M.V. Lomonosov Moscow State University, Moscow, Russia

Laichuan Shen School of Science and Engineering, The Chinese University of Hong Kong, Shenzhen, Guangdong, China

Sunae So Department of Mechanical Engineering, Pohang University of Science and Technology (POSTECH), Pohang, Republic of Korea

Robert L. Stamps Department of Physics & Astronomy, University of Manitoba, Winnipeg, Canada

Oleg A. Tretiakov School of Physics, The University of New South Wales, Sydney, Australia

L. F. Urrutia Instituto de Ciencias Nucleares, Universidad Nacional Autónoma de México, Ciudad de México, Mexico

Frederic Vanderveken Departement Materiaalkunde, KU Leuven, Leuven, Belgium;
Imec, Leuven, Belgium

J. Enrique Vázquez-Lozano Nanophotonics Technology Center, Universitat Politècnica de València, Valencia, Spain

X. R. Wang Physics Department, The Hong Kong University of Science and Technology, Clear Water Bay, Kowloon, Hong Kong;
HKUST Shenzhen Research Institute, Shenzhen, China

X. S. Wang School of Physics and Electronics, Hunan University, Changsha, China

Jing Xia School of Science and Engineering, The Chinese University of Hong Kong, Shenzhen, Guangdong, China

Hidemasa Yamane Department of Physics and Electronics, Osaka Prefecture University, Sakai, Osaka, Japan

Peng Yan School of Electronic Science and Engineering, University of Electronic Science and Technology of China, Chengdu, China

Younghwan Yang Department of Mechanical Engineering, Pohang University of Science and Technology (POSTECH), Pohang, Republic of Korea

Nobuhiko Yokoshi Department of Physics and Electronics, Osaka Prefecture University, Sakai, Osaka, Japan

Tao Yu Max Planck Institute for the Structure and Dynamics of Matter, Hamburg, Germany

Xichao Zhang School of Science and Engineering, The Chinese University of Hong Kong, Shenzhen, Guangdong, China

Yan Zhou School of Science and Engineering, The Chinese University of Hong Kong, Shenzhen, Guangdong, China

Chapter 1
Chiral Coupling to Magnetodipolar Radiation

Tao Yu and Gerrit E. W. Bauer

Abstract We review and extend the theory of chiral pumping of spin waves by magnetodipolar stray fields that generate unidirectional spin currents and asymmetric magnon densities. We illustrate the physical principles by two kinds of chiral excitations of magnetic films, i.e., by the evanescent Oersted field of a narrow metallic stripline with an AC current bias and a magnetic nanowire under ferromagnetic resonance.

1.1 Introduction

"Handedness" or "chirality" of wave propagation is a lively research topic in optics, acoustics, and condensed matter physics. The "spin" of magnons is rooted in the time-reversal symmetry breaking of the magnetic order and leads to chiral coupling with other excitations when locked to the momentum. This phenomenon is governed by non-universal selection rules. This chapter clarifies a specific mechanism, viz. evanescent microwaves that can efficiently generate chiral dynamics.

Magnonics and magnon spintronics [1–4] are emergent fields that hold the promise of a next-generation low-power and scalable information processing and communication technology. The generation of coherent and propagating spin waves is a crucial ingredient, which can be realized by magnetic fields generated by microwave antennas such as current-biased metallic striplines. In order to generate stray fields with

T. Yu (✉)
Max Planck Institute for the Structure and Dynamics of Matter,
Luruper Chaussee 149, 22761 Hamburg, Germany
e-mail: tao.yu@mpsd.mpg.de

G. E. W. Bauer
Institute for Materials Research & WPI-AIMR & CSRN, Tohoku University,
Sendai 980-8577, Japan
e-mail: G.E.W.Bauer@tudelft.nl

Kavli Institute of Nanoscience, Delft University of Technology, 2628 CJ Delft,
The Netherlands

E. Kamenetskii (ed.), *Chirality, Magnetism and Magnetoelectricity*,
Topics in Applied Physics 138,
https://doi.org/10.1007/978-3-030-62844-4_1

high-momentum Fourier components these must be small in size and placed close to the magnetic medium. Not only the amplitude, but also the direction of the excited spin waves depend on the excitation conditions that obey right-hand rules and are therefore *chiral*.

In this chapter, we focus on the chirality of the dipolar coupling between the magnetization dynamics in ferromagnetic heterostructures [5–10], while those in optics [11–17], plasmonics [18, 19], and magnetic structures with Dzyaloshinskii–Moriya interaction are treated in other chapters. We focus on the favorite material of magnonics, viz. the ferrimagnetic insulator yttrium iron garnet (YIG) with high Curie temperature and outstanding magnetic and acoustic quality [20]. Its magnons can be excited electrically by heavy metal contacts [21], acoustically [22], as well as by a large spectrum of electromagnetic waves from gigahertz (microwaves) to petahertz (light). Magnonic transducers with spatially separated contact that excite and detect magnons [5, 21, 23–29] are sensitive probes to study magnon transport. We illustrate the chiral physics for thin YIG films with in-plane magnetizations, but other materials and configurations can be treated by changing the model parameters.

The spin waves of in-plane magnetized films can be classified by the interaction that governs their dispersion as a function of wave vector, into the dipolar, dipolar-exchange and exchange type with energies ranging from a few gigahertz to many terahertz [1–4, 23]. The long-wavelength modes are dipolar, whereas the short-wavelength ones are exchange. Bulk volume modes and surface (Damon–Eshbach) modes propagate along or perpendicular to the magnetization direction with different dispersion relations [30–33]. Moreover, the surface modes are chiral: their propagation direction (linear momentum) is fixed by the outer product of surface normal and magnetization direction, allowing unidirectional spin current generation by dominantly exciting one surface of a magnetic film [34–37]. However, Damon–Eshbach spin waves are not well suited for applications—their group velocity tends to be zero when the linear momentum is larger than the inverse of film thickness, leading to a small spin conductivity. They are also very sensitive to dephasing by surface roughness [38], and do not exist in sufficiently thin films.

An alternative to intrinsically chiral spin waves is the chiral excitation of non-chiral ones. Micromagnetic simulations [5] revealed that the AC dipolar field emitted by a magnetic nanowire on top of an in-plane magnetized film with magnetization normal to the wire can excite unidirectional spin waves. We have been motivated by experiments on an array of magnetic nanowires on top of an ultrathin YIG film that generated unidirectional spin waves parallel to the surface and perpendicular to the nanowires [7] to develop a general theory of coherent and incoherent chiral excitation of magnons [6, 8] by the dipolar interaction between the dynamics of a magnetic film and a magnetic transducer. The chirality can be traced to the different stray fields generated by spin waves with opposite polarization and propagation. By angular momentum conservation electromagnetic waves with particular polarization emitted by a magnetic transducer couple only the circularly polarized component of a spin wave with a certain propagation direction [11]. When dipolar or crystal anisotropy mixes the right and left circularly polarized components, magnons are still excited preferentially, but not exclusively, in one direction. Finally, a (short-range) exchange coupling between film and transducer is not sensitive to the propagation direction, and reduces the chirality.

Recently, two remarkable experiments confirmed our predictions. By NV magnetometry Bertelli et al. [39] observed chiral pumping of spin waves by a stripline antenna. Wang et al. [40] measured unidirectional microwave transmission mediated by two magnetic wires on top of a thin magnetic film, i.e. chiral magnon-magnon coupling.

The chiral coupling to spin waves enables the generation and control of spin currents [6–8] or spin accumulations [9, 10] in ferromagnetic insulators, which is beneficial for spintronic devices. In this short review, we comprehensively illustrate two kinds of chiral coupling to the magneto-dipolar radiation, including the evanescent field of a thin stripline that carries an AC current (Sect. 1.2) and that of a magnetic wire under resonant excitation (Sect. 1.3).

1.2 Chiral Excitation of Spin Waves by Metallic Stripline

We call a wave "chiral" when it propagates with handedness, i.e. in a certain direction that is determined by two other control vectors, such as surface normal and magnetic field. A rotating *electrical* dipole [41, 42] excites surface plasmon polaritons in one direction only [18, 19, 41], while a precessing *magnetic* dipole excite magnons unidirectionally [43, 44]. Here we analyze solutions of the combined Maxwell and Landau–Lifshitz–Gilbert equations that explain the available experimental evidence. We analyze the near microwave field from a normal metal strip line in Sect. 1.2.1 and its effect on a thin magnetic film in Sect. 1.2.2 (see Fig. 1.1). We focus for simplicity on a configuration in which the film normal is along the x-direction, $\hat{\mathbf{z}}$ is parallel to a stripline that is assumed to be very long, and the excited spin waves propagate in the y-direction.

1.2.1 Oersted Magnetic Fields

We first demonstrate that even though the magnetic field of a stripline is linearly-polarized in real space (see Fig. 1.1), it is chiral in momentum space. Ampere's Law states that the current density $\mathbf{J}(\mathbf{r})$ generates the vector potential [42]

$$\mathbf{A}(\mathbf{r}, t) = \frac{\mu_0}{4\pi} \int d\mathbf{r}' dt' \frac{\mathbf{J}(\mathbf{r}', t')}{|\mathbf{r} - \mathbf{r}'|} \delta \left(t' + \frac{|\mathbf{r} - \mathbf{r}'|}{c} - t \right), \tag{1.1}$$

where μ_0 is the vacuum permeability and the delta-function represents (non-relativistic) retardation. For a harmonic source $\mathbf{J}(t) \sim \mathbf{J}(\omega)e^{-i\omega t}$,

$$\mathbf{A}(\mathbf{r},\omega)=\frac{\mu_0}{4\pi}\int d\mathbf{r}'\mathbf{J}(\mathbf{r}',\omega)\frac{e^{ik|\mathbf{r}-\mathbf{r}'|}}{|\mathbf{r}-\mathbf{r}'|}, \tag{1.2}$$

where $k=\omega/c$. The current in the stripline is uniform over the cross section of width w and thickness t as well as length $L \ll c/\omega$. In the long wavelength limit and square cross section $\mathbf{J}(\mathbf{r},\omega)\simeq\delta(x)\delta(y)\mathcal{J}(\omega)\hat{\mathbf{z}}$, where $\mathcal{J}$ is the total electric current, leading to

$$\mathbf{A}(\mathbf{r},\omega)=\frac{\mu_0}{4\pi}\mathcal{J}(\omega)\hat{\mathbf{z}}\int_{-\infty}^{\infty}dz'\frac{e^{ik\sqrt{x^2+y^2+z'^2}}}{\sqrt{x^2+y^2+z'^2}}, \tag{1.3}$$

which does not depend on z. Substituting the Weyl identity [41]

$$\frac{e^{ik\sqrt{x^2+y^2+z^2}}}{\sqrt{x^2+y^2+z^2}}=\frac{i}{2\pi}\int dk_y dk_z\frac{e^{ik_x|x|+ik_y y+ik_z z}}{k_x}, \tag{1.4}$$

where $k=\sqrt{k_x^2+k_y^2+k_z^2}$ and $k_x=|a|+i|b|$ is complex, into (1.3) yields

$$\mathbf{A}(x,y,\omega)=\frac{i\mu_0}{4\pi}\mathcal{J}(\omega)\hat{\mathbf{z}}\int dk_y\frac{e^{ik_x|x|+ik_y y}}{k_x}. \tag{1.5}$$

The magnetic field $\mathbf{H}(\mathbf{r})=\nabla\times\mathbf{A}(\mathbf{r})/\mu_0=\left(\partial_y A_z,-\partial_x A_z,0\right)/\mu_0$ is transverse to the wire, see Fig. 1.1. Below the stripline ($x<0$),

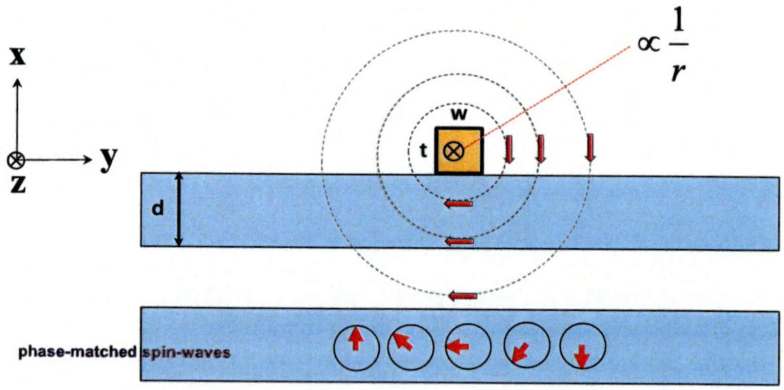

Fig. 1.1 (Color online) Chiral excitation of spin waves in a magnetic thin film by the near field of a stripline antenna. The ac magnetic field is axially symmetric with an oscillating modulus and in the film a position-dependent linear polarization. It excites spin waves with the same frequency and phase-matched spatial amplitude. The film magnetization direction (here parallel to the stripline) can be tuned by a static magnetic field

$$H_x(x, y, \omega) \equiv \int dk_y e^{ik_y y} H_x(x, k_y) = -\int dk_y e^{ik_y y} \frac{\mathcal{J}(\omega)}{4\pi} \frac{k_y}{k_x} e^{-ik_x x},$$
$$H_y(x, y, \omega) \equiv \int dk_y e^{ik_y y} H_y(x, k_y) = -\int dk_y e^{ik_y y} \frac{\mathcal{J}(\omega)}{4\pi} e^{-ik_x x}, \tag{1.6}$$

where $k_x = \sqrt{(\omega/c)^2 - k_y^2}$. Directly above or below the wire $H_x\,(x, y = 0, \omega) = 0$, i.e. the magnetic field is linearly-polarized along y. The polarization rotates as a function of y until $H_y\,(0, y \to \infty, \omega) = 0$. Surprisingly, a circular polarization emerges in the Fourier components

$$H_x(x, k_y, \omega) = -\frac{\mathcal{J}(\omega)}{4\pi} \frac{k_y}{k_x} e^{-ik_x x},$$
$$H_y(x, k_y, \omega) = -\frac{\mathcal{J}(\omega)}{4\pi} e^{-ik_x x}. \tag{1.7}$$

For an evanescent field with $k_y > \omega/c \equiv k$, $k_x = i\sqrt{k_y^2 - k^2}$ and $x < 0$

$$H_x(x, k_y, \omega) = \frac{i\mathcal{J}(\omega)}{4\pi} \frac{k_y}{\sqrt{k_y^2 - k^2}} e^{\sqrt{k_y^2 - k^2}x},$$
$$H_y(x, k_y, \omega) = -\frac{\mathcal{J}(\omega)}{4\pi} e^{\sqrt{k_y^2 - k^2}x}. \tag{1.8}$$

At microwave frequencies $\omega/(2\pi) \sim 10\,\text{GHz}$, $k \equiv \omega/c \sim 200\,\text{m}^{-1}$ and wavelength $\lambda = 2\pi/k \sim 3\,\text{cm}$. The spin wavelength at the same frequency is much smaller with $\sqrt{k_y^2 + k_z^2} \gg \omega/c$, so we are in the near-field limit. The magnetic field component $H_x \to i\,\text{sgn}(k_y) H_y$ is then circularly polarized with a sign locked to its linear momentum.

For a finite rectangular cross section with $0 < x < t$ and $-w/2 < y < w/2$ the Fourier components of the magnetic field read

$$H_x(x, k_y, \omega) = i\frac{J(\omega)}{4\pi} \mathcal{F}(t, w) \frac{k_y}{\sqrt{k_y^2 - k^2}} e^{\sqrt{k_y^2 - k^2}x},$$
$$H_y(x, k_y, \omega) = -\frac{J(\omega)}{4\pi} \mathcal{F}(t, w) e^{\sqrt{k_y^2 - k^2}x}, \tag{1.9}$$

which differ from the previous results only by the form factor

$$\mathcal{F}(t, w) = \frac{4}{k_x k_y} e^{ik_x \frac{t}{2}} \sin\left(k_x \frac{t}{2}\right) \sin\left(k_y \frac{w}{2}\right). \tag{1.10}$$

Irrespective to the shape of the stripline, the magnetic field components are circularly polarized when $|k_y| \gg \omega/c$ but oscillate now as function of the wave vector.

1.2.2 Chiral Excitation of Spin Waves

We focus here on thin YIG films with thickness $d \sim \mathcal{O}(10\,\text{nm})$, which allows an analytical treatment of the dispersion and spin wave amplitudes in the dipolar-exchange regime [6]. An applied magnetic field $H_{\text{app}}\hat{\mathbf{z}}$ parallel to the stripline corresponds to the Damon–Eshbach configuration, but we stress that for ultrathin films there are no Damon–Eshbach surface modes. The spin wave energy dispersion [6]

$$\omega_{\mathbf{k}} = \mu_0\gamma M_s\sqrt{\left[\Omega_H + \alpha_{\text{ex}}k^2 + 1 - f(|k_y|)\right]\left[\Omega_H + \alpha_{\text{ex}}k^2 + (k_y^2/k^2)f(|k_y|)\right]}, \tag{1.11}$$

where $-\gamma$ is the electron gyromagnetic ratio, M_s denotes the saturated magnetization, α_{ex} is the exchange stiffness, $\Omega_H \equiv H_{\text{app}}/M_s$, and

$$f(|k_y|) = 1 - \frac{1}{|k_y|d} + \frac{1}{|k_y|d}\exp(-|k_y|d), \tag{1.12}$$

is highly anisotropic. The spin waves amplitudes across sufficiently thin films are constant [6]:

$$m_x = \sqrt{\frac{B+1}{4d(B-1)}},\ m_y = i\sqrt{\frac{B-1}{4d(B+1)}}, \tag{1.13}$$

where we chose the normalization [38, 45, 46]

$$\int d\mathbf{r}\left[m_x(\mathbf{r})m_y^*(\mathbf{r}) - m_x^*(\mathbf{r})m_y(\mathbf{r})\right] = -i/2, \tag{1.14}$$

and

$$B = \frac{1/2 - (1/2)\left(1 + k_y^2/k^2\right)f(|k_y|)}{\omega_{\mathbf{k}}/(\mu_0\gamma M_s) - \left(\Omega_H + \alpha_{\text{ex}}k_y^2 + 1/2\right) + (1/2)\left(1 - k_y^2/k^2\right)f(|k_y|)}. \tag{1.15}$$

When $k_y \to 0$: $f(|k_y|) = 0$, $\lim_{k_y\to 0}\omega_{\mathbf{k}} = \mu_0\gamma M_s\sqrt{\Omega_H(\Omega_H+1)}$, $B \to -1 - 2\Omega_H - 2\sqrt{\Omega_H(\Omega_H+1)}$. When $\Omega_H \to 0$ with a small static magnetic field, $B \to -1 - 2\sqrt{\Omega_H}$, $|m_y| \gg |m_x|$, so the Kittel mode is (nearly) linearly polarized. In the opposite (exchange) limit of $|k_y|d \gg 1$ and $\alpha_{\text{ex}}k^2 \gg 1$, $f(k_y) \to 1$, $|B| \gg 1$, and the spin waves are right-circularly polarized with $m_y = im_x$.

The Oersted magnetic fields from the stripline interact with spin waves by the Zeeman interaction [47]

$$\hat{H}_{\text{int}} = -\mu_0\int \mathbf{M}(\mathbf{r})\cdot\mathbf{H}(\mathbf{r})dV. \tag{1.16}$$

The excited magnetization in the film can be expressed by time-dependent perturbation theory [48]

$$M_\alpha(x,\rho,t) = -i\int_{-\infty}^{t} dt' \left\langle \left[\hat{\mathbf{M}}_\alpha(x,\rho,t), \hat{H}_{\rm int}(t') \right] \right\rangle . \tag{1.17}$$

in terms of the retarded spin susceptibility tensor

$$\chi_{\alpha\delta}(x,x';\rho-\rho';t-t') = i\Theta(t-t')\left\langle \left[\hat{\mathbf{S}}_\alpha(x,\rho,t), \hat{\mathbf{S}}_\delta(x',\rho',t') \right] \right\rangle , \tag{1.18}$$

where $\hat{\mathbf{S}}_\alpha = -\hat{\mathbf{M}}_\alpha/(\gamma\hbar)$is the spin operator and a sum over repeated indices is implied. Hence [6–8],

$$\mathbf{M}_\alpha(x,k_y,\omega) = \mu_0(\gamma\hbar)^2\int_{-d}^{0} dx' \chi_{\alpha\beta}(x,x',k_y,\omega)H_\beta(x',k_y,\omega), \tag{1.19}$$

where

$$\chi_{\alpha\beta}(x,x',\mathbf{k},\omega) = -\frac{2M_s}{\gamma\hbar} m_\alpha^{(\mathbf{k})}(x) m_\beta^{(\mathbf{k})*}(x') \frac{1}{\omega-\omega_{\mathbf{k}}+i\Gamma_{\mathbf{k}}} . \tag{1.20}$$

Here, $\Gamma_{\mathbf{k}} = 2\alpha\omega_{\mathbf{k}}$ is the reciprocal lifetime in terms of the Gilbert damping constant α. The excitation efficiency is determined by $m_\beta^{(k_y)*}(x')H_\beta(x',k_y,\omega)$, so the excitation of circularly polarized spin waves is chiral (or unidirectional) by the polarization-momentum locking with the stripline magnetic field. Since the amplitudes across thin films are constant for $kd \ll 1$, the excited magnetization in time domain and position space is the real part of the inverse Fourier transform ($q \equiv k_y$),

$$\begin{aligned} \mathbf{M}_\alpha(x,y,t) &= \sum_q e^{iqy-i\omega t}\mathbf{M}_\alpha(x,q) \\ &\approx 2i\mu_0\gamma\hbar d M_s m_\alpha^{(q_\omega)} m_\beta^{(q_\omega)*}\frac{1}{v_{q_\omega}} e^{-i\omega t} \begin{cases} e^{iq_\omega y-\delta_\omega y}H_\beta(q_\omega,\omega) & \text{for } y>0 \\ e^{-iq_\omega y+\delta_\omega y}H_\beta(-q_\omega,\omega) & \phantom{\text{for }} y<0 \end{cases} , \end{aligned} \tag{1.21}$$

where $q_\omega + i\delta_\omega$ is the positive root of $\omega_q = \omega + i\Gamma_q$, and v_{q_ω} is the modulus of the group velocity $|\partial\omega_q/\partial q|_{q_\omega}$. The polarization-momentum locking of the stripline field generates two different magnetization dynamics. When the excited spin waves are circularly polarized, they not only propagate in one direction only, but the excitation is also spatially limited to half of the film, i.e. the chirality is perfect. We can understand this phenomenon in terms of the interference between the spin waves and the stripline magnetic field that is constructive and destructive on opposite sides, as illustrated in Fig. 1.1.

The dominant excitation direction can be switched with the film magnetization. For a finite angle θ between the saturated magnetization and the stripline, the situation becomes complicated by the reduced symmetry. It is advantageous to transform (1.21) following the Supplements of [7, 8]:

$$m_x^{(k_y)} \to m_x^{(\mathbf{l})}, \quad m_y^{(k_y)} \to \cos\theta\, m_y^{(\mathbf{l})}$$

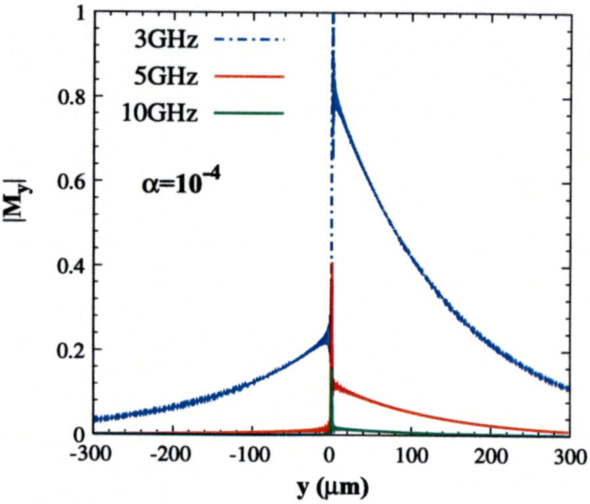

Fig. 1.2 Calculated magnetization amplitude profile $|M_y(y)|$ of a YIG film with ground state magnetization along z, $d = 20\,\text{nm}$, and $\alpha = 10^{-4}$, excited by a metal stripline with $t = 100\,\text{nm}$, $w = 1\,\mu\text{m}$, carrying an AC current with excitation frequencies $\omega/(2\pi) = 3$, 5, and 10 GHz. $|M_y(y)|$ is proportional to the current density, which is here normalized by its maximum value for $\omega/(2\pi) = 3\,\text{GHz}$

where $\mathbf{l} = (0, q\cos\theta, q\sin\theta)$ and q is determined by $\omega_{\mathbf{l}} + i2\alpha\omega_{\mathbf{l}} = \omega$. Even for circularly polarized spin waves, the chirality is not perfect anymore while situation is complicated for elliptical spin waves since their polarization depends on the wave vector. For $\theta_c = \pi/2$ the chirality always vanishes. Since mirror symmetry is broken, the two roots $|q_\omega^{(+)} + i\delta_\omega^{(+)}| \neq |-q_\omega^{(-)} - i\delta_\omega^{(-)}|$ for $\theta \neq 0, \theta_c$. The wavelength and propagation direction of the excited spin waves may therefore be different on the two sides of the stripline.

Figure 1.2 is a plot of the calculated excited magnetization profile for a YIG magnetic film for constant current density but different excitation frequencies $\omega/(2\pi)$. At low frequencies the excitation efficiency is high, but since the dipolar interaction renders the spin wave precession elliptical, the chirality is relatively weak. At high frequencies the chirality improves, but the magnetization amplitude is suppressed by the form factor $\sin(k_y w/2)$ that favors spin waves with wavelengths around w. A narrower stripline helps to excite spin waves with short wavelengths and higher chirality. The spatial decay on both sides of the stripline is governed by the Gilbert damping. Chiral spin waves can also be generated by magnetic striplines with high coercivity that allow efficient excitation and almost perfect chirality at frequencies $>10\,\text{GHz}$. The physics is quite different, however, and explained in the following section.

1.3 Chiral Spin Wave Excitation and Absorption by a Magnetic Transducer

Coherent exchange-dipolar spin waves with short wavelengths $\lambda < 100\,\text{nm}$ are attractive information carriers by their long lifetime and high group velocity. According to the discussion above their excitation is difficult because striplines cannot be fabri-

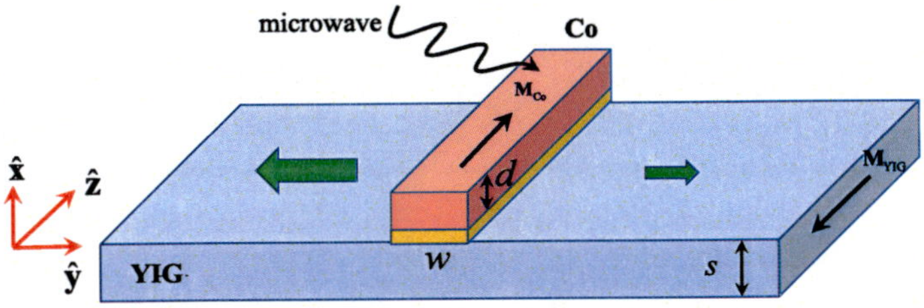

Fig. 1.3 A magnetic (Co) nanowire transducer separated by a non-magnetic spacer (optional) from a YIG film. The dipolar coupling is maximized for the antiparallel magnetization. The direction of the magnon spin currents pumped into the $\pm\hat{\mathbf{y}}$-directions is indicated by the green arrows, whose size indicates the magnitude of the magnon currents. The black arrows indicate the (nearly uniform) microwave input to the magnetic nanowire

cated much finer than this wave length. A small stripline cross section also increases Joule heating and thereby limits the maximum applicable currents. A new strategy is to use magnetic nanowires with high coercivity and resonance frequencies that can be fabricated with the same feature sizes as normal metal ones. Rather than applying an AC current directly, magnetic nanowires can be used as "antennas" that are excited by proximity coplanar wave guides [5, 25–29]. A direct contact between film and nanowires can suppress chirality by the interface exchange interaction and associated spin transfer [6], but an insulating spacer of a few atomic monolayers strongly suppresses exchange without much affecting the dipolar interaction. Figure 1.3 shows a typical configuration with a Co nanowire on top of the YIG film.

1.3.1 Chiral Magnetodipolar Field

The dipolar field from the magnetic nanowire fundamentally differs from the Oersted field of the AC current-biased normal metal wire discussed above. The precessing magnetization is a magnetic dipole and generates a rotating dipolar field rather than the oscillating axially symmetric field of the normal metal wire sketched in Fig. 1.1. The amplitudes of dipolar waves decay faster than that of (monopolar) current-induced ones, but are still long-ranged compared to e.g. the exchange interaction. The nanowire and its equilibrium magnetization are parallel to the z-direction as shown in Fig. 1.3. When driven with a frequency ω, the macrospin (Kittel) magnetization dynamics of a wire with thickness d and width w is the real part of

$$\tilde{M}_{x,y}(\mathbf{r},t) = \tilde{m}_{x,y}\Theta(x)\Theta(-x+d)\Theta(y+w/2)\Theta(-y+w/2)e^{-i\omega t}, \qquad (1.22)$$

where $\Theta(x)$ is the Heaviside step function and $\tilde{m}_{x,y}$ are constant amplitudes that depend on the geometry and the excitation power. The corresponding dipolar magnetic field [47]

$$\begin{aligned}\tilde{h}_\beta(\mathbf{r},t) &= \frac{1}{4\pi}\partial_\beta\partial_\alpha\int\frac{\tilde{M}_\alpha(\mathbf{r}',t)}{|\mathbf{r}-\mathbf{r}'|}\mathrm{d}\mathbf{r}' \\ &= \frac{1}{4\pi}\partial_\beta\partial_\alpha\int \mathrm{d}z'\int_0^{\mathrm{d}}\mathrm{d}x'\int_{-\frac{w}{2}}^{\frac{w}{2}}\mathrm{d}y'\frac{\tilde{m}_\alpha e^{-i\omega t}}{\sqrt{z'^2+(x-x')^2+(y-y')^2}}. \end{aligned} \tag{1.23}$$

We use the Coulomb integral [6, 41]

$$\frac{1}{\sqrt{z'^2+(x-x')^2+(y-y')^2}} = \frac{1}{2\pi}\int \mathrm{d}k_x\mathrm{d}k_y\frac{e^{-|z'|\sqrt{k_x^2+k_y^2}}}{\sqrt{k_x^2+k_y^2}}e^{ik_x(x-x')+ik_y(y-y')}, \tag{1.24}$$

a variation of the Weyl identity used in (1.4), to express the magnetic field below the nanowire ($x < 0$) with partial Fourier components k_y

$$\begin{aligned}\tilde{h}_\beta(k_y,x,t) &= \int h_\beta(\mathbf{r},t)e^{-ik_y y}\mathrm{d}y \\ &= \frac{1}{\pi}\int \mathrm{d}k_x(k_x\tilde{m}_x+k_y\tilde{m}_y)k_\beta e^{ik_x x-i\omega t}\frac{1}{k_x^2+k_y^2}\frac{1-e^{-ik_x\mathrm{d}}}{ik_x}\frac{\sin(k_y w/2)}{k_y}. \end{aligned} \tag{1.25}$$

Closing the contour of the k_x integral in the lower half of the complex plane yields

$$\begin{pmatrix}\tilde{h}_x(k_y,x,t)\\ \tilde{h}_y(k_y,x,t)\end{pmatrix} = -\frac{i}{4\pi}e^{|k_y|x}(1-e^{-|k_y|d})\frac{2\sin(k_y w/2)}{k_y|k_y|}\begin{pmatrix}|k_y| & ik_y\\ ik_y & -|k_y|\end{pmatrix}\begin{pmatrix}\tilde{m}_x\\ \tilde{m}_y\end{pmatrix}e^{-i\omega t}. \tag{1.26}$$

The perfectly right-circularly polarized wire dynamics of the Kittel mode in rectangular wires ($\tilde{m}_y = i\tilde{m}_x$ when $w = d$) implies that the Fourier components of $\tilde{\mathbf{h}}$ with $k_y > 0$ vanish. The Fourier component with $k_y < 0$ is then perfectly left circularly polarized $\left(\tilde{h}_y = -i\tilde{h}_x\right)$. Above the nanowire, the magnetic field direction and polarization are reversed, as sketched in Fig. 1.4. The elliptical polarization of the Kittel mode in rectangular nanowires breaks the perfect chirality. Analogous expressions can be derived for arbitrarily shaped magnetic transducers such as discs, but analytical expressions become complex or may not exist when the symmetry is reduced.

Equation (1.21) can be used also for magnetic fields $\tilde{\mathbf{h}}$ generated by a magnetic transducer, i.e. (1.26), a left-circularly polarized dipolar field that propagates to the left. An ellipticity of the spin waves in the film does not affect the chirality since the excited magnetization still propagates to the left and lives only in the left half-space, but it reduces the excitation efficiency. The same holds when the Kittel mode in a rectangular nanowire is elliptical and the spin waves in the film are circularly polarized. We illustrate these conclusion below from different viewpoint.

Let us compare the dipolar stray fields $\tilde{\mathbf{h}}$ emitted by the excited magnetic wire and $\mathbf{H}$ generated by a stripline as discussed in the previous section. The main dif-

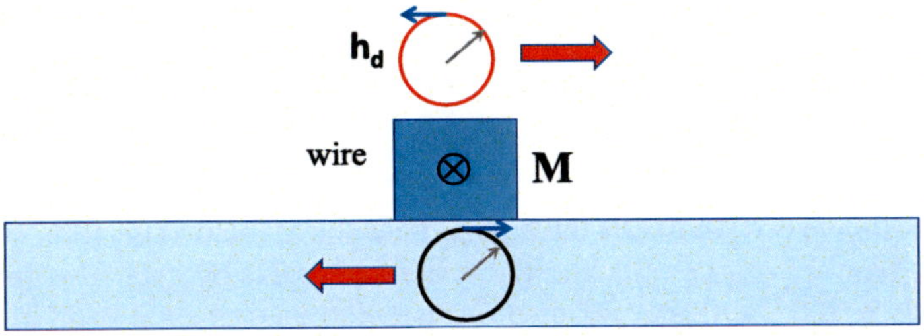

Fig. 1.4 Dipolar magnetic field $\tilde{\mathbf{h}}$ generated by a Kittel mode excitation of a magnetic nanowire ($\parallel \hat{\mathbf{z}}$). The thick red and thin blue arrows indicate the propagation and precession directions of $\tilde{\mathbf{h}}$, respectively, both above and below the wire

ference between these "Oersted" versus "dipolar" radiation is that the latter has additional chirality that induces a circularly-polarized magnetic field in real space, in contrast to the linearly-polarized magnetic field of the former. Equation (1.26) can be summarized as $\tilde{h}_x \propto |k_y|(\tilde{m}_x + i\mathrm{sgn}(k_y)\tilde{m}_y)$ and $\tilde{h}_y \propto ik_y(\tilde{m}_x + i\mathrm{sgn}(k_y)\tilde{m}_y)$. $\tilde{h}_y = i\mathrm{sgn}(k_y)\tilde{h}_x$ is the polarization-momentum locking in reciprocal space, which is the same as that of the evanescent Oersted field. However, the magnetic chirality affects $\tilde{m}_x + i\mathrm{sgn}(k_y)\tilde{m}_y$: for right circularly-polarized (when $w = d$) $\tilde{m}_y = i\tilde{m}_x$, $\tilde{\mathbf{h}}$ simply vanishes for positive k_y. Thus, the magnetic field is unidirectional with linear momentum components normal to the wire that are negative, which is more than just a locking between polarization and momentum. $\tilde{\mathbf{h}}$ therefore couples chirally to spins with arbitrary polarizations.

The Zeeman interaction $\sim\mathbf{M}\cdot\tilde{\mathbf{H}}$ between the wire and film is governed as used above is completely equivalent to the interaction $\sim\tilde{\mathbf{M}}\cdot\mathbf{h}$, where $\tilde{\mathbf{M}}$ is the wire magnetization and $\mathbf{h}$ the dipolar field generated by the spin waves in the film. It is instructive to discuss the physics from this second viewpoint. We assume again that the equilibrium wire magnetization is fixed by the form anisotropy to the z-direction. A sufficiently soft film magnetization can be rotated in the x-z plane by an applied magnetic field, but we address here only (anti)parallel magnetizations but general wave propagation direction [7, 8]. We allow for the elliptical spin wave polarization in the magnetostatic regime. At frequency ω and in the coordinate system defined in Fig. 1.3 with in-plane wave vector $\mathbf{k} = k_y\hat{\mathbf{y}} + k_z\hat{\mathbf{z}}$, we define $M_x(\mathbf{r},t) = m_R^{(\mathbf{k})}(x)\cos(\mathbf{k}\cdot\boldsymbol{\rho} - \omega t)$ and $M_y(\mathbf{r},t) \equiv -m_R^{(\mathbf{k})}(x)\sin(\mathbf{k}\cdot\boldsymbol{\rho} - \omega t)$, where $m_R^{(\mathbf{k})}(x)$ is the time-independent amplitude into the film and $\boldsymbol{\rho} = y\hat{\mathbf{y}} + z\hat{\mathbf{z}}$. The dipolar field outside the film with $\alpha, \beta = \{x, y, z\}$ [47],

$$h_\beta(\mathbf{r},t) = \frac{1}{4\pi}\partial_\beta\partial_\alpha \int d\mathbf{r}' \frac{M_\alpha(\mathbf{r}',t)}{|\mathbf{r}-\mathbf{r}'|}, \tag{1.27}$$

then reads

$$\begin{pmatrix} h_x(\mathbf{r},t) \\ h_y(\mathbf{r},t) \\ h_z(\mathbf{r},t) \end{pmatrix} = \begin{pmatrix} \left(k+\eta k_y\right)\cos\left(\mathbf{k}\cdot\boldsymbol{\rho}-\omega t\right) \\ \left(\frac{k_y^2}{k}+\eta k_y\right)\sin\left(\mathbf{k}\cdot\boldsymbol{\rho}-\omega t\right) \\ k_z\left(\frac{k_y}{k}+\eta\right)\sin\left(\mathbf{k}\cdot\boldsymbol{\rho}-\omega t\right) \end{pmatrix} \frac{1}{2}e^{-\eta kx}\int dx' m_R^{\mathbf{k}}\left(x'\right)e^{\eta kx'}, \tag{1.28}$$

where $x > 0$ ($x < -s$) indicates the dipolar field above (below) the film, $\eta = 1$ (-1) when $x > 0$ ($x < -s$), $k = |\mathbf{k}|$, and the spatial integral is over the film thickness.

When $k_z = 0$, $k_y \neq 0$ spin waves propagate normal to the wire and $h_z = 0$. The distribution of the dipolar field above and below the film then strongly depends on the sign of k_y: the dipolar field generated by the right (left) moving spin waves only appears above (beneath) the film [6–8] and precesses in the opposite direction of the magnetization. These features provide an alternative explanation of the chiral coupling between these spin waves and *any* magnet close to the film surface [6, 7]. The chiral dipolar coupling is most pronounced when the magnetizations of the film and wire are antiparallel [6–8].

When the film magnetization is rotated by 90° in perpendicular to the wire, the wire magnetization excites spin waves that propagate *parallel* to the magnetization ($k_y = 0$, $h_y = 0$), which for thick films correspond to the backward moving bulk modes. Surprisingly, these also couple chirally to the wire dynamics, but by a different mechanism. According to (1.4), $h_x \propto |k_z| \cos(k_z z - \omega t)$ and $h_z \propto \eta k_z \sin(k_z z - \omega t)$. The dipolar fields generated by spin waves with positive (negative) k_z are left (right) circularly polarized, respectively, while below the film, the polarizations are reversed. These spin waves chirally interact with the transducer magnet since the polarization of the transverse magnetization dynamics of the latter has to match that of the stray field $\mathbf{h}$ [5].

Therefore, two mechanisms contribute to the chiral excitation, depending on the magnetic configuration. When spin waves propagate perpendicular to the magnetization with opposite momenta, their dipolar fields vanish on opposite sides of the film; when propagating parallel to the magnetization, their dipolar field is chiral, i.e., polarization-momentum locked. Purely chiral coupling between magnons can be achieved in the former case without constraints on the polarization of the local magnet, but in the latter case elliptical polarization of the wire leads to partial chirality.

The resonance frequency of a magnetic nanowire can be tuned by an applied magnetic field and excites spin waves in a frequency window that is governed by the wire form factor. The magnetodipolar field emitted by a coherently excited magnetic nanowire array can also be chiral [6, 7]. However, such a nanowire grating with period a and translational symmetry $na\hat{\mathbf{y}}$ excites discrete spin waves with momenta $(m\pi/a)\hat{\mathbf{y}}$, where $\{m, n\} \in Z_0$ that are observable as sharp and intense feature in the microwave transmission (more details are shown below).

1.3.2 Non-local Detection

Here we illustrate the principle of non-local excitation and detection of magnons by a device consisting of two magnetic nanowires on top of a YIG film. The generation of DC currents by AC forces in the absence of a DC bias is generally referred to as "pumping" [49]. Spin pumping is the injection of a spin current by the magnetization dynamics of a magnet into a normal metal contact by the interface exchange interaction [50, 51]. *Chiral spin pumping* is the generation of unidirectional spin waves by the dynamics of a proximity magnetic wire as discussed above. Its inverse is the *chiral spin absorption,* i.e. the wire dynamics induced by the stray fields caused by spin waves in the film. We develop below a semi-analytic theory of chiral spin pumping/absorption for antiparallel magnetic configurations and describe two effects—non-reciprocal microwave transmission and chiral spin Seebeck effect. Whereas the former is due to coherent pumping by applied microwaves, the latter represents the incoherent (thermal) pumping by a temperature difference [52–55]. Both effects can be observed in terms of the magnon population or temperature in the detector, e.g., inductively or by light scattering.

We switch from a purely classical picture of previous sections to a quantum description of the chiral coupling in terms of Hamiltonian matrix elements between generalized harmonic oscillators. This does not introduce new physics since we can simply replace operators by classical amplitudes, but it provides a compact formalism used in many other fields such as nanomechanical systems and optics, and prepares the stage for the treatment of real quantum problems. For simplicity, we focus on the antiparallel magnetic configuration with maximized dipolar coupling (for arbitrary magnetization directions see [8]). The dipolar coupling of the wire magnetization $\tilde{\mathbf{M}}$ with that of a film $\mathbf{M}$ is governed by the Zeeman interaction with the respective stray magnetic fields $\mathbf{h}$ and $\tilde{\mathbf{h}}$ [47]

$$\hat{H}_{\rm int}/\mu_0 = -\int \tilde{\mathbf{M}}(\mathbf{r},t)\cdot\mathbf{h}(\mathbf{r},t)d\mathbf{r} = -\int \mathbf{M}(\mathbf{r},t)\cdot\tilde{\mathbf{h}}(\mathbf{r},t)d\mathbf{r}, \tag{1.29}$$

where $\mathbf{h}$ and $\tilde{\mathbf{h}}$ have been introduced in (1.28) and (1.26). The magnetization dynamics of film ($\hat{\mathbf{M}}$) and nanowire ($\hat{\tilde{\mathbf{M}}}$) are now interpreted as operators with Cartesian components $\beta \in \{x, y\}$. To leading order of the expansion in magnon creation and annihilation operators [38, 45, 46],

$$\begin{aligned}\hat{M}_\beta(\mathbf{r}) &= -\sqrt{2M_s\gamma\hbar}\sum_{\mathbf{k}}\left[m_\beta^{(\mathbf{k})}(x)e^{i\mathbf{k}\cdot\boldsymbol{\rho}}\hat{\alpha}_{\mathbf{k}} + \text{H.c.}\right],\\ \hat{\tilde{M}}_\beta(\mathbf{r}) &= -\sqrt{2\tilde{M}_s\gamma\hbar}\sum_{k_z}\left[\tilde{m}_\beta^{(k_z)}(x,y)e^{ik_z z}\hat{\beta}_{k_z} + \text{H.c.}\right],\end{aligned} \tag{1.30}$$

where M_s and $\tilde{M}_s$ are the respective saturation magnetizations, $m_\beta^{(\mathbf{k})}(x)$ and $\tilde{m}_\beta^{(k_z)}(x, y)$ are the spin wave amplitudes across the film and nanowire, and $\hat{\alpha}_{\mathbf{k}}$ and $\hat{\beta}_{k_z}$ denote the magnon (annihilation) operator in the film and nanowire, respectively. The total system Hamiltonian then reads

$$\hat{H}/\hbar = \sum_{\mathbf{k}} \omega_{\mathbf{k}} \hat{\alpha}_{\mathbf{k}}^{\dagger} \hat{\alpha}_{\mathbf{k}} + \sum_{k_z} \tilde{\omega}_{k_z} \hat{\beta}_{k_z}^{\dagger} \hat{\beta}_{k_z} + \sum_{\mathbf{k}} \left(g_{\mathbf{k}} e^{-ik_y y_0} \hat{\alpha}_{\mathbf{k}}^{\dagger} \hat{\beta}_{k_z} + g_{\mathbf{k}}^{*} e^{ik_y y_0} \hat{\beta}_{k_z}^{\dagger} \hat{\alpha}_{\mathbf{k}} \right), \tag{1.31}$$

where $\omega_{\mathbf{k}}$ and $\tilde{\omega}_{k_z}$ are the frequencies of spin waves in the film and nanowire and the coupling

$$g_{\mathbf{k}} = F(\mathbf{k}) \left(m_x^{(\mathbf{k})*}, m_y^{(\mathbf{k})*} \right) \begin{pmatrix} |\mathbf{k}| & ik_y \\ ik_y & -k_y^2/|\mathbf{k}| \end{pmatrix} \begin{pmatrix} \tilde{m}_x^{(k_z)} \\ \tilde{m}_y^{(k_z)} \end{pmatrix}, \tag{1.32}$$

with $F(\mathbf{k}) = -\mu_0 \gamma \sqrt{M_s \tilde{M}_s / L} \phi(\mathbf{k})$. The form factor $\phi(\mathbf{k}) = 2\sin(k_y w/2)(1 - e^{-kd})(1 - e^{-ks})/(k_y k^2)$ couples spin waves with wavelengths of the order of the nanowire width (mode selection) and $\lim_{\mathbf{k}\to 0} \phi(\mathbf{k}) = wsd$. Pure exchange waves are right-circularly polarized with $m_y^{(k_y)} = i m_x^{(k_y)}$ and their coupling is perfectly chiral since $g_{-|k_y|} = 0$ and $g_{|k_y|} \neq 0$.

Equations (1.11) and (1.13) give the spin-wave dispersion and amplitudes in the thin film. The spin waves propagate in the nanowire along $\hat{\mathbf{z}}$ with amplitudes [6, 8]

$$\tilde{m}_x^{k_z} = \sqrt{\frac{1}{4\mathcal{D}(k_z) w d}}, \quad \tilde{m}_y^{k_z} = i \sqrt{\frac{\mathcal{D}(k_z)}{4wd}}, \tag{1.33}$$

where

$$\mathcal{D}(k_z) = \sqrt{\frac{H_{\text{app}} + N_{xx} \tilde{M}_s + \tilde{\lambda}_{\text{ex}} k_z^2 \tilde{M}_s}{H_{\text{app}} + N_{yy} \tilde{M}_s + \tilde{\lambda}_{\text{ex}} k_z^2 \tilde{M}_s}}. \tag{1.34}$$

H_{app} and $\tilde{\lambda}_{\text{ex}}$ are the applied magnetic field and the exchange stiffness of the nanowire, respectively. The demagnetization factors $N_{xx} \simeq w/(d+w)$ and $N_{yy} = d/(d+w)$ [6] also govern the spin waves frequency

$$\tilde{\omega}_{k_z} = \mu_0 \gamma \sqrt{(H_{\text{app}} + N_{yy} \tilde{M}_s + \tilde{\lambda}_{\text{ex}} k_z^2 \tilde{M}_s)(H_{\text{app}} + N_{xx} \tilde{M}_s + \tilde{\lambda}_{\text{ex}} k_z^2 \tilde{M}_s)}. \tag{1.35}$$

When the magnetic field is antiparallel to the nanowire magnetization we require $|H_{\text{app}}| < \min\{N_{yy}\tilde{M}_s, N_{xx}\tilde{M}_s\}$. The ellipticity of the Kittel mode with $k_z = 0$ is strongly affected by the shape anisotropy when the applied field is sufficiently small and the aspect ratio large: when $d \ll w$, $N_{xx} \to 1$, $N_{yy} \to 0$, $\mathcal{D}$ is large and the mode is nearly linearly-polarized. On the other hand, when $d \approx w$, $\mathcal{D} \to 1$, and the Kittel mode is circularly polarized. When $d \lesssim w$, and the Kittel mode traces an elliptical orbit. Figure 1.5 illustrates the chirality of the coupling parameter $g_{\mathbf{k}}$ of the k_z-Kittel

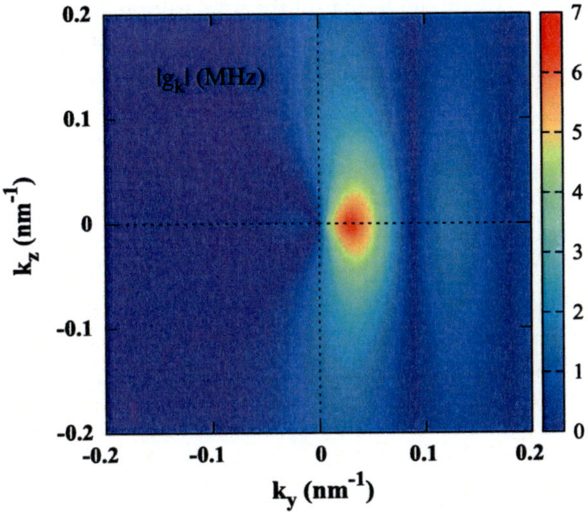

Fig. 1.5 Momentum dependence of the dipolar coupling strength $|g_{\mathbf{k}}|$ of a magnetic nanowire and film (parameters in the text) [8]

mode in a nanowire of dimensions $w = 70\,\text{nm}$ and $d = 20\,\text{nm}$ and magnons in a film of thickness $s = 20\,\text{nm}$ with wave vector $\mathbf{k} = (0, k_y, k_z)$ [8]. The coupling maximum can be shifted to larger momenta by a smaller feature size of the wire. The excitation of such short-wavelength spin waves is possible with a magnetically hard transducer that has a high ferromagnetic resonance frequency [5, 25–29].

1.3.3 Coherent Chiral Spin Wave Transmission

The quantum description leads to expressions that are fully equivalent with (1.21) obtained from the classical description [56, 57]. The excitation of magnons saps nanowire energy and angular momentum, thereby contributing to the magnetization damping, which can be observed as an increased linewidth of the ferromagnetic resonance spectrum. In the quantum description, this broadening is determined by the imaginary part of the magnetic self-energy, which in the first Born approximation or the Fermi-golden rule reads

$$\delta\tilde{\kappa}_{k_z} = 2\pi \sum_{k_y} |g_{\mathbf{k}}|^2 \delta(\tilde{\omega}_{k_z} - \omega_{\mathbf{k}}). \tag{1.36}$$

We predict a very significant additional damping for a Co nanowire with width $w = 70\,\text{nm}$, thickness $d = 20\,\text{nm}$, magnetization $\mu_0 \tilde{M}_s = 1.1\,\text{T}$ [7, 29], and exchange stiffness $\tilde{\lambda}_{\text{ex}} = 3.1 \times 10^{-13}\,\text{cm}^2$ [58]. We adopt a YIG film $s = 20\,\text{nm}$ with magnetization $\mu_0 M_s = 0.177\,\text{T}$ and exchange stiffness $\lambda_{\text{ex}} = 3.0 \times 10^{-12}\,\text{cm}^2$ [7, 29, 38]. A magnetic field $\mu_0 H_{\text{app}} = 0.05\,\text{T}$ is sufficient to switch the film magnetizations antiparallel to that of the wire to maximize the effect [28, 29]. The calculated additional damping of nanowire Kittel dynamics is then $\delta\alpha_{\mathrm{Co}} = \delta\tilde{\kappa}_{k_z=0}/(2\tilde{\omega}_{k_z=0}) = 3.1 \times 10^{-2}$, which is one order of magnitude larger than the intrinsic Gilbert damping coefficient $\alpha_{\mathrm{Co}} = 2.4 \times 10^{-3}$ [59].

Almost perfect chiral pumping by a nanowire *array* has been observed by the microwave transmission and Brillouin light scattering in [7]. We here focus on the new features in the broadband non-local excitation-detection by *two nanowires*. The magnetic order in two nanowires located at $\mathbf{r}_1 = R_1\hat{\mathbf{y}}$ and $\mathbf{r}_2 = R_2\hat{\mathbf{y}}$ act as transducers for microwaves that are emitted or absorbed by local microwave (normal metal) antennas such as coplanar wave guides. The observable is the scattering matrix of the microwaves with excitation (input) at R_1 and the detection (output) at R_2, which can be formulated by the input-output theory [56, 57]. The equation of motion of magnons localized at R_1 and R_2 with operators $\hat{m}_L$ and $\hat{m}_R$ and coupled by the film magnons with operators $\hat{\alpha}_q$ (not to be confused with the Gilbert damping constant) read

$$\begin{aligned}
\frac{d\hat{m}_L}{dt} &= -i\omega_{\mathrm{K}}\hat{m}_L(t) - i\sum_q g_q e^{iqR_1}\hat{\alpha}_q(t) - \left(\frac{\kappa_L}{2} + \frac{\kappa_{p,L}}{2}\right)\hat{m}_L(t) - \sqrt{\kappa_{p,L}}\hat{p}_{\mathrm{in}}^{(L)}(t),\\
\frac{d\hat{m}_R}{dt} &= -i\omega_{\mathrm{K}}\hat{m}_R(t) - i\sum_q g_q e^{iqR_2}\hat{\alpha}_q(t) - \frac{\kappa_R}{2}\hat{m}_R(t),\\
\frac{d\hat{\alpha}_q}{dt} &= -i\omega_q\hat{\alpha}_q(t) - ig_q e^{-iqR_1}\hat{m}_L(t) - ig_q e^{-iqR_2}\hat{m}_R(t) - \frac{\kappa_q}{2}\hat{\alpha}_q(t).
\end{aligned} \tag{1.37}$$

Here, κ_L and κ_R are the intrinsic damping of the Kittel modes in the left and right nanowires, respectively, $\kappa_{p,L}$ is the additional radiative damping induced by the microwave photons $\hat{p}_{\mathrm{in}}^{(L)}$, i.e. the coupling of the left nanowire with the microwave source, and κ_q denotes the intrinsic (Gilbert) damping of magnons in the films. In frequency space:

$$\begin{aligned}
\hat{\alpha}_q(\omega) &= g_q G_q(\omega)\left[e^{-iqR_1}\hat{m}_L(\omega) + e^{-iqR_2}\hat{m}_R(\omega)\right],\\
\hat{m}_R(\omega) &= \frac{-i\sum_q g_q^2 G_q(\omega)\, e^{iq(R_2-R_1)}}{-i(\omega-\omega_{\mathrm{K}}) + \kappa_R/2 + i\sum_q g_q^2 G_q(\omega)}\hat{m}_L(\omega),\\
\hat{m}_L(\omega) &= \frac{-\sqrt{\kappa_{p,L}}}{-i(\omega-\omega_{\mathrm{K}}) + (\kappa_L+\kappa_{p,L})/2 + i\sum_q g_q^2 G_q(\omega) - f(\omega)}\hat{p}_{\mathrm{in}}^{(L)}(\omega),
\end{aligned} \tag{1.38}$$

with spin wave propagator $G_q(\omega) = \left[(\omega-\omega_q) + i\kappa_q/2\right]^{-1}$ and

$$f(\omega) \equiv -\frac{\left(\sum_q g_q^2 G_q(\omega)\, e^{iq(R_1-R_2)}\right)\left(\sum_q g_q^2 G_q(\omega)\, e^{iq(R_2-R_1)}\right)}{-i(\omega-\omega_{\mathrm{K}}) + \kappa_R/2 + i\sum_q g_q^2 G_q(\omega)}. \tag{1.39}$$

The excitation of the left nanowire propagates to the right nanowire by the spin waves in the film. When chiral coupling is perfect, $f(\omega)$ vanishes without the back-action. The microwave output of both left and right nanowires as inductively detected by coplanar wave guides are denoted $\hat{p}_{\rm out}^{(L)}(\omega)$ and $\hat{p}_{\rm out}^{(R)}(\omega)$ with input-output relations [56, 57]

$$\begin{aligned}\hat{p}_{\rm out}^{(L)}(\omega) &= p_{\rm in}^{(L)}(\omega) + \sqrt{\kappa_{p,L}}\hat{m}_L(\omega),\\ \hat{p}_{\rm out}^{(R)}(\omega) &= \sqrt{\kappa_{p,R}}\hat{m}_R(\omega),\end{aligned} \tag{1.40}$$

where $\kappa_{p,R}$ is the additional radiative damping induced by the detector. Therefore, the elements in the microwave scattering matrix describing reflection (S_{11})and transmission (S_{21}) amplitudes become

$$\begin{aligned}S_{11}(\omega) &\equiv \frac{\hat{p}_{\rm out}^{(L)}}{\hat{p}_{\rm in}^{(L)}} = 1 - \frac{\kappa_{p,L}}{-i(\omega-\omega_{\rm K}) + (\kappa_L + \kappa_{p,L})/2 + i\sum_q g_q^2 G_q(\omega) - f(\omega)},\\ S_{21}(\omega) &\equiv \frac{\hat{p}_{\rm out}^{(R)}}{\hat{p}_{\rm in}^{(L)}} = [1 - S_{11}(\omega)]\sqrt{\frac{\kappa_{p,R}}{\kappa_{p,L}}}\frac{i\sum_q g_q^2 G_q(\omega)\, e^{iq(R_2-R_1)}}{-i(\omega-\omega_{\rm K}) + \kappa_R/2 + i\sum_q g_q^2 G_q(\omega)}.\end{aligned} \tag{1.41}$$

The real parts of S_{11} and S_{12} at different magnetic fields and microwave frequencies are illustrated in Fig. 1.6 for antiparallel magnetizations. The frequency of the Co Kittel mode decreases with increasing magnetic field until its direction is reversed to the magnetic-field direction (here $|H_{\rm app}| \lesssim 200\,\mathrm{mT}$). The interference pattern on the Kittel resonance in Fig. 1.6b reflects the transmission phase delay $e^{ik(R_1-R_2)}$ in (1.41). We note that in our model the nanowires do not reflect spin waves, the features should therefore not be interpreted in terms of standing spin waves.

1.3.4 Incoherent Chiral Pumping

A temperature gradient between the magnetic nanowire and film also injects unidirectional magnon currents, i.e., causes a chiral spin Seebeck effect [52–55] . Here we consider again two identical transducers, i.e., a magnetic nanowire at $\mathbf{r}_2 = R_2\hat{\mathbf{y}}$ that detects magnons, which are now thermally injected by the nanowire at $\mathbf{r}_1 = R_1\hat{\mathbf{y}}$ and $R_1 < R_2$. This is the configuration of the non-local spin Seebeck effect as detected electrically in many experiments starting with [21]. The magnons in those experiments are believed to be injected by the interface exchange interaction or generated by a temperature gradient in the bulk and results are interpreted by spin diffusion models. Here we consider the regime in which the exchange effect is suppressed, magnon propagation is ballistic and we disregard the bulk spin Seebeck effect due to possible temperature gradients. We predict a spin non-local spin Seebeck effect that is caused exclusively by dipolar fields and carried by magnons with long wave

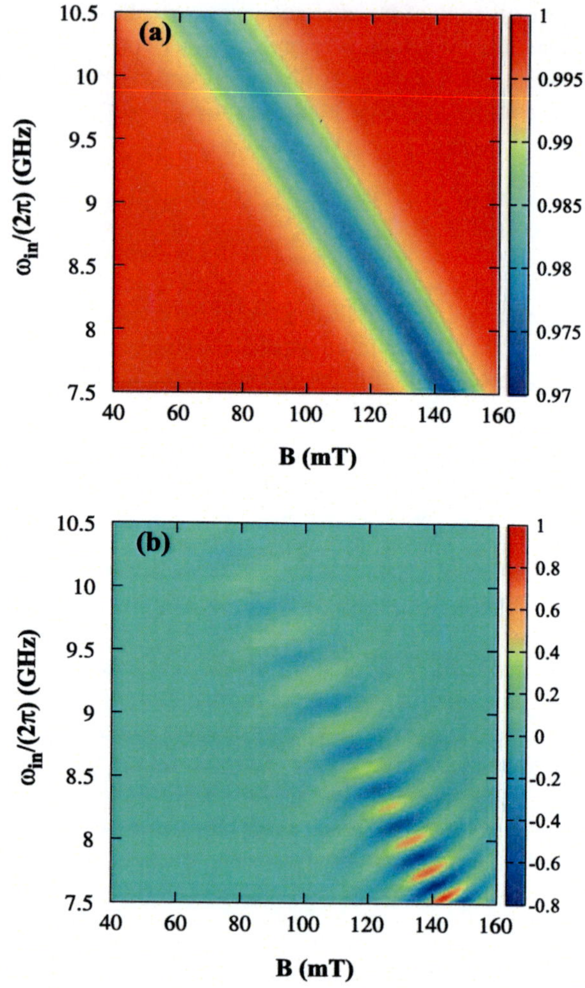

Fig. 1.6 Microwave reflection ReS_{11} (**a**) and transmission ReS_{12} (**b**) amplitudes, (1.41), for a system of two Co nanowires on a YIG film as a function of frequency $\omega_{\rm in}$. The radiative damping of both nanowires is $\kappa_p/(2\pi) = 10\,\text{MHz}$ and other parameters are given in the text

lengths and lifetimes. We focus on the Kittel magnons in the wires since the dipolar coupling between the film and higher bands in the nanowire is very small. The coupling strength $|g_{\mathbf{k}}|$ in Fig. 1.5 illustrates that magnons with wavelength around half of the nanowire width (here $\pi/w = 0.045\,\text{nm}^{-1}$) dominate the coupling. Pumping from other than the those modes can therefore be disregarded even at elevated temperatures. Furthermore, the spin current in the film is dominated by spin waves with small momentum and long mean-free paths, so the effects of magnon-magnon and magnon-phonon interactions that otherwise render magnon transport phenomena diffuse [21] should be negligibly small. The narrow-band thermal injection requires an inductive (or optical) detection of the magnons accumulated in the detector contact, since the inverse spin Hall effect with heavy metal contacts is very inefficient.

The equation of motions of the Kittel modes in the nanowire and film spin waves with momentum q in the coupled system read

$$\frac{d\hat{m}_L}{dt} = -i\omega_K\hat{m}_L - \sum_q ig_q^* e^{iqR_1}\hat{\alpha}_q - \frac{\kappa}{2}\hat{m}_L - \sqrt{\kappa}\hat{N}_L,$$
$$\frac{d\hat{m}_R}{dt} = -i\omega_K\hat{m}_R - \sum_q ig_q^* e^{iqR_2}\hat{\alpha}_q - \frac{\kappa}{2}\hat{m}_R - \sqrt{\kappa}\hat{N}_R,$$
$$\frac{d\hat{\alpha}_q}{dt} = -i\omega_q\hat{\alpha}_q - ig_q e^{-iqR_1}\hat{m}_L - ig_q e^{-iqR_2}\hat{m}_R - \frac{\kappa_q}{2}\hat{\alpha}_q - \sqrt{\kappa_q}\hat{N}_q, \quad (1.42)$$

where κ is caused by the same Gilbert damping in both nanowires, and $\hat{N}_L$ and $\hat{N}_R$ represent the thermal noise in the left and right nanowires, with $\langle \hat{N}_\eta^\dagger(t)\hat{N}_{\eta'}(t')\rangle = n_\eta\delta(t-t')\delta_{\eta\eta'}$. Here, $\eta \in \{L, R\}$ and $n_\eta = 1/\left\{\exp\left[\hbar\tilde{\omega}_K/(k_B T_\eta)\right] - 1\right\}$ and T_R is also the film temperature. Integrating out the spin-wave modes in the film, we obtain equations for dissipatively coupled [60, 61] nanowires. In frequency space,

$$\left(-i(\omega-\omega_K) + \frac{\kappa}{2} + \frac{\Gamma_1+\Gamma_2}{2}\right)\hat{m}_L(\omega) + \Gamma_2 e^{iq_*|R_2-R_1|}\hat{m}_R(\omega) \quad (1.43)$$
$$= \sum_q ig_q^* e^{iqR_1}\sqrt{\kappa_q}G_q(\omega)\hat{N}_q(\omega) - \sqrt{\kappa}\hat{N}_L(\omega),$$
$$\left(-i(\omega-\omega_K) + \frac{\kappa}{2} + \frac{\Gamma_1+\Gamma_2}{2}\right)\hat{m}_R(\omega) + \Gamma_1 e^{iq_*|R_2-R_1|}\hat{m}_L(\omega)$$
$$= \sum_q ig_q^* e^{iqR_2}\sqrt{\kappa_q}G_q(\omega)\hat{N}_q(\omega) - \sqrt{\kappa}\hat{N}_R(\omega), \quad (1.44)$$

where $\Gamma_1 = |g_{q_*}|^2/v_{q_*}$ and $\Gamma_2 = |g_{-q_*}|^2/v_{q_*}$ are assumed constant (for the Kittel mode). Here, q_* is the positive root of $\omega_{q_*} = \tilde{\omega}_K$.

For perfectly chiral coupling with $\Gamma_2 = 0$ the solutions of (1.44) read

$$\hat{m}_L(\omega) = \frac{\sum_q ig_q^* e^{iqR_1}\sqrt{\kappa_q}G_q(\omega)\hat{N}_q(\omega) - \sqrt{\kappa}\hat{N}_L(\omega)}{-i(\omega-\omega_K) + \frac{\kappa}{2} + \frac{\Gamma_1}{2}},$$
$$\hat{m}_R(\omega) = \frac{\sum_q ig_q^* e^{iqR_2}\sqrt{\kappa_q}G_q(\omega)\hat{N}_q(\omega) - \sqrt{\kappa}\hat{N}_R(\omega) - \Gamma_1 e^{q_*(R_2-R_1)}\hat{m}_L(\omega)}{-i(\omega-\omega_K) + \frac{\kappa}{2} + \frac{\Gamma_1}{2}}. \quad (1.45)$$

With $\hat{m}_{L,R}(t) = \int e^{-i\omega t}\hat{m}_{L,R}(\omega)d\omega/(2\pi)$, the Kittel modes are occupied according to

$$\rho_L \equiv \langle \hat{m}_L^\dagger(t)\hat{m}_L(t)\rangle = n_L + \int \frac{d\omega}{2\pi} \frac{\kappa}{(\omega-\omega_K)^2 + (\kappa/2+\Gamma_1/2)^2}(n_{q_*} - n_L), \tag{1.46}$$

$$\rho_R \equiv \langle \hat{m}_R^\dagger(t)\hat{m}_R(t)\rangle = n_R + \int \frac{d\omega}{2\pi} \frac{\Gamma_1^2\kappa}{\left[(\omega-\omega_K)^2 + (\kappa/2+\Gamma_1/2)^2\right]^2}(n_L - n_{q_*}), \tag{1.47}$$

where the damping in the high-quality film has been disregarded $(\kappa_q \to 0)$. In the linear regime the non-local thermal injection of magnons into the right transducer by the left one then reads

$$\delta\rho_R = \begin{cases} \mathcal{S}_{\mathrm{CSSE}}(T_L - T_R) \\ 0 \end{cases} \text{when} \begin{array}{c} T_L > T_R \\ T_L \leq T_R \end{array},$$
$$\mathcal{S}_{\mathrm{CSSE}} = \int \frac{d\omega}{2\pi} \frac{\Gamma_1^2\kappa}{\left[(\omega-\omega_K)^2 + (\kappa/2+\Gamma_1/2)^2\right]^2} \left.\frac{dn_L}{dT}\right|_{T=(T_L+T_R)/2}. \tag{1.48}$$

where we defined the chiral (or dipolar) spin Seebeck coefficient $\mathcal{S}_{\mathrm{CSSE}}$.

The device therefore operates as a heat diode, apparently acting as a "Maxwell demon" that rectifies the thermal fluctuations at equilibrium. However, in thermal equilibrium all right and left moving magnons are eventually connected by reflection of spin waves at the edges and absorption and re-emission by connected heat baths. The Second Law of thermodynamics is therefore safe, but it might be interesting to search for chirality-induced transient effects.

1.4 Conclusion and Outlook

Handedness or chirality of wave propagation is a popular research topic in optics, acoustics, and condensed matter physics. Here we contribute by a theory for the coherent and incoherent chiral pumping of spin waves into thin magnetic films through the chiral magnetodipolar radiation generated by the Oersted field of metallic striplines and dipolar field of magnetic nanostructures. Spin waves excited coherently in the film under magnetic resonance of the nanowire are unidirectional, generating a non-equilibrium magnetization in only half of the film. A temperature gradient between a local magnet and a film leads to the unidirectional excitation of incoherent magnons, i.e., a chiral spin Seebeck effect.

$\mathcal{PT}$ symmetry has been predicted to amplify unidirectional response [62–64]. Even though our system is dissipative and therefore not PT symmetric, the nonreciprocal coupling of the two wires still allows directional amplification [9, 10]. It would be interesting to introduce $\mathcal{PT}$ symmetry into our system via gain in one wire that compensates the damping in the other, possibly leading to enhanced effects.

Magnons can interact remotely by their chiral dipolar magnetic fields with other quasiparticles including other magnons, photons, phonons, and conduction electron spins. Strong chiral coupling between magnons and photons exist, e.g., in microwave waveguides or cavities that contain chains of small magnets on special lines [9, 10]. Large magnon numbers accumulate at one edge of a chain of magnets when excited by local antennas [9, 10]. Spin currents by electrons or phonons may be generated by the chiral magnetodipolar radiation as well. Chirality is a functionality that has not yet been employed much in spintronics, but could be the basis for a new generation of spin-based devices made from conventional materials.

Acknowledgements This work is financially supported by the Nederlandse Organisatie voor Wetenschappelijk Onderzoek (NWO) as well as JSPS KAKENHI Grant No. 19H006450. We thank Yaroslav M. Blanter, Haiming Yu, Bi-Mu Yao, Toeno van der Sar, Sanchar Sharma, Yu-Xiang Zhang, Weichao Yu, and Xiang Zhang for helpful discussions.

References

1. B. Lenk, H. Ulrichs, F. Garbs, M. Muenzenberg, Phys. Rep. **507**, 107 (2011)
2. A.V. Chumak, V.I. Vasyuchka, A.A. Serga, B. Hillebrands, Nat. Phys. **11**, 453 (2015)
3. D. Grundler, Nat. Nanotechnol. **11**, 407 (2016)
4. V.E. Demidov, S. Urazhdin, G. de Loubens, O. Klein, V. Cros, A. Anane, S.O. Demokritov, Phys. Rep. **673**, 1 (2017)
5. Y. Au, E. Ahmad, O. Dmytriiev, M. Dvornik, T. Davison, V.V. Kruglyak, Appl. Phys. Lett. **100**, 182404 (2012)
6. T. Yu, C.P. Liu, H.M. Yu, Y.M. Blanter, G.E.W. Bauer, Phys. Rev. B **99**, 134424 (2019)
7. J.L. Chen, T. Yu, C.P. Liu, T. Liu, M. Madami, K. Shen, J.Y. Zhang, S. Tu, M.S. Alam, K. Xia, M.Z. Wu, G. Gubbiotti, Y.M. Blanter, G.E.W. Bauer, H.M. Yu, Phys. Rev. B **100**, 104427 (2019)
8. T. Yu, Y.M. Blanter, G.E.W. Bauer, Phys. Rev. Lett. **123**, 247202 (2019)
9. T. Yu, Y.-X. Zhang, S. Sharma, Y.M. Blanter, G.E.W. Bauer, Phys. Rev. Lett. **124**, 107202 (2020)
10. T. Yu, X. Zhang, S. Sharma, Y.M. Blanter, G.E.W. Bauer, Phys. Rev. B **101**, 094414 (2020)
11. P. Lodahl, S. Mahmoodian, S. Stobbe, A. Rauschenbeutel, P. Schneeweiss, J. Volz, H. Pichler, P. Zoller, Nature (London) **541**, 473 (2017)
12. F. Le Kien, S.D. Gupta, K.P. Nayak, K. Hakuta, Phys. Rev. A **72**, 063815 (2005)
13. M. Scheucher, A. Hilico, E. Will, J. Volz, A. Rauschenbeutel, Science **354**, 1577 (2016)
14. P. Lodahl, S. Mahmoodian, S. Stobbe, A. Rauschenbeutel, P. Schneeweiss, J. Volz, H. Pichler, P. Zoller, Nature **541**, 473 (2017)
15. B. Vermersch, P.-O. Guimond, H. Pichler, P. Zoller, Phys. Rev. Lett. **118**, 133601 (2017)
16. P. Schneeweiss, S. Zeiger, T. Hoinkes, A. Rauschenbeutel, J. Volz, Opt. Lett. **42**, 85 (2017)
17. C.A. Downing, J.C. López Carreño, F.P. Laussy, E. del Valle, A.I. Fernández-Domínguez, Phys. Rev. Lett. **122**, 057401 (2019)
18. F.J. Rodríguez-Fortuño, G. Marino, P. Ginzburg, D. O'Connor, A. Martínez, G.A. Wurtz, A.V. Zayats, Science **340**, 328 (2013)
19. J. Petersen, J. Volz, A. Rauschenbeutel, Science **346**, 67 (2014)
20. H. Chang, P. Li, W. Zhang, T. Liu, A. Hoffmann, L. Deng, M. Wu, IEEE Magn. Lett. **5**, 6700104 (2014)
21. L.J. Cornelissen, J. Liu, R.A. Duine, J. Ben Youssef, B.J. van Wees, Nat. Phys. **11**, 1022 (2015)

22. K. Uchida, H. Adachi, T. An, T. Ota, M. Toda, B. Hillebrands, S. Maekawa, E. Saitoh, Nat. Mater. **10**, 737 (2011)
23. A.A. Serga, A.V. Chumak, B. Hillebrands, J. Phys. D **43**, 264002 (2010)
24. T. Shinjo (ed.), *Nanomagnetism and Spintronics* (Elsevier, Oxford, 2009)
25. H. Yu, G. Duerr, R. Huber, M. Bahr, T. Schwarze, F. Brandl, D. Grundler, Nat. Commun. **4**, 2702 (2013)
26. H. Qin, S.J. Hämäläinen, S. van Dijken, Sci. Rep. **8**, 5755 (2018)
27. S. Klingler, V. Amin, S. Geprägs, K. Ganzhorn, H. Maier-Flaig, M. Althammer, H. Huebl, R. Gross, R.D. McMichael, M.D. Stiles, S.T.B. Goennenwein, M. Weiler, Phys. Rev. Lett. **120**, 127201 (2018)
28. C.P. Liu, J.L. Chen, T. Liu, F. Heimbach, H.M. Yu, Y. Xiao, J.F. Hu, M.C. Liu, H.C. Chang, T. Stueckler, S. Tu, Y.G. Zhang, Y. Zhang, P. Gao, Z.M. Liao, D.P. Yu, K. Xia, N. Lei, W.S. Zhao, M.Z. Wu, Nat. Commun. **9**, 738 (2018)
29. J.L. Chen, C.P. Liu, T. Liu, Y. Xiao, K. Xia, G.E.W. Bauer, M.Z. Wu, H.M. Yu, Phys. Rev. Lett. **120**, 217202 (2018)
30. L.R. Walker, Phys. Rev. **105**, 390 (1957)
31. R.W. Damon, J.R. Eshbach, J. Phys. Chem. Solids **19**, 308 (1961)
32. A. Akhiezer, V. Baríakhtar, S. Peletminski, *Spin Waves* (North-Holland, Amsterdam, 1968)
33. D.D. Stancil, A. Prabhakar, *Spin Waves-Theory and Applications* (Springer, New York, 2009)
34. T. An, V.I. Vasyuchka, K. Uchida, A.V. Chumak, K. Yamaguchi, K. Harii, J. Ohe, M.B. Jungfleisch, Y. Kajiwara, H. Adachi, B. Hillebrands, S. Maekawa, E. Saitoh, Nat. Mater. **12**, 549 (2013)
35. O. Wid, J. Bauer, A. Müller, O. Breitenstein, S.S.P. Parkin, G. Schmidt, Sci. Rep. **6**, 28233 (2016)
36. E. Shigematsu, Y. Ando, S. Dushenko, T. Shinjo, M. Shiraishi, Appl. Phys. Lett. **112**, 212401 (2018)
37. P. Wang, L.F. Zhou, S.W. Jiang, Z.Z. Luan, D.J. Shu, H.F. Ding, D. Wu, Phys. Rev. Lett. **120**, 047201 (2018)
38. T. Yu, S. Sharma, Y.M. Blanter, G.E.W. Bauer, Phys. Rev. B **99**, 174402 (2019)
39. I. Bertelli, J.J. Carmiggelt, T. Yu, B.G. Simon, C.C. Pothoven, G.E.W. Bauer, Y.M. Blanter, J. Aarts, T. van der Sar, Sci. Adv. **6**, eabd3556 (2020)
40. H.C. Wang, J.L. Chen, T. Yu, C.P. Liu, C.Y. Guo, H. Jia, S. Liu, K. Shen, T. Liu, J.Y. Zhang, M.A. Cabero Z, Q.M. Song, S. Tu, M.Z. Wu, X.F. Han, K. Xia, D.P. Yu, G.E.W. Bauer, H.M. Yu, Nano Res. https://doi.org/10.1007/s12274-020-3251-5
41. L. Novotny, B. Hecht, *Principles of Nano-Optics* (Cambridge University Press, Cambridge, 2006)
42. J.D. Jackson, *Classical Electrodynamics* (Wiley, New York, 1998)
43. T. Schneider, A.A. Serga, T. Neumann, B. Hillebrands, M.P. Kostylev, Phys. Rev. B **77**, 214411 (2008)
44. V.E. Demidov, M.P. Kostylev, K. Rott, P. Krzysteczko, G. Reiss, S.O. Demokritov, Appl. Phys. Lett. **95**, 2509 (2009)
45. C. Kittel, *Quantum Theory of Solids* (Wiley, New York, 1963)
46. T. Holstein, H. Primakoff, Phys. Rev. **58**, 1098 (1940)
47. L.D. Landau, E.M. Lifshitz, *Electrodynamics of Continuous Media*, 2nd edn. (Butterworth-Heinenann, Oxford, 1984)
48. G.D. Mahan, *Many Particle Physics* (Plenum, New York, 1990)
49. M. Büttiker, H. Thomas, A. Prêtre, Z. Phys. B **94**, 133 (1994)
50. Y. Tserkovnyak, A. Brataas, G.E.W. Bauer, Phys. Rev. Lett. **88**, 117601 (2002)
51. Y. Tserkovnyak, A. Brataas, G.E.W. Bauer, B.I. Halperin, Rev. Mod. Phys. **77**, 1375 (2005)
52. K. Uchida, J. Xiao, H. Adachi, J. Ohe, S. Takahashi, J. Ieda, T. Ota, Y. Kajiwara, H. Umezawa, H. Kawai, G.E.W. Bauer, S. Maekawa, E. Saitoh, Nat. Mater. **9**, 894 (2010)
53. J. Xiao, G.E.W. Bauer, K. Uchida, E. Saitoh, S. Maekawa, Phys. Rev. B **81**, 214418 (2010)
54. H. Adachi, J. Ohe, S. Takahashi, S. Maekawa, Phys. Rev. B **83**, 094410 (2011)
55. G.E.W. Bauer, E. Saitoh, B.J. van Wees, Nat. Mat. **11**, 391 (2012)

56. C.W. Gardiner, M.J. Collett, Phys. Rev. A **31**, 3761 (1985)
57. A.A. Clerk, M.H. Devoret, S.M. Girvin, F. Marquardt, R.J. Schoelkopf, Rev. Mod. Phys. **82**, 1155 (2010)
58. R. Moreno, R.F.L. Evans, S. Khmelevskyi, M.C. Muñoz, R.W. Chantrell, O. Chubykalo-Fesenko, Phys. Rev. B **94**, 104433 (2016)
59. M.A.W. Schoen, D. Thonig, M.L. Schneider, T.J. Silva, H.T. Nembach, O. Eriksson, O. Karis, J.M. Shaw, Nat. Phys. **12**, 839 (2016)
60. B.M. Yao, T. Yu, X. Zhang, W. Lu, Y.S. Gui, C.-M. Hu, Y.M. Blanter, Phys. Rev. B **100**, 214426 (2019)
61. B.M. Yao, T. Yu, Y.S. Gui, J.W. Rao, Y.T. Zhao, W. Lu, C.-M. Hu, Commun. Phys. **2**, 161 (2019)
62. A. Mostafazadeh, J. Phys. A: Math. Theor. **47**, 505303 (2014)
63. L. Ge, Y.D. Chong, A.D. Stone, Phys. Rev. A **85**, 023802 (2012)
64. A. Galda, V.M. Vinokur, Phys. Rev. B **94**, 020408(R) (2016)

Chapter 2
Surface Plasmons for Chiral Sensing

Sotiris Droulias and Lykourgos Bougas

Abstract Chiral sensitive techniques have been used to probe the fundamental symmetries of the universe, study biomolecular structures, and even develop safe drugs. The traditional method for the measurement of chirality is through optical activity, however, chiroptical signals are inherently weak and often suppressed by large backgrounds. Different techniques have been proposed to overcome the limitations of traditionally used optical polarimetry, such as cavity- and/or nanophotonic-based schemes. In this chapter we demonstrate how surface plasmon resonance can be employed as a new research tool for chiral sensing, which we term here as CHIral Surface Plasmon Resonance (CHISPR). We present how surface plasmons at a metal-chiral interface are sensitive to the chirality parameter of the chiral medium and how their properties can be exploited to reveal information not easily accessible using standard polarimetric/nanophotonic approaches. We then present an experimental realisation of CHISPR, an angle-resolved measurement scheme, and demonstrate how can one detect the complete chirality (handedness and magnitude) of a chiral sample while being also sensitive to both the real and imaginary part of a chiral sample's refractive index. We present analytical results and numerical simulations of CHISPR measurements, predicting signals in the mdeg range for chiral samples of <100 nm thickness at visible wavelengths. Finally, we present a theoretical analysis that clarifies the underlying physics of the near-field chiral interactions and their far-field manifestation. In overall, CHISPR builds upon the strengths of standard SPR: does not require elaborate fabrication and has the advantage of being directly implementable on existing SPR instrumentation, making it, thus, an ideal modality for studying chirality dynamics on surfaces.

S. Droulias (✉)
Institute of Electronic Structure and Laser, FORTH, 71110 Heraklion, Crete, Greece
e-mail: sdroulias@iesl.forth.gr

L. Bougas (✉)
Institut für Physik, Johannes Gutenberg-Universität Mainz, 55128 Mainz, Germany
e-mail: lybougas@uni-mainz.de

E. Kamenetskii (ed.), *Chirality, Magnetism and Magnetoelectricity*,
Topics in Applied Physics 138,
https://doi.org/10.1007/978-3-030-62844-4_2

2.1 Introduction

Chirality, the geometric property of an object that is non-superimposable with its mirror image, is a foundational property of life: the weak interaction between fundamental particles violates parity [1]; biomolecular structures fundamental to life, such as amino acids, sugars, RNA and DNA, are both chiral and single-handed (i.e. homochiral) [1]; the chemistry of life and the functionalities of its building blocks are largely stereospecific [2]; organisms ranging from protists to plants and animals possess morphological asymmetries with respect to their left-right axis [3]. The development of sensitive chiral sensitive techniques has, therefore, been vital for this wide range of scientific disciplines, and has enabled the study of fundamental symmetries of the universe [4], determination of biomolecular structures [5–7], and even the development of safe and effective drugs [8, 9], to name few of its most prominent applications.[1]

2.1.1 *Chirality and Optical Activity*

Starting from the observation by Arago in 1811 of colours in the sunlight as seen through the optic axis of a quartz crystal placed between crossed polarizers, and the observation of optical rotation in organic liquids such as lemon oils and turpentine by Biot in 1815, the polarimetric techniques of optical rotatory dispersion (ORD) and circular dichroism (CD) have remained as the most widely used research tools in science for chiral sensing [11]. By 1825, Fresnel had discovered that linearly polarized light can be regarded as a superposition of the two possible forms of circularly polarized light [right (RCP) and left (LCP)], which lead to his proposal of the first phenomenological theory about optical activity, i.e. the ability of a chiral medium to rotate linearly polarized light travelling through it. His theory correctly attributed this effect to the propagation at different speeds in the *optically active* medium of the left- and right-circularly polarized components of the incident linearly polarized light. The expression for optical activity as proposed by Fresnel has the general form:

$$\varphi = \frac{\pi l}{\lambda}(n_- - n_+), \tag{2.1}$$

where $n_\pm$ are the (complex) indices of refraction of a chiral medium for RCP and LCP light, respectively, λ is the vacuum wavelength of light, and l the length of the medium.

With the advent of electromagnetism, a description of natural optically active materials on the macroscopic level became possible and is now contained in the

[1]Portions of this chapter have been reprinted with permission from [10]. Copyright 2019 American Chemical Society.

material parameters of the constitutive relations. In particular, these, along with Maxwell's equations, can be written as (according to Condon [12, 13], in the $e^{-i\omega t}$ convention):

$$\nabla \times \mathbf{E} = i\omega\mathbf{B}, \tag{2.2}$$

$$\nabla \times \mathbf{H} = -i\omega\mathbf{D}, \tag{2.3}$$

$$\mathbf{D} = \epsilon_0\epsilon_r\mathbf{E} + i\,(\kappa/c)\mathbf{H}, \tag{2.4}$$

$$\mathbf{B} = \mu_0\mu_r\mathbf{H} - i\,(\kappa/c)\mathbf{E}, \tag{2.5}$$

where ϵ_0, μ_0 are the vacuum permittivity and permeability, ϵ_r, μ_r the relative material permittivity and permeability, respectively, and c is the vacuum speed of light; κ is the chirality (also known as 'Pasteur') parameter, which expresses the chiral molecular response. The eigenwaves of such a medium are RCP/LCP (or $\pm$) waves, which propagate with refractive indices $n_\pm = n_c \pm \kappa$, respectively, where $(n_+ + n_-)/2 = n_c \equiv \sqrt{\epsilon_r\mu_r}$ is the average (background) refractive index.

Considering all the above, we can now see that $\varphi = (2\pi l/\lambda) \cdot \kappa$, and that the real part of κ, i.e. Re(κ), is associated with effects of circular birefringence, while the imaginary part of κ, Im(κ), with circular dichroism.

2.1.2 Chiral Sensing Techniques

The polarimetric techniques of ORD and CD allow for the direct detection of the real part and the imaginary part of φ, and, thus, of the chirality parameter κ of a natural optically active medium. A typical ORD/CD spectro-polarimeter generally consists of a light source, a set of polarizers for state preparation and analysis, and a spectral analysis and detection stage. Yet, two separate instruments are actually designed to perform, *separately*, ORD and CD measurements. Only in the recent years, new approaches towards generalized polarimetry, such as the technique of Mueller Matrix polarimetry, have been demonstrated, which enable a complete characterization of the optical properties of a medium and the simultaneous detection of both its circular birefringence and dichroism [14]. Notwithstanding, spectropolarimetry is being extensively used in basic research and remains the established analytical technique for quality and process control in the pharmaceutical, chemical, and agricultural industries. Despite their extensive use, though, the sensitivity limits of commercially available optical spectro-polarimeters, circular dichroism and optical rotation modules, are at the ~10–100 μdeg levels corresponding to analyte concentration detection limits at the (sub)-micromolar levels, constraining, thus, the extension of polarimetry to a wide range of important research and industrial applications that require improved sensitivity levels (e.g. sub-nM sensitivities). To overcome the limitations of traditional polarimetry in chiral sensing, different techniques have been proposed in the recent years. These techniques, which aim to enhance the matter-wave chiral interaction, can in principle be arranged into two main categories, as they rely mainly on either *(a)* path-length enhancement or *(b)* chiral-field enhancement.

Path-length enhancement techniques typically rely on the use of optical mirrors to create either multipass cells or optical cavities, where in both cases spectro-polarimetric signals are enhanced by the average number of passes through the medium. Multipass cells [15, 16] are technically easy to construct and implement, and optical path-length enhancements as large as $\times 500$ have been demonstrated [17, 18]; crucially though, multipass techniques cannot be, in principle, employed for the measurement of optical activity (owing to the effective round-trip cancellation of the polarization effects [19]). The solution is to use stable optical cavities [20–24], with which one can achieve path-length enhancements of up to 10^5 using state-of-the-art high quality mirrors, with effective path-lengths of up to several hundred kilometers (one needs to compare this to the 10 cm-long sample cell in a single-pass commercial polarimeter), enabling record sensitivities for measurements of absorption and birefringence (CD and ORD, respectively). Importantly, optical cavities can easily be made compact and allow for wave-matter interactions in small volumes. However, cavity-enhanced techniques become inadequate in systems with losses originating from absorption and/or scattering (e.g. chiral molecules within complex matrices, thin films, liquid and/or solid systems), because losses hinder the path-length enhancement. For the case of CD, in particular, path-length enhancement techniques can be mainly used to probe weak molecular transitions [20, 22, 25].

Chiral-field-enhancement techniques rely primarily on generating probing electromagnetic fields with chiral densities higher than circularly polarized plane waves, i.e. *superchiral* fields [26–32]. Chiral and/or achiral nanophotonic systems, such as plasmonic/dielectric nanostructures and metamaterials, can generate contorted intense near-fields with high chiral densities around a resonance frequency of the nanosystem, thus, amplifying the chiral-chiral interactions between them and a molecule. In general, nanophotonic approaches have proven to be a powerful means for granting access to weak chiroptical signals not previously attainable with traditional polarimetric techniques, however, the general principle of operation behind (almost) all contemporary nanophotonic chiral-sensing approaches primarily relies on the detection of enhanced CD signals in the presence of an optically active chiral substance. To achieve this, right- and left-circularly polarized waves are used to excite the system and generate these superchiral fields, and enable the ability to perform CD measurements in transmission. While several works have attributed the resulting CD signal to be proportional only to the imaginary part of the chirality parameter, i.e. $\mathrm{Im}(\kappa)$ (see, e.g., [31–33]), in reality, as supported by past and recent experimental and theoretical results [34–37] the observed CD signals depend on both the real and the imaginary part of κ [$\mathrm{Re}(\kappa)$ and $\mathrm{Im}(\kappa)$, respectively]. Thus, with most contemporary nanophotonic approaches, sensing of the magnitude and sign of both the real and imaginary part of the chirality parameter of a natural optically active substance (*complete* measurement) has not been possible, while elaborate fabrication is required and the employed nanosystems typically have intrinsic chiroptical responses that contribute to the total signal, often precluding direct quantitative measurement of chirality [26–31].

Considering, therefore, the importance of chiral sensing in research, it is vital to develop alternative schemes that overcome the above-mentioned limitations of path-length enhancement and/or chiral-field enhancement techniques. In particular, it is apparent that a general chiral sensing scheme should be able to discriminate between the contributions of $\mathrm{Re}(\kappa)$ and $\mathrm{Im}(\kappa)$, especially far from the chiral molecular resonances, where $\mathrm{Im}(\kappa)$ is weak and $\mathrm{Re}(\kappa)$ is dominant. Additionally, it should be sensitive to both the magnitude of the chirality parameter, $|\kappa|$, and its sign $\mathrm{sgn}(\kappa)$, as $|\kappa|$ is a function of the molecular properties (i.e. polarizability) and its concentration, while $\mathrm{sgn}(\kappa)$ depends on the handedness of the medium.

In this chapter we discuss how surface plasmon resonance (SPR) allows for the complete measurement of chirality (handedness and magnitude) of a chiral system. SPR has become an important technology in the areas of biochemistry, biology, and medical sciences because of its real-time, label-free, and noninvasive nature (see, e.g., [38, 39]). We demonstrate how chiral-sensitive SPR, i.e. CHISPR, is able to quantitatively detect both the real and imaginary part of the refractive index of a chiral substance (responsible for the refraction and absorption, respectively), contrary to most demonstrations that employ nanophotonic structures. We show that CHISPR is particularly suitable for chiral sensing from thin samples which are not easily measurable using alternative polarimetric techniques, and that it makes use of the advantage of being applicable directly on existing SPR instrumentation without the need for additional elaborate fabrication.

2.2 Surface Plasmon Resonance (SPR)

Surface plasmon resonance (SPR) refers to the resonant excitation of a surface plasmon polariton (SPP) at the interface between a metal and a dielectric (or, in general, between two materials, one with negative and one with positive permittivity). SPPs, in particular, are electromagnetic excitations that propagate along the metal-dielectric interface and are evanescently confined in the perpendicular direction. They involve the collective oscillation of conduction electrons at the surface of the metal (hence the term '*surface plasmon*') and they arise via the coupling of the electromagnetic fields to the surface plasmon (hence the term '*polariton*'). SPPs are TM (or '*p*')-polarized waves, i.e., their magnetic field lies entirely on the metal-dielectric interface.

SPR is the basis for many biosensor applications and different lab-on-a-chip sensors, owing to the sensitive dependence of the SPP characteristics on the permittivity of the dielectric region extending over the metal. In essence, slight changes in the permittivity of the dielectric lead to different propagation characteristics in the SPP and therefore a frequency shift in its resonance. By measuring such frequency shifts, one can detect the material changes occurring at the metal-dielectric interface.

Typical SPR setups involve a thin metallic layer (usually Ag or Au in the order of 50 nm) placed directly on a glass substrate, as shown in Fig. 2.1a in the well-known Kretschmann configuration (see [40, 41] for other customary setups). In this configuration the metallic layer extends along the xy-plane and SPP propagates along

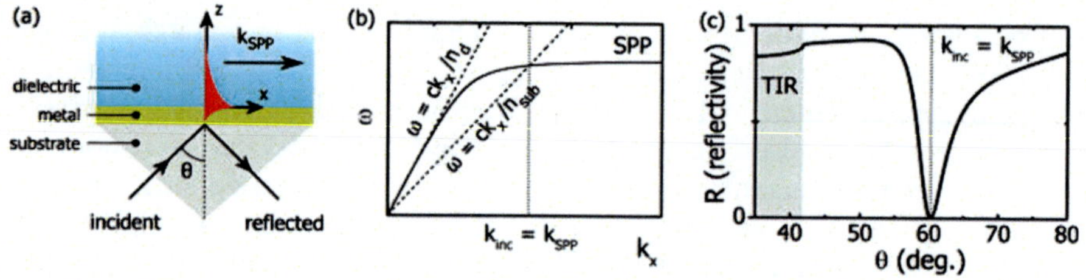

Fig. 2.1 SPR principle of operation. **a** Experimental implementation for SPR measurements (Kretschmann configuration). **b** Explanation of SPP excitation. The glass substrate is used to provide the necessary tangential wavenumber (k_{inc}) to match that of the SPP (k_{SPP}). This is shown as a crossing between the SPP dispersion (solid) and the light line in glass (dashed) (the light line in the dielectric is also shown). **c** Excitation of surface plasmons in the Kretschmann configuration for a Au-H_2O interface at 633 nm. The shaded area marked with 'TIR' (Total Internal Reflection) denotes the region below the critical angle (41.8°)

the x-direction at the interface between the metal and the dielectric medium located above; the evanescent field is confined in the z-direction, as also shown schematically in Fig. 2.1a. To excite the SPP wave, first an incident wave must match the polarization properties of the SPP and, therefore, a TM(p)-polarized wave is required (components E_x, H_y, E_z). Second, the incident wave must also match the tangential wavenumber of the SPP, k_{SPP}; this is provided by the substrate, as shown schematically in Fig. 2.1a and explained in Fig. 2.1b. There it is shown that the dispersion of the SPP lies below the lightline in the dielectric ($k_{SPP} > k_d$; k_d: wavenumber in the dielectric) and, hence, the incident wave must have a high tangential wavenumber k_{inc} to match k_{SPP} and achieve efficient power transfer to the SPP wave. This can be achieved, for example, via the incident angle θ in an angle-resolved experiment, as shown in Fig. 2.1c. For this example the wavelength of the incident wave is 633 nm, a typical wavelength employed in SPR spectroscopy, and the calculations have been performed for H_2O on a 50 nm Au layer and a prism of refractive index $n_{sub} = 1.5$ used as substrate. By expressing the tangential wavenumber of the incident wave as $k_{inc} = k_0 \cdot n_{sub} \cdot \sin\theta$, where k_0 is the free-space wavenumber, it becomes clear that k_{inc} can be controlled by both by θ and n_{sub}. Therefore, as θ is scanned, maximum power transfer from the incident wave to the SPP wave can be achieved at a certain angle where the condition $k_{inc} = k_{SPP}$ is met. At this angle, the excitation of the SPP wave becomes the most efficient and the reflected optical power is therefore minimized. This is manifested as a dip in the angle-resolved measured reflection.

The CHISPR sensing scheme is an extension of the typical SPR configuration, in which the dielectric layer is replaced by a chiral medium. Therefore, to understand the principles of CHISPR it is instructive first to examine the properties of SPPs at a metal-dielectric interface and how these are modified when the dielectric layer becomes chiral.

2.2.1 SPPs at a Metal-Dielectric Interface

The propagation characteristics of SPPs and their associated effects on the SPR have been widely discussed in many textbooks [41–43] and papers [42, 44–52]. To find the analytical form of the SPP, one starts with solving Maxwell's equations for the simple case of a flat interface between two semi-infinite spaces (conductor and dielectric), i.e. (2.2)–(2.5) with $\kappa = 0$. Although realistic systems do not involve semi-infinite material regions, the results are directly applicable to metallic films of finite thickness, because the field penetration inside the metallic region is usually much smaller than the thickness of the finite metallic layer (i.e., the film is is seen by the wave as having effectively infinite thickness).

Let us assume that the metallic layer extends along the xy-plane and SPP propagation occurs along the x-direction, with the evanescent field being confined in the z-direction, as shown schematically in Fig. 2.2. Both regions are non-magnetic ($\mu_r = 1$) and we may denote the relative permittivity ϵ_r of the metal and dielectric as ϵ_m and ϵ_d, respectively.

Due to the homogeneity of the geometry along the y-direction ($\partial/\partial y = 0$), Maxwell's equations are decomposed in two sets, one involving only H_x, E_y, H_z components (TE mode) and one involving only E_x, H_y, E_z components (TM mode). As also shown in [42], the SPP is a TM mode (it cannot exist for TE polarization) and, assuming propagation along the x-direction with propagation constant k_{SPP}, $H_y \sim e^{ik_{\text{SPP}}x}$, ($\partial/\partial x = ik_{\text{SPP}}$), the y-component satisfies the wave equation in both regions:

$$\frac{\partial^2 H_y}{\partial z^2} + (k_0^2 \epsilon_{m,d} - k_{\text{SPP}}^2) H_y = 0, \tag{2.6}$$

where $\epsilon_{m,d}$ denotes ϵ_m or ϵ_d, depending on the material region this equation refers to. The solution for both sub-spaces is written in the form:

Fig. 2.2 Properties of SPPs at a metal-dielectric interface. **a** Geometry for SPP propagation **b** Field components of SPP wave at a Au-H_2O interface at 633 nm. The electric field components are normalized with ζ the wave impedance in H_2O and the penetration distance z is normalized with $\lambda_{\text{SPP}} = 2\pi/\text{Re}(k_{\text{SPP}})$, the SPP wavelength. The SPP is studied analytically in the configuration shown in **a**, however the results are directly applicable to real SPR experiments with metallic films of finite size, as the field penetration inside the metal is usually much smaller than its thickness

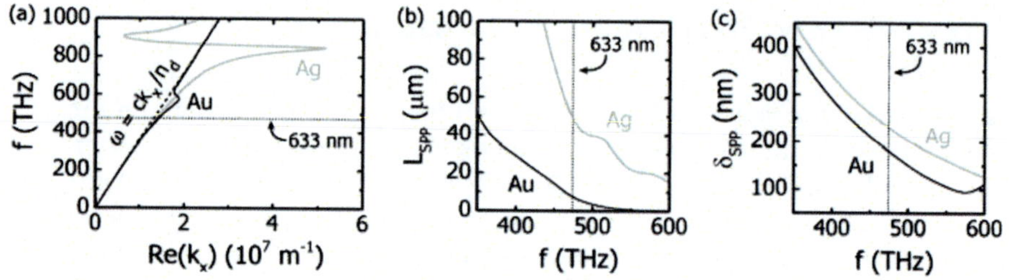

Fig. 2.3 SPP properties at a single interface between Ag-H_2O (grey lines) or Au-H_2O (black lines). **a** Dispersion relation. **b** Propagation length, L_{SPP}. **c** Penetration depth, δ_{SPP}. In all cases the dotted line marks the operation wavelength of 633 nm

$$H_y(z) = \begin{cases} H_0 e^{ik_{SPP}x} e^{-k_d z}, \ z \geq 0 \\ H_0 e^{ik_{SPP}x} e^{+k_m z}, \ z < 0 \end{cases}, \tag{2.7}$$

with

$$k_{m,d} = \sqrt{k_{SPP}^2 - k_0^2 \epsilon_{m,d}}, \tag{2.8}$$

where H_0 is a complex constant common to both branches in (2.7)—satisfying the continuity of H_y at the interface—and $k_{m,d}$ is the wavenumber in the perpendicular direction that expresses the confinement in the metal (subscript 'm') or dielectric (subscript 'd') (the remaining E_x, E_z components are directly calculated from H_y using Maxwell's equations). The continuity of the tangential electric field, E_x, dictates that $k_d/k_m = -\epsilon_d/\epsilon_m$, which in combination with (2.8) leads to the dispersion relation of SPPs propagating at the interface between the two half-spaces:

$$k_{SPP} = k_0 \sqrt{\frac{\epsilon_d \epsilon_m}{\epsilon_d + \epsilon_m}}. \tag{2.9}$$

This dispersion relation is associated with several important properties, see discussions in [44–46]. As an example, in Fig. 2.3a we plot (2.9) for a Ag-H_2O (grey line) and a Au-H_2O (black-line) interface (the material parameters for Ag and Au taken from [53], and for H_2O we use a constant $n_d = \sqrt{\epsilon_d} = 1.33$, the refractive index of water in the visible). The qualitative difference between the dispersion in Fig. 2.1b and both curves in Fig. 2.3a (i.e., the absence of a horizontal asymptote) is due to the high metallic losses of Ag and Au, which damp the propagating SPPs (see [45, 50, 54] for further details). The propagation distance of a SPP, L_{SPP}, is usually defined as the distance over which the mode can propagate along the supporting interface until the field amplitudes drop to $1/e$ of their initial magnitude, i.e. $L_{SPP} = 1/\text{Im}[k_{SPP}]$ [50]. This is shown in Fig. 2.3b for the two cases considered here (L_{SPP} is typically between 10 and 100 μm in the visible regime, depending on the particular metal/dielectric material properties). In relevance to the SPR experiments that we are concerned here, the most important feature of (2.9) is that the dispersion branch related to the propagating SPP *(a)* lies below the lightline (i.e. the SPP is confined to the interface) and *(b)* acquires high k-values with increasing frequency (i.e. the confinement becomes stronger); note that the SPP fields in the dielectric fall off as $e^{-|k_d||z|}$ with $k_d = \sqrt{k_{\text{SPP}}^2 - k_0^2\epsilon_d}$ [see (2.7), (2.8)]. The field-penetration depth $\delta_{\text{SPP}} = 1/\text{Im}[\text{k}_\text{d}]$, i.e. the distance over which the field amplitudes drop to $1/e$ of their initial magnitude, is shown in Fig. 2.3c for both cases. As shown, δ_{SPP} is in the range of a few nm and, therefore, SPPs are ideal for sensing material changes particularly close to the metal-dielectric interface, such as those from thin-subwavelength in size-films.

2.2.2 SPPs at a Metal-Chiral Interface

The introduction of chirality in the dielectric medium that comprises the metal-dielectric system imposes changes in the features of the SPP, as examined in detail in the work of Mi and Van [51]. With chirality, SPPs are still supported at a metal-chiral interface, however both their field components and their dispersion change. Let us consider such an interface, where the chiral medium is characterized by permittivity ϵ_c, permeability μ_c, and chirality (Pasteur) parameter κ. Chirality introduces magneto-electric coupling by means of κ and, as a result, in the chiral medium the fields satisfy the coupled wave equations:

$$\nabla^2 \begin{bmatrix} \mathbf{E} \\ \mathbf{H} \end{bmatrix} + k_0^2(n_c^2 + \kappa^2) \begin{bmatrix} \mathbf{E} \\ \mathbf{H} \end{bmatrix} + 2k_0\kappa \begin{bmatrix} +i\omega\mu_c\mathbf{H} \\ -i\omega\epsilon_c\mathbf{E} \end{bmatrix} = \begin{bmatrix} 0 \\ 0 \end{bmatrix}, \tag{2.10}$$

where $n_c = \sqrt{\epsilon_c\mu_c}$ is the average (background) refractive index of the chiral medium. For propagation along the x-direction [$\mathbf{E}, \mathbf{H} \sim e^{ik_{\text{SPP}}x}$ $(\partial/\partial x = ik_{\text{SPP}})$] and homogeneity along the y-direction $(\partial/\partial y = 0)$, this system simplifies to:

$$\frac{\partial^2}{\partial z^2} \begin{bmatrix} \mathbf{E} \\ \mathbf{H} \end{bmatrix} + (k_0^2(n_c^2 + \kappa^2) - k_{\text{SPP}}^2) \begin{bmatrix} \mathbf{E} \\ \mathbf{H} \end{bmatrix} + 2k_0\kappa \begin{bmatrix} +i\omega\mu_c\mathbf{H} \\ -i\omega\epsilon_c\mathbf{E} \end{bmatrix} = \begin{bmatrix} 0 \\ 0 \end{bmatrix}. \tag{2.11}$$

From the form of (2.11) it is evident that due to κ, each E-field component now couples with the respective H-field component and, therefore, all three components of the electric and magnetic field exist in both the metal and the chiral medium. The requirement for continuity of the tangential field components leads to the dispersion relation:

$$\left(\frac{\zeta_c k_{z,c+}}{k_{c+}} + \frac{\zeta_m k_{z,m}}{k_m}\right)\left(\frac{k_{z,c-}}{\zeta_c k_{c-}} + \frac{k_{z,m}}{\zeta_m k_m}\right) + \left(\frac{\zeta_c k_{z,c-}}{k_{c-}} + \frac{\zeta_m k_{z,m}}{k_m}\right)\left(\frac{k_{z,c+}}{\zeta_c k_{c+}} + \frac{k_{z,m}}{\zeta_m k_m}\right) = 0, \tag{2.12}$$

where $\zeta_c = \sqrt{\mu_c/\epsilon_c}$, $\zeta_m = \sqrt{\mu_m/\epsilon_m}$ is the wave impedance in the chiral medium and the metallic region, respectively and

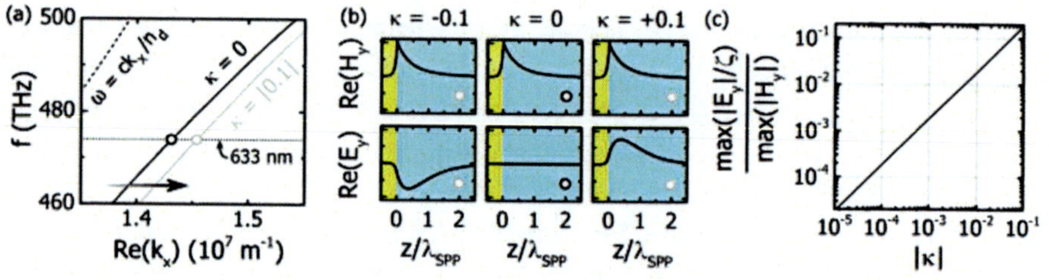

Fig. 2.4 Chiral SPP properties at a single Au-H_2O interface. **a** Dispersion relation for $\kappa = 0$ (black line) and $\kappa = \pm 0.1$ (grey line). Chirality shifts the dispersion to higher in-plane wavenumbers, independent of the sign of κ. This is also seen at the points marked with open circles, denoting operation at 633 nm. **b** Plots of H_y and E_y for $\kappa = -0.1$, $\kappa = 0$ and $\kappa = +0.1$. Chirality induces a non-vanishing E_y-component, with phase depending on the sign of κ. The open dots correspond to the dispersion points shown in **a**. **c** Magnitude of chiral-induced E_y-component in terms of $|\kappa|$, normalized with the magnitude of H_y-component (ζ: wave impedance in water)

$$k_{z,c\pm} = \sqrt{k_{\text{SPP}}^2 - k_{c\pm}^2},\, k_{c\pm} = k_0 n_\pm, \tag{2.13}$$

$$k_{z,m} = \sqrt{k_{\text{SPP}}^2 - k_m^2},\, k_m = k_0 n_m, \tag{2.14}$$

with $n_\pm = \sqrt{\epsilon_c \mu_c} \pm \kappa$, $n_m = \sqrt{\epsilon_m \mu_m}$ being the refractive index in the chiral medium and the metallic region, respectively.

This is the dispersion relation of SPPs propagating at a chiral-metal interface. Note that, in the limit $\kappa \to 0$, $k_{z,c+} \to k_{z,c-} \equiv k_{z,c}$, and the above relation reduces to the familiar dispersion of SPPs at an achiral dielectric-metal interface, which supports only the TM polarization (in this case ϵ_c is simply ϵ_d, as considered previously for the typical metal-dielectric interface). The chiral dispersion relation is also presented in the work of Mi and Van [51], however in a slightly different notation, due to the different formulation of the constitutive relations [52] (there, the chirality parameter is denoted as ξ and is related to κ as $\kappa = \zeta_0 \xi$, where ζ_0 is the vacuum wave impedance).

Using this dispersion relation, in Fig. 2.4 we examine the SPP properties at a single Au-H_2O interface, where a chiral medium is dispersed in the water region. To emphasize our findings, we assume a large chirality parameter ($\kappa \sim 0.1$). In Fig. 2.4a we present plots of the dispersion relation for $\kappa = 0$ (black line) and $\kappa = \pm 0.1$ (grey line), which demonstrate how chirality induces a shift to higher in-plane wavenumbers (k). This is also seen at the points marked with open circles, denoting operation at 633 nm; therefore, in an angle-resolved SPR experiment, chirality is expected to manifest as a shift of the reflectance dip at higher angles. Note that the k-shift does not depend on the sign of κ and, therefore, a typical SPR measurement cannot distinguish between left and right enantiomers. For the dispersion points marked with the open circles in Fig. 2.4a, in Fig. 2.4b we plot Re(H_y) and Re(E_y). We present individually three cases, namely for $\kappa = -0.1$, $\kappa = 0$ and $\kappa = +0.1$. From this plot it is evident that chirality modifies the pure-TM character of the SPP, thereby inducing a non-vanishing E_y-component. Most importantly, the phase of E_y depends on the sign of κ; as we will see in the next section, this is a key element to our CHISPR protocol. Last, in Fig. 2.4c we plot the magnitude of E_y in terms of the magnitude of κ. To quantify the results with respect to the magnitude of H_y, we normalize E_y with the wave impedance ζ in water. The plot is shown in logarithmic scale, starting from realistic values of $|\kappa|(\sim 10^{-5})$ up to $|\kappa| = 0.1$, the value used in the examples of Fig. 2.4a, b. The linearity of this scaling is a second key element to our CHISPR protocol, as we will discuss in Sect. 2.3.

2.3 CHISPR

To demonstrate the principles of CHISPR we start by considering a standard angle-resolved SPR setup in the Kretschmann configuration [40] where a Au layer is deposited on a prism surface upon which a chiral substance dispersed in water is placed. A realistic rendering of such a setup is shown in Fig. 2.5.

To excite the SPP wave at the metal-chiral interface, a TM (p)-polarized wave is incident from the prism side (Figs. 2.1 and 2.5). For the case of a single metal-chiral interface we saw previously that the presence of the chiral layer qualitatively changes the SPP wave by generating an s-wave and by shifting the SPP dispersion to higher in-plane wavenumbers. Therefore, the properties of the chiral layer itself should be observable through angle-resolved SPR measurements; as discussed in the previous section, these are expected as an angular shift in the reflectance dip. In Fig. 2.6 we present the results of a simulated angle-resolved SPR experiment in the presence

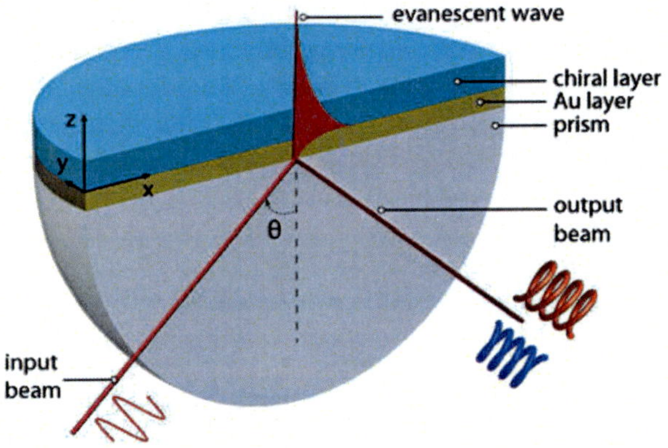

Fig. 2.5 CHISPR experimental setup (Kretschmann configuration) for the detection of chirality from thin (sub-wavelength) chiral layers: A linearly TM(p) polarized beam incident on a thin gold layer (Au layer thickness ~50 nm) excites a surface plasmon polariton (SPP) (indicated by the evanescent wave) at a particular angle, θ, which propagates along the metal-chiral interface. The SPP wave is modified by the chiral environment, resulting in an outgoing optical chirality flux which can be used to infer the properties of the chiral layer (see text for details). Figure reprinted with permission from [10]. Copyright 2020 American Chemical Society

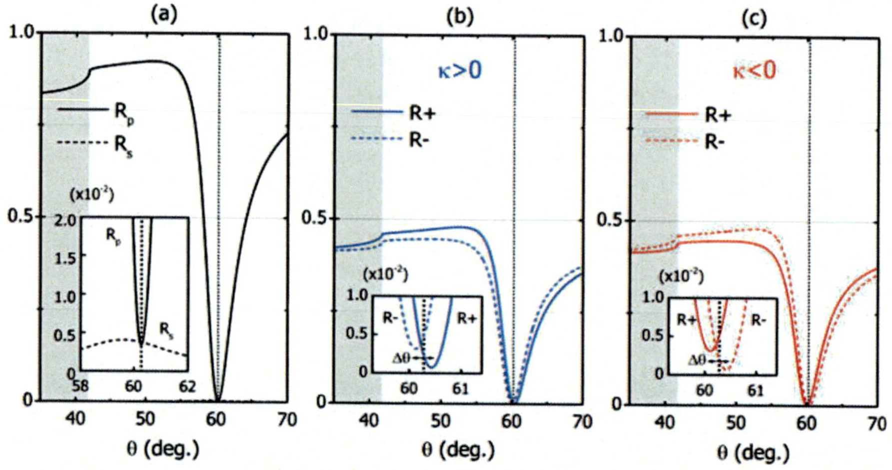

Fig. 2.6 SPR reflectance under the presence of a chiral layer ($n_c = 1.33$, 100 nm thickness). The SPR is excited with a TM(p)-polarized wave and the reflected power is analyzed into **a** p- and s-components, R_p, R_s, respectively (same for $\kappa = \pm 0.1$), and RCP (+) and LCP (-) components, R_+, R_-, respectively for **b** $\kappa = +0.1$ and **c** $\kappa = -0.1$. The effect of chirality appears in (a) as a enantio-independent angular shift of R_p, accompanied by nonzero R_s and in (b) & (c) as a chiral-dependent angular split ($\Delta\theta = \theta_+ - \theta_-$) between R_+ and R_-. The magnitude and sign of $\Delta\theta$ depends on $|\kappa|$ and sgn(κ), respectively. In all subplots, the vertical dashed lines denote the angle of minimum R_p, i.e. the SPR angle, and the shaded areas denote the region below the critical angle (41.8 deg)

of a thin chiral layer, where, to clarify our findings, we again use a large chirality parameter κ and consider both possibilities for the sign, i.e. $\kappa = \pm 0.1$.

We analyze the reflected wave in terms of s and p components and calculate the power at each polarization, namely R_s, R_p. Additionally, we analyze the total reflected power $R_s + R_p$ in terms of $+/-$ components, which we denote as $R_\pm = |r_\pm|^2$, where r_+ (r_-) is the complex amplitude of the RCP (LCP) wave (that is, $R_+ + R_- = R_s + R_p$). In an actual experiment, measurement of $R_{s/p}$ and $R_{+/-}$ can be easily performed with the incorporation of a Stokes polarimeter at the analysis stage.

In Fig. 2.6a we show the reflected power measured in terms of s/p waves, as is typically performed and presented in SPR experiments. The R_p curve has a pronounced reflection-dip at 60.3 deg, indicating the excitation of a SPP wave, while we also observe a nonzero R_s peaking at 59.5 deg [Fig. 2.6a, inset], as now part of the p-wave is transferred to the s-wave due to the presence of the chiral layer. We note here that, in accord with our analysis in Sect. 2.2.2, κ induces a shift on R_p towards larger angles and this shift is identical for both $\kappa = \pm 0.1$ [for $\kappa = 0$, the R_p reflection-dip is located at 60.1 deg, while $R_s = 0$, as also shown in Fig. 2.1c]. Thus, *measurement analysis based on the s/p waves cannot differentiate between left-handed and right-handed chiral substances*. We also note that this measurement modality has been used in previous works discussing the possibility of detecting chirality using SPR measurements [55, 56], but it is effectively insensitive to the sign of the chiral parameter κ.

When we now analyze the reflected wave in terms of RCP/LCP (+/−) components, we observe that the minima of the R_+, R_- reflectances do not coincide, but are separated by an angle $\Delta\theta \equiv \theta_+ - \theta_-$, where θ_+ (θ_-) denotes the angle of the R_+ (R_-) minimum [Fig. 2.6b, c]. Moreover, we observe that for $\kappa > 0$ ($\kappa < 0$), $\Delta\theta > 0$ ($\Delta\theta < 0$). Thus, the presence of a thin chiral layer results in a *chiral-dependent angular split* $\Delta\theta$ between the measured reflectances of R_+ and R_-, which has a distinct behaviour depending on the sign and magnitude of κ. We wish to note here that our polarimetric measurement scheme, where the reflected wave is analyzed in terms of +/− components, is sensitive only to chiral effects and not to general spectral shifts (possible if variations to the host refractive index are present), and moreover it is equivalent to the chirality flux spectroscopy used to probe the chiral near-fields of chiral nanosystems, as presented in [57].

2.3.1 Mechanism of Chiral-Dependent SPR-Reflectance Angular Split

To understand the mechanism behind the chiral-dependent SPR-reflectance angular split, we examine how the near-field properties of the SPP wave are associated with the properties of the reflected wave in the far-field. We start by analyzing the SPP wave along its propagation direction (x) into +/− components, i.e. $\mathbf{A}_{\mathrm{SPP}} = A_y\hat{y} + A_z\hat{z} = A^+_{\mathrm{SPP}}(\hat{y} + i\hat{z}) + A^-_{\mathrm{SPP}}(\hat{y} - i\hat{z})$, where $A^{\pm}_{\mathrm{SPP}} = (A_y \mp A_z)/2$ and $\mathbf{A}$ is any of the electromagnetic field quantities $\mathbf{E}$, $\mathbf{H}$, $\mathbf{B}$, $\mathbf{D}$; then, we calculate the electric and magnetic energy densities $w^{\pm}_e = (1/4)\mathbf{E}^{\pm}_{\mathrm{SPP}}(\mathbf{D}^{\pm}_{\mathrm{SPP}})^*$ and $w^{\pm}_m = (1/4)\mathbf{B}^{\pm}_{\mathrm{SPP}}(\mathbf{H}^{\pm}_{\mathrm{SPP}})^*$, respectively, which we integrate to find the total energy stored in each of the two (+/−) components, namely $W_{\pm} = \int_V (w^{\pm}_e + w^{\pm}_m)\,\mathrm{d}^3x$ [Fig. 2.7a]. Here, the integration volume V is the entire SPP volume extending above the metal (where the chiral layer is to be probed). For $\kappa = 0$ we obtain $W_+ = W_-$, as the SPP wave has only an E_z-component on the yz-plane, which is equally distributed between the two +/− components (typical nonchiral SPR case). This is shown in Fig. 2.7b, where the energy difference $W_+ - W_-$ is normalized to the incident energy $S_{\mathrm{inc}}/2\omega$ (ω is the angular frequency and S_{inc} is the magnitude of the time-averaged Poynting vector). However, the onset of chirality causes the emergence of an E_y-component [51] and, hence, an unbalanced storage of the optical energy between the +/− components of the SPP. In fact, for $\kappa > 0$ ($\kappa < 0$), RCP (LCP) components are favoured and, therefore, $W_+ > W_-$ ($W_+ < W_-$) [Fig. 2.7b]. This stored energy excess between +/− SPP components in the near-field results in nonzero R_s reflectance in the far-field; this is apparent in the fact that the peak of R_s coincides with the peak of $W_+ - W_-$ at 59.5 deg which differs from the R_p minimum at 60.3 deg [Figs. 2.6a and 2.7b]. In other words, the E_y-component that emerges in the near-field due to chirality, is identified in the far-field as well, as power transfer from the outgoing p-wave

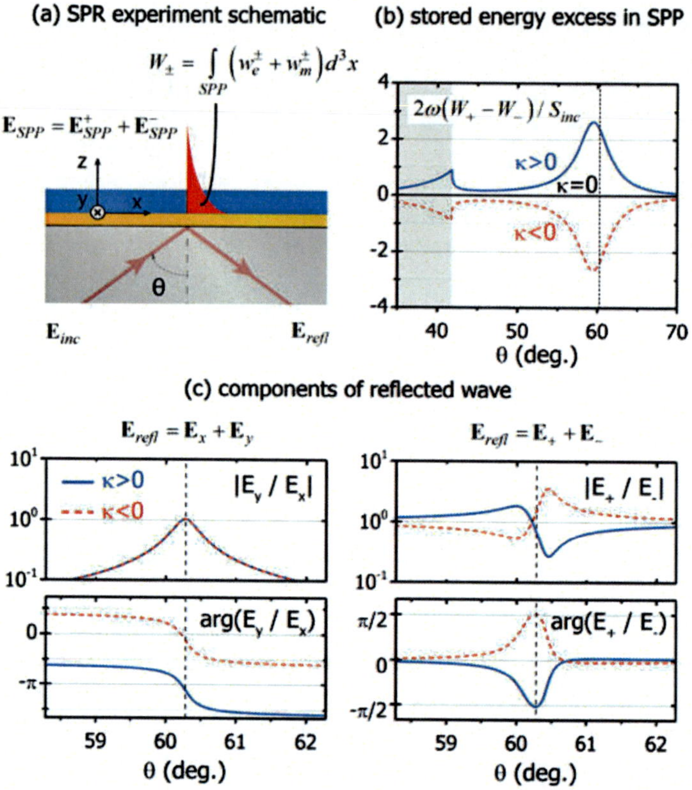

Fig. 2.7 Mechanism of the R_+, R_- angular split. **a** The incident wave excites the SPP, which we analyze in RCP/LCP (+/−) components along its propagation direction. The total energy stored in each component is denoted as $W_\pm$ (see also text for details). **b** Stored energy difference between RCP/LCP components of the SPP wave (normalized over the incident energy $S_{\text{inc}}/2\omega$). Chirality causes energy excess between RCP/LCP waves, which changes sign upon sign change of κ ($\kappa = \pm 0.1$). **c** Ratio of amplitude and phase of the reflected wave components for $\kappa = +0.1$ (solid blue lines), $\kappa = -0.1$ (dashed red lines). We analyze the wave in terms of s/p components (left) and RCP/LCP components (right). We present the amplitude in logarithmic scale to emphasize the inversion symmetry of the ratio $|E_+/E_-|$ upon sign change in κ. The vertical dashed line denotes the angle of minimum R_p, i.e. the SPR angle. Figure adapted with permission from [10]. Copyright 2020 American Chemical Society

to the s-wave. When the total reflected wave in the far-field is analyzed in +/− components as well, the reflectance splits into two parts, which have their minima at different angles [Fig. 2.6b, c]. This angular split is mediated by the resonance of the surface plasmon; the amplitude of the E_y/E_x ratio is symmetric around the SPR angle [Fig. 2.7c], however, the phase $\arg(E_y/E_x)$ undergoes a π-shift, favouring the advance of either the E_x or the E_y component, depending on whether the angle of incidence is below or above the SPR angle [Fig. 2.7c]. Consequently, the mixture of the reflected RCP and LCP wave-components is weighted differently, resulting in an excess of either RCP or LCP waves below or above the SPR angle and, hence, a reflectance split between R_+ and R_- waves. As for the magnitude of κ, it does not significantly affect the $\arg(E_y/E_x)$ (it induces a slight angular shift), however, it notably changes the amplitude of E_y/E_x, which increases with increasing κ. Changing the sign of κ induces a π-shift of the E_y phase, without affecting E_x [Fig. 2.7c]. Hence, the sign of κ does not affect the amplitude of the ratio E_y/E_x but causes the interchange between RCP/LCP components.

2.3.2 Sensitivity of Chiral-Dependent SPR-reflectance Angular Split

Because the resonance of the SPP wave depends strongly on the material parameters at the metal-chiral interface, it is expected that the strength of the SPR reflectance angular split $\Delta\theta$ will depend, besides the chirality parameter κ, also on the chiral substance's refractive index, n_c.

In Fig. 2.8 we show this dependence for the case of real κ (we discuss the case of imaginary κ in a following section). Overall, we observe *(a)* a linear dependence between the magnitudes of $\Delta\theta$ and κ, *(b)* a distinct correspondence between the signs of $\Delta\theta$ and κ, and *(c)* a non-monotonic dependence of $\Delta\theta$ on n_c. This non-monotonic dependence is related to the interplay between the coupling strength of the incident wave to the SPP wave and the interaction strength of the SPP wave with the chiral layer. In particular, as n_c increases, the SPR dispersion changes and the reflection-dip is shifted to higher angles due to higher k_{SPP} [Fig. 2.8a]. In turn, the coupling of the incident wave to the SPP becomes stronger, leading to higher $\Delta\theta$ and, hence, in increased sensitivity [Fig. 2.8b]. Eventually, for very high incident angles the coupling of the incident wave to the SPP becomes weaker, leading to weaker $\Delta\theta/\Delta\kappa$ accordingly. In addition, for angles close to the critical angle the effect becomes the weakest. In reflection-based polarimetric measurements, the optical rotation signals scale as $\sim\sqrt{\text{Re}(\kappa)}$ when approaching the critical angle (see [23, 24, 58, 59]), which is not the case here [Fig. 2.8b]. This is a consequence of the fact that the measurement is mediated entirely by the SPP wave and is not associated with direct polarimetric signals from the chiral layer. Thus, by measuring the magnitude and sign of this chiral-dependent angular split, we obtain information about the magnitude and sign of κ.

Furthermore, because the thickness of the chiral film is finite, the SPP interacts both with the chiral film and the dielectric region above (air). In essence, the SPP experiences an effective index in the chiral-air region, which depends on the chiral film thickness; with increasing film thickness, the evanescent tails of the SPP interact less and less with the air above, until this effective index converges to n_c. In our simulations so far we considered a chiral layer of 100 nm thickness, which is in the order of the calculated field penetration depth [see Fig. 2.3b]. For this reason, in Fig. 2.9 we repeat the calculations of the measurement sensitivity ($\Delta\theta/\Delta\kappa$) pre-

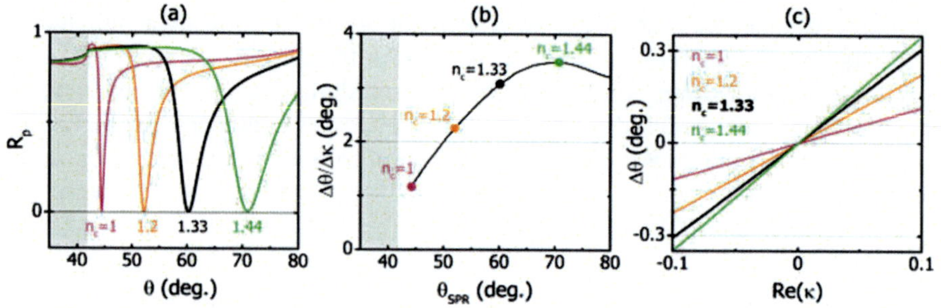

Fig. 2.8 Measurement sensitivity of $\Delta\theta$ on the chirality parameter κ. **a** Angle-resolved R_p reflectance denoting the SPR angle ($\theta_{\rm SPR}$) for the shown selected values of n_c with $\kappa = 0$. **b** $\Delta\theta/\Delta\kappa$ vs SPR angle. The solid black line represents a multitude of individual calculations, on which the cases for n_c shown in (a) are marked with dots of the same colour. **c** $\Delta\theta$ vs κ for the selected values of the host index n_c. The solid black line denotes the system of Fig. 2.6 with $n_c = 1.33$

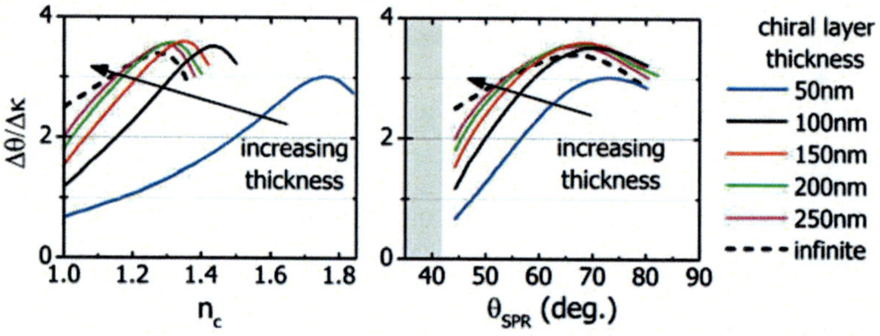

Fig. 2.9 $\Delta\theta/\Delta\kappa$ as a function of the chiral-layer thickness, for $\kappa = 0.1$. The dashed black line represents the results shown in Fig. 2.8

sented in Fig. 2.8b, for chiral layers of variable thickness. We see that, with increasing chiral layer thickness, the measurement sensitivity converges to the limit of chiral substances of theoretically infinite extent (practically referring to electrically thick samples). In addition, we observe that for small SPR angles this increase is monotonic, but for large SPR angles the measurement sensitivity reaches a maximum level for a thickness of ~100–150 nm, beyond which it gradually drops until convergence. Therefore, we see that due to the evanescent character of the SPP wave inside the chiral region, one can achieve through measurements of $\Delta\theta$ similar levels of sensitivity for a large range of chiral-layer thicknesses.

2.3.3 Differential Measurements

In the previous subsections we demonstrate how the presence of a thin chiral layer results in a *chiral-dependent angular split* between the measured reflectances of R_+

and R_{-}, which has a distinct behaviour depending on the sign and magnitude of κ. However, as we show in Fig. 2.8, for small values of κ ($\kappa < 10^{-3}$), this angular split, $\Delta\theta$, becomes similarly small and its detection can be hindered by background noise sources. Despite this, there exist alternative measurements one can consider performing in a CHISPR scheme, thanks to the ability to obtain distinct chiral-dependent signals. In particular, we can consider measurement configurations based on differential signals which are largely immune to signal fluctuations and drifts (a direct analogy is the case of CD measurements, which are differential type of measurements). Specifically, we consider two relevant quantities associated with the reflected (outgoing) RCP/LCP waves: the amplitude and phase differential reflectances (DR), namely ρ_{DR} and ϕ_{DR}, respectively. We define the DR signals as,

$$\rho_{\mathrm{DR}} = \frac{|r_{+}|^2 - |r_{-}|^2}{|r_{+}|^2 + |r_{-}|^2}, \text{ and } \phi_{\mathrm{DR}} = \mathrm{Arg}\left[\frac{\mathrm{r}_{+}}{\mathrm{r}_{-}}\right]. \tag{2.15}$$

In Fig. 2.10 we present the DR signals, ρ_{DR} and ϕ_{DR}, respectively, for a value of $\kappa = \pm 10^{-5}$ as a function of the background index of the chiral layer, n_c. We

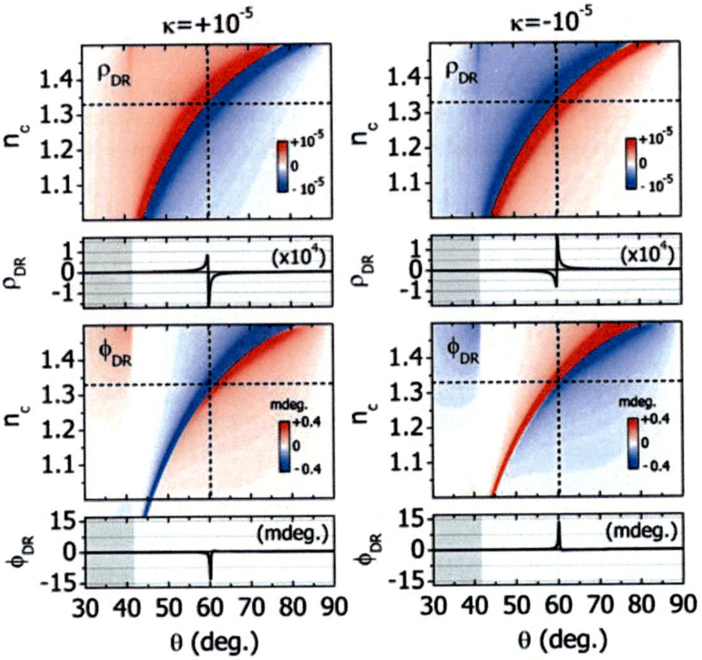

Fig. 2.10 Differential reflectance (DR) signals for a 100 nm thin chiral layer with $\kappa = \pm 10^{-5}$, as a function of the background index of the chiral layer (n_c). The shown range is $\sim(-6.3 \ldots + 6.3) \times 10^{-2} \times \max(\rho_{\mathrm{DR}})$ in the ρ_{DR} plots and $\sim(-2.7 \ldots + 2.7) \times 10^{-2} \times \max(\phi_{\mathrm{DR}})$ in the ϕ_{DR} plots to emphasize the broadening with increasing SPR angle and the distinct association with the sign of κ. Specific examples for $n_c = 1.33$ are marked with horizontal dashed lines and are shown separately below each panel. The vertical dashed line marks the SPR angle for the chosen value of n_c

emphasize here that this is a realistic value for the chirality parameter, corresponding, for example, to the case of aqueous solutions of monosaccharides [23, 24] or biomolecules [34–36]. For such a realistic value of κ, therefore, we observe ρ_{DR} signals of the order of $\sim 10^{-4}$ and ϕ_{DR} signals of the order of a few $\sim$mdegs, both within the sensitivity range of SPR instruments [60–62]. As a comparison, we note that the optical rotation signal from a transmission measurement of a 100 nm chiral layer with $\kappa = +10^{-5}$ at 633 nm, is $\varphi \simeq 6 \times 10^{-4}$ deg (2.1). Furthermore, we observe that $\rho_{\mathrm{DR}}(-\kappa) = -\rho_{\mathrm{DR}}(\kappa)$ and $\phi_{\mathrm{DR}}(-\kappa) = -\phi_{\mathrm{DR}}(\kappa)$. Thus, using the DR-dependent signals, we are able to quantify $\mathrm{Re}(\kappa)$ (magnitude and sign) with increased sensitivity compared to measurements of the angular split, $\Delta\theta$ [61, 63, 64].

Another important feature of the differential signals, ρ_{DR} and ϕ_{DR}, is that these decrease in amplitude as the plasmon resonance moves away from the critical angle, contrary to $\Delta\theta/\Delta\kappa$ which we observe to increase (Fig. 2.8). This decrease is related to the broadening of the SPR feature due to increased losses for higher k_{SPP}, and to the reduction of the R_s/R_p ratio, which expresses the strength of the p- to s-wave conversion. In particular, in Fig. 2.11 we show the change in the R_s/R_p ratio with increasing SPR angle (due to increasing n_c). We observe that the R_s/R_p ratio decreases while simultaneously broadening, which yields, thus, reduced differential signals. Moreover, the peak-to-peak values of ρ_{DR} and ϕ_{DR} [$\Delta\rho_{\mathrm{DR}}$ and $\Delta\phi_{\mathrm{DR}}$, respectively; Fig. 2.11b], qualitatively follow a similar trend indicating a strong connection with the strength of R_s/R_p. Furthermore, we observe that the variation of R_s/R_p (and consequently of $\Delta\rho_{\mathrm{DR}}$ and $\Delta\phi_{\mathrm{DR}}$) is non-monotonic and it generally depends on the properties of the particular metal. Despite these, it is apparent that regardless of the exact variation, the differential signals of ρ_{DR} and ϕ_{DR} allow for unambiguous determination of κ.

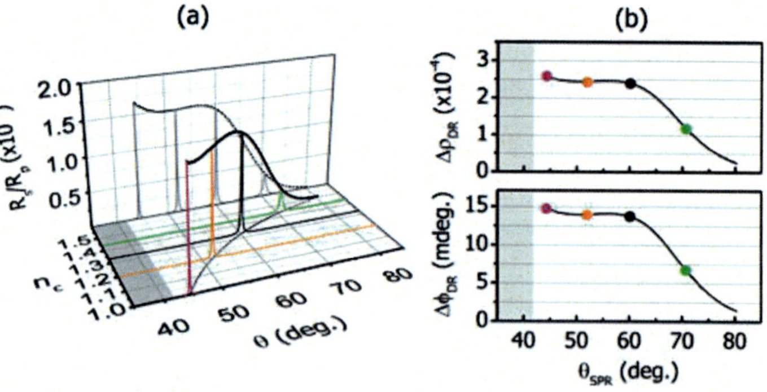

Fig. 2.11 Effect of coupling strength between $s-$ and $p-$waves due to chirality on differential reflection measurements. **a** R_s/R_p ratio as function of n_c for $\kappa = \pm 10^{-5}$. **b** Peak-to-peak values of the differential signals ρ_{DR} and ϕ_{DR} [$\Delta\rho_{\mathrm{DR}}$ (top) and $\Delta\phi_{\mathrm{DR}}$ (bottom), respectively]. The solid black lines represent a multitude of individual calculations, on which the cases for n_c shown in **a** are marked with dots of the same colour. In addition, the marked cases in **a**, **b** correspond to the cases shown in Fig. 2.8 using the same colour-code

2.4 Complete Measurement of Chirality

So far we have examined CHISPR signals assuming only a real-valued chirality parameter κ. Under realistic experimental conditions, this is a valid approximation for interpreting the results when one performs measurements at optical frequencies far detuned from any molecular resonances where chiral-dependent differential absorption (i.e. circular dichroism), which is proportional to Im(κ), is negligible [11]. However, when the optical frequency of an SPR instrument is near a molecular transition, both circular birefringence and circular dichroism [proportional to Re(κ) and Im(κ), respectively] become substantial and the proposed measurements should be interpreted with care.

To examine circular dichroism effects we introduce a nonzero imaginary part in the chirality parameter κ and demonstrate its effect on the CHISPR signals. However, to appropriately examine the case of circular dichroism (by introducing an imaginary part in the chirality parameter κ) without violating the passivity, we must ensure that $\text{Im}(n_c \pm \kappa) > 0$. For this reason, we introduce artificial loss in the average refractive index of the chiral layer, i.e. $n_c = 1.33 + 0.01\,i$.

We start by demonstrating how each CHISPR signal ($\Delta\theta$ and DR signals) change under the influence of an imaginary-valued chirality parameter. In Fig. 2.12a we present the chiral-dependent angular split $\Delta\theta$ as a function of Re(κ) for $\text{Im}(\kappa) = -10^{-3}, 0, 10^{-3}$. We see that in the presence of absorption [$\text{Im}(\kappa) \neq 0$] the angular split $\Delta\theta$ obtains a *linear chiral-dependent offset*, and the effects of Re(κ) and Im(κ) appear as linear superpositions in the total $\Delta\theta$. Moreover, in Fig. 2.12b we present the change in the differential signals, ρ_{DR} and ϕ_{DR}, in the presence of absorption. In particular, we calculate ρ_{DR} and ϕ_{DR} for $\kappa = 10^{-5}$, $\kappa = 10^{-5}\,i$, and the sum of the two individual signals. We also present the same signals for the case of $\kappa = 10^{-5} + 10^{-5}\,i$, which we show to coincide with the sums of the individual signals [Fig. 2.12b]. In overall, we see that in the presence of absorption and birefringence, the effects of the real and imaginary parts of the chirality parameter appear as linear superpositions in the final CHISPR signals ($\Delta\theta$ and DR signals).

In Fig. 2.13 we examine the resulting DR signals for the cases of both chiral-dependent refraction, $\kappa = \pm 10^{-5}$, and absorption, $\kappa = \pm 10^{-5}\,i$. From the individual simulations we observe a clear distinction in the DR signals between the four cases, that is, CHISPR enables the detection of the magnitude and sign of both the real and imaginary part of the chirality parameter, and discrimination of their contribution through the distinct CHISPR signals. In combination with the results shown in Fig. 2.12, it becomes apparent that in the case where the SPR operational wavelength is near the vicinity of a molecular resonance, where circular dichroism is accompanied by an dispersive circular birefringence, i.e Re(κ) & Im(κ) $\neq 0$ (Cotton effect [11]), the resulting DR signals will be the result of a linear superposition of the individual signals for the real and the imaginary part of the total chirality parameter (as these are presented in Fig. 2.13).

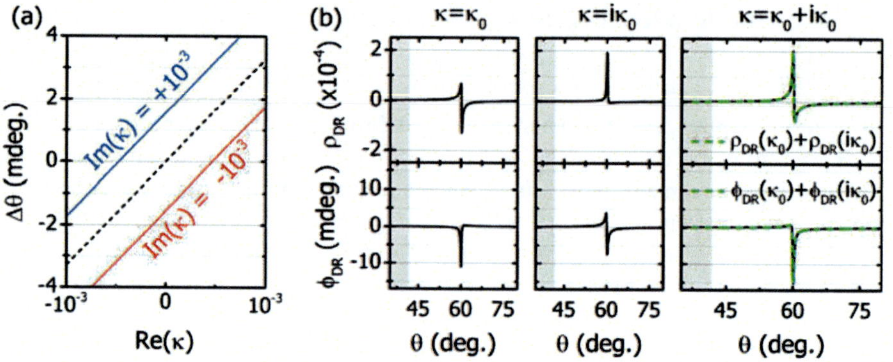

Fig. 2.12 CHISPR measurements in the presence of molecular absorption, i.e. $\mathrm{Im}(\kappa) \neq 0$. **a** $\Delta\theta$ as a function of $\mathrm{Re}(\kappa)$ for $\mathrm{Im}(\kappa) = \pm 0.001i$ [dashed line corresponds to $\mathrm{Im}(\kappa) = 0$]. To maintain the passivity of the system, we add artificial loss to the chiral layer index, which now is $n_c = 1.33 + 0.01i$. **b** Demonstration of linearity of the effects of $\mathrm{Re}(\kappa)$ and $\mathrm{Im}(\kappa)$ on the differential signals ρ_{DR} and ϕ_{DR}. The calculations have been performed for $\kappa = \kappa_0$ (left) and $\kappa = i\,\kappa_0$ (middle) separately, and for $\kappa = \kappa_0 + i\,\kappa_0$ (right), with $\kappa_0 = 10^{-5}$. Here we use: $n_c = 1.33 + 10^{-3}i$. Figure adapted with permission from [10]. Copyright 2020 American Chemical Society

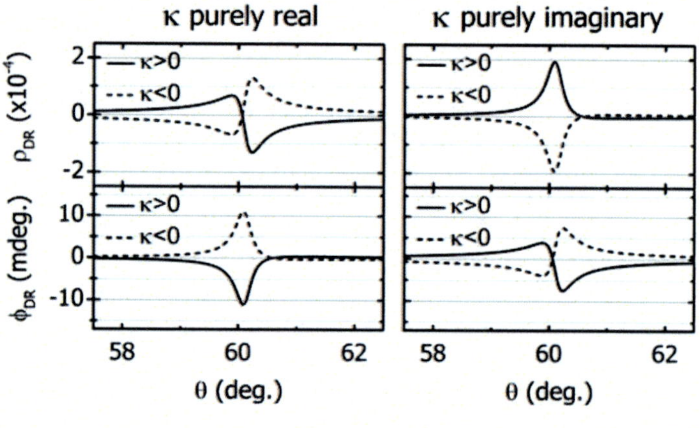

Fig. 2.13 Complete sensing of total chirality via measurements of the differential signals ρ_{DR} and ϕ_{DR}. *Left column*: $\kappa = \pm 10^{-5}$ (purely real). *Right column*: $\kappa = \pm 10^{-5}i$ (purely imaginary). Here we use: $n_c = 1.33 + 10^{-3}i$

2.5 Optical Chirality Conservation

The fact that the presence of the chiral layer modifies the evanescent field of the SPP, naturally raises the question of whether the reflectance measurements in the far-field can provide information about the chiral near-field features. To relate the two quantities, we utilize the conservation law of optical chirality density [65–70], which in the time-averaged, time-harmonic case is written as:

$$- 2\omega \int_V \mathrm{Im}(\chi_e - \chi_m)\, \mathrm{d}^3x + \int_V \mathrm{Re}(\nabla \cdot \mathbf{F})\, \mathrm{d}^3x = 0, \tag{2.16}$$

where χ_e and χ_m are the electric and magnetic optical chirality densities, respectively, and $\mathbf{F}$ is the corresponding chirality flux:

$$\chi_e = \frac{1}{8}\Big[\mathbf{D}^* \cdot (\nabla \times \mathbf{E}) + \mathbf{E} \cdot (\nabla \times \mathbf{D}^*)\Big], \tag{2.17}$$

$$\chi_m = \frac{1}{8}\Big[\mathbf{H}^* \cdot (\nabla \times \mathbf{B}) + \mathbf{B} \cdot (\nabla \times \mathbf{H}^*)\Big], \tag{2.18}$$

$$\mathbf{F} = \frac{1}{4}\Big[\mathbf{E} \times (\nabla \times \mathbf{H}^*) - \mathbf{H}^* \times (\nabla \times \mathbf{E})\Big]. \tag{2.19}$$

Similarly to as we performed in Sect. 2.3.1, here we again analyze the SPP wave along its propagation direction (x) into $+/-$ components. For $\kappa \neq 0$ the integral of the total chirality density $\chi = \chi_e + \chi_m$ across the SPP volume, X, is unequally stored between the $+/-$ SPP components, i.e. $|X_+| \neq |X_-|$, where $X_\pm = \int_V (\chi_e^\pm + \chi_m^\pm)\, \mathrm{d}^3x$. Because X_+ and X_- are associated with waves of the opposite handedness, they possess opposite sign and, hence, the total X, which is the sum $X_+ + X_-$, is written as $|X_+| - |X_-|$, which expresses a chirality excess between the $+/-$ components. In Fig. 2.14 we plot the integrated optical-chirality-density excess $|X_+| - |X_-|$, and to directly relate it with the simulations we present in Fig. 2.6 we choose $\kappa = 0, \pm 0.1$. We see that the result is qualitatively similar to $W_+ - W_-$, as shown in Fig. 2.7b. Due to the chirality conservation law (2.16), this unbalance results in a chirality flux $\mathbf{F}$ in the far-field, manifested as unequal RCP and LCP components and observed through the angular split or the DR signals. In fact, as shown in [70] the chirality flux $\mathbf{F}$ of a certain propagating wave is proportional to the third Stokes parameter, and also related to the Poynting vector $\mathbf{S} = \frac{1}{2}(\mathbf{E} \times \mathbf{H}^*)$, via $F_\pm = \mp(\omega/c)S_\pm$ (where c is the speed of light in the medium, and $F_\pm$ and $S_\pm$ are the magnitudes of $\mathbf{F}$ and $\mathbf{S}$ with the signs $+/-$ corresponding to RCP/LCP waves, respectively). Therefore, in the CHISPR measurement protocol, there must be a connection between the measurable reflectances R_+, R_- and the far-field chiral quantities.

To find this connection, we start by analyzing the incident wave in $+/-$ components. We calculate the magnitudes of the power flux $S_{\mathrm{inc}}^\pm$ and chiral flux $F_{\mathrm{inc}}^\pm$ for each component and, because the incident wave is linearly polarized (p-wave), we find that these quantities are equally distributed between the $+/-$ components, i.e. $S_{\mathrm{inc}}^+ = S_{\mathrm{inc}}^- \equiv S_{\mathrm{inc}}/2$ and $F_{\mathrm{inc}}^+ = F_{\mathrm{inc}}^- \equiv F_{\mathrm{inc}}/2$, where S_{inc} and F_{inc} are the magnitudes of the total incident power and chiral flux, respectively. In fact, because the incident wave is linearly polarized, the total incident chirality flux is zero, i.e. $\mathbf{F}_{\mathrm{inc}} = \mathbf{F}_{\mathrm{inc}}^+ + \mathbf{F}_{\mathrm{inc}}^- = 0$. However, because the individual fluxes have nonzero magnitude (and equal; they correspond to circularly polarized waves of equal amplitude), we define the incident flux magnitude as $F_{\mathrm{inc}} = |\mathbf{F}_{\mathrm{inc}}^+| + |\mathbf{F}_{\mathrm{inc}}^-| = 2|\mathbf{F}_{\mathrm{inc}}^\pm| \equiv 2F_{\mathrm{inc}}^\pm$.

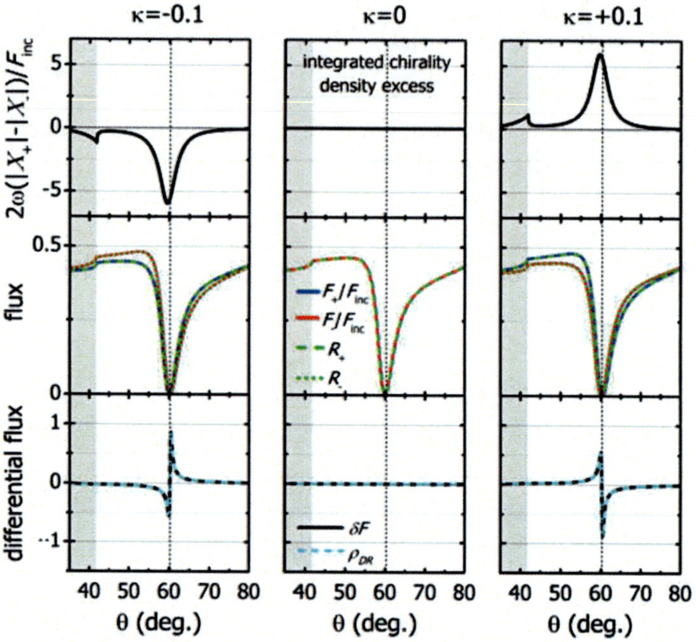

Fig. 2.14 Top row: Integrated optical chirality density excess $X_+ - X_-$ between RCP/LCP $(+/-)$ components of the SPP wave, normalized with the incident $F_{\text{inc}}/2\omega$. Middle row: Reflected chirality flux $F_\pm$ (solid blue/red lines), normalized with the incident chirality flux F_{inc}, and reflectance $R_\pm$ (dashed/dotted lines). Bottom row: differential flux δF and differential reflectance amplitude ρ_{DR}. For these simulations we use a 100 nm thin chiral layer with $n_c = 1.33$ and $\kappa = -0.1$ (left column), $\kappa = 0$ (middle column) and $\kappa = +0.1$ (right column). In all panels, the SPR angle (angle of minimum R_p) is marked with a vertical dashed line. Figure adapted with permission from [10]. Copyright 2020 American Chemical Society

Next, after following a similar analysis for the reflected wave, we associate the quantities related to both the incident and reflected $+/-$ components as,

$$F_{\text{inc}}^{\pm} = \mp\frac{\omega}{c}S_{\text{inc}}^{\pm}, \quad F_{\text{refl}}^{\pm} = \mp\frac{\omega}{c}S_{\text{refl}}^{\pm}, \tag{2.20}$$

and hence

$$\frac{F_{\text{refl}}^{\pm}}{F_{\text{inc}}^{\pm}} = \frac{S_{\text{refl}}^{\pm}}{S_{\text{inc}}^{\pm}} \Rightarrow \frac{F_{\text{refl}}^{\pm}}{F_{\text{inc}}} = \frac{S_{\text{refl}}^{\pm}}{S_{\text{inc}}} \equiv R_\pm, \tag{2.21}$$

or simply $F_\pm/F_{\text{inc}} = R_\pm$. As a result, we find that the reflected chirality flux $F_\pm$, normalized by the incident optical chirality flux F_{inc}, is equal to the reflectance $R_\pm$ in the CHISPR measurement protocol. To emphasize this equivalence, in Fig. 2.14 we present calculations of $F_\pm/F_{\text{inc}}$ and $R_\pm$ for the three cases considered, namely $\kappa = 0$, $\kappa = \pm 0.1$. Therefore, we see that our proposed measurement scheme results in a direct measurement of the optical chirality flux, which is directly connected with the near-field optical chirality density [65–70]. Moreover, using $F_{\pm}/F_{\text{inc}} = R_{\pm}$ it follows that:

$$\delta F = \frac{F_{+} - F_{-}}{F_{+} + F_{-}} = \frac{R_{+}F_{\text{inc}} - R_{-}F_{\text{inc}}}{R_{+}F_{\text{inc}} + R_{-}F_{\text{inc}}} = \frac{R_{+} - R_{-}}{R_{+} + R_{-}} = \rho_{\text{DR}}, \tag{2.22}$$

i.e. the differential chirality flux δF is equal to ρ_{DR}, as we also demonstrate in Fig. 2.14.

2.6 Discussion and Conclusions

In this chapter, we start from the observation that plasmonic near fields at metal-chiral interfaces exhibit non-vanishing optical chirality density, which generates a far-field chirality flow. By measuring this chirality flow, which we show to be simply the reflectance in our measurement protocol, we are able to detect changes in the near-field chirality and we demonstrate how this can be exploited for chiral-sensitive measurements using SPR instrumentations. From this, we proceed to demonstrate the following: *(a)* how chiral-sensitive SPR measurements allow for the complete determination of chirality of a natural optically active substance (handedness and magnitude) and for the detection of the near-field wave at the metal-chiral interface; *(b)* how CHISPR is able to detect both the real and imaginary part of the chiral index of refraction, i.e. detect both circular birefringent and circular dichroism effects; and *(c)* how CHISPR is particularly sensitive for the case of sub-wavelength chiral layers, for which traditional (commercial) polarimetry has typically insufficient sensitivities to detect. A particularly significant advantage of the angle-resolved CHISPR protocol we present here, is that it can be employed for spectroscopy of a molecule simply by tuning the frequency of the incident laser radiation over a molecular absorption line [71, 72]. In this case, one can record either the output of an angular split (Figs. 2.6 and 2.8) or a differential measurement (Fig. 2.10) as a function of frequency, and the outcome of such an experiment would be the molecular spectrum of a chiral molecule through angle-resolved CHISPR measurements. Furthermore, in a SPR-based chiral-sensing scheme the whole evanescent-wave volume is sensitive to the probed chiral substance (due to the mobility of the propagating SPPs), contrary to contemporary chiral-sensing nanophotonic schemes that typically rely on localised surface plasmons, where the sign and magnitude of the chiroptical response can possess a complex dependence on sample geometry [26–31, 73]. Crucially, the CHISPR signals we demonstrate are within the sensitivity of current SPR instrumentation for the realistic values of the chirality parameter we consider and, therefore, CHISPR measurements can be directly realized on existing SPR measurement instrumentations with slight modifications on the analysis stage. As a final remark, we wish to emphasize here that since the observed relationships between κ and the measured quantities ($\Delta\theta$, ρ_{DR} and ϕ_{DR}) will effectively impose a lower limit of chiral detection, we expect that this can be further improved by enhancing the local fields, for example via modification of the thin metal layer, e.g. via perforation, in order to take advantage of the strong evanescent fields at the gaps.

In overall, we expect our findings to be of great interest for chiral-biosensing applications, considering also that an angle-resolved CHISPR sensing scheme is a surface-sensitive measurement and differs from conventional chiral-sensing techniques based on transmission measurements. Crucially, the CHISPR scheme and its predicted signals we demonstrate are within, respectively, the capabilities and sensitivity of current SPR instrumentation and, therefore, CHISPR measurements it can be directly realized on existing SPR measurement instrumentations with slight modifications on the analysis stage. Furthermore, CHISR has also the potential for miniaturization and portable design [74–76] (we note here that recent developments in nanophotonics have unveiled novel ways to develop nanoscale Stokes polarimeters [77–79]), and such a possibility could enable compact devices for real-time sensing of biological processes that occur in limited regions of space

Acknowledgements We acknowledge the European Commission Horizon 2020, ULTRA-CHIRAL Project (grant no. FETOPEN-737071) for the financial support.

References

1. S.F. Mason, From Pasteur to parity violation: cosmic dissymmetry and the origins of biomolecular handedness. Ambix **38**(2), 85–108 (1991). https://doi.org/10.1179/amb.1991.38.2.85
2. S.F. Mason, Optical activity and molecular dissymmetry. Contemp. Phys. **9**(3), 239–256 (1968)
3. M. Levin, A.J.S. Klar, A.F. Ramsdell, Introduction to provocative questions in left–right asymmetry. Philos. Trans. R. Soc. B: Biol. Sci. **371**(1710), 20150399 (2016). https://doi.org/10.1098/rstb.2015.0399, https://royalsocietypublishing.org/doi/abs/10.1098/rstb.2015.0399
4. E.N. Fortson, L.L. Lewis, Atomic parity nonconservation experiments. Phys. Rep. **113(5)**(5), 289–344 (1984). https://doi.org/10.1016/0370-1573(84)90005-X, http://www.sciencedirect.com/science/article/pii/037015738490005X
5. G.D. Fasman, *Circular Dichroism and the Conformational Analysis of Biomolecules* (Springer US, New York, 2010)
6. S.M. Kelly, T.J. Jess, N.C. Price, How to study proteins by circular dichroism. Biochim. Biophys. Acta **1751**(2), 119–139 (2005). https://doi.org/10.1016/j.bbapap.2005.06.005, http://www.sciencedirect.com/science/article/pii/S1570963905001792
7. B. Nordén, A. Rodger, T. Dafforn, *Linear Dichroism and Circular Dichroism* (The Royal Society of Chemistry, London, 2010)
8. A.J. Hutt, S.C. Tan, Drug chirality and its clinical significance. Drugs **52**(5), 1–12 (1996). https://doi.org/10.2165/00003495-199600525-00003
9. Nguyen, L.A., He, H., Pham-Huy, C.: Chiral drugs: an overview. Int. J. Biomed. Sci. (IJBS) **2**(2), 85–100 (2006). https://www.ncbi.nlm.nih.gov/pubmed/23674971, https://www.ncbi.nlm.nih.gov/pmc/PMC3614593/
10. S. Droulias, L. Bougas, Surface plasmon platform for angle-resolved chiral sensing. ACS Photonics **6**(6), 1485–1492 (2019). https://doi.org/10.1021/acsphotonics.9b00137
11. L.D. Barron, *Molecular Light Scattering and Optical Activity*, 2 edn. (Cambridge University Press, Cambridge, 2004). https://doi.org/10.1017/CBO9780511535468
12. E.U. Condon, Theories of optical rotatory power. Rev. Mod. Phys. **9**, 432–457 (1937). https://doi.org/10.1103/RevModPhys.9.432, https://link.aps.org/doi/10.1103/RevModPhys.9.432

13. I.V. Lindell, A.H. Sihvola, S.A. Tretyakov, A.J. Viitanen, *Electromagnetic Waves in Chiral and Bi-isotropic Media* (Artech House, Norwood, 1994)
14. O. Arteaga, B. Kahr, Mueller matrix polarimetry of bianisotropic materials. J. Opt. Soc. Am. B **36**(8), F72–F83 (2019). https://doi.org/10.1364/JOSAB.36.000F72, http://josab.osa.org/abstract.cfm?URI=josab-36-8-F72
15. J.U. White, Long optical paths of large aperture. J. Opt. Soc. Am. **32**(5), 285–288 (1942). https://doi.org/10.1364/JOSA.32.000285, http://www.osapublishing.org/abstract.cfm?URI=josa-32-5-285
16. D. Herriott, H. Kogelnik, R. Kompfner, Off-axis paths in spherical mirror interferometers. Appl. Opt. **3**(4), 523–526 (1964). https://doi.org/10.1364/AO.3.000523, http://ao.osa.org/abstract.cfm?URI=ao-3-4-523
17. D. Das, A.C. Wilson, Very long optical path-length from a compact multi-pass cell. Appl. Phys. B **103**(3), 749–754 (2011). https://doi.org/10.1007/s00340-010-4337-7
18. K. Krzempek, M. Jahjah, R. Lewicki, P. Stefański, S. So, D. Thomazy, F.K. Tittel, CW DFB RT diode laser-based sensor for trace-gas detection of ethane using a novel compact multipass gas absorption cell. Appl. Phys. B **112**(4), 461–465 (2013). https://doi.org/10.1007/s00340-013-5544-9
19. T. Müller, K.B. Wiberg, P.H. Vaccaro, Cavity ring-down polarimetry (CRDP): a new scheme for probing circular birefringence and circular dichroism in the gas phase. J. Phys. Chem. A **104**(25), 5959–5968 (2000). https://doi.org/10.1021/jp000705n
20. T. Müller, K.B. Wiberg, P.H. Vaccaro, Cavity ring-down polarimetry (CRDP): a new scheme for probing circular birefringence and circular dichroism in the gas phase. J. Phys. Chem. A **104**(25), 5959–5968 (2000). https://doi.org/10.1021/jp000705n
21. T. Müller, K.B. Wiberg, P.H. Vaccaro, J.R. Cheeseman, M.J. Frisch, Cavity ring-down polarimetry (CRDP): theoretical and experimental characterization. J. Opt. Soc. Am. B **19**(1), 125–141 (2002). https://doi.org/10.1364/JOSAB.19.000125, http://josab.osa.org/abstract.cfm?URI=josab-19-1-125
22. L. Bougas, G.E. Katsoprinakis, W. von Klitzing, J. Sapirstein, T.P. Rakitzis, Cavity-enhanced parity-nonconserving optical rotation in metastable Xe and Hg. Phys. Rev. Lett. **108**, 210801 (2012). https://doi.org/10.1103/PhysRevLett.108.210801, https://link.aps.org/doi/10.1103/PhysRevLett.108.210801
23. D. Sofikitis, L. Bougas, G.E. Katsoprinakis, A.K. Spiliotis, B. Loppinet, T.P. Rakitzis, Evanescent-wave and ambient chiral sensing by signal-reversing cavity ringd-own polarimetry. Nature **514**, 76 (2014). https://doi.org/10.1038/nature13680, http://10.0.4.14/nature13680
24. L. Bougas, D. Sofikitis, G.E. Katsoprinakis, A.K. Spiliotis, P. Tzallas, B. Loppinet, T.P. Rakitzis, Chiral cavity ring down polarimetry: chirality and magnetometry measurements using signal reversals. J. Chem. Phys. **143**(10), 104202 (2015). https://doi.org/10.1063/1.4930109
25. G.E. Katsoprinakis, L. Bougas, T.P. Rakitzis, V.A. Dzuba, V.V. Flambaum, Calculation of parity-nonconserving optical rotation in iodine at 1315 nm. Phys. Rev. A **87**, 040101 (2013). https://doi.org/10.1103/PhysRevA.87.040101, https://link.aps.org/doi/10.1103/PhysRevA.87.040101
26. E. Hendry, T. Carpy, J. Johnston, M. Popland, R.V. Mikhaylovskiy, A.J. Lapthorn, S.M. Kelly, L.D. Barron, N. Gadegaard, M. Kadodwala, Ultrasensitive detection and characterization of biomolecules using superchiral fields. Nat. Nanotechnol. **5**, 783 (2010)
27. Y. Tang, A.E. Cohen, Enhanced enantioselectivity in excitation of chiral molecules by superchiral light. Science **332**(6027), 333–336 (2011). https://doi.org/10.1126/science.1202817
28. T.J. Davis, E. Hendry, Superchiral electromagnetic fields created by surface plasmons in nonchiral metallic nanostructures. Phys. Rev. B **87**, 085405 (2013). https://doi.org/10.1103/PhysRevB.87.085405
29. A.S. Karimullah, C. Jack, R. Tullius, V.M. Rotello, G. Cooke, N. Gadegaard, L.D. Barron, M. Kadodwala, Disposable plasmonics: plastic templated plasmonic metamaterials with tunable chirality. Adv. Mater. **27**(37), 5610–5616 (2015). https://doi.org/10.1002/adma.201501816
30. Y. Luo, C. Chi, M. Jiang, R. Li, S. Zu, Y. Li, Z. Fang, Plasmonic chiral nanostructures: chiroptical effects and applications. Adv. Opt. Mater. **5**(16), 1700040 (2017). https://doi.org/10.1002/adom.201700040

31. Y. Zhao, A.N. Askarpour, L. Sun, J. Shi, X. Li, A. Alù, Chirality detection of enantiomers using twisted optical metamaterials. Nat. Commun. **8**, 14180 (2017)
32. E. Mohammadi, K.L. Tsakmakidis, A.N. Askarpour, P. Dehkhoda, A. Tavakoli, H. Altug, Nanophotonic platforms for enhanced chiral sensing. ACS Photonics **5**(7), 2669–2675 (2018). https://doi.org/10.1021/acsphotonics.8b00270
33. M.L. Solomon, J. Hu, M. Lawrence, A. García-Etxarri, J.A. Dionne, Enantiospecific optical enhancement of chiral sensing and separation with dielectric metasurfaces. ACS Photonics **6**(1), 43–49 (2019). https://doi.org/10.1021/acsphotonics.8b01365
34. N.A. Abdulrahman, Z. Fan, T. Tonooka, S.M. Kelly, N. Gadegaard, E. Hendry, A.O. Govorov, M. Kadodwala, Induced chirality through electromagnetic coupling between chiral molecular layers and plasmonic nanostructures. Nano Lett. **12**(2), 977–983 (2012). https://doi.org/10.1021/nl204055r
35. C. Kelly, L. Khosravi Khorashad, N. Gadegaard, L.D. Barron, A.O. Govorov, A.S. Karimullah, M. Kadodwala, Controlling metamaterial transparency with superchiral fields. ACS Photonics **5**(2), 535–543 (2018).https://doi.org/10.1021/acsphotonics.7b01071
36. J. García-Guirado, M. Svedendahl, J. Puigdollers, R. Quidant, Enhanced chiral sensing with dielectric nanoresonators. Nano Lett. **20**(1), 585–591 (2020). https://doi.org/10.1021/acs.nanolett.9b04334. PMID: 31851826
37. S. Droulias, Chiral sensing with achiral isotropic metasurfaces. Phys. Rev. B **102**, 075119 (2020). https://doi.org/10.1103/PhysRevB.102.075119, https://link.aps.org/doi/10.1103/PhysRevB.102.075119
38. H.H. Nguyen, J. Park, S. Kang, M. Kim, Surface plasmon resonance: a versatile technique for biosensor applications. Sensors **15**(5), 10481–10510 (2015). https://doi.org/10.3390/s150510481
39. R.B.M. Schasfoort (ed.), *Handbook of Surface Plasmon Resonance* (The Royal Society of Chemistry, London, 2017). https://doi.org/10.1039/9781788010283
40. E. Kretschmann, Decay of non radiative surface plasmons into light on rough silver films. comparison of experimental and theoretical results. Opt. Commun. **6**(2), 185–187 (1972). https://doi.org/10.1016/0030-4018(72)90224-6, http://www.sciencedirect.com/science/article/pii/0030401872902246
41. H. Raether, *Surface Plasmons on Smooth and Rough Surfaces and on Gratings*. Springer Tracts in Modern Physics (Springer, Berlin, 1988). https://doi.org/10.1007/BFb0048317, https://www.springer.com/gp/book/9783662151242
42. S.A. Maier, *Plasmonics: Fundamentals and Applications* (Springer US, New York, 2007). https://doi.org/10.1007/0-387-37825-1, https://www.springer.com/gp/book/9780387331508
43. L. Novotny, B. Hecht, *Principles of Nano-Optics*, 2 edn. (Cambridge University Press, Cambridge, 2012). https://doi.org/10.1017/CBO9780511794193, https://www.cambridge.org/core/books/principles-of-nanooptics/E884E5F4AA76DF179A1ECFDF77436452
44. E.N. Economou, Surface plasmons in thin films. Phys. Rev. **182**(2), 539–554 (1969). https://doi.org/10.1103/PhysRev.182.539
45. J.J. Burke, G.I. Stegeman, T. Tamir, Surface-polariton-like waves guided by thin, lossy metal films. Phys. Rev. B **33**(8), 5186–5201 (1986). https://doi.org/10.1103/PhysRevB.33.5186
46. P. Berini, Plasmon-polariton waves guided by thin lossy metal films of finite width: bound modes of symmetric structures. Phys. Rev. B **61**(15), 10484–10503 (2000). https://doi.org/10.1103/PhysRevB.61.10484
47. W.L. Barnes, A. Dereux, T.W. Ebbesen, Surface plasmon subwavelength optics. Nature **424**(6950), 824–830 (2003). https://doi.org/10.1038/nature01937
48. W.L. Barnes, Surface plasmon-polariton length scales: a route to sub-wavelength optics. J. Opt. A: Pure Appl. Opt. **8**(4), S87–S93 (2006). https://doi.org/10.1088/1464-4258/8/4/S06
49. J.M. Pitarke, V.M. Silkin, E.V. Chulkov, P.M. Echenique, Theory of surface plasmons and surface-plasmon polaritons. Rep. Prog. Phys. **70**(1), 1–87 (2006). https://doi.org/10.1088/0034-4885/70/1/R01
50. B. Dastmalchi, P. Tassin, T. Koschny, C.M. Soukoulis, A new perspective on plasmonics: confinement and propagation length of surface plasmons for different materials and geometries. Adv. Opt. Mater. **4**(1), 177–184 (2016). https://doi.org/10.1002/adom.201500446

51. G. Mi, V. Van, Characteristics of surface plasmon polaritons at a chiral-metal interface. Opt. Lett. **39**(7), 2028–2031 (2014). https://doi.org/10.1364/OL.39.002028
52. P. Pelet, N. Engheta, The theory of chirowaveguides. IEEE Trans. Antennas Propag. **38**(1), 90–98 (1990). https://doi.org/10.1109/8.43593
53. P.B. Johnson, R.W. Christy, Optical constants of the noble metals. Phys. Rev. B **6**, 4370–4379 (1972). https://doi.org/10.1103/PhysRevB.6.4370
54. S.A. Maier, H.A. Atwater, Plasmonics: localization and guiding of electromagnetic energy in metal/dielectric structures. J. Appl. Phys. **98**(1), 011101 (2005). https://doi.org/10.1063/1.1951057
55. K. Zhelyazkova, M. Petrov, B. Katranchev, G. Dyankov, Surface plasmon resonance on the surface: metal - liquid crystal layer. J. Phys.: Conf. Ser. **558**, 012023 (2014). https://doi.org/10.1088/1742-6596/558/1/012023
56. M. Wang, H. Li, T. Xu, H. Zheng, M. Yu, G. Li, J. Xu, J. Wu, Probing bianisotropic biomolecules via a surface plasmon resonance sensor. Opt. Express **26**(22), 28277–28287 (2018). https://doi.org/10.1364/OE.26.028277
57. L.V. Poulikakos, P. Thureja, A. Stollmann, E. De Leo, D.J. Norris, Chiral light design and detection inspired by optical antenna theory. Nano Lett. **18**(8), 4633–4640 (2018). https://doi.org/10.1021/acs.nanolett.8b00083. PMID: 29533637
58. M.P. Silverman, J. Badoz, B. Briat, Chiral reflection from a naturally optically active medium. Opt. Lett. **17**(12), 886–888 (1992). https://doi.org/10.1364/OL.17.000886
59. M. Silverman, J. Badoz, Large enhancement of chiral asymmetry in light reflection near critical angle. Opt. Commun. **74**(3), 129–133 (1989)
60. M. Piliarik, J. Homola, Surface plasmon resonance (SPR) sensors: approaching their limits? Opt. Express **17**(19), 16505–16517 (2009). https://doi.org/10.1364/OE.17.016505
61. X. Wang, M. Jefferson, P.C.D. Hobbs, W.P. Risk, B.E. Feller, R.D. Miller, A. Knoesen, Shot-noise limited detection for surface plasmon sensing. Opt. Express **19**(1), 107–117 (2011). https://doi.org/10.1364/OE.19.000107
62. R.C. Pooser, B. Lawrie, Plasmonic trace sensing below the photon shot noise limit. ACS Photonics **3**(1), 8–13 (2016). https://doi.org/10.1021/acsphotonics.5b00501
63. A.V. Kabashin, S. Patskovsky, A.N. Grigorenko, Phase and amplitude sensitivities in surface plasmon resonance bio and chemical sensing. Opt. Express **17**(23), 21191–21204 (2009). https://doi.org/10.1364/OE.17.021191
64. S. Patskovsky, M. Meunier, P.N. Prasad, A.V. Kabashin, Self-noise-filtering phase-sensitive surface plasmon resonance biosensing. Opt. Express **18**(14), 14353–14358 (2010). https://doi.org/10.1364/OE.18.014353
65. D.M. Lipkin, Existence of a new conservation law in electromagnetic theory. J. Math. Phys. **5**(5), 696–700 (1964). https://doi.org/10.1063/1.1704165
66. Y. Tang, A.E. Cohen, Optical chirality and its interaction with matter. Phys. Rev. Lett. **104**, 163901 (2010). https://doi.org/10.1103/PhysRevLett.104.163901, https://link.aps.org/doi/10.1103/PhysRevLett.104.163901
67. K.Y. Bliokh, F. Nori, Characterizing optical chirality. Phys. Rev. A **83**, 021803 (2011). https://doi.org/10.1103/PhysRevA.83.021803
68. M.M. Coles, D.L. Andrews, Chirality and angular momentum in optical radiation. Phys. Rev. A **85**, 063810 (2012). https://doi.org/10.1103/PhysRevA.85.063810
69. T.G. Philbin, Lipkin's conservation law, Noether's theorem, and the relation to optical helicity. Phys. Rev. A **87**, 043843 (2013). https://doi.org/10.1103/PhysRevA.87.043843, https://link.aps.org/doi/10.1103/PhysRevA.87.043843
70. L.V. Poulikakos, P. Gutsche, K.M. McPeak, S. Burger, J. Niegemann, C. Hafner, D.J. Norris, Optical chirality flux as a useful far-field probe of chiral near fields. ACS Photonics **3**(9), 1619–1625 (2016). https://doi.org/10.1021/acsphotonics.6b00201
71. A. Ikehata, T. Itoh, Y. Ozaki, Surface plasmon resonance near-infrared spectroscopy. Anal. Chem. **76**(21), 6461–6469 (2004). https://doi.org/10.1021/ac049003a
72. R. Zektzer, L. Stern, N. Mazurski, U. Levy, Enhanced light-matter interactions in plasmonic-molecular gas hybrid system. Optica **5**(4), 486–494 (2018). https://doi.org/10.1364/OPTICA.5.000486, http://www.osapublishing.org/optica/abstract.cfm?URI=optica-5-4-486

73. Y. Tang, L. Sun, A.E. Cohen, Chiroptical hot spots in twisted nanowire plasmonic oscillators. Appl. Phys. Lett. **102**(4), 043103 (2013). https://doi.org/10.1063/1.4789529
74. P. Preechaburana, M.C. Gonzalez, A. Suska, D. Filippini, Surface plasmon resonance chemical sensing on cell phones. Angew. Chem. Int. Ed. **51**(46), 11585–11588 (2012). https://doi.org/10.1002/anie.201206804, https://onlinelibrary.wiley.com/doi/abs/10.1002/anie.201206804
75. Y. Liu, Q. Liu, S. Chen, F. Cheng, H. Wang, W. Peng, Surface plasmon resonance biosensor based on smart phone platforms. Sci. Rep. **5**, 12864 (2015). https://doi.org/10.1038/srep12864, http://10.0.4.14/srep12864
76. H. Guner, E. Ozgur, G. Kokturk, M. Celik, E. Esen, A.E. Topal, S. Ayas, Y. Uludag, C. Elbuken, A. Dana, A smartphone based surface plasmon resonance imaging (SPRI) platform for on-site biodetection. Sens. Actuators B: Chem. **239**, 571–577 (2017). https://doi.org/10.1016/j.snb.2016.08.061, http://www.sciencedirect.com/science/article/pii/S092540051631293X
77. J.P.B. Mueller, K. Leosson, F. Capasso, Ultracompact metasurface in-line polarimeter. Optica **3**(1), 42–47 (2016). https://doi.org/10.1364/OPTICA.3.000042, http://www.osapublishing.org/optica/abstract.cfm?URI=optica-3-1-42
78. A. Espinosa-Soria, F.J. Rodríguez-Fortuño, A. Griol, A. Martínez, On-chip optimal stokes nanopolarimetry based on spin-orbit interaction of light. Nano Lett. **17**, 3139–3144 (2017). https://doi.org/10.1021/acs.nanolett.7b00564
79. P.C. Wu, J.W. Chen, C.W. Yin, Y.C. Lai, T.L. Chung, C.Y. Liao, B.H. Chen, K.W. Lee, C.J. Chuang, C.M. Wang, D.P. Tsai, Visible metasurfaces for on-chip polarimetry. ACS Photonics **5**(7), 2568–2573 (2018). https://doi.org/10.1021/acsphotonics.7b01527

Chapter 3
Spin-Polarized Plasmonics: Fresh View on Magnetic Nanoparticles

Vladimir P. Drachev, Maria Pogodaeva, Sergey V. Levchenko, and Ali E. Aliev

Abstract Here we discuss effect of spin-polarization on plasmon excitation in deep ultra-violet spectral range for Co nanoparticles with a single-domain magnetic structure. Structural, magnetic, and optical characterizations of Co nanoparticles shine a light on a mechanism of the magneto-plasmonic response.

3.1 Introduction

It is a common belief that the quality of the plasmon resonance of magnetic nanoparticles such as Co is quite low, which follows, in particular, from the experimental data for permittivity of bulk cobalt by Johnson and Christy (J&C) [1]. Our recent paper shows that for single-domain magnetic nanoparticles the usual approach, based on bulk permittivity, does not work, while used to work perfectly for nonmagnetic nanoparticles [2]. Indeed, our experiments prove that Co nanoparticles with a single-domain magnetic structure support a sharp plasmon resonance at about 280 nm with the resonance quality comparable to gold nanoparticles. This type of plasmons is quite different from known plasmons in noble metals. Note that the plasmon resonance of Co is in the deep ultraviolet spectral range, which is the range for bio-molecule resonances [3], as it is shown in Fig. 3.1, and, therefore, attractive for bio-medical applications in addition to its magnetic nature.

Deep ultraviolet (DUV) Raman spectroscopy selectively visualizes nucleotide bases, monomeric units of deoxyribonucleic acids (DNA) and aromatic amino acids, monomeric units of proteins, in cells due to the resonant effect (spectra are presented

V. P. Drachev (✉) · M. Pogodaeva · S. V. Levchenko
Skolkovo Institute of Science and Technology, Moscow, Russia
e-mail: v.drachev@skoltech.ru

V. P. Drachev
University of North Texas, Denton, TX, USA

A. E. Aliev
A. G. MacDiarmid NanoTech Institute, University of Texas at Dallas, Richardson, TX, USA

E. Kamenetskii (ed.), *Chirality, Magnetism and Magnetoelectricity*,
Topics in Applied Physics 138,
https://doi.org/10.1007/978-3-030-62844-4_3

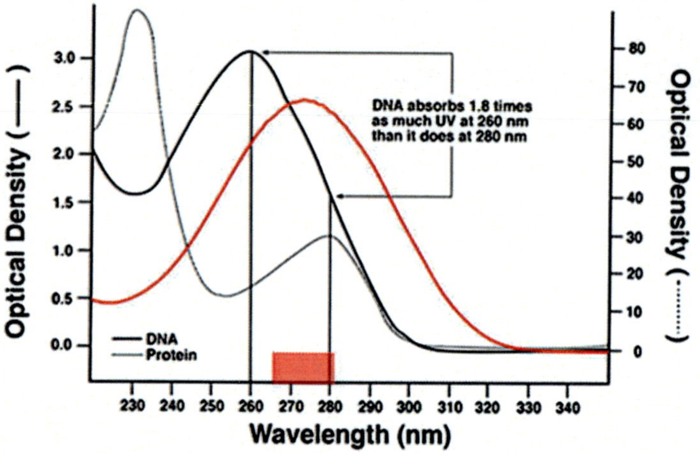

Fig. 3.1 Absorption spectra of DNA (black) and aromatic proteins (gray) (from [3] © Institute of Physics and Engineering in Medicine. Reproduced by permission of IOP Publishing. All rights reserved). The CoNP plasmon resonance in absorption is shown in red, optical density in arbitrary units (from [2] with permission licensed under CC BY 4.0 https://creativecommons.org). The red bar shows the Raman shift range at the laser excitation wavelength 266 nm. The enhancement of the CoNPs will cover this range

in Fig. 3.1, black and gray lines correspondingly) [4–7]. In the resonant Raman condition, where the electronic transition energy of a molecule corresponds to photon energy of Raman excitation light, the Raman scattering intensity of the molecule is enhanced by as much as 10^6 compared to the non-resonant Raman scattering [5]. Because other biological compounds in cells, such as lipids and sugars, are not in the resonant condition at the DUV, Raman scattering from nucleotide bases and aromatic amino acids [4–7] are selectively enhanced in the DUV resonant Raman spectroscopy of cells. Main obstacle of the current resonance Raman approach is that the UV light overdose is harmful to the cells viability and biomolecules functionality [5]. Surface-enhanced resonance Raman scattering (SERRS) will strongly reduce the required energy density for robust detection. Unique combination of plasmonic and magnetic properties makes this platform appropriate for a combined approach of diagnostics and therapy (theranostics). Magnetic nanoparticles represent an attractive tool for medical applications based on their ability to be simultaneously functionalized and guided by an external magnetic field [8–17]. Various biomedical applications of magnetic nanoparticles include enhancing and targeting gene delivery by magnetic force in vitro and in vivo [8, 9], magnetic fluid hyperthermia and cancer therapy [9, 10], cells separation [12], magnetic resonance imaging [13–15]. Biocompatibility of magnetic nanoparticles is under extensive studies and can be achieved by an appropriate coating [16, 17].

Currently plasmonic applications in bio-sensing involve noble metals, Ag or Au, since the quality of their plasmon resonance is highest [18–20]. The SERS protocols based on Ag and Au nanoparticles are demonstrated for tag free protein-protein binding detection, the feasibility of using SERS to distinguish protein conformational states, which was shown for human insulin and its analog insulin lispro [21–24]. A protocol has been developed to detect cell surface markers, CD44 and CD24, in three breast cancer cell lines [25]. The dielectric functions Ag and Au for this type of nanostructures were carefully studied as well [26–28].

However, Co nanoparticles (Co-NPs), under certain conditions can support an excellent plasmon resonance at about 280 nm with a quality factor greater than Al, In, and comparable to Au in the visible [2]. Importantly, this Co platform is comprising both magnetic and plasmonic properties. A long lasting search for plasmonic materials in the ultraviolet spectral range does not consider Co as a promising candidate [29–32]. One of the criteria for a high quality plasmonic material is that the number of electrons involved in interband transitions must be low, and at the highest possible frequency. These criteria significantly reduces the number of materials that are likely to have favorable optical properties, by the simple fact that all materials with partially occupied d or f states are going to perform poorly across the visible due to interband transitions [33]. Recent works pushed the plasmonics to high-energy range using Al [33–35] and In [36].

The Mott model [37, 38] of conductivity in magnetic metals helps to qualitatively explain observed phenomena for Co nanoparticles. Indeed, the electrical conductivity in metals can be described in terms of two largely independent conducting channels, corresponding to the spin-up and spin-down electrons. Importantly, the probability of spin-flip scattering processes in metals is normally small as compared to the probability of the scattering processes in which the spin is conserved. This means that the spin-up and spin-down electrons do not mix over long distances and, therefore, the electrical conduction occurs in parallel for the two spin channels. *Also*, the scattering rates in ferromagnetic metals of the spin-up and spin-down electrons are quite different, whatever the nature of the scattering centers is. These two channels of conductivity with a distinct spin-dependent scattering is the primary origin of giant magnetoresistance [39].

Here, we discuss the effect of spin polarization on plasmon oscillations of the free electrons in nanoparticles, which is, crucial in many envisioned applications at the cross road of magnetism and plasmonics.

3.2 Spin Polarization in Co Nanoparticles

A new type of plasmons is specific for spin-polarized magnetic nanoparticles. One can expect two independent plasmons which co-exist in a spin-polarized metal nanoparticle following Mott's model. These two plasmons coexist in a particle at the same frequency and polarizations of excitation, but for electrons of opposite spin. Inter-nanoparticle interactions completely demolish plasmon quality resonance, which is the probable reason why it was not observed previously and why the results for bulk films [1] cannot be used for single domain nanoparticles evaluations. It is known that the exchange interaction of electrons splits the energy bands between spin-up (majority) electrons and spin-down (minority) electrons. We suggest that a low quality of the plasmon resonance for spin-down electrons is due to the large relaxation rate of the conduction electrons caused by high density of empty states in a partially populated d-band. However, the majority electrons with a completely filled d-band does not affect the relaxation rate and plasmon resonance of the conduction spin-up electrons within magnetic nanoparticles.

Figure 3.2 shows spin polarization for bulk Co and Co nanocluster calculated using density functional theory (DFT) simulations. Transition metals are challenging for DFT, since standard exchange-correlation (XC) functional approximations, local density approximation (LDA) and generalized gradient approximation (GGA) underestimate localization of valence d-electrons. This problem is commonly addressed via ad hoc inclusion of a Hubbard correction to e.g. GGA with an effective U term (GGA+U). The resulting GGA+U method has the same low computational cost as GGA. U is a parameter that can be tuned to reproduce experimental results, in particular lattice parameters and magnetic moments. Although optimal U have been suggested in the literature for various metals including Co, [40, 41] not all properties can be reproduced with good accuracy at the same time. Moreover, the value of the U parameter depends on the coordination of metal atoms. Therefore, a different value of U may be required to describe metal bulk and clusters.

In Fig. 3.2 projected density of states (DOS) for the bulk hexagonal close packed (hcp) cobalt (panel b) and a 48-atom cluster (panel a) are shown. DOS for face-centered cubic (fcc) crystal structure looks qualitatively similar. For this comparison we used DFT with the Perdew–Burke-Ernzerhof (PBE) exchange-correlation functional [42]. All calculations are done with all-electron full-potential electronic-structure package FHI-aims [43–46] with “tight” numerical settings. The hcp lattice constants were optimized with PBE. The initial atomic structure of the cluster is obtained using Wulff construction for hcp Co, [47, 48] and then fully relaxed. Projected DOS for the cluster is calculated using Gaussian width of 0.01 eV.

As can be seen in Fig. 3.2, the general shapes of DOS for bulk and particle for both spin channels are similar, but there are also important differences. In particular,

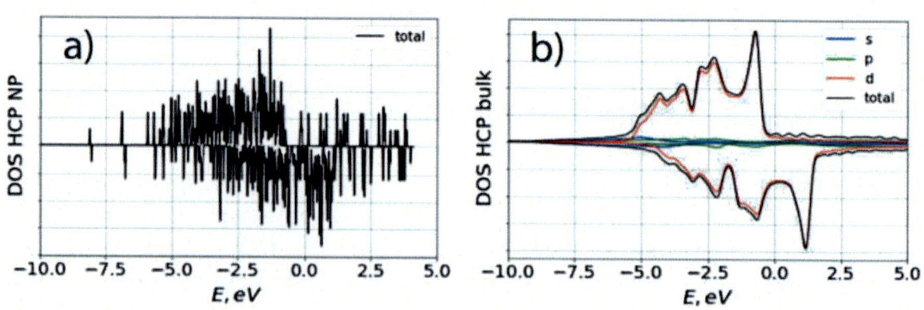

Fig. 3.2 Projected density of states (DOS) for **a** 48-atom Co cluster and **b** bulk hexagonal close packed (hcp) cobalt. Projections on valence *s*, *p*, and *d* orbitals of Co are shown. Spin-minority DOS is shown with negative sign. Zero on the energy axis corresponds to the Fermi level

more structure is observed for the particle both below and above the Fermi level (corresponding to 0 eV). This is not surprising, since the particle has additional states due to the presence of the surface. Notably, the pronounced peaks around the Fermi level in both spin channels are discrete and broadened in the cluster compared to bulk.

3.3 Methods

In this work, the oleic acid (OA) coated cobalt nanoparticles were fabricated by the high temperature reduction of cobalt salt in the presence of trioctylphosphine (TOP) as a surfactant and lithium triethylborohydride as a reducing reagent [49–51]. Cobalt nanoparticles were synthesized using a method similar to that of Sun and Murray [52]. The reduction of cobalt nanoparticles was conducted under inert atmosphere. At room temperature, 0.13 g (0.019 M) of anhydrous cobalt chloride, 0.3 mL (0.018 M) of oleic acid and 30 mL (1.87 M) of dioctyl ether were mixed together under purged nitrogen gas in the three-necked flask containing magnetic stir bar and heated to 100 °C. Then 1.5 mL (0.063 M) of trioctylphosphine, which was injected via syringe and the temperature raised to 205 °C. At this temperature, a strong reducing reagent, 1.5 mL (0.236 M) lithium triethylborohydride, was added in solution and the cobalt nanoparticles begin to grow immediately. The blue colour of the solution turns to black upon nucleation and growth of cobalt nanoparticles. The reaction was terminated by cooling the solution to room temperature and 20 mL (4.8 M) of anhydrous ethanol was added to precipitate the particles. The solution was aged overnight at room temperature in order to attach cobalt nanoparticles to the magnetic stir bar in the flask. The cobalt nanoparticles are removed from a magnetic stir bar and washed several times with ethanol by centrifugation. Finally, oleic acid coated cobalt nanoparticles were suspended in 8 mL of hexane.

To address the mechanism of new type of plasmons specific for magnetic nanoparticles our work involves the structural electron microscopy, superconducting quantum interference device (SQUID) magnetometry, dynamic light scattering (DLS), and spectroscopy of Co nanoparticles. The structural and magnetic characterizations prove the single-domain and superparamagnetic properties of nanoparticles required for spin dependent channels of plasmon oscillations. The magnetic field induced aggregation of nanoparticles in our experiments results in the suppression of the resonance quality.

3.4 Structural Properties

The scanning electron microscopy (SEM) energy dispersive X-ray (EDX) *image* of the cobalt nanoparticles synthesized by the high temperature decomposition of cobalt salt (Fig. 3.3).

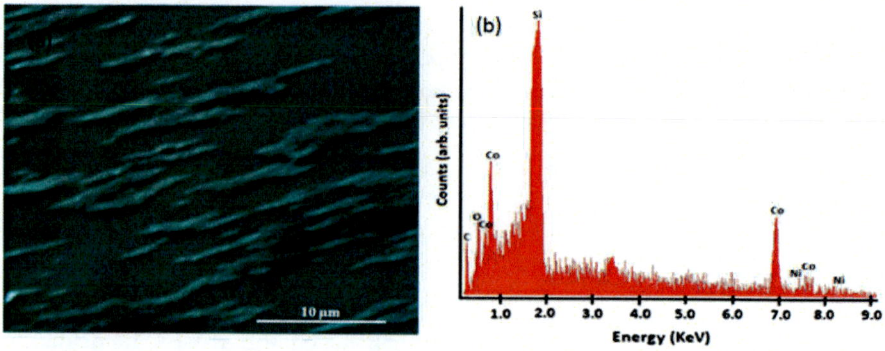

Fig. 3.3 **a** SEM image of the Co nanoparticles. **b** Corresponding EDX spectrum. From [2] with permission licensed under CC BY 4.0 https://creativecommons.org

We observed the large aggregates of cobalt nanoparticles instead of isolated particles because the circular magnet placed underneath the silicon substrate attracted the magnetic nanoparticles from the solution. An energy dispersive X-ray (EDX) spectrum analysis has been obtained from the cobalt nanoparticle sample. This spectrum clearly shows the presence of cobalt peaks. In addition, EDX spectrum also shows the presence of nickel peaks because the sample was made conducting by coating with nickel.

High resolution transmission electron microscopy (TEM) images were obtained with the FEI Tecnai G2 F20 S-Twin 200 keV field emission scanning transmission electron microscope (S/TEM). The high magnification TEM image of spherical cobalt nanoparticles shows the size distribution of cobalt nanoparticles ranges from 6 to 12 nm with average particle diameter of 8.7 nm. High resolution TEM images show that our particles form both hcp and fcc crystal structure. Figure 3.4a shows

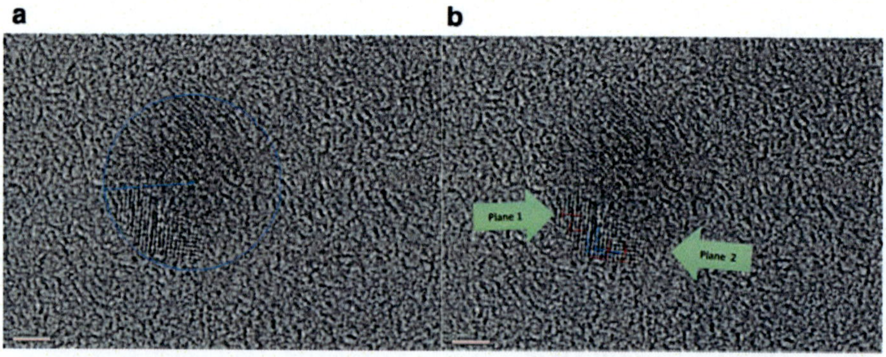

Fig. 3.4 **a** High resolution TEM pattern of [100] hcp crystal structure of Co nanoparticle. The radius here measures 4.59 ±0.05 nm. **b** High resolution TEM pattern of [100]—[001] hcp crystal structure of Co nanoparticle. Plane 1: 0.216 nm ±0.05 nm, Plane 2: 0.223 nm ±0.05 nm, interplanar angle 91.8° ±3°. Calculated: [100] 0.217113, [001] 0.217113, [100]—[001] angle 90°

TEM of [100] crystal structure of some cobalt nanoparticles have lattice spacing 2.102 Å and inter-planar angle 83.6° typical for hcp. Figure 3.4b shows high resolution TEM image of [111] crystal structure of other cobalt nanoparticles has lattice spacing 1.95 Å and inter-planar angle 25.63°, typical for fcc.

For calculation expressions explaining the basic crystallography of cobalt was used [53]. Following by definition, if two planes have indices (hkl) the distance perpendicularly between them is shown by the inverse of the magnitude of the lattice vector.

$$d_{hkl} = \frac{1}{\left|h\boldsymbol{a}^{\bullet} + k\boldsymbol{b}^{*} + l\boldsymbol{c}^{\bullet}\right|} \tag{3.1}$$

Then:

$$\left|h\boldsymbol{a}^{*} + k\boldsymbol{b}^{*} + l\boldsymbol{c}^{*}\right|^{2} = \left(h\boldsymbol{a}^{*} + k\boldsymbol{b}^{*} + l\boldsymbol{c}^{*}\right)\cdot\left(h\boldsymbol{a}^{*} + k\boldsymbol{b}^{*} + l\boldsymbol{c}^{*}\right)$$

$$= h^{2}\boldsymbol{a}^{*2} + k^{2}\boldsymbol{b}^{*2} + l^{2}\boldsymbol{c}^{*2} + 2kl\boldsymbol{b}^{*}\cdot\boldsymbol{c}^{*} + 2lh\boldsymbol{c}^{*}\cdot\boldsymbol{a}^{*} + 2hk\boldsymbol{a}^{*}\cdot\boldsymbol{b}^{*}$$

$$= h^{2}a^{*2} + k^{2}b^{*2} + l^{2}c^{*2} + 2klb^{*}c^{*}\cos\alpha^{*} + 2lhc^{*}a^{*}\cos\beta^{*} + 2hka^{*}b^{*}\cos\gamma^{*}$$

here, $\boldsymbol{a}^{\bullet}$,$\boldsymbol{b}^{*}$, and $\boldsymbol{c}^{\bullet}$ are lattice vectors, α^{*} is the angle between lattice vectors $\boldsymbol{b}^{*}$ and $\boldsymbol{c}^{\bullet}$, β^{*} is the angle between $\boldsymbol{c}^{\bullet}$ and $\boldsymbol{a}^{\bullet}$. Knowing that cobalt is found primarily in two phases, hexagonal close packed and face-centered cubic further expansion of this formula was used only for those two overall structures. The lattice constants used for cobalt are found in a variety of literatures. For hcp lattice constants of $a = b = 0.2507$ nm, $c = 0.4069$ nm. For fcc lattice constants of $a = b = c = 0.35446$ nm. First calculations were done to families of standard planes for both fcc and hcp structures. Hexagonal first:

$$\alpha^{*} = \beta^{*} = 90^{\circ},\ \gamma^{*} = 60^{\circ}\ \text{and}\ a^{*} = b^{*},\ d_{hkl}^{2} = \frac{1}{(h^{2} + k^{2} + hk)a^{*2} + l^{2}c^{*2}} \tag{3.2}$$

$$\cos\phi = d_{hkl}d_{h'k'l'}\left[\left\{hh' + kk' + \frac{1}{2}(hk' + kh')\right\}a^{*2} + ll'c^{*2}\right]$$

In which:

$$a^{\bullet} = \frac{2}{a\sqrt{3}},\ c^{\bullet} = \frac{1}{c}$$

Next for cubic:

$$\alpha^{*} = \beta^{*} = \gamma^{*} = 90^{\circ}\ \text{and}\ a^{*} = b^{*} = c^{*},\ d_{hkl}^{2} = \frac{1}{(h^{2} + k^{2} + l^{2})a^{*2}} \tag{3.3}$$

$$\cos\phi = \frac{hh' + kk' + ll'}{\sqrt{h^2 + k^2 + l^2}\sqrt{h'^2 + k'^2 + l'^2}} \tag{3.4}$$

In which:

$$a^* = \frac{1}{a}$$

Using these equations [53] the calculated interplanar distances and then angles were able to be deduced. These equations were used to code a simple Mathematica program that would output both interplanar distances as well as angles.

The analysis of the TEM images showed and confirmed the likely structure of cobalt. Though, a large amount of the images had planes and angles that were inconclusive the data definitely points most directly at hexagonal close packed to be the average crystalline phase of the particles. The large amount of inconclusive interplanar angles are likely due to surface deformities, as well as, small and unpredictable tilts in the particles. This would allow for the interplanar angles and distances to be skew, reduced in the case of the interplanar distances and increased up to 180° in the case of the interplanar angles. In general, cobalt should not be found in face-centered cubic crystalline phase below 400 °C. However, some nanoparticles in our case did show possible indications of being in this phase.

3.5 Magnetic Response

Magnetic properties of Co-NP embedded into poly(methyl methacrylate) (PMMA) host matrix were measured using direct current (DC) option of 7 Tesla SQUID magnetometer (Magnetic Property Measurement Device, Quantum Design Inc.). Figure 3.5 shows the step-by-step sample preparation for magnetic measurements: Co-NP collected from magnetic stirring bar were dispersed in PMMA, deposited on a substrate, dried, and then the PMMA film with embedded Co-NP were packed into a gelatine capsule. To reduce the influence of the sample shape (demagnetization effect), all Co-NP embedded sheets were placed in gelatine capsule parallel to the applied magnetic field.

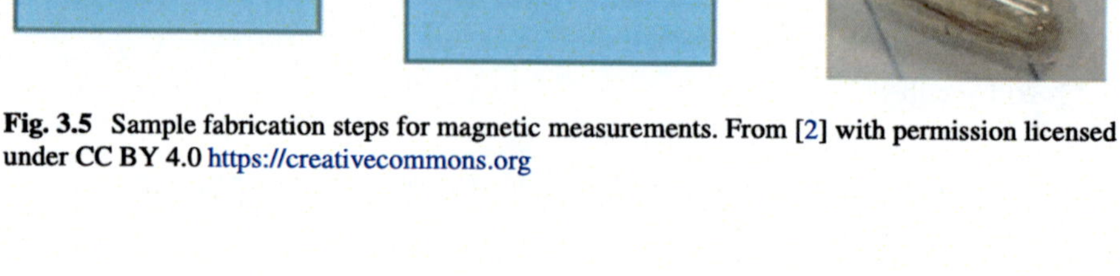

Fig. 3.5 Sample fabrication steps for magnetic measurements. From [2] with permission licensed under CC BY 4.0 https://creativecommons.org

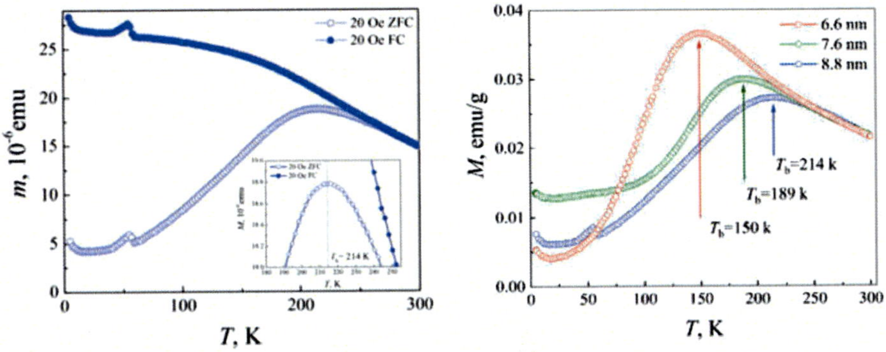

(a) (b)

Fig. 3.6 **a** Temperature dependence of magnetic moment for 8.8 nm Co-NP measured in FC (solid blue circles) and ZFC (open blue circles) regimes at applied field of 20 Oe (sample #25). The inset shows an expanded view at a peak of ZFC magnetization. The small bump at 52 K indicates an oxygen contents in the sealed capsule. **b** Shift of ZFC magnetization peaks towards lower temperatures for decreasing particle sizes: 8.8 nm for sample #25; 7.6 nm for sample #28; and 6.6 nm for sample #33, respectively. From [2] with permission licensed under CC BY 4.0 https://creativecommons.org

The temperature dependence of magnetization in field cooling (FC) and zero field cooling (ZFC) regimes exploited to determine the blocking temperature (T_b) at which the ZFC magnetization shows a pronounced peak. T_b is the temperature below which the magnetization curve shows the hysteresis and above superparamagnetic properties [54]. The observed curves for FC and ZFC regimes shown in Fig. 3.6a, b are typical for single-domain ferromagnetic nanoparticles.

For very small particles at finite temperatures the magnetic anisotropy energy, K_u, becomes comparable to the thermal energy resulting in random flip of the magnetization direction and in superparamagnetic (SP) relaxation. Thus, the T_b is defined as the temperature at which the SP relaxation time (response of magnetic dipole), equals the timescale of the experimental technique used to study the magnetic properties, $\omega\tau = 1$. The SP relaxation time τ, also called the Neel relaxation time, τ_N [54], given by [55],

$$\tau = \tau_0 \exp\left(\frac{K_u V}{k_b T}\right). \tag{3.5}$$

was measured using ACMS option of Physical Property Measurement Device (Quantum Design Inc.) in the frequency range of $10 \leq f \leq 10^4$ Hz with alternating current (AC) magnetic field amplitude of ± 10 Oe. Here, K_u is the magnetic anisotropy energy, V is the particle volume, k_B is Boltzmann's constant and T is the temperature. The value of τ_0 extracted from the linear extrapolation of τ to zero $1000/T$ for Co-NP with 6.6 nm in diameter is 4.1×10^{-14} s (see Fig. 3.7). Here we ignore the temperature dependence of τ_0, as it is small compared to the effect of the temperature

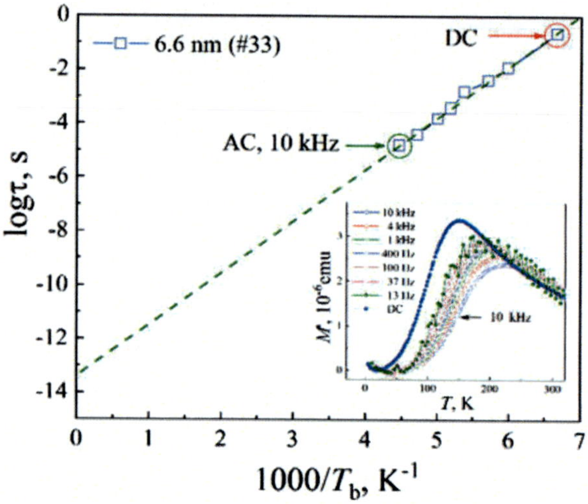

Fig. 3.7 Thermal variation of a relaxation time versus reciprocal temperature for $d = 6.6$ nm (sample #33). The green dashed line is an extrapolation of relaxation time to $1000/T_b = 0$. The inset shows the ZFC magnetization curves measured for each frequency ($f = 1/2\pi\tau$) shown in the main panel. From [2] with permission licensed under CC BY 4.0 https://creativecommons.org

through the exponential [56]. Hence, for DC measurements, where $\ln(\tau/\tau_0) \approx 29$, the blocking temperature, above which the single domain Co-NP starts randomly flip its magnetic moment and is small enough to display superparamagnetism, should roughly satisfy the relationship

$$T_b \approx \frac{K_u V}{29 k_B}. \tag{3.6}$$

With a knowledge of average particle diameter from the precise analysis of TEM images and dynamic light scattering measurements ($d_{\#25} = 8.8$ nm, $d_{\#28} = 7.6$ nm, $d_{\#33} = 6.6$ nm,) the magnetic anisotropy energy extracted from Fig. 3.6b ($K_u^{\#25} = 2.74 \times 10^6$ erg/cm^3, $K_u^{\#28} = 3.73 \times 10^6$ erg/cm^3, $K_u^{\#33} = 3.9 \times 10^6$ erg/cm^3) falls between bulk fcc and hcp structures (2.7×10^6 erg/cm^3 for fcc and 4.7×10^6 erg/cm^3 for hcp), respectively.

The increase in anisotropy constant (energy) for small Co-NP and clusters is resulted from strong contribution of surface atoms. The fraction of Co atoms on the surface of nanoparticles increases with decrease in particle size, which results in decrease of coordination number of surface atoms, increased spin and orbital magnetic moments towards free atoms [57]. The anisotropy energy, K_u, is also increases due to the reduction of spherical symmetry of small nanoparticles [58].

Above the blocking temperature the susceptibility is precisely follows the Curie law, $\chi \sim C/T$, where $C = n(\mu_0\mu)^2/k_B$ is the Curie constant, n is number of particles, μ_0 is magnetic moment of vacuum, and μ is the relative magnetic moment of Co-NP. The linearity of $\chi \sim T_c/T$ plot observed for studied Co-NP (see Fig. 3.8) also implies a low interaction between nanoparticles. The interaction between nanoparticles is a crucial parameter determining the strength of plasmon resonance in Co-NP.

The correctly measured T_b should increase as a cub of the particle diameter, $T_b \sim V \sim D^3$. The deviation of T_b dependence from cubic behaviour in Fig. 3.9a implies

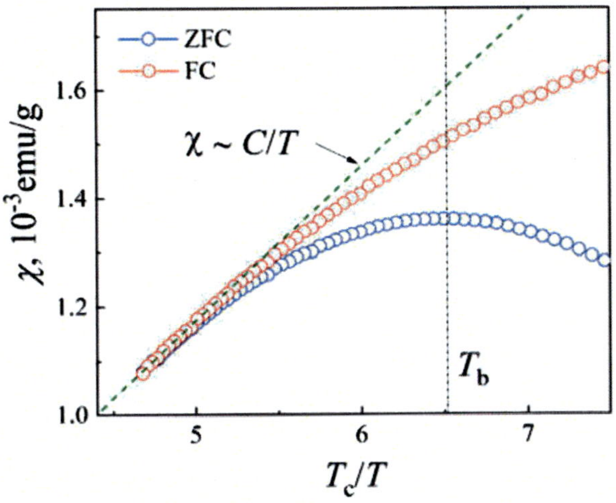

Fig. 3.8 ZFC and FC susceptibility ($\chi = M/H$) for 8.8 nm Co-NP (#25) measured at applied magnetic field of 20 Oe plotted versus reciprocal temperature, $T_c = 1394$ K is the Curie temperature for bulk Co. From [2] with permission licensed under CC BY 4.0 https://creativecommons.org

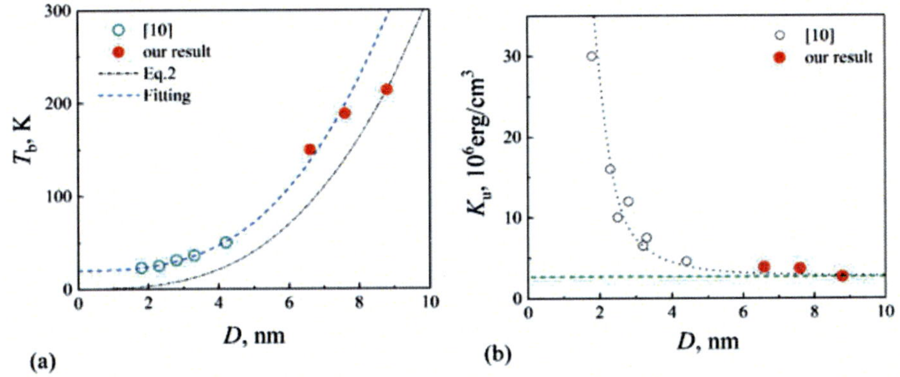

Fig. 3.9 **a** The particle diameter dependence of the blocking temperature, T_b. Dashed and dash-dot curves show a fitting to the cub dependence, $T_b \sim V$. **b** The particle diameter dependence of the anisotropy energy. The dot line is a fitting of reciprocal-cub dependence for [58] and our result. The bulk value for fcc cobalt is shown by dashed (green) line. From [2] with permission licensed under CC BY 4.0 https://creativecommons.org

the decrease of anisotropy energy. On the other hand, the increase of particle size above single-domain size should bring the anisotropy constant towards the balk value [58]. Figure 3.9b shows that for particles larger than 8 nm the anisotropy energy is close to the bulk fcc value, 2.7×10^6 erg/cm^3.

Co nanoparticles coated with CoO (or partially oxidized during fabrication) should exhibit exchange anisotropy due to an interfacial interaction between ferromagnetic

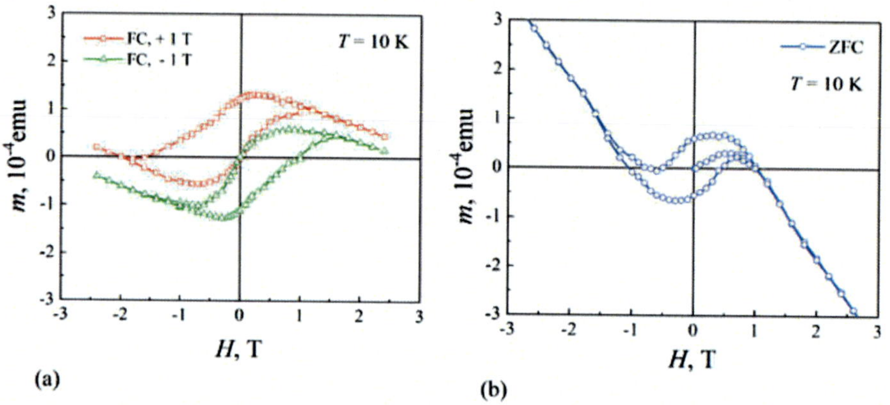

Fig. 3.10 **a** The shift of magnetization loops for sample #25 cooled in field +1 T (open red squares) and −1 T (open green triangles), $T = 10$ K. **b** The magnetization loop for the same sample cooled in zero field is symmetrical. The diamagnetic slope of $m(H)$ curves beyond the hysteresis loop comes from a PMMA matrix. From [2] with permission licensed under CC BY 4.0 https://creativecommons.org

Co metal and antiferromagnetic CoO. The shift of the symmetry of hysteresis loop measured at low temperatures (10 K), after the sample was cooled in a magnetic field of 1 T (FC), implies that CoO shell was formed on the particle surfaces (see Fig. 3.10a). At the same time, the ZFC cooled sample shows quite symmetrical hysteresis loop at $T = 10$ K (see Fig. 3.10b). These measurements were taken for samples #25 (8.8 nm) four months later from preparation date. Since the samples were embedded into PMMA matrix, the oxidation, perhaps, comes from the surfactant shell on the surface of Co-NP. However, as prepared samples measured within 2–3 weeks do not show any shift of the hysteresis loop.

The high-field magnetic moment for 8.8 nm Co-NP (#25) measured at 5, 100 and 298 K are shown in Fig. 3.11a. The analysis of saturation magnetization (M_s) at 5 K (the diamagnetic contribution of PMMA matrix was subtracted) shows that the Co-NP exhibit no-saturation behaviour up to highest available field of 7 T. The saturation magnetization obtained from the extrapolation of M versus $1/H$ line to 0 gives ~10 emu/g, which is close to the value obtained for 7.8 nm Co-NP having fcc structure produced by the Kraschmer carbon arc process [59]. At the same time, this value is one order lower than for bulk fcc Co (162 emu/g, or 175 emu/g, see in [60]). The reduced remanence, $M_r/M_s = 0.02$, is far below of theoretical values for nanocrystals having uniaxial anisotropy such as cobalt in the hcp form (0.5) or in the fcc form (0.8) [60].

Among the possible explanations for the reduced saturation magnetization and remanence is the multiphase (fcc–hcp) crystalline domain structure of single particle separated by amorphous cobalt and the exchange coupling between adjacent Co-NP. The dipole coupling enhancements are attributed to the long-range order of the 2D lattice (particles are embedded into thin PMMA film) and collective “flips” of the

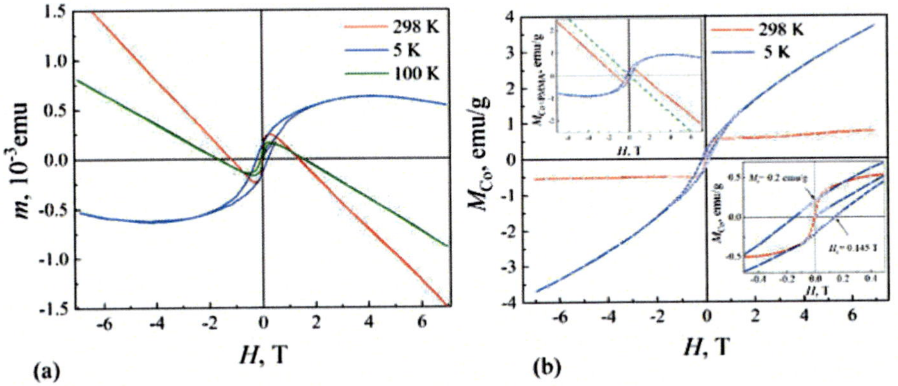

Fig. 3.11 **a** The field dependence of magnetic moment of Co nanoparticles (0.69 mg) dispersed in 10.31 g PMMA matrix taken at 5 K (blue line), 100 K (green line) and 298 K (red line). **b** The field dependence of magnetization for 5 K and 298 K. Main panel shows magnetization of Co-nanoparticles with subtracted diamagnetic contribution of PMMA host material. Top-left inset shows as-measured magnetization at 5 and 298 K. Green dashed line shows the reference line for diamagnetic contribution of host material. Bottom-right inset shows expanded view of magnetization at low fields. From [2] with permission licensed under CC BY 4.0 https://creativecommons.org

magnetic dipoles. On the other hand, the coercive field at $T = 5$ K, $H_c = 0.145$ T, is comparable to reported fcc CoNP [60].

The magnetic moment per particle was calculated for 8.8 nm Co-NP (#25) from the susceptibility χ above the blocking temperature at low field using following equation [58]

$$\chi = \frac{M_s \mu}{3k_b T}, \quad (3.8)$$

where μ is the magnetic moment per particle, M_s is the saturation magnetization, and k_b is Boltzmann's constant. The diamagnetic susceptibility of the PMMA matrix was measured and subtracted to obtain these results: At 214 K contribution of diamagnetic part (PMMA) for sample #25 is 0.45×10^{-6} emu; Magnetic moment, $m = (18.8–0.45) \times 10^{-6}$ emu $= 18.35 \times 10^{-6}$ emu; Saturation magnetization M_s at $T =$ 298 K (the same at 214 K): $M_s = 0.57$ emu/g. Susceptibility χ measured for applied field $H = 20$ Oe at 214 K: $\chi = m/(m_\rho \cdot H) = 18.35 \times 10^{-6}/(0.69 \times 10^{-3} \times 20\ \text{Oe})$ $= 1.33 \times 10^{-3}$ emu/g Oe. Here $m_\rho = 0.69$ mg is the mass of Co nanoparticles in PMMA host material (10.31 mg). Thus, the magnetic moment per particle above T $= 214$ K:

$$\mu = \frac{3k_B T \chi}{M_s} = \frac{3 \cdot 1.38 \cdot 10^{-16} \cdot 214 \cdot 1.33 \cdot 10^{-3}}{0.57} \cong 2.07 \cdot 10^{-16}\ \text{Erg/G} \quad (3.9)$$

Number of Bohr magnetons: $n = \mu/\mu_b = 2.07 \times 10^{-16}/9.27 \times 10^{-21} = 22300$, which is ~$0.687\mu_b$ per atom. Number of Co atoms in single particle, n_a, was calculated as follows: $m_{(\text{Co atom})}$ = mol. weight/N_A = 58.933 g/mol/6.022 × 10^{-23} atom/mol = 9.786 × 10^{-23} g/atom. Thus, $n_a = m_{\text{particle}}/m_{(\text{Co atom})} = 3.175 \times 10^{-18}/9.786 \times 10^{-23} = 32444$.

The above number of Bohr magnetons per atom in single 8.8 nm nanoparticle is much lower than the number of Bohr magnetons per single atom in bulk cobalt (300 K), $1.67\mu_b$ (fcc) and $1.73\mu_b$ (hcp). Partially it is caused by demagnetization factor, $N_d = 1/3$, for spherical, non-interacting particles. Note, the demagnetization does not affect the M_s in (3.8), but only χ measured at low fields, $\chi = \chi_{\text{eff}}\,(1+N_d)$. Thus, taking into account the demagnetization factor results in $0.96\mu_b$ per atom.

3.6 Optical Resonance in Spin-Polarized Co Nanoparticles

Good quality plasmon resonance in absorption is proven below to be the property of isolated Co nanoparticles. Indeed, we observe a complete suppression of sharp plasmon resonance for aggregated Co nanoparticles, probably due to the interparticle interaction inducing a spin-flip electron scattering at the particle surface. This behavior is reversible, i.e., the sharp resonance is totally restored for separated nanoparticles after sonication, as shown below. Note, that the absence of cobalt oxide shell, which could introduce an antiferromagnetic response, is controlled with the low temperature SQUID measurements.

The ab-initio simulations of the relaxation constants performed for the giant magnetoresistance show big difference for spin-up and spin-down electrons [61, 62]. Susceptibility of Co nanoparticles can be expressed as a sum of two terms coming from two independent group of electrons, thus the total polarizability is given by:

$$\alpha = r^3(\chi_\uparrow + \chi_\downarrow) = r^3\left(\frac{1}{X_\uparrow + i\delta_\uparrow} + \frac{1}{X_\downarrow + i\delta_\downarrow}\right) \tag{3.10}$$

Here we use the spectral representation of the Drude-Sommerfield model [63, 64].

$$\chi_i = \frac{\varepsilon_h - \varepsilon_{mi}}{2\varepsilon_h + \varepsilon_{mi}} = \frac{1}{X_i + i\delta_i},\ \varepsilon_{mi} = \varepsilon_{0i} - \frac{\omega_p^2}{\omega(\omega + i2\Gamma)},\ \varepsilon_h \approx \varepsilon_0,\ \omega_{sp}^2 = \frac{\omega_p^2}{\varepsilon_0 + 2\varepsilon_h} \tag{3.11}$$

$X_i = \frac{\omega_{sp}^2 - \omega^2}{\omega_{sp}^2}$, $\delta_i = \frac{\omega 2\Gamma_i}{\omega_{sp}^2}$, $2\Gamma_\uparrow = \upsilon_F/\lambda_\uparrow$ and $2\Gamma_\downarrow = \upsilon_F/\lambda_\downarrow$, where $\lambda_\uparrow$ = 12 nm, $\lambda_\downarrow$ = 0.6 nm, Fermi velocity $\upsilon_F = 2.1 \times 10^5$ m/s, thus $2\Gamma_\uparrow \approx 72.4$ meV and $2\Gamma_\downarrow \approx 1448$ meV [61, 62]. Extinction cross-section is $k\text{Im}\alpha$. Thus, the absorption spectra should look like a sharp resonance, due to spin-up electrons, plus a

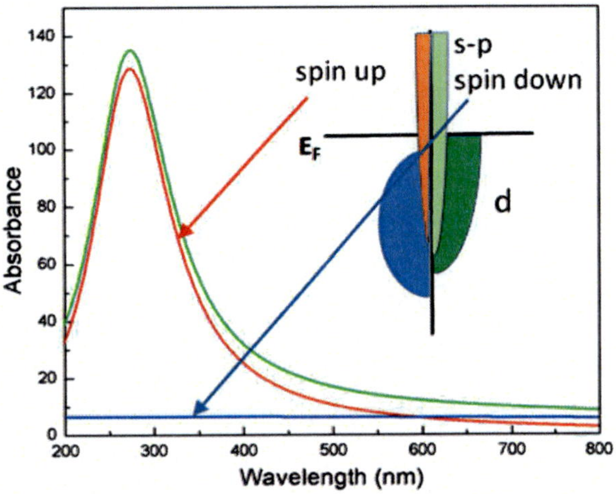

Fig. 3.12 Two plasmon model for Co nanoparticles absorbance. ω_{sp} is taken 280 nm. Green is the sum of red (spin up) and blue (spin down). Insert: a cartoon of the projected density of states typical for Co. From [2] with permission licensed under CC BY 4.0 https://creativecommons.org

broad background coming from spin-down electrons (Fig. 3.12). Thus, as soon as all possible electron scattering processes go without spin-flip, meaning that two group of electrons are independent, one should expect sharp plasmon resonance. In particular, it requires single domain nanoparticles, since inter-domain walls increase probability of spin flip electron scattering and thus two group of electrons are not independent anymore.

3.7 Effect of Dimers

Experiments show sharp plasmon resonance for isolated, single-domain Co nanoparticles (Co NPs). However, the plasmon resonance disappears, if small, two-three particles aggregates were formed. The magnetization measurements by SQUID system show superparamagnetic properties of the Co NPs at room temperature, which indicates the single-domain structure. The temperature dependence of the magnetization gives blocking temperature, which corresponds to the particle volume of this size. Below the blocking temperature field dependence of the magnetization has hysteresis behavior. The shift of the hysteresis loop cooled to 10 K at field +1 T and opposite shift for the sample cooled at −1 T allows to control the oxidation level of nanoparticles. All the results below correspond to the particles without oxide shell. Figure 13a demonstrates remarkable resonance quality of the representative spectrum for Co NPs in hexane solution shown in red. The plasmon resonance quality is about the same as for gold nanoparticles, which have resonance in the green spectral range. Co-NPs are isolated due to surfactants, trioctylphosphine and oleic acid. Dynamic light scattering data show an average size close to the mean size from TEM images.

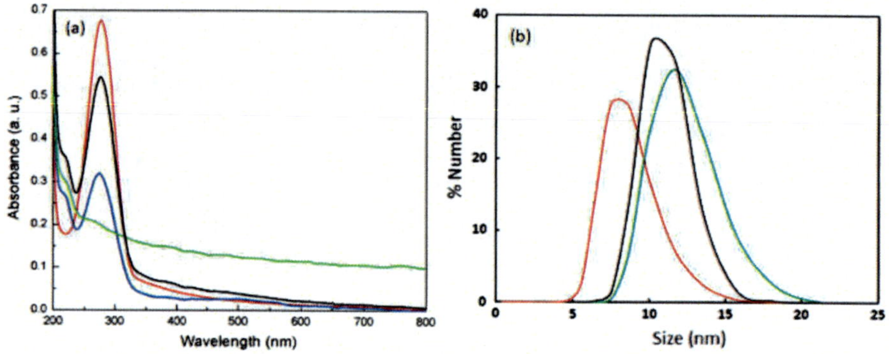

Fig. 3.13 **a** Experimental absorbance of Co-NPs in hexane. As grown Co-NPs (red), after 1 h sonication with external 130 mT DC magnetic field (blue), after 2.5 h sonication with external 130 mT DC magnetic field (green, plasmon peak is demolished), magnetic field is off and 1 h sonication (black). **b** Co-NPs size distribution measured with dynamic light scattering (red)-as grown; (green) -2.5 h sonication with external 130 mT DC magnetic field; (black) after 1 h sonication without magnetic field. From [2] with permission licensed under CC BY 4.0 https://creativecommons.org

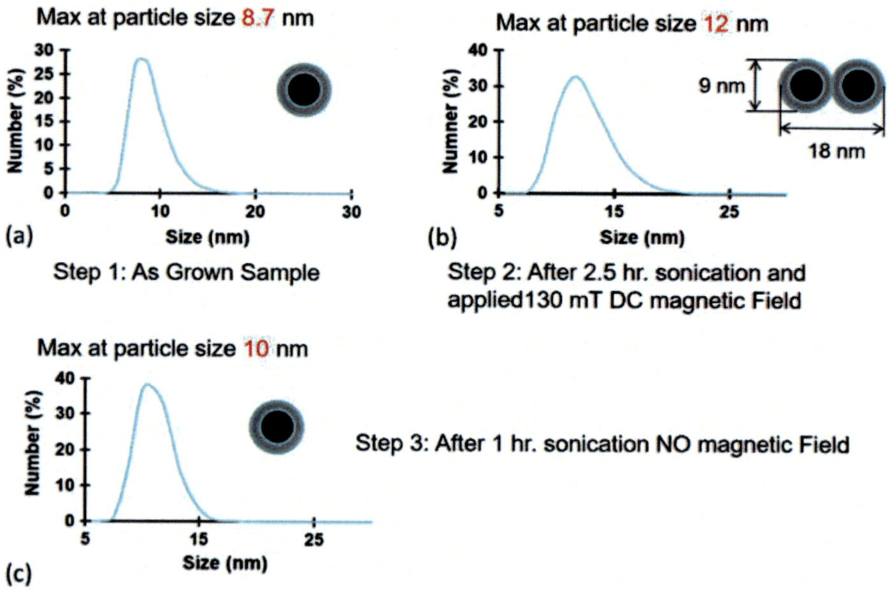

Fig. 3.14 Dynamic light scattering (DLS) size distribution of Co nanoparticles (**a**), as grown sample; (**b**), after 2.5 h sonication with external 130 mT DC magnetic field; and (**c**), after 1 h sonication without external magnetic field. From [2] with permission licensed under CC BY 4.0 https://creativecommons.org

Shown in Fig. 3.14 are dynamic light scattering (DLS) results on particle size distribution of undiluted Co nanoparticles samples. As grown Co nanoparticles have DLS average particle size 8.7 nm. Then, the solution was sonicated in the presence of 130 mT external DC magnetic field. The effective particles size has been increased from 8.7 nm to 12 nm. This increase in particle size is due to agglomeration of Co nanoparticles in the presence of magnetic field. To reverse the agglomeration, the sample was sonicated for 1 h in the absence of an external DC magnetic field. The particle size has been reduced down to 10 nm. When an external field is removed, sonication isolates the nanoparticles and results in decreasing of the effective particle size. The DLS measurements were performed at the same time as collected the optical spectra and prepared sample for magnetometry like in Fig. 3.13.

The following experiment illustrates interaction of Co nanoparticles separated by thin surfactant shell. To initiate aggregation, the 130 mT DC magnetic field together with sonication were applied to the Co NP hexane suspension in a quartz cuvette. After 1 h of "aggregation" the dynamic light scattering and absorption spectra were collected. Figure 3.3 shows reduced plasmon peak (shown in blue). After 2.5 h sonication in magnetic field plasmon peak disappeared (shown in green). The dynamic light scattering shown in the Fig. 3.3b gives increase in the hydrodynamic particles size from 8.7 to 12–13 nm corresponding to small, two-three particles aggregates. Remarkable, that the following up sonication, without external magnetic field, separates aggregated particles and the plasmon resonance is restored. Thus, this magnetic/sonication induced aggregation is a reversible process.

The magnetization of as grown, after aggregation, and after sonication without magnetic field samples, shown in Fig. 3.15, also demonstrate a reversible behavior. It first decreases after 2.5 h of sonication in magnetic field (Fig. 3.15 blue line), then return to the initial value after sonication without magnetic field (Fig. 3.15 black line).

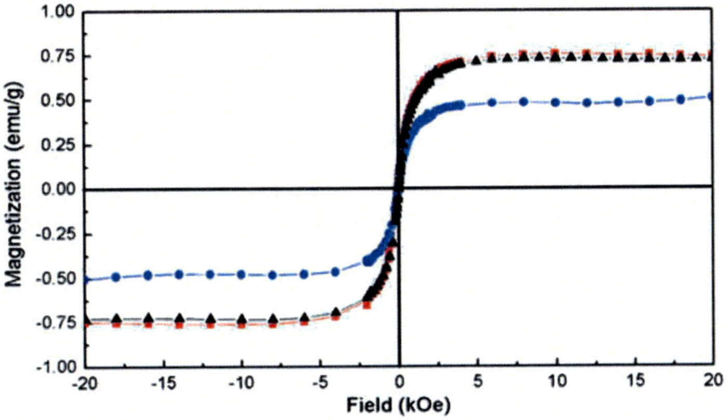

Fig. 3.15 Magnetization of Co-NPs embedded in PMMA. As grown Co-NPs (red), after 2.5 h sonication with external 130 mT DC magnetic field (blue), and 1 h sonication with magnetic field off (black). From [2] with permission licensed under CC BY 4.0 https://creativecommons.org

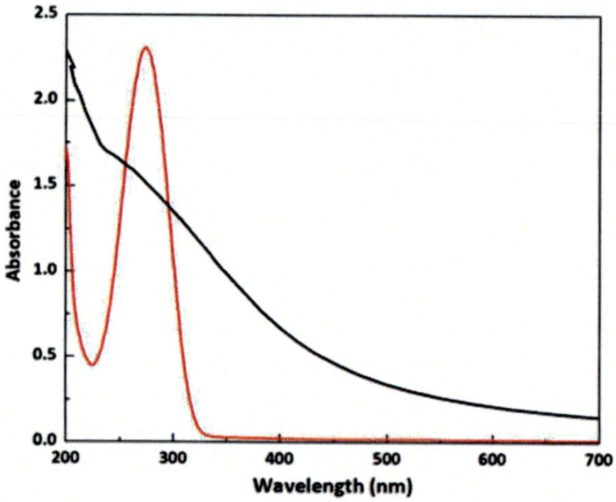

Fig. 3.16 Absorbance of the Co NPs in hexane solution: experiment (red) and calculated (black) using J&C data. [1] From [2] with permission licensed under CC BY 4.0 https://creativecommons.org

Note that calculated absorption spectrum for Co nanoparticles using J&C permittivity for Co films does not show a pronounced resonance (Fig. 3.16 black line) in contrast to the experimental spectrum for single-domain Co nanoparticles (Fig. 3.16 red line). For the nanoparticles with substantially sub-wavelength size the dipole approximation reduces Mie's theory to the following expression for the extinction cross-section [29]:

$$\sigma_{ext} = 9\frac{\omega\varepsilon_h^{1/2}}{c}V\frac{\varepsilon_h\varepsilon_2(\omega)}{[\varepsilon_1(\omega)+2\varepsilon_h]^2+\varepsilon_2^2(\omega)} \tag{3.12}$$

where ω is the light frequency, V is the volume of the spherical particle, ε_h is the dielectric permittivity of the surrounding (host) medium, and c is the speed of light. The spectrum of nanoparticles was calculated using bulk material complex permittivity $\varepsilon(\omega) = \varepsilon_1 + i\varepsilon_2$ from J&C [1]. Note that, this approach for modeling nanoparticles spectra works for nonmagnetic metals like Au, Ag , but cannot be used for Co. One can see that the calculated spectrum using permittivity measured for Co films has no good resonance. That is the reason why Co was not consider as a promising candidate so far. Indeed, if the film has multi-domain structure, where neighbour domains are typically disoriented, the electron scattering easily changes the spin polarization. Thus, electrons with spin-up become with spin-down and immediately got huge increase in relaxation rate due to available empty states in the d-band.

3.8 Conclusions

Spin-polarized metals have two channels of conductivity resulted in interesting applications including the most known giant magnetoresistance. Magnetic nanoparticles have unusual features compared to bulk materials. They manifest superparamagnetic properties in case of single-domain size. They have discrete density of states due to quantum size effect. Our experiments with Co nanoparticles clearly show a new type of plasmon excitation. This type of plasmon has unusual properties due to existence of two independent groups of electrons with opposite spins providing weak interaction so that all electron scattering processes occur without spin flip. Magnetic response of the nanoparticles enables controlled and reversible aggregation accompanied by the tailoring of optical absorption. Magnetic nature of the nanoparticles suggests a new type of these plasmons. Magnetic response of Co nanoparticles shows less magnetic moments per atom relative to the bulk value, namely 0.68 μ_b and taking into account the demagnetization factor 0.96 μ_b per atom versus 1.7 μ_b. The exchange interaction of electrons splits the energy bands between spin-up electrons and spin-down electrons. It makes possible to coexist two independent channels of conductivity as well as two independent plasmons in the same nanoparticle with very different electron relaxation. Indeed, the density of empty states in a partially populated d-band is high, resulting in a large relaxation rate of the spin-down conduction electrons and consequently in low quality of the plasmon resonance. In contrast, the majority electrons with a completely filled d-band do not provide final states for the scattering processes of the conduction spin-up electrons, therefore supporting a good quality plasmon resonance. The scattering without spin flip is required to keep these two plasmons independent.

References

1. P.B. Johnson, R.W. Christy, Optical constants of transition metals: Ti, V, Cr, Mn, Fe Co, Ni, and Pd. Phys. Rev. **9**, 5056 (1974). https://doi.org/10.1103/PhysRevB.9.5056
2. H. Bhata, A.E. Aliev, V.P. Drachev, New mechanism of plasmons specific for spin-polarized nanoparticles. Sci. Rep. **9**, 2019 (2019). https://doi.org/10.1038/s41598-019-38657-w
3. A.R. Young, Chromophores in human skin. Phys. Med. Biol. **42**, 789–802 (1997)
4. G.J. Thomas, Raman spectroscopy of protein and nucleic acid assemblies. Annu. Rev. Biophys. Biomol. Struct. **28**, 1–27 (1999)
5. Y. Kumamoto, A. Taguchi, N.I. Smith, S. Kawata, Deep UV resonant Raman spectroscopy for photodamage characterization in cells. Biomed. Optics Express **2**, 927–936 (2011)
6. H. Takeuchi, UV Raman markers for structural analysis of aromatic side chains in proteins. Anal. Sci. **27**, 1077–1086 (2011)
7. L. Ashton, C.E.M. Hogwood, A.S. Tait, J. Kuligowski, C.M. Smales, D.G. Bracewell, A.J. Dicksone, R. Goodacre, UV resonance Raman spectroscopy: a process analytical tool for host cell DNA and RNA dynamics in mammalian cell lines. J. Chem. Technol. Biotechnol. **90**, 237–243 (2015)
8. F. Scherer, M. Anton, U. Schillinge, J. Henke, C. Bergemann, A. Kruger, B. Gansbacher, C. Plank, Magnetofection: enhancing and targeting gene delivery by magnetic force in vitro and in vivo. Gene Ther. **9**, 102–109 (2002)

9. A.S. Lübe, C. Alexiou, C. Bergemann, Clinical applications of magnetic drug targeting. J. Surg. Res. **95**, 200 (2001)
10. N.A. Brusentsov, L.V. Nikitin, T.N. Brusentsova, F.S. Bayburtskiy, L.I. Shumakov, N.Y. Jurchenko, Magnetic fluid hyperthermia of the mouse experimental tumor. J. Magn. Magn. Mater. **252**, 378 (2002)
11. V.I. Shubayev, T.R. Pisanic, S. Jin, Magnetic nanoparticles for theragnostics. Adv. Drug Deliv. Rev. **61**, 467–477 (2009)
12. C. Jiang, J.B. Wechuck, W.F. Goins, D.M. Krisky, D. Wolfe, M.M. Ataai, J.C. Glorioso, Immobilized cobalt affinity chromatography provides a novel, efficient method for herpes simplex virus type 1 gene vector purification. J. Virol. **78**, 8994–9006 (2004)
13. J.H. Lee, Y.M. Huh, Y.W. Jun, J.W. Seo, J.T. Jang, H.T. Song, S. Kim, Cho, H.G. Yoon, J.S. Suh, J. Cheon, Artificially engineered magnetic nanoparticles for ultra-sensitive molecular imaging. Nat. Med. **13**, 95–99 (2007)
14. L.M. Parkes, R. Hodgson, L.T. Lu, L.D. Tung, I. Robinson, D.G. Fernig, N.T.K. Thanl, Cobalt nanaoprticles as a novel magnetic resonance contrast agent-relaxivities at 1.5 and 3 Tesla. Contrast Media Mol. Imag. **3**, 150–156 (2008)
15. N.A. Frey, S. Peng, K. Cheng, S. Sun, Magnetic nanoparticles: synthesis, functionalization, and applications in bioimaging and magnetic energy storage. Chem. Soc. Rev. **38**, 2532–2542 (2009)
16. L.H. Reddy, J.L. Arias, J. Nicolas, P. Couvreur, Magnetic nanoparticles: design and characterization, toxicity and biocompatibility, pharmaceutical and biomedical applications. Chem. Rev. **112**, 5818–5878 (2012)
17. B. Issa, I.M. Obaidat, B.A. Albiss, Y. Haik, Magnetic nanoparticles: surface effects and properties related to biomedicine applications. Int. J. Mol. Sci. **14**, 21266–21305 (2013)
18. S. Lal, S. Link, N.J. Halas, Nano-optics from sensing to waveguiding. Nat. Photon. **1**, 641–648 (2007)
19. K.A. Willets, R.P. Van Duyne, Localized surface plasmon resonance spectroscopy and sensing Annu. Rev. Phys. Chem. **58**, 267–297 (2007)
20. M. Stockman, Nanophotonics: past, present, and glimpse into future. Opt. Express **19**, 22029–22106 (2011)
21. V.P. Drachev, V.M. Shalaev, Biomolecule sensing with adaptive plasmonic nanostructure, a chapter in a "Surface Enhanced Raman Scattering - Physics and Applications", ed. by K. Kneipp, M. Moskovits, and H. Kneipp, Springer Verlag, Topics in applied physics (2006)
22. V.P. Drachev, V. Nashine, M.D. Thoreson, D. Ben-Amotz, V.J. Davisson, V.M. Shalaev, Adaptive silver films for detection of antibody-antigen binding. Langmuir **21**, 8368–8373 (2005)
23. V.P. Drachev, V.C. Nashine, M.D. Thoreson, D. Ben-Amotz, V.J. Davisson, V.M. Shalaev, Surface-enhanced Raman scattering of biomolecules with adaptive nanostructures. J. Raman Spectrosc. **36**, special issue on SERS, 648–656 (2005)
24. V.P. Drachev, M.D. Thoreson, E.N. Khaliullin, V.J. Davisson, V.M. Shalaev, Surface-enhanced raman difference between human insulin and insulin lispro detected with adaptive nanostructures. J. Phys. Chem. **108**, 18046–18052 (2004)
25. K. Lee, V.P. Drachev, J. Irudayaraj, DNA-Gold nanoparticle reversible networks grown on cell surface marker sites: application in diagnostics. ACS Nano **5**(3), 2109–2117 (2011)
26. V.P. Drachev, A.K. Buin, H. Nakotte, V.M. Shalaev, Size dependent $\chi(3)$ for conduction electrons in Ag nanoparticles. Nano Lett. **4**, 1535 (2004)
27. K.P. Chen, V.P. Drachev, J.D. Borneman, A.V. Kildishev, V.M. Shalaev, Drude relaxation rate in grained gold nanoantennas. Nano Lett. **10**, 916–922 (2010)
28. V.P. Drachev, U.K. Chettiar, A.V. Kildishev, W. Cai, H.-K. Yuan, V.M. Shalaev, The Ag dielectric function in plasmonic metamaterials. Opt. Express **16**, 1186–1195 (2008)
29. U. Kreibig, M. Volmer, Optical properties of metal clusters. Springer-Verlag, Berlin (1995), 532p. https://doi.org/10.1007/978-3-662-09109-8
30. J.M. McMahon, G.C. Schatz, S.K. Gray, Plasmonics in the ultraviolet with the poor metals Al, Ga, In, Sn, Tl, Pb, and Bi. Phys. Chem. Chem. Phys. **15**(15), 5415–5423 (2013)

31. M.G. Blaber, C.J. Engel, S.R.C. Vivekchand, S.M. Lubin, T.W. Odom, G.C. Schatz, Eutectic liquid alloys for plasmonics: theory and experiment. Nano Lett. **12**, 4324–4328 (2012)
32. M.G. Blaber, M.D. Arnold, M.J. Ford, A review of the optical properties of alloys and intermetallics for plasmonics J. Phys.: Condens. Matter **22**, 143201–143215 (2010)
33. G. Maidecchi, G. Gonella, R.P. Zaccaria, R. Moroni, L. Anghinolfi, A. Giglia, S. Nannarone, L. Mattera, H.-L. Dai, M. Canepa, F. Bisio, Deep ultraviolet plasmon resonance in aluminum nanoparticle arrays. ACS Nano **7**, 5834 (2013)
34. M.W. Knight, N.S. King, L. Liu, H.O. Everitt, P. Nordlander, N.J. Halas, Aluminum for plasmonics. ACS Nano **8**, 834–840 (2014)
35. T. Ding, D.O. Sigle, L.O. Herrmann, D. Wolverson, J. J. Baumberg, Nanoimprint Lithography of Al Nanovoids for Deep-UV SERS. ACS Appl. Mater. Interf. https://doi.org/10.1021/am505511v
36. Y. Kumamoto, A. Taguchi, M. Honda, K. Watanabe, Y. Saito, S. Kawata, Indium for deep ultraviolet surface-enhanced resonance raman scattering. ACS Photon. **1**, 598–603 (2014)
37. N.F. Mott, The resistance and thermoelectric properties of the transition metals. Proc. Royal Soc. **156**, 368 (1936). https://doi.org/10.1098/rspa.1936.0154
38. N.F. Mott, Electrons in transition metals. Adv. Phys. **13**, 325 (1964). https://doi.org/10.1080/00018736400101041
39. M.N. Baibich, J.M. Broto, A. Fert, N. Van Dau, F. Petroff, P. Etienne, G. Creuzet, A. Friederich, Chazelas, Giant magnetoresistance of (001) Fe/(001) Cr magnetic superlattices. J. Phys. Rev. Lett. **61**, 2472 (1988). https://doi.org/10.1103/PhysRevLett.61.2472
40. V.A. de la Peña O'Shea, I. de P. R. Moreira, A. Roldán, A. Illas, Electronic and magnetic structure of bulk cobalt: The αα, ββ, and εε-phases from density functional theory calculations.J. Chem. Phys. **133**, 024701 (2010)
41. D. Gull, Electron mean free path in elemental metals. J. Appl. Phys. **119**, 085101 (2016)
42. P. Perdew, K. Burke, M. Ernzerhof, Phys. Rev. Lett. **77**, 3865 (1996)
43. V. Blum, R. Gehrke, F. Hanke, P. Havu, V. Havu, X. Ren, K. Reuter, M. Scheffler, Comput. Phys. Commun. **180**, 2175–2196 (2009)
44. V. Havu, V. Blum, P. Havu, M. Scheffler, J. Comput. Phys. **228**, 8367–8379 (2009)
45. A. Marek, V. Blum, R. Johanni, V. Havu, B. Lang, T. Auckenthaler, A. Heinecke, H.-J. Bungartz, H. Lederer, J. Phys. Condens. Matter **26**, 213201 (2014)
46. S. Levchenko, X. Ren, J. Wieferink, R. Johanni, P. Rinke, V. Blum, M. Scheffler, Comput. Phys. Commun. **192**, 60–69 (2015)
47. J.-X. Liu, H.-Y. Su, D.-P. Sun, B.-Y. Zhang, W.-X. Li, J. Am. Chem. Soc. **135**, 16284 (2013)
48. J.M. Rahm, P. Erhart, WulffPack: a python package for Wulff constructions. J. Open Source Softw. 5.45 (2020): 1944
49. V.F. Puntes, K. Krishnan, A.P. Alivisatos, Colloidal nanocrystal shape and size control: the case of cobalt. Science **291**, 2115–2117 (2001)
50. H.T. Yang, C.M. Shen, Y.G. Wang, Y.K. Su, T.Z. Yang, H.J. Gao, Stable cobalt nanoparticles passivated with oleic acid and trioctylphosphine. Nanotechnology **15**, 70–74 (2004)
51. Y.K. Su, C.M. Shen, T.Z. Yang, H.T. Yang, H.J. Gao, H.L. Li, The dependence of Co nanoaprticle sizes on the ratio of surfactants and the influence of different crystal sizes on magnetic properties. Appl. Phys. A **81**, 569–572 (2005)
52. S.H. Sun, C.B. Murray, Synthesis of monodisperse cobalt nanoparticles and their assemble into magnetic superlattices. J. Appl. Phys. **85**, 4325 (1999). https://doi.org/10.1063/1.370357
53. A. Kelly, K. Knowles (2012) *Crystallography and Crystal Defects* 2nd Ed. (Wiley, 2011)
54. L. Néel, Théorie du traînage magnétique des ferromagnétiques en grains fins avec application aux terres 5 cuites. Ann. Géophys. **5**, 99–136 (1949)
55. W.T. Coffey, D.S.F. Crothers, Yu.P. Kalmykov, E.S. Massawe, J.T. Waldron, Exact analytic formula for the correlation time of a single-domain ferromagnetic particle. Phys. Rev. E **49**, 1869 (1994)
56. J.L. Dormann; F. D'Orazio, F. Lucari, E. Tronc, P. Prene, J.P. Jolivet, D. Fiorani, R. Cherkaoui, M. Nogue's, Thermal variation of the relaxation time of the magnetic moment of γ-Fe_2O_3 nanoparticles with interparticle interactions of various strengths. Phys. Rev. B **53**, 14291–14297 (1996)

57. P. Gambardella, M. Rusponi, M. Veronese, S.S. Dhesi, C. Grazioli, A. Dallmeyer, I. Cabria, R. Zeller, P.H. Dederichs, K. Kern, C. Carbone, H. Brune, Giant magnetic anisotropu of single atoms and nanoparticles. Science **300**, 1130–1133 (2003)
58. J.P. Chen, C.M. Sorensen, K.J. Klabund, Enhanced magnetization of nanoscale colloidal cobalt particles. Phys. Rev. B **51**, 11527 (1995)
59. M.E. Mohenry, S.A. Majetich, J.O. Artman, M. DeGraef, S.W. Staley, Superparamagnetism in carbon-coated Co particles produced by the Kratschmer carbon arc process. Phys. Rev. B **49**, 11358 (1994)
60. C. Petit, Z.L. Wang, M.P. Pileni, Seven-nanometer hexagonal close packed cobalt nanocrystals for high-temperature magnetic applications through a novel annealing process. J. Phys. Chem. B **109**, 15309–15316 (2005)
61. E.Y. Tsymbal, D.G. Pettifor, Perspectives of giant magnetoresistance. in *Solid state Physics*, ed. by H. Ehrenreich, F. Spaepen (Eds), (Academic Press, Boston), 56, pp. 113–237 (2001). https://digitalcommons.unl.edu/physicstsymbal/50
62. D.J. Gall, App. Phys. **119**, 085101–085101–085105 (2016). https://doi.org/10.1063/1.4942216
63. D.J. Bergman, D. Stroud, Properties of macroscopically ingomogeneous media, in *Solid State Physics*, ed. by H. Ehrenreich, D. Turnbull (Academic, Boston), 46, pp. 149–270 (1992)
64. V.A. Markel, L.S. Muratov, M.I. Stokman, Optical properties of fractals: theory and numerical simulation. Sov. Phys. JETP **71**(3), 455–464 (1990)

Chapter 4
Chirality and Antiferromagnetism in Optical Metasurfaces

Kun Huang

Abstract In this chapter, we will introduce optical chiral metasurfaces that modulate the phase and amplitude of the circularly polarized (CP) light by using orientation-rotated artificial subwavelength structures. For high-efficiency dielectric chiral metasurfaces, the nanostructures work as nanoscale half-waveplates that hold the physical origins of antiferromagnetism by inducing multiple anti-parallel magnetic dipoles. The interaction between metasurfaces and chiral CP light (carrying the spin angular momentum) enables polarization meta-optics for the applications such as lens, grating and hologram. In addition, the optical vortex beams carrying the orbital angular momentum of light is introduced as another chiral light because the helical wavefronts with opposite handednesses exhibit the mirror symmetry. Such a chiral feature of optical vortex beam is used to probe the chirality of micro-structures, i.e., helical dichroism (HD), which works as a counterpart of circular dichroism (CD) created by the interaction between CP light and chiral molecules. The concluding remarks about the interaction between chiral light and nano-/microstructures are made at the end of this chapter.

4.1 Introduction

Light is a kind of electromagnetic radiation that carries the detectable energy with high oscillating frequencies of hundreds of terahertzs for various usages such as animal vision, illumination and display, information transform, material processing, optical imaging, energy transfer and storage in plants, heating, nano-fabrication, remote sensing and medical surgery. Due to the rich diversity in these applications, the requirement of manipulating light in a highly customized and arbitrary way increases rapidly with the development of modern technology. The fundamental physics of tailoring light is the interaction of light with natural or artificial materials

K. Huang (✉)
Department of Optics and Optical Engineering, University of Science and Technology of China, Hefei, China
e-mail: huangk17@ustc.edu.cn

E. Kamenetskii (ed.), *Chirality, Magnetism and Magnetoelectricity*, Topics in Applied Physics 138,
https://doi.org/10.1007/978-3-030-62844-4_4

at the different scales ranging from macroscopic bulky crystals, to man-made nanostructures, to microscopic molecules and atoms. Recently, optical metasurfaces made of artificial subwavelength structures has drawn great attention due to its capacity of tailoring the amplitude, phase and polarization of light in a subwavelength thickness. It therefore offers the opportunity to integrate optical elements (Sect. 4.1.1) within a compact volume. The rapid development of optical metasurfaces (Sect. 4.1.2) benefits from matured nanotechnology, which enables the fabrication of optical subwavelength structures. These nanostructures could confine nanoscale resonating modes that behave like electric and magnetic dipoles or multipoles. The electromagnetic resonances of the confined nanomodes are modified through changing the dimension or geometry of the nanostructures, hereby realizing the required modulation of light. Based on these fundamental origins, metasurfaces can be mainly categorized into shape-varied metasurfaces, geometric metasurfaces, Huygens metasurfaces. In this chapter, we only focus on geometric metasurfaces composed of size-fixed orientation-rotated nanostructures that control the phase of circularly polarized light, which carries the spin angular momentum (SAM) of light and exhibits the polarization chirality of light (Sect. 4.2.1). In addition, the orbital angular momentum (OAM) of light carried by optical vortices beams with a helical wavefront (Sect. 4.2.2) characterizes the phase chirality of light. Due to the strong response to the polarization chirality of light, the geometric metasurfaces can also be called as chiral metasurfaces (Sect. 4.3). According to their material platforms, chiral metasurfaces are discussed in terms of plasmonic chiral metasurfaces (Sect. 4.3.1), chiral nanosieves (Sect. 4.3.2), and dielectric chiral metasurfaces (Sect. 4.3.3). In the transmission mode, the dielectric chiral metasurfaces have much higher efficiency than plasmonic metasurfacea and chiral nanosieves, due to the low absorption of dielectrics. In the configuration of dielectric metasurfaces, the nanostructures could support the induced antiparallel magnetic dipoles, implying the antiferromagnetic resonances. Such a concept of antiferromagnetism is used to explain the physics governing dielectric nano-halfwaveplates, which is the functionality of nano-structures in chiral metasurfaces. Finally, the applications of chiral light and chiral metasurfaces (Sect. 4.4) are introduced by addressing the exciting topics such as circular dichroism and helical dichroism (Sect. 4.4.1), and chiral meta-optics (Sect. 4.4.2), followed by a brief conclusion in Sect. 4.5.

4.1.1 Optical Elements

Before 1900, the optical elements such as lenses, glasses, gratings, birefringent crystals, polarizers, mirrors and filters have been well developed with the help of traditional technology of cold machining, thus forming the cornerstone of geometry optics where the dimension is much larger than the operating wavelengths. Light in geometry optics is simplified into a ray of light so that the intensity of an optical beam can be taken as the density of the rays at its cross section. The manipulation of light originates mainly from the refraction and refraction of the ray at the interface between two media with different refractive indices, where the curvature of the interface and the index difference between two media are the main parameters to tailor the ray of light. The disadvantage of geometry optics comes from the fact that the other variables such as phase, polarization, coherence and diffraction of light are ignored completely due to the model of the ray, so that its applied territory is quite limited.

When the optical elements have the feature size at the level of wavelength, the diffraction and interference dominate its propagation of light and the light-matter interaction due to the wave properties of light. The well-developed lithography enables us to acquire the micro- or nano-scale elements such as optical fibers [1], holograms [2], photonic crystals [3], plasmonic devices [4], optical metamaterials [5, 6] and metasurfaces [7–11], quantum devices, where their feature sizes range from tens of microns to several nanometers. The diffraction effects in micro-devices is weak, which is a good issue for optical fibers during the long-distance propagation. But, the weak diffraction makes micro-pixelated traditional holograms with small field-of-view, leading to the limited applications. Meanwhile, the micro-pixel pitch in traditional holograms allows only the phase and amplitude modulation of light.

When the pixels of devices are below the subwavelength, the polarization effect is important because the subwavelength structures have different electromagnetic resonances for transverse electric and magnetic fields. The photonic crystals with periodic subwavelength pixels could prohibit the motion of photons through confining the electromagnetic modes within the index-contrast structures, hereby creating the forbidden states at the designed frequencies. A well-designed photonic crystal could support the simultaneous oscillation of multiple modes with different polarizations, which is useful to generate the cylindrical vector beams with spatially variant states of polarization in active nanolasers [12]. As its killer-man application, the photonic crystal fibers offer the well-confined modes that are impossible in conventional fibers. However, the periodic properties make photonic crystals incompetent at manipulating the phase of light at the transvers plane vertical to the propagation direction. Such a situation also exists at the periodic meta-materials with the pixel pitches much smaller than one wavelength, where the effective medium theory can be used to approximate its optical properties. For visible light with the wavelengths ranging from 400 nm to 800 nm, the pixel pitches in metamaterials is too small for the state-of-art nanolithography to fabricate a three-dimensional structure, so that the novel properties of metamaterials are frequently demonstrated at the micro-wave and radio frequencies [13].

Fortunately, we can realize two-dimensional meta-materials made of high-refraction-index dielectric or metal-based plasmonic nano-structures with the pixel pitches around a half of wavelength [14], facilitating the fabrication issue by using electron-beam or focused ion lithography. Instead of effective-medium approach in three-dimensional metamaterials, two-dimensional metamaterials have the distinct working principle of controlling the polarization, amplitude, phase and wavelength of light through the electromagnetic responses of the confined nanomodes, which oscillate between two lower-refraction-index media (i.e., substrate and surrounding medium) [15]. Due to its planarity and ultrathin thickness of smaller than one wavelength, two-dimensional meta-materials are usually phrased as "metasurfaces", which can be realized easily by using one-step top-down lithography that is a matured technology of manufacturing mass-product integrated circuits in semiconductor industry. Since it could realize full control of light arbitrarily in a compact volume, metasurface-based devices are drawing the increasing attention in integrated opto-electronics and portable devices. Therefore, optical metasurfaces are the main topic that we will focus in this chapter.

From the viewpoint of functionality, the main contribution of optical metasurfaces to optics community is to introduce arbitrary polarization manipulation of light, in addition to amplitude and phase modulation. It allows us to manipulate the local states of polarization at any pixel of interest across an entire meta-device, and simultaneously customize the corresponding amplitude and phase in a high accuracy, which is not possible for all the previous optical elements. For a polarized light, two commonly used orthogonal bases such as linear polarization (i.e., $\mathbf{e}_x$ and $\mathbf{e}_y$, where x and y are the transverse coordinates that are perpendicular to the propagating direction of light) and circular polarization (i.e., $\mathbf{e}_x \pm i\mathbf{e}_y$) can be used to characterize its polarization property. The circular polarization is a simple combination of two orthogonal linear polarizations with an additional phase delay of $\pm\pi/2$, which leads to the rotating vector of electric field in circularly polarized light. Determined by the sign of delayed phase, the rotation is along the clockwise or anti-clockwise direction, which is related with the chirality of light. Optical geometric metasurfaces could transfer a circularly polarized light into its cross-polarization, meanwhile imprint an additional phase that is two times of the rotating angle of the size-fixed rectangle-shape nanorodes. The rectangle nanorodes operate as a miniaturized half-waveplate that provides a phase delay of π between two orthogonal components of electric fields. The mechanism of these half-waveplates depends on the materials: the metal-based plasmonic nano-halfwaveplates hold the physical origin of electric dipole resonances [16], while the high-refraction-index dielectric nano-halfwaveplates originate from the antiferromagnetic resonances [15]. The interaction between chiral light and geometric metasurfaces increases significantly the full manipulation of circular polarized light, and therefore results in many intriguing applications such as polarization meta-optics, spin Hall effects of light [17, 18], beam shaping of chiral light, detection of chirality, which forms the primary configuration of this chapter.

4.1.2 *History of Optical Metasurfaces*

Ultrathin metasurfaces could full control of electromagnetic waves by using the spatially varied subwavelength structures [11, 13, 14, 19–23] that supports the local responses of well-confined modes, which are commonly used elements at its long-wavelength spectrum. For example, the antennas at the radio and microwave frequencies are designed at the scale of subwavelength for sending or receiving the signals. However, optical metasurfaces have much shorter history because the fabrication and simulation issue of subwavelength structures at the visible or infrared wavelengths cannot be well-solved until the personal computers with powerful scientific computing capacities are popularized to assist the design and manufacture of optical nano-structures. The first optical metasurfaces were reported in 1998 by Simion Astilean, Philippe Lalanne and their collaborators to demonstrate the blazed gratings with spatially varied subwavelength features of TiO_2 on a quartz substrate at the working wavelength of 633 nm [7, 8]. The phase modulation is controlled by the filling factor of size-varied subwavelength structures in a pixel pitch, with the guidance of effective medium theory. The measured efficiency of 1-order diffraction is 83%, which is slightly larger than the theoretical prediction and indicates a fundamental breakthrough as a novel element. Although further demonstration of meta-lenses is also done with good agreement, the potential of these subwavelength structures has not been fully recognized due to the rough design tools of effective medium theory that ignores the local response of confined nanomodes and the unpolarized illumination that excludes the deep investigation into polarization effects. In addition, the product yield of these devices is lower due to the challenging fabrication at that moment. As a result, these excellent works at their group terminate so that they are not widely exposed among the optics community until recently.

Since 2000, Hasman group reported a series of works that utilized the rotating subwavelength metal or dielectric nanorods to control the phase and polarization of light by introducing the important concept of geometric phase or Berry phase [9, 10, 24–26]. Compared with the size-varied subwavelength structures developed by Philippe Lalanne et al., geometric metasurfaces with size-fixed nanorodes offer much higher level of phase modulation because the geometric phase is determined by the rotating angle of nanorodes without changing their dimension, hereby facilitating the fabrication due to uniformly distributed nanostructures. With this powerful platform of geometric metasurfaces, Hasman group demonstrated many novel devices such as cylindrical vector beams, polarization-dependent hologram and lenses at the infrared wavelengths. These works highlight the significance of subwavelength devices and therefore play an important role in developing and popularizing various optical meta-devices with polarization features. Considering that the dominated electromagnetic resonances existing in geometric metasurfaces were not unveiled clearly at that moment, these metal or dielectric geometric metasurfaces, operating in a transmission mode that is preferred in most optical equipment and systems, were not designed with the suitable geometry and materials, leading to low efficiency for practical applications.

With the rapid development of metamaterials for applications such as perfect lens and electromagnetic cloaking, the interest of utilizing subwavelength structures to control the waves increases quickly among the entire electromagnetic community. The related computing algorithms such as finite-element method (FEM) and time-domain finite-difference (FDTD) method are available at low cost and carried out even at a personal computer, which extremely decreases the threshold of simulating the electromagnetic responses of optical nano-structures. In 2011, Federico Capasso et al. developed size-varied V-shape structures with deep-subwavelength features to tailor the discrete phase of a linearly polarized light (perpendicular to the incident polarization) for blazed gratings and vortex generation at the mid-infrared wavelengths [11]. It offers the first time demonstration of manipulating the phase and amplitude of scattering light with linear co-polarization, and completely releases the ability of subwavelength structures in shaping the polarized light arbitrarily. The successive works by Capasso group refresh the concept of traditional optics [20, 27–32], which leads to the persistent burst of optical metasurfaces from infrared and visible to ultraviolet regions.

After the consequent development around 10 years, optical metasurfaces, nowadays, have the clear physics that shapes light through the electromagnetic interaction with subwavelength structure in terms of the interference of the confined nanomodes. In theory, we can categorize optical metasurfaces into gradient metasurfaces with sized-varied geometry [11], geometric metasurfaces [16, 33], Huygens metasurfaces [34–36], nanosieve metasurfaces with amplitude modulation [37–40]. The working principle of gradient and geometric metasurfaces have been introduced above. Huygens metasurfaces support the simultaneous resonances of two transversely located electric and magnetic dipoles having the crossed directions and the equal amplitudes, so that every meta-atom or nanostructure in Huygens metasurfaces can be taken as an ideal Huygens point source that has the forward transmission of 1 but the backward scattering of 0, thus offering another strategy of high-efficiency metasurfaces. Photon nanosieves are one kind of amplitude metasurfaces that shapes the binary transmission of light with the etched holes on an opaque thin film. They become the only choice of meta-devices at the extremely short wavelengths such as extreme ultra-violet light and soft-X ray, where all the materials are absorbing [41] so that the electromagnetic resonances of nanomodes cannot exist.

From the viewpoint of structure, optical metasurfaces could utilize the unit cell of any geometric shape for controlling light. The size-varied circular, V-shape and rectangle nano-structures are frequently used in gradient metasurfaces [11, 42], while the orientation-rotated rectangle structures are much preferred in geometric metasurfaces [16, 33] due to the limitation of its mechanism. Optical Huygens metasurfaces frequently adopt the circular dielectric nanodisks to excite the electric and magnetic dipoles [34–36]. Nanosieve metasurfaces generally use the circular or rectangle apertures to modulate the amplitude or geometric phase of light [37–40].

Efficiency of optical metasurfaces is a key parameter for their practical applications in industry. Due to the intrinsic properties of electromagnetic resonances, optical metasurfaces have the efficiency determined mainly by the material platform. The reflective metasurfaces could exhibit the total efficiency of ~90% by using metals or a combination of metallic and dielectric materials with low absorption at the wavelengths of interest [15, 42, 43]. However, the metals are not preferred in an optical transmissive metasurface because of their strong ohmic loss and low polarization conversion [44]. High-efficiency optical metasurfaces in the transmission mode usually uses the refraction-index and large-bandgap dieletric nano-structures sitting on a low-index and transparent substrate, creating a well-confined dielectric nano-cavity that has much higher refractive index than the surrounding environment. This conclusion is valid for all gradient, geometric and Huygens metasurfaces in transmission. Since most of optical equipment and systems work in a transmission mode, optical dielectric metasurfaces with high efficiency are the hotspot among the

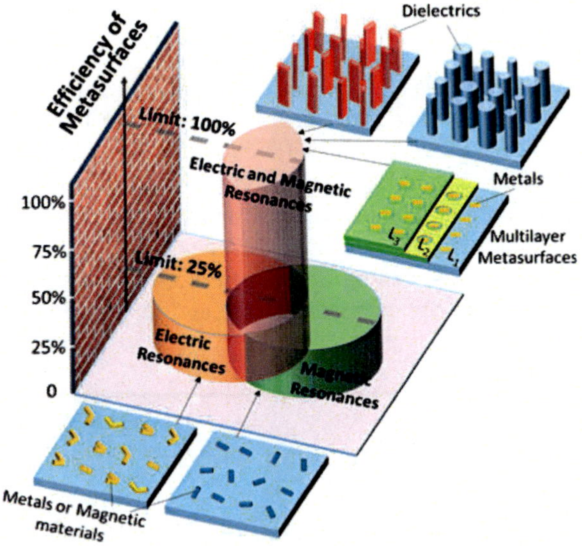

Fig. 4.1 The summary about the efficiency limits of optical metasurfaces. The limit for metasurfaces supporting either electric or magnetic resonance is 25%, while the high aspect-ratio dielectric and multilayer metasurfaces have the efficiency limit of 100%

current metasurface community. To achieve high efficiency, optical reflective metasurfacecs employ the gold (Au) at the near-infrared frequencies, silver (Ag) at the visible wavelengths and aluminum (Al) at the ultrav-violet spectrum, while transmission metasurfaces prefer the silicon (Si) for the near-infrared light, titanium dioxide (TiO_2), gallium nitride (GaN), silicon nitride (Si_3N_4) and gallium phosphide (GaP) for visible light, niobium pentoxide (Ni_2O_5), hafnium dioxide (HfO_2) and diamond for ultra-violet light. Figure 4.1 shows a brief summary about the efficiency limit of optical metasurfaces with electric and magnetic resonances.

Optical devices demonstrated by metasurfaces are frequently called as meta-optics. Compared with traditional optics, meta-optics such as metalenses [33, 45, 46], meta-holograms [37, 38, 43, 44, 47–50], meta-gratings [51–54], beam meta-splitters [55, 56], vortex meta-generators [57, 58], meta-retroreflectors [59], meta-biosensors [60] and image meta-processors [61] have the advantages of compact volume, polarization sensitivity, multiplexing functionality, large degree of freedom, high integration and light weight for portable devices. Benefitting from these novel properties, meta-optics could be used in high-resolution imaging, display, nano-lasing, beam shaping, information recognition, nonlinear optics, quantum physics, bio-sensing, spectroscopy and augmented reality. Note that, this technology of optical metasurfaces is hunting for its killer-man applications where only meta-optics can solve the relative problems so that the meta-devices cannot be substituted by others. Before reaching this goal, we speculate that optical reconfigurable metasurfaces might be the most important issue that should be steadily solved, although many intriguing proposals [62–69] have been reported in a limited or imperfect functionality for the practical usage.

4.2 Chirality of Light

Chirality refers to an object or system cannot be superposed or brought to coincide with itself by its mirror image. It exists in our daily life (left and right hands), physics (spin of a particle), optics (circularly polarized light, optical vortices with opposite topological charge) and chemistry (molecules). The interaction between chiral light and matter dominates the dichromic responses of microscopic particles, molecules and nano-structures, which becomes an efficient tool to detect or characterize the properties of interested objects in both physics and chemistry. Here, the chirality of light and optical metasurfaces and their interactions will be highlighted in this chapter.

4.2.1 Spin of a Photon and Spin Angular Momentum

A circularly polarized (CP) monochromatic light contains two orthogonal components in its vector potential that can be expressed as [70]

$$\mathbf{A} = u(x, y, z)\mathbf{e}^{i(kz-\omega t)} \cdot (\mathbf{e}_x + \sigma i \mathbf{e}_y), \tag{4.1}$$

where the item $\sigma = \pm 1$ denotes the spin or chirality of light, the wave number is $k = \omega/c$, ω is the angular frequency and c is the speed of light in vacuum, $\mathbf{e}_x$ and $\mathbf{e}_y$ are the unit vectors along x and y directions, respectively. Its vector rotates in an anti-clockwise or clockwise way, which is determined by σ. Generally, $\sigma = 1$ indicates the left-handed circular polarized light while $\sigma = -1$ for the right-handed CP light, as sketched in Fig. 4.2. The vectors show the mirror symmetry, implying the feature of chirality. Under the Lorentz gauge, we have the electric and magnetic fields [71]

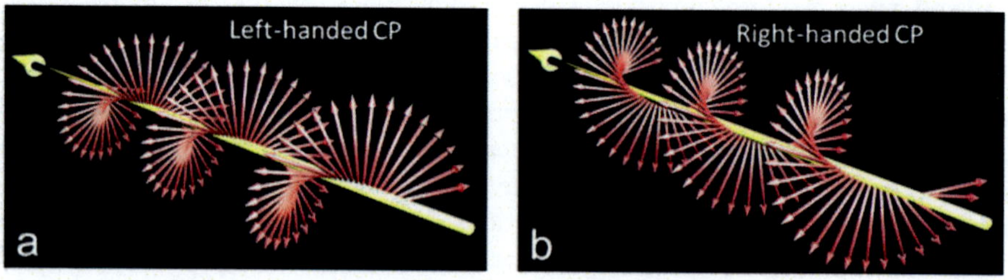

Fig. 4.2 Sketched vectors of electric fields in left- (**a**) /right- (**b**) handed CP light

$$\mathbf{E} = i\omega\mathbf{A} + \nabla\left(\frac{c^2}{i\omega}\nabla\cdot\mathbf{A}\right) = \left[i\omega u\mathbf{e}_x - \sigma\omega u\mathbf{e}_y + c\left(\frac{\partial u}{\partial x} + \sigma i\frac{\partial u}{\partial y}\right)\mathbf{e}_z\right] \exp[i(kz - \omega t), \tag{4.2}$$

and

$$\mathbf{B} = \nabla\times\mathbf{A} = \left[-i\sigma\left(\frac{\partial u}{\partial z} + iku\right)\mathbf{e}_x + \left(\frac{\partial u}{\partial z} + iku\right)\mathbf{e}_y + \left(\sigma i\frac{\partial u}{\partial x} - \frac{\partial u}{\partial y}\right)\mathbf{e}_z\right] \exp[i(kz - \omega t), \tag{4.3}$$

where the paraxial approximation is employed by ignoring the two-order derivations. As suggested by Allen et al., the spin angular momentum (SAM) of a CP Gaussian beam can be evaluated by using [72]

$$\frac{J_z}{W} = \frac{\int\int(\mathbf{r}\times\mathbf{E}\times\mathbf{B})_z r dr d\varphi}{\int\int c\mathbf{E}\times\mathbf{B}_z r dr d\varphi} = \frac{\sigma}{\omega}, \tag{4.4}$$

where only the SAM of light is considered because the Gaussian beam takes no orbital angular momentum (OAM). Equation (4.4) means that the SAM of a photon orientates in the positive or negative z direction, which is labelled by the sign of σ. Light carrying the SAM can be used to control the rotation of an object such as birefringent crystals. More frequently, the CP light with SAMs is used to interact with microscopic chiral molecules that have the distinguished absorption of the left-handed and right-handed light, the phenomena of which is named as "circular dichroism (CD)'. By using circular dichroism, one can distinguish the chirality of molecules that cannot be resolved by using the traditional microscopy due to the diffraction limit. Currently, the CD spectroscopy has been widely used in physics, chemistry and biology.

Generally, the CP light is taken to be chiral because of its unique response to chiral objects. However, the chirality of CP light has not been well understood from the viewpoint of electromagnetic waves until the definition of chirality for electromagnetic waves is defined elegantly by Tang and Cohen [73]. For an electromagnetic wave with the fields **E** and **B**, its chirality is proposed to be expressed by [73, 74]

$$C \equiv \frac{\varepsilon}{2}\mathbf{E}\cdot\nabla\times\mathbf{E} + \frac{1}{2\mu}\mathbf{B}\cdot\nabla\times\mathbf{B}, \tag{4.5}$$

where ε and μ are the permittivity and permeability of the surrounding medium. After introducing (4.2) and (4.3) into (4.5), we have its chirality $C = \sigma\varepsilon k\omega^2 u^2$, which is related with the spin σ and the intensity density u^2 at the position of interest. Therefore, (4.5) is also referred as the local chirality density. Once the CP light interacts with the plasmonic structures, the local intensity might be much higher than the incident light due to the enhancement of local electric fields, leading to a larger C than $\sigma\varepsilon k\omega^2 u^2$. Such a phenomena is named as superchiral, which is also found in the standing waves generated by the interference between two beams with the equal amplitude and opposite propagating directions. Although the superchiral phenomena can enhance the CD response, the value of CD is still located at the level of micro-degrees when the CP light is used to detect the molecules. The possible reason might come from the mismatch between the dimension of molecules and the operating wavelength of light. In contrast, the interaction between the CP light and the subwavelength structures in optical metasurfaces is quite strong so that it can be used to induce the giant spin Hall effects of light [13] or demonstrate the spin-dependent meta-optics [44] (see more details in Sect. 4.2). In addition, it has been recently discovered that the topological structure of the CP light plays an important role in enhancing the CD responses because the symmetry of the beam is changed when the optical vortices with helical wavefronts are introduced [75].

4.2.2 *Optical Vortices and Orbital Angular Momentum*

Beams carrying a helical wavefront exhibit the vortex-like energy flow, alike the case of vortices in water, and therefore are called optical vortex beam. The paraxial vortex beams such as Laguerre-Gaussian and Bessel beams have the singularities at their centers. Along the arbitrary closed curve containing the singularity, the accumulated phase difference divided by 2π is an integer of l, which is the topological charge of optical vortices and corresponds to the topological feature such as the self-healing effects. Thus, an optical vortex beam takes an angle-dependent phase profile of $e^{il\varphi}$, which generates the helical wavefront and the axis-symmetry doughnut-shape intensity profile (see Fig. 4.3). In 1992, Allen et al. found that such kind of optical vortex beams under the paraxial approximation have the orbital angular momentum

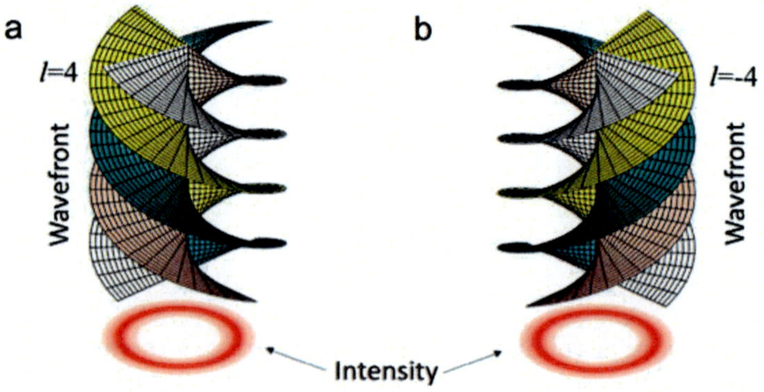

Fig. 4.3 Intensity and phase profiles of optical vortex beams with the opposite-handedness wavefronts: $l = 4$ (**a**) and $l = -4$ (**b**)

of $l\hbar$ per photon, where $\hbar$ is the reduced Planck constant [72]. It thus connects the microscopic quantity of OAM with the macroscopic phase profile of $e^{il\varphi}$. The mathematical proof of the OAM carried in optical vortex beam can be found easily by using a circularly polarized Laguerre-Gaussian beam with its vector potential of $\mathbf{A} = u(r,z)\mathrm{e}^{il\varphi} \bullet \mathrm{e}^{i(kz-\omega t)} \bullet (\mathbf{e}_x + \sigma i\, \mathbf{e}_y)$, where the amplitude part is angle-independent, r and φ are the polar coordinates. The Lorentz gauge yields the similar electric and magnetic fields in (4.2) and (4.3) after $u(x, y, z)$ is substituted by $u(r, z)\mathrm{e}^{il\varphi}$. By using (4.4) under the paraxial approximation, we have the total angular momentum

$$\frac{J_z}{W} = \frac{l + \sigma}{\omega}, \tag{4.6}$$

where the spin item σ refers to the SAM of the circular polarized light and l is related with the OAM of light. For a linearly polarized vortex beam, its SAM is zero, leaving the OAM of $l\hbar$ per photon. Considering that the OAM of light is a microscopic variable, optical vortices with the OAM has been intensely investigated in quantum physics. From the helical phase profiles, one can find that the orthogonality between two arbitrary OAM states is extremely good, so that it can be used for the purpose of communication. In quantum entanglement [76] and optical communication [77–79], the OAM channel could provide the low crosstalk for high-fidelity data transmission. Meanwhile, because the integer l can take any value, the number of OAM channels in communication is unlimited, which therefore could increase the capacity of data transmission.

As denoted in (4.6), the OAM in an optical vortex beam can take the positive or negative value, which presents its direction of z or –z, respectively. It means that the energy flow in the vortex beam moves clockwise or anti-clockwise, behaving like the spin of a photon in a CP beam. When a tightly focused OAM is used to trap the micro-particles within the ring-shape intensity, the OAM of the photons is transferred to micro-particles, leaving a clockwise or anti-clockwise moving trajectory that is determined by the sign of l. All these phenomena suggest the chiral behavior of optical vortices, which is therefore expected to have the strong interaction with chiral objects. However, a plethora of theoretical and experimental works show no obvious interaction with chiral molecules [80]. The rigorous theory proves that the OAM states can respond to the electric quadrupole excited in molecules [81], but its corresponding excitation rate is extremely low in real world. With the help of unique nanoparticle aggregates that could excite the electric quadrupole efficiently, a recent work has reported the first observation of discriminating two chiral enantiomers by the optical vortices with opposite topological charges [82]. Note that, since the electric quadrupoles are excited by the third part, the resulting scattering induced by two vortices has the little difference, leading to a quite weak helical dichroism (HD) of 0.8% [82]. From the viewpoint o practical applications, such a tiny HD cannot be used to detect the chirality of objects in a steady way. More efforts should be made to enhance the HD for the wide usage. An experimental attempt to solve the problem [83] is provided under the theoretical consideration in Sect. 4.4.

4.3 Optical Chiral Metasurfaces

As introduced in Sect. 2., optical metasurfaces contain many types of structures that are employed to control light with different polarizations. Considering the limitation of the topic about chirality, we constrain ourselves to discuss the chiral metasurfaces that manipulate the circularly polarized light with the geometric phase. All the chiral metasurfaces follow the same mechanism of realizing the geometric phase. We start by giving a universal introduction about geometric metasurfaces containing the birefringent nanostructures that have different responses to two orthogonal components of an incident circularly polarized light. Without the loss of generality, we define the scattering Jones matrix of the non-rotating nanostructure by using $S(\varphi = 0) = \begin{bmatrix} s_x & 0 \\ 0 & s_y \end{bmatrix}$, where the scattering factors $s_x$and s_y stand for the complex modulation of x- and y-components of the incident polarized light, respectively. The anti-diagonal elements are fixed to zeros under the assumption that the optical scattering by the nanostructures will not generate a new electric component orthogonal to the incident one, which is usually valid for rectangle-shape nanostructures. If the nanostructure has a rotating angle of φ with the x-axis, its scattering matrix is

$$\mathbf{S}(\varphi) = \mathbf{R}(-\varphi)\begin{bmatrix} s_x & 0 \\ 0 & s_y \end{bmatrix}\mathbf{R}(\varphi) \tag{4.7}$$

where the rotation matrix $\mathbf{R}(\varphi) = \begin{bmatrix} \cos\varphi & \sin\varphi \\ -\sin\varphi & \cos\varphi \end{bmatrix}$. When a circularly polarized beam with the electric field of $\mathbf{E}^{\sigma} = E\bullet(\mathbf{e}_x{+}\sigma i\mathbf{e}_y)$ is incident on the nanostructure, the scattering light has the electric field

$$\begin{aligned} \mathbf{E}_s = \mathbf{S}(\varphi)\mathbf{E}^{\sigma} &= E\begin{bmatrix} \cos\varphi & -\sin\varphi \\ \sin\varphi & \cos\varphi \end{bmatrix}\begin{bmatrix} s_x & 0 \\ 0 & s_y \end{bmatrix}\begin{bmatrix} \cos\varphi & \sin\varphi \\ -\sin\varphi & \cos\varphi \end{bmatrix}\begin{bmatrix} 1 \\ \sigma i \end{bmatrix} \\ &= \frac{s_x + s_y}{2}\mathbf{E}^{\sigma} + \frac{s_x - s_y}{2}e^{2i\sigma\varphi}\mathbf{E}^{-\sigma}, \end{aligned} \tag{4.8}$$

where $\mathbf{E}^{\pm\sigma} = E\bullet(\mathbf{e}_x{\pm}\sigma i\mathbf{e}_y)$. In (4.8), the left item refers to the co-polarized part that is usually taken as the background in the scattering light, while the right one implies that the scattering light contains a cross-polarized part with an additional phase modulation of $e^{2i\sigma\varphi}$. Such geometric phase depends on both the chirality of incident light and the rotating angle φ of the nanostructures. It means that the geometric metasurfaces respond to the chiral part in a polarized light, which is therefore named as chiral metasurfaces in this chapter. In addition, the geometric phase arises from the rotation of spatial coordinates, which is thus valid for any birefringent nanostructures (such as chiral metasurfaces and the molecules of liquid crystal) or bulky crystals (such as half-waveplates).

For optical chiral metasurfaces, the total scattering $\mathbf{E}_s$ can be tuned by controlling the parameters s_x and s_y, which are determined by the geometry of nanostructures, materials and working mode. In (4.8), the item $|s_x - s_y|/2$ is related with the efficiency of metasurfaces. For the passive nanostructures without gain, we have $|s_x| \leq 1$ and $|s_y| \leq 1$. To achieve high efficiency, one should maximize $|s_x\text{-}s_y|/2$ and minimize the co-polarization part $|s_x+s_y|/2$. Mathematically, the highest conversion efficiency is 1 by setting $s_x = -s_y=1$, which means that the nanostructures have the equal amplitude of 1 and the phase delay of π between two orthogonal components of the scattering light. Correspondingly, the conversion efficiency is 0 when $s_x = s_y$, implying that the nanostructures have the isotropic responses for the E_x and $E_y$polarized light. The zero-efficiency case usually appears in the circular and square nanostructures. Therefore, most of the demonstrated geometric metasurfaces adopt the nanostructures with anisotropic geometry, such as rectangle shape. On the basis of working principles, they are classified into plasmonic chiral metasurfaces, helical nanosieves, and dielectric chiral metasurfaces, where the nano-structures work as miniaturized half-waveplates that holds the physical origins of electric dipole resonances, birefringent transmission and antiferromagnetic resonances, respectively. In practical design of nanostructures, one tune the parameters s_x and s_y that are controlled by the geometry of nanostructure for a give material platform and numerically simulated with the help of electromagnetic numerical calculation methods such as finite-element method (FEM) and finite-difference time-domain (FDTD) methods. Both electromagnetic simulation methods are important in designing optical metasurfaces and the relative sources can be found easily, so that we will not introduce them here.

4.3.1 Plasmonic Chiral Metasurfaces

Figure 4.4 sketches plasmonic chiral metasurfaces in a transmission and reflective mode. The working principle is the anisotropic resonances of the electric dipoles induced by the circularly polarized incident light. With the illumination by the circular polarized light, the metal nanorodes have the plasmonic resonances at the designed wavelengths. Considering that the feature size of nanorode is smaller than one wavelength, the plasmonic resonances induced by E_x and E_y components could be approximated theoretically as optical responses of two electric dipoles orientated along x and y directions, respectively. For the transmission mode as shown in Fig. 4.4a, the single-layer nanostructures can excite the transversely located electric dipoles that have both the forward and backward radiation. But, only the forward radiation is useful for controlling the s_x and s_y parameters, which refer to the transmittance r_x and r_y in the transmission mode. In addition, due to small filling factor of nanostructures, most of the incident light has no any interaction with nano-structures and directly passed through the substrate, leaving the background with co-polarization in the transmitted light. Therefore, the transmissive plasmonic chiral metasurfaces have the low conversion efficiency. The rigorous electromagnetic theory predicts that the highest conversion efficiency is 25% for metasurfaces that support only electric

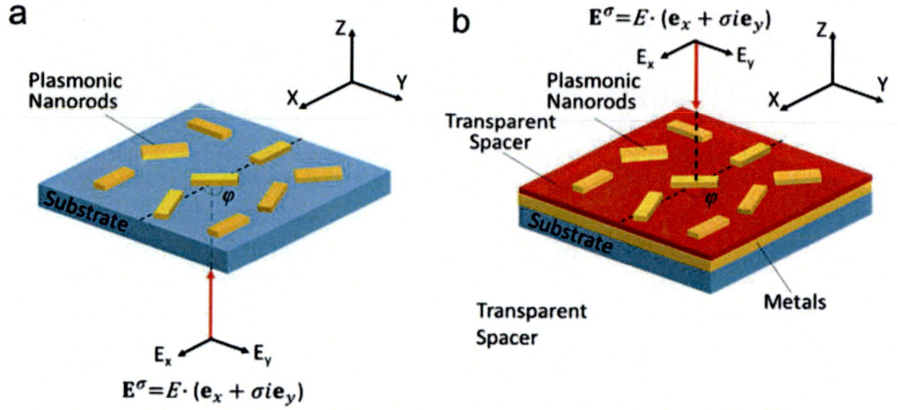

Fig. 4.4 Sketch of plasmonic chiral metasurfaces working in the transmission (**a**) and reflective (**b**) mode

or magnetic resonances [84], while the efficiency limit is 100% if both electric and magnetic resonances exist in the nanostructures [85]. In this transmissive plasmonic chiral metasurfaces, only the electric dipoles are excited, implying its theoretical limitation of 25% in the transverse efficiency.

To enhance the efficiency, the reflective plasmonic chiral metasurfaces have been proposed as shown in Fig. 4.4. In this configuration, the reflective metasurfaces are composed of a single-layer nanostructures, a transparent spacer and a single-layer high-reflectivity metal film on the substrate. Although the plasmonic resonances are similar with the transmission case, the reflective architecture allows two-time interaction between light and nanostructures during the incidence and reflectivity of light. Equivalently, the reflective metasurfaces behave like two-layer meta-materials with the same top and bottom nano-structures, which are separated by the transparent spacer with twofold thickness. The area of transparent spacer sandwiched by the metallic nanostructures and films allows for the magnetic resonances that are induced by the electric responses. Figure 4.5 shows the simulated intensity profiles of such reflective metasurfaces, where the metal is gold and the transparent spacer is made of silicon dioxide (SiO_2). One can clearly observe that both electric and magnetic dipoles oscillate in a unit cell, which offer much higher conversion efficiency of 80% at the visible wavelengths [43]. Such a conclusion implies that, the chiral metasufaces working in a transmission mode could also provide high conversion efficiency if the magnetic dipoles are also induced by constructing a three-dimensional metal-ring structure, which is, however, challenging to be fabricated due to the small footprint of hundreds of nanometers for optical metasurfaces.

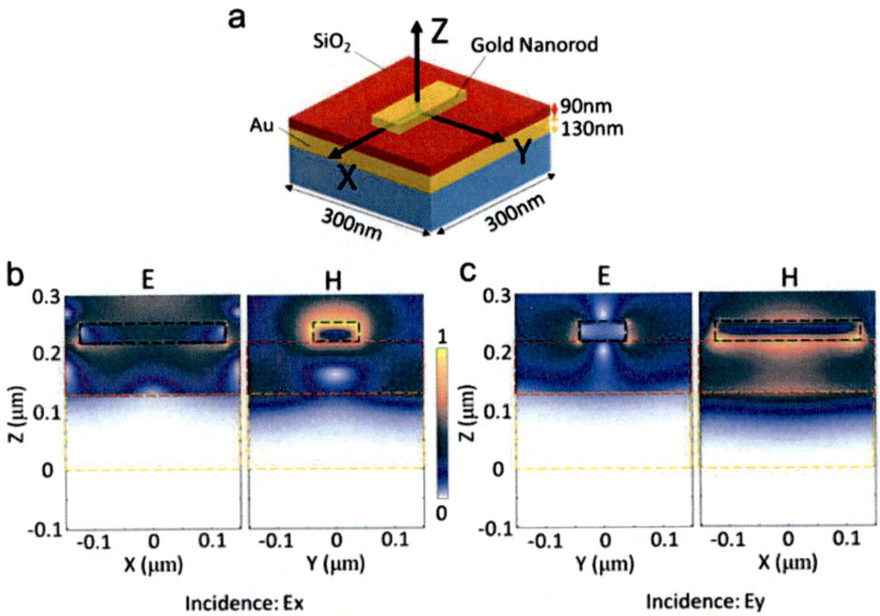

Fig. 4.5 Electric and magnetic fields of scattering light by the reflective plasmonic chiral metasurfaces. The reflective chrial metasurfaces are composed of an 80 nm * 250 nm * 30 nm gold nanostructure, a single-layer SiO_2 film of 90 nm thickness and a reflective gold film of 130 nm on a quartz substrate, as sketched in (**a**). The simulated electric and magnetic fields under the illumination of E_x and E_y polarization are provided in (**b**) and (**c**), respectively

4.3.2 Chiral Nanosieves

Photon sieves are the etched micro-apertures on an opaque film and have been proposed to control light in 2001 due to their advantages of suppressing the side-lobes of a focal spot in optical focusing [86]. In 2015, this concept of photon sieves was introduced into nano-optics by using the nano-apertures, i.e., nanosieves [37], which were used as one kind of typical amplitude metasurfaces to generate a tight spot beyond the diffraction limit and construct a high-uniformity ultra-broadband holographic image [38]. The interference between the diffracted light from these nanosieves will yield the expected intensity profiles by optimizing the locations of nanosieves. Since these nano-apertures is subwavelength, the vector features of diffracting light by them must be taken by using vector electromagnetic analysis that combines the coupled-mode theory and the multipole expansion, yielding an analytical solution of optical diffraction by a circular hole [37]. Due to the axis-symmetry feature, the circular nanosieves have no any dependence on the polarization of incident light, which is good for robust manipulation of arbitrarily polarized light. On the other hand, the circular nanoiseves lose the polarization and phased degree of freedom during the manipulating of light. So, the rectangle nanosieves are proposed

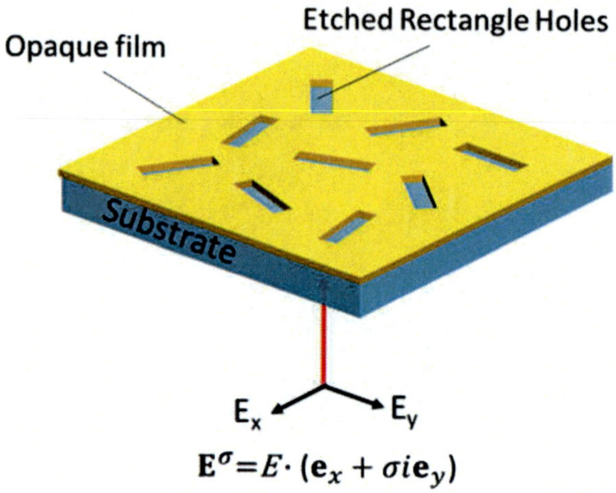

Fig. 4.6 Sketch of chiral nanosieves that contains the rectangle holes etched on an opaque film

to shape the amplitude, phase and polarization of light simultaneously [39, 87]. Here, we just discuss the chiral nanosieves with rotating rectangular apertures, which can be used to control the chiral light that is focused in this chapter.

Figure 4.6 shows the sketch of chiral nanosieves that work in a transmission mode. In appearance, the chiral nanosieves are the complementary structures of transmissive plasmonic chiral metasurfaces (Fig. 4.4a). But, the working principle of chiral nanosieves depends on the anisotropic transmission of x- and y-polarized light, that is, the parameters s_x and s_y refer to t_x and t_y that are mainly determined by the dimension of the hole. It means that the plasmonic resonances are not mandatorily required in chiral nanosieves, which holds the significant difference from the transmissive plasmonic chiral metasurfaces. Such a fundamental difference makes the chiral nanosieves more useful than plasmonic chiral metasurfaces because the plasmonic resonances are absent at the short-wavelength electromagnetic waves such as extreme ultraviolet light and soft-X rays, where all the materials are absorbing [40]. Thus, the chiral nanosieves become the good candidate to shaping the spin of EUV and soft-X photons in an easier way than the traditional reflective [41] and refractive [88] elements. The recent experimental results have revealed that the phase modulation of chiral nanosieves is valid in the absence of plasmonic resonances [40]. The conversion efficiency of chiral nanosieves is around 0.4% without plasmonic resonances, which confirms the feasibility of manipulating the EUV and soft-X photons by using chiral nanoisieves. In comparison with some high-efficiency elements, the chiral nanosieves are incompetent at the visible frequencies.

4.3.3 Dielectric Chiral Metasurfaces and Anti-ferromagnetic Resonances

Both transmissive chiral metasurfaces mentioned above cannot obtain high conversion efficiency. According to (4.8), high-efficiency chiral metasurfaces should maximize s_x and s_y, which is determined by the simultaneous oscillation of electric and magnetic nanomodes. For the plasmonic chiral metasurfaces and chiral nanosieves, both resonances cannot be satisfied due to either the single-layer induced electric responses or the strong ohmic loss of metallic materials [44]. In addition, the phase delay of π between s_x and s_y should also be met to realize the polarization conversion, and any deviation from π will lead to the existence of co-polarized light even if both s_x and s_y approaches 1. Although the reflective plasmonic chiral metasurfaces [43] offers high conversion efficiency that can only be checked by numerical simulations, it is difficult to explain the fundamental reason of generating the phase delay of π, which is usually ignored in most literatures. In addition to the reflective plasmonic chiral metasurfaces, dielectric chiral metasurfaces could also generate high conversion efficiency by employing the low-loss and high-refractive-index high-aspect-ratio dielectric nanostructures, which support the simultaneous electric and magnetic dipoles [35]. The magnetic dipoles are generated by the induced circle-shape electric displacements with alternating handedness. Thus, both resonances of electric and magnetic dipoles promise the large s_x and s_y that approach 1. Due to the high refractive index, the effective wavelength in dielectric nanostructures is much smaller than the vacuum wavelength, thus allowing multiple electric and magnetic dipoles in nanostructures. It leads to the existence of antiferromagnetic modes that contain a series of vertically locate antiparallel magnetic dipoles [15], which holds the fundamental physics of generating the phase delay of π.

By following our previous work about ultra-violet dielectric chiral metasurfaces, we will discuss the roles of antiferromagnetic nanomodes [15]. Figure 4.7 plots the sketch of the chiral metasurfaces and their optical performance with the simulated efficiency as high as 80% at the designed wavelength of 355 nm. The material of niobium pentoxide (Nb_2O_5) offer high refractive index of 2.2 and low absorption at the wavelengths ranging from 350 nm to 400 nm, hereby enabling the ultraviolet metasurfaces. To reveal the underlying physics, we investigate the detailed electromagnetic responses of dielectric chiral metasurfaces with the maximum conversion efficiency of 80% when the nanostructure has the dimension of L = 150 nm, W = 70 nm and H = 430 nm.

Figures 4.7d, e show the simulated electric and magnetic fields induced by $E_x$and E_y components of the incident polarized light. For E_x component in Fig. 4.7d, the induced electric displacements contain four alternative circular currents in clockwise and anti-clockwise directions, which could generate the relative antiparallel magnetic dipoles (AMPs) sitting vertically along z axis. This staggered magnetization has the similar behavior with one-dimensional antiferromagnetic chain, which has recently been reported in plasmonic nano-disks and hybrid metamaterials. The antiferromagnetic modes have the even electric circle-shape currents with alternative directions,

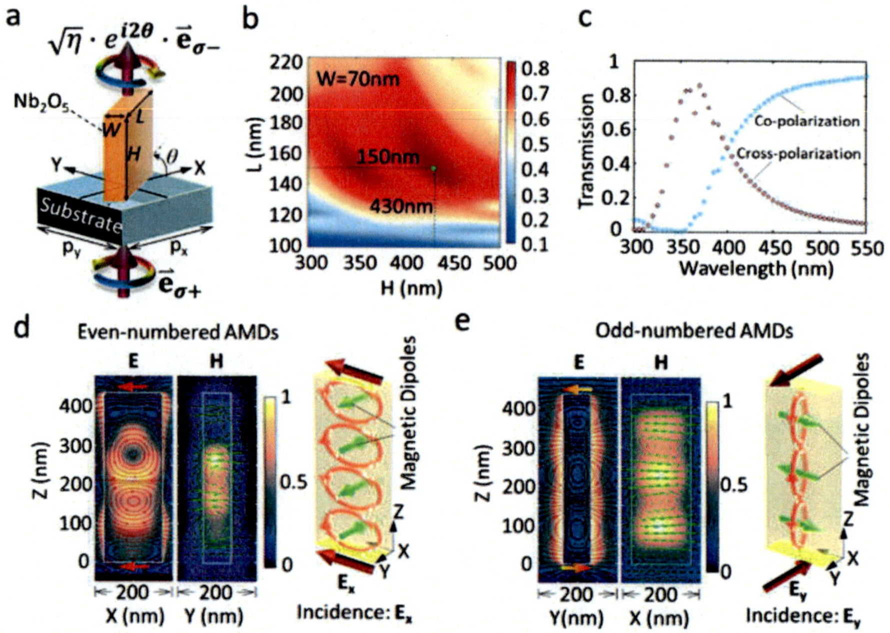

Fig. 4.7 **a** Sketch of dielectric chiral metasurfaces that use Nb_2O_5 to manipulating the ultraviolet light. **b** The simulated conversion efficiency of Nb_2O_5-base chiral metasurfaces at the interested wavelength of 355 nm. Considering the limitation of fabrication issue, the width is set to be 70 nm. The simulation is carried out by scanning the length and height of nanostructures with the help of FDTD methods. **c** The broadband response of the optimized nanostructures with W = 70 nm, L = 150 nm and H = 430 nm. **d–e** The simulated electric and magnetic fields under the illumination of E_x and E_y-polarization light. The inserts show the simplified sketches of magnetic dipoles. Reproduced with the permission from 2019 WILEY-VCH Verlag GmbH [15]

which thus enable the same orientation of both the electric vectors at the bottom and top ends of nanostructure (see Fig. 4.7d). It means that the x-component of CP light is not changed after the electromagnetic interaction with nanostructures. However, for E_y component, the antiferromagnetic mode is induced with three AMPs, leading to the odd circular currents that could reverse the electric vector of light at the terminal of transmission, as observed in Fig. 4.7e. It indicates a phase delay of π, which is desired for realizing high conversion efficiency. As a result, the transmitted light has the orthogonal polarization to the incident light, which is highly deserved in high-efficiency meta-devices [15]. Figure 4.8 sketches the details during the entire optical process intermediated by antiferromagnetic modes during the realization of circular polarization conversion into its cross-polarization.

Two coupling effects dominates the electromagnetic interaction between the oscillating nanomodes and the designed nanostructures [15]. Firstly, the number of the reduced magnetic dipoles in the antiferromagnetic nanomode is determined by the coupling between the volume modes and the dimension of nanostructures. Along the long axis (i.e., x axis) of nanostructures, the volume of magnetic dipole moments is

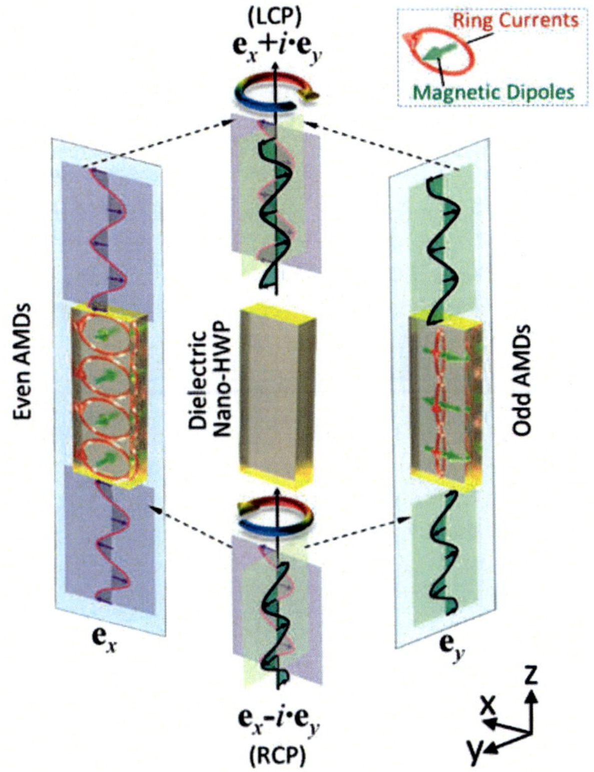

Fig. 4.8 Working principle of antiferromagnetic modes in realizing the polarization conversion

larger than the *y*-orientated magnetic dipoles. The limited volume of a nanostructure leads to the fact that, y-orientated magnetic dipoles are larger than the x-orientated ones, leading to the even and odd AMPs for realizing the phase delay of π. Secondly, every magnetic dipole coupled in an identical antiferromagnetic mode will compete, leading to the different magnetic momenta that influence the total magnetic field [15].

4.4 Applications of Chiral Light and Metasurfaces

In this section, we will discuss the applications of chiral light and chiral metasurfaces in various fields such as detecting the chirality of microscopic objects, optical spin Hall effects and polarization meta-optics.

4.4.1 Circular Dichroism and Helical Dichroism

In chemistry and biology, many molecules have the same compound of different atoms but are assembled with different chirality, which can be distinguished by their circular dichroism (CD) spectroscopy that measures the absorption difference of the right-handed and left-handed circularly polarized light over a wide range of wavelengths. The fundamental physics is that, the chiral enantiomers have the different extinction coefficients for right- and left-handed circular-polarization light. The CD topic has been discussed in many literatures and therefore is ignored here, so that we can focus the detection of chirality by using another chiral light with orbital angular momentum as introduced in Sect. 3. The OAM-based chiral detection draws the increasing attention because the traditional CD spectroscopy has the disadvantages of the weak CD responses at the level of micro-degrees and the broadband illumination, which imposes the requirement on the optical detectors and the lasers.

One of our recent works has addressed the issue about OAM-based chiral detection of micro-objects [83], as sketched in Fig. 4.9. Since the beams carrying the OAM of light have the doughnut-shape intensity profile, the light-matter interaction happens in a ring of intensity profile, where the energy of photons flows along a helical trajectory of micro-sized radius under the tightly focusing condition. It determines the interacting dimension of OAM at the level of microns. It is different from the spin of a circularly polarized light which the chirality of the photon is local at the cross section of entire beam, implying that a spinning photon has strong interaction dimension at the level of operating wavelengths. Therefore, we infer that the OAM can be used to detect the chirality of micro-sized objects. Figure 4.9 sketches the interaction between optical vortices and chiral micro-objects. Due to the helical wavefront of optical vortices, the incident angle generated by two optical vortices with opposite topological charges is different as illustrated in Fig. 4.9, leading to the differential reflectivity between these two optical vortices. During this procedure, the polarization issue is ignored so that we can define the differential scattering induced by OAM as [83]

$$\mathbf{VDS} = \frac{R_{l,s} - R_{-l,s}}{(R_{l,s} + R_{-l,s})/2}, \tag{4.9}$$

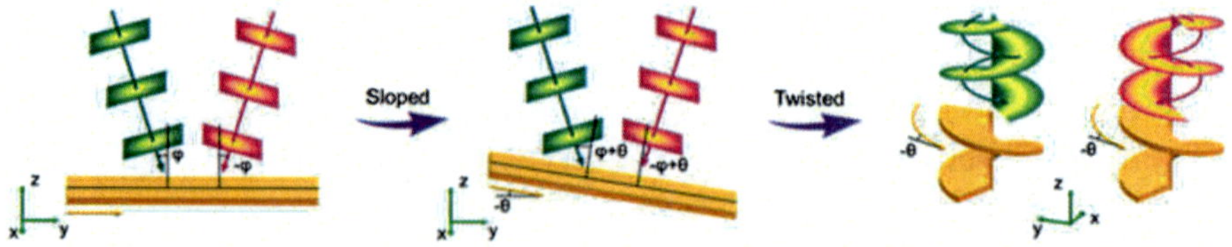

Fig. 4.9 Working principle of OAM-based chirality detection

where R is the optical reflectivity, l denotes the topological charge of optical vortices, the spin s of the incident beam is set to be identical for avoiding the CD effect. The definition of vortical differential scattering (VDS) emphasizes the importance of optical vortices by evaluating the OAM-induced dichroism between the topological-charge-opposite vortices. We have to emphasize that, the VDS has the fundamental difference from CD, because the VDS refers to the scattering difference between two chirality-opposite vortices while the CD comes from the difference of absorption to both circularly polarized light [83].

Based on the definition of VDS, the experimental setup is proposed to characterize the reflectivity difference induced by the chirality-opposite vortices [83]. A femtosecond laser with linear polarization is used here to generate the chirality-opposite optical vortices with a spatial light modulator. The optical vortices are focused by a high numerical aperture objective lens and then illuminate the chiral microstructures. The scattering signals that reflect from the chiral microstructures are collected by the focusing objective lens and later recorded by a CCD camera. The achiral structure of circular disk is used as a control of experiment. The recorded signals at the different topological charges of $|l|$ clearly shows the difference between the optical vortices with the opposite wavefronts (RHW and LHW) for the chiral micro-structures. In comparison, the controlled experiment gives nearly the same reflectivity from both optical vortices. All these experimental observations have suggested that the optical vortices with opposite topological charges can be used to distinguish the chirality of micro-structures due to the difference of the scattering light. To check its VDS further, we carried out the numerical calculation by using the experimental data. It exhibits that the VDS is positive for left-handed structures, negative for right-handed structures and zero for achiral structures. It gives the proof that the chirality of microstructures has been discriminated by using VDS, which therefore provides another method of detecting the chirality of objects in addition to the traditional CD spectroscopy [83].

Compared with CD method, the VDS has the following properties. Firstly, VDS operates at a single wavelength, and therefore releases the requirement of broadband sources in CD spectroscopy. Secondly, the strong VDS happens when the micro-structures have the dimension larger than the operating wavelength. Such a feature is determined by the large interacting area of optical vortices due to the doughnut-shape intensity profiles. Thirdly, the topological charges of optical vortices are unlimited so that it provides the infinite dimension of chiral OAMs, implying that the VDS has much larger degree of freedom than the CD with two spins of photons in a circularly polarized light [83].

4.4.2 *Chiral Meta-Optics*

By using (4.8), one can find that the arbitrary manipulation of chiral light could be realized by customizing the spatially varied phase profiles to meet the special requirement in practical applications. Here, we introduce three kinds of phase profiles to demonstrate lenses, holograms and gratings for optical imaging, vectorial holographic anticounterfeiting and beam splitter. In addition, when illuminated by a linearly polarized beam, the chiral metasurfaces will induce the spin-dependent shifts of beam centroids, which is usually related with the spin Hall effects of light [17].

I. Metalenses and imaging

To realize a focusing spherical-aberration-free lens by using chiral metasurfaces, the phase profiles have the function of [33, 46, 89]

$$\phi(x, y) = \frac{2\pi}{\lambda}\left(f - \sqrt{r^2 + f^2}\right), \tag{4.10}$$

where f is the focal length of the lens, $r^2 = x^2 + y^2$, λ is the wavelength. Correspondingly, the rotation angle of every nanostructure is $\varphi(x, y) = \phi(x, y)/2$. From (4.8), we can see that the real focusing length depends on the spin of the incident circularly polarized light. It means that a metalens with the phase profile in (4.10) can offer the convergent or divergent functionality, which can be switched by changing the spin of incident light.

The focusing spot generated by such a metalens gives the Airy-like intensity profile that obeys the diffraction limit of $0.515\lambda/\mathrm{NA}$ [46], where NA is the numerical aperture of the metalens. Therefore, the metalens can give the diffraction-limited imaging, which, however, suffers from the off-axis and achromatic aberrations. The off-axis aberration has been solved by a meta-doublet. One singlet is responsible for the aberration correction [90] and the other is used to realize the functionality of focusing. To solve the achromatic aberration, the hybrid nanostructure [91] or ultrahigh aspect-ratio dielectric nanorod [89] has been proposed and demonstrated experimentally in a low-NA lens, which behaves better than the uncorrected metalens. However, it should be noted that the aberration is much more severe in a high-NA lens than the low-NA one. Especially, a high-NA metalens has a phase cycle of 2π over the radial distance of one wavelength at the outmost rings. Since the pixel pitch of current metalens is around half a wavelength [46], there is only two nanostructures along the radial direction of the metalens so that the aberration cannot be eliminated completely.

II. Meta-gratings and beam splitters

Benefitting from the subwavelength pixel pitch, the chiral meta-gratings usually have the form of blazed gratings with a linearly varied phase. Similarly, a circularly polarized light passing through the metagratings deflects in one direction, while the opposite-spin light runs out in the other direction. When a linearly polarized light works as the incident light, these both deflecting beams can be generated with different spins, which refers to the spin Hall effects of light [92]. It means that the chiral meta-gratings can be used to filter out the spin components of a polarized light. If the chiral meta-gratings are combined with the linear-polarization metagratings, the diffracted beam could contain four parts: two orthogonal linear-polarization components and two spin components, which can be used to reconstruct the state of incident polarization. It holds the working principle of a metasurface-based polarimetry [54].

III. Chiral meta-holograms and vectorial anticounterfeiting

If the encoded phase profile in (4.8) is used to generate a holographic image, such a metasurface is called as meta-hologram that is different from the traditional holograms. Due to the distinct polarization response in a subwavelength pixel, the chiral meta-holograms could generate the spin-dependent or vectorial images in a large angle-of-view, without the high-order diffraction and twin-image issues that exists in the traditional holograms. For a chiral meta-hologram, the generated image also has a spin-dependent behavior. If the chiral meta-hologram works in a Fresnel distance, it will generate a real holographic image along the propagating direction of the circularly polarized incident light with one spin but form a virtual image for the other spin [44]. However, a Fraunhofer metahologram illuminated by the circularly polarized light with both spins will create the real images with center-symmetry properties [49]. Both Frensnel and Fraunhofer metaholograms have been experimentally demonstrated with the good agreement.

In fact, both cases can be mathematically derived. For a Fresnel metahologram with a phase profile of $\phi(x, y)$ for the circularly polarized light with its spin $\sigma = 1$, it generates a holographic image at the $z = z'$, where the electric field with the paraxial approximation can be expressed as [44, 93]

$$E_\sigma\left(x, y, z'\right) = \frac{\exp(ikz)}{i\lambda z} \int\int A(x_0, y_0) e^{i\phi(x_0, y_0)} \exp\left\{i\frac{k}{2z'}\left[(x-x_0)^2 + (y-y_0)^2\right]\right\} dx_0 dy_0, \tag{4.11}$$

where the amplitude A of incident light is considered without the phase (thus $A = A^*$), $k = 2\pi/\lambda$, λ is the operating wavelength. If the same hologram is illuminated by a circularly polarized light with the other spin $\sigma = -1$, we have the electric field as

$$E_{-\sigma}\left(x, y, z'\right) = \frac{\exp\left(ikz'\right)}{i\lambda z'} \int\int A(x_0, y_0) e^{-i\phi(x_0, y_0)} \exp\left\{i\frac{k}{2z'}\left[(x-x_0)^2 + (y-y_0)^2\right]\right\} dx_0 dy_0. \tag{4.12}$$

Its relative intensity profile has the form of

$$
\begin{aligned}
I_{-\sigma}(x,y,z) &= \left|E_{-\sigma}(x,y,z')\right|^2 = \left|E_{-\sigma}(x,y,z')^*\right|^2 \\
&= \left|\frac{\exp(-ikz')}{i\lambda(-z')}\int\int A^*(x_0,y_0)e^{i\phi(x_0,y_0)}\exp\left\{i\frac{k}{2(-z')}\left[(x-x_0)^2+(y-y_0)^2\right]\right\}dx_0dy_0\right|^2 \\
&= \left|E_\sigma\left(x,y,-z'\right)\right|^2 = I_{-\sigma}\left(x,y,-z'\right)
\end{aligned}
\tag{4.13}
$$

which means that the holographic image for the inverse spin is located at z = −z′, i.e., a virtual image.

For a Fraunhofer meta-hologram illuminated by a circularly polarized light with $\sigma = 1$, its intensity at the target plane can be expressed as [49]

$$
I_\sigma\left(x,y,z'\right) = \frac{1}{\left(\lambda z'\right)^2}\left|\int\int A(x_0,y_0)e^{i\phi(x_0,y_0)}\exp\left[-i\frac{k}{z'}(xx_0+yy_0)\right]dx_0dy_0\right|^2. \tag{4.14}
$$

When it is illuminated by a circularly polarized light with $\sigma = -1$, the intensity profile is

$$
\begin{aligned}
I_{-\sigma}\left(x,y,z'\right) &= \frac{1}{(\lambda z')^2}\left|\int\int A(x_0,y_0)e^{-i\phi(x_0,y_0)}\exp\left[-i\frac{k}{z'}(xx_0+yy_0)\right]dx_0dy_0\right|^2 \\
&= \frac{1}{(\lambda z')^2}\left|\int\int A^*(x_0,y_0)e^{i\phi(x_0,y_0)}\exp\left[-i\frac{k}{z'}(-xx_0-yy_0)\right]dx_0dy_0\right|^2 = I_\sigma\left(-x,-y,z'\right) \\
&= \frac{1}{(\lambda z')^2}\left|\int\int A^*(x_0,y_0)e^{i\phi(x_0,y_0)}\exp\left[-i\frac{k}{-z'}(xx_0+yy_0)\right]dx_0dy_0\right|^2 = I_\sigma\left(x,y,-z'\right),
\end{aligned}
\tag{4.15}
$$

which means that the Fraunhofer metahologram could give two images. One is the same with the Fresnel metahologram and the other refers to a real image with centrosymmetric feature, although the Fraunhofer metahologram is observed by the second case.

In addition, when the sparse meta-holograms are introduced [15], one can realize a vectorial holographic image by combining two sparse metaholograms that generate two complementary images with the dependence of spin. Such a concept has been demonstrated at the ultra-violet region for the applications of optical vectorial anticounterfeiting [15].

4.5 Conclusions

In summary, we have discussed the chirality of light and optical metasurfaces and introduced their properties, working principles, the spin-structure interaction and applications in various fields. The involved antiferromagnetism existing in dielectric nanostructures helps us to explain the underlying physics of nanoscale halfwaveplates in terms of the well-confined electromagnetic modes, which is a good attempt to investigate and unreal the novel phenomena in nanophotonics by using classical electrodynamics. The chirality detection by using the orbital angular momentum of light is also discussed with the experimental proofs, which, we believe, will excite more interesting researches in the near future. Furthermore, the interaction between chiral light, chiral nanostructures, and macroscopic external electric and magnetic fields has not been well investigated with many unanswered problems, which might stimulate the potential applications.

References

1. G. Keiser, Optical fiber communications. Wiley Encyclopedia of Telecommunications (2003)
2. D. Gabor, A new microscopic principle. Nature **161**, 777–778 (1948)
3. E. Yablonovitch, Inhibited Spontaneous Emission in Solid-State Physics and Electronics. Phys Rev Lett **58**, 2059–2062 (1987)
4. T.W. Ebbesen, Extraordinary optical transmission through sub-wavelength hole arrays. Nature **391**, 3 (1998)
5. J.B. Pendry, Negative refraction makes a perfect lens. Phys. Rev. Lett. **85**, 3966 (2000)
6. D.R. Smith, W.J. Padilla, D.C. Vier, S.C. Nemat-Nasser, S. Schultz, Composite medium with simultaneously negative permeability and permittivity. Phys. Rev. Lett. **84**, 4184–4187 (2000)
7. S. Astilean, P. Lalanne, P. Chavel, E. Cambril, H. Launois, High-efficiency subwavelength diffractive element patterned in a high-refractive-index material for 633nm. Opt. Lett. **23**, 552–554 (1998)
8. P. Lalanne, S. Astilean, P. Chavel, E. Cambril, H. Launois, Blazed binary subwavelength gratings with efficiencies larger than those of conventional échelette gratings. Opt. Lett. **23**, 1081–1083 (1998)
9. Z. Bomzon, G. Biener, V. Kleiner, E. Hasman, Spatial Fourier-transform polarimetry using space-variant subwavelength metal-stripe polarizers. Optics Lett. **26**, 1711–1713 (2001)
10. Z. Bomzon, V. Kleiner, E. Hasman, Pancharatnam-Berry phase in space-variant polarization-state manipulations with subwavelength gratings. Opt. Lett. **26**, 1424–1426 (2001)
11. N. Yu, P. Genevet, M.A. Kats, F. Aieta, J.-P. Tetienne, F. Capasso, Z. Gaburro, Light propagation with phase discontinuities: generalized laws of reflection and refraction. Science **334**, 333–337 (2011)
12. S. Iwahashi, Y. Kurosaka, K. Sakai, K. Kitamura, N. Takayama, S. Noda, Higher-order vector beams produced by photonic-crystal lasers. Opt. Express **19**, 11963–11968 (2011)
13. L. Zhang, S. Mei, K. Huang, C.-W. Qiu, Advances in full control of electromagnetic waves with metasurfaces. Adv. Opt. Mater. **4**, 818–833 (2016)
14. H.-T. Chen, A.J. Taylor, N. Yu, A review of metasurfaces: physics and applications. Rep. Prog. Phys. **79**, 076401 (2016)
15. K. Huang, J. Deng, H.S. Leong, S.L.K. Yap, R.B. Yang, J. Teng, H. Liu, Ultraviolet metasurfaces of ≈80% efficiency with antiferromagnetic resonances for optical vectorial anti-counterfeiting. Laser & Photonics Rev. **13**, 1800289 (2019)

16. L. Huang, X. Chen, H. Mühlenbernd, G. Li, B. Bai, Q. Tan, G. Jin, T. Zentgraf, S. Zhang, Dispersionless phase discontinuities for controlling light propagation. Nano Lett. **12**, 5750–5755 (2012)
17. X. Ling, X. Zhou, K. Huang, Y. Liu, C.-W. Qiu, H. Luo, S. Wen, Recent advances in the spin Hall effect of light. Rep. Prog. Phys. **80**, 066401 (2017)
18. K.Y. Bliokh, F.J. Rodriguez-Fortuno, F. Nori, A.V. Zayats, Spin-orbit interactions of light. Nat. Photon **9**, 796–808 (2015)
19. P. Genevet, F. Capasso, Holographic optical metasurfaces: a review of current progress. Rep. Prog. Phys. **78**, 024401 (2015)
20. N. Yu, F. Capasso, Flat optics with designer metasurfaces. Nat. Mater. **13**, 139–150 (2014)
21. S.B. Glybovski, S.A. Tretyakov, P.A. Belov, Y.S. Kivshar, C.R. Simovski, Metasurfaces: From microwaves to visible. Phys. Rep. **634**, 1–72 (2016)
22. S. Kruk, Y. Kivshar, Functional meta-optics and nanophotonics governed by mie resonances. ACS Photon. **4**, 2638–2649 (2017)
23. A.E. Minovich, A.E. Miroshnichenko, A.Y. Bykov, T.V. Murzina, D.N. Neshev, Y.S. Kivshar, Functional and nonlinear optical metasurfaces. Laser Photon. Rev. **9**, 195–213 (2015)
24. Z.E. Bomzon, G. Biener, V. Kleiner, E. Hasman, Radially and azimuthally polarized beams generated by space-variant dielectric subwavelength gratings. Opt. Lett. **27**, 285–287 (2002)
25. Z. Bomzon, V. Kleiner, F. Hasman, Computer-generated space-variant polarization elements with subwavelength metal stripes. Opt. Lett. **26**, 33–35 (2001)
26. A. Niv, G. Biener, V. Kleiner, E. Hasman, Formation of linearly polarized light with axial symmetry by use of space-variant subwavelength gratings. Opt. Lett. **28**, 510–512 (2003)
27. N.A. Rubin, G. D'Aversa, P. Chevalier, Z. Shi, W.T. Chen, F. Capasso, Matrix Fourier optics enables a compact full-Stokes polarization camera. Science **365**, eaax1839 (2019)
28. W.T. Chen, A.Y. Zhu, J. Sisler, Z. Bharwani, F. Capasso, A broadband achromatic polarization-insensitive metalens consisting of anisotropic nanostructures. Nat. Commun. **10**, 355 (2019)
29. W.T. Chen, A.Y. Zhu, J. Sisler, Y.-W. Huang, K.M.A. Yousef, E. Lee, C.-W. Qiu, F. Capasso, Broadband achromatic metasurface-refractive optics. Nano Lett. **18**, 7801–7808 (2018)
30. R.C. Devlin, F. Capasso, Broadband high-efficiency dielectric metasurfaces for the visible spectrum. Proc. Natl. Acad. Sci. USA **113**, 10473 (2016)
31. F. Aieta, P. Genevet, M. Kats, F. Capasso, Aberrations of flat lenses and aplanatic metasurfaces. Opt. Express **21**, 31530–31539 (2013)
32. P. Genevet, N. Yu, F. Aieta, J. Lin, M.A. Kats, R. Blanchard, M.O. Scully, Z. Gaburro, F. Capasso, Ultra-thin plasmonic optical vortex plate based on phase discontinuities. Appl. Phys. Lett. **100**, 013101 (2012)
33. X. Chen, L. Huang, H. Muhlenbernd, G. Li, B. Bai, Q. Tan, G. Jin, C.W. Qiu, S. Zhang, T. Zentgraf, Dual-polarity plasmonic metalens for visible light. Nat. Commun. **3**, 1198 (2012)
34. M. Kim, A.M. Wong, G.V. Eleftheriades, Optical Huygens' metasurfaces with independent control of the magnitude and phase of the local reflection coefficients. Phys. Rev. X **4**, 041042 (2014)
35. M. Decker, I. Staude, M. Falkner, J. Dominguez, D.N. Neshev, I. Brener, T. Pertsch, Y.S. Kivshar, High-efficiency dielectric huygens' surfaces. Adv. Opt. Mater. **3**, 813–820 (2015)
36. C. Pfeiffer, N.K. Emani, A.M. Shaltout, A. Boltasseva, V.M. Shalaev, A. Grbic, Efficient light bending with isotropic metamaterial Huygens' surfaces. Nano Lett. **14**, 2491–2497 (2014)
37. K. Huang, H. Liu, F.J. Garcia-Vidal, M. Hong, B. Luk'yanchuk, J. Teng, C.-W. Qiu, Ultrahigh-capacity non-periodic photon sieves operating in visible light. Nat. Commun. **6**, 7059 (2015)
38. K. Huang, H. Liu, G. Si, Q. Wang, J. Lin, J. Teng, Photon-nanosieve for ultrabroadband and large-angle-of-view holograms. Laser Photon. Rev. **11**, 1700025 (2017)
39. S. Mei, M.Q. Mehmood, S. Hussain, K. Huang, X. Ling, S.Y. Siew, H. Liu, J. Teng, A. Danner, C.-W. Qiu, Flat helical nanosieves. Adv. Fun. Mater. **26**, 5255–5262 (2016)
40. J. Li, G. Si, H. Liu, J. Lin, J. Teng, K. Huang, Resonance-free ultraviolet metaoptics via photon nanosieves. Opt. Lett. **44**, 3418–3421 (2019)
41. D. Attwood, A. Sakdinawat, *X-rays and Extreme Ultraviolet Radiation: Principles and Applications* (Cambridge University Press, 2017)

42. A. Arbabi, Y. Horie, M. Bagheri, A. Faraon, Dielectric metasurfaces for complete control of phase and polarization with subwavelength spatial resolution and high transmission. Nat. Nanotechnol. **10**, 937–943 (2015)
43. G. Zheng, H. Mühlenbernd, M. Kenney, G. Li, T. Zentgraf, S. Zhang, Metasurface holograms reaching 80% efficiency. Nat. Nanotechnol. **10**, 308–312 (2015)
44. K. Huang, Z. Dong, S. Mei, L. Zhang, Y. Liu, H. Liu, H. Zhu, J. Teng, B. Luk'yanchuk, J.K. Yang, C. Qiu, Silicon multi-meta-holograms for the broadband visible light. Laser Photon. Rev. **10**, 500–509 (2016)
45. M. Khorasaninejad, W.T. Chen, R.C. Devlin, J. Oh, A.Y. Zhu, F. Capasso, Metalenses at visible wavelengths: Diffraction-limited focusing and subwavelength resolution imaging. Science **352**, 1190–1194 (2016)
46. K. Huang, F. Qin, H. Liu, H. Ye, C.W. Qiu, M. Hong, B. Luk'yanchuk, J. Teng, Planar diffractive lenses: fundamentals, functionalities, and applications. Adv. Mater. **30**, 1704556 (2018)
47. L. Huang, X. Chen, H. Mühlenbernd, H. Zhang, S. Chen, B. Bai, Q. Tan, G. Jin, K.-W. Cheah, C.-W. Qiu, J. Li, T. Zentgraf, S. Zhang, Three-dimensional optical holography using a plasmonic metasurface. Nat. Commun. **4**, 2808 (2013)
48. X. Ni, A.V. Kildishev, V.M. Shalaev, Metasurface holograms for visible light. Nat. Commun. **4** (2013)
49. D. Wen, F. Yue, G. Li, G. Zheng, K. Chan, S. Chen, M. Chen, K.F. Li, P.W.H. Wong, K.W. Cheah, E. Yue Bun Pun, S. Zhang, X. Chen, Helicity multiplexed broadband metasurface holograms. Nat. Commun. **6**, 8241 (2015)
50. W. Ye, F. Zeuner, X. Li, B. Reineke, S. He, C.-W. Qiu, J. Liu, Y. Wang, S. Zhang, T. Zentgraf, Spin and wavelength multiplexed nonlinear metasurface holography. Nat. Commun. **7**, 11930 (2016)
51. K. Huang, H. Liu, S. Restuccia, M.Q. Mehmood, S. Mei, D, Giovannini, A. Danner, M.J. Padgett, J. Teng, C.-W. Qiu, Spiniform phase-encoded metagratings entangling arbitrary rational-order orbital angular momentum. Light: Sci. Appl. **7**, 17156 (2018)
52. D. Sell, J. Yang, S. Doshay, R. Yang, J.A. Fan, Large-angle, multifunctional metagratings based on freeform multimode geometries. Nano Lett. **17**, 3752–3757 (2017)
53. Y. Ra'di, D.L. Sounas, A. Alu, Metagratings: Beyond the limits of graded metasurfaces for wave front control. Phys. Rev. Lett. **119**, 067404 (2017)
54. A. Pors, M.G. Nielsen, S.I. Bozhevolnyi, Plasmonic metagratings for simultaneous determination of Stokes parameters. Optica **2**, 716–723 (2015)
55. M. Khorasaninejad, K.B. Crozier, Silicon nanofin grating as a miniature chirality-distinguishing beam-splitter. Nat. Commun. **5**, 5386 (2014)
56. M. Papaioannou, E. Plum, N.I. Zheludev, All-optical pattern recognition and image processing on a metamaterial beam splitter. ACS Photon. **4**, 217–222 (2017)
57. C. Huang, C. Zhang, S. Xiao, Y. Wang, Y. Fan, Y. Liu, N. Zhang, G. Qu, H. Ji, J. Han, L. Ge, Y. Kivshar, Q. Song, Ultrafast control of vortex microlasers. Science **367**, 1018–1021 (2020)
58. Y. Yang, W. Wang, P. Moitra, I.I. Kravchenko, D.P. Briggs, J. Valentine, Dielectric meta-reflectarray for broadband linear polarization conversion and optical vortex generation. Nano Lett. **14**, 1394–1399 (2014)
59. A. Arbabi, E. Arbabi, Y. Horie, S.M. Kamali, A. Faraon, Planar metasurface retroreflector. Nat. Photon. **11**, 415–420 (2017)
60. F. Yesilkoy, E.R. Arvelo, Y. Jahani, M. Liu, A. Tittl, V. Cevher, Y. Kivshar, H. Altug, Ultrasensitive hyperspectral imaging and biodetection enabled by dielectric metasurfaces. Nat. Photon. **13**, 390–396 (2019)
61. J. Zhou, H. Qian, C.-F. Chen, J. Zhao, G. Li, Q. Wu, H. Luo, S. Wen, Z. Liu, Optical edge detection based on high-efficiency dielectric metasurface. Proc. Natl. Acad. Sci. 201820636 (2019)
62. P. Berto, L. Philippet, J. Osmond, C.F. Liu, A. Afridi, M. Montagut Marques, B. Molero Agudo, G. Tessier, R. Quidant, Tunable and free-form planar optics. Nat. Photon. **13**, 649–656 (2019)
63. S.-Q. Li, X. Xu, R. Maruthiyodan Veetil, V. Valuckas, R. Paniagua-Domínguez, A.I. Kuznetsov, Phase-only transmissive spatial light modulator based on tunable dielectric metasurface. Science **364**, 1087–1090 (2019)

64. C. Zou, A. Komar, S. Fasold, J. Bohn, A.A. Muravsky, A.A. Murauski, T. Pertsch, D.N. Neshev, I. Staude, Electrically tunable transparent displays for visible light based on dielectric metasurfaces. ACS Photon. (2019)
65. R.A. Maniyara, D. Rodrigo, R. Yu, J. Canet-Ferrer, D.S. Ghosh, R. Yongsunthon, D.E. Baker, A. Rezikyan, F.J. García de Abajo, V. Pruneri, Tunable plasmons in ultrathin metal films. Nat. Photon. (2019)
66. E. Arbabi, A. Arbabi, S.M. Kamali, Y. Horie, M. Faraji-Dana, A. Faraon, MEMS-tunable dielectric metasurface lens. Nat. Commun. **9**, 812 (2018)
67. Y.-W. Huang, H.W.H. Lee, R. Sokhoyan, R.A. Pala, K. Thyagarajan, S. Han, D.P. Tsai, H.A. Atwater, Gate-tunable conducting oxide metasurfaces. Nano Lett. **16**, 5319–5325 (2016)
68. D. Franklin, Y. Chen, A. Vazquez-Guardado, S. Modak, J. Boroumand, D. Xu, S.-T. Wu, D. Chanda, Polarization-independent actively tunable colour generation on imprinted plasmonic surfaces. Nat. Commun. **6** (2015)
69. J. Lin, J.P.B. Mueller, Q. Wang, G. Yuan, N. Antoniou, X.-C. Yuan, F. Capasso, Polarization-controlled tunable directional coupling of surface plasmon polaritons. Science **340**, 331–334 (2013)
70. L. Allen, M. Padgett, M. Babiker, The orbital angular momentum of light. Progress Optics: Vol XXXIX **39**, 291–372 (1999)
71. H.A. Haus, *Waves and Fields in Optoelectronics*, vol. 1 (Prentice-Hall Englewood Cliffs, NJ, 1984).
72. L. Allen, M. Beijersbergen, R. Spreeuw, J. Woerdman, Orbital angular momentum of light and the transformation of Laguerre-Gaussian laser modes. Phys. Rev. A **45**, 8185–8189 (1992)
73. Y. Tang, A.E. Cohen, Optical chirality and its interaction with matter. Phys. Rev. Lett. **104**, 163901 (2010)
74. K.Y. Bliokh, F. Nori, Characterizing optical chirality. Phys. Rev. A **83**, 021803 (2011)
75. X. Zambrana-Puyalto, X. Vidal, G. Molina-Terriza, Angular momentum-induced circular dichroism in non-chiral nanostructures. Nat. Commun. **5**, 1–7 (2014)
76. A. Mair, A. Vaziri, G. Weihs, A. Zeilinger, Entanglement of the orbital angular momentum states of photons. Nature **412**, 313–316 (2001)
77. J. Wang, J.-Y. Yang, I.M. Fazal, N. Ahmed, Y. Yan, H. Huang, Y. Ren, Y. Yue, S. Dolinar, M. Tur, Terabit free-space data transmission employing orbital angular momentum multiplexing. Nat. Photon. **6**, 488–496 (2012)
78. N. Bozinovic, Y. Yue, Y. Ren, M. Tur, P. Kristensen, H. Huang, A.E. Willner, S. Ramachandran, Terabit-scale orbital angular momentum mode division multiplexing in fibers. Science **340**, 1545–1548 (2013)
79. L. Gong, Q. Zhao, H. Zhang, X.-Y. Hu, K. Huang, J.-M. Yang, Y.-M. Li, Optical orbital-angular-momentum-multiplexed data transmission under high scattering. Light: Sci. Appl. **8**, 27 (2019)
80. W. Löffler, D. Broer, J. Woerdman, Circular dichroism of cholesteric polymers and the orbital angular momentum of light. Phys. Rev. A **83**, 065801 (2011)
81. M. Babiker, C.R. Bennett, D.L. Andrews, L.C. Dávila Romero, Orbital angular momentum exchange in the interaction of twisted light with molecules. Phys. Rev. Lett. **89**, 143601 (2002)
82. W. Brullot, M.K. Vanbel, T. Swusten, T. Verbiest, Resolving enantiomers using the optical angular momentum of twisted light. Sci. Adv. **2** (2016)
83. J. Ni, Z. Wang, Z. Lao, Y. Hu, K. Huang, S. Ji, J. Li., Z. Huang. D. Wu, C. Qiu, Ultra-sensitive experimental detection of chiral meso-structures by orbital angular momentum of light. arXiv preprint arXiv:1809.02754 (2018)
84. F. Monticone, N.M. Estakhri, A. Alù, Full control of nanoscale optical transmission with a composite metascreen. Phys. Rev. Lett. **110**, 203903 (2013)
85. T. Cai, G. Wang, S. Tang, H. Xu, J. Duan, H. Guo, F. Guan, S. Sun, Q. He, L. Zhou, High-efficiency and full-space manipulation of electromagnetic wave fronts with metasurfaces. Phys. Rev. Appl. **8**, 034033 (2017)
86. L. Kipp, M. Skibowski, R.L. Johnson, R. Berndt, R. Adelung, S. Harm, R. Seemann, Sharper images by focusing soft X-rays with photon sieves. Nature **414**, 184–188 (2001)

87. M. Mehmood, S. Mei, S. Hussain, K. Huang, S. Siew, L. Zhang, T. Zhang, X. Ling, H. Liu, J. Teng, Visible-frequency metasurface for structuring and spatially multiplexing optical vortices. Adv. Mater. **28**, 2533–2539 (2016)
88. L. Drescher, O. Kornilov, T. Witting, G. Reitsma, N. Monserud, A. Rouzée, J. Mikosch, M.J.J. Vrakking, B. Schütte, Extreme-ultraviolet refractive optics. Nature **564**, 91–94 (2018)
89. S. Wang, P.C. Wu, V.-C. Su, Y.-C. Lai, M.-K. Chen, H.Y. Kuo, B.H. Chen, Y.H. Chen, T.-T. Huang, J.-H. Wang, R.-M. Lin, C.-H. Kuan, T. Li, Z. Wang, S. Zhu, D.P. Tsai, A broadband achromatic metalens in the visible. Nat. Nanotechnol. **13**, 227–232 (2018)
90. A. Arbabi, E. Arbabi, S.M. Kamali, Y. Horie, S. Han, A. Faraon, Miniature optical planar camera based on a wide-angle metasurface doublet corrected for monochromatic aberrations. Nat. Commun. **7**, 13682 (2016)
91. W.T. Chen, A.Y. Zhu, V. Sanjeev, M. Khorasaninejad, Z. Shi, E. Lee, F. Capasso, A broadband achromatic metalens for focusing and imaging in the visible. Nat. Nanotechnol. **13**, 220–226 (2018)
92. X. Ling, X. Zhou, X. Yi, W. Shu, Y. Liu, S. Chen, H. Luo, S. Wen, Fan, D.: Giant photonic spin Hall effect in momentum space in a structured metamaterial with spatially varying birefringence. Light: Sci. Appl. **4**, e290 (2015)
93. K. Huang, H.F. Gao, G.W. Cao, P. Shi, X.B. Zhang, Y. Li , Design of DPE for modulating the electric field at the out-of-focus plane in a lens system. Appl. Opt. **51**, 5149–5153 (2012)

Chapter 5
Light-Nanomatter Chiral Interaction in Optical-Force Effects

Hajime Ishihara, Masayuki Hoshina, Hidemasa Yamane, and Nobuhiko Yokoshi

Abstract The interaction between chiral objects and chiral environment is a fundamental topic in various research domains such as chemistry, material physics, optics, and bioscience. Particularly, the chiral interplay between light and nanostructures has been a fascinating topic because of its potential applications in chemical analyses, molecular sensing, novel light sources, and optical manipulations with high degree of freedom. Further, the study of relevant subjects will shed light on the unconventional fundamentals of light–matter interaction where the nanoscale spatial correlation between the light and matter geometries plays an essential role therein, which is beyond the conventional picture of optical response based on the long wavelength approximation of light or dipole approximation of matter systems. In this chapter, we introduce the recent theoretical studies on light–nanomatter chiral interactions, focusing on two topics in optical-force effects. The first topic is a scheme to measure the circular dichroism (CD) of the chiral near field in the vicinity of metallic nanostructures. This scheme evaluates the CD by measuring the optical force that acts on the probe tip with the circularly polarized lights irradiated. The second topic is the proposal of the rotational-motion control of nanoparticles in a nanoscale area by using the optical force generated through the interaction between a chiral light field (optical vortex) and a metallic nanocomplex. The results revealed the unique role of the optical force in the studies of the light–nanomatter chiral interaction.

H. Ishihara (✉)
Department of Materials Engineering Science, Osaka University,
1-3 Machikaneyama-cho, Toyonaka, Osaka 560-8531, Japan
e-mail: ishi@mp.es.osaka-u.ac.jp

Department of Physics and Electronics, Osaka Prefecture University,
1-1 Gakuen-cho, Naka-ku, Sakai, Osaka 599-8531, Japan

Quantum Information and Quantum Biology Division, Institute for Open and Transdisciplinary Research Initiatives, Osaka University, 1-3 Machikaneyama-cho, Toyonaka, Osaka 560-8531, Japan

M. Hoshina · H. Yamane · N. Yokoshi
Department of Physics and Electronics, Osaka Prefecture University,
1-1 Gakuen-cho, Naka-ku, Sakai, Osaka 599-8531, Japan

E. Kamenetskii (ed.), *Chirality, Magnetism and Magnetoelectricity*,
Topics in Applied Physics 138,
https://doi.org/10.1007/978-3-030-62844-4_5

5.1 Introduction

Chirality is the property that the spatial geometries of matters, fields, and motions cannot be identical to their mirror images. There exist various kinds of chiral structures or phenomena for different objects such as nucleons, amino acids, and liquid crystals. The chiral interaction between light and matter, especially, has been an important subject in various research domains. For example, the circular dichroism (CD) of materials, which is a difference between the absorptions of left- and right-circularly polarized lights, has been a standard measure of matter chirality for a long time. However, the chirality of light is also affected by the interaction with matter. Optical rotation and circularly polarized luminescence are typical examples. Recently, chiral interaction has been applied to the generation of light with orbital angular momentum, sometimes referred to as optical vortex. Accordingly, chiral interactions have been one of the fundamental principles in germinating, probing, and designing the nature and the function of matter and light.

Recently, the research on chiral interactions has taken a new step because of the rapid development of nanofabrication technologies and characterization techniques of few-molecule systems. For example, the high-accuracy fabrication of metallic structures enables us to control the localized electric field in nanoscale areas owing to localized surface plasmon (LSP). By using LSP effect, we can realize circularly polarized fields with significantly reduced pitch of polarization rotation, i.e., the so-called superchiral fields [1]. In the past 20 years, various nanostructures have been reported to generate superchiral fields [2–9]. In addition, the CD signal of the plasmonic near field was measured via photon scanning tunneling microscopy or near-field scanning optical microscopy (SNOM) [10–13]. Superchiral fields boost the molecular CD signal, although the signals of individual molecules, such as proteins, DNA, and carbon nanotubes, are generally small. The superchiral field is expected to be a promising tool for performing the sensitive enantioselective detections of chiral molecules [14–17]. Another interesting application of LSP is the conversion between the spin angular momentum and orbital angular momentum. A plane wave light with circularly polarization has spin angular momentum, but not orbital angular momentum. However, when it is radiated onto metallic nanocomplexes, the induced plasmonic near field can exhibit optical current with orbital angular momentum [18]. The optical response of a matter manifests not only as optical outputs but also as a mechanical force exerted on the matter system. Therefore, the plasmonic near field with the nanoscale radius of gyration is expected to enhance the degree of freedom to manipulate the center-of-mass motion of nanoobjects.

For the analyses of light–mater chiral interactions, we should note that the conventional scheme of light–mater interaction based on the long wavelength approximation (LWA) of light or dipole approximation (DA) of matter does not work because the chiral nanostructures have a nanoscale polarization configuration. Moreover, the chiral light induces non-dipole polarization structures of matter systems. Thus, the nonlocal response is important in nanoscale chiral light–matter interactions, in which the nanoscale spatial correlation between light and matter plays an essential role.

Anomalous optical responses of nanostructures due to the nonlocal characteristics of the response were theoretically proposed [19, 20], and have been experimentally demonstrated as the giant nonlinear response of quadrupole excitons [21], the optical forbidden electronic transition of a single-wall carbon nanotube [22], and the ultra-short radiative decay time in the femtosecond regime [23, 24], etc. Chiral interaction of light–nanomaterials is another frontier wherein the nonlocal optical response induces anomalous effects beyond LWA or DA.

Based on the nonlocal scheme, we have focused on the mechanical force (optical force) associated with the chiral interaction between plasmonic near-field and matter. Specifically, the following two issues are introduced. The first issue is the three-dimensional (3D) near-field CD during the optical-force measurement. As aforementioned, the aperture-type SNOM is a powerful tool to unveil the chiral near field [13]. However, it is difficult to elucidate the 3D structure of a superchiral field, especially around the edges of metallic structures, as the longitudinal component is dominant there. We show that if we measure the optical-force between the dipoles induced on the sample and the probe tip irradiated by light, we can observe the 3D distribution of the electric-field intensity to evaluate the 3D near-field CD (3D NF-CD). The other issue is the optical manipulation of nanoparticles (NPs) near metallic nanocomplexes with a high degree of freedom. We show that the flexible rotational optical manipulation, such as the rotation control in nanoscale area and switching of the rotational direction, of nanoobjects can be achieved. This means that all the basic elements of nanoobject-motion control (i.e., pushing, pulling, and rotating) are realized in principle.

The remainder of this chapter is organized as follows: In Sect. 5.2, we discuss the manner in which we can observe the chiral near field in the vicinity of metallic chiral structures. In addition, we propose a scheme to measure the 3D NF-CD using the optical force that visualizes CD with nm resolution by numerically demonstrating the CD map of the observed force on the gammadion metallic structures. In Sect. 5.3, we demonstrate that the NPs can be mechanically rotated and manipulated using the designed chiral field by plasmonic structures [25]. In this demonstration, we specifically consider that utilizing the nonlinear optical response considerably enhances the degree of freedom to manipulate NPs.

5.2 3D Near-Field CD by Optical-Force Measurement

One of the goals of analytical chemistry is to determine the structures of isolated single molecules. Accordingly, the single-molecule chiral analysis gains importance. To that end, the superchiral field is a promising tool [14–17], and its electromagnetic-field analysis is essential. However, in the aforementioned studies, the far-field CD (FF-CD) was evaluated based on the extinction of the light propagating through the target. The FF-CD signal is a convenient indicator that provides macroscopic information on the integrated target ensemble. However, nanoscale structures such as single molecules are averaged together with the information of the microscopic surrounding environment. Therefore, to maximize the potential of superchiral fields for measuring single molecules, we must develop a measuring device for the in-situ evaluation of the near field.

Recent studies reported the NF-CDs for various nanostructures and obtained characteristic signals that reflected their local geometry [10–13]. However, these studies measured the 2D projections of the NF-CD, which are the images of the propagated component of the scattered light in the vicinity of the target. Therefore, it is difficult to obtain the information regarding the localized longitudinal field, which is a significant component of the near field. Hence, we propose another scheme that uses optical force under laser illumination. If the targeted sample and the probe tip of the atomic force microscope are irradiated by light, dipole-dipole interaction occurs between them, and the near-field profile around the sample can be obtained by detecting the induced force, which is proportional to the gradient of the field intensity. This technique is called photoinduced force microscopy (PiFM) [26]. Because this scheme is sensitive to the longitudinal component of the polarization of the chiral near field, it can simultaneously acquire the 3D profile of the near field, i.e., the transverse and localized longitudinal components. To examine the validity of the scheme, for the superchiral field near the specific metallic structure, the force-measurement calculation results are compared with the electromagnetic-field calculation results.

5.2.1 *Model and Method*

The system under consideration is depicted in Fig. 5.1. In our numerical demonstration, the metallic structure, which comprises four gold gammadions, is set on the dielectric substrate. This structure exhibits large FF-CD [4] and boosts the FF-CD of a few molecules [14, 15]. The tip probe is modeled using a gold hemisphere and is scanned at 5 nm above the metallic structure. The circularly polarized plane wave lights are illuminated from below the substrate and propagate along the z-axis. We calculated the total electric field near the metallic gammadion and probe tip to detect the optical force.

Based on linear response theory, we can express the induced polarization $\mathbf{P}(\mathbf{r}_i, \omega) = \chi(\mathbf{r}_i, \omega)\mathbf{E}(\mathbf{r}_i, \omega)$ at the position $\mathbf{r}_i$, i.e., the i-th cubic cell of the volume V_c, where $\mathbf{E}(\mathbf{r}_i, \omega)$ denotes the total electric field. We set the local susceptibility in the substrate to $\chi(\mathbf{r}, \omega) = \varepsilon_s - 1$ with the relative permittivity set as $\varepsilon_s = 2.25$. On the other hand, the gold gammadion structures and the probe tip are composed of cells, whose dielectric constant is represented using the Drude model [27]. The dielectric function here is Drude-type, i.e.,

$$\chi_{\rm metal}(\omega) = \epsilon_{\rm metal}(\omega) - 1 = \epsilon_{\rm b} - 1 - \frac{(\hbar\Omega^{\rm pl})^2}{\hbar^2\omega^2 + i\hbar\omega(\hbar\Gamma^{\rm bulk} + \frac{\hbar v_{\rm f}}{L_{\rm eff}})}, \tag{5.1}$$

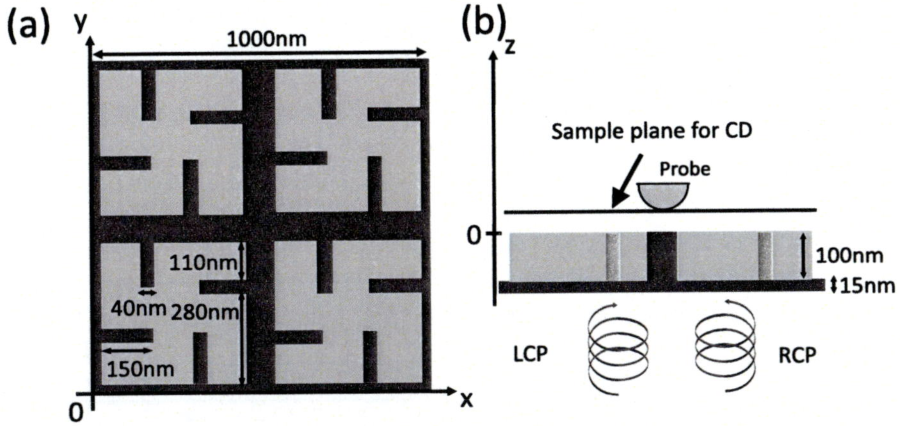

Fig. 5.1 Incident light condition and metallic structure on the dielectric substrate. The incident light is a circularly polarized plane wave and propagates along the z-axis. We allocate a calculation space of 1000 × 1000 × 250nm^3 and discretize the space using cubes of the volume V_c 5 × 5 × 5 nm^3. The CD signal is the summation of the field intensity of the sample surface. The top view (**a**) and side view (**b**) of the gammadion structure on the substrate. The length of each gammadion arm is illustrated, and the thickness is 100 nm. The relative permittivity and thickness of the substrate are set to 2.25 and 15 nm, respectively. The gammadion structure is aligned at the intervals of 500 nm. The scanning probe illustrated in (**b**) is a gold hemisphere of radius 50 nm. The size of the center gap is 75 × 75 nm^2. The scanning area is 90 × 90 nm^2

where ϵ_b denotes the background dielectric constant of the metal, Ω^{pl} the bulk plasma frequency, Γ^{bulk} the electron-relaxation constant of the metal, V_f the Fermi velocity, and L_{eff} the effective mean free path of the electrons. We have used the following parameters for the gold cells: $\epsilon_b = 12.0$, $\hbar\Omega^{pl} = 8.958$ eV, $\hbar\Gamma^{bulk} = 72.3$ meV, $\hbar v_f = 0.9215$ eV·nm, and $L_{eff} = 20$ nm [27]. In the calculations, we have used the discrete dipole approximation (DDA) method [28] to solve Maxwell's equation. We solved the following discretized integral equation:

$$\mathbf{E}(\mathbf{r}_i, \omega) = \mathbf{E}_0(\mathbf{r}_i, \omega) + \sum_j \mathbf{G}_0(\mathbf{r}_i, \mathbf{r}_j, \omega)\mathbf{P}_{metal}(\mathbf{r}_j, \omega) V_c, \tag{5.2}$$

where $\mathbf{E}_0$ is the electric field of the incident light, and $\mathbf{G}_0$ the Green's function of the electric field vector in vacuum. In the preceding equation, we take the sum for all the cells including the dielectric substrate, gammadion structures, and probe tip. Therefore, the calculated response field takes into account the nonlocal response, and the influence of the shape of the metals.

Notably, the definition of the circularly polarized light is not clear in case of electric fields that contain much localized components. To evaluate the circularly polarized components along the z-, y- and x-directions, the following projections were employed:

$$E_{xy\mp}(\mathbf{r}_i, \omega) = \frac{1}{\sqrt{2}}(1, \pm i, 0)\mathbf{E}(\mathbf{r}_i, \omega), \tag{5.3}$$

$$E_{xz\mp}(\mathbf{r}_i, \omega) = \frac{1}{\sqrt{2}}(1, 0, \pm i)\mathbf{E}(\mathbf{r}_i, \omega), \tag{5.4}$$

$$E_{yz\mp}(\mathbf{r}_i, \omega) = \frac{1}{\sqrt{2}}(0, \pm i, 1)\mathbf{E}(\mathbf{r}_i, \omega). \tag{5.5}$$

The "-" and "+" signs denote the left-handed circularly polarized (LCP) component and right-handed circularly polarized (RCP) component, respectively, of the electric field. In Fig. 5.2, we have mapped each electric-field polarization component directed along each direction just above the metallic structures (i.e., at $z = 5$ nm). Here, the incident light is the RCP light, and the electric-field intensity is normalized using the incident-light intensity $|E_0|^2$. The LCP component of the near field in the xy-plane is small at the center gap of the gammadion metallic structure, while the RCP component is remarkably enhanced there (see Fig. 5.2 a, b). In Fig. 5.2 c, d, we show the circularly polarized components in the xz-plane; these components include the longitudinal component of the near field. Compared with the case in the xy-plane, the longitudinal component appears more locally at the edges of the metallic structures. In the yz-plane, similar maps are obtained (no figure). Notably, a considerable part of the electric field is occupied by the longitudinal near-field component, and the polarizations do not change with propagation unlike those in the case of the circularly polarized plane wave. This implies that the longitudinal component of the near field in NF-CD must be a significant component of the superchiral field.

The proposed measurement scheme uses the time-averaged optical force on the scanning probe tip, as depicted in Fig. 5.1 b. We derive the optical force on the probe tip as follows [29]:

$$\langle \mathbf{F}(\omega) \rangle = \frac{1}{2}\mathrm{Re}\Big[\int_{V_{\mathrm{tip}}} d\mathbf{r}[\nabla \mathbf{E}(\mathbf{r}, \omega)^*] \cdot \mathbf{P}_{\mathrm{probe}}(\mathbf{r}, \omega)]\Big] = \mathbf{F}_z(\omega) + \mathbf{F}_{xy}(\omega), \tag{5.6}$$

where the integration range is within the volume of the probe tip V_{tip}. In the present setup, a strong and steep localized field gradient appears between the tip and the metal. Thus, the gradient force is dominant in the optical force, where the vertical force $F_z(\omega)$ is proportional to the gradient in the z-direction of the probe-target potential $U(\mathbf{r})$, and proportional to $|\mathbf{E}(\mathbf{r}, \omega)|^2$. In addition, the lateral photo-induced force $\mathbf{F}_{xy}$ is approximated by the differentiation of $U(\mathbf{r})$ in the lateral xy-plane [30]. Here, the electric field $\mathbf{E}(\mathbf{r}, \omega)$ includes the field scattered by the gold probe tip, and it considerably differs from the one that is observed when only the metallic structures are set. The numerical simulations presented in the following subsections show that the optical force well reproduces the CD of the superchiral field near the metal structure, despite the change in the electric field profile with and without the probe tip.

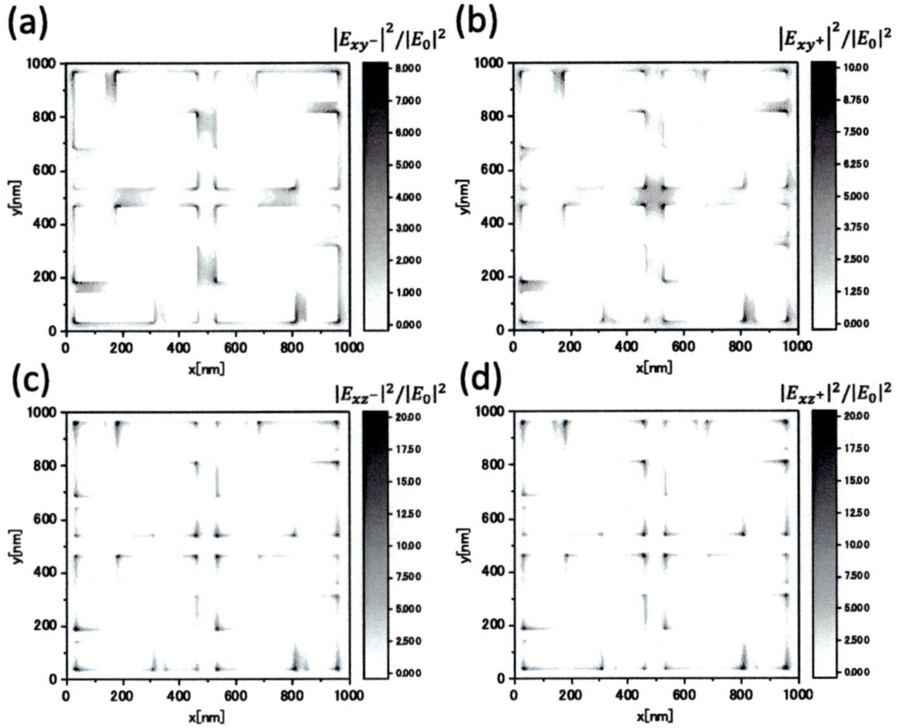

Fig. 5.2 Circularly polarized components are mapped along each direction at 5 nm above the top plane of the metallic structure. As the incident field, the RCP light at 1.65 eV is employed. **a, b** The intensity map of the LCP component (**a**) and RCP component (**b**) in the xy-plane. **c, d** The intensity map of the LCP component (**c**) and RCP component (**d**) in the xz-plane. The color bars indicate the intensity of each circularly polarized component normalized using the incident-light intensity

5.2.2 CD Spectra and NF-CD Maps

Prior to calculating the optical force, we analyzed the CD of the electric field to be referenced. We allocated a computational space of $1000 \times 1000 \times 1125\text{nm}^3$ and discretized it into the cubes of $5 \times 5 \times 5\,\text{nm}^3$. Since the localized field is considerably attenuated at $z \sim 1000\,\text{nm}$, the influence of the boundary of the computational space on the superchiral field is negligibly small. Throughout this work, we defined the NF-CD signal as $\Delta T = T_{\text{RCP}} - T_{\text{LCP}}$, where T_{RCP} and T_{LCP} denote the transmitted field intensities of the RCP and LCP optical signals, respectively. This is because we considered the in-situ CD of the superchiral field, and could not evaluate the CD from the difference in the extinction far from the metal structures.

As illustrated in Fig. 5.3 a, b, we investigated the CD spectra both at 5 and 1000 nm above the top plane of the metallic structure, respectively. We have plotted $\overline{\Delta T}$, which is the average of ΔT in each z over an area of $1000 \times 1000\,\text{nm}^2$. To better examine the effect of the longitudinal field on the CD, we decomposed the CD into

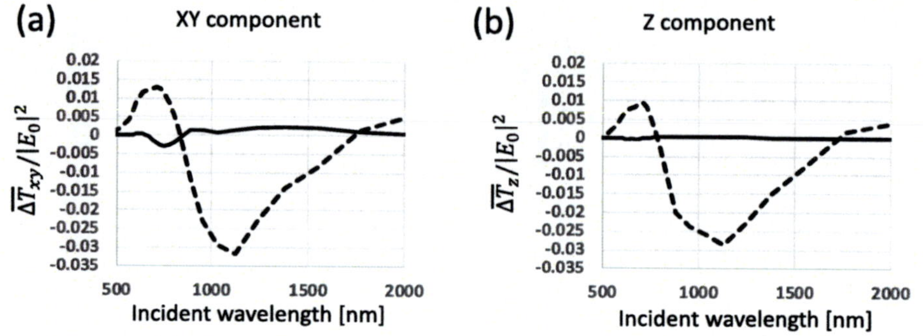

Fig. 5.3 CD spectra of the **a** xy component and **b** z component. The CD spectra at 5 nm (broken line) and 1000 nm (solid line) above the top plane of the metallic structure

the in-plane component (normal to the incident propagation) ΔT_{xy} and z component ΔT_z, and both are compared with each other. Here, ΔT_i denotes the intensity of the i-component of the transmitted light, and $\Delta T_{xy} = \Delta T_x + \Delta T_y$. At a considerably distant region above the metal (i.e., $z = 1000$ nm), the intensity of the longitudinal z-component becomes approximately 10 times smaller than that of the in-plane one. In the conventional FF-CD measurement, the distance between the probe and sample is more than 1 mm; hence, the longitudinal component can be safely neglected. However, at z =5 nm above the sample, the longitudinal CD intensity is of the same order of magnitude as the in-plane one. Moreover, even for the in-plane component, the NF-CD spectra (z =5 nm) are fairly different from the those at z =1000 nm. This clearly indicates that we cannot estimate the NF-CD spectra from the FF-CD ones, as confirmed by the experiment [13].

In Fig. 5.4a–d, we present the NF-CD contour maps of the transmitted in-plane intensity ΔT_{xy} and the vertical longitudinal one ΔT_z at z =5 nm. The incident RCP light has the energy of 1.65 eV (wavelength of approximately λ =750 nm), where we observe the CD peak in Fig. 5.3. Because half of the wavelength is as long as each gammadion, the NF-CD maps should reflect well the geometrical figure of the system. Notably, the in-plane NF-CD distributes differently than the longitudinal one. The in-plane components of the NF-CD appear mainly in the gap between the metallic structures, while the longitudinal components are concentrated on the gammadion edges. Therefore, while examining the spatial structure of the NF-CD, it is difficult to estimate the structure of its longitudinal component by using the structure of the in-plane component.

5.2.3 *CD of Optical Force*

As can be seen in the previous subsection, observing the 3D figure, as well as the longitudinal component of the superchiral field, for individual metallic structures

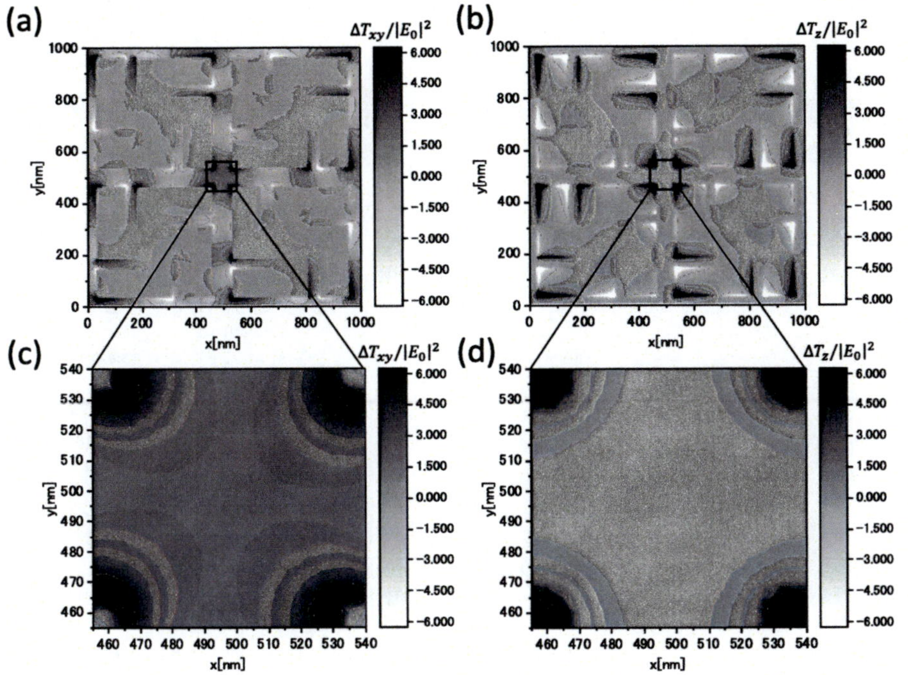

Fig. 5.4 NF-CD maps about the xy- and z-components on sample planes. **a**, **b** NF-CD maps about the xy- and z-component at z =5 nm. The grayscale bar indicates the CD intensity normalized using the incident-light intensity. **c**, **d** Enlarged view near the center gap of the metallic structure in (**a**, **b**). The grayscale bar indicates the CD intensity normalized using the incident-light intensity $|E_0|^2$

is fairly important. We simulated the 3D NF-CD measurement by calculating the time-averaged optical force that acts on the metallic probe tip [29, 31]. Using a state-of-the-art technique of PiFM, the spatial resolution of 1 nm was achieved [32]. In addition, the technique enables 3D vector force imaging [30].

When the probe tip is scanned on the gap area of the four gammadions in Fig. 5.1 a at z = 5 nm, we calculated the difference between the optical forces when illuminated with the LCP and RCP lights. We set the incident light intensity to 1 kW/cm^2 and the energy is the same as that in the previous subsection. The computational space is $1000 \times 1000 \times 250\,\text{nm}^3$, which is smaller in the z-direction compared to the space defined in the previous subsection. Since the localized field between the probe tip and metal structure is the dominant contributor to the photo-induced force, we can ignore the influence due to the space boundary. The difference of the z-component of the optical force is plotted in Fig. 5.5 a. The gradients along the z-direction of ΔT_{xy} and ΔT_z are depicted in Fig. 5.5 b, c. If the vertical force $F_z(\mathbf{r})$ on the probe is integrated over z, we can obtain the field intensity and the probe-target potential $U(\mathbf{r})$. Instead of performing the aforementioned integration, we have depicted the

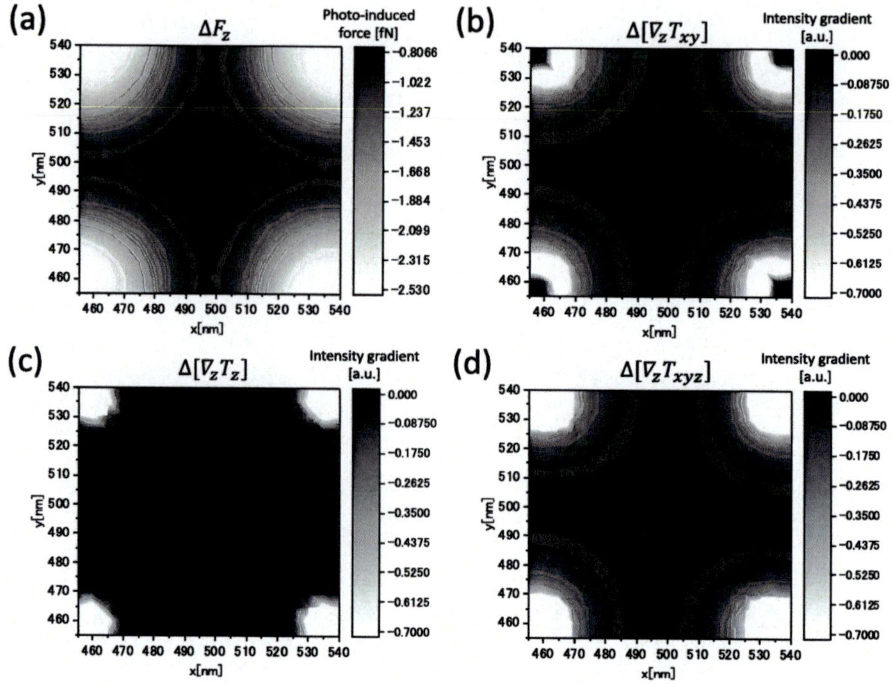

Fig. 5.5 **a** Contour map of the difference between the z-directed photo-induced forces on the probe under the RCP illumination and those under the LCP illumination. **b–d** The differentiations in the vertical direction ∇_z of the in-plane CD, longitudinal CD and their summation are mapped

gradient of the field intensities in Fig. 5.5b, c and compared them with the forces depicted in Fig. 5.5a.

Notably, the map in Fig. 5.5a resembles that in Fig. 5.5b inside the gap area, as the longitudinal component is weak in this region. However, at the corner regions, the gradient of the longitudinal component becomes remarkably large, as depicted in Fig. 5.5c. This situation is well reflected in the corner regions in Fig. 5.5a. Specifically, the field gradient at the corner regions is weak for the xy-component, as depicted in Fig. 5.5b, whereas the induced force at the corner regions becomes strong, as depicted in Fig. 5.5a. This situation becomes clearer if we note that the map of the gradient of sum $\Delta T_{xyz} = \Delta T_{xy} + \Delta T_z$ shown in Fig. 5.5d is very similar to the map in Fig. 5.5a. Therefore, the force map successfully provides information including the contribution of the longitudinal component that cannot be observed using the aperture-type scanning near-field optical microscope.

Finally, the feasibility of this proposal warrants a mention. The sensitivity of the optical force enables the application of the present scheme in the state-of-the-art technology of optical-force microscope with reasonable incident intensity [32]. However, the question may arise whether the probe tip might change the image from the actual field strength profile in our proposal. In practice, the force map blurs slightly, as the probe-tip diameter, i.e., 100 nm, is greater than the gap spacing between the gammadions in Fig. 5.5 b, c. However, when the spacing between the gammadions and probe tip is small, the electric-field enhancement is determined only by the structure near the probe tip. Therefore, the difference between the profile of the electric field and the one measured using the force is not significant. In actual experiments using optical-force microscope, the tip-sample spacing is less than 1 nm. Therefore, we can obtain satisfactory correspondence between the optical force and field gradients in actual experiments, as well as obtain further enhanced optical force. The information obtained provides significant insight into the spatial structures of the 3D NF-CD. For the cases in the presence of targeted molecules on the metallic structures, more detailed analyses of the force map to obtain the 3D NF-CD information are desired.

5.3 Optical Force to Rotate Nano-Particles in Nanoscale Area

According to Maxwell's theory of electromagnetism, both traveling and standing light waves can exert mechanical force on targets because of scattering and absorption. This force is sometimes called the "optical force". The optical force is classified as a dissipative force, which arises from the transfer of optical momentum to a substance by absorption and scattering, and a gradient force due to the electromagnetic interaction between the induced polarization and the incident light. The dissipative force usually pushes and transports particles, and the gradient force can be used to attract and trap particles. One of the most impressive applications of this force is an optical tweezer with a focused laser, as proposed by Ashkin et al. [33]. This optical force is attributed to the transfer of momentum from light to the matter target. Similarly, the spin and orbital angular momenta of light also can be transferred to the mechanical motion of small particles. The light with orbital angular momentum, such as a Laguerre–Gaussian (LG) beam [34], is considered to result in the orbital rotation of targets. Notably, micro-particles are swirled along a ring-shaped region, where the field intensity of the LG beam is strong [35]. However, presently, it is not known how the optical manipulation for the rotational control in a nanoscale area can be performed.

This section is devoted to discussing the chiral interaction between light and metallic nanocomplexes, wherein the chiral interaction induces rotational motion of nano-particles (NPs) in a nanoscale region. In recent years, the target of optical manipulation has shifted to the nanometer-scale. However, within the Rayleigh scattering regime, the optical force is approximately proportional to the volume of the object, hence, the induced force is quite weak. To enhance the force to overcome the disturbance due to the environment of NPs, the use of the evanescent field with a steep gradient of the electric field has been proposed [36, 37], and recently, the trapping of NPs associated with LSP resonance has been extensively studied [38–44]. Another approach is the use of resonance with transitions between the electronic levels in

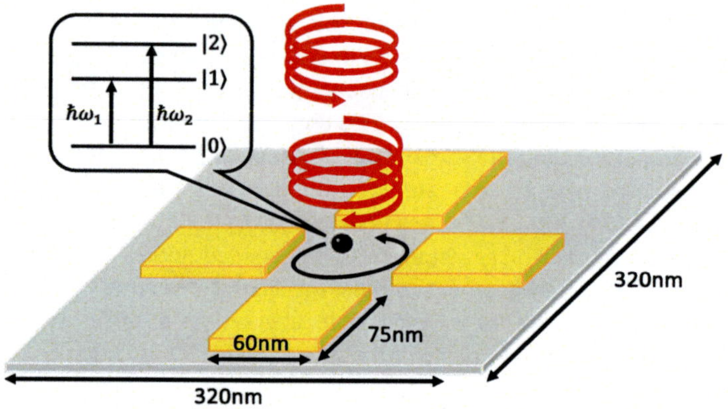

Fig. 5.6 System setup. The image of super-resolution rotational optical manipulation inside the tetramer metallic structures with a circularly polarized plane wave. The nanostructures are assumed to be four gold panels that form nanogaps on the glass substrate. The size of each panel is $75{\times}60{\times}20\,\mathrm{nm}^3$, and the gaps are $\sqrt{10^2+10^2} \simeq 14$ nm. The NP is assumed to have three levels as depicted in the inset. The size of the NP is set to be $5{\times}5{\times}5\,\mathrm{nm}^3$. (The entire system is divided into cells of size $5{\times}5{\times}5\,\mathrm{nm}^3$ for the DDA calculation.) In both the cases, the incident lights propagate along the negative z-axis. See, [25]

NPs [31, 45–49]. In general, nanostructures have quantized electronic levels. Thus, the optical force can be resonantly enhanced if the frequency of the incident light coincides with their transition energies. Successful trapping and transport of NPs using resonant laser light have been previously reported [50–54]. Herein, we have demonstrated a scheme to realize rotation and the switching direction of NPs in a nanoscale region using metallic nanocomplexes with LSP resonance and the resonant optical response of NPs [25]. In particular, optical nonlinearity is key to realizing the rotational motion of NPs. Using the resonant optical response often results in optical nonlinearities, thereby enhancing both the trapping efficiency and the degree of freedom of NP manipulation [49, 55, 56]. We have shown that by introducing the optical nonlinearity of NPs due to LSP resonance, nanoscale NP rotation, and the switching of the direction is possible. The transfer of the orbital angular momentum of light to a small object has potential applications in various technologies such as nanoelectromechanical systems and chiral sensing [57, 58].

5.3.1 *Model and Method*

We assumed a metallic nanocomplex as a platform for rotating the NP within the nanoscale region, as illustrated in Fig. 5.6. When it is irradiated using a circularly polarized light with the spin angular momentum of $s = \pm 1$, a nano-optical vortex, which is connected to the LSPs near the metal, is generated. Recent theoretical works revealed that the excited plasmon modes inside the gap of the metallic nanocomplex

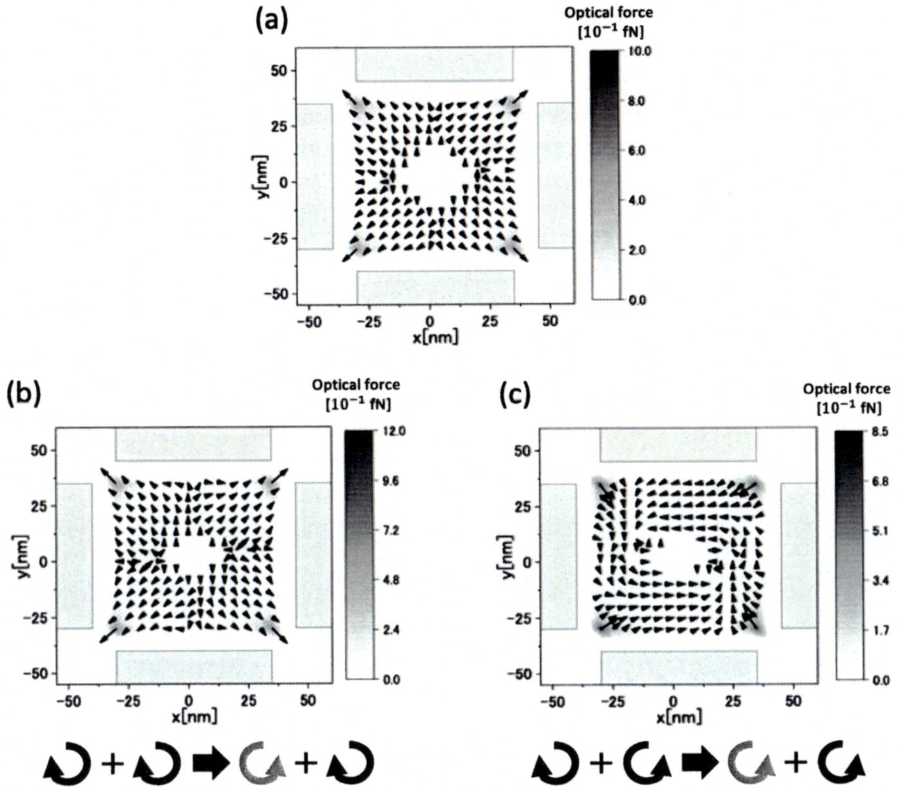

Fig. 5.7 Optical-force map inside the metallic nanocomplex. The black arrows indicate the force vectors in the x-y plane. The grayscale bars represent the magnitude of the optical force. The manipulation light and pump light energies are resonant to the 0–1 and 0–2 transitions, respectively. **a** The case that contains only the RCP manipulation light with the spin angular momentum of $s = +1$. The intensity is 100 kW/cm^2. **b** The case that contains both the RCP pump light and RCP manipulation light with the intensities of 100 kW/cm^2. **c** The case that contains both the LCP pump light and RCP manipulation light with the intensities of 100 kW/cm^2. In this case, the manipulation light energy is slightly red-detuned (1.798 eV). The arrows that draw circle below **b**, **c** show the rotation direction of the radiation force due to the pump light and manipulation light, respectively. (Reprinted with permission from [25] ©The Optical Society.)

can be determined using the symmetry of the metallic structures and orbital angular momentum of the incident light [18]. Notably, we employed the tetramer structure because of its satisfactory matching with the modeling in the DDA method [28].

To calculate the optical force that acts on the NP, we considered two features. One is to assume a specific metal structure using the DDA method, as done in the previous section. The other is to assume a three-level NP with levels {0, 1, 2} as depicted in Fig. 5.6 and to incorporate the nonlinear optical response there. Further, we ignored the nonlinear effect and temperature dependence of the dielectric constant in the metal. This is because the effects of possible saturation or broadening do not change the nature of the results of the intensity regions considered, and only slight quantitative corrections are expected [59, 60].

The expression of the time-averaged optical force is the same as (5.6), where $\mathbf{P}_{\text{probe}}(\mathbf{r}, \omega)$ should be replaced by $\mathbf{P}_{\text{NP}}(\mathbf{r}, \omega)$, which denotes the induced polarization of the NP. The integration was performed over the volume of the NP. We calculated the electric field $\mathbf{E}$ and polarization $\mathbf{P}_{\text{NP}}$ according to the following process. We obtained the simultaneous master equations of the three-level NP and Maxwell's equations. To solve the Maxwell's equations, we solved the same equation as (5.2) with (5.1) by using the same parameters of gold used in the previous section. The formal solution of the total electric field in the presence of the NP is expressed as the following integral equation:

$$\mathbf{E}(\mathbf{r}_i, \omega) = \mathbf{E}_b(\mathbf{r}_i, \omega) + \sum_j^{\text{NP}} \mathbf{G}(\mathbf{r}_i, \mathbf{r}_j, \omega)\mathbf{P}_{\text{NP}}(\mathbf{r}_j, \omega)V_j, \tag{5.7}$$

where $\mathbf{G}$ denotes the renormalized Green's function that includes the geometrical information of the metallic structures. To derive the renormalized Green's function of arbitrary-shaped metallic structures, we solved the following integral equation:

$$\mathbf{G}(\mathbf{r}_i, \mathbf{r}_j, \omega) = \mathbf{G}_0(\mathbf{r}_i, \mathbf{r}_j, \omega) + \sum_k^{\text{metal}} \mathbf{G}_0(\mathbf{r}_i, \mathbf{r}_k, \omega)\chi_{\text{metal}}(\omega)\mathbf{G}(\mathbf{r}_k, \mathbf{r}_j, \omega)V_k, \tag{5.8}$$

where $\mathbf{G}_0$ denotes the free-space Green's function.

The NP polarization should be determined using the total electric field. Accordingly, we assumed the following Hamiltonian of NP with isotropic dipole moments:

$$\hat{H}(t) = \sum_{a=1,2} \hbar\omega_a \hat{\sigma}_{aa} - \int_V d\mathbf{r} \hat{P}_{\text{NP}}(\mathbf{r})|\mathbf{E}(\mathbf{r}, t)|, \tag{5.9}$$

where index a represents the excited levels of the NP, $\hbar\omega_a$ the transition energy between the ground state 0 and state a of the NP, and $\hat{\sigma}_{aa}$ the population of state a. Further, we described the induced polarization as follows:

$$\hat{P}_{\text{NP}}(\mathbf{r}) = \sum_{k<l} d_{kl}\hat{\sigma}_{kl}\delta(\mathbf{r} - \mathbf{r}_p) + \text{h.c.}, \tag{5.10}$$

where d_{kl} denotes the matrix element of the dipole moment. In addition, $\hat{\sigma}$ denotes the dimensionless polarization operator, $\mathbf{r}_p$ its position, and indices $k = \{0, 1\}$ and $l = \{1, 2\}$ its energy levels. The Markovian master equation for the three-level NP is described using the following equation [61]:

$$\frac{d}{dt}\rho(t) = -\frac{i}{\hbar}[\hat{H}(t), \rho(t)] + \sum_{k<l} \frac{\gamma_{kl}}{2}[2\hat{\sigma}_{kl}\rho(t)\hat{\sigma}_{lk} - \{\hat{\sigma}_{lk}\hat{\sigma}_{kl}, \rho(t)\}]$$
$$+ \sum_{(l \neq m)} \frac{\gamma_{p_l}}{2}\left[(\hat{\sigma}_{ll} - \hat{\sigma}_{mm} - \hat{\sigma}_{00})\rho(t)(\hat{\sigma}_{ll} - \hat{\sigma}_{mm} - \hat{\sigma}_{00}) - \rho(t)\right], \quad (5.11)$$

where ρ denotes the density matrix of the NP and γ (γ_p) a nonradiative population damping constant (pure dephasing constant). Notably, the radiative decay rate of the NP is automatically incorporated into the calculation by using the renormalized Green's function. Using (5.11), we derived the equations of motion for the polarization operator $\langle\sigma_{kl}\rangle$, and expanded via the Fourier components with respect to the incident frequencies.

Using the mean-field approximation, we solved $\langle\sigma_{kl}\rangle$ and $\mathbf{E}(\mathbf{r}_i, \omega)$ in a self-consistent manner; thus, we obtained the NP polarization and total electric field. In the calculation, the Green's function was expanded into the Fourier components, as well as the polarization, and we considered only the frequency components near the plasmon resonance. In addition, we assumed that the directions of the dipole moments of the NP coincide with those of the background electric-field resonant to each transition. In this solution, if we consider up to the higher-order correlations in the master equation, we can take into account the effects of the nonlinear optical response in the NP beyond the perturbation regime, so that the population inversion can be treated. Substituting the obtained **E** and **P** in (5.6), we obtained the optical force, as well as the nonlinear effect of the NP.

5.3.2 Optical Force to Rotate the NP

In the numerical simulation, we employed the parameters by considering fluorescent dyes; the resonance energies for the 0–1 and 0–2 transitions were 1.80 and 1.85 eV, respectively. The dipole moments of the NP were set to 10 Debye for the 1–0 and 2–0 transitions, and this is realistic for the molecular aggregate of size 5^3 nm^3. The nonradiative population decay constants for the 1–0 and 2–1 transitions were set to 1μeV and 20 meV, respectively. The pure dephasing constant for the excited levels was 2 meV. As for the incident lasers, we considered two plane waves with circular polarization, and these waves were irradiated from above. The incident waves, hereinafter referred to as manipulation light and pump light, had spin angular momentum $s = \pm 1$ and energies resonant to the 0–1 and 0–2 transitions, respectively. The rotation direction of the NP depends on its spin angular momentum. Notably, the spin angular momentum of light can result in the orbital motion of the NP by utilizing the metallic nanocomplex.

In Fig. 5.7a, we depict the force map inside the tetramer structure in the x-y plane, where only the RCP manipulation light is irradiated with the spin angular momentum $s = +1$, and the intensity is 100 kW/cm^2. In this case, the dissipative force to rotate the NP saturates because of the nonlinearity, and the gradient force becomes the

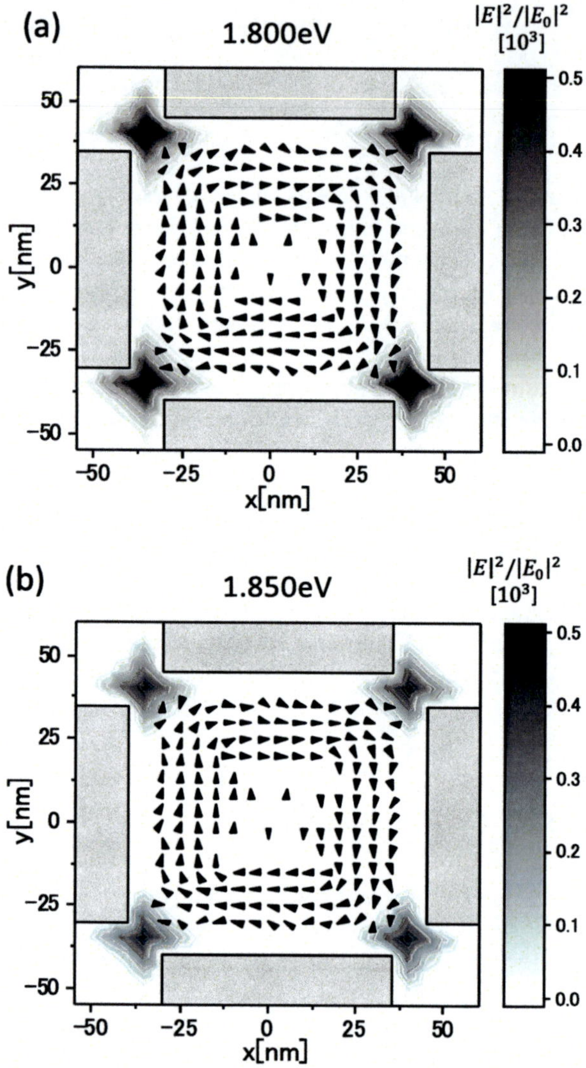

Fig. 5.8 Optical current map inside the metallic nanocomplex. The black arrows denote the optical current vectors. The grayscale bars show the field intensity. **a** Incident energy is 1.80 eV, and the spin angular momentum is $s = +1$. **b** Incident energy is 1.85 eV, and the spin angular momentum is $s = +1$. (Reprinted with permission from [25] ©The Optical Society.)

dominant force to move the NP. Therefore, the rotational force is suppressed, and the NP is attracted toward the gaps where the field intensity is strong. However, when the pump light is switched on, the pump intensity enables us to adjust the ratio of the dissipative force to gradient force. The criterion to adjust this ratio by using nonlinearity is explained in Appendix 1.

In Fig. 5.7b, we depict a force map, in which the pump light has the same spin angular momentum and intensity as those of the manipulation light. In this case, the 0–2 transition by the strong LSP field easily makes the population of the 1 state exceed 0.5; i.e., population inversion occurs. This inversion reverses the force vector induced by the manipulation light [49]. Consequently, although both the incident lights have the same spin angular momentum, the rotation direction of the manipulation light becomes opposite to that of the pump light. Therefore, their rotational forces (dissipative forces) eliminate each other and, hence, the optical force to rotate the NP is not induced, although the spin directions of both the lights are parallel. Subsequently, we examined the case in which the LCP pump light has different spin angular momenta. In this case, the rotational forces reinforce each other. Therefore, the rotational force of the NP was realized as depicted in Fig. 5.7c. Notably, we can avoid pointing the optical force outward by slightly red-detuning the manipulation-light energy to the resonance of the 0–1 transition (see Appendix 2). This rotational force is sufficiently strong because of the LSP. Therefore, one can realize the rotational manipulation at the nanometer scale and can selectively control the rotational direction by switching the pump light.

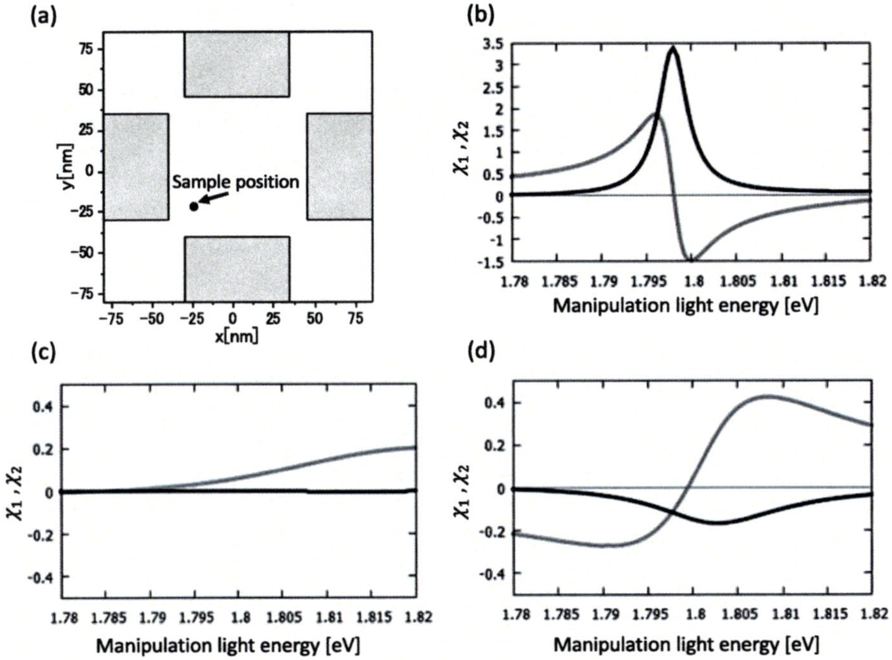

Fig. 5.9 **a** Sample position of the NP inside the tetramer structure in calculating susceptibility $\chi_{1(2)}$, which is induced by the manipulation light. The coordinates of the position are set to be (x, y, z)=(−20 nm, −20 nm, 25 nm). **b–d** Spectra of susceptibility χ_1 (gray line) and χ_2 (black line) as the functions of the manipulation light energy. **b** Case with weak excitation. The intensity of the manipulation light is 1 W/cm^2. **c** Case with strong excitation. The intensity of the manipulation light is 100 kW/cm^2. **d** Case with stimulated emission. The pump light energy and intensity are 1.85 eV and 100 kW/cm^2, respectively. The intensity of the manipulation light is 100 kW/cm^2. (Reprinted with permission from [25] ©The Optical Society.)

5.3.3 Optical Current

To gain a better understanding of the optical force that rotates the NP inside the metallic nanocomplex, we considered the x-y component of the Poynting vector, which is zero for plane wave light. This vector, which is also referred to as optical current, contributes to the dissipative force exerted on the NP [62, 63]. The optical current is classified into orbital and spin parts [63]. In Fig. 5.8 a, b, we depict the orbital part of the time-averaged optical current inside the metallic nanocomplex when the circularly polarized light with $s = +1$ is irradiated. Whether irradiating only the manipulation light (resonant to the 0–1 transition) or only the pump light (resonant to the 0–2 transition), the optical current flows clockwise inside the metallic nanocomplex. These results are consistent with the rotational force discussed in Fig. 5.7c, and also with the relation between population inversion and the rotational direction of the NP. In addition, it is observed that the direction of the rotational flow in the metallic nanocomplex depends on the direction of the optical current at the four gap positions. If the manipulation scheme suggested here is realized, the basic elements of manipulating the NP (pushing, pulling, and rotating) are also achieved. This will introduce new technologies not only for nanofabrication but also for nano-optomechanics that involve chiral materials and highly sensitive and selective chiral sensing.

5.4 Summary

Light–nanomatter chiral interaction has been a central subject of research in various domains for a long time. The development of nanofabrication technologies and single-molecular detection techniques take the study of this domain to a new stage. Particularly, the chiral interaction between light and metallic nanostructures has garnered attention because of its wide potential applications. Further, this phenomenon is the basis of our understanding of the nonlocal optical response, which is beyond the conventional model based on LWA or DA. For example, a strong CD of the localized field appears due to the geometric effect of the entire structure of the samples. The conversion of the spin angular momentum of light to orbital angular momentum via multipole excitation of nanoscale metallic complexes is also a peculiar manifestation of nonlocalty.

One of the interesting aspects of optical response is the fact that its manifestation appears not only as optical signals but also as mechanical force induced on matter systems. Accordingly, in this chapter, we discussed the chiral interaction between light and metallic structures visualized and appearing in optical-force effects.

The first topic is the manner in which we can visualize the 3D NF-CD that appears in the vicinity of chiral metallic structures. The aperture-type scanning near-field optical microscope is a powerful tool to unveil the chiral near field. However, it is difficult to elucidate the 3D structure of superchiral field, especially around the edges of metallic structures, as the longitudinal localized component is dominant there. However, optical force is significantly sensitive to the spatial distribution of the field, which includes both the transverse traveling component and longitudinal localized one. This sensitivity feature can be utilized to observe the 3D field distribution by using a microscope. Our numerical simulations showed that the optical-force microscope satisfactorily visualized the 3D NF-CD.

The second topic is the rotational-motion control of NPs in the nanoscale area, and this control is realized using the chiral interaction between the circularly polarized light and metallic nanocomplex. We theoretically demonstrated that the conversion between the spin angular momentum and orbital angular momentum using the aforementioned interaction induced the rotational motion of NPs. Furthermore, it was shown that by performing the balance control between the dissipation force and gradient force via nonlinear optical responses, we could switch the rotation direction of NPs.

The studies introduced in this chapter confirmed the essential roles of optical-force effects to research the chiral interactions at the metallic nanostructures that sustain LSP. It is important for various kinds of technologies to analyze molecular substances and control the mechanical motion of nanostructures via chiral interactions. The present results might stimulate further study of microscopic chiral interactions in various research domains.

Acknowledgements The authors thank Y. Togawa, S. Hashiyada, H. Okamoto and K. Sasaki for their fruitful discussions. This work was supported in part by JSPS KAKENHI Grant Number JP16H06504 for Scientific Research on Innovative Areas "Nano-Material Optical-Manipulation".

Appendix 1

Balance Between Dissipative Force and Gradient Force Under a Strong Near Field [25]

To understand the balance between the dissipative force and gradient force exerted on an NP under a strong near field, we considered the time-averaged optical force as follows:

$$\langle \mathbf{F}(\omega) \rangle = \frac{1}{2}\mathrm{Re}\Big[\int_{\mathrm{V}} \mathbf{dr}[\nabla \mathbf{E}(\mathbf{r},\omega)^{*}] \cdot \mathbf{P}_{\mathrm{NP}}(\mathbf{r},\omega)\Big]. \quad (5.12)$$

Here, we consider an evanescent field with a simple profile as follows:

$$\mathbf{E}(\mathbf{r},\omega) = \mathbf{E}(x,y)e^{i(k_1+ik_2)z}, \quad (5.13)$$

where k_1 and k_2 denote the wavenumber and extinction coefficients along the z-direction, respectively. The induced polarization is represented using complex susceptibility as follows:

$$\mathbf{P}_{\mathrm{NP}}(\mathbf{r},\omega)=(\chi_1+i\chi_2)\mathbf{E}(\mathbf{r},\omega)=(\chi_1+i\chi_2)\mathbf{E}(x,y)e^{i(k_1+ik_2)z}. \quad (5.14)$$

The real (imaginary) part of NP's susceptibility is given by $\chi_{1(2)}$. Substituting (5.13) and (5.14) in (5.12) and regarding the induced polarization as the point dipole, the optical force is written as follows:

$$\langle\mathbf{F}(\omega)\rangle=\frac{1}{2}V_{\mathrm{NP}}\left[\frac{1}{2}\chi_1\nabla_{x,y}|\mathbf{E}|^2-\chi_1k_2|\mathbf{E}|^2\mathbf{n}_z+\chi_2k_1|\mathbf{E}|^2\mathbf{n}_z\right]e^{-2k_2z}, \quad (5.15)$$

where V_{NP} denotes the volume of the NP, $\nabla_{x,y}$ a 2D gradient, and $\mathbf{n}_z$ a unit vector along the z-axis.

We classified the terms of (5.15) into dissipative and gradient forces. The first and third terms in the right hand side denote the conventional gradient force and dissipative force, respectively. In the presence of a finite k_2, we should also regard the second terms as the gradient force. Notably, under the LSP near-field, the gradient and dissipative forces depend on χ_1 and χ_2, respectively. This means that one can control over the trapping/exclusion of the NP by adjusting the balance between χ_1 and χ_2 and by tuning the laser frequencies.

Appendix 2

Dissipative and Gradient Forces Induced by the Detuned Light [25]
When the non-resonant light is radiated, χ_1 and χ_2 are constant for light energy. However, the resonant polarization in the linear optical response results in dispersion-type χ_1 and Lorentz-type χ_2. Therefore, their ratio drastically changes near the resonant energy of NP. Moreover, strong nonlinear optical effects such as absorption saturation and population inversion can result in the sign inversion of χ_1 and χ_2. Here, we have regarded the ratio of χ_1 to χ_2 as that of gradient force to dissipative force, and demonstrated it for the following three cases: weak excitation, strong excitation, and stimulated emission.

In Fig. 5.9b–d, we depict the real and imaginary parts of the induced polarization of the NP at the sample position indicated in Fig. 5.9a. The polarization includes the resonant and non-resonant elements for the 0–1 and 0–2 transitions that are induced by the manipulation light. The parameter of the NP is the same as that in the main text. Under the weak excitation, the nonlinear optical effects are negligible, and the polarization obeys the linear optical response. Notably, the resonant energy (e.g.., the peak energy of χ_2) is slightly red-shifted because of the metal-NP interaction. Under strong excitation, χ_2 is approximately zero, as the absorption and emission balance with each other owing to the absorption saturation. With respect to χ_1, it becomes large as the manipulation light energy increases. This is because the absorption saturation effect suppresses the polarization with the 0–1 transition, and the polarization with the 0–2 transition starts to appear. In addition, in the presence of the pump light, the sign of the susceptibility is inverted, compared with the weakly excited case. The negative value of χ_2 does not indicate reduction but gain. In other words, the negative value indicates the generation of the stimulated emission. The stimulated emission also results in the sign inversion of χ_1. Inside the tetramer structure, the dissipative force induces orbital motion, and the gradient force points toward the hotspots at gap areas. In Fig. 5.7b, we have employed the red-detuned manipulation light and utilize the repulsive force (negative value of χ_1) from the hotspots, as depicted in Fig. 5.9d.

References

1. Y. Tang, A.E. Cohen, Phys. Rev. Lett. **104**(16), 163901 (2010)
2. A. Papakostas, A. Potts, D. Bagnall, S. Prosvirnin, H. Coles, N. Zheludev, Phys. Rev. Lett. **90**(10), 107404 (2003)
3. T. Vallius, K. Jefimovs, J. Turunen, P. Vahimaa, Y. Svirko, Appl. Phys. Lett. **83**(2), 234 (2003)
4. M. Kuwata-Gonokami, N. Saito, Y. Ino, M. Kauranen, K. Jefimovs, T. Vallius, J. Turunen, Y. Svirko, Phys. Rev. Lett. **95**(22), 227401 (2005)
5. B. Bai, Y. Svirko, J. Turunen, T. Vallius, Phys. Rev. A **76**(2), 023811 (2007)
6. X. Yin, M. Schäferling, B. Metzger, H. Giessen, Nano Lett. **13**(12), 6238 (2013)
7. X. Yin, M. Schäferling, A.K.U. Michel, A. Tittl, M. Wuttig, T. Taubner, H. Giessen, Nano Lett. **15**(7), 4255 (2015)
8. X. Duan, S. Yue, N. Liu, Nanoscale **7**(41), 17237 (2015)
9. S. Zu, Y. Bao, Z. Fang, Nanoscale **8**(7), 3900 (2016)
10. V.P. Drachev, W.D. Bragg, V.A. Podolskiy, V.P. Safonov, W.T. Kim, Z.C. Ying, R.L. Armstrong, V.M. Shalaev, J. Opt. Soc. Am. B **18**(12), 1896 (2001)
11. T. Narushima, H. Okamoto, J. Phys. Chem. C **117**(45), 23964 (2013)
12. T. Narushima, S. Hashiyada, H. Okamoto, ACS Photonics **1**(8), 732 (2014)
13. S. Hashiyada, T. Narushima, H. Okamoto, J. Phys. Chem. C **118**(38), 22229 (2014)
14. E. Hendry, T. Carpy, J. Johnston, M. Popland, R. Mikhaylovskiy, A. Lapthorn, S. Kelly, L. Barron, N. Gadegaard, M. Kadodwala, Nat. Nanotechnol. **5**(11), 783 (2010)
15. Y. Tang, A.E. Cohen, Science **332**(6027), 333 (2011)
16. A. Kuzyk, R. Schreiber, Z. Fan, G. Pardatscher, E.M. Roller, A. Högele, F.C. Simmel, A.O. Govorov, T. Liedl, Nature **483**(7389), 311 (2012)
17. J. Kumar, H. Eraña, E. López-Martínez, N. Claes, V.F. Martín, D.M. Solís, S. Bals, A.L. Cortajarena, J. Castilla, L.M. Liz-Marzán, Proceedings of the National Academy of Sciences, p. 201721690 (2018)
18. K. Sakai, T. Yamamoto, K. Sasaki, Sci. Rep. **8**(1), 7746 (2018)
19. H. Ishihara, K. Cho, Phys. Rev. B **48**(11), 7960 (1993)
20. H. Ishihara, K. Cho, Solid State Commun. **89**(10), 837 (1994)
21. H. Ishihara, K. Cho, K. Akiyama, N. Tomita, Y. Nomura, T. Isu, Phys. Rev. Lett. **89**(1), 017402 (2002)
22. M. Takase, H. Ajiki, Y. Mizumoto, K. Komeda, M. Nara, H. Nabika, S. Yasuda, H. Ishihara, K. Murakoshi, Nat. Photonics **7**(7), 550 (2013)
23. M. Ichimiya, M. Ashida, H. Yasuda, H. Ishihara, T. Itoh, Phys. Rev. Lett. **103**(25), 257401 (2009)
24. T. Kinoshita, T. Matsuda, T. Takahashi, M. Ichimiya, M. Ashida, Y. Furukawa, M. Nakayama, H. Ishihara, Phys. Rev. Lett. **122**(15), 157401 (2019)
25. M. Hoshina, N. Yokoshi, H. Ishihara, Opt. Express **28**(10), 14980 (2020)
26. D. Nowak, W. Morrison, H.K. Wickramasinghe, J. Jahng, E. Potma, L. Wan, R. Ruiz, T.R. Albrecht, K. Schmidt, J. Frommer et al., Sci. Adv. **2**(3), e1501571 (2016)
27. P.B. Johnson, R.W. Christy, Phys. Rev. B **6**(12), 4370 (1972)
28. E.M. Purcell, C.R. Pennypacker, Astrophys. J. **186**, 705 (1973)

29. T. Iida, H. Ishihara, Phys. Rev. B **77**(24), 245319 (2008)
30. M. Ternes, C.P. Lutz, C.F. Hirjibehedin, F.J. Giessibl, A.J. Heinrich, Science **319**(5866), 1066 (2008)
31. T. Iida, H. Ishihara, Phys. Rev. Lett. **97**(11), 117402 (2006)
32. J. Yamanishi, Y. Naitoh, Y. Li, Y. Sugawara, Phys. Rev. Appl. **9**(2), 024031 (2018)
33. A. Ashkin, J.M. Dziedzic, J. Bjorkholm, S. Chu, Opt. Lett. **11**(5), 288 (1986)
34. L. Allen, M.W. Beijersbergen, R. Spreeuw, J. Woerdman, Phys. Rev. A **45**(11), 8185 (1992)
35. A. O'neil, I. MacVicar, L. Allen, M. Padgett, Phys. Rev. Lett. **88**(5), 053601 (2002)
36. T. Sugiura, S. Kawata, Bioimaging **1**(1), 1 (1993)
37. L. Novotny, R.X. Bian, X.S. Xie, Phys. Rev. Lett. **79**(4), 645 (1997)
38. M. Righini, A.S. Zelenina, C. Girard, R. Quidant, Nat. Phys. **3**(7), 477 (2007)
39. M.L. Juan, M. Righini, R. Quidant, Nat. Photonics **5**(6), 349 (2011)
40. K. Wang, E. Schonbrun, P. Steinvurzel, K.B. Crozier, Nat. Commun. **2**(1), 1 (2011)
41. T. Shoji, Y. Tsuboi, J. Phys. Chem. Lett. **5**(17), 2957 (2014)
42. P.R. Huft, J.D. Kolbow, J.T. Thweatt, N.C. Lindquist, Nano Lett. **17**(12), 7920 (2017)
43. K. Liu, N. Maccaferri, Y. Shen, X. Li, R.P. Zaccaria, X. Zhang, Y. Gorodetski, D. Garoli, Opt. Lett. **45**(4), 823 (2020)
44. A.N. Koya, J. Cunha, T.L. Guo, A. Toma, D. Garoli, T. Wang, S. Juodkazis, D. Cojoc, R. Proietti Zaccaria, Advanced Optical Materials, p. 1901481 (2020)
45. T. Iida, H. Ishihara, Opt. Lett. **27**(9), 754 (2002)
46. R.R. Agayan, F. Gittes, R. Kopelman, C.F. Schmidt, Appl. Opt. **41**(12), 2318 (2002)
47. T. Iida, H. Ishihara, Phys. Rev. Lett. **90**(5), 057403 (2003). See also, Phys. Rev. Focus 11, Story 6, 11 Feb. (2003)
48. H. Ajiki, T. Iida, T. Ishikawa, S. Uryu, H. Ishihara, Phys. Rev. B **80**(11), 115437 (2009)
49. T. Kudo, H. Ishihara, Phys. Rev. Lett. **109**(8), 087402 (2012)
50. K. Inaba, K. Imaizumi, K. Katayama, M. Ichimiya, M. Ashida, T. Iida, H. Ishihara, T. Itoh, physica status solidi (b) **243**(14), 3829 (2006)
51. H. Li, D. Zhou, H. Browne, D. Klenerman, J. Am. Chem. Soc. **128**(17), 5711 (2006)
52. C. Hosokawa, H. Yoshikawa, H. Masuhara, Jpn. J. Appl. Phys. **45**(4L), L453 (2006)
53. T. Shoji, N. Kitamura, Y. Tsuboi, J. Phys. Chem. C **117**(20), 10691 (2013)
54. S.E.S. Spesyvtseva, S. Shoji, S. Kawata, Phys. Rev. Appl. **3**(4), 044003 (2015)
55. Y. Jiang, T. Narushima, H. Okamoto, Nat. Phys. **6**(12), 1005 (2010)
56. T. Kudo, H. Ishihara, H. Masuhara, Opt. Express **25**(5), 4655 (2017)
57. H. He, M. Friese, N. Heckenberg, H. Rubinsztein-Dunlop, Phys. Rev. Lett. **75**(5), 826 (1995)
58. N. Simpson, K. Dholakia, L. Allen, M. Padgett, Opt. Lett. **22**(1), 52 (1997)
59. A. Alabastri, S. Tuccio, A. Giugni, A. Toma, C. Liberale, G. Das, F.D. Angelis, E.D. Fabrizio, R.P. Zaccaria, Materials **6**(11), 4879 (2013)
60. S. Chu, T. Su, R. Oketani, Y. Huang, H. Wu, Y. Yonemaru, M. Yamanaka, H. Lee, G. Zhuo, M. Lee et al., Phys. Rev. Lett. **112**(1), 017402 (2014)
61. H.J. Carmichael, *Statistical Methods in Quantum Optics 1: Master Equations and Fokker-Planck Equations* (Springer Science & Business Media, 2013)
62. S. Albaladejo, M.I. Marqués, M. Laroche, J.J. Sáenz, Phys. Rev. Lett. **102**(11), 113602 (2009)
63. M.V. Berry, J. Opt. A: Pure Appl. Opt. **11**(9), 094001 (2009)

Chapter 6
Magnetoelectricity of Chiral Micromagnetic Structures

A. P. Pyatakov, T. T. Gareev, A. S. Kaminskiy, K. S. Antipin, E. P. Nikolaeva, D. P. Kulikova, A. S. Sergeev, and A. V. Nikolaev

Abstract The concept of chirality has profound implications throughout science, from elementary particle physics to biology. In this review, we will refer to chirality as a rotational sense of spin structures such as spin cycloid in spiral magnets, micromagnetic structure in curved magnetic film, twisted spin space of 2D electron gas in an ultrathin magnetic metal film, and micromagnetic structures observed in iron garnet samples. It will be shown that chirality plays the key role in magnetoelectric phenomena observed in iron garnet films: the electric field-induced generation, motion and annihilation of magnetic topological defects such as magnetic bubble domains, domain walls and vertical Bloch lines. As a new degree of freedom that can be controlled by electric and magnetic field, the chirality is an important issue for spintronic applications.

The whole of my physics is nothing other than geometry
R. Decartes

6.1 Introduction. Chiral Structures of an Order Parameter

The idea of electromagnetic properties defined by the geometry of structure rather than its chemical composition is a cornerstone of the concept of metamaterial. However the notion of geometry-defined physical properties can be also useful while considering many other problems in condensed matter science, in particular, dealing with noncollinear ordering in magnets [1, 2] ferroelectrics [3], and liquid crystals [4, 5].

A. P. Pyatakov (✉) · T. T. Gareev · A. S. Kaminskiy · K. S. Antipin · E. P. Nikolaeva · D. P. Kulikova · A. S. Sergeev · A. V. Nikolaev
M.V. Lomonosov Moscow State University, Leninskie Gori, Moscow 119991, Russia
e-mail: pyatakov@physics.msu.ru

A. V. Nikolaev
Skolkovo Institute of Science and Technology, Bolshoy Boulevard 30, Bld. 1, Moscow 121205, Russia

E. Kamenetskii (ed.), *Chirality, Magnetism and Magnetoelectricity*, Topics in Applied Physics 138,
https://doi.org/10.1007/978-3-030-62844-4_6

In the case of magnetically ordered media the order parameters can be magnetization (for ferro- and ferrimagnets), antiferromagnetic vector (for antiferromagnets) and toroidal moment (for ferrotoroidal media). In ferroelectrics the analogue of magnetization is electric polarization. In the case of liquid crystals the role of the order parameter is played by the director, i.e. the average molecular orientation.

The regular structures with spatially varying order parameter direction can be represented as a superposition of two basic classes: the cycloidal and helicoidal ones (Fig. 6.1). The former has the axis of order parameter rotation that is perpendicular to the direction of spin modulation **k** (Fig. 6.1a) while the latter is a proper screw with the rotation axis parallel to the wave vector **k** (Fig. 6.1b).

From the standpoints of symmetry, the cycloid is analogous to the bending of the crystal (compare Fig. 6.1a, c). The bending results in the strain gradient (upper layers are stretched and the bottom layers are shrunk) pointing a polar direction in the crystal that is a prerequisite for the appearance of electric polarization (Fig. 6.1c). This phenomenon of the electric polarization induced by strain gradient is called the flexoelectric effect (from Latin *flexura*, meaning "bending") [6, 7]. Another type of spatially modulated structure, the helicoid, is symmetrically equivalent to the twisting deformation (compare Fig. 6.1b, d). In contrast to cycloid, the helicoid does

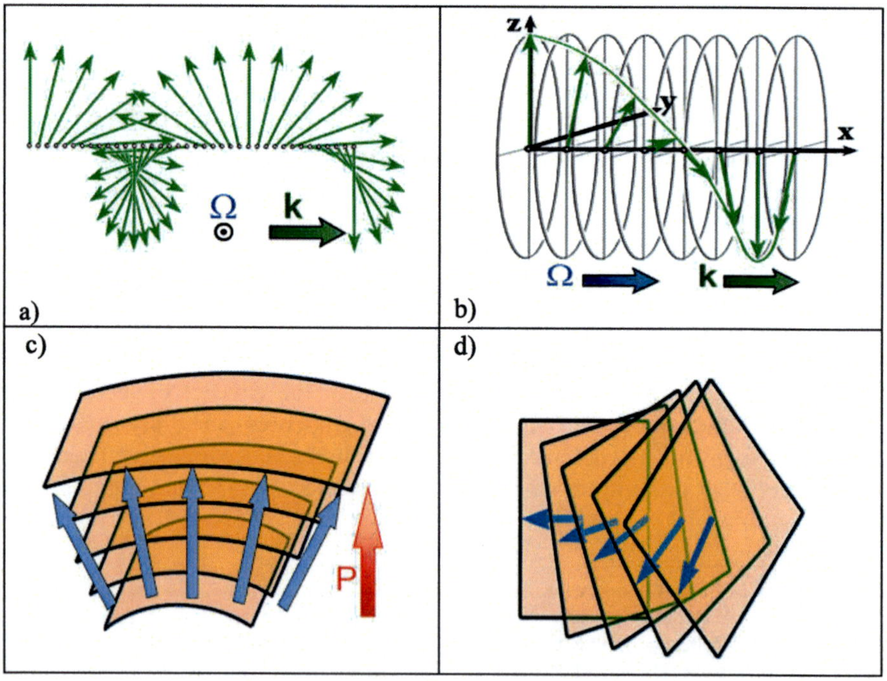

Fig. 6.1 The analogy between the spatially modulated order parameter structures and mechanical deformations. **a** the cycloid, **b** the helicoid, **c** flexural distortion, **d** twisting distortion

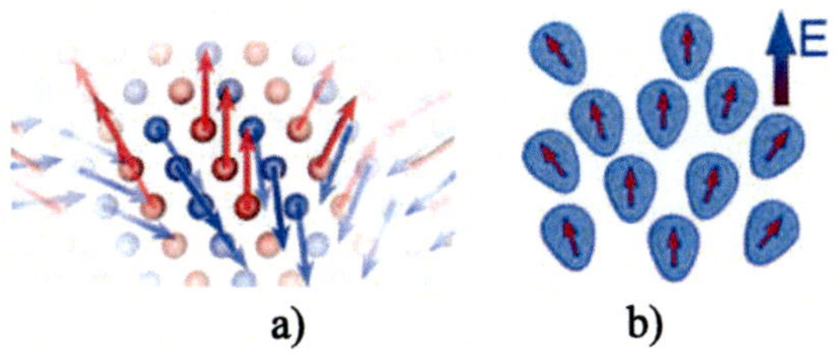

Fig. 6.2 The analogy between the structures of various nature: **a** spin cycloidal ordering of ion spins in antiferromagnet; **b** molecular structures in nematic liquid crystals

not single out the polar direction in the crystal and cannot be the single cause of the electric polarization.

The sense of rotation of order parameter in these spatially modulated structures is often called *chirality* [8]. Chirality is a general term for asymmetry with respect to the mirror symmetry and plays an important role in natural science from biology to optics of metamaterials, in particular, the magnetoelectric ones (see, for example [9, 10]). The chirality of spatially modulated structures can be controlled either through interface engineering [11] or by external field: magnetic [12–14] and electric one [15–17]. As will be shown below chirality is a key feature that determines the magnetoelectric properties of the magnet, i.e. the cross-coupling between magnetic and ferroelectric subsystems in crystal.

The certain chirality of a spatially modulated spin structure lowers the symmetry of a magnet. The spin cycloid is symmetrically equivalent to the fan-shaped molecular pattern that is formed in nematic liquid crystals in response to the electric field (see Fig. 6.2) [4]. In both cases, there is a kind of "bending" of order parameter distribution (the magnetization or director, respectively) whose symmetry is equivalent to the flexural strain. Conversely, the flexural deformation in solids results in appearance of electric polarization in the crystal (Fig. 6.1c). That is why all these phenomena are described by an umbrella term of *flexo*-effects (Fig. 6.3) [4].

In the case of flexoelectric effect, the strain gradient ∇U_{ij} induces the electric polarization $\boldsymbol{P}$. In a similar way the spatial modulation of magnetization generates the ferroelectricity in spiral multiferroics due to the *flexomagnetoelectric* effect [1]. The modulated spin structures like domain walls [18, 19], magnetic vortices [20], magnetic skyrmions [2] and other magnetic topological defects [21], can be the sources of local ferroelectricity. The converse flexomagnetoelectric effect [22] implies the modification of magnetic state due to the presence of electric field gradient (for example, from the charged tip of scanning probe microscope), in the same way as the polarization gradient induces the strain in the case of converse flexoelectric effect (Fig. 6.3) [6].

Flexomagnetoelectric effects can be mediated by mechanical deformation but the real bending of the crystal lattice is not a prerequisite of their existence. The bending in the magnetic subsystem, i.e. the spin cycloid (Fig. 6.1) also leads to the inversion symmetry breaking and the onset of polar direction in crystal. The flexomagnetoelectric interaction is described by the contribution to the free energy in the form of Lifshitz-type invariant [4]:

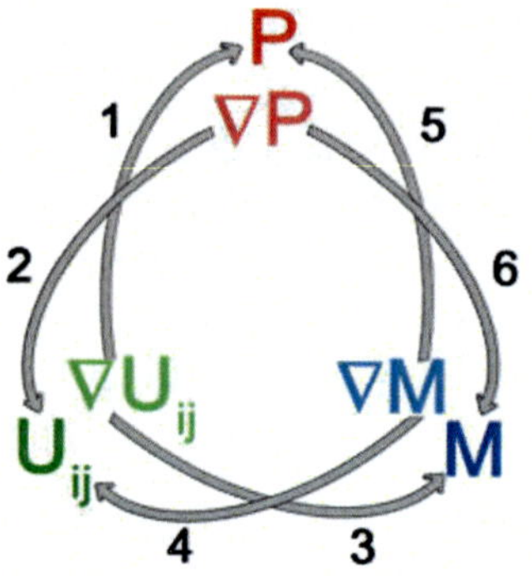

Fig. 6.3 Cross-type effects that couple the mechanical, magnetic and electric subsystems in magnetic crystals: 1—direct flexoelectric effect and 2 is converse flexoelectric effect; 3,4 are the direct and converse flexomagnetic effects; 5, 6 are direct and converse flexomagnetoelectric effects, respectively. **M**, **P**, u_{ij} stand for magnetization, electric polarization and strain tensor components, respectively

$$F_{FlexoME} = \gamma \mathbf{P} \cdot (\mathbf{n} div \mathbf{n} - (\mathbf{n} \cdot \nabla)\mathbf{n}), \tag{1}$$

where γ is the constant of flexo-type interaction, **n** is the unit vector of the order parameter, **P** is electric polarization or another polar vector (in the case of the thin film with chiral spin structure [23, 24] it is directed along the normal to the plane). In the case of liquid crystals **n** stands for director [4], in magnets **n** is magnetization or antiferromagnetic vector. The free energy term (1) is universal for multiferroic antiferromagnets, thin films of ferromagnets or liquid crystals emphasizes the profound analogy between various types of flexoelectric phenomena.

The switching of the chirality of the spatially modulated structure (the change of the signs of spatial derivatives of the order parameter **n** in (1)) results in the reversal of electric polarization **P**. This mechanism of switchable polarization is inherent to the multiferroics like manganites [25–27], tungstates [28, 29], and hexaferrites [12, 30–32] whose ferroelectricity is induced by cycloidal magnetic order.

Spin flexoelectricity should not be confused with *flexomagnetism* [33–38], that relates the strain gradient ∇u with magnetization **M** (Fig. 6.4) and, conversely, the inhomogeneous magnetization distribution with homogeneous strain. For example, the chirality of flux-closure domain structures in curved magnetic films [36] is determined by the sign of curvature (and consequently the sign of magnetization divergence that influences on the magnetostatic energy). The magnetism of curved surfaces is considered in special review [39].

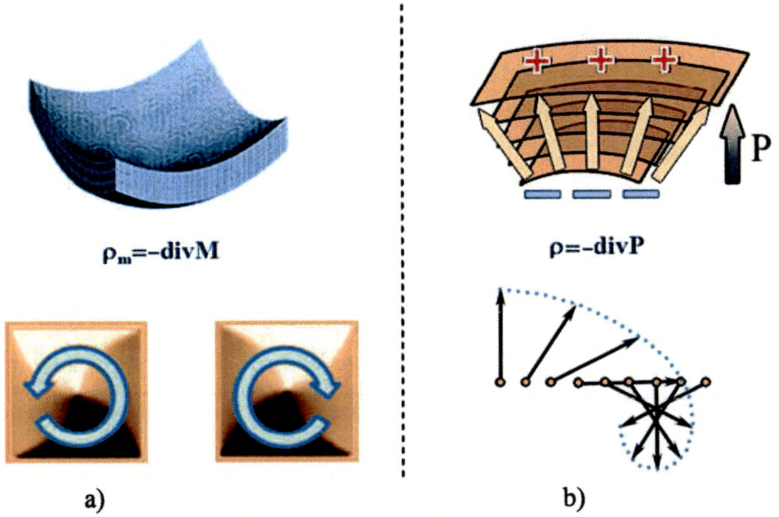

Fig. 6.4 Flexomagnetism and spin flexoelectricity (flexomagnetoelectricity): **a** the bending of the magnetic plate results in the vortex-like magnetization distribution whose chirality is determined by the direction of bending; **b** "the bending" in magnetization distribution (being translated at a large distance it converts to the spin cycloid in the bottom right corner) results in electric polarization in analogy to the electric polarization induced by mechanical bending due to the flexoelectric effect

6.2 Microscopic Mechanisms of Spin Flexoelectricity

On the microscopic level, the spatial derivatives in Lifshitz-type invariant (1) correspond to the cross product of spins of neighboring magnetic ions $[\mathbf{S}_1 \times \mathbf{S}_2]$. This type of relativistic and antisymmetric exchange interaction that is proportional to the cross product of the localized spins is called Dzyaloshinskii-Moriya interaction (DMI):

$$H_{DM} = \mathbf{D} \cdot [\mathbf{s}_1 \times \mathbf{s}_2], \quad (2)$$

where $\mathbf{s}_1$, $\mathbf{s}_2$ are unit vectors of the magnetic moments of exchange coupled ions, $\mathbf{D}$ is the Dzyaloshinskii vector.

In antiferromagnets, the DMI can have two macroscopic manifestations:

(i) a long-range chiral spin structure
(ii) a weak ferromagnetism: homogeneous magnetic state with non-zero net magnetization.

In the three-site indirect exchange model of DMI [40, 41] these two cases can be explained by Keffer formula for Dzyaloshinskii vector [42]:

$$\mathbf{D} = V_0[\mathbf{r}_1 \times \mathbf{r}_2], \quad (3)$$

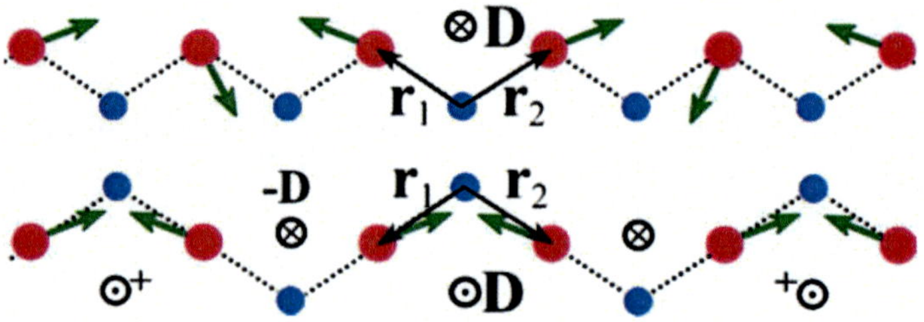

Fig. 6.5 (color online). The three-site ion model for DMI. **a** the unidirectional displacement of ligand ions (blue balls) leads to the constant angle between the spins of magnetic ions (brown balls). **b** the staggered displacements result in spin canting of antiferromagnetic sublattices

where V_0 is microscopic constant, $\mathbf{r}_1$, $\mathbf{r}_2$ are the magnetic ions position vectors directed from ligand ion to the nearest magnetic ions (Fig. 6.5).

The existence of electric polarization in the crystal implies the uniform displacement of the ligands that results in spin cycloid structure (Fig. 6.5a), at the same time the staggered displacement of ligand ions corresponds to sign-alternating Dzyaloshinskii vector **D** that results in antiferromagnetic ordering with non-zero net magnetization (Fig. 6.5b).

The DMI interaction (2) can be represented in terms of small displacement of the ligand ions **p** and spin canting $\delta\mathbf{s} = \mathbf{s}_1 - \mathbf{s}_2$ in the following way:

$$H_{DM} = V_0\big([\mathbf{p} \times \mathbf{a}] \cdot [\mathbf{s}_1 \times \delta\mathbf{s}]\big), \tag{3}$$

where **a** is the primitive vector of crystal lattice and **p** is the polar displacement of the ligand. The linear term in Taylor series expansion for the spin canting $\delta\mathbf{s}$ in (3) gives the (1) (for details, see [43]).

It is noteworthy that the antisymmetric exchange expressed by (2) is not always realized as superexchange interaction described by the three-ion model. Superexchange is specific for magnetic dielectrics (oxides and fluorites) while in magnetic metal films another mechanism of relativistic indirect exchange based on the Ruderman-Kittel-Kasuya-Yosida (RKKY) model is possible.

RKKY interaction between localized spins is mediated by the conduction electrons. In two-dimensional electron gas systems (2DEG) there is a precession of spins due to the spin-orbit Rashba interaction. The electron propagation with precessing spin is equivalent to motion in the "curved spin space" with spin quantization axis changing its orientation in a way the normal to some curved surface changes its direction (Fig. 6.6 inset).

In ultrathin metal films, Rashba interaction is caused by interfacial electric field [44] while in bent magnetic nanostructures it is related to curvature-induced quantum effects [45]. This is another manifestation of the profound analogy between magnetic media with chiral spin structures and bent surfaces.

The RKKY interaction in 2DEG with Rashba interaction is modified to the so-called *twisted RKKY interaction* where the conventional scalar product of localized

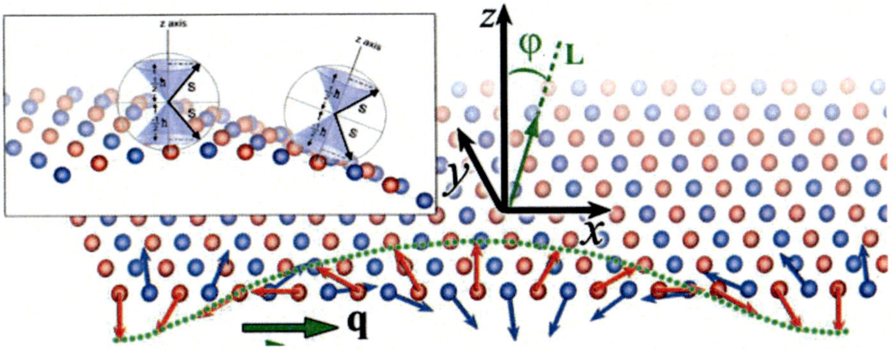

Fig. 6.6 (color online). The twisted RKKY interaction model. The cycloid structure in the single-atom layer of magnetic metal atoms (the ions of two antiferromagnetic sublattices are shown with different colors). The spatial rotation of spins and antiferromagnetic vector **L** takes place in the xz plane. The φ is the angle antiferromagnetic vector spatial rotation, **q** is the wave vector of spin cycloid. In the inset, there is an art representation of the curved spin space

spins $\mathbf{S}_1$, $\mathbf{S}_2$ is augmented with the cross product [46]:

$$H_{RKKY}^{twisted} = F(R)\left\{\cos\theta(\mathbf{S}_1\mathbf{S}_2) + \sin\theta[\mathbf{S}_1 \times \mathbf{S}_2]_y\right\}, \tag{4}$$

where *F(R)* is an oscillating range function of RKKY interaction, and the angle $\theta = 2k_R(x_1 - x_2)$(where x_1 and x_2 are the positions of interacting magnetic ions [46]), the k_R is the Rashba splitting of spin bands. For the case of 2D array of atoms, the cross product of spins in the second term of (4) takes the form of the mixed product of (3). The polar vector **p** in this case is directed along the surface normal and related to the surface electric field. This fact implies that the voltage tuning of Rashba splitting is possible [47].

6.3 Chirality Dependent Domain Wall Motion

An interface between the domains in magnets (a domain wall) is a region of gradual reorientation of magnetic moments and, in this context, it can be considered as a spin cycloid or a spin helicoid fragment that corresponds to the magnetization rotation angle equal to 90° or 180° (depending on the domain structure). Being the mobile interface, a domain wall is an ideal object to investigate the behavior under the influence of various types of stimuli like magnetic field, spin polarized current and electric field. The chirality of the domain wall plays a key role in the latter case since it determines the direction of the domain wall motion.

The examples of two types of 180-degree domain wall: the Neel domain wall and the Bloch one are shown in Fig. 6.7. They are analogous to the spin cycloid and the helicoid, respectively. According to (1) the Neel type domain wall can host electric

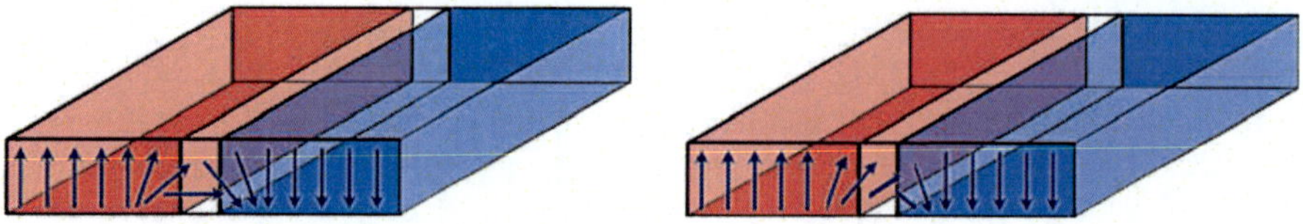

Fig. 6.7 Domain walls in magnets: a Neel-type domain wall; b Bloch-type domain wall

polarization due to nonzero divergence of magnetization while for Bloch wall the free energy term (1) remains zero.

The existence of domain wall magnetoelectricity was experimentally proven on the domain structure in iron garnet films [48, 49] that was subjected to the influence of gradient electric field from the tip electrode (Fig. 6.8). When the voltage was applied between the tip electrode and the film substrate, the displacement of the domain walls was observed. As soon as the voltage was turned off the domain wall came back to the initial equilibrium position. Reversing the polarity of the voltage at the tip resulted in the change of the direction of the displacement to the opposite (Fig. 6.8a). The attraction/repulsion of the domain wall with respect to the tip depended only on the electric polarity and was independent of the tip electrode position with respect to the domain wall.

The electric field-induced magnetic domain wall motion has been proved to be chirality dependent [13]: whether the domain wall attracts to or repels from the tip depends on the sense of spatial rotation of magnetization across the wall. The sense of rotation determines the sign of the spatial derivatives of magnetization in (1) and therefore the electric polarity of the wall.

Fig. 6.8 The electric-field induced magnetic domain wall displacement at various electric polarities: 1 is the tip electrode, 2 is the domain wall right (blue) is the domain wall position at the negatively charged tip, red (left) is for positive potential. The neutral position the wall is shown with a dashed line (Sample 1 from the table in the Appedix is used). b, c) illustrate the influence of electrically charged tip electrode on the stripe domain structure subjected to the in-plane magnetic bias field perpendicular to the domain walls. The sample 2 from the table b) is the initial unperturbed stripe domain structure c) represent the distortion of the structure: the nearest left with respect to the tip domain wall repulse from the tip, the right one attracts to it, the attraction of the next to near left domain wall is also seen

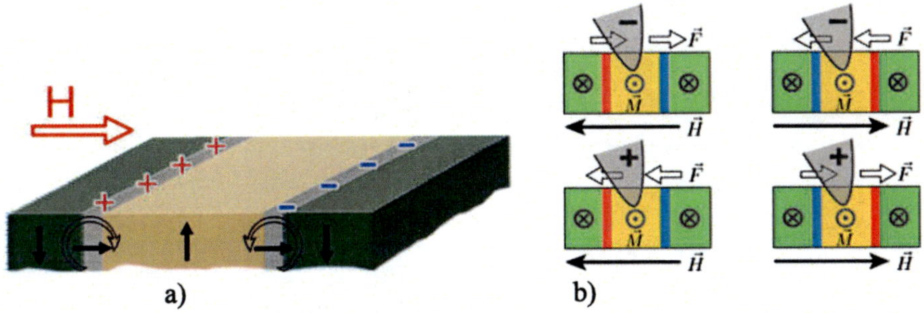

Fig. 6.9 The chirality dependent electric field-induced domain wall motion **a** if one moves from left to right the neighboring domain walls have an opposite sense of magnetization rotation: clockwise and counterclockwise (due to external in-plane magnetic field that orients the magnetization in domain walls), the surface charges are shown with «+» and «-»; **b** the top view of the domain structure for both polarities of the electric and magnetic field: force acting on the domain wall depends on the charge of the tip and the chirality of the wall (opposite chiralities shown with red/blue colors)

In the absence of magnetic bias field, the domain walls have a built-in chirality. In the external magnetic field, the magnetization direction in the center of the domain wall tends to orient along the field, and as a result, the neighboring domain walls have opposite chiralities (Fig. 6.9). In experiment these domain walls move in opposite directions with respect to the tip: if one attracts to the electrode then the other repels from it (Fig. 6.8c). The magnetic field reversal leads to the switching of the domain wall chirality, as well as its electric polarity and the direction of displacement (Fig. 6.9b) [13].

Remarkably, the domain wall chirality also determines the current-driven domain wall motion in heavy-metal/ferromagnet multilayers [50, 51]. In this case, domain wall motion is induced by spin-transfer torque produced by spin-Hall current from the non-magnetic heavy metal substrate (Pt, Ta, Ir) [52], as shown in Fig. 6.10a. In external in-plane magnetic field every second domain wall moves against an electron flow (Fig. 6.10b). This counterintuitive behavior had been remained puzzling until the chirality issue was considered [50]: since the chiralities of the neighboring domain wall in the external magnetic field are opposite (Fig. 6.9a) the directions the chirality-dependent forces acting on the domain walls are opposite as well (Fig. 6.10b).

6.4 Chirality Dependent Bubble Domain Generation

The chirality of the domain walls plays a key role in another type of magnetoelectric phenomena: the electric field-induced bubble domain generation. It should be noted that in the sample with stripe domain structure (samples 1,2 from table in the Appedix) the magnetic bubble state is unstable and cannot be induced solely with magnetic fields. The bubble domain is nucleated at the electrically charged tip electrode in

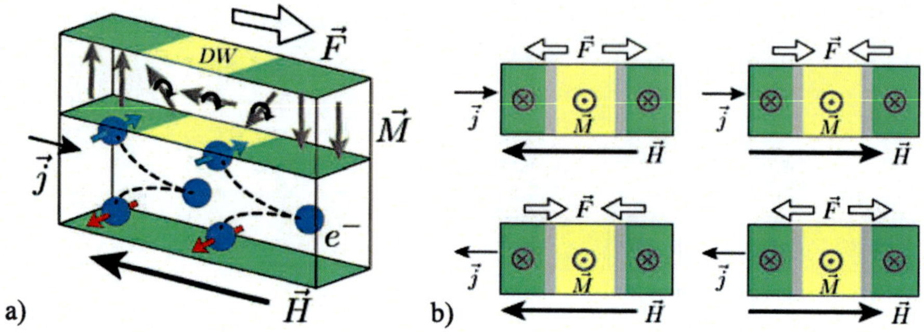

Fig. 6.10 (Color online) The chirality dependent current-induced domain wall motion: **a** Spin Hall effect in the bilayer of magnetic conducting material on non-magnetic heavy metal substrate: the spin-Hall current injected from the substrate acts with a force F on the domain wall due to the spin-transfer torque **b** the top view of the domain structure for various directions of current and magnetic field: as in the case of Fig. 6.9 the neighboring domain walls have opposite chiralities

the presence of constant in-plane and out-of-plane magnetic fields near the phase transition to the single domain state (Fig. 6.11b).

The role of the tip is not limited to just being the nucleation site of the bubble: the increase of voltage at the tip leads to the inflating of the bubble (Fig. 6.12). This effect may be interpreted in the light of the experiment with the in-plane magnetic field (Fig. 6.8b, c). Indeed, the neighboring domain walls experience the opposite electrostatic forces, therefore the bubbles in Figs. 6.11 and 6.12 are inflated by the pair of forces acting on the opposite edges of the bubble: the boundary nearest to the tip is attracted to it, the farthest one repels from the tip.

The mechanism of electric field induced magnetic bubble domain generation is illustrated schematically in Fig. 6.13. The chirality is shown with circle arrows.

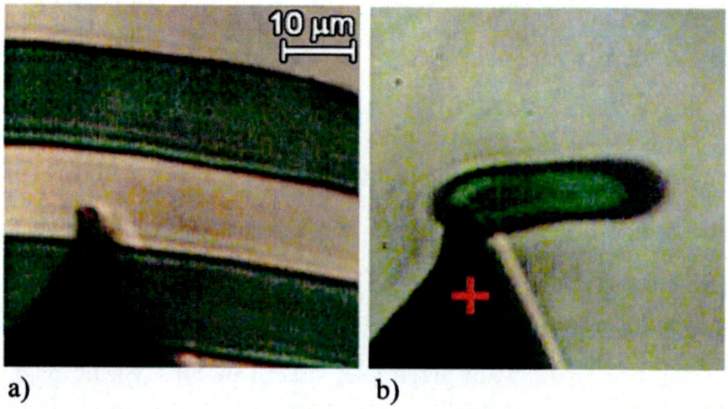

Fig. 6.11 The electric field induced magnetic bubble domain generation: **a** the original stripe domain structure in spontaneous state **b** The bubble domain nucleation at the electrically biased AFM cantilever tip in single domain state. Sample 3 from the table in the Appedix is used

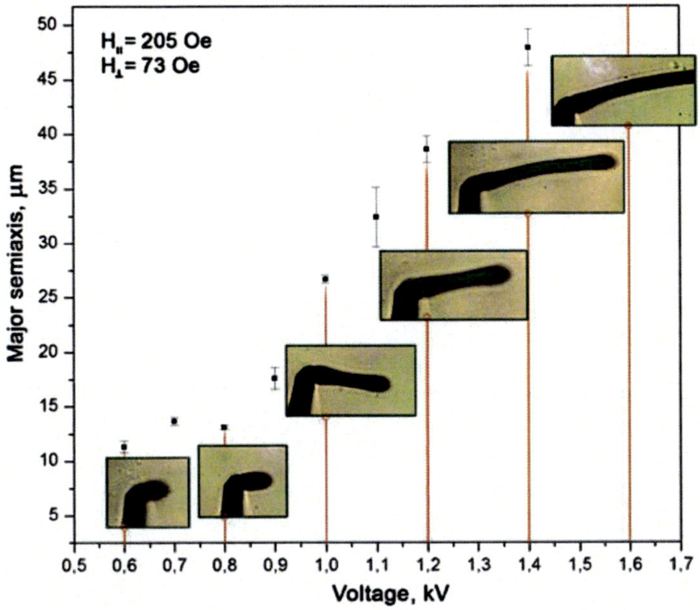

Fig. 6.12 The dependence of the electric field generated domain size (the major semi-axis for elliptical form) on the voltage applied to the tip electrode. Sample 3 from table in the Appedix is used

Various configurations of in-plane and out-of-plane magnetic fields as well as the tip electrode position with respect to the bubble are shown. It is noteworthy that the chirality of the domain wall located under the tip remains the same in all these situations.

In (210) iron garnet film (sample 2 from the Table) the bubble domain nucleation was observed only at positively biased tip electrode while application of negative voltage led to the shrinkage or even collapse of the nucleated bubble [53]. This can be explained in terms of the "built-in chirality" of domain walls in a spontaneous state that makes all the domain walls attracting to the positively charged tip (Fig. 6.8a). To generate the magnetic bubble domain with a negatively charged tip one needs to overcome this inherent tendency of nucleation of the domain wall with this preferred chirality that was presumably related to the magnetic film growth conditions.

In fact, it will be possible to achieve the bipolar nucleation of bubble domain walls at accessible voltage value if the energy cost of domain wall generation is lower. For example, this happens in the case of a magnetic domain with 90-degree domain walls. The bubble generation induced by negatively biased tip was observed in (110) iron garnet films (Fig. 6.14). According to the theory of spin flexoelectric interaction, the switching of electric polarity of the tip results in the chirality reversal (the spatial derivatives in (1) change signs). Indeed, according to Fig. 6.14, the chirality of the domain wall located under the negatively biased tip electrode is always opposite to the chirality of the domain wall under the positively charged tip.

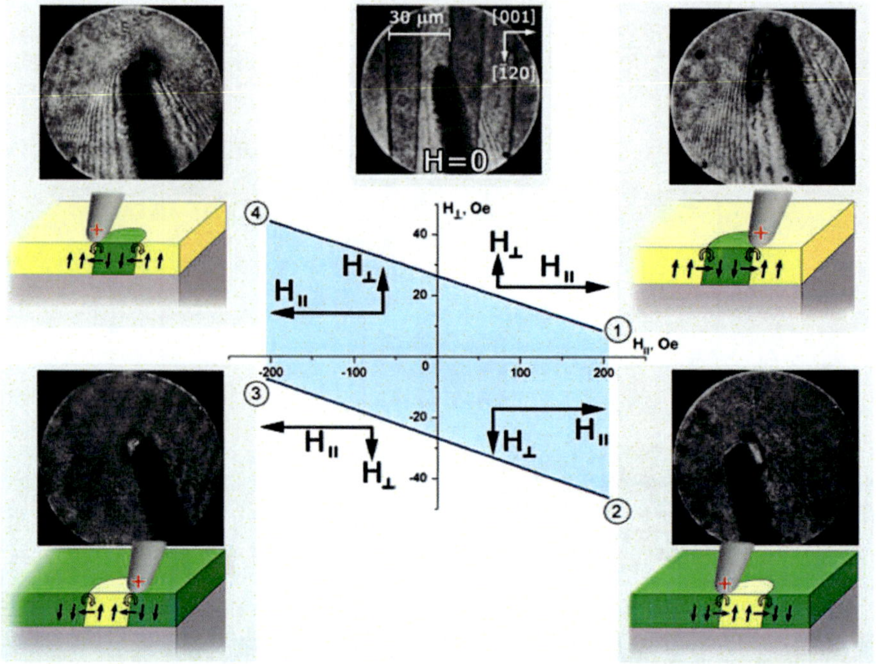

Fig. 6.13 Various scenarios of bubble domain nucleation corresponding to four points on magnetic phase diagram with coordinates ($H_\parallel$ stands for in plane magnetic field, and $H_\perp$ stands for out-of-plane field). The magneto-optical images of the bubble domain nucleated at the tip as well as the schematic pictures of micromagnetic configurations are shown in the insets. Sample 3 from table in the Appedix is used

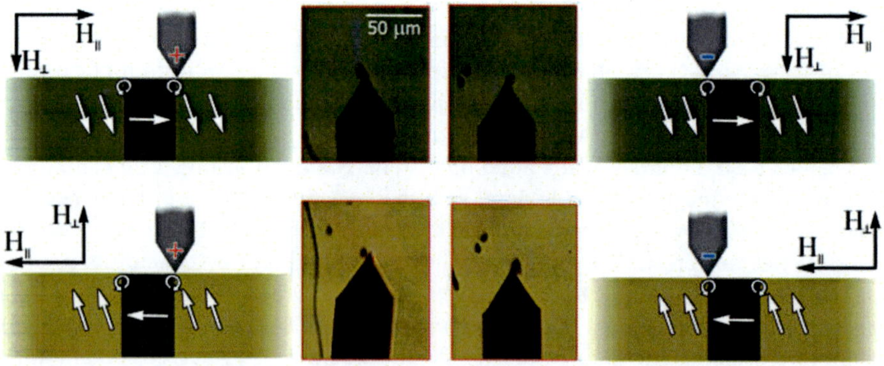

Fig. 6.14 The electric-field-induced domain nucleation with two electric polarities of the tip: in the red insets the top-view magnetooptical images of the iron garnet film are shown. The corresponding schematics of the magnetization distribution in the cross-section of the film and magnetic bias fields are shown. $H_\parallel = 170$ Oe, $H_\perp = 16$ Oe, the electric voltage at the tip is 500 V. The cantilever width is 50 μm. Sample 2 from table in the Appedix is used

Fig. 6.15 The reconstruction of the domain structure by electrically biased cantilever tips. The images are taken in the absence of external magnetic field, the electric voltage between the tip and the substrate is 500 V. Sample 4 from table in the Appedix. The white scale bar in the first frame corresponds to the 50mum

The electric field-induced generation of magnetic inhomogeneities can be observed in the spontaneous state with no magnetic bias as well (Fig. 6.15). The newborn inhomogeneity modifies the domain structure pushing aside the neighboring domain wall (the second frame in the series). The in-plane movement of the cantilever tip leads to the merging of the inhomogeneity with the domain wall (two central frames), and finally, to the radical reconstruction of the structure (the appearance of a new dislocation captured in the last two frames).

6.5 Spin Flexoelectricity of Bloch Lines, Vortexes and Skyrmions

The electric field can modify not only the shape and position of the domain wall but also its internal micromagnetic structure. The evidence for that is the influence of the electric field on the vertical Bloch lines. Bloch line is a kind of "boundary in the boundary": the region where the segments of the wall with the clockwise and counter-clockwise rotation of the magnetization meet each other. The pair of vertical Bloch lines (VBL) oriented along the normal to the plane are shown in the figure (Fig. 6.16a). They have opposite polarities (the so-called σ-charges): one line

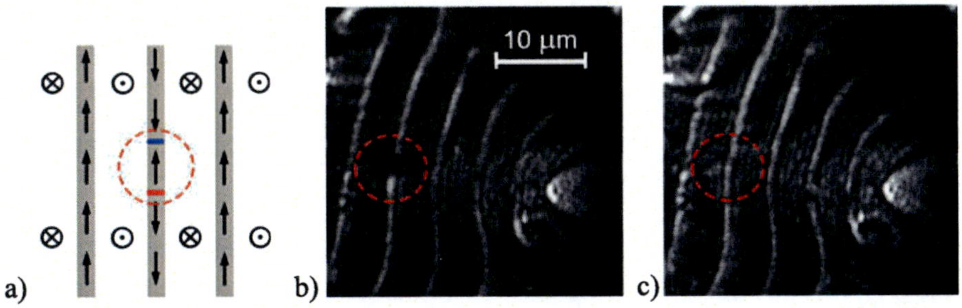

Fig. 6.16 Bloch line optical detection and electric control: **a** the schematic picture of a pair of vertical Bloch lines with the opposite σ-charges. **b** the magneto-optical image of a pair of VBL. **c** the image of the same region after the sweeping of the selected area in (b) with electrically charged tip. Sample 5 from table in the Appedix is used. The tip voltage is 1kV

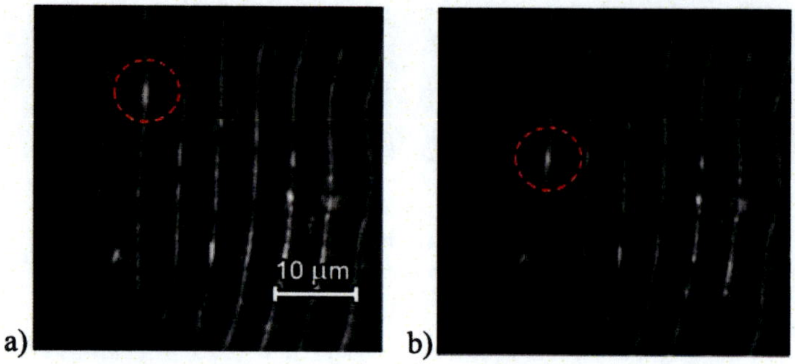

Fig. 6.17 Electric field-induced Bloch line motion: **a** the initial position of the VBL, **b** the displacement of VBL after the influence of electrically charged tip. Sample 5 from table in the Appedix is used. The tip voltage is 1kV

corresponds to the head-to-head magnetization orientation while the other is tail-to-tail VBL. Due to magnetostatic interaction of the VBL with the magnetic domain stray fields the segment of the wall where the VBL is located is slightly tilted (by an angle of several degrees) and the direction of the tilt depends on the σ-charge. This enables visualization the VBL with a dark field technique [54] since the tilted domain wall scatters light in a different way compared to the regular segment of the wall (see Apendix for details of the anisotropic dark field observation used in the experiments). Moreover, the lines with the opposite σ-charges due to the different orientation of the domain wall plane with respect to the direction of the light ray can be detected as a brighter or darker segment of the domain wall (Fig. 6.16b).

The sweeping of the domain wall that host VBL with electrically charged tip leads to the changes in the micromagnetic structure of the wall: the annihilation (Fig. 6.16c) of the pair of VBL with the opposite σ-charges or the displacement (Fig. 6.17) of the single VBL [55]. There were also reports on the broadening of VBL dark-field images in the electric field of flat electrodes [56].

The domain wall is often considered as a magnetic topological soliton with the magnetization modulated in one direction normal to the plane. The skyrmion [57, 58] is another type of topological soliton with axisymmetric geometry: its crossection in every radial direction mimics the magnetization distribution in the domain wall provided that the full angle of rotation is 360°. In analogy to domain walls, the skyrmions can be classified into two types: the Bloch and the Neel ones. Following the same logic as in the case of Neel domain wall, one can conclude that the Neel-type skyrmion should be electrically polarized due to spin flexoelectricity. More accurate consideration with the use of simulated annealing technique [2] shows that there are two components of electric polarization: along the normal to the film and the radial one (Fig. 6.18a). First one corresponds to the polarization of the Neel-like structure, the second one arises due to the in-plane curvature of the skyrmion. The estimate of the critical value of electric field strength required for skyrmion creation using the parameters of the iron garnet film [2] gives the value ~1 MV/cm comparable to the one used in our experiments with the bubble domain nucleation [53, 59] The

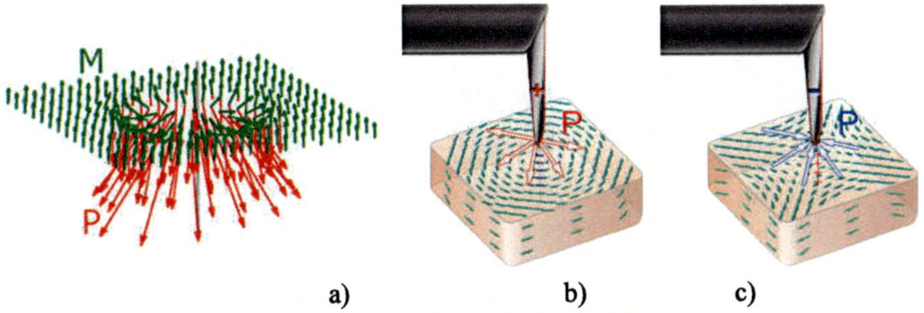

Fig. 6.18 The vortex-like micromagnetic structures and corresponding electric polarization distribution: **a** the skyrmion, **b** the magnetic vortex, **c** the magnetic antivortex

electric field-induced nucleation and annihilation of skyrmion were actually observed in experiment with the use of scanning tunneling microscopy probe [60] though its mechanism is still under discussion.

Finally, in the case of the magnetic vortex structure there is the only radial component of electric polarization (Fig. 6.18b). Due to the divergence of electric polarization the vortex core hosts a bound electric charge [20]. Its polarity does not depend on whether the vortex has a clockwise or counter-clockwise rotation of magnetization. The vortex-like structure of opposite polarity corresponds to an antivortex (Fig. 6.18c) [61]. Multiple vortex and antivortex generation should lead to the formation of a "magnetic atom" structure with a "nucleus" consisting of densely packed vortices and "a shell" of distant antivortices [62].

6.6 Conclusion

All the magnetic textures considered above can serve as illustrations of the general geometrical idea: the flexural deformation of order parameter distribution singles out the polar direction in media. There are various physical consequences of this fact including flexomagnetic phenomena, spin torque-driven domain wall dynamics and spin flexoelectricity that the paper is focused on. The chirality of the micromagnetic structure plays the key role in spin flexoelectricity: the electric polarity of the magnetic topological defect and the direction of its motion changes to the opposite upon the chirality switching. The magnetoelectric phenomena demonstrated, e.g. field-induced generation, motion and annihilations of magnetic topological defects, are interesting both from fundamental and applied points of view.

Acknowledgements Authors are grateful to Prof. A.M. Balbashov and Prof. F.V. Lisovskii for samples provided and Z.A. Pyatakova for image processing and figures design. The support from RFBR grant Support by Russian Foundation for Basic Research RFBR #19-02-00828 is acknowledged.

Appendix: Experimental and Calculation Details

In our experiments, the Bi-substituted rare earth iron garnet films $(BiR)_3(FeGa)_5O_{12}$ (R stands for rare earth) were used that were epitaxially grown on $Gd_3Ga_5O_{12}$ substrate by liquid-phase epitaxy (for details see elsewhere [63]). Bi substitution allows us to obtain a clear magneto-optical image of the domain structure.

The films grown on (111) substrate (sample 6 from the table) in spontaneous state demonstrate labyrinth-type domain structure (Fig. A.1a) while (110) and (210) films host stripe domain structure due to the in-plane magnetic anisotropy (Fig. A.1b).

No. of the sample		Chemical composition	h (μm)	p (μm)	$4\pi M_{s}$, G	K_u (erg/cm^3)	K_{orth} (erg/cm^3)	K_c (erg/cm^3)
1	(210)	$(BiLu)_3(FeGa)_5O_{12}$	7.4	44	77	732	5333	3208
2	(110)	$(BiLu)_3(FeGa)_5O_{12}$	11	40	–	–	–	–
3	(210)	$(BiLu)_3(FeGa)_5O_{12}$	10	28	62	–498	3398	1813
4	(210)	$(BiLu)_3(FeGa)_5O_{12}$	11	35	44	1054	1830	1016
5	(111)	$(BiTm)_3(FeGa)_5O_{12}$	10	8,7	144	5400	NA*	NA*
6	(111)	$(BiLu)_3(FeGa)_5O_{12}$	19	39	78	–	NA*	NA*

The table of samples parameters: h is the thickness of the film, p is the period of the domain structure in the spontaneous state, M_s is the saturation magnetization, K_u, K_{orth}, K_c are the constants of uniaxial, orthorhombic and cubic anisotropies, respectively

*The easy directions of magnetization for uniaxial and cubic anisotropies are the same

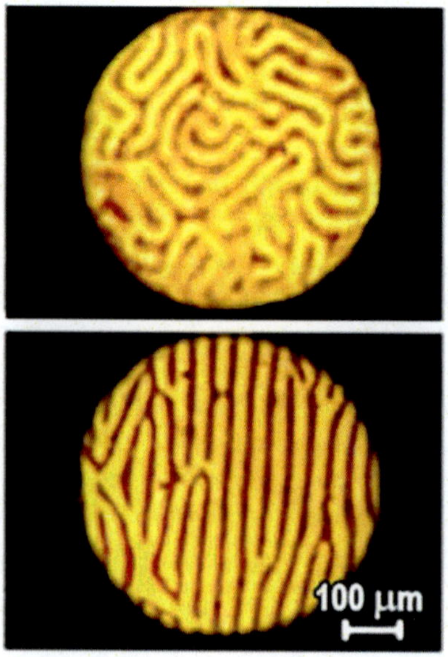

Fig. A.1 Magneto-optical images of domain structure: **a** labyrinth type, **b** stripe type

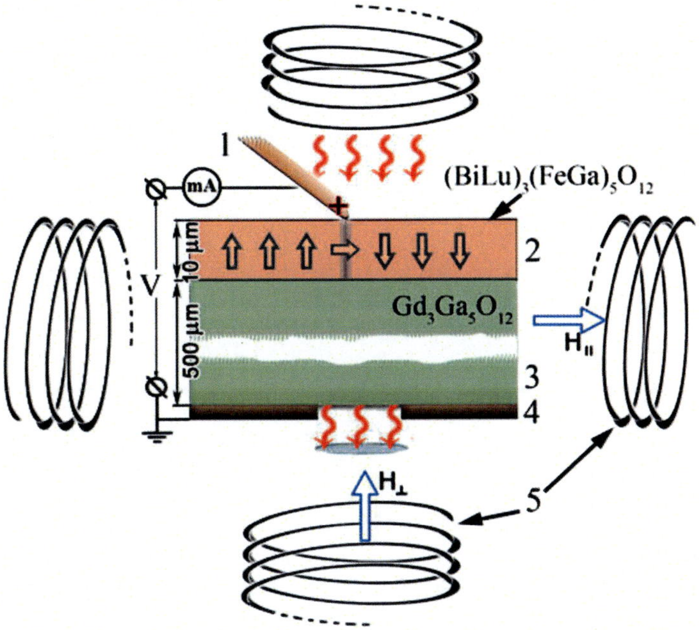

Fig. A.2 Schematic representation of the geometry of the experiment and the configurations of the magnetic field, electric field and magnetization. The electric field is generated by the tip electrode (1) in the iron garnet film (2) and the GGG substrate (3). (4) is the grounding electrode. The inductive coils (5) produce the in-plane $H_{||}$ and out-of-plane $H_{\perp}$ magnetic field

The schematics of the experiment with the electric field applied to the iron garnet film is shown in the Fig. A.2. The visual contact silicon cantilevers VIT_P_C-A (NT-MDT production) were used as tip electrodes (Figs. 6.8b, c, 6.11, 6.14 and 6.15). In some cases (Figs. 6.12 and 6.13) the 10μm-diameter molybdenum wire or 50μm-diameter copper wire (Figs. 6.8a, 6.16 and 6.17) were used.

To visualize the VBL the polarized anisotropic dark field observation (PADO) technique was used [64, 65]. The idea of PADO is to illuminate the domain wall at some angle to the film normal and to collect the light scattered on it (Fig. A.3). In the case of Figs. 6.15 and 6.16 the plane of laser light illumination was perpendicular to the plane of the domain wall.

For calculations of Fig. 6.17 the exchange stiffness constant $A = 10^{-7}$ erg/cm, and spin flexoelectricity constant $\gamma = 10^{-6}$ (erg/cm)$^{1/2}$ (the material parameters typical for iron garnet films [66]), as well as magnetization and magnetic anisotropy values from table in the Appedix, were used.

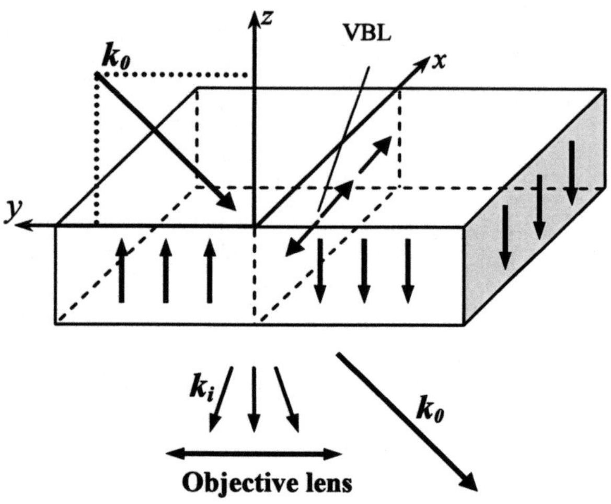

Fig. A.3 Schematics of PADO (polarized anisotropic dark field observation). VBL is a vertical Bloch line. k_0 is the wave vector of the incident light, the k_i are wave vectors of scattered light

References

1. S.-W. Cheong, M. Mostovoy, Nat. Mater. **6**, 13 (2007)
2. A.P. Pyatakov, A.S. Sergeev, F.A. Mikailzade, A.K. Zvezdin, J. Magn. Magn. Mater. **383**, 255 (2015)
3. J. Y. Chauleau, T. Chirac, S. Fusil, V. Garcia, W. Akhtar, J. Tranchida, P. Thibaudeau, I. Gross, C. Blouzon, A. Finco, M. Bibes, B. Dkhil, D. D. Khalyavin, P. Manuel, V. Jacques, N. Jaouen, M. Viret, Nat. Mater. **19**, 386 (2020)
4. A. Sparavigna, A. Strigazzi, A. Zvezdin, Phys. Rev. B **50**, 2953 (1994)
5. H. Ueda, T. Akita, Y. Uchida, T. Kimura, Appl. Phys. Lett. **111**, 262901 (2017)
6. P. Zubko, G. Catalan, A.K. Tagantsev, Annu. Rev. Mater. Res. **43**, 387 (2013)
7. B. Wang, Y. Gu, S. Zhang, and L. Q. Chen, Prog. Mater. Sci. **106**,100570 (2019)
8. S. Cheong, D. Talbayev, V. Kiryukhin, A. Saxena, Npj. Quantum. Materials. **3**, 19 (2018)
9. E.O. Kamenetskii, Ann. Phys. **532**, 1900423 (2020)
10. E.O. Kamenetskii, J. Mod. Opt. **66**, 909 (2019)
11. G. Chen, T. Ma, A.T. N'Diaye, H. Kwon, C. Won, Y. Wu, A.K. Schmid, Nat. Commun. **4**, 2671 (2013)
12. S. Ishiwata, Y. Taguchi, H. Murakawa, Y. Onose, Y. Tokura, Science **319**, 1643 (2008)
13. A.P. Pyatakov, D.A. Sechin, A.S. Sergeev, A.V. Nikolaev, E.P. Nikolaeva, A.S. Logginov, A.K. Zvezdin, EPL **93**, 17001 (2011)
14. V.Y. Ivanov, A.M. Balbashov, A.A. Mukhin, L.D. Iskhakova, M.E. Voronchikhina, J. Exp. Theor. Phys. **124**, 604 (2017)
15. Y. Yamasaki, H. Sagayama, T. Goto, M. Matsuura, K. Hirota, T. Arima, Y. Tokura, Phys. Rev. Lett. **98**, 147204 (2007)
16. T. Srivastava, M. Schott, R. Juge, V. Křižáková, M. Belmeguenai, Y. Roussigné, A. Bernand-Mantel, L. Ranno, S. Pizzini, S.M. Chérif, A. Stashkevich, S. Auffret, O. Boulle, G. Gaudin, M. Chshiev, C. Baraduc, H. Béa, Nano Lett. **18**, 4871 (2018)
17. J.A. Rodríguez-Velamazán, O. Fabelo, J. Campo, J. Rodríguez-Carvajal, N. Qureshi, L.C. Chapon, Sci. Rep. **8**, 10665 (2018)
18. V.G. Bar'yakhtar, V.A. L'vov, D. A. Yablonskii, JETP Lett. **37**, 673 (1983)
19. A.P. Pyatakov, A.S. Sergeev, E.P. Nikolaeva, T.B. Kosykh, A.V. Nikolaev, K.a. Zvezdin, A. K. Zvezdin, Uspekhi Fiz. Nauk **185**, 981 (2015)
20. A.P. Pyatakov, G.A. Meshkov, A.K. Zvezdin, J. Magn. Magn. Mater. **324**, 3551 (2012)

21. A.P. Pyatakov, Phys. B Condens. Matter **542**, 59 (2018)
22. A.F. Kabychenkov, F.V. Lisovskii, E.G. Mansvetova, JETP Lett. **97**, 265 (2013)
23. A. K. Zvezdin, Bull. Lebedev, Phys. Inst. **29**, 7 (2002)
24. M. Bode, M. Heide, K. von Bergmann, P. Ferriani, S. Heinze, G. Bihlmayer, a Kubetzka, O. Pietzsch, S. Blügel, R. Wiesendanger, Nature **447**, 190 (2007)
25. T. Kimura, T. Goto, H. Shintani, K. Ishizaka, T. Arima, Y. Tokura, Nature **426**, 55 (2003)
26. Y. Tokura, S. Seki, Adv. Mater. **22**, 1554 (2010)
27. A.P. Pyatakov, A.M. Kadomtseva, G.P. Vorob'ev, Y.F. Popov, S.S. Krotov, A.K. Zvezdin, M.M. Lukina, J. Magn. Magn. Mater. **321**, 858 (2009)
28. D. Meier, M. Maringer, T. Lottermoser, P. Becker, L. Bohatý, M. Fiebig, Phys. Rev. Lett. **102**, 107202 (2009)
29. T. Finger, D. Senff, K. Schmalzl, W. Schmidt, L. P. Regnault, P. Becker, L. Bohat\'y,M. Braden, Phys. Rev. B **81**, 54430 (2010)
30. S.H. Chun, Y.S. Chai, Y.S. Oh, D. Jaiswal-Nagar, S.Y. Haam, I. Kim, B. Lee, D.H. Nam, K.-T. Ko, J.-H. Park, J.-H. Chung, K.H. Kim, Phys. Rev. Lett. **104**, 37204 (2010)
31. Y. Kitagawa, Y. Hiraoka, T. Honda, T. Ishikura, H. Nakamura, T. Kimura, Nat. Mater. **9**, 797 (2010)
32. M. Soda, T. Ishikura, H. Nakamura, Y. Wakabayashi, T. Kimura, Phys. Rev. Lett. **106**, 87201 (2011)
33. E.A. Eliseev, A.N. Morozovska, M.D. Glinchuk, R. Blinc, Phys. Rev. B **79**, 165433 (2009)
34. P. Lukashev, R.F. Sabirianov, Phys. Rev. B **82**, 94417 (2010)
35. E. Eliseev, M. Glinchuk, V. Khist, V. Skorokhod, R. Blinc, A. Morozovska, Phys. Rev. B **84**, 174112 (2011)
36. R. Hertel, Spin **3**, 1340009 (2013)
37. J.H. Lee, K. Kim, B. Jang, A.A. Ünal, S. Valencia, F. Kronast, K. Ko, S. Kowarik, J. Seidel, C. Yang, Phys. Rev. B **96**, 64402 (2017)
38. B.A. Belyaev, A.V. Izotov, P.N. Solovev, N.M. Boev, Phys. Status Solidi RRL **2019**, 1900467 (2019)
39. R. Streubel, P. Fischer, F. Kronast, V.P. Kravchuk, D.D. Sheka, Y. Gaididei, O.G. Schmidt, D. Makarov, J. Phys. D Appl. Phys. **49**, 363001 (2016)
40. T. Moriya, Phys. Rev. Lett. **4**, 228 (1960)
41. A. Fert, P.M. Levy, Phys. Rev. Lett. **44**, 1538 (1980)
42. F. Keffer, Phys. Rev. **126**, 896 (1962)
43. A.K. Zvezdin, A.P. Pyatakov, EPL **99**, 57003 (2012)
44. A.P. Pyatakov, A.K. Zvezdin, EPL (Europhysics Lett. **107**, 67002 (2014)
45. P. Gentile, M. Cuoco, C. Ortix, Spin **3**, 1340002 (2013)
46. H. Imamura, P. Bruno, Y. Utsumi, Phys. Rev. B **69**, 121303(R) (2004)
47. S.E. Barnes, J. Ieda, S. Maekawa, Sci. Rep. **4**, 4105 (2014)
48. A.S. Logginov, G.A. Meshkov, A.V. Nikolaev, A.P. Pyatakov, JETP Lett. **86**, 115 (2007)
49. A.S. Logginov, G.A. Meshkov, A.V. Nikolaev, E.P. Nikolaeva, A.P. Pyatakov, A.K. Zvezdin, Appl. Phys. Lett. **93**, 182510 (2008)
50. S. Emori, U. Bauer, S. Ahn, E. Martinez, G.S.D. Beach, Nat. Mater. **12**, 611 (2013)
51. P.P.J. Haazen, E. Murè, J.H. Franken, R. Lavrijsen, H.J.M. Swagten, B. Koopmans, Nat. Mater. **12**, 299 (2013)
52. L. Liu, O.J. Lee, T.J. Gudmundsen, D.C. Ralph, R.A. Buhrman, Phys. Rev. Lett. **109**, 96602 (2012)
53. D.P. Kulikova, T.T. Gareev, E.P. Nikolaeva, TB. Kosykh, A.V. Nikolaev, Z.A. Pyatakova, A.K. Zvezdin, A.P. Pyatakov, Phys. Status Solidi - Rapid Res. Lett. **12**, 1800066 (2018)
54. A. Thiaville, J. Miltat, J. Appl. Phys. **68**, 2883 (1990)
55. A.S. Logginov, G.A. Meshkov, A.V. Nikolaev, A.P. Pyatakov, V.A. Shust, A.G. Zhdanov, A.K. Zvezdin, J. Magn. Magn. Mater. **310**, 2569 (2007)
56. V.E. Koronovskyy, Y.A. Vakyla, Phys. B Condens. Matter **547**, 79 (2018)
57. A. Bogdanov A.N., Yablonskii, JETP **68**, 101 (1989)
58. U.K. Rössler, a N. Bogdanov, C. Pfleiderer, Nature **442**, 797 (2006)

59. D.P. Kulikova, AP. Pyatakov, E.P. Nikolaeva, A.S. Sergeev, T.B. Kosykh, Z.A. Pyatakova, A.V. Nikolaev, A.K. Zvezdin, JETP Lett. **104**, 197–200 (2016)
60. P. Hsu, A. Kubetzka, A. Finco, N. Romming, K. von Bergmann, R. Wiesendanger, Nat. Nanotechnol. **12**, 123 (2016)
61. A.P. Pyatakov, G.A. Meshkov, A.S. Logginov, Moscow Univ. Phys. Bull. **65**, 329 (2010)
62. P.I. Karpov, S.I. Mukhin, Phys. Rev. B **95**, 195136 (2017)
63. G.V. Arzamastseva, A.M. Balbashov, F.V. Lisovskii, E.G. Mansvetova, A.G. Temiryazev, M.P. Temiryazeva, J. Exp. Theor. Phys. **120**, 687 (2015)
64. A. Thiaville, F. Boileau, J. Miltat, L. Arnaud, J. Appl. Phys. **63**, 3153 (1988)
65. A.S. Logginov, A.V. Nikolaev, V.V. Dobrovitski, IEEE Trans. Magn. **29**, 2590 (1993)
66. A.P. Pyatakov, D.A. Sechin, A.S. Sergeev, A.V. Nikolaev, E.P. Nikolaeva, A.S. Logginov, A. K. Zvezdin, EPL (Europhysics Lett. **93**, 17001 (2011))

Chapter 7
Current-Induced Dynamics of Chiral Magnetic Structures: Creation, Motion, and Applications

Jan Masell and Karin Everschor-Sitte

Abstract Magnetic textures can be manipulated by electric currents via the mechanisms of spin-transfer and spin-orbit-torques. We review how these torques can be exploited to create chiral magnetic textures in magnets with broken inversion symmetries, including domain walls and skyrmions. These chiral textures can also be moved by (electric) currents and obey very rich dynamics. For example, magnetic domain walls feature the famous Walker breakdown, and magnetic whirls are subject to the skyrmion Hall effect, which is rooted in their real-space topology. These properties led to a variety of potential novel applications which we briefly overview.

7.1 Introduction

Magnetic materials have been studied over the centuries for various prospects, in particular yielding the fundamental building blocks in computers that enable us to store tremendous amounts of data and transcending our culture to the age of information technology. Permanent magnetism as a key feature in these devices which offers not only fundamentally interesting, but also application-wise impressive and practical phenomena. The fact that magnets can be strongly influenced by external magnetic fields is both, a blessing and a curse. On the one hand, localized magnetic fields can be used to easily manipulate magnetic states of matter. On the other hand, magnetic devices are sensitive to invasive, external stimuli. Even nowadays, where magnetic mass storages in the form of rotating hard disc drives are steadily replaced by all-electric devices, magnetic recording media still appears throughout our everyday lives. Besides their intrinsic advantage of being non-volatile, magnetic recording media have to overcome some challenges such as increasing the speed

J. Masell
RIKEN Center for Emergent Matter Science, 2-1 Hirosawa, Wako, Saitama 351-0198, Japan
e-mail: jan.masell@riken.jp

K. Everschor-Sitte (✉)
Johannes Gutenberg University, Institute of Physics, Staudingerweg 7, 55128 Mainz, Germany
e-mail: kaeversc@uni-mainz.de

E. Kamenetskii (ed.), *Chirality, Magnetism and Magnetoelectricity*,
Topics in Applied Physics 138,
https://doi.org/10.1007/978-3-030-62844-4_7

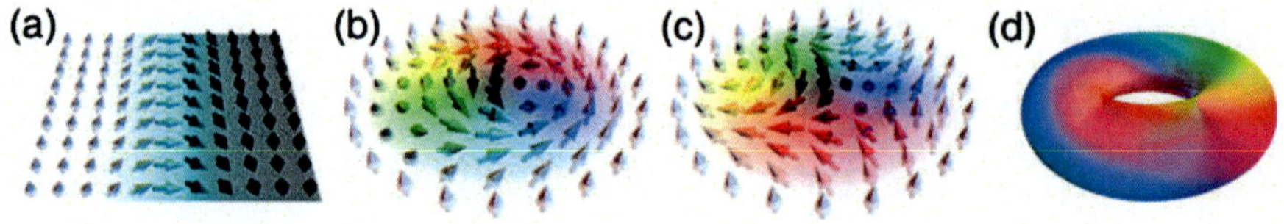

Fig. 7.1 Schematic figures of a **a** Néel-type domain wall, **b** Bloch-type skyrmion, **c** antiskymion, and **d** hopfion. The color code represents the direction of the normalized local magnetization. For the hopfion, we sketch an isosurface of the magnetization, using the same color code as in (**a**–**c**)

for reading and writing information and reducing the energy consumption to be competitive with nowadays all-electric information technology. For example, in the 1970s, some memories and computers used magnetic bubbles as mobile information carriers which, however, by the 1980s were completely replaced by magnetic hard drives or transistor-based controllers which turned out to be faster and better scalable. However, since the 1980s, research has unveiled a number of new effects and novel ways to control the static and dynamic properties of magnetic materials. These include most importantly (i) chiral magnetic systems and the ability to control their relativistic asymmetric exchange interaction—the Dzyaloshinskii-Moriya interaction (DMI) [1, 2]—and (ii) the ability to generate current-induced spin-torques, in particular spin-transfer torques (STTs) [3, 4] and spin-orbit torques (SOTs) [5, 6]. These spin-torques can be used to manipulate the magnetization directly, providing a new toolbox for potentially more competitive magnetic applications and opening the door to a whole range of interesting new physical phenomena.

This book chapter is intended to serve as an overview over the basic theoretical concepts in the context of chiral magnetic textures and their dynamics, in particular, when subject to spin-torques. Those spin textures which are stabilized, e.g., in systems with DMI or in systems with strong frustration comprise the well-studied magnetic domain walls, [7] but also the miniaturized versions of magnetic bubbles, i.e., magnetic skyrmions and antiskyrmions, [8–12] and magnetic hopfions [13, 14]. Representatives of such structures are shown in Fig. 7.1. We first review in Sect. 7.2 the description of magnetic textures within a continuum (micromagnetic) model, discussing their energy functional and their effective dynamic equation—the Landau-Lifshitz-Gilbert-Slonczewski (LLGS) equation. In this part, we also address the interaction of magnetic textures with electric currents, focusing on the origin and effects of spin-torques. In Sect. 7.3 we review the most common magnetic textures. In Sect. 7.4 we address how to create magnetic textures focusing on all-electrical methods. In Sect. 7.5 we review the recent progress made in the analysis of the motion of spin textures subject to spin-torques. In particular, we provide a detailed review on one of the most important and yet simple theoretical concepts for the motion of magnetic textures—the Thiele equation in its generalized form. We demonstrate how to apply it to the dynamics of magnetic textures such as domain walls, skyrmions, and hopfions. Finally, in Sect. 7.6, we give a brief overview over the plethora of suggested possible applications for chiral magnetic textures.

7.2 Continuum Model for the Magnetization

In this section, we present the continuum description of magnets and their interplay with electric currents, which in a simplified form is known as the micromagnetic model.

7.2.1 *Magnetization Statics*

The static properties of any magnet are well determined by an energy functional whose form depends strongly on the symmetries of the system. The precise determination of this energy functional in all its components is a very hard task. For sufficiently simple systems, the spin wave dispersion can be calculated with ab initio methods and then be fitted to a model of localized magnetic moments $\{\boldsymbol{S}_i\}$. Such treatments are very successful in describing magnetism on the atomic scale, which often requires exchange interactions $\boldsymbol{S}_i \cdot \boldsymbol{S}_j$ beyond nearest neighbors and, potentially, also more exotic interactions between multiple spins [15].

The magnetization in most chiral ferromagnets is, however, smooth, i.e., it is polarized on the length scale of the atomic lattice and varies only on much larger length scales. In this limit, the magnetic system can be well described by a phenomenological Ginzburg-Landau theory where an effective energy functional for the magnetization $\boldsymbol{M}$ is derived as a series expansion in powers of $\boldsymbol{M}$ and spatial derivatives ∂_α. Moreover, for temperatures far below the Curie temperature the magnetic system is in an ordered state and the local magnitude of the magnetization corresponds to the saturation magnetization M_s. The resulting energy functional can then be expressed in terms of the normalized magnetization $\boldsymbol{m} = \boldsymbol{M}/M_s$ in very general terms as

$$\begin{aligned} E[\boldsymbol{m}] = \int \mathrm{d}\boldsymbol{r}\ \big[&- B_i \boldsymbol{m}_i - K_{ij}\, \boldsymbol{m}_i \boldsymbol{m}_j - K_{ijkl}\, \boldsymbol{m}_i \boldsymbol{m}_j \boldsymbol{m}_k \boldsymbol{m}_l \\ &- D_{ij}^{\alpha}\, \boldsymbol{m}_i \partial_\alpha \boldsymbol{m}_j + A_{ij}^{\alpha\beta}\, \partial_\alpha \boldsymbol{m}_i \partial_\beta \boldsymbol{m}_j - Q_{ijk}^{\alpha\beta}\, \boldsymbol{m}_i \partial_\alpha \boldsymbol{m}_j \partial_\beta \boldsymbol{m}_k \\ &+ A_{ijkl}^{\alpha\beta}\, \boldsymbol{m}_i \boldsymbol{m}_j \partial_\alpha \boldsymbol{m}_k \partial_\beta \boldsymbol{m}_l + A_{ij}^{\alpha\beta\gamma\delta}\, \partial_\alpha \partial_\beta \boldsymbol{m}_i \partial_\gamma \partial_\delta \boldsymbol{m}_j - \ldots \big] \end{aligned} \tag{7.1}$$

where we implicitly sum over all spatial indices α, β and magnetization indices i, j. The first term is usually written explicitly as $\boldsymbol{B} = \mu_0 M_s(\frac{1}{2}\boldsymbol{H}_\mathrm{d} + \boldsymbol{H})$ where $\boldsymbol{H}_\mathrm{d}$ is the demagnetizing field and $\boldsymbol{H}$ is the externally applied magnetic field. All other interaction tensors are material specific and their tensorial structure is determined by the point group symmetry of the system. In principle, they can be completely anisotropic and even non-local, similar to the demagnetizing field. For an effective description of the low energy physics on large length scales, the infinite series in (7.1) is restricted to only the most relevant terms. Higher order interaction processes are usually small which suppresses terms which are higher order in the magnetization. Higher orders of derivatives, moreover, are suppressed as they become increasingly irrelevant on larger scales. Other terms, such as the DMI term with D^{α}_{ij}, are only non-vanishing because of the finite spin-orbit coupling, which is usually also small. Here we list the most common and relevant examples focusing on magnetic systems with their dominant lowest order chiral interaction

- For a *time-reversal invariant system* all terms with an odd power of $\boldsymbol{m}$ vanish.
- For an *inversion symmetric system* there is no chiral interaction, i.e. the DMI term vanishes, $D^{\alpha}_{ij} = 0$, and so do all terms with an odd number of derivatives.
- For a *bulk chiral magnet with a cubic unit cell* and a three-fold screw axis in the [109] direction, like the prototypical chiral magnets MnSi or FeGe, the DMI tensor simplifies to what is denoted as Bloch-type DMI in the literature, i.e., $D^{\alpha}_{ij} = D\epsilon_{i\alpha j}$, with ϵ_{ijk} being the Levi-Civita symbol. The exchange interaction becomes $A^{\alpha\beta}_{ij} = A\delta_{ij}\delta_{\alpha\beta} + A'\delta_{ij}\delta_{\alpha\beta}\delta_{i\alpha}$ with the Kronecker delta δ_{ij}. The last term proportional to A' reflects an anisotropic exchange coupling which can be present in cubic systems, but for MnSi and FeGe it turns out to be negligible [16].
- In *thin films or monolayers*, the inversion symmetry along the film normal (e.g., the $\hat{z}$-direction) is explicitly broken by the sandwich structure of the material or the substrate, but is usually preserved in the other directions. In such a setup the DMI tensor simplifies to what is known as Néel-type DMI, i.e. $D^{\alpha}_{ij} = D(\delta_{i\alpha}\delta_{jz} - \delta_{iz}\delta_{j\alpha})$. The exchange interaction simplifies to $A^{\alpha\beta}_{ij} = A\delta_{ij}\delta_{\alpha\beta} + A^{z}\delta_{ij}\delta_{iz}\delta_{\alpha\beta}$. Besides exchange and DMI, the term that is often relevant in such systems is the uniaxial anisotropy $K_{ij} = K\delta_{ij}\delta_{iz}$. In combination with the demagnetizing field, it can lead to the stabilization of magnetic bubbles.
- For systems with lower symmetry, the emerging terms and the corresponding tensor entries become more and more complex. We still would like to highlight *systems with C_{2v} symmetry*, where the two-fold rotational symmetry allows not only to realize magnetic skyrmions but also antiskyrmions [17], see Fig. 7.1. In a basis where the $\hat{z}$-axis is the two-fold rotational symmetry and the $\hat{x}$ and $\hat{y}$-axes are defined to be along the two reflection planes of the C_{2v} point group [18], the exchange parameters are $A^{\alpha\beta}_{ij}\,\partial_{\alpha}\boldsymbol{m}_i\partial_{\beta}\boldsymbol{m}_j = A_i\delta_{ij}\delta_{\alpha\beta}$ and there are seven independent DMI tensor components given by D^{x}_{xz}, D^{x}_{zx}, D^{y}_{yz}, D^{y}_{zy}, D^{z}_{zz}, D^{z}_{xx}, and D^{z}_{yy}. For further interesting systems we refer to [19, 20].

To summarize, the specific systems determine which magnetic interaction scales are relevant and which magnetic structures can be realized as (meta-)stable states. Over the past century, magnets with strong uniaxial anisotropy have been in the focus of material research, mostly application-oriented. With the advances made over the past decades, more detailed engineering of the properties of magnetic materials became possible and experimental techniques were developed that enable the observation of magnetic structures on the nanometer scale. With these new techniques at hand, more exotic materials can be studied where other interactions are dominant and stabilize new forms of magnetic textures.

7.2.2 Magnetization Dynamics in the Presence of Spin-Torques

The interplay of magnetism and currents is very complex and they mutually influence each other. For example, upon traversing a topologically non-trivial magnetic structure, the electrons pick up a Berry phase [21] which then leads to a topological Hall effect [12, 22–24] in addition to other Hall contributions such as the anomalous Hall effect. In this part, we will focus on the effects that an electric current has on the magnetization.

Within the micromagnetic framework, where the local magnitude of the magnetization is constant, the slow and smooth magnetization dynamics can be described effectively within the LLGS equation [3]

$$d_t \boldsymbol{m} = -\gamma\, \boldsymbol{m} \times \boldsymbol{B}_{\text{eff}} + \alpha\, \boldsymbol{m} \times d_t \boldsymbol{m} + \tau, \tag{7.2}$$

where γ is the (positive) gyromagnetic ratio, α is the dimensionless Gilbert damping parameter, and $\boldsymbol{B}_{\text{eff}} = -\delta E[\boldsymbol{m}]/(M_s \delta \boldsymbol{m})$ is the effective magnetic field due to interactions in the magnetization. τ represents the current-induced magnetic torques. It comprises STTs as well as SOTs, $\tau = \tau_{STT} + \tau_{\text{SOT}}$. Their lowest order terms comprise each a field- and damping-like term [25]

$$\tau_{\text{STT}} = -\,(\boldsymbol{v}_e \cdot \nabla)\, \boldsymbol{m} + \beta\, \boldsymbol{m} \times (\boldsymbol{v}_e \cdot \nabla) \boldsymbol{m} \tag{7.3a}$$

$$\tau_{\text{SOT}} = -\tau_{\text{FL}}\, \boldsymbol{m} \times \boldsymbol{\sigma} - \tau_{\text{DL}}\, \boldsymbol{m} \times (\boldsymbol{m} \times \boldsymbol{\sigma}), \tag{7.3b}$$

where $\boldsymbol{v}_e = -[P\mu_{\text{B}}/eM_s(1+\beta^2)]\boldsymbol{j}_e$ is the effective spin velocity [26] with $\boldsymbol{j}_e$ the electric current density, P the polarization, μ_{B} the Bohr magneton, and $e > 0$ the electron charge. β is the non-adiabatic damping parameter. $\boldsymbol{\sigma}$ encodes the spin polarized current: For the typical situation where the SOTs [5, 6] are generated by the spin Hall effect at an interface between a ferromagnet and a heavy metal, it is $\boldsymbol{\sigma} = \hat{\boldsymbol{n}} \times \boldsymbol{j}_e$ where $\hat{\boldsymbol{n}}$ is the normal direction of the interface between the materials. The strengths τ_{FL} and τ_{DL} for the field-like and damping-like terms are material dependent.

Note that the two torque expressions are not uniquely linked to STTs and SOTs, respectively, and we use these labels mostly for a better distinction of the two mathematically different expressions. For example, Eq. (7.3b) also describes STTs in layered systems where a current perpendicular to one layer is σ-polarized according to the magnetization in the adjacent layer.

7.3 Magnetic Solitons

In this part we review the most common magnetic structures focusing on chiral solitons, shown in Fig. 7.1.

Magnetic domain walls are rather ubiquitous one-dimensional textures that connect two distinctly polarized phases. The reason for this is that they do not require any particular stabilization mechanisms; the two distinct ferromagnetic ordered phases can be fixed by the boundary conditions. Therefore, magnetic domain walls have been observed and studied already long ago and can be found in many different samples with different properties. By continuation in further dimensions, domain walls can also be hosted in two-dimensional or three-dimensional systems. For example, in symmetric thin films Bloch- or Néel-type domain walls can be stabilized. Their helicity is determined by magnetostatic interactions and therefore depends on the film geometry. For very thin films, Néel type domain walls are formed, where the magnetization winds from one out-of-plane polarized state to the oppositely polarized state in the plane spanned by the out-of-plane state and the direction of rotation, as shown in Fig. 7.1. For thicker films mainly Bloch type domain walls are realized. In two-dimensional systems, domain walls can be effectively described as strings [27, 28] and closing this string leads to structures that are called magnetic bubbles. Furthermore, domain walls can obey localized defects for example in the version of Bloch lines, i.e., localized windings in the domain wall where the helicity switches from one Bloch handedness to the other handedness.

Magnetic skyrmions are localized whirls in two dimensions which can be viewed as a closed magnetic domain wall, embedded as defects in a surrounding background phase or they can be ordered in a lattice. In three-dimensional systems, skyrmions form extended strings. Skyrmions received lots of attention in particular due to their non-trivial real-space topology. The two-dimensional winding number for skyrmions (located in the xy-plane)

$$Q = \frac{1}{4\pi} \int_{\Omega} \mathrm{d}\boldsymbol{r}\ \mathbf{m} \cdot (\partial_x \boldsymbol{m} \times \partial_y \boldsymbol{m}) = \frac{1}{4\pi} \int_{\Omega} \mathrm{d}\boldsymbol{r}\ F_z \in \mathbb{Z} \tag{7.4}$$

evaluates to $Q = -1$ for the skyrmion and to $Q = +1$ for the antiskyrmion shown in Fig. 7.1, when integrating over the open area Ω of the skyrmion. Note that Q only evaluates to an integer if Ω is a closed surface, i.e. $\partial\Omega = 0$, which can, however, be mapped to an open area Ω with a topologically trivial boundary $\partial\Omega$. In the second equality we have introduced the solenoidal gyro-vector field $\boldsymbol{F}$ as

$$F_\alpha = \frac{1}{2} \epsilon_{\alpha\beta\gamma} \mathbf{m} \cdot \left(\frac{\boldsymbol{m}}{\partial r_\beta} \times \frac{\boldsymbol{m}}{\partial r_\gamma} \right). \tag{7.5}$$

While for the skyrmion only one component of this vector field is important, the topological index in 3D—the Hopf invariant—involves all components, see below.

Note that from a topological point of view, skyrmions and the earlier studied magnetic bubbles are equivalent. Even, given the various systems where such magnetic whirl-like textures with a winding number of ± 1 occur, a clear definition and full disentanglement might not be possible. Here, we will refer to magnetic bubbles when the domain wall width of the topological whirl-like structure is small compared to its the center area and to a skyrmion otherwise. While a strict differentiation between the two is not possible, the static and dynamic properties of magnetic whirls do depend on their detailed energy scales, and can be very different. In particular, skyrmions are typically smaller and more stable such that they are potentially interesting for future technological applications, see Sect. 7.6.

Roughly speaking, skyrmions occur in systems with competing interactions, of which some favor the alignment of magnetic moments, and others prefer their twisting. In most systems, however, it is a more complicated interplay that finally stabilizes the topological magnetic whirls. Experimentally, skyrmions have first been observed in bulk crystals with broken inversion symmetry as a result of a competition between a uniform stiffness A, DMI strength D, an applied magnetic field B and strong thermal fluctuations at temperatures slightly below the critical temperature [29]. By now several other systems have been identified to host skyrmions, revealing alternative stabilization mechanisms such as spatial confinement and frustrated exchange, e.g., via RKKY [30, 31]. Moreover, materials have been tailored to exhibit a strong interfacial DMI to host skyrmions at room temperature [32, 33]. For an overview of different material systems we refer to [11].

Typically, when discussing magnetic skyrmions, it is assumed that these are whirls in an out-of-plane polarized background. However, just as domain walls, skyrmions can be hosted by in-plane polarized backgrounds [34, 35] or even more complex background phases such as conical backgrounds in 3d [36], or embedded inside a helical phase [37]. While skyrmions are effectively two-dimensional structures, there is an ongoing search to find three-dimensional magnetic solitons.

A bit in the middle of two or three-dimensional structures are *magnetic bobbers*, [38] which, for example, occur in extended films. They look like a skyrmion on the top surface and then turn into a Bloch point within the material. Chiral bobbers are metastable states which are stabilized by the interplay of DMI and the boundary condition. The DMI induces a repulsive force between the skyrmion at the surface and the the Bloch point, wherefore the remaining skyrmion string is not expelled from the material. Similar surface effects have been known to occur due to demagnetization effects [39].

Magnetic hopfions are three-dimensional topological objects which, similar to the relation between skyrmions and domain walls, can be viewed as a closed skyrmion string, see Fig. 7.1. They can be characterized by the Hopf index H, which can be calculated by the Whitehead formula [40]

$$H = -\frac{1}{4\pi^2} \int_{\mathbb{R}^3} (\boldsymbol{F} \cdot \boldsymbol{A})\, d\boldsymbol{r} \tag{7.6}$$

with the vector field $\boldsymbol{F}$ defined above in (7.5) and $\boldsymbol{A}$ being an appropriate vector potential $\nabla \times \boldsymbol{A} = \boldsymbol{F}$. They have been predicted to occur in magnetic systems [13, 14] and recent progress in the development of new experimental techniques led to the first experimental signatures of hopfions in magnetic multilayers [114].

Other topological magnetic textures apart from the above mentioned ones, are predicted including those which have a more complex order parameter than just the normalized magnetization. Several of them have not yet been observed experimentally. However, the vast progress in recent years, allowing to engineer coupling strengths and image magnetizations in more and more detail, might reveal more exotic states in the future.

7.4 Creation of Magnetic Solitons

In this section we discuss, from a theoretical point of view, how to create magnetic solitons in different dimensions. These solitons comprise domain walls and skyrmions, see Sect. 7.3, and can be introduced into a given magnetic background, such as the ferromagnetic or helical state. Before discussing specific properties of different creation mechanisms, we first comment on a few very generic principles.

A soliton is stable and does not decay into magnons if it is protected by a (free) energy barrier. Vice versa, the creation of a soliton is also associated with an energy barrier, otherwise the solitons would just spontaneously proliferate and trigger a phase transition. As an example, Fig. 7.2 shows two possible mechanisms for the creation/annihilation of a skyrmion in a finite-size two-dimensional system. On the blue path, the magnetization twists at the edge and a skyrmion enters from outside

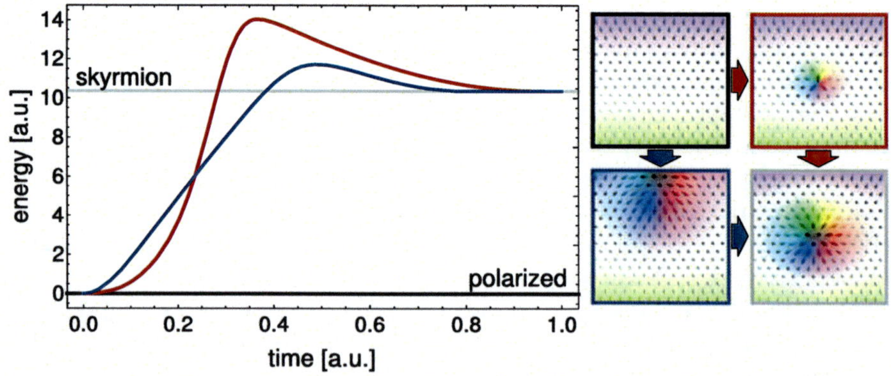

Fig. 7.2 Energetics of skyrmion creation in a finite size system with interfacial DMI. The energy barrier depends on details of the interactions but also on the creation process which can, for example, involve a continuous change of the winding number via the edge of the system (blue path) or a discontinuous change via the creation of a skyrmion in the bulk (red path)

the sample, while on the red path the skyrmion emerges between lattice points within the sample and then grows. In either case, the energy (as function of time in arbitrary units) has to rise above the bare energy difference between the initial and the final state but the absolute height of the barrier depends on how the soliton is introduced. Furthermore, introducing a soliton into the system requires to "twist" some parts of the current magnetization state, i.e., exerting local torques on the magnetization structure which are also very different for the two distinct paths shown in Fig. 7.2. Thus, the different creation mechanisms can be classified by the effective dimensionality of the magnetic soliton, its embedding background, and the origin of the acting torques. While creating one and two-dimensional textures is explored quite well, the controlled creation of three-dimensional magnetic structures is subject to current and future research.

7.4.1 Creation of One-Dimensional Solitons

A magnetic domain wall is an (effectively) *one-dimensional* magnetic soliton which usually connects two oppositely polarized phases, see Sect. 7.3. Within a nanowire with uniform magnetization, domain walls can only be created pairwise, as an odd number of domain walls naturally leads to opposite background orientations on both ends. To create such a pair of domain walls, one somehow has to locally flip the orientation of the magnetization. The most naive way is to locally apply a magnetic field in the desired direction, see Fig. 7.3a. An alternative is to switch the magnetization by means of locally applied spin-currents.

Single domain walls can be created at the edge of the sample. One can employ similar techniques as mentioned above, but at the edge the restrictive condition of having the same ferromagnetic state on both sides of the created magnetic texture does not apply. Alternatively, one can utilize magnetic inhomogeneities in the sample which effectively act as the edge of a smaller subsample. When an inhomogeneity

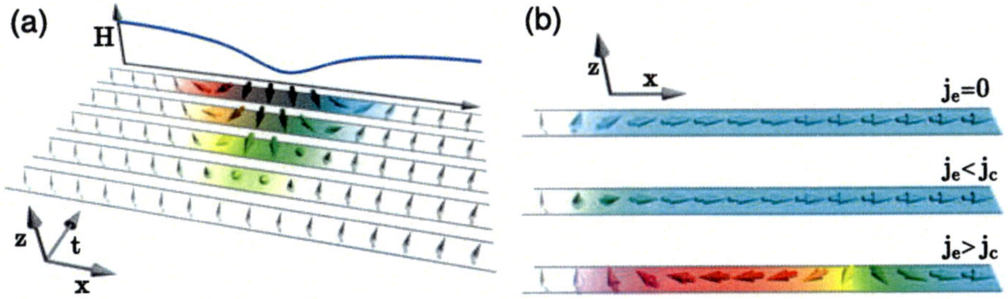

Fig. 7.3 Possible mechanisms to create magnetic domain walls. **a** Pairwise creation in the middle of a nanowire by a local magnetic field H or spin-currents (not shown). **b** Insertion of individual domain walls at the end of the wire via the interplay of spin-torques and an inhomogeneity (white spin fixed e.g. by strong perpendicular anisotropy)

alters the local magnetization direction, the generation of domain walls is not necessarily pairwise, see Fig. 7.3b. One idea is to exploit that the magnetic profile around an inhomogeneity is twisted and, therefore, spin-torques can act on this part by both, further twisting and pulling on the magnetic texture [41]. Increasing the applied current will enhance the twisting until a domain wall structure is built, that at the critical current density j_e^c rips off and travels along the system. Such a creation mechanism also works in a minimal model consisting of exchange and anisotropy interaction and basic STTs [41]. In this setup, domain walls are created periodically with a period T that depends on the applied current strengths j_e, or respectively on the effective spin velocity v_e as

$$T \sim (v_e - v_e^c)^{-1/2} \sim (j_e - j_e^c)^{-1/2}, \tag{7.7}$$

where the exponent is independent of the microscopic details. This universal behavior of the shedding period T can be proven by explicitly solving for the magnetic profile and its shedding period in the one-dimensional model including only exchange and anisotropy interactions. Furthermore, it is valid for a large class of magnetic systems independent of the details of the microscopic Hamiltonian, including the applicability for higher dimensions [42]. The required assumptions are (i) presuming a translationally invariant model away from the inhomogeneity and (ii) neglecting non-adiabatic spin-torque terms. The argument for the universal exponent in the shedding period is based on combining three ingredients:

(1) the postulate of a critical current density j_e^c above which there will be no statically stable solution and the created magnetic texture rips off the inhomogeneity,
(2) the behavior of the magnetic structure in the "just still static limit", i.e., for $j_e \lesssim j_e^c$ and
(3) the "just dynamic limit", i.e. for $j_e \gtrsim j_e^c$.

For the last two, one employs that the magnetic profile at the critical point will not differ too much in these two limits. The main influence on the magnetic structure will be a (time-dependent) shift in the position x_0 where the structure is centered in combination with a mild perturbation on the profile. Solving the LLGS equation in these two limits, yields for the "just still static" limit the relation $j_e^c - j_e^s \sim x_0^2$ and for the "just dynamic limit" $\partial_t x_0 = j_e^d - j_e^s$, where j_e^s is the current strength in the just still static limit and j_e^d in the just dynamic limit. These relations are the simplest, that satisfy the expected behavior: (i) the velocity of the domain wall depends linearly on the current strength beyond the threshold value and (ii) inverting the direction of the current should, in principle, create the domain wall structure in the opposite direction. Eliminating j_e^s allows to calculate the period of the magnetic texture formation $T \sim (j_e - j_e^c)^{-1/2}$ and thus explains the universal dependence.

Note that this universal behavior holds independent of the dimension, provided the above mentioned assumptions are satisfied. In dimensions higher than one the precise shape of the created magnetic texture cannot be calculated analytically. Based on topology, one can, however, conclude that the winding number during the production process must be conserved, opening up the possibility to shed more complex topological structures and their anti-particles.

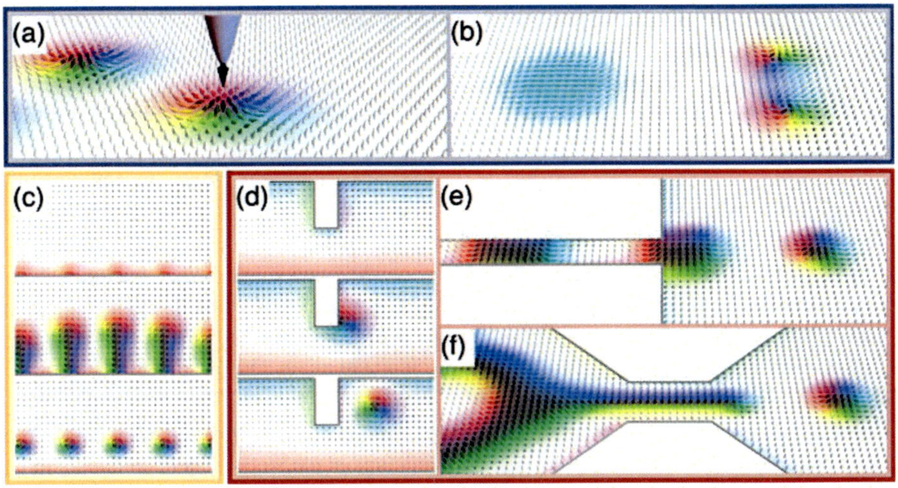

Fig. 7.4 Possible mechanisms to create magnetic skyrmions in ferromagnetic background by creating them within the sample (blue box), at the boundary (orange box) or via an engineered geometry (red box): **a** writing them locally, e.g. with spin-polarized electric currents, **b** creation of skyrmion-antiskyrmion pairs due to the interplay of magnetic inhomogeneities and spin-torques. **c** a train of skyrmions is created due to a particular sequence of applied magnetic fields. **d** a skyrmion is created at a notch via spin currents. **e** current-driven domain wall pairs are fusing to skyrmions, and **f** at the end of a constriction domain walls are chopped off to form skyrmions

7.4.2 Creation of Two-Dimensional Solitons

Examples of (effectively) *two-dimensional* magnetic solitons are skyrmions and antiskyrmions, see Sect. 7.3. To create skyrmions numerous methods exist, see for example [11] for an overview. Similar to domain walls, their creation mechanisms can be categorized by (i) being created within the sample, (ii) at the boundary, or (iii) because of a specialized geometry, see Fig. 7.4.

To create skyrmions within the sample in a ferromagnetic background one has to invert the magnetization in a small region. This can be done for example by local magnetic fields [43, 44], by local spin currents flowing perpendicular to the material [25, 45] by electric fields induced, e.g., by spin-polarized STM [46, 47], by effective local heating [48] or spontaneously by fluctuations [49]. Furthermore, one can generate skyrmions dynamically by means of the interplay of spin-currents and some inhomogeneity or defect, as indicated above when discussing domain wall production. Increasing the spin-current density above a critical values allows to produce skyrmion-antiskyrmion pairs dynamically by means of STTs [41, 42, 50, 51]. While the creation mechanism itself is independent of micromagnetic details and, in principle no twisting-like interactions such as DMI are required, in the subsequent dynamics, only the (meta-)stable solutions will continue to exist. For example, in a material with Bloch DMI, the antiskyrmion will annihilate and only a Bloch skyrmion will remain. Similarly, skyrmions can be created by SOTs [52–54].

An alternative is to create skyrmions via exploiting the tailored geometry of the material, see Fig. 7.4 for different options. For example, one can convert a domain wall pair into a skyrmion [55] or one can generate skyrmions through what has become known as "blowing bubble" technique [56] where a worm domain is sent through constrictions and "chopped" into pieces, i.e. skyrmions, by means of the diverging current upon leaving the constriction. Or skyrmions can be produced at a notch [57]. The latter technique leads over to the another principal option, i.e. to create skyrmions at the boundaries of a sample. Here the effect of the chiral surface states are helpful in pre-twisting the magnetic configurations. By means of a properly chosen protocol of an applied field strength one can even generate a whole train of skyrmions at the boundary [58].

While in a ferromagnetic background single magnetic skyrmions are (meta-)stable states, magnetic skyrmions can be the ground state of a chiral magnet in the form of skyrmion lattices under certain conditions [29]. To switch from the competing stripe domain phase into the skyrmion phase several methods exist, including triggering the magnetic material by means of AC field excitations [59].

7.5 Motion of Magnetic Solitons

The micromagnetic dynamics of the magnetization are mainly governed by the LLGS equation (7.2). This equation describes the local precession and relaxation of the magnetization, formulated in terms of a local effective magnetic field which accounts for the interaction of the magnetization with itself and its environment and, moreover, additional torques due to current-induced STTs and/or SOTs. In general, these non-linear dynamics lead to a complicated dynamical behavior which can even trigger the creation of magnetic solitons as described in Sect. 7.4 and can usually only be solved numerically.

However, once the solitons are created, they are influenced by the applied spin-torques and other external forces, e.g., due to field gradients. The reaction of the magnetization is most strongly expressed in the low energy degrees of freedom. An effective and potentially more efficient description of the soliton dynamics can therefore be formulated by taking only a few collective coordinates into account. We will review the derivation of these effective equations of motion, known as (generalized) *Thiele equations*,[1] in the following. We then show examples for their application when we use them as the starting point for the discussion of the dynamics of current-driven magnetic solitons.

---

[1]Note that the original equation that Thiele derived in his seminal works [60, 61] refers to the steady-state motion of domain walls. By now the concept that Thiele used to obtain his equation of motion for the domain wall has been generalized for any structure described by a finite set of collective coordinates.

7.5.1 A Collective Coordinate Approximation: Thiele Equations of Motion

The main step to obtain the Thiele equations for a given magnetic structure is to project the LLGS equation onto the corresponding collective coordinates. This said, the first step is to select suitable collective coordinates for a given magnetic structure. In principle, these collective coordinates can represent any property of the quasiparticle. To achieve an accurate description of the system with only a few coordinates it makes sense to choose coordinates which are related to zero modes or low energy modes as these are most easily activated, and thus most relevant for the low-energy physics of the system. A suitable choice of coordinates should therefore depend on the symmetries of the entire setup: the quasi-particle itself, the energy landscape, and the acting spin-torques.

To give an example for an appropriate collective coordinate, let us consider the standard assumption of the standard Thiele approach, i.e., a translational invariant model with a *rigid* magnetic texture. This means that the magnetic texture does not change its shape when driven by an electric current. In this situation, the position of the quasi-particle $\boldsymbol{R}(t)$ is a proper collective coordinate (or more generally, *any* position of the rigid magnetic structure) and the magnetization behaves as $\boldsymbol{m}(\boldsymbol{r}, t) = \boldsymbol{m}(\boldsymbol{r} - \boldsymbol{R}(t), 0)$.

For the derivation of the generalized Thiele equations, suppose that the time-dependence of the magnetic texture $\boldsymbol{m}(\boldsymbol{r}, t)$ is described by N collective coordinates $\boldsymbol{q}(t) = \{q_i(t)\}_{i=1,...,N}$. We first isolate the expression for the effective magnetic field $\boldsymbol{B}_{\text{eff}}$ by multiplying the LLGS equation (7.2), by $\boldsymbol{m}\times$ from the left.[2] Next, we project the LLGS equation onto the translational mode $\frac{\mathrm{d}\boldsymbol{m}}{\mathrm{d}\boldsymbol{q}_i}$ of the i-th collective coordinate $\boldsymbol{q}_i$, where the projection $\mathcal{P}(\boldsymbol{q}_i)$ is implemented by the scalar product $\mathcal{P}(\boldsymbol{q}_i) = \langle\frac{\mathrm{d}\boldsymbol{m}}{\mathrm{d}\boldsymbol{q}_i}|\,.\,\rangle = \int \mathrm{d}\boldsymbol{r}(\frac{\mathrm{d}\boldsymbol{m}}{\mathrm{d}\boldsymbol{q}_i}\cdot .)$. Moreover, we explicitly use that all time-dependence is now expressed in the collective coordinates to replace $d_t\boldsymbol{m} = \sum_{j=1}^{N} \dot{\boldsymbol{q}}_j \frac{\mathrm{d}\boldsymbol{m}}{\mathrm{d}\boldsymbol{q}_j}$ where $\dot{\boldsymbol{q}}_j = d_t\boldsymbol{q}_j$. A compact form of the $i = 1, ..., N$ Thiele equations for an arbitrary magnetic texture with both STTs and SOTs then reads

$$F_i(\boldsymbol{q}) = \mathcal{G}_{ij}\dot{\boldsymbol{q}}_j + \alpha\mathcal{D}_{ij}\dot{\boldsymbol{q}}_j + \mathcal{G}_{i\mu}^{\text{STT}} v_{e,\mu} + \beta\mathcal{D}_{i\mu}^{\text{STT}} v_{e,\mu} + \tau_{\text{FL}}\mathcal{G}_{i\mu}^{\text{SOT}}\boldsymbol{\sigma}_{\mu} + \tau_{\text{DL}}\mathcal{D}_{i\mu}^{\text{SOT}}\boldsymbol{\sigma}_{\mu} \tag{7.8}$$

with implicit summation over both, the collective coordinates $j = 1, ..., N$ and the spatial dimensions $\mu = x, y, z$. The projection of the effective magnetic field $\boldsymbol{B}_{\text{eff}} = -\delta E[\boldsymbol{m}]/(M_s\delta\boldsymbol{m})$ can be interpreted as a force

$$F_i(\boldsymbol{q}) = -\frac{\gamma}{M_s}\frac{\mathrm{d}E[\boldsymbol{q}]}{\mathrm{d}q_i} = -\frac{\gamma}{M_s}\int \frac{\mathrm{d}\boldsymbol{m}}{\mathrm{d}q_i}\cdot\frac{\delta E}{\delta\boldsymbol{m}}\mathrm{d}\boldsymbol{r} \tag{7.9}$$

[2] We exploit that the magnetization is a normalized vector field with $|\boldsymbol{m}(\boldsymbol{r})| = 1$. Thus, $\boldsymbol{m} \perp \partial_i\boldsymbol{m}$ and $\boldsymbol{m} \times (\boldsymbol{m} \times \partial_i\boldsymbol{m}) = -\partial_i\boldsymbol{m}$ for all coordinates $i = x, y, z, t$. Moreover, $\boldsymbol{m} \perp \boldsymbol{B}_{\text{eff}}$ is always achieved by adding a term $\lambda(\boldsymbol{r})(1 - \boldsymbol{m}^2) = 0$ to the energy functional which does not change the energy but cancels all components of $\boldsymbol{B}_{\text{eff}}$ that are parallel to $\boldsymbol{m}(\boldsymbol{r})$.

acting on the i-th collective coordinate q_i. Moreover (7.8) is implicitly non-linear as, in general, all the matrices on the right hand side depend on $q(t)$ and explicitly read

$$\begin{aligned}
\mathcal{G}_{ij} &= -\int \boldsymbol{m}\cdot\left(\frac{\mathrm{d}\boldsymbol{m}}{\mathrm{d}q_i}\times\frac{\mathrm{d}\boldsymbol{m}}{\mathrm{d}q_j}\right)\mathrm{d}\boldsymbol{r}, & \mathcal{D}_{ij} &= \int\left(\frac{\mathrm{d}\boldsymbol{m}}{\mathrm{d}q_i}\cdot\frac{\mathrm{d}\boldsymbol{m}}{\mathrm{d}q_j}\right)\mathrm{d}\boldsymbol{r},\\
\mathcal{G}^{\mathrm{STT}}_{i\mu} &= -\int \boldsymbol{m}\cdot\left(\frac{\mathrm{d}\boldsymbol{m}}{\mathrm{d}q_i}\times\frac{\mathrm{d}\boldsymbol{m}}{\mathrm{d}x_\mu}\right)\mathrm{d}\boldsymbol{r}, & \mathcal{D}^{\mathrm{STT}}_{i\mu} &= \int\left(\frac{\mathrm{d}\boldsymbol{m}}{\mathrm{d}q_i}\cdot\frac{\mathrm{d}\boldsymbol{m}}{\mathrm{d}x_\mu}\right)\mathrm{d}\boldsymbol{r},\\
\mathcal{G}^{\mathrm{SOT}}_{i\mu} &= -\int \boldsymbol{m}\cdot\left(\frac{\mathrm{d}\boldsymbol{m}}{\mathrm{d}q_i}\times(\boldsymbol{m}\times\hat{x}_\mu)\right)\mathrm{d}\boldsymbol{r}, & \mathcal{D}^{\mathrm{SOT}}_{i\mu} &= \int\left(\frac{\mathrm{d}\boldsymbol{m}}{\mathrm{d}q_i}\cdot(\boldsymbol{m}\times\hat{x}_\mu)\right)\mathrm{d}\boldsymbol{r}.
\end{aligned} \tag{7.10}$$

Here x_μ is the coordinate in the spatial direction μ and the corresponding unit vector is $\hat{x}_\mu$. We would like to emphasise that the Thiele approach is only a good approximation if sufficiently many relevant coordinates are considered. Furthermore, it is only of practical quantitative use if the matrix elements can be computed with a reasonable effort, which can also involve numerical simulations [62].

Note that for the above example of a translationally invariant system with a rigid magnetic texture with $\boldsymbol{q} = \boldsymbol{R}$ one obtains $\frac{\mathrm{d}\boldsymbol{m}}{\mathrm{d}R_i} = -\frac{\mathrm{d}\boldsymbol{m}}{\mathrm{d}x_i}$. Hence, the gyro-matrix $\mathcal{G}$ and the STT-coupling matrix $\mathcal{G}^{\mathrm{STT}}$ are directly related via $\mathcal{G}_{XY} = -\mathcal{G}^{\mathrm{STT}}_{Xy} = -4\pi Q$, where Q is the skyrmion winding number, see (7.4). Similarly, in this standard Thiele approach, the dissipation matrix $\mathcal{D}$ and the dissipative STT-coupling matrix $\mathcal{D}^{\mathrm{STT}}$ are related via $\mathcal{D}_{ij} = -\mathcal{D}^{\mathrm{STT}}_{ij}$ and their components resemble the magnetic stiffness in the energy functional, see (7.1).

In the following, we apply the generalized Thiele equations to describe the motion of magnetic solitons focusing on domain wall and skyrmion dynamics.

7.5.2 *Magnetization Dynamics of Domain Walls in Nanowires*

Magnetic domain walls can be moved by various sources, including, in particular, magnetic fields and spin-currents. The details of the motion as well as their possible maximal velocity typically depend on details of the system and the relevant magnetic interactions. In systems without DMI, for example, the plane in which the magnetization rotates when passing through the domain wall, i.e., *domain wall angle* or *helicity*, is determined by magnetostatic interactions, which are a rather weak effect. When increasing the driving magnetic field above a certain threshold value, the helicity unpins and the magnetization inside the domain wall precesses. This effect, known as the *Walker breakdown* [63], leads to a reduced domain wall speed and is therefore detrimental for the application in information technology, as discussed in Sect. 7.6. Nowadays, it is possible to design materials which have a strong DMI that more strongly pins the helicity and, consequently, raises the barrier for the activation of the Walker breakdown.

The magnetic field-driven dynamics of one-dimensional magnetic domain walls have been extensively studied over many decades and can be well described in the Thiele framework. Also magnetic domain walls in higher dimensions can be well described by this simple technique. Here, additionally to the one-dimensional case, the position of the domain wall is not only a one-dimensional parameter, but characterized by an extended line or surface. The additional degrees of freedom that then typically become relevant is the tilting [39] or bending of the hyperplanes of the domain walls.

To demonstrate the Thiele approach, let us consider a domain wall in an effectively one-dimensional system. This means we assume that a domain wall is located in a nanowire which is narrow compared to the length scale of the variations of the magnetic texture. In such a system, a simple ansatz for the domain wall profile can be written as

$$\boldsymbol{m}\,(x - X, \psi) = (\cos\psi \sin\theta\,(x - X)\,, \sin\psi \sin\theta\,(x - X)\,, \cos\theta\,(x - X))\,. \tag{7.11}$$

where X and ψ are the position and the helicity of the domain wall, respectively, where $\psi = \pm\pi/2$ describes a Bloch type wall and $\psi = 0$ or π describes a Néel type wall. $\theta(x)$ is the azimuthal angle of the magnetization. Here we assumed that the nanowire is along the $\hat{x}$-direction and that the helicity is not spatially dependent.

Consider now the standard thin film setup as introduced in Sect. 7.2, where the DMI is the relevant source of the twisting of the magnetization and magnetostatic interactions only enters on the level of a modified uniaxial anisotropy. In its simplified form, the only parameters that enter the energy functional (7.1) for a low energy description are the uniform exchange A, interfacial DMI $D > 0$, and the easy-axis anisotropy $K > 0$. In one spatial dimension, the energy functional then explicitly takes the form

$$E[\boldsymbol{m}] = \int A\left(\frac{\mathrm{d}\boldsymbol{m}}{\mathrm{d}x}\right)^2 - D\left(m_x\frac{\mathrm{d}m_z}{\mathrm{d}x} - m_z\frac{\mathrm{d}m_x}{\mathrm{d}x}\right) - K\,m_z^2\,\mathrm{d}x\,. \tag{7.12}$$

A domain wall which connects two polarized phases $\boldsymbol{m}(-\infty) = -\hat{z}$ and $\boldsymbol{m}(\infty) = \hat{z}$ minimizes this energy functional for the profile

$$\theta(x) = -2\arctan\left(e^{-\sqrt{K/A}\,x}\right) \quad \text{and} \quad \psi = \pi\,. \tag{7.13}$$

Here, the DMI term fixes the helicity $\psi = \pi$ while the other terms are independent of ψ. In the following, we will use the ansatz, (7.11), and the profile, (7.13), to discuss the current-driven motion of domain walls on the Thiele level.

Note that, in broader nanowires, the additional spatial dimension can allow for more complex domain wall profiles and also dynamics. In particular, domain walls in finite-width systems with DMI show a tilting of the domain wall normal [64] which can be explained by the interaction with the edges of the system [65]. The dynamics of Bloch lines are also known to lead to more complex behavior [27]. However, these effects go beyond the scope of this introduction.

7.5.2.1 Domain Wall Motion Due to Spin-Transfer Torques

In the continuum limit, without inhomogeneities the system is translationally invariant, i.e., $F_X = 0$. We will first consider the case of a small driving current which can only activate the zero mode, i.e., the translational mode. Next we will discuss the case of stronger driving which leads to the activation of the helicity degree of freedom and, finally, to the Walker breakdown under STTs.

Pinned helicity. In the limit of a small STT $\boldsymbol{v}_e = v_e\,\hat{\boldsymbol{x}}$, only the true zero modes are activated. Therefore, for the one-dimensional domain wall, the only relevant collective coordinate is the position X. Due to the lack of further spatial dimensions, the only terms which contribute from (7.8) are the dissipation terms. Since $\mathcal{D}_{XX} = -\mathcal{D}^{\mathrm{STT}}_{Xx}$, however, the Thiele equation reduces to the simple expression

$$\dot{X} = \frac{\beta}{\alpha} v_e. \tag{7.14}$$

In the limit of a rigid texture, the velocity $\dot{X}$ is directly proportional to the effective spin velocity v_e, and it is completely independent of details of the domain wall shape, see Fig. 7.5.

Unpinned helicity. In a next step, we consider the role of collective coordinates beyond the translational zero mode. The position X is still a zero mode with $F_X(X, \psi) = 0$ and, moreover, the off-diagonal dissipation matrix elements vanish, i.e., $\mathcal{D}_{X\psi} = \mathcal{D}^{\mathrm{STT}}_{\psi x} = 0$. Thus, the two coupled Thiele equations read

$$\mathcal{G}_{X\psi}\dot{\psi} \qquad + \mathcal{D}_{XX}(\alpha\dot{X} - \beta v_e) = 0\ , \tag{7.15a}$$

$$\mathcal{G}_{\psi X}(\dot{X} - v_e) + \mathcal{D}_{\psi\psi}\,\alpha\,\dot{\psi} \qquad = F_\psi(X, \psi)\ . \tag{7.15b}$$

With the ansatz from (7.11) and the solution for the profile in (7.13), the gyro-coupling and dissipation matrices of the Thiele equations evaluate to

$$\mathcal{G}_{\psi X} = -\mathcal{G}_{X\psi} = m_z(\infty) - m_z(-\infty) = 2\ , \tag{7.16a}$$

$$\mathcal{D}_{XX} \qquad = \int_{-\infty}^{\infty} (\theta'(x))^2\,\mathrm{d}x \qquad = 2\sqrt{K/A}\ , \tag{7.16b}$$

$$\mathcal{D}_{\psi\psi} \qquad = \int_{-\infty}^{\infty} \sin^2\theta(x)\,\mathrm{d}x \qquad = 2\sqrt{A/K}\ . \tag{7.16c}$$

For a non-equilibrium helicity, i.e. $\psi \neq \pi$, the DMI term yields a positive energy contribution while the other terms remain unaffected. Relative to the energy of the domain

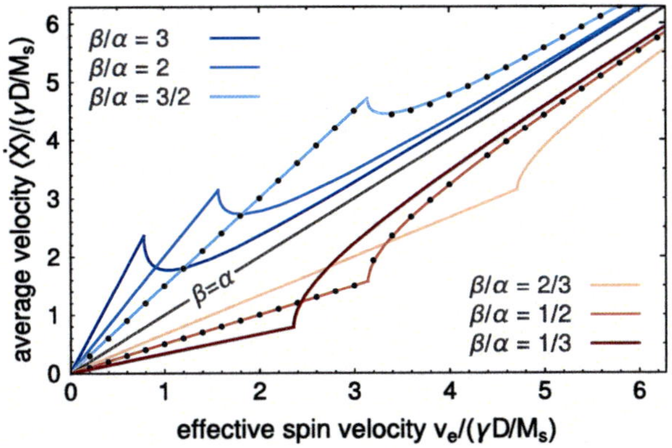

Fig. 7.5 STT-driven domain wall motion. Shown is the average domain wall velocity $\langle \dot{X} \rangle$ as function of the effective spin velocity v_e. Solid lines show Thiele results, see (7.14) and (7.21), respectively. The dots are LLGS simulation results, see (7.2). We fixed $\alpha = 0.1$ for various β, i.e., $\beta = 3\alpha, 2\alpha, \frac{3}{2}\alpha$ (dark to light blue), $\beta = \alpha$ (gray dashed), and $\beta = \frac{2}{3}\alpha, \frac{1}{2}\alpha, \frac{1}{3}\alpha$ (light to dark red)

wall in equilibrium, we therefore obtain the energy $E(X, \psi)$ and force $F_\psi(X, \psi)$

$$E(X, \psi) = \pi D(1 + \cos\psi) \quad \Rightarrow \quad F_\psi(X, \psi) = -\frac{\gamma}{M_s}\partial_\psi E(\psi) = \frac{\pi\gamma D}{M_s}\sin\psi \tag{7.17}$$

which completes the constituents of Eqs. (7.15a) and (7.15b). This set of coupled non-linear differential equations can be solved analytically, in both cases, (i) below and (ii) above the Walker-like breakdown.

Below the Walker breakdown, the helicity rotates away from its equilibrium position and, in the long-time limit, assumes a constant value, i.e., $\dot{\psi} = 0$. In this limit (7.15a) reduces to the simplified case (7.14) where the helicity dynamics are absent and the velocity $\dot{X}$ is independent of details of the domain wall texture. From (7.15b), we obtain the current-dependent helicity $\psi(v_e)$ of the driven domain wall which gives

$$\psi(v_e) = \pi + \arcsin\left(\frac{\alpha - \beta}{\alpha}\frac{2M_s v_e}{\pi\gamma D}\right) \quad \text{for } |v_e| \le v_e^c = \frac{\alpha}{|\alpha - \beta|}\frac{\pi\gamma D}{2M_s}\ . \tag{7.18}$$

For currents above the critical current v_e^c the restoring force $F_\psi(X, \psi)$ cannot compensate for the velocity anymore and, therefore, solutions with $\dot{\psi} = 0$ can no longer be obtained. Consequently, v_e^c marks the onset of the Walker breakdown.

Above the Walker breakdown, we can solve (7.15a) for $\dot{X}$ and make (7.15b) an equation of only ψ and $\dot{\psi}$. This differential equation can be solved exactly for a constant current density v_e and the solutions can be written in the form

$$\dot{X}(v_e, t) = \frac{\beta}{\alpha} v_e + \frac{\sqrt{A/K}}{\alpha} \dot{\psi}(v_e, t) , \tag{7.19a}$$

$$\psi(v_e, t) = -2 \operatorname{arccot}\left(\frac{u \operatorname{sign}(\alpha - \beta)}{1 - \sqrt{u^2 - 1} \tan(\omega_\psi t)} \right) \quad \text{with } u = \frac{v_e}{v_e^c} \geq 1. \tag{7.19b}$$

Here $T = \pi/\omega_\psi$ is the period of one helicity rotation and ω_ψ is the frequency given by

$$\omega_\psi = \frac{\alpha}{1+\alpha^2} \frac{\pi \gamma D}{4M_s} \sqrt{\frac{K}{A}} \sqrt{\left(\frac{v_e}{v_e^c}\right)^2 - 1} . \tag{7.20}$$

As can be seen from (7.19a), the velocity $\dot{X}$ of the domain wall is also periodic with the frequency ω_ψ and shows a very complicated behavior as function of time. The average velocity $\langle \dot{X} \rangle$, however, can be obtained from the time-average of (7.19a) where we can exploit the relation $\langle \dot{\psi} \rangle = (2\pi/T) \operatorname{sign}(\alpha - \beta)$. This yields

$$\langle \dot{X} \rangle = \frac{\beta}{\alpha} v_e + \frac{\operatorname{sign}(\alpha - \beta)}{1+\alpha^2} \frac{\pi \gamma D}{2M_s} \sqrt{\left(\frac{v_e}{v_e^c}\right)^2 - 1} \quad \text{for } |v_e| \geq v_e^c \tag{7.21}$$

as the average velocity of the domain wall above the Walker breakdown, $v_e > v_e^c$, in the Thiele framework. Interestingly, it turns out that above the Walker breakdown, for $\beta < \alpha$ the domain wall speed does not get reduced but boosted instead. In Fig. 7.5, we illustrate these different behaviors obtained from the Thiele approach, see (7.14) and (7.21). For comparison, we also show data obtained from numerical simulations of the full LLGS equation (7.2).

Despite the Walker breakdown there are other interesting effects for field-driven domain walls and magnetic bubbles [39, 66]. For example, for a time-dependent current, the coupled dynamics of the position and helicity degree of freedom (7.15a) and (7.15b), lead to an effective mass similar to the *Döring mass* [67].

7.5.2.2 Domain Wall Motion Due to Spin-Orbit Torques

Spin-transfer torques act via gradients $\mathbf{v}_e \cdot \nabla$ only on local changes of the magnetization. This is very different for SOTs being characterized by a spin polarization $\boldsymbol{\sigma}$, where $\boldsymbol{\sigma}$ couples explicitly to the local direction of the magnetization. Therefore, these can apply a torque also on a uniform magnetization and, moreover, induce a helicity-dependence of the total forces. Upon including the helicity degree of freedom ψ as a collective coordinate for the description of the domain wall dynamics, and using the ansatz from (7.11), we obtain the following Thiele equations for the SOT-driven domain wall

$$\mathcal{G}_{X\psi}\dot{\psi} + \tau_{\mathrm{FL}}\boldsymbol{\sigma}\cdot\boldsymbol{G}_X^{\mathrm{SOT}}(\psi) + \alpha\mathcal{D}_{XX}\dot{X} + \tau_{\mathrm{DL}}\boldsymbol{\sigma}\cdot\boldsymbol{D}_X^{\mathrm{SOT}}(\psi) = 0\ , \tag{7.22a}$$

$$\mathcal{G}_{\psi X}\dot{X} + \tau_{\mathrm{FL}}\boldsymbol{\sigma}\cdot\boldsymbol{G}_\psi^{\mathrm{SOT}}(\psi) + \alpha\mathcal{D}_{\psi\psi}\dot{\psi} + \tau_{\mathrm{DL}}\boldsymbol{\sigma}\cdot\boldsymbol{D}_\psi^{\mathrm{SOT}}(\psi) = F_\psi(X,\psi)\ . \tag{7.22b}$$

Here, for better readability, we have summarized the matrix products from (7.8) into scalar products of $\boldsymbol{\sigma}$ with SOT gyro or dissipation vectors. For the SOT-independent terms we refer to the previous section. For the domain wall ansatz in (7.11) together with the solution in (7.13), these SOT-specific vectors read

$$\boldsymbol{G}_X^{\mathrm{SOT}}(\psi) = \boldsymbol{\psi}\sin\theta + \hat{z}\cos\theta\big|_{\theta(-\infty)}^{\theta(+\infty)} \ = \ 2\hat{z}\ , \tag{7.23a}$$

$$\boldsymbol{G}_\psi^{\mathrm{SOT}}(\psi) = \int_{-\infty}^{\infty} -(\hat{z}\times\boldsymbol{\psi})\sin\theta(x)\,\mathrm{d}x \ = \ \pi\sqrt{A/K}\,(\hat{z}\times\boldsymbol{\psi})\ , \tag{7.23b}$$

$$\boldsymbol{D}_X^{\mathrm{SOT}}(\psi) = \int_{-\infty}^{\infty} (\hat{z}\times\boldsymbol{\psi})\,\theta'(x)\,\mathrm{d}x \ = \ \pi\,(\hat{z}\times\boldsymbol{\psi})\ , \tag{7.23c}$$

$$\boldsymbol{D}_\psi^{\mathrm{SOT}}(\psi) = \int_{-\infty}^{\infty} \boldsymbol{\psi}\cos\theta(x)\sin\theta(x) - \hat{z}\sin^2\theta(x)\,\mathrm{d}x \ = \ -2\sqrt{A/K}\,\hat{z}\ , \tag{7.23d}$$

where we defined the helicity vector $\boldsymbol{\psi} = (\cos\psi, \sin\psi, 0)$ for a more compact notation. For the standard setup with $\boldsymbol{j}_e = j_e\hat{x}$ and where the spin-polarization is determined by the spin Hall effect, i.e., $\boldsymbol{\sigma} = \hat{z}\times\boldsymbol{j}_e = j_e\hat{y}$, the contributions of gyro and dissipation vectors $\boldsymbol{G}_X^{\mathrm{SOT}}(\psi)$ and $\boldsymbol{D}_\psi^{\mathrm{SOT}}(\psi)$ vanish. Moreover, for a Bloch-type domain wall with $\psi = \pm\pi/2$ the other scalar products also vanish and, hence, the Bloch wall remains unaffected by the SOT. In contrast, for a Néel type domain wall, the scalar products are maximized. The Néel wall with $\psi = \pi$ then moves in the direction of the current $\boldsymbol{j}_e$ with additional dynamics of ψ whereas it moves in the opposite direction for $\psi = 0$. For additional information on SOT-induced dynamics we refer to [68].

7.5.3 *Magnetization Dynamics of Two-Dimensional Solitons*

Skyrmions and related magnetic textures, see Sect. 7.3, are not only thought to have an enhanced (topological) stability, but their non-zero winding number Q, see (7.4), also leads to a gyromagnetic tensor element $\mathcal{G}_{XY} = -4\pi Q$ in the Thiele equation, (7.8). This gyromagnetic coupling induces a force, similar to the Magnus force in Newtonian mechanics which acts on rotating bodies, leading to very particular dynamics. These include a response perpendicular to extrinsic forces and an intrinsic *skyrmion Hall effect*, which we both discuss in the following.

As for domain walls in one spatial dimension, to demonstrate the Thiele approach, let us introduce an ansatz for a skyrmion-like profile in an out-of-plane polarized background of the form

$$m(r - R, \psi) = (\cos(\phi + \psi) \sin\theta, \sin(\phi + \psi) \sin\theta, \cos\theta), \tag{7.24}$$

where $\phi = \phi(r - R)$ sets the inplane magnetic profile and $\theta = \theta(r - R)$ determines the m_z profile. R is the position of the skyrmion and ψ is the helicity. For circular skyrmions R is usually their center position, the profile only depends on the radial coordinate $\rho = |r - R|$ and ϕ depends only the axial coordinate χ of the cylindrical coordinate system centered at R. In this convention, the Bloch-type skyrmion shown in Fig. 7.1b is described by $\phi = \chi$, $\psi = -\pi/2$, and $\theta = \theta(\rho)$ with $\theta(0) = \pi$ and $\theta(\infty) = 0$. The antiskyrmion in Fig. 7.1c is described by $\phi = -\chi$, $\psi = \pi/3$. Other skyrmion-like structures, e.g., higher order skyrmions with $Q = -N$ can be described by setting $\phi = N\chi$, and the topologically trivial *skyrmionium* is characterized by $\theta(0) = 2\pi$ and $\theta(\infty) = 0$.

7.5.3.1 Pinning and Deformation

At ultra low current densities all magnetic solitons are pinned by material defects. For skyrmion lattices, it has been shown experimentally that the critical current density for depinning is very low [69, 70]. Theoretically, the influence of disorder on skyrmion lattices was studied in various micromagnetic simulations [71]. The micromagnetic results agree well with particle model simulations which are based on the Thiele equation of motion of skyrmions which interact with each other and with random pinning sites [72]. Also for isolated skyrmions in the presence of defects, the pinning, depinning, and motion has been studied experimentally [73] and can be described in a generalized Thiele equation when taking deformations due to defects into account [62]. For simplicity, however, we will neglect pinning effects in the following.

Deformations of moving solitons also occur in the absence of impurities, for example, due to internal dynamics, as has been shown already in early studies in magnetic bubble dynamics with Bloch lines [27]. For skyrmions, which do not have Bloch lines, deformations can also arise due to spin-torques. In this case, the matrix elements of the Thiele equation (7.10), become dependent on the current strength which leads to non-trivial corrections of the particle-like motion [74]. These effects, as well as deformations due to thermal fluctuations, interactions with defects or other magnetic textures can induce an effective mass for two-dimensional solitons which might potentially be described by the broader term *automotion* [27]. In the limit where skyrmions can be treated as rigid objects, i.e., when the bound state excitation gap of the skyrmion is large, deformation effects can be neglected, as we will assume in the following.

7.5.3.2 Skyrmion Motion Due to External Forces

Historically, before spin-torques became an active research field, the motion of magnetic bubbles was studied intensively, for example, with pulsed field gradients. It was found that the bubbles do not move along the direction of the external force, but along a deflected direction that depends on the winding number Q [27]. For both, skyrmions and magnetic bubbles, the side-drift response can be understood within the Thiele approach, (7.8). Moreover, this effect occurs not only for field gradients but for all forces $\boldsymbol{F}(\boldsymbol{q})$ in the Thiele equation, e.g., due to field gradients or the interaction with defects and other magnetic structures. It is also the source of the unusual Brownian motion of skyrmions which in two dimensions diffuse less if the Gilbert damping α is reduced [37, 75, 76]. In the following, analog to the domain wall case, we will first discuss the limit of a pinned helicity and then consider what happens beyond this limit.

Pinned helicity. Consider a skyrmion with winding number Q and the position $\boldsymbol{R} = (X, Y)$ as the only collective coordinates for the Thiele approach. In a spatially dependent energy landscape $E(\boldsymbol{R})$, e.g., due to anisotropy gradients, magnetic field gradients, defects, or other magnetic textures, $\boldsymbol{R}$ is not a true zero mode but can still be a good collective coordinate. As the system is dissipative, the skyrmion will at some point be trapped in a local minimum of $E(\boldsymbol{R})$. An elegant form of the resulting Thiele equation then reads

$$\boldsymbol{G} \times \dot{\boldsymbol{R}} + \alpha \mathcal{D}(\boldsymbol{R})\dot{\boldsymbol{R}} = \boldsymbol{F}(\boldsymbol{R}) \tag{7.25}$$

where $\boldsymbol{F}(\boldsymbol{R}) = -(\gamma/M_s)\nabla_{\boldsymbol{R}} E(\boldsymbol{R})$ is the force on the skyrmion, $\boldsymbol{G} = 4\pi Q\hat{z}$ is the gyro-vector, and $\mathcal{D}(\boldsymbol{R})$ is the dissipation matrix. The gyro-vector $\boldsymbol{G}$ couples the motion of the X and Y coordinates and leads to the side-deflection in the motion of two-dimensional solitons with a finite topological charge. For a circular skyrmion, using the notation in (7.24) and the angular dependence $\phi(\chi) = N\chi$, $N \in \mathbb{Z}$, the dissipation matrix reduces to a scalar with

$$\mathcal{D}(\boldsymbol{R}) = D_s(\boldsymbol{R}) = \int_0^\infty \pi\rho \left(\frac{N^2}{\rho^2} \sin^2\theta(\rho) + \left(\theta'(\rho)\right)^2 \right) \mathrm{d}\rho \tag{7.26}$$

where ρ is the distance to $\boldsymbol{R}$, i.e., the center of the skyrmion. Usually, it is assumed that the texture of the skyrmion does not change much with the position such that $D_s(\boldsymbol{R}) \approx D_s$ is a good approximation.

The Thiele equation, (7.25), can be solved for the skyrmion velocity $\dot{\boldsymbol{R}}$. Its absolute value $|\dot{\boldsymbol{R}}|$ and the direction relative to the force $\boldsymbol{F}$, parameterized by the deflection angle θ_d, then read

$$|\dot{\boldsymbol{R}}| = \frac{|\boldsymbol{F}(\boldsymbol{R})|}{\sqrt{(4\pi Q)^2 + \alpha^2 D_s^2}} \quad \text{and} \quad \theta_\mathrm{d} = -\arctan\left(\frac{4\pi Q}{\alpha D_s}\right) . \tag{7.27}$$

In this formulation, the real-space topological nature of the side-deflection can be identified as $\theta_d \neq 0$ only for $Q \neq 0$. The deflection angle $\theta_d \neq 0$ is schematically illustrated in Fig. 7.6 for various skyrmion-like textures and Gilbert dampings α. Moreover, (7.27) reveals that a finite charge Q reduces the (absolute) velocity $|\dot{\boldsymbol{R}}|$. The dependence on Q should, however, be investigated more thoroughly as the dissipation scalar D_s depends explicitly on the vorticity N, see (7.26), and, hence, also on the winding number. Furthermore, magnetic textures with different winding numbers usually relax to different magnetization profiles and, thus, different values of D_s [27].

Unpinned helicity. Let us assume now that the skyrmion is stabilized in a system where the helicity ψ is a zero mode and can, in principle, be activated. However, this activation is not straightforward. In the Thiele equation, (7.8), all matrix elements $\mathcal{G}_{X\psi}, \mathcal{G}_{Y\psi}, \mathcal{D}_{X\psi}$, and $\mathcal{D}_{Y\psi}$ vanish for circular solitons. Therefore, ψ does not couple to the position $\boldsymbol{R}$ or derivatives thereof, which seemingly suggests that the helicity does not show any dynamics. However, this conclusion is wrong as, for example, simulations with a magnetic field gradient show a steady rotation of the helicity while the skyrmion moves towards the direction of the smaller field [77, 78]. In the following, we discuss this example in more detail and show how to resolve the apparent contradiction.

Consider a magnetic field of the form $\boldsymbol{B}(\boldsymbol{r}) = (B_0 + x\,\delta B)\,\hat{z}$. Let us assume, moreover, that the field gradient δB is a sufficiently small so that the skyrmion profile is still approximately circular and the above arguments still hold. Due to the field gradient, the position $\boldsymbol{R}$ is not a zero mode but still a good collective coordinate which is subject to a force $\boldsymbol{F_R}$ which drags the skyrmion towards regions with lower field. While moving there, however, the skyrmion profile has to adapt to the local magnetic field $\boldsymbol{B}(\boldsymbol{r})$, leading to an inflation of the skyrmion size ξ. Unlike $\boldsymbol{R}$, the collective coordinate ξ couples directly to the helicity ψ via the matrices in the Thiele equation but, due to the circular shape, ξ does not couple to $\boldsymbol{R}$. In a compact form, the four Thiele equations then read

$$\boldsymbol{G} \times \dot{\boldsymbol{R}} + \alpha D_s \dot{\boldsymbol{R}} = \boldsymbol{F_R}(\boldsymbol{R}, \xi)\ , \tag{7.28a}$$

$$\mathcal{G}_{\psi\xi}\,\dot{\xi} + \alpha \mathcal{D}_{\psi\psi}\,\dot{\psi} = 0\ , \tag{7.28b}$$

$$\mathcal{G}_{\xi\psi}\,\dot{\psi} + \alpha \mathcal{D}_{\xi\xi}\,\dot{\xi} = F_\xi(\boldsymbol{R}, \xi)\ . \tag{7.28c}$$

All matrix elements with indices ψ or ξ are, in principle, dependent on ξ. This dependence can be neglected on small time scales. The force $F_\xi(\boldsymbol{R}, \xi)$ ensures that the skyrmion size adapts to the local magnetic field. In a small field gradient δB, the skyrmion moves slow enough that we can assume ξ to be close to the energetically optimal value. Then its contribution to the force $\boldsymbol{F}_{\boldsymbol{R},\xi}$ can be neglected and, to lowest order in δB, this force is $\boldsymbol{F_R} \propto -\delta B\,\hat{x}$. Now, (7.28a) is decoupled from the other two equations of motion and the skyrmion moves according to the results of the previous section, (7.27). In particular, the parallel velocity is $\dot{X} \propto -\delta B$ and, therefore, $\dot{\xi} \propto \dot{B}(\boldsymbol{R}) = \dot{X}\delta B \propto \delta B^2$. Equation (7.28b) then yields the velocity of the helicity $\dot{\psi} \propto \delta B^2/\alpha$ which continuously rotates while the skyrmion moves in magnetic field gradient [77, 78], similar to the domain wall above the Walker breakdown [63].

We would like to point out that $\dot{\psi} \propto \delta B^2/\alpha$ is also the consequence of another effect which we did not capture in the above discussion: So far, we assumed that the skyrmion maintains its circular shape. In the field gradient $\boldsymbol{B}(\boldsymbol{R})$, however, the skyrmion becomes slightly non-circular which adds a finite direct coupling $\mathcal{D}_{X\psi} \propto$

δB between the velocity $\dot{X}$ and $\dot{\psi}$ to (7.28a) and (7.28b). For the full dynamics, therefore, both the change of the skyrmion size and the its non-circular distortion contribute.

In the following sections, we review the dynamics of current-driven instead of force-driven skyrmions which follow the same basic concepts.

7.5.3.3 Skyrmion Motion Due to Spin-Transfer Torques

A standard example for the application of a Thiele equation is to model the motion of spin-tranfer torque-driven skyrmions in chiral magnets. The Thiele formalism, for example, provides a direct mean to explain the *skyrmion Hall effect*, where the skyrmions move at an angle relative to the direction of the applied current density $\boldsymbol{j}_e$. We will discuss this in the following. Analog to the previous chapters we will first discuss the limit of a pinned helicity and then explain briefly what happens beyond this regime.

Pinned helicity. Let us consider a frequently used assumption for skyrmions in chiral magnets, namely that the helicity is pinned by DMI to a fixed value ψ and does not contribute to the dynamics. Consider, moreover, that the system is translation invariant, i.e., the position $\boldsymbol{R}$ is a zero mode, and that the skyrmion can be described by the ansatz in (7.24). The Thiele equation then reads

$$\boldsymbol{G} \times (\dot{\boldsymbol{R}} - \boldsymbol{v}_e) + D_s(\alpha \dot{\boldsymbol{R}} - \beta \boldsymbol{v}_e) = 0 \tag{7.29}$$

where $\boldsymbol{G} = 4\pi Q\hat{z}$ is the gyro-vector, and the dissipation matrix $\mathcal{D}$ reduces to a scalar D_s as in (7.26). In principle, this equation of motion can be solved for $\dot{\boldsymbol{R}}$ which yields the skyrmion Hall effect. Alternatively, we can interpret the effect of the STTs from a different perspective. By isolating all terms which originate from the STT on the right hand side of (7.29), we effectively recover the Thiele equation for a skyrmion driven by an external force, (7.25), with

$$\boldsymbol{F}^{\mathrm{STT}} = \boldsymbol{G} \times \boldsymbol{v}_e + \beta D_s \boldsymbol{v}_e \,. \tag{7.30}$$

The skyrmion Hall angle θ_d^{STT} is then the sum of (i) angle θ_F^{STT} between the effective STT-force $\boldsymbol{F}^{\mathrm{STT}}$ and the direction of the current $\boldsymbol{v}_e$ and (ii) the deflection angle θ_d for a force-driven skyrmion, see (7.27), and reads

$$\theta_d^{\mathrm{STT}} = \arctan\left(\frac{4\pi Q}{\beta D_s}\right) - \arctan\left(\frac{4\pi Q}{\alpha D_s}\right) = \arctan\left(\frac{4\pi Q D_s(\alpha - \beta)}{(4\pi Q)^2 + \alpha\beta D_s^2}\right) \,. \tag{7.31}$$

The result reflects the trivial cases $\theta_d^{\mathrm{STT}} = 0$ for $\alpha = \beta$ or for $Q = 0$ where the magnetic texture just moves along with the current. In contrast to the deflection angle θ_d in the previous section (7.27), the skyrmion Hall angle θ_d^{STT} shrinks for increasing Q and, for typical values of parameters, the maximal θ_d^{STT} is at $Q =$

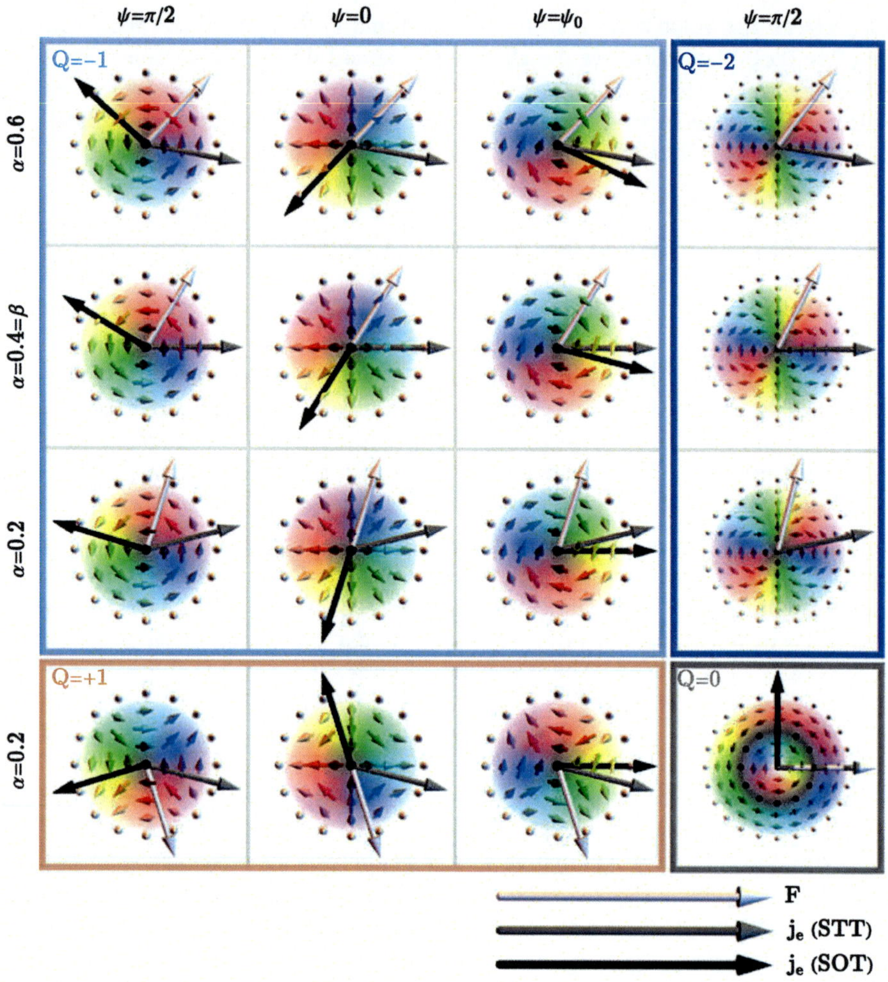

Fig. 7.6 Deflection angle for force-driven and skyrmion Hall angles for STT-driven and SOT-driven two-dimensional solitons. The force $\boldsymbol{F} = -(\gamma/M_s)\partial_{\boldsymbol{R}} E$ and the electric currents $\boldsymbol{j_e}$ for both STTs and SOTs point to the right, as indicated. For STTs we use $\beta = 0.4$ in all panels and for SOTs we assume $\boldsymbol{\sigma} = \hat{z} \times \boldsymbol{j_e}$. The direction of the velocity of the reacting soliton is indicated by arrows in each panel, illustrating the Thiele results of (7.27), (7.31), and (7.34). A skyrmion with $Q = -1$ and an antiskyrmion with $Q = 1$ are shown, both for three different helicities $\psi = \pi/2$, 0, and ψ_0. The compensation helicity ψ_0 is chosen such that the skyrmion Hall angle with SOT vanishes for $\alpha = 0.2$. The impact of the damping α is shown by various $\alpha = 0.6$, 0.4, and 0.2, where the skyrmion Hall angle in the second row vanishes ($\alpha = \beta$). Moreover, we show a higher order $Q = -2$ skyrmion and a $Q = 0$ skyrmionium, both with $\psi = \pi/2$. The $Q = -2$ skyrmion is unaffected by SOTs and has slightly modified responses to forces and STTs, compared to the skyrmion with $Q = -1$. The skyrmionium moves precisely in the direction of the force and STT, while its reaction to SOTs is solely determined by its helicity

± 1. The properties of θ_d^{STT} for different skyrmion-like solitons are schematically summarized in Fig. 7.6.

The skyrmion Hall angle can change dramatically, for example, when considering small random defects which additionally reduce the velocity [62, 72]. Extended defects, such as the DMI-induced twisting at the edge of a sample, in turn, can speed up the skyrmion motion $\propto |Q|/\alpha$ when STTs push the skyrmion into nonequilibrium positions [71]. To accommodate the translationally non-invariant case in the Thiele formalism one has to take a spatially dependent force in (7.29) into account.

Moreover, the STT-induced torques can distort the skyrmion profile on the level of the LLGS equation which eventually leads to strong corrections to the skyrmion Hall effect and, even more importantly, a speed limit above which the STTs destroy the skyrmion [74]. The latter cannot be derived from a simple Thiele ansatz and requires more rigorous models or numerical simulations of the LLGS equation.

Unpinned helicity. As discussed above in Sect. 7.5.3.2, for a circular skyrmion, the coupling between the collective coordinates $\boldsymbol{R}$ and the helicity ψ is absent. Similarly, because of $\partial_X \boldsymbol{m} = -\partial_x \boldsymbol{m}$, STTs do not directly couple to the helicity. However, STT-driven skyrmions can still show dynamics of the helicity, e.g., in an energy landscape $E(\boldsymbol{R}, \psi)$ where the position and helicity are coupled. In this case, the helicity of the moving skyrmion shows dynamics around the local optimum $\psi_0(\boldsymbol{R})$, potentially showing features of an effective helicity mass [51]. Moreover, STTs can deform the skyrmion which enables the coupling in the Thiele equation, leading to a steady rotation of the helicity.

7.5.3.4 Skyrmion Motion Due to Spin-Orbit Torques

In contrast to STTs, SOTs couple directly to the magnetization texture and not derivates thereof. Therefore, SOT induced dynamics are sensitive to the helicity ψ of the magnetic soliton as we will discuss in more detail in the following. A skyrmion Hall effect can also be derived for skyrmions with SOTs and was recently also confirmed experimentally [73, 79].

Pinned helicity. In monolayers on heavy metal substrates or thin films of stacked heterostructures, skyrmions can be stabilized by a strong DMI with extra stabilizing support from dipolar interactions. These interactions usually pin the helicity such that it does not contribute to the dynamics.

For the axially symmetric soliton with $\boldsymbol{m}(\infty) = \hat{z}$ and $\phi = N\chi$ in the ansatz in (7.24), the gyro-coupling SOT matrix evalutes to zero whereas the dissipative SOT coupling matrix elements only vanish for $|N| \neq 1$. Note that in this convention, the relation $Q = N(1 - m_z(0))/2$ implies that not only the skyrmion and the anti-skyrmion, but also the topologically trivial skyrmionium can be driven by SOTs. In turn, higher order skyrmions with $|Q| > 1$ do not react to SOTs within these approximations. The Thiele equation for these $|N| \neq 1$ objects can be written as

$$\boldsymbol{G} \times \dot{\boldsymbol{R}} + \alpha D_s \dot{\boldsymbol{R}} + \tau_{\mathrm{DL}} \boldsymbol{\sigma} \cdot \mathcal{D}_{\boldsymbol{R}}^{\mathrm{SOT}}(\psi) = 0\,. \tag{7.32}$$

Here, we have introduced the 3×2 dissipation tensor $(\mathcal{D}_{\boldsymbol{R}}^{\mathrm{SOT}})_{\mu i}$, $\mu = x, y, z$, and $i = X, Y$, for the SOT-induces torque which reads

$$\mathcal{D}_{\boldsymbol{R}}^{\mathrm{SOT}}(\psi) = \left(\hat{z} \times \boldsymbol{\psi}, -N\boldsymbol{\psi}\right)\pi\,\delta_{|N|,1} \int_0^\infty \cos\theta(\rho)\sin\theta(\rho) + \rho\theta'(\rho)\,\mathrm{d}\rho \qquad (7.33)$$

with $\boldsymbol{\psi} = (\cos\psi, \sin\psi, 0)$. The Kronecker delta $\delta_{|N|,1}$ indicates that only solitons with $N = \pm 1$ give finite contributions. The asymmetric N-dependence in only the second column of $\mathcal{D}_{\boldsymbol{R}}^{\mathrm{SOT}}$ is an artefact of the ansatz (7.24), which flips $m_y \to -m_y$ for $N \to -N$.

Similar to the discussion in Sect. 7.5.3.3, we can interpret the SOT-induced terms as an external force $\boldsymbol{F}^{\mathrm{SOT}}$ and derive the skyrmion Hall angle from the direction of this effective force. Assuming again the standard spin Hall setup for the SOT with $\boldsymbol{j_e} = j_e\hat{x}$ and $\boldsymbol{\sigma} = \hat{z} \times \boldsymbol{j_e} = j_e\hat{y}$, the skyrmion Hall angle θ_d^{SOT} becomes

$$\theta_d^{\mathrm{SOT}}(\psi) = -N\left(\psi + \pi - \arctan\left(\frac{4\pi Q}{\alpha D_s}\right)\right) \quad \text{with} \quad |N| = 1\,, \qquad (7.34)$$

which is only well-defined for $|N| = 1$ as otherwise the soliton does not move. Note that the skyrmion Hall angle θ_d^{SOT} is a function of the helicity ψ and can result in a motion in arbitrary directions, including parallel to the current, by fine-tuning the DMI [80]. This angular dependence is also schematically summarized in Fig. 7.6 for various soliton configurations and parameters.

Unpinned helicity. For circular two-dimensional solitons driven by SOTs, the same physics arises as for the other driving mechanisms, namely that neither the collective coordinates $\boldsymbol{R}$ couple directly to ψ nor do the Thiele matrices $\mathcal{G}_{\psi i}^{\mathrm{SOT}}$ and $\mathcal{D}_{\psi i}^{\mathrm{SOT}}$ yield a finite coupling between $\boldsymbol{\sigma}$ and ψ (except for $\mathcal{D}_{\psi z}^{\mathrm{SOT}}$, which is usually not relevant as $\sigma_z = 0$).

A distinguished feature of SOTs is that they tilt the background magnetization. This naturally leads to deformations of the soliton, breaking the axial symmetry and enabling a finite coupling of ψ and $\dot{\boldsymbol{R}}$, see Sect. 7.5.3.2. Thus, while moving with velocity $\dot{\boldsymbol{R}}$ at a skyrmion Hall angle $\theta_d^{\mathrm{SOT}}(\psi)$, see (7.34), the helicity ψ changes which feeds back on $\theta_d^{\mathrm{SOT}}(\psi)$. Consequently, the skyrmion with an activated helicity degree of freedom can end up orbiting around a fixed point [81] or, for sufficiently asymmetric energy landscapes $E(\psi)$, perform a trochoidal motion [82] which is a combination of translation and orbiting. Moreover, once the helicity becomes dynamical, it can also lead to an effective mass in the Thiele equation [51, 81].

7.5.4 *Magnetization Dynamics of Three-Dimensional Hopfions*

As magnetic hopfions in chiral magnets have only recently been proposed theoretically, see Sect. 7.3, their dynamics are a field that is still much under investigation.

In thin films of chiral magnets with perpendicularly magnetized surfaces, hopfions are predicted to be stabilized due to geometric confinement. The magnetic texture of the hopfion is then fixed by the DMI such that only translational modes can be activated easily. For such a setup, it was shown theoretically that the STT-driven $H = 1$ hopfion behaves like a skyrmionium, i.e., it moves like a two-dimensional soliton, straight along the applied current without any Hall angle [83].

More complex dynamics are predicted for three-dimensional frustrated magnets: Here, the translation in all spatial dimensions and rotation around all axes are zero modes. It was shown in a theoretical study by Liu et al. [84] that the STT-driven $H = 1$ hopfion indeed rotates while moving with the current, adjusting such that its skyrmionium-like cross-section aligns perpendicular to the current. Moreover, inside the hopfion, regions with positive and negative skyrmion charge Q are present which are subject to opposite skyrmion Hall angles. As a consequence, the STT-driven hopfion either inflates or deflates, dependent on the direction of the current. For a detailed description of the dynamics, featuring also a discussion in the Thiele framework, we refer to [84].

7.6 Potential Applications

Based on the very rich playground of spintronics with chiral magnetic structures, several potential applications have been proposed over the recent years. In the following we will briefly introduce some of them.

7.6.1 *Storage and Logic Technologies*

Magnetic racetrack. The central idea behind the racetrack is that information is encoded by magnetic bits which are placed in a one-dimensional shift register device. Data can be accessed or written at a particular point of the nanowire. It has the great advantage that instead of moving mechanical parts, only the magnetic bits are moved, e.g., by spin-currents. In the classically suggested version [85, 86], the bits are magnetic domains, separated by domain walls. For the racetrack based on magnetic skyrmions [87], the state of a bit can be represented by the presence or absence of a skyrmion. The latter has the advantage to circumvent the impact of edge roughness in the nanowire, as skyrmions opposed to domain walls do not necessarily touch the edge. However, it has also some disadvantages. In particular, the skyrmion Hall effect hinders the straight motion of the skyrmion through the nanowire. To enhance the speed of the magnetic data, (synthetic) antiferromagnetic instead of ferromagnetic materials have been studied within the recent years. An antiferromagnetic coupling would also resolve the problem with the skyrmion Hall effect as in this case the forces in the direction perpendicular to the current direction typically cancel [115, 116]. Moreover, similar devices with closely packed skyrmions or other similar solitons have been suggested as the information encoded in not well-defined inter-skyrmion distances is very fragile [88].

Bubble memory. In the 1970s and 1980s, before magnetic racetracks were discussed, memory devices exploiting magnetic bubbles have been commercially available. These are non-volatile two-dimensional shift register memories that exploit the magnetic field-driven motion of small magnetized areas—the bubbles [89, 90].

Magnetic transistor. Transistors as key elements for controlling integrated circuits and logic devices have also been proposed to be implemented based on magnetic textures, such as a domain wall based transistor [91] or a skyrmion based transistor [92] These exploit the gate-voltage controlled motion of the magnetic nanostructures.

Magnetic logic. Another key field in spintronics is the idea to create magnetic-based logic gates [93]. This is, on the one hand, done by studying nano-magnetic logic, where nano-magnetic islands with a uniaxial-anisotropy represent the zero and one state based on their orientation with respect to this anisotropy direction, e.g., "up" and "down". The other idea is to send signals through an appropriately shaped device, which represent the logical gates. This includes magnonic logic [94] as well as logic based on chiral magnetic states such as skyrmions. An example of the latter was suggested by Zhang et al. [95] which exploits the possibility to convert spin-torque driven skyrmions into domain walls in narrow wires. In a convention where a logical 1 or 0 is represented by the presence or absence of a skyrmion, respectively, an OR gate and AND gate have been simulated by properly designing the width of the narrow wires.

Magnetic nano-oscillators. Oscillators exploit the system's natural time scale and responses to external sources to provide a tunable frequency source. In magnetic texture based systems, these oscillators are naturally on the nano-scale and exploit, for example, the current-driven self oscillation of domain walls [96] or skyrmions [97]. While for ferromagnets the frequencies are in the GHz regime and can be tuned, e.g., by an externally applied magnetic field, they can be in the THz regime for antiferromagnetic materials, thereby bridging the THz gap.

7.6.2 Unconventional Spintronics-Based Computing Schemes

Within the recent years, more and more unconventional computational paradigms are being explored. Based on their low-energy consumption, compact nanometer size scale, and manipulability, magnetic textures could play an important role in the development of such novel computational technologies [98, 99].

Magnetic artificial neural networks. The vast progress within the field of artificial intelligence is mainly based on the widely enhanced available hardware power, while most of the concepts have been suggested already a few years ago. So as with deep artificial neural networks, which nowadays are widely used for different types of AI applications. However, so far they are mostly performed on the existing hardware which, due to the classical segmentation in computational units and storage, are not optimally suited for these types of applications as their power consumption shows. Instead, alternative architectures which adjust to the deep neural network structure are proposed, with a focus of creating their central components, i.e., artificial synapses and neurons, in hardware. There are also several suggestions for magnetic neuromorphic computing [99, 100] and how to implement artificial *neurons* [101] and *synapses* [102, 103]. In particular, *memristors*, [104] i.e., devices whose resistance depends on the previous state, are suggested to function as a basis for synaptic applications.

Spintronics based reservoir computing. Reservoir computing has the goal to exploit the response of a reservoir to simplify, for example, spatial-temporal recognition tasks. The reservoir itself projects the input into a higher dimensional space, where it becomes easier to classify. For this concept to work, the reservoir needs to be a non-linear, complex system with a short-term memory, which is fulfilled by several physical systems opening up the path for in-materio computing [105]. As spintronics systems often naturally fulfill these criteria for the reservoir and additionally provide a lot of tune-ability as well as complexity, together with their low energy consumption, they do provide a promising hardware-based solution for reservoir computing [106]. It has been proposed that skyrmion fabrics are very well suited for reservoir computing applications [107].

Stochastic computing. The ansatz of stochastic computing is to trade speed for accuracy, exploiting the law of large numbers where upon enhancing the number of experiments the result converges to the expectation value. For example, one can stochastically multiply two numbers in-between zero and one, when interpreting them as a probability of having a one in a bit-string. For uncorrelated bit-strings the multiplication of these two numbers can then be efficiently calculated as sending the two bit-strings through an AND gate. Spintronics offers a potential ansatz with respect to stochastic computing, as spintronics systems can naturally exhibit stochastic behavior. Furthermore, recently a device which allows to reshuffle bit-strings based on magnetic skyrmions has been realized [76, 108]. Such a skyrmion reshuffler allows to restore the decoherence between signals which possibly synchronized. A similar suggestion is to encode the information in probabilistic bits, also called p-bits. These are bits that fluctuate between 0 and 1 and, in this sense, interpolate between a classical bit and a q-bit. It has been suggested that magnetic states naturally provide a realization for such p-bits [109].

Topological quantum computing. Even more exotically, chiral magnetic states could contribute to topological quantum computing. It has been suggested that Majorana modes localize at skyrmions [110] or compound structures of superconducting vortices and skyrmions [111–113]. This might provide a path to perform the key operation of topological quantum computing, i.e., braiding of the localized modes with a non-Abelian statistics, via the manipulation of magnetic textures.

7.7 Conclusion

This book chapter presented an introduction to current-induced dynamics of chiral magnetic structures. We briefly summarized the basic concepts for deriving a continuum theory of magnetization dynamics in Sect. 7.2 and introduced *domain walls*, *(anti-)skyrmions*, and *hopfions* as examples for magnetic solitons in Sect. 7.3. In the main part of this chapter, we focused on the manipulation of magnetic textures by spin-troques, both due to spin-transfer and spin-orbit mechanisms. We reviewed (i) selected creation processes for domain walls and skyrmions in Sect. 7.4 and (ii) the motion of the above solitons in Sect. 7.5 with a particular focus on the generalized Thiele method. Finally, in Sect. 7.6 we summarized already implemented or theoretically suggested applications of magnetic textures which are manipulated by electric currents.

The field of *spintronics*, which explores the interplay of electric currents and the magnetization, has shown an enormous theoretical and experimental progress in the past years and a vast variety of possible new routes have emerged, including antiferromagnetic materials which are not discussed in this chapter. We can look forward with excitement to the future of current-induced magnetization dynamics, what new physics and which new quasi-particles will be revealed in the future, and how spintronics might continue contributing to our everyday life.

Acknowledgements We thank D. R. Rodrigues for fruitful discussions. JM is supported as an Alexander von Humboldt/JSPS International Research Fellow (Project No. 19F19815). KES acknowledges funding from the German Research Foundation (DFG) Project No. 320163632 (Emmy Noether), No. 403233384 (SPP 2137 Skyrmionics), No. 268565370 (TRR 173, B12), as well as from the Emergent AI Center funded by the Carl-Zeiss-Stiftung.

References

1. I.E. Dzyaloshinskii, A thermodynamic theory of "weak" ferromagnetism of antiferromagnetics. J. Phys. Chem. Solids **4**, 241–255 (1958)
2. T. Moriya, Anisotropic Superexchange Interaction and Weak Ferromagnetism. Phys. Rev. **120**, 91 (1960)
3. J.C. Slonczewski, Current-driven excitation of magnetic multilayers. J. Magn. Magn. Mater. **159**, L1–L7 (1996)
4. L. Berger, Emission of spin waves by a magnetic multilayer traversed by a current. Phys. Rev. B **54**, 9353 (1996)

5. I.M. Miron, K. Garello, G. Gaudin, P.J. Zermatten, M.V. Costache, S. Auffret, S. Bandiera, B. Rodmacq, A. Schuhl, P. Gambardella, Perpendicular switching of a single ferromagnetic layer induced by in-plane current injection. Nature **476**, 189–193 (2011)
6. L. Liu, C.-F. Pai, Y. Li, H.W. Tseng, D.C. Ralph, R.A. Buhrman, Spin-Torque Switching with the Giant Spin Hall Effect of Tantalum. Science **336**, 555–558 (2012)
7. G. Catalan, J. Seidel, R. Ramesh, J.F. Scott, Domain wall nanoelectronics. Rev. Mod. Phys. **84**, 119 (2012)
8. A.N. Bogdanov, D.A. Yablonskii, Thermodynamically stable "vortices" in magnetically ordered crystals. The mixed state of magnets. Zh. Eksp. Teor. Fiz. **95**, 178 (1989) [JETP **68**, 101 (1989)]
9. A. Fert, N. Reyren, V. Cros, Magnetic skyrmions: advances in physics and potential applications. Nat. Rev. Mater. **2**, 17031 (2017)
10. W. Jiang, G. Chen, K. Liu, J. Zang, S.G. te Velthuis, A. Hoffmann, Skyrmions in magnetic multilayers. Phys. Rep. **704**, 1–49 (2017)
11. K. Everschor-Sitte, J. Masell, R.M. Reeve, M. Kläui, Perspective: Magnetic skyrmions - Overview of recent progress in an active research field. J. Appl. Phys. **124**, 240901 (2018)
12. C. Back, V. Cros, H. Ebert, K. Everschor-Sitte, A. Fert, M. Garst, T. Ma, S. Mankovsky, T.L. Monchesky, M. Mostovoy, N. Nagaosa, S.S.P. Parkin, C. Pfleiderer, N. Reyren, A. Rosch, Y. Taguchi, Y. Tokura, K. von Bergmann, J. Zang, The 2020 Skyrmionics Roadmap. J. Phys. D: Appl. Phys. **53**, 363001 (2020)
13. I.L. Bogolubsky, Three-dimensional topological solitons in the lattice model of a magnet with competing interactions. Phys. Lett. A **126**, 511–514 (1988)
14. P. Sutcliffe, Skyrmion knots in frustrated magnets. Phys. Rev. Lett. **118**, 247203 (2017)
15. S. Heinze, K. von Bergmann, M. Menzel, J. Brede, A. Kubetzka, R. Wiesendanger, G. Bihlmayer, S. Blügel, Spontaneous atomic-scale magnetic skyrmion lattice in two dimensions. Nat. Phys. **7**, 713–718 (2011)
16. P. Bak, M.H. Jensen, Theory of helical magnetic structures and phase transitions in MnSi and FeGe. J. Phys. C: Solid State Phys. **13**, L881 (1980)
17. M. Hoffmann, B. Zimmermann, G.P. Müller, D. Schürhoff, N.S. Kiselev, C. Melcher, S. Blügel, Antiskyrmions stabilized at interfaces by anisotropic Dzyaloshinskii-Moriya interactions. Nat. Commun. **8**, 308 (2017)
18. R.R. Birss, *Symmetry and magnetism* (North-Holland Pub, Co, 1964)
19. K.M.D. Hals, K. Everschor-Sitte, Twists in ferromagnetic monolayers with trigonal prismatic symmetry. Phys. Rev. B **99**, 104422 (2019)
20. I.A. Ado, A. Qaiumzadeh, A. Brataas, M. Titov, Chiral ferromagnetism beyond Lifshitz invariants. Phys. Rev. B **101**, 161403(R) (2020)
21. M.V. Berry, Quantal phase factors accompanying adiabatic changes. Proc. R. Soc. Lond. A **392**, 45–57 (1984)
22. A. Neubauer, C. Pfleiderer, B. Binz, A. Rosch, R. Ritz, P.G. Niklowitz, P. Böni, Topological Hall Effect in the A Phase of MnSi. Phys. Rev. Lett. **102**, 186602 (2009)
23. M. Lee, W. Kang, Y. Onose, Y. Tokura, N.P. Ong, Unusual Hall Effect Anomaly in MnSi under Pressure. Phys. Rev. Lett. **102**, 186601 (2009)
24. K. Everschor-Sitte, M. Sitte, Real-space Berry phases: Skyrmion soccer (invited). J. Appl. Phys. **115**, 172602 (2014)
25. J. Sampaio, V. Cros, S. Rohart, A. Thiaville, A. Fert, Nucleation, stability and current-induced motion of isolated magnetic skyrmions in nanostructures. Nat. Nanotechnol. **8**, 839–844 (2013)
26. S. Zhang, Z. Li, Roles of nonequilibrium conduction electrons on the magnetization dynamics of ferromagnets. Phys. Rev. Lett. **93**, 127204 (2004)
27. A.P. Malozemoff, Mobility of bubbles with small numbers of Bloch lines. J. Appl. Phys. **44**, 5080 (1973)
28. D.R. Rodrigues, Ar. Abanov, J. Sinova, K. Everschor-Sitte, Effective description of domain wall strings. Phys. Rev. B **97**, 134414 (2018)

29. S. Mühlbauer, B. Binz, F. Jonietz, C. Pfleiderer, A. Rosch, A. Neubauer, R. Georgii, P. Böni, Skyrmion Lattice in a Chiral Magnet. Science **323**, 915–919 (2009)
30. A.V. Bezvershenko, A.K. Kolezhuk, B.A. Ivanov, Stabilization of magnetic skyrmions by RKKY interactions. Phys. Rev. B **97**, 054408 (2018)
31. T. Kurumaji, T. Nakajima, M. Hirschberger, A. Kikkawa, Y. Yamasaki, H. Sagayama, H. Nakao, Y. Taguchi, T. Arima, Y. Tokura, Skyrmion lattice with a giant topological Hall effect in a frustrated triangular-lattice magnet. Science **365**, 914–918 (2019)
32. O. Boulle, J. Vogel, H. Yang, S. Pizzini, D. de Souza Chaves, A. Locatelli, T.O. Menteş, A. Sala, L.D. Buda-Prejbeanu, O. Klein, M. Belmeguenai, Y. Roussigné, A. Stashkevich, S.M. Chérif, L. Aballe, M. Foerster, M. Chshiev, S. Auffret, I.M. Miron, G. Gaudin, Room-temperature chiral magnetic skyrmions in ultrathin magnetic nanostructures. Nat. Nanotechnol. **11**, 449–454 (2016)
33. C. Moreau-Luchaire, C. Moutafis, N. Reyren, J. Sampaio, C.A.F. Vaz, N. Van Horne, K. Bouzehouane, K. Garcia, C. Deranlot, P. Warnicke, P. Wohlhüter, J.M. George, M. Weigand, J. Raabe, V. Cros, A. Fert, Additive interfacial chiral interaction in multilayers for stabilization of small individual skyrmions at room temperature. Nat. Nanotechnol. **11**, 444–448 (2016)
34. B. Göbel, A. Mook, J. Henk, I. Mertig, O.A. Tretiakov, Magnetic bimerons as skyrmion analogues in in-plane magnets. Phys. Rev. B **99**, 060407 (2019)
35. R. Zarzuela, V.K. Bharadwaj, K.-W. Kim, J. Sinova, K. Everschor-Sitte, Stability and dynamics of in-plane skyrmions in collinear ferromagnets. Phys. Rev. B **101**, 054405 (2020)
36. A.O. Leonov, T.L. Monchesky, J.C. Loudon, A.N. Bogdanov, Three-dimensional chiral skyrmions with attractive interparticle interactions. J. Phys.: Condens. Matter **28**, 35LT01 (2016)
37. J. Müller, J. Rajeswari, P. Huang, Y. Murooka, H.M. Rønnow, F. Carbone, A. Rosch, Magnetic Skyrmions and Skyrmion Clusters in the Helical Phase of Cu_2OSeO_3. Phys. Rev. Lett. **119**, 137201 (2017)
38. F.N. Rybakov, A.B. Borisov, S. Blügel, N.S. Kiselev, New Type of Stable Particlelike States in Chiral Magnets. Phys. Rev. Lett. **115**, 117201 (2015)
39. A.P. Malozemoff, J.C. Slonczewski, Magnetic Domain Walls in Bubble Materials. Applied Solid State Science: Supplement 1. Elsevier (1979)
40. J.H.C. Whitehead, An Expression of Hopf's Invariant as an Integral. Proc. Natl. Acad. Sci. **33**, 117–123 (1947)
41. M. Sitte, K. Everschor-Sitte, T. Valet, D.R. Rodrigues, J. Sinova, Ar. Abanov, Current-driven periodic domain wall creation in ferromagnetic nanowires. Phys. Rev. B **94**, 064422 (2016)
42. K. Everschor-Sitte, M. Sitte, T. Valet, Ar. Abanov, J. Sinova, Skyrmion production on demand by homogeneous DC currents. New J. Phys. **19**, 092001 (2017)
43. V. Flovik, A. Qaiumzadeh, A.K. Nandy, C. Heo, T. Rasing, Generation of single skyrmions by picosecond magnetic field pulses. Phys. Rev. B **96**, 140411(R) (2017)
44. S. Zhang, J. Zhang, Q. Zhang, C. Barton, V. Neu, Y. Zhao, Z. Hou, Y. Wen, C. Gong, O. Kazakova, W. Wang, Y. Peng, D.A. Garanin, E.M. Chudnovsky, X. Zhang, Direct writing of room temperature and zero field skyrmion lattices by a scanning local magnetic field. Appl. Phys. Lett. **112**, 132405 (2018)
45. P. Dürrenfeld, Y. Xu, J. Åkerman, Y. Zhou, Controlled skyrmion nucleation in extended magnetic layers using a nanocontact geometry. Phys. Rev. B **96**, 054430 (2017)
46. N. Romming, C. Hanneken, M. Menzel, J.E. Bickel, B. Wolter, K. von Bergmann, A. Kubetzka, R. Wiesendanger, Writing and Deleting Single Magnetic Skyrmions. Science **341**, 636–639 (2013)
47. P.-J. Hsu, A. Kubetzka, A. Finco, N. Romming, K. von Bergmann, R. Wiesendanger, Electric-field-driven switching of individual magnetic skyrmions. Nat. Nanotechnol. **12**, 123–126 (2017)
48. W. Koshibae, N. Nagaosa, Creation of skyrmions and antiskyrmions by local heating. Nat. Commun. **5**, 5148 (2014)
49. F. Rendell-Bhatti, R.J. Lamb, J.W. van der Jagt, G.W. Paterson, H.J.M. Swagten, D. McGrouther, Spontaneous creation and annihilation dynamics and strain-limited stability of magnetic skyrmions. Nat. Commun. **11**, 3536 (2020)

50. M. Stier, W. Häusler, T. Posske, G. Gurski, M. Thorwart, Skyrmion-Anti-Skyrmion Pair Creation by in-Plane Currents. Phys. Rev. Lett. **118**, 267203 (2017)
51. A.O. Leonov, M. Mostovoy, Edge states and skyrmion dynamics in nanostripes of frustrated magnets. Nat. Commun. **8**, 14394 (2017)
52. A. Hrabec, J. Sampaio, M. Belmeguenai, I. Gross, R. Weil, S.M. Chérif, A. Stashkevich, V. Jacques, A. Thiaville, S. Rohart, Current-induced skyrmion generation and dynamics in symmetric bilayers. Nat. Commun. **8**, 15765 (2017)
53. W. Legrand, D. Maccariello, N. Reyren, K. Garcia, C. Moutafis, C. Moreau-Luchaire, S. Collin, K. Bouzehouane, V. Cros, A. Fert, Room-Temperature Current-Induced Generation and Motion of sub-100 nm Skyrmions. Nano Lett. **17**, 2703–2712 (2017)
54. F. Büttner, I. Lemesh, M. Schneider, B. Pfau, C.M. Günther, P. Hessing, J. Geilhufe, L. Caretta, D. Engel, B. Krüger, J. Viefhaus, S. Eisebitt, G.S.D. Beach, Field-free deterministic ultrafast creation of magnetic skyrmions by spin-orbit torques. Nat. Nanotechnol. **12**, 1040–1044 (2017)
55. Y. Zhou, M. Ezawa, A reversible conversion between a skyrmion and a domain-wall pair in a junction geometry. Nat. Commun. **5**, 4652 (2014)
56. W. Jiang, P. Upadhyaya, W. Zhang, G. Yu, M.B. Jungfleisch, F.Y. Fradin, J.E. Pearson, Y. Tserkovnyak, K.L. Wang, O. Heinonen, S.G.E. te Velthuis, A. Hoffmann, Blowing magnetic skyrmion bubbles. Science **349**, 283–286 (2015)
57. J. Iwasaki, M. Mochizuki, N. Nagaosa, Current-induced skyrmion dynamics in constricted geometries. Nat. Nanotech. **8**, 742–747 (2013)
58. J. Müller, A. Rosch, M. Garst, Edge instabilities and skyrmion creation in magnetic layers. New J. Phys. **18**, 065006 (2016)
59. S. Woo, K. Litzius, B. Krüger, M.-Y. Im, L. Caretta, K. Richter, M. Mann, A. Krone, R.M. Reeve, M. Weigand, P. Agrawal, I. Lemesh, M.-A. Mawass, P. Fischer, M. Kläui, G.S.D. Beach, Observation of room-temperature magnetic skyrmions and their current-driven dynamics in ultrathin metallic ferromagnets. Nat. Mater. **15**, 501–506 (2016)
60. A.A. Thiele, Steady-State Motion of Magnetic Domains. Phys. Rev. Lett. **30**, 230 (1973)
61. A.A. Thiele, Applications of the gyrocoupling vector and dissipation dyadic in the dynamics of magnetic domains. J. Appl. Phys. **45**, 377 (1974)
62. J. Müller, A. Rosch, Capturing of a magnetic skyrmion with a hole. Phys. Rev. B **91**, 054410 (2015)
63. N.L. Schryer, L.R. Walker, The motion of 180° domain walls in uniform dc magnetic fields. J. Appl. Phys. **45**, 5406 (1974)
64. O. Boulle, S. Rohart, L.D. Buda-Prejbeanu, E. Jué, I.M. Miron, S. Pizzini, J. Vogel, G. Gaudin, A. Thiaville, Domain Wall Tilting in the Presence of the Dzyaloshinskii-Moriya Interaction in Out-of-Plane Magnetized Magnetic Nanotracks. Phys. Rev. Lett. **111**, 217203 (2013)
65. C.B. Muratov, V.V. Slastikov, A.G. Kolesnikov, O.A. Tretiakov, Theory of the Dzyaloshinskii domain-wall tilt in ferromagnetic nanostrips. Phys. Rev. B **96**, 134417 (2017)
66. V.V. Slastikov, C.B. Muratov, J.M. Robbins, O.A. Tretiakov, Walker solution for Dzyaloshinskii domain wall in ultrathin ferromagnetic films. Phys. Rev. B **99**, 100403(R) (2019)
67. A. Hubert, Statics and dynamics of domain walls in bubble materials. J. Appl. Phys. **46**, 2276 (1975)
68. A. Manchon, J. Železný, I.M. Miron, T. Jungwirth, J. Sinova, A. Thiaville, K. Garello, P. Gambardella, Current-induced spin-orbit torques in ferromagnetic and antiferromagnetic systems. Rev. Mod. Phys. **91**, 035004 (2019)
69. F. Jonietz, S. Mühlbauer, C. Pfleiderer, A. Neubauer, W. Münzer, A. Bauer, T. Adams, R. Georgii, P. Böni, R.A. Duine, K. Everschor, M. Garst, A. Rosch, Spin Transfer Torques in MnSi at Ultralow Current Densities. Science **330**, 1648–1651 (2010)
70. T. Schulz, R. Ritz, A. Bauer, M. Halder, M. Wagner, C. Franz, C. Pfleiderer, K. Everschor, M. Garst, A. Rosch, Emergent electrodynamics of skyrmions in a chiral magnet. Nat. Phys. **8**, 301–304 (2012)
71. J. Iwasaki, M. Mochizuki, N. Nagaosa, Universal current-velocity relation of skyrmion motion in chiral magnets. Nat. Commun. **4**, 1463 (2013)

72. C. Reichhardt, D. Ray, C.J. Olsen Reichhardt, Collective Transport Properties of Driven Skyrmions with Random Disorder. Phys. Rev. Lett. **114**, 217202 (2015)
73. K. Litzius, J. Leliaert, P. Bassirian, D. Rodrigues, S. Kromin, I. Lemesh, J. Zazvorka, K.-J. Lee, J. Mulkers, N. Kerber, D.N. Heinze, R.M. Reeve, M. Weigand, B. Van Waeyenberge, G. Schütz, K. Everschor-Sitte, G.S.D. Beach, M. Kläui, The role of temperature and drive current in skyrmion dynamics. Nat. Electron. **3**, 30–36 (2020)
74. J. Masell, D.R. Rodrigues, B.F. McKeever, K. Everschor-Sitte, Spin-transfer torque driven motion, deformation, and instabilities of magnetic skyrmions at high currents. Phys. Rev. B **101**, 214428 (2020)
75. C. Schütte, J. Iwasaki, A. Rosch, N. Nagaosa, Inertia, diffusion, and dynamics of a driven skyrmion. Phys. Rev. B **90**, 174434 (2014)
76. J. Zázvorka, F. Jakobs, D. Heinze, N. Keil, S. Kromin, S. Jaiswal, K. Litzius, G. Jakob, P. Virnau, D. Pinna, K. Everschor-Sitte, L. Rózsa, A. Donges, U. Nowak, M. Kläui, Thermal skyrmion diffusion used in a reshuffler device. Nat. Nanotechnol. 14, 658–661 (2019)
77. J.J. Liang, J.H. Yu, J. Chen, M.H. Qin, M. Zeng, X.B. Lu, X.S. Gao, J.-M. Liu, Magnetic field gradient driven dynamics of isolated skyrmions and antiskyrmions in frustrated magnets. New J. Phys. **20**, 053037 (2018)
78. V. Lohani, C. Hickey, J. Masell, A. Rosch, Quantum Skyrmions in Frustrated Ferromagnets. Phys. Rev. X **9**, 041063 (2019)
79. K. Litzius, I. Lemesh, B. Krüger, P. Bassirian, L. Caretta, K. Richter, F. Büttner, K. Sato, O.A. Tretiakov, J. Förster, R.M. Reeve, M. Weigand, I. Bykova, H. Stoll, G. Schütz, G.S.D. Beach, M. Kläui, Skyrmion Hall effect revealed by direct time-resolved X-ray microscopy. Nat. Phys. **13**, 170–175 (2017)
80. K.-W. Kim, K.-W. Moon, N. Kerber, J. Nothhelfer, K. Everschor-Sitte, Asymmetric skyrmion Hall effect in systems with a hybrid Dzyaloshinskii-Moriya interaction. Phys. Rev. B **97**, 224427 (2018)
81. X. Zhang, J. Xia, Y. Zhou, X. Liu, H. Zhang, M. Ezawa, Skyrmion dynamics in a frustrated ferromagnetic film and current-induced helicity locking-unlocking transition. Nat. Commun. **8**, 1717 (2017)
82. U. Ritzmann, S. von Malottki, J.-V. Kim, S. Heinze, J. Sinova, B. Dupé, Trochoidal motion and pair generation in skyrmion and antiskyrmion dynamics under spin-orbit torques. Nat. Electron. **1**, 451–457 (2018)
83. X.S. Wang, A. Qaiumzadeh, A. Brataas, Current-Driven Dynamics of Magnetic Hopfions. Phys. Rev. Lett. **123**, 147203 (2019)
84. Y. Liu, W. Hou, X. Han, J. Zang, Three-Dimensional Dynamics of a Magnetic Hopfion Driven by Spin Transfer Torque. Phys. Rev. Lett. **124**, 127204 (2020)
85. S.S.P. Parkin, M. Hayashi, L. Thomas, Magnetic Domain-Wall Racetrack Memory. Science **320**, 190–194 (2008)
86. M. Hayashi, L. Thomas, R. Moriya, C. Rettner, S.S.P. Parkin, Current-Controlled Magnetic Domain-Wall Nanowire Shift Register. Science **320**, 209–211 (2008)
87. A. Fert, V. Cros, J. Sampaio, Skyrmions on the track. Nat. Nanotechnol. **8**, 152–156 (2013)
88. J. Müller, Magnetic skyrmions on a two-lane racetrack. New J. Phys. **19**, 025002 (2017)
89. A.H. Bobeck, P.I. Bonyhard, J.E. Geusic, Magnetic bubbles - An emerging new memory technology. Proc. IEEE **63**, 1176–1195 (1975)
90. T.H. O'Dell, Magnetic bubble domain devices. Rep. Prog. Phys. **49**, 589 (1986)
91. C. Ma, T. Jin, X. Liu, S.N. Piramanayagam, Switching domain wall motion on and off using a gate voltage for domain wall transistor applications. Appl. Phys. Lett. **113**, 232401 (2018)
92. X. Zhang, Y. Zhou, M. Ezawa, G.P. Zhao, W. Zhao, Magnetic skyrmion transistor: Skyrmion motion in a voltage-gated nanotrack. Sci. Rep. **5**, 11369 (2015)
93. D.A. Allwood, G. Xiong, C.C. Faulkner, D. Atkinson, D. Petit, R.P. Cowburn, Magnetic domain-wall logic. Science **309**, 1688–1692 (2005)
94. K. Wagner, A. Kàkay, K. Schultheiss, A. Henschke, T. Sebastian, H. Schultheiss, Magnetic domain walls as reconfigurable spin-wave nanochannels. Nat. Nanotechnol. **11**, 432–436 (2016)

95. X. Zhang, M. Ezawa, Y. Zhou, Magnetic skyrmion logic gates: conversion, duplication and merging of skyrmions. Sci. Rep. **5**, 9400 (2015)
96. S. Sharma, B. Muralidharan, A. Tulapurkar, Proposal for a Domain Wall Nano-Oscillator driven by Non-uniform Spin Currents. Sci. Rep. **5**, 14647 (2015)
97. F. Garcia-Sanchez, J. Sampaio, N. Reyren, V. Cros, J.-V. Kim, A skyrmion-based spin-torque nano-oscillator. New J. Phys. **18**, 075011 (2016)
98. G. Finocchio, M. Di Ventra, K.Y. Camsari, K. Everschor-Sitte, P.K. Amiri, Z. Zeng, The promise of spintronics for unconventional computing. J. Magn. Magn. Mater. **521**, 167506 (2021)
99. J. Grollier, D. Querlioz, K.Y. Camsari, K. Everschor-Sitte, S. Fukami, M.D. Stiles, Neuromorphic spintronics. Nat. Electron. **3**, 360–370 (2020)
100. Li, S., Kang, W., Huang, Y., Zhang, X., Zhou, Y., Zhao, W.: Magnetic skyrmion-based artificial neuron device. Nanotechnology **28**, 31LT01 (2017)
101. A. Sengupta, Y. Shim, K. Roy, Proposal for an All-Spin Artificial Neural Network: Emulating Neural and Synaptic Functionalities Through Domain Wall Motion in Ferromagnets. IEEE Trans. Biomed. Circuits Syst. **10**, 1152–1160 (2016)
102. M. Sharad, C. Augustine, G. Panagopoulos, K. Roy, Spin-Based Neuron Model With Domain-Wall Magnets as Synapse. IEEE Trans. Nanotechnol. **11**, 843–853 (2012)
103. K.M. Song, J.-S. Jeong, B. Pan, X. Zhang, J. Xia, S. Cha, T.-E. Park, K. Kim, S. Finizio, J. Raabe, J. Chang, Y. Zhou, W. Zhao, W. Kang, H. Ju, S. Woo, Skyrmion-based artificial synapses for neuromorphic computing. Nat. Electron. **3**, 148–155 (2020)
104. X. Wang, Y. Chen, H. Xi, H. Li, D. Dimitrov, Spintronic Memristor Through Spin-Torque-Induced Magnetization Motion. IEEE Electron Device Lett. **30**, 294–297 (2009)
105. L. Appeltant, M.C. Soriano, G. Van der Sande, J. Danckaert, S. Massar, J. Dambre, B. Schrauwen, C.R. Mirasso, I. Fischer, Information processing using a single dynamical node as complex system. Nat. Commun. **2**, 468 (2011)
106. G. Tanaka, T. Yamane, J.B. Héroux, R. Nakane, N. Kanazawa, S. Takeda, H. Numata, D. Nakano, A. Hirose, Recent advances in physical reservoir computing: A review. Neural Netw. **115**, 100–123 (2019)
107. D. Prychynenko, M. Sitte, K. Litzius, B. Krüger, G. Bourianoff, M. Kläui, J. Sinova, K. Everschor-Sitte, Magnetic Skyrmion as a Nonlinear Resistive Element: A Potential Building Block for Reservoir Computing. Phys. Rev. Appl. **9**, 014034 (2018)
108. D. Pinna, F. Abreu Araujo, J.-V. Kim, V. Cros, D. Querlioz, P. Bessiere, J. Droulez, J. Grollier, Skyrmion Gas Manipulation for Probabilistic Computing. Phys. Rev. Appl. **9**, 064018 (2018)
109. K.Y. Camsari, B.M. Sutton, S. Datta, p-bits for probabilistic spin logic. Appl. Phys. Rev. **6**, 011305 (2019)
110. G. Yang, P. Stano, J. Klinovaja, D. Loss, Majorana bound states in magnetic skyrmions. Phys. Rev. B **93**, 224505 (2016)
111. K.M.D. Hals, M. Schecter, M.S. Rudner, Composite Topological Excitations in Ferromagnet-Superconductor Heterostructures. Phys. Rev. Lett. **117**, 017001 (2016)
112. J. Nothhelfer, Localized Majorana modes in heterostructures - A path towards topological quantum computation. Master Thesis, JGU Mainz, Germany (2019)
113. S. Rex, I.V. Gornyi, A.D. Mirlin, Majorana bound states in magnetic skyrmions imposed onto a superconductor. Phys. Rev. B **100**, 064504 (2019)
114. N. Kent, N. Reynolds, D. Raftrey, I.T.G. Campbell, S. Virasawmy, S. Dhuey, R.V. Chopdekar, A. Hierro-Rodriguez, A. Sorrentino, E. Pereiro, S. Ferrer, F. Hellman, P. Sutcliffe, P. Fischer, Creation and confirmation of Hopfions in magnetic multilayer systems. ArXiv preprint, 2010.08674 (2020)
115. J. Barker, O.A. Tretiakov, Static and Dynamical properties of antiferromagnetic skyrmions in the presence of applied current and temperature. Phys. Rev. Lett. 116, 147203 (2016)
116. X. Zhang, Y. Zhou, M. Ezawa, Antiferromagnetic Skyrmion: Stability, Creation and Manipulation. Sci. Rep. 6, 24795 (2016)

Chapter 8
Microwave-Driven Dynamics of Magnetic Skyrmions Under a Tilted Magnetic Field: Magnetic Resonances, Translational Motions, and Spin-Motive Forces

Masahito Mochizuki

Abstract Magnetic skyrmions, particle-like magnetic textures characterized by a quantized topological invariant in magnets with broken spatial inversion symmetry, are currently attracting enormous research interest in the field of spintronics. Recent intensive studies have uncovered that magnetic skyrmions exhibit rich physical phenomena and device functions originating from their topological nature. In this chapter, we discuss microwave-induced dynamical phenomena of magnetic skyrmions associated with their peculiar spin-wave modes. In particular, we focus on a situation that the magnetic skyrmions are confined in a quasi-two-dimensional thin-plate magnet under application of a static magnetic field tilted from the perpendicular direction. It is theoretically demonstrated that the spin-wave excitations of these magnetic skyrmions by microwave irradiation give rise to translational motion of the skyrmions and generation of the DC electric voltages. These phenomena indicate that rich physics and functionalities are hidden in the microwave-driven skyrmion dynamics under a tilted magnetic field.

8.1 Introduction

Magnetic skyrmions, topological magnetic entities carrying a quantized topological invariant in magnets without spatial inversion symmetry, are currently attracting a great deal of research interest [1–7]. In the early 1960s, an English theoretical physicist, Tony Hilton Royle Skyrme, proposed a theoretical concept of Skyrmion as a soliton solution of the nonlinear sigma model to account for the stability of hadron in the nuclear physics [8, 9]. This originally proposed skyrmion is a hedgehog-like particle composed of field vectors pointing in every direction wrapping a sphere

M. Mochizuki (✉)
Department of Applied Physics, Waseda University, 3-4-1 Okubo, Shinjuku-ku, Tokyo 169-8555, Japan
e-mail: masa_mochizuki@waseda.jp

E. Kamenetskii (ed.), *Chirality, Magnetism and Magnetoelectricity*, Topics in Applied Physics 138,
https://doi.org/10.1007/978-3-030-62844-4_8

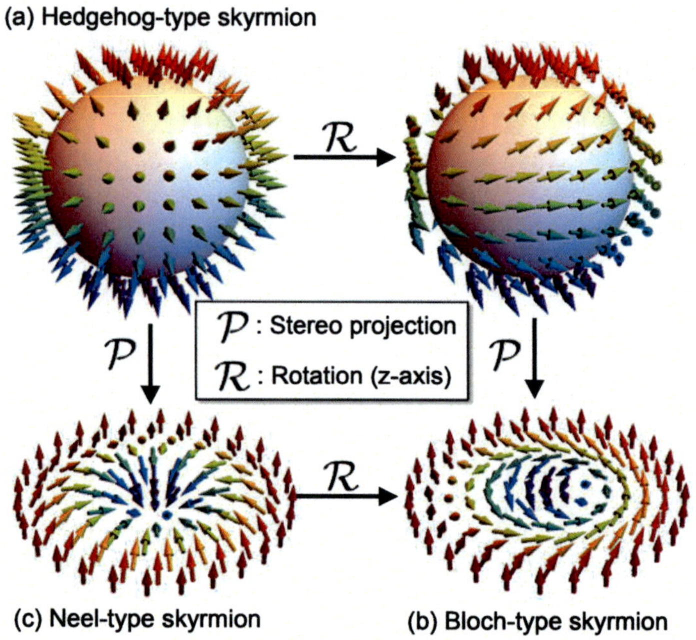

Fig. 8.1 **a** Hedgehog-type skyrmion originally proposed by Tony Skyrme in early 1960s. **b** Bloch-type skyrmion with magnetizations rotating within a plane normal to the radial direction. **c** Neel-type skyrmion with magnetizations rotating within a plane parallel to the radial direction. These three types of skyrmions are mutually related via a stereo projection and a rotation with respect to the z axis

(Fig. 8.1a). About 30 years later, Bogdanov and his coworkers theoretically predicted realization of these conceptual objects in magnets as textures of magnetizations [10, 11]. They proposed that the magnetic skyrmions can emerge in magnets with spatial inversion asymmetry as vortex-like (Fig. 8.1b) or fountain-like (Fig. 8.1c) magnetic textures. These magnetic textures can be reproduced by a stereo projection of the original hedgehog-type skyrmion onto a two-dimensional plane. This theoretical prediction was indeed confirmed by experimental discoveries of magnetic skyrmions in B20-type alloys [12–14]. In 2009, a small-angle neutron scattering experiment reported a discovery of skyrmion crystals in which numerous magnetic skyrmions are crystallized into a hexagonal form in a chiral-lattice magnet MnSi [12]. In 2010, a Lorentz transmission electron microscopy reported an observation of real-space images of skyrmion crystals in $Fe_{1-x}Co_xSi$ [13]. This experiment also revealed that magnetic skyrmions emerge not only as the crystallized form but also as isolated defects in the ferromagnetic phase. Subsequently, the magnetic skyrmions have been discovered in various magnetic systems such as other chiral-lattice magnets (FeGe [14], Cu_2OSeO_3 [15], and βMn-type Co-Mn-Zn alloys [16]), polar magnets (GaV_4S_8 [17], GaV_4Se_8 [18, 19], $Mn_{1.4}Pt_{0.9}Pd_{0.1}Sn$ [20], $VOSe_2O_5$ [21]), and magnetic heterostructures [3, 22–24].

All the above-mentioned magnets and magnetic systems have structures with broken spatial inversion symmetry. Consequently, the Dzyaloshinskii-Moriya interaction becomes active, which favors rotating alignment of magnetizations with a pitch angle of 90° [25, 26]. This interaction strongly competes with the ferromagnetic exchange interaction, which favors parallel alignment of magnetizations. The keen competition between these two interactions results in the formation of helical magnetic order with a moderate pitch angle. When an external magnetic field is applied to this helical magnetic state, a skyrmion crystal appears on a plane normal to the applied magnetic field. In each skyrmion constituting the skyrmion crystal, the magnetizations at periphery are oriented along the external magnetic field so as to maximize an energy gain of the Zeeman interaction. The magnetizations gradually rotate when we move from the periphery to the center along the radial direction. Eventually, the magnetizations are oriented antiparallel to the external magnetic field at the center. In bulk magnets, these two-dimensional skyrmions are stacked along the three-dimensional direction to form tubular structures. It should also be mentioned that the magnetic skyrmions can be classified into several types according to the way of the magnetization rotation. Skyrmions in which the magnetizations rotate within planes perpendicular (parallel) to the radial direction are called Bloch type (Neel type). The three types of skyrmions, i.e., the hedgehog-type, the Bloch-type, and the Neel-type, are mutually related via the rotational operation with respect to the z axis and the stereo projection (see Fig. 8.1).

Skyrmions are characterized by a quantized topological invariant, which represents a sum of the solid angles spanned by three neighboring normalized magnetization vectors $\boldsymbol{m}(\boldsymbol{r})$,

$$\int_{\mathrm{UC}} dxdy \left(\frac{\partial \boldsymbol{m}}{\partial x} \times \frac{\partial \boldsymbol{m}}{\partial y} \right) \cdot \boldsymbol{m} = 4\pi Q \qquad (Q = \pm 1). \tag{8.1}$$

Here the integration is taken over the area of the unit cell (UC). Because the magnetization vectors constituting a skyrmion point in every direction wrapping a sphere, the sum of the solid angles is identical to the surface area of a unit sphere ($\mp 4\pi$). Consequently, the quantity Q, which is called the topological charge or the skyrmion number, becomes $+1$ or -1. The sign is positive (negative) when the magnetization at the core points upward (downward). This number does not change upon continuous variation of the magnetization alignment. Because the value of Q is zero for ferromagnetic and spiral magnetic orders, magnetic skyrmions belong to a different topological class from these magnetic structures. This means that magnetic skyrmions cannot be created or erased in a uniform ferromagnetic state by continuous variation of the magnetization alignment. Instead, a local flop of the magnetization is needed to create and erase them, which necessarily requires a large cost of energy, magnitude of which is comparable to the energy scale of the ferromagnetic exchange interaction. The topological protection by this large energy barrier makes magnetic skyrmions robust against external assitations and perturbations.

This significant stability is one of the advantageous properties of magnetic skyrmions for spintronics application to ubiquitous magnetic devices in the coming

IoT era. Future electronic devices sustaining the IoT society such as sensor, logic, and storage devices are demanded to keep working normally with a little energy supply in harsh environments such as weather-beaten places, extremely hot or cold areas, places suffered from continuous mechanical shocks and vibrations, and outer spaces exposed by cosmic rays. Therefore, the devices should have the resistance against thermal assitations, the durability against radioactive rays, and the sustainability with conserved electric power. From this viewpoint, the magnetic skyrmions have high potentials as building blocks of the next-generation magnetic devices because they have not only the topologically protected stability but also the nanoscale ultrasmall size and the operability with ultralow energy consumption.

Aiming at the technical applications, magnetic skyrmions have attracted a great deal of interest recently as a target of the spintronics research, and several intriguing phenomena and useful devices functions have been discovered or proposed successively. In particular, the dynamical magnetoelectric phenomena and the microwave devices functions associated with their peculiar microwave-active spin-wave modes [27] are important examples [6, 7]. In this chapter, we discuss recent theoretical studys on the microwave-induced physical phenomena and device functions of magnetic skyrmions by particularly focusing on theoretical proposals of the microwave-driven translational motion [28–30] and the efficient conversion of microwaves to DC electric voltages [31] expected in the spin-wave excitations of magnetic skyrmions in a quasi-two-dimensional magnet under a tilted external magnetic field.

8.2 Spin Model of the Skyrmion-Hosting Magnets

In this chapter, we deal with magnetic skyrmions confined in a quasi-two-dimensional thin-plate magnet with broken spatial inversion symmetry. The simplest spin model for such quasi-two-dimensional skyrmion-hosting magnets is given in a continuum form as [32],

$$\mathcal{H}_0 = \int d\boldsymbol{r} \left[\frac{J}{2a}(\nabla \boldsymbol{m})^2 + \frac{D}{a^2}\boldsymbol{m} \cdot (\nabla \times \boldsymbol{m}) - \frac{1}{a^3}\boldsymbol{H} \cdot \boldsymbol{m} \right] \tag{8.2}$$

By dividing the continuum space into cubic cells, the above continuum model is reduced to a classical Heisenberg model on a square lattice as [33],

$$\mathcal{H}_0 = -J \sum_{i,\mu} \boldsymbol{m}_i \cdot \boldsymbol{m}_{i+\hat{\mu}} - \sum_{i,\mu} \boldsymbol{D}_\mu \cdot (\boldsymbol{m}_i \times \boldsymbol{m}_{i+\hat{\mu}}) - \boldsymbol{H}_{\text{ex}} \cdot \sum_i \boldsymbol{m}_i. \tag{8.3}$$

Here $\boldsymbol{m}_i$ represents a normalized magnetization vector on the ith lattice site. The first term represents the ferromagnetic exchange interaction, whereas the second term represents the Dzyaloshinskii-Moriya interaction. The third term depicts the Zeeman coupling associated with an external magnetic field $\boldsymbol{H}_{\text{ex}}$. Types of the skyrmions are determined by the structure of the Moriya vectors $\boldsymbol{D}_\mu$ ($\mu = x, y$). The Bloch-type

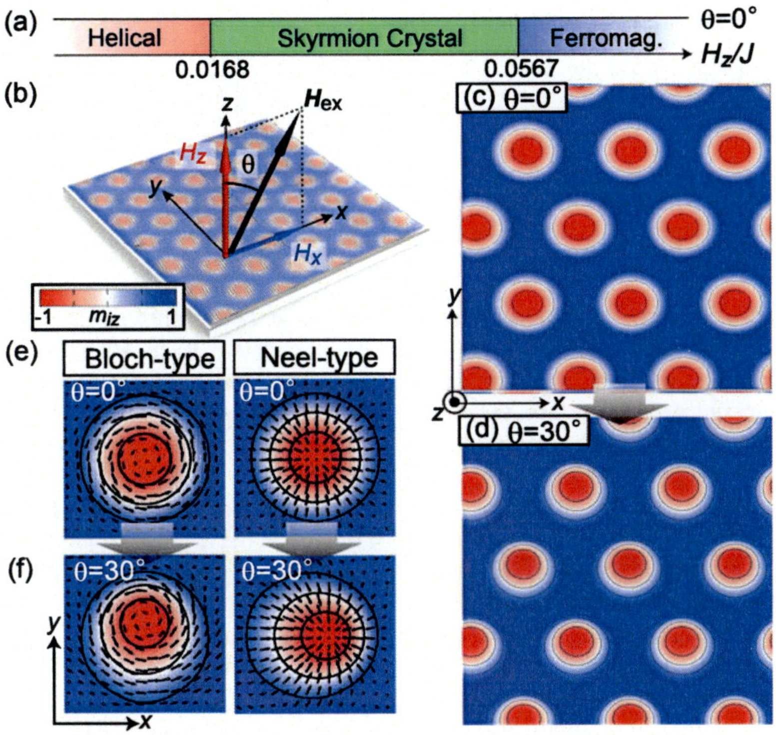

Fig. 8.2 **a** Theoretical phase diagram of the spin model in (8.3) on a square lattice as a function of the magnetic-field strength H_z when $\boldsymbol{H}_{\text{ex}} = (0, 0, H_z)$ is applied normal to the plane. Here the strength of the DM interaction is set to be $D/J = 0.27$. **b** Thin-plate specimen of chiral-lattice magnet hosting a skyrmion crystal under a magnetic field $\boldsymbol{H}_{\text{ex}} = (H_z \tan\theta, 0, H_z)$ with a tilting angle of θ. **c, d** Skyrmion crystal under a perpendicular [tilted] $\boldsymbol{H}_{\text{ex}}$ field with $\theta = 0°$ [$\theta = 30°$]. **e, f** Magnetization configurations of Bloch-type and the Neel-type skyrmions under a perpendicular [tilted] $\boldsymbol{H}_{\text{ex}}$ field (Reproduced from [29].)

skyrmions are produced by $\boldsymbol{D}_x = (D, 0, 0)$ and $\boldsymbol{D}_y = (0, D, 0)$, whereas the Neel-type skyrmions are produced by $\boldsymbol{D}_x = (0, D, 0)$ and $\boldsymbol{D}_y = (-D, 0, 0)$. Figure 8.2a shows a theoretical phase diagram of this spin model for $D/J = 0.27$ at $T = 0$ as a function of the magnetic-field strength H_z when $\boldsymbol{H}_{\text{ex}} = (0, 0, H_z)$ is applied perpendicular to the two-dimensional plane. This phase diagram exhibits the skyrmion-crystal phase in a region of moderate field strength sandwiched by the helical phase and the field-polarized ferromagnetic phase.

The lattice spacing of skyrmion crystal, i.e., the distance between cores of neighboring skyrmions in the skyrmion crystal is determined by competition between the Dzyaloshinskii-Moriya interaction and the ferromagnetic exchange interaction, which favor rotating and parallel magnetization alignments, respectively. A stronger Dzyaloshinskii-Moriya interaction causes more rapid rotation of the magnetizations, which results in a smaller skyrmion size. Because $\phi \sim D/(\sqrt{2}J)$ holds for the

Table 8.1 Unit conversion table when $J = 1\,\text{meV}$

	Dimensionless quantity	Corresponding value with units
Exchange int.	$J = 1$	$J = 1\,\text{meV}$
Magnetic field	$H = 1$	$J/g\mu_B = 8.64\,\text{T}$
Time	$t = 1$	$\hbar/J = 0.66\,\text{ps}$
Frequency $f = \omega/2\pi$	$\omega = 1$	$J/h = 241\,\text{GHz}$

magnetization rotation angle ϕ, the spatial period in the skyrmion crystal becomes $\lambda_m \sim 2\pi a/\phi$ when $\boldsymbol{H}_{ex} = 0$ where a is the lattice constant. Even when $\boldsymbol{H}_{ex}$ is finite, this spatial period does not change so much although the magnetization rotation is no longer uniform. Therefore, the ratio $D/J = 0.27$ gives $\lambda_m \sim 18\,\text{nm}$ if we assume a typical lattice constant of $a = 0.5\,\text{nm}$, which corresponds to the experimentally observed skyrmion size in MnSi [12, 34]. When $J = 1\,\text{meV}$ is the energy units, the dimensionless field strength $H_z = 1$ corresponds to ~8.64 T. Therefore, the threshold magnetic fields of $H_{c1} = 0.0168$ and $H_{c2} = 0.0567$ in the theoretical phase diagram of Fig. 8.2a correspond to 0.145 T and 0.49 T, respectively. These values again coincide well with the experimentally observed threshold magnetic fields of ~0.15 T and ~0.45 T for MnSi at low temperatures [34]. The unit conversions when $J = 1\,\text{meV}$ are summarized in Table 8.1.

We examine the cases in which the $\boldsymbol{H}_{ex}$ field is tilted from the normal direction ($\parallel z$) of the quasi-two-dimensional system towards the x direction. The magnetic field is given in the form,

$$\boldsymbol{H}_{ex} = (H_x, 0, H_z), \tag{8.4}$$

with $H_x = H_z \tan\theta$, where the angle θ describes to what extent the $\boldsymbol{H}_{ex}$ field is tilted towards the x direction (Fig. 8.2b). In bulk materials, the magnetic skyrmions usually appear on a plane normal to the $\boldsymbol{H}_{ex}$ field and have a circularly symmetric shape. This circular symmetry is kept even when a direction of the $\boldsymbol{H}_{ex}$ field is changed because the stacked magnetic skyrmions can change their tubular orientations to keep the skyrmion plane normal to the $\boldsymbol{H}_{ex}$ field in the three-dimensional systems so as to maximize the energy gain of the Zeeman interaction. On the other hand, when the skyrmions are confined in a system of strong two-dimensionality, the situation is no longer the same. Although the skyrmions have a circularly symmetric shape when the $\boldsymbol{H}_{ex}$ field applied perpendicular to the two-dimensional plane (Fig. 8.2c, e), they become to have a disproportionate weight in the magnetization distribution and lose their circular symmetry when the $\boldsymbol{H}_{ex}$ field is tilted (Fig. 8.2d, f). In fact, these deformed skyrmions under a tilted $\boldsymbol{H}_{ex}$ field turn out to exhibit intriguing microwave-related physical phenomena and device functions.

The microwave-induced dynamics of magnetic skyrmions are investigated by the micromagnetic simulations based on the Landau–Lifshitz–Gilbert equation. The equation is given by,

$$\frac{d\boldsymbol{m}_i}{dt} = -\gamma \boldsymbol{m}_i \times \boldsymbol{H}_i^{\rm eff} + \frac{\alpha_{\rm G}}{m}\boldsymbol{m}_i \times \frac{d\boldsymbol{m}_i}{dt}. \tag{8.5}$$

The first term of the right-hand side is the gyrotropic term where $\gamma = g\mu_{\rm B}/\hbar$ is the gyromagnetic ratio. This term describes the precessional motion of magnetizations $\boldsymbol{m}_i$ around the effective local magnetic field $\boldsymbol{H}_i^{\rm eff}$. The second term is the Gilbert-damping term introduced phenomenologically to depicts the dissipation of gyration energy. The effective magnetic field $\boldsymbol{H}_i^{\rm eff}$, which acts on the local magnetization $\boldsymbol{m}_i$, is calculated from the Hamiltonian $\mathcal{H} = \mathcal{H}_0 + \mathcal{H}'(t)$ as,

$$\boldsymbol{H}_i^{\rm eff} = -\frac{1}{\gamma\hbar}\frac{\partial\mathcal{H}}{\partial\boldsymbol{m}_i}, \tag{8.6}$$

where $\mathcal{H}_0$ is the model Hamiltonian given by (8.3). Here an additional term of the Hamiltonian $\mathcal{H}'(t)$ is introduced, which describes the dynamical coupling between the skyrmion magnetizations and a time-dependent magnetic field or a microwave magnetic field in the form,

$$\mathcal{H}'(t) = -\boldsymbol{H}(t) \cdot \sum_i \boldsymbol{m}_i. \tag{8.7}$$

The initial magnetic configuration of a skyrmion crystal is prepared by the Monte-Carlo thermalization at low temperatures and by further relaxing it in the micromagnetic simulation without applying a microwave magnetic field. The microwave-induced dynamics are simulated by applying a microwave field to thus obtained sufficiently converged skyrmion-crystal configuration.

8.3 Microwave-Active Spin-Wave Modes

A skyrmion crystal confined in a quasi-two-dimensional magnet exhibits characteristic microwave-active spin-wave modes (Fig. 8.3a–c) [27]. An in-plane polarized microwave magnetic field $\boldsymbol{H}^{\omega} \| \boldsymbol{x}, \boldsymbol{y}$ activates a pair of rotation modes, in which all the skyrmions constituting the skyrmion crystal uniformly circulate in counterclockwise and clockwise ways. The counterclockwise rotation mode has a lower resonance frequency and a larger intensity than the clockwise rotation mode. On the other hand, an out-of-plane polarized microwave magnetic field $\boldsymbol{H}^{\omega} \| z$ activates the breathing mode in which all the skyrmions in the skyrmion crystal uniformly expand and shrink in an oscillatory manner. Then how are these spin-wave modes modulated when the $\boldsymbol{H}_{\rm ex}$ field is tilted from the perpendicular direction?

The spin-wave modes and their resonance frequencies can be identified by calculating the dynamical magnetic susceptibility,

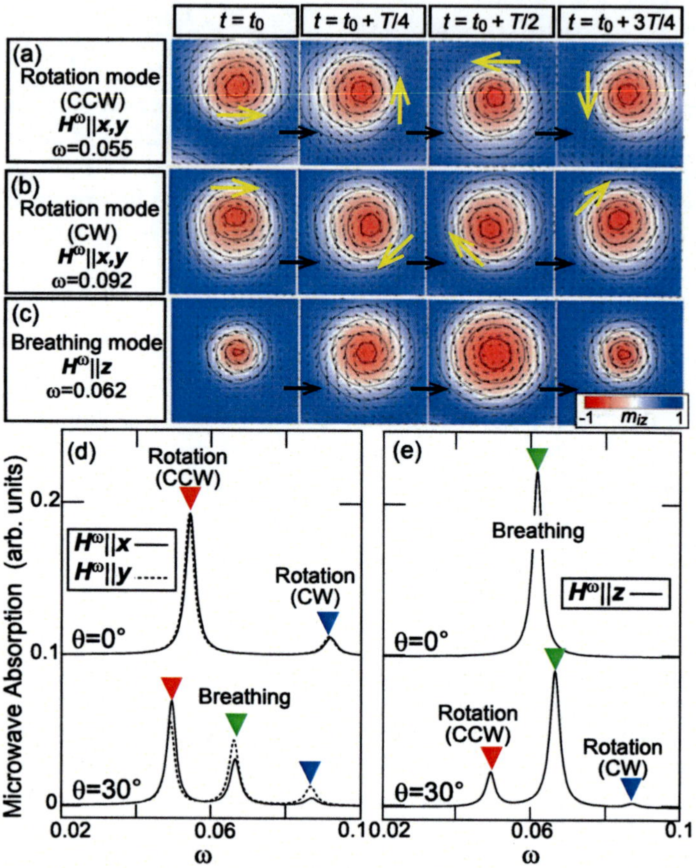

Fig. 8.3 **a–c** Simulated snapshots of the three spin-wave modes of a skyrmion crystal confined in a quasi-two-dimensional thin-plate magnet under a perpendicular $\boldsymbol{H}_{\rm ex}$ field. Here one skyrmion in the skyrmion crystal is focused on because all the skyrmions oscillate uniformly in each mode. The two rotation modes with counterclockwise (CCW) and clockwise (CW) rotation senses are active to an in-plane microwave magnetic field $\boldsymbol{H}^{\omega}\|\boldsymbol{x}, \boldsymbol{y}$, whereas the breathing mode is active to an out-of-plane microwave magnetic field $\boldsymbol{H}^{\omega}\|z$. **d** Calculated microwave absorption spectra under perpendicular ($\theta = 0°$) and tilted ($\theta = 30°$) $\boldsymbol{H}_{\rm ex}$ fields as functions of the microwave angular frequency ω for the in-plane microwave field $\boldsymbol{H}^{\omega}\|\boldsymbol{x}, \boldsymbol{y}$. **e** Those for the out-of-plane microwave field $\boldsymbol{H}^{\omega}\|z$. Here the simulations are performed for $J = 1$, $D/J = 0.27$, $H_z = 0.036$, $H^{\omega} = 0.0018$, and $\alpha_{\rm G} = 0.02$ (Reproduced from [29].)

$$\chi_\mu(\omega) = \frac{\Delta M_\mu(\omega)}{\mu_0 H_\mu(\omega)} \quad (\mu = x, y, z), \tag{8.8}$$

where $H_\mu(\omega)$ and $\Delta M_\mu(\omega)$ are the Fourier transforms of the time-dependent magnetic field $\boldsymbol{H}(t)$ and the simulated time-profile of the net magnetization $\Delta \boldsymbol{M}(t) = \boldsymbol{M}(t) - \boldsymbol{M}(0)$ with $\boldsymbol{M}(t) = \frac{1}{N}\sum_{i=1}^{N} \boldsymbol{m}_i(t)$. In the calculations, a short rectangular pulse is used for $\boldsymbol{H}(t)$ whose component is given by,

$$H_\mu(t) = \begin{cases} H_{\text{pulse}} & 0 \le t \le 1 \\ 0 & \text{others} \end{cases} \tag{8.9}$$

where $t = (J/\hbar)\tau$ is the dimensionless time with τ being the real time. An advantage of using the short pulse is that the Fourier component $H_\mu(\omega)$ becomes constant being independent of ω up to first order in $\omega\Delta t$ for a sufficiently short duration Δt with $\omega\Delta t \ll 1$. The Fourier component is

$$H_\mu(\omega) = \int_0^{\Delta t} H_{\text{pulse}} e^{i\omega t} dt = \frac{H_{\text{pulse}}}{i\omega}\left(e^{i\omega\Delta t} - 1\right) \sim H_{\text{pulse}}\Delta t. \tag{8.10}$$

Consequently, we obtain the relationship $\chi_\mu(\omega) \propto \Delta M_\mu(\omega)$. The imaginary part of thus calculated $\chi_\mu(\omega)$ corresponds to the microwave absorption spectrum. Figure 8.3d shows the spectra for the in-plane polarized microwave field $\boldsymbol{H}^\omega \| \boldsymbol{x}, \boldsymbol{y}$ under perpendicular ($\theta = 0°$) and tilted ($\theta = 30°$) $\boldsymbol{H}_{\text{ext}}$ fields, whereas Fig. 8.3e shows the spectra for the out-of-plane polarized microwave field $\boldsymbol{H}^\omega \| z$ [29]. When the $\boldsymbol{H}_{\text{ext}}$ field is perpendicular ($\theta = 0°$), two spectral peaks appear in Fig. 8.3d originating from the two rotation modes, whereas a single peak appears in Fig. 8.3e originating from the breathing mode, indicating that the in-plane (out-of-plane) microwave field can activate the rotation modes (the breathing mode) only when the $\boldsymbol{H}_{\text{ext}}$ field is perpendicular. On the other hand, when the $\boldsymbol{H}_{\text{ext}}$ field is tilted ($\theta = 30°$), three spectral peaks appear in both Fig. 8.3d, e, indicating that all the three spin-wave modes become active to both microwave polarization under the tilted $\boldsymbol{H}_{\text{ext}}$ field.

8.4 Microwave-Magnetic-Field-Driven Translational Motion of Skyrmion Crystal

The continuous spin-wave excitation by microwave irradiation under a tilted $\boldsymbol{H}_{\text{ext}}$ field can induce translational motion of a skyrmion crystal [28, 29]. Figure 8.4 shows simulated snapshots of a skyrmion crystal driven by an in-plane microwave field $\boldsymbol{H}^\omega \| \boldsymbol{x}$ (right upper panel) and the same skyrmion crystal driven by an out-of-plane microwave field $\boldsymbol{H}^\omega \| z$ (right lower panel) at $t = 400$ ns after the microwave irradiation commences. The figure also shows the initial configuration of the skyrmion crystal at $t = 0$ (left panel) under application of the tilted magnetic field $\boldsymbol{H}_{\text{ex}} =$

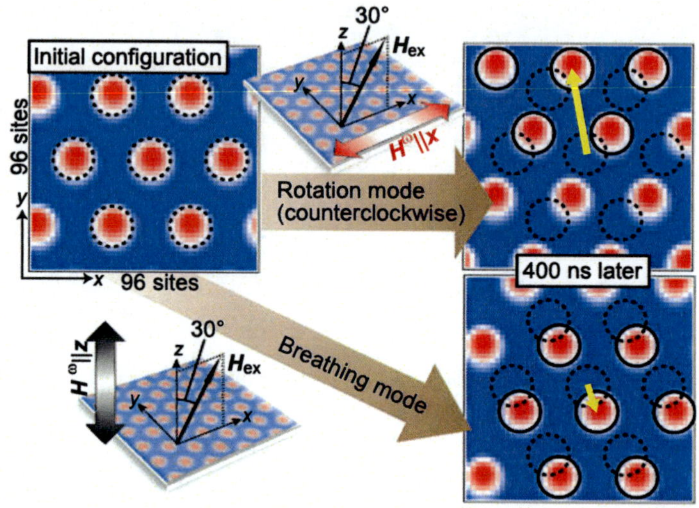

Fig. 8.4 Microwave-driven translational motion of a skyrmion crystal in a quasi-two-dimensional system under a tilted $\boldsymbol{H}_{\text{ext}}$ field with $\theta = 30°$. The system is irradiated by a microwave field $H_{\mu}^{\omega} \sin \omega t$ ($\mu = x, y, z$) with $H_{\mu}^{\omega} = 0.0006$. The skyrmion crystal moves approximately towards the positive (negative) y direction when a microwave field $\boldsymbol{H}^{\omega}\|\boldsymbol{x}$ ($\boldsymbol{H}^{\omega}\|z$) with $\omega = 0.0494$ ($\omega = 0.0666$) activates the counterclockwise rotation (breathing) mode. Displacement vectors connecting the original position (dashed circles) and the position after 400-ns duration (solid circles) are indicated by the thick arrows in the right panels. The simulations are performed for $J = 1$, $D/J = 0.27$, $H_z = 0.036$, and $\alpha_G = 0.04$ (Reproduced from [29].)

($H_z \tan\theta, 0, H_z$) with $H_z = 0.036$ and $\theta = 30°$. Here, the microwave field is given by $H_{\mu}^{\omega} \sin \omega t$ ($\mu = x, y, z$) with $H_{\mu}^{\omega} = 0.0006$. The displacement vectors connecting the original position and the position at $t = 400$ ns are indicated by thick arrows shown in the right-side panels. When the microwave field $\boldsymbol{H}^{\omega}\|\boldsymbol{x}$ with $\omega = 0.0494$ activates the counterclockwise rotation mode, the skyrmion crystal propagates in a direction close to the positive y direction, whereas the same skyrmion crystal propagates in a direction close to the negative y direction when $\boldsymbol{H}^{\omega}\|z$ with $\omega = 0.0666$ activates the breathing mode. It is also found that the travel distance in the former case is much longer than that in the latter case, which indicates that the in-plane microwave field $\boldsymbol{H}^{\omega}\|x$ drives much faster motion of the skyrmion crystal than the out-of-plane microwave field $\boldsymbol{H}^{\omega}\|z$.

Figure 8.5a, b display the simulated microwave frequency dependence of the drift velocity $v = (v_x, v_y)$ of the driven skyrmion crystal under a tilted $\boldsymbol{H}_{\text{ex}}$ field for different microwave polarizations. Apparently, the velocities are enhanced to have peaks at the resonance frequencies that correspond to the spin-wave modes, indicating that the resonant spin-wave excitations of skyrmion crystal indeed drive its quick translational motion. The velocity is highest with $v_x \sim 0.04$ m/s when the in-plane

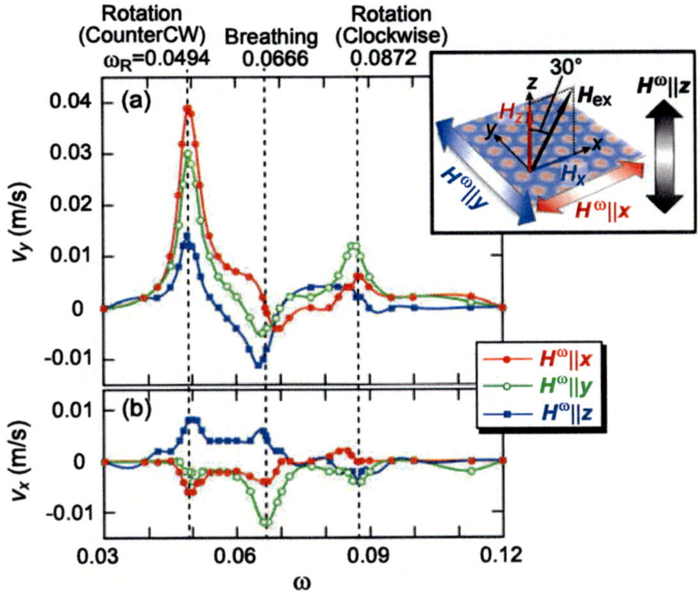

Fig. 8.5 Calculated microwave frequency dependence of the velocity $\boldsymbol{v} = (v_x, v_y)$ for translational motion of the skyrmion crystal induced by a microwave field $\boldsymbol{H}^\omega$ under a tilted $\boldsymbol{H}_{\rm ext}$ field $\boldsymbol{H}_{\rm ex} = (H_z \tan\theta, 0, H_z)$ with $H_z = 0.036$ and $\theta = 30°$. The system is irradiated by a microwave field $H_\mu^\omega \sin\omega t$ ($\mu = x, y, z$) with $H_\alpha^\omega = 0.0006$ where $\omega = 2\pi f$ is its angular frequency. The velocities show peaks at the resonant frequencies of the spin-wave modes, while their signs vary depending on the mode and the microwave polarization (Reproduced from [29].)

microwave field $\boldsymbol{H}^\omega \parallel \boldsymbol{x}$ activates the counterclockwise rotation mode. It is also found that the speed and the direction of this translational motion sensitively depend on the excited spin-wave mode and the microwave polarization.

8.5 Microwave-Electric-Field-Driven Translational Motion of Isolated Skyrmions

In this section, we discuss an efficient method to drive isolated magnetic skyrmions embedded in ferromagnetic environment with microwave *electric* fields instead of microwave *magnetic* fields [30]. As mentioned in the introduction section, the magnetic skyrmions appear not only as a crystallized form but also as isolated defects in the ferromagnetic phase. Recently possible application of magnetic skyrmions to high-performance memory devices are studied intensively. For the usage as information carriers in memory devices, isolated magnetic skyrmions rather than crystallized ones are recognized to be convenient. However, when a microwave magnetic field is applied to a device to activate the isolated skyrmion defects, it is unavoidable to excite a vast majority of background ferromagnetic magnetizations, which hinders the resonance modes of the skyrmion defects.

To solve this problem, it was recently proposed theoretically that isolated magnetic skyrmions can be selectively activated using microwave electric fields without exciting ferromagnetic resonances, in contrast to conventional methods using microwave magnetic fields. It was also demonstrated by numerical simulations that the selective activation of a skyrmion can efficiently drive its translational motion in a ferromagnetic nanotrack under application of a tilted $\boldsymbol{H}_{\rm ext}$ field.

We consider a magnetic bilayer system with a ferromagnetic layer fabricated on top of a heavy-metal layer with strong spin-orbit interaction, where the Dzyaloshinskii-Moriya interaction becomes active at their interface due to the spatial inversion asymmetry to form Neel-type skyrmions. A tilted magnetic field $\boldsymbol{H}_{\rm ex} = (H_x, 0, H_z)$ with $H_x = H_z \tan\theta$ is applied to this bilayer system. To investigate microwave-driven phenomena of skyrmions in this system, the following classical Heisenberg model on a square lattice was employed,

$$\mathcal{H} = -J \sum_{<i,j>} \boldsymbol{m}_i \cdot \boldsymbol{m}_j - [\boldsymbol{H}_{\rm ex} + \boldsymbol{H}(t)] \cdot \sum_i \boldsymbol{m}_i + D(t) \sum_i [(\boldsymbol{m}_i \times \boldsymbol{m}_{i+\hat{x}}) \cdot \hat{\boldsymbol{y}} - (\boldsymbol{m}_i \times \boldsymbol{m}_{i+\hat{y}}) \cdot \hat{\boldsymbol{x}}], \tag{8.11}$$

where $\boldsymbol{m}_i$ is the normalized magnetization vector. In this Hamiltonian, a time-dependent magnetic field or a microwave magnetic field $\boldsymbol{H}(t) = (0, 0, H_z(t))$ with $H_z(t) = H_z^\omega \sin(\omega t)$ is taken into account via the Zeeman coupling term. On the other hand, a time-dependent electric field $\boldsymbol{E}(t) = (0, 0, E_z(t))$ with $E_z(t) = E_z^\omega \sin(\omega t)$ applied perpendicular to the sample plane is incorporated via the time-dependent interfacial Dzyaloshinskii-Moriya interaction. The strength of this interaction can be tuned by applying a gate electric field normal to the plane via varying the extent of the spatial inversion asymmetry [35–37]. Importantly, the coefficient $D(t) = D_0 + \Delta D(t)$ has two components, specifically, a steady component D_0 and a $\boldsymbol{E}(t)$-dependent component $\Delta D(t) = \kappa E_z(t)$ with κ being the coupling constant.

A time profile of the net magnetization $M_z(t) = (1/N) \sum_i \boldsymbol{m}_{zi}(t)$ and $\Delta M_z(t) = M_z(t) - M_z(0)$ are simulated by numerically solving the Landau–Lifshitz–Gilbert equation after application of a short pulse $H_z(t)$ or $E_z(t)$ with duration of $\Delta t = 1$. From its Fourier transform ΔM_z^ω, the dynamical magnetic and electromagnetic susceptibilities $\chi^{\rm mm}$ and $\chi^{\rm em}$ are calculated as,

$$\chi^{\rm mm}(\omega) = \frac{\Delta M_z^\omega}{\mu_0 H_{\rm pulse}}, \quad \chi^{\rm em}(\omega) = \sqrt{\frac{\mu_0}{\epsilon_0}} \frac{\Delta M_z^\omega}{E_{\rm pulse}} \tag{8.12}$$

Note that the magnetic susceptibility $\chi^{\rm mm}$ represents the response of the magnetizations to the microwave magnetic field $\boldsymbol{H}(t)$, whereas the electromagnetic susceptibility $\chi^{\rm em}$ represents the response of the magnetizations to the microwave electric field $\boldsymbol{E}(t)$.

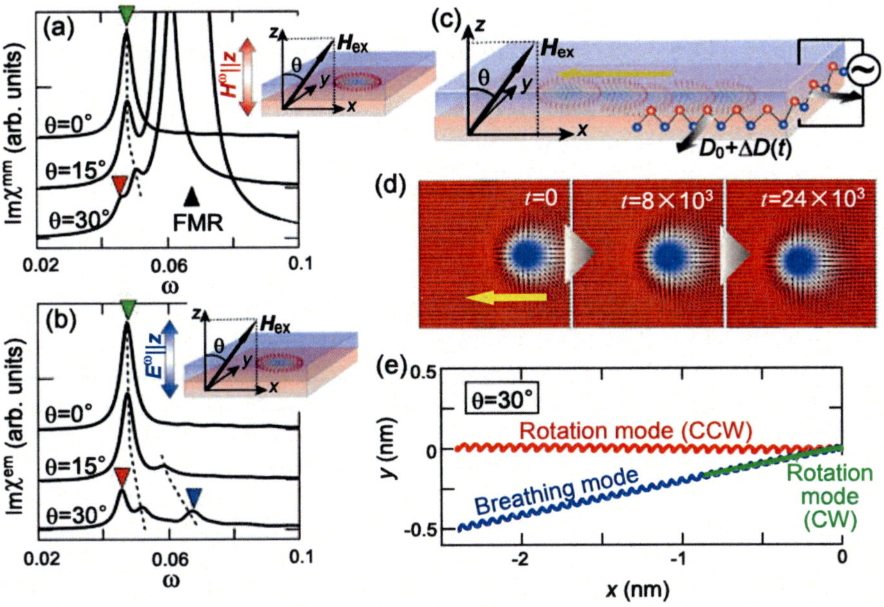

Fig. 8.6 **a, b** Microwave absorption spectra $\mathrm{Im}\chi^{\mathrm{mm}}(\omega)$ [$\mathrm{Im}\chi^{\mathrm{em}}(\omega)$] when a quasi-two-dimensional ferromagnet with a skyrmion defect under perpendicular ($\theta = 0°$) and tilted ($\theta \neq 0°$) $\boldsymbol{H}_{\mathrm{ext}}$ field is irradiated with an out-of-plane polarized microwave magnetic [electric] field $\boldsymbol{H}^{\omega} \| z$ [$\boldsymbol{E}^{\omega} \| z$]. The spectra $\mathrm{Im}\chi^{\mathrm{mm}}(\omega)$ for various angles θ indicate that the ferromagnetic resonance mode becomes active to the microwave magnetic field $\boldsymbol{H}^{\omega} \| z$ when the $\boldsymbol{H}_{\mathrm{ext}}$ field is tilted, and its intensity becomes rapidly enhanced as θ increases, although it is silent when the $\boldsymbol{H}_{\mathrm{ext}}$ field is perpendicular. In contrast, the spectra $\mathrm{Im}\chi^{\mathrm{em}}(\omega)$ indicate that the microwave electric field $\boldsymbol{E}^{\omega} \| z$ can selectively activate resonance modes of the skyrmion defect without exciting the background ferromagnetic magnetizations. **c** Schematic illustration of an experiment for the temporal variation of the interfacial Dzyaloshinskii-Moriya interaction by application of a microwave electric field to the magnetic bilayer system under a tilted $\boldsymbol{H}_{\mathrm{ext}}$ field. **d** Simulated snapshots of the translational motion of the driven skyrmion defect. **e** Trajectories of a driven skyrmion defect for three different spin modes activated by the microwave electric field. The simulations for **d** and **e** are performed for $J = 1$, $D_0/J = 0.27$, $\kappa E_z^{\omega} = 0.05 D_0$, $H_z = 0.057$, $\theta = 30°$, and $\alpha_{\mathrm{G}} = 0.04$ (Reproduced from [30].)

Figure 8.6a displays imaginary parts of the calculated dynamical magnetic susceptibilities $\mathrm{Im}\chi^{\mathrm{mm}}$ of a ferromagnetic system with a single skyrmion defect under application of the out-of-plane microwave magnetic field ($\boldsymbol{H}^{\omega} \| z$) for various values θ. The calculations are performed for a ferromagnetic system of 160×160 sites with periodic boundary conditions, in which one skyrmion exists as a defect. When $\theta = 0°$, only a single peak appears in the spectrum indicating that only the breathing mode of the skyrmion defect is activated without exciting the background ferromagnetic magnetizations. However, as θ increases, the intensity of the breathing mode decreases, and, alternatively, a large spectral peak due to the ferromagnetic resonance mode appears in the higher frequency regime. Namely, under the tilted $\boldsymbol{H}_{\mathrm{ext}}$ field, the microwave magnetic field cannot avoid activating the intense ferromagnetic resonance mode, which inevitably results in large loss of energy and significant rise of temperature.

To drive isolated skyrmions in the ferromagnetic phase efficiently, it is necessary to activate the skyrmions selectively avoiding the situation that the weak skyrmion resonance modes are masked by the intense ferromagnetic resonance mode. This seemingly impossible operation can be achieved by taking advantage of the microwave electric field. More concretely, when a microwave electric field is applied to a magnetic bilayer system or a thin-film specimen of chiral-lattice magnet fabricated on insulating substrate, the isolated skyrmions can be selectively activated through temporal variation of the extent of spatial inversion asymmetry and resulting temporal oscillation of the Dzyaloshinskii-Moriya interaction. The calculated microwave absorption spectra (the imaginary parts of the electromagnetic susceptibilities $\mathrm{Im}\chi^{\mathrm{em}}$) in this case are displayed in Fig. 8.6b, which clearly show selective activations of the resonance modes of the skyrmion defect.

When the spin-wave modes of isolated skyrmions are activated in a system shown in Fig. 8.6c with a microwave electric field, their translational motion can be driven as shown in Fig. 8.6d. Figure 8.6e displays trajectories of the skyrmion translational motion when a microwave electric field with each resonance frequency is applied for a certain duration (Note that the length scale is different between the horizontal and vertical axes). Interestingly, the breathing mode turns out to drive the skyrmion most quickly although the intensity of this mode is not so large.

Another interesting aspect to be mentioned is the trajectory is straight and exactly parallel to the x axis (direction toward which the $\boldsymbol{H}_{\mathrm{ex}}$ field is tilted) when the counterclockwise rotation mode is excited. It is recognized that one of the most promising forms of the skyrmion-based magnetic memories is the skyrmion race track memory based on the skyrmion motion in magnetic nanowires driven by a spin-polarized electric current, which can be regarded as the race track memory with its ferromagnetic domains being replaced with magnetic skyrmions. However, one of the critical problems that hinders its realization is the skyrmion Hall effect. The current-driven skyrmions have not only a velocity component parallel to the electric current but also that perpendicular to it. Due to this effect, the skyrmion cannot avoid colliding to the horizontal edges of devices, which results in absorption and pinning of the skyrmions. In contrast to the current-driven case, we can achieve the translational motion of skyrmions exactly parallel to the nanowire and thus can avoid this kind of problem when we drive them by microwave irradiation under a tilted $\boldsymbol{H}_{\mathrm{ext}}$ field. For the breathing mode and the clockwise rotation mode, the skyrmion moves in a direction slanted from the tilting direction of the $\boldsymbol{H}_{\mathrm{ex}}$ field. Even in these cases, we can achieve the skyrmion motion exactly parallel to the nanowire by tuning the tilting direction of $\boldsymbol{H}_{\mathrm{ex}}$. It is always possible to realize the straight and parallel skyrmion motion in a nanowire without collision to the edges in this way because the $\boldsymbol{H}_{\mathrm{ex}}$ field can be oriented in an arbitrary direction in contrast to an electric current flowing always along the nanowire.

8.6 Electrically Driven Spin Torque and Dynamical Dzyaloshinskii-Moriya Interaction

In the above section, we argued that the application of microwave electric field can drives translational motion of isolated skyrmions embedded in a ferromagnetic background of magnetic bilayer system through inducing the temporal variation of the interfacial Dzyaloshinskii-Moriya interaction. In this section, we discuss a theoretical formulation of this electrically induced time-dependent Dzyaloshinskii-Moriya interaction [35]. From a theoretical perspective, we demonstrate that the spin torques can be exerted into magnetic bilayer systems via the Rashba spin-orbit interaction by application of an AC electric voltage. The exerted spin torques turn out to resemble the well-known electric-current-induced torques, i.e., the spin-transfer torque and the nonadiabatic torque, providing similar controllability of magnetism with microwave electric fields. The spin torques also turn out to work as an interfacial Dzyaloshinskii-Moriya interaction, which contains both steady and oscillating components and enables us to create and activate noncollinear magnetism like magnetic skyrmions by application of a microwave electric field.

In the magnetic bilayer system with broken spatial inversion symmetry, the Rashba spin-orbit interaction becomes active. This interaction works as an effective magnetic field acting on the conduction-electron spins, through mediating mutual coupling between spins and orbital momenta of the electrons [38, 39]. Importantly, strength and direction of the effective magnetic field are determined by the momentum of the electron. Therefore, the Rashba spin-orbit interaction can induce nontrivial spin torques acting on the magnetizations through controlling the spin polarizations of the conduction electrons which couple to the magnetizations via the exchange interaction. The strength of the Rashba spin-orbit interaction can be tuned by application of a gate electric voltage normal to the interfacial plane [40], through modulating the extent of the spatial inversion asymmetry. This suggests that an AC gate voltage produces nontrivial Rashba-mediated dynamical spin torques.

We consider a magnetic bilayer system with a ferromagnet/heavy-metal interface (Fig. 8.7), which is fabricated on an insulating substrate. The insulating substrate prevents the electric-current flow and thus enhances the effects of gate electric voltage acting on the ferromagnet/heavy-metal interface. The Hamiltonian for this system has four terms as

$$H = H_{\rm K} + H_{\rm R} + H_{\rm ex} + H_{\rm imp} \tag{8.13}$$

with

$$H_{\rm K} = \frac{1}{2m_{\rm e}} \int d^2r \, |\boldsymbol{p}\psi(\boldsymbol{r},t)|^2 - E_{\rm F} \int d^2r \, \psi^\dagger(\boldsymbol{r},t)\psi(\boldsymbol{r},t), \tag{8.14}$$

$$H_{\rm R} = -\frac{\alpha_{\rm R}(t)}{\hbar} \int d^2r \, \psi^\dagger(\boldsymbol{r},t)(\boldsymbol{p}\times\boldsymbol{\sigma})_z\psi(\boldsymbol{r},t), \tag{8.15}$$

$$H_{\rm ex} = J_{\rm ex} \int d^2r \, \boldsymbol{m}(\boldsymbol{r}) \cdot \psi^\dagger(\boldsymbol{r},t)\boldsymbol{\sigma}\psi(\boldsymbol{r},t), \tag{8.16}$$

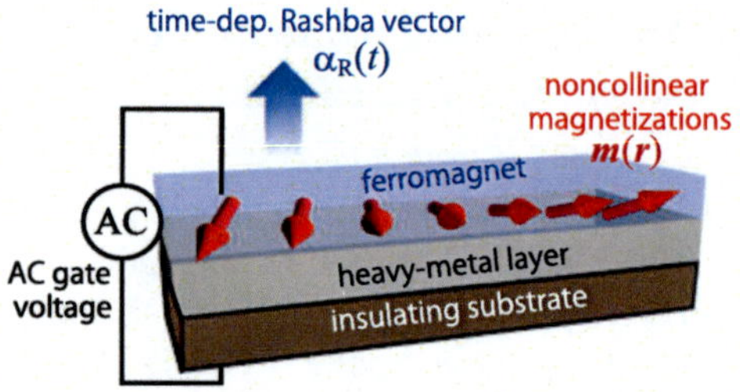

Fig. 8.7 Schematic illustration of a magnetic bilayer system in which local magnetizations $\boldsymbol{m}(\boldsymbol{r})$ couple to a conduction-electron system with the time-dependent Rashba spin-orbit interaction. The strength of the Rashba interaction $\alpha_R(t)$ is temporally varying under application of an AC gate electric voltage. The insulating substrate prevents electric-current flows to enhance the effects of electric voltage acting on the interface hosting the Rashba spin-orbit interaction (Reproduced from [35].)

$$H_{\rm imp} = \int d^2r\, v_{\rm imp}(\boldsymbol{r})\psi^\dagger(\boldsymbol{r},t)\psi(\boldsymbol{r},t), \tag{8.17}$$

where m_e and $\boldsymbol{p}$ denote, respectively, the mass and momentum of a conduction electron, E_F the Fermi energy, $\boldsymbol{\sigma}$ the Pauli matrices, and $\psi^\dagger$ (ψ) the creation (annihilation) operator of a conduction electron. The term H_K represents the kinetic energies of the conduction electrons, while the term H_R describes the time-varying Rashba spin-orbit interaction where $\alpha_R(t)$ is the time-dependent coupling coefficient. The term H_{ex} represents the exchange interaction between the conduction-electron spins and the local magnetization where J_{ex} and $\boldsymbol{m}$ are the coupling constant and the normalized local magnetization vector, respectively. The term $H_{\rm imp}$ depicts the scattering potentials from spatially distributed nonmagnetic impurities, which determine the relaxation time τ of the conduction electrons.

The impurity potential is given by,

$$v_{\rm imp}(\boldsymbol{r}) = u_{\rm imp}\sum_i \delta(\boldsymbol{r}-\boldsymbol{r}_i) \tag{8.18}$$

where $u_{\rm imp}$ is the strength of the impurity scattering, $\boldsymbol{r}_i$ denotes positions of the impurities, and $\delta(\boldsymbol{r})$ is the Dirac delta function. Taking averages over the impurity positions as

$$\overline{v_{\rm imp}(\boldsymbol{r})} = 0, \quad \overline{v_{\rm imp}(\boldsymbol{r})v_{\rm imp}(\boldsymbol{r}')} = n_{\rm imp}u_{\rm imp}^2\delta(\boldsymbol{r}-\boldsymbol{r}'), \tag{8.19}$$

the relaxation time of the conduction electrons is given by

$$\tau = \hbar/2\pi\nu_e n_{\rm imp}u_{\rm imp}^2 \tag{8.20}$$

in the first Born approximation. Here, $n_{\rm imp}$ denotes the concentration of impurities and $\nu_e = m_e/2\pi\hbar^2$ is the density of state.

The spin torque induced by the conduction-electron spins via the exchange interaction is defined as

$$\boldsymbol{T} = \frac{J_{\rm ex}a^2}{\hbar}\boldsymbol{m}\times\boldsymbol{s}, \tag{8.21}$$

where a is the lattice constant, $\boldsymbol{s} = \langle\psi^\dagger\boldsymbol{\sigma}\psi\rangle$ is the conduction-electron spin density, and the brackets denote the quantum expectation value. The analytical formula of the spin torque is given in the form,

$$\begin{aligned}\boldsymbol{T} &= \boldsymbol{T}_1 + \boldsymbol{T}_2 + \boldsymbol{T}_3 \\ &= -\frac{a}{\hbar}D_1(\boldsymbol{m}\times\nabla)_z\boldsymbol{m} + \frac{a}{\hbar}D_2(\boldsymbol{m}\times\nabla)_z\boldsymbol{m} \\ &\quad -\frac{a}{\hbar}\beta_{\rm R}D_2\boldsymbol{m}\times\Big[(\boldsymbol{m}\times\nabla)_z\boldsymbol{m}\Big],\end{aligned} \tag{8.22}$$

where

$$D_1(t) = \frac{\hbar\nu_{\rm e}a}{2\pi\tau}\left[\frac{E_{\rm F}}{J_{\rm ex}}\ln\left(\frac{E_{\rm F}+J_{\rm ex}}{E_{\rm F}-J_{\rm ex}}\right)-2\right]\alpha_{\rm R}(t), \tag{8.23}$$

$$D_2(t) = \nu_{\rm e}aE_{\rm F}\tau\frac{J_{\rm ex}^2(J_{\rm ex}^2-\eta^2)}{(J_{\rm ex}^2+\eta^2)^2}\frac{d\alpha_{\rm R}(t)}{dt}, \tag{8.24}$$

$$\beta_{\rm R} = \frac{2J_{\rm ex}\eta}{J_{\rm ex}^2-\eta^2}, \tag{8.25}$$

with $\eta = \hbar/2\tau$.

The above formula is derived from perturbation calculations based on some assumptions summarized below:

- Metallic bilayer systems with $J_{\rm ex} < E_{\rm F}$,
- Slowly varying magnetizations with $q \ll k_{\rm F}$,
- Weak magnitudes $\alpha_{\rm R}$ with $\alpha_{\rm R}k_{\rm F} \ll E_{\rm F}$,
- Low frequencies Ω for $\alpha_{\rm R}$ with $\hbar\Omega \ll E_{\rm F}$,

where q is the wavenumber of local magnetization and $k_{\rm F}$ is the Fermi wavenumber. The coefficients $D_1(t)$, $D_2(t)$, and $\beta_{\rm R}$ are well defined when the relaxation time is sufficiently long to satisfy the conditions $E_{\rm F} \gg \hbar/\tau$, $J_{\rm e} \gg \hbar/\tau$ and $E_{\rm F} - J_{\rm ex} \gg \hbar/\tau$. Note that D_1 vanishes in the clean limit with $\tau \to \infty$ for the present quasi-two-dimensional metallic system with $J_{\rm ex} < E_{\rm F}$, whereas it is known to survive in the three-dimensional systems or in the half-metallic systems with $J_{\rm ex} > E_{\rm F}$ even in the clean limit. For details of the derivation, see [35].

The first two contributions in (8.22), $\boldsymbol{T}_1 + \boldsymbol{T}_2$, describe an effective Dzyaloshinskii-Moriya interaction, which is given in the continuum form as

$$\mathcal{H}_{\rm DMI} = \frac{D_1 - D_2}{a}\epsilon_{\alpha\beta z}\int d^2r\,(\boldsymbol{m}\times\nabla_\alpha\boldsymbol{m})^\beta. \tag{8.26}$$

Table 8.2 Typical material parameters for magnetic bilayer systems

	Metallic systems	Semiconducting systems
Lattice constant a	5 Å	5 Å
Fermi energy E_F	4 eV	10 meV
Fermi wavenumber k_F	1 $Å^{-1}$	0.01 $Å^{-1}$
Exchange int. J_{ex}/E_F	0.25	0.5
Relaxation time τ	10^{-14} s	10^{-12} s
Rashba parameter 1: α_0	2 eV·Å	0.07 eV·Å
Rashba parameter 2: α_{ext}/α_0	0.1	0.1
Frequency $\Omega/2\pi$	1 GHz	1 GHz

(Note that $\boldsymbol{m} \times (a^2/\hbar)\delta\mathcal{H}_{DMI}/\delta\boldsymbol{m}$ leads to $\boldsymbol{T}_1 + \boldsymbol{T}_2$). The contribution D_1, which is proportional to α_R, appears even with a steady Rashba spin-orbit interaction. In contrast, the contribution D_2, which is proportional to $\partial_t\alpha_R$, appears only in the presence of a time-dependent Rashba spin-orbit interaction. This interfacial Dzyaloshinskii-Moriya interaction is expected to be tuned by applying an electric gate voltage. More interestingly, application of an AC voltage is expected to induce an oscillating component of the Dzyaloshinskii-Moriya interaction. The Rashba parameter $\alpha_R(t)$ in the driven Rashba electron system is composed of both steady and time-dependent components as $\alpha_R(t) = \alpha_0 + \alpha_{ext}(t)$ with $\alpha_{ext}(t) = \alpha_{ext}\sin(\Omega t)$. Using typical material parameters summarized in Table 8.2 [40–42], the strength of this Rashba-induced Dzyaloshinskii-Moriya interaction is roughly estimated as $D_1 \sim 0.1$ meV and $D_2 \sim 5 \times 10^{-3}$ meV for metallic bilayer systems, while $D_1 \sim 6 \times 10^{-6}$ meV and $D_2 \sim 2 \times 10^{-6}$ meV for semiconducting bilayer systems. The strength of the Rashba-induced Dzyaloshinskii-Moriya interaction is relatively strong in metallic bilayer systems, whereas it is rather weak in the semiconducting bilayer systems.

It should be mentioned that the magnitude of D_2 being proportional to $\partial_t\alpha_{ext}(t)$ can be tuned by varying the amplitude and frequency of the AC gate voltage, although it is usually small as compared to the magnitude of D_1. The ratio D_2/D_1 is approximately given by $\Omega\varepsilon_F\tau^2/2\pi\hbar$, which takes $\sim 10^{-4}$ (10^{-2}) for metallic (semiconducting) bilayer systems when a typical frequency of $\Omega = 1$ GHz is assumed. The ratio D_2/D_1 tends to be larger for the semiconducting system, whereas the absolute value of D_2 tends to be larger for the metallic system. An appropriate system should be chosen depending on the purpose.

The last two terms in (8.22), both of which are proportional to D_2, can be rewritten as

$$\boldsymbol{T}_2 + \boldsymbol{T}_3 \propto (\boldsymbol{j}_s \cdot \nabla)\boldsymbol{m} - \beta_R \boldsymbol{m} \times (\boldsymbol{j}_s \cdot \nabla)\boldsymbol{m}. \tag{8.27}$$

where

$$\boldsymbol{j}_s \equiv (e/\hbar a) D_2 \boldsymbol{z} \times \boldsymbol{m}. \tag{8.28}$$

When the time-dependent Rashba spin-orbit interaction $\alpha_R(t)$ is induced by the AC gate voltage, the vector quantity $\boldsymbol{j}_s \propto \partial_t \alpha_R(t) z \times \boldsymbol{m}$ gives rise to AC spin torques. In the clean limit with a long relaxation time ($\hbar/J_{ex}\tau \ll 1$), the coefficient β_R in the second term is reduced to $\hbar/J_{ex}\tau$. Interestingly, the vector quantity $\boldsymbol{j}_s$ defined here can be regarded as a fictitious spin current because the above expressions of $\boldsymbol{T}_2$ and $\boldsymbol{T}_3$ have equivalent forms with those of the spin-transfer torque and the nonadiabatic torque in the presence of the real spin current j_s, respectively. It should be noted, however, the present relaxation time τ corresponds to a different time scale. Specifically, the relaxation time is governed by the coherence of the conduction-electron momenta in the present case, while in the current-induced case, it is governed by that of the conduction-electron spins.

Assuming the material parameters for the metallic bilayer systems in Table 8.2, the values of $j_s = (e/\hbar a)D_2$ and β_R are evaluated as ~2 A/m and ~0.07, respectively. These values are large enough to induce the magnetization dynamics. Numerical simulations in [35] indeed demonstrated that not only a skyrmion crystal with hexagonally packed magnetic skyrmions but also isolated skyrmions embedded in a ferromagnetic background can be excited resonantly through temporal variation of the Dzyaloshinskii-Moriya interaction achieved by application of a microwave electric voltage. This technique provides a means to drive magnetic skyrmions electrically with a low energy consumption. Recently, a lot of ferromagnet/heavy-metal bilayer systems hosting magnetic skyrmions have been reported. The above theoretical proposals are anticipated to be realized by future experiments on these magnetic bilayer systems.

8.7 Microwave-Induced DC Spin-Motive Force

We next discuss a proposed method to generate DC electric voltages by exploiting the spin-wave excitations of magnetic skyrmions under a tilted $\boldsymbol{H}_{ext}$ field [31]. It is well known that injection of spin-polarized electric currents can drive noncollinear skyrmion textures in metallic magnets via the spin-transfer torque mechanism, whereas the noncollinear skyrmion magnetizations inversely affect transport properties of conduction electrons as exemplified by the topological Hall effect. The spin-driven electromotive force (i.e., an emergent electric field induced by magnetization dynamics) is another important example of the latter kinds of phenomena [43, 44]. It was proposed theoretically that spatially modulated magnetic textures such as magnetic skyrmions, magnetic helices and ferromagnetic domain walls produce effective vector potential acting on the conduction electrons via exchange coupling with the conduction-electron spins, which is called non-Abelian gauge field. When these magnetic textures are temporally varied by an applied time-periodic field such as microwave electromagnetic fields, this effective vector potential also changes temporally. This temporal variation of the vector potential gives rise to an effective electric field which acts on the conduction electrons. The electric motive force due to this effective electric field is referred to as the spin-motive force, which is one of

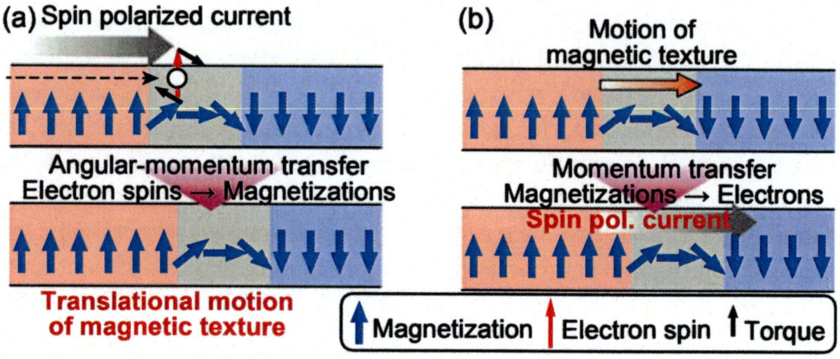

Fig. 8.8 Relationship between the spin-transfer torque mechanism and the spin-motive force. **a** Schematic illustrations of the spin-transfer torque mechanism. Translational motion of a magnetic texture is driven by the angular-momentum transfer from conduction-electron spins of injected spin-polarized currents to the noncollinear magnetizations. **b** Schematic illustrations of the spin-motive force. Effective electromotive force acting on the conduction electrons is induced by the momentum transfer from driven noncollinear magnetic texture to the conduction electrons via exchange coupling, resulting in the generation of electric currents (Reproduced from [31].)

the important subjects of the recent spintronics research. This phenomenon can be interpreted as the inverse effect of the spin-transfer torque mechanism (see Fig. 8.8).

An expression of this spin-induced effective electric field is given by,

$$E_\mu(t) = \frac{\hbar}{2e}\boldsymbol{m} \cdot (\partial_\mu \boldsymbol{m} \times \partial_t \boldsymbol{m}) \qquad (\mu = x, y, z), \tag{8.29}$$

where $\boldsymbol{m}(\boldsymbol{r}, t)$ is the normalized classical magnetization vector. This formula explicitly indicates that both temporal and spatial variations of magnetizations are required to generate the spin-motive force. Several experimental reports have discussed the generation and observation of the spin-motive force in ferromagnetic domain walls and magnetic vortices activated by microwave fields [45, 46].

There has been a theoretical proposal that the microwave activation of skyrmion crystal under a perpendicular $\boldsymbol{H}_{\text{ex}}$ field gives rise to an enhanced spin-motive force [47, 48]. However, the spin-motive force available in this way is a pure AC voltage with an average of zero. In fact, there have been several theoretical proposals and experimental demonstrations of the generation of the AC spin-motive force. However, a method to generate a stationary DC spin-motive force has long been missing. One possible way to obtain a DC voltage is to use an AC-DC transducer to convert the AC voltage to a DC voltage. But it is difficult to fabricate such a precise device in nanometric systems. Moreover, significant reduction of the voltage cannot be avoided in the conversion process, which can be a critical problem because the spin voltage is originally very tiny. Therefore, it is highly desired to establish a simple technique to generate a DC spin voltage for spintronics applications.

To solve this problem, it was recently proposed theoretically that an oscillating spin voltage with a large DC component can be generated by exciting the microwave-active spin-wave modes of skyrmion crystal on a quasi-two-dimensional thin-plate magnet under a tilted $\boldsymbol{H}_{\text{ex}}$ field in [31]. In this study, micromagnetic simulations based on the Landau–Lifshitz–Gilbert equation were performed to trace the magnetization dynamics of a skyrmion crystal activated by a microwave magnetic field $\boldsymbol{H}^{\omega}$. Using the simulated data of the magnetization dynamics, the spatiotemporal profiles of the spin-motive force $\boldsymbol{E}(\boldsymbol{r}, t)$ were numerically calculated. For the numerical calculations, it is convenient to rewrite (8.29) in discretized form as

$$E_{\mu,i}(t) = \frac{\hbar}{2e}\boldsymbol{m}_i(t) \cdot \left[\frac{\boldsymbol{m}_{i+\hat{\mu}}(t) - \boldsymbol{m}_{i-\hat{\mu}}(t)}{2a} \times \frac{\boldsymbol{m}_i(t+\Delta t) - \boldsymbol{m}_i(t-\Delta t)}{2\Delta t}\right], \tag{8.30}$$

where $\mu = x, y$ and a(=5 Å) is the lattice constant. Time profiles of the spin voltage were calculated by numerically solving the Poisson equation using the spatial distribution data of the spin-motive force $\boldsymbol{E}(\boldsymbol{r}, t)$ at each moment t.

The spin-motive force that contains a large DC component can indeed be generated by activation of the spin-wave modes of magnetic skyrmions under a tilted $\boldsymbol{H}_{\text{ext}}$ field. Figure 8.9a–d show time profiles of the spin-motive force simulated for a 50 nm × 50 nm squared system which contains a skyrmion crystal composed of twelve skyrmions. When the $\boldsymbol{H}_{\text{ex}}$ field is perpendicular, the generated spin-motive force is of pure AC for the counterclockwise rotation mode (Fig. 8.9a) or constantly zero for the breathing mode (Fig. 8.9b). On the contrary, a spin-motive force with a large DC component of 0.5-1 μV is generated when the $\boldsymbol{H}_{\text{ex}}$ field is tilted by $\theta = 30°$ (Fig. 8.9c, d).

The simulated time profiles of spin voltages turn out to be well fitted by an approximate formula of the forced oscillation with a damping.

$$V_\mu = V_\mu^{\text{DC}} + V_\mu^{\text{AC}}(1 - e^{-t/\tau}) \sin \omega t, \tag{8.31}$$

with $\mu = x, y$. Here V_μ^{DC}, V_μ^{AC}, $\omega(= 2\pi f)$, and τ are the DC component, the AC amplitude, the angular frequency, and the decay rate of the induced temporally oscillating spin voltage, respectively. Fig. 8.9e–f show the microwave-frequency dependence of the DC component V_μ^{DC} ($\mu = x, y$) for different microwave polarizations, which are evaluated by the fitting. Specifically, Fig. 8.9e shows V_x^{DC} for $\boldsymbol{H}^{\omega} \| \boldsymbol{x}, \boldsymbol{y}$, Fig. 8.9f shows V_y^{DC} for $\boldsymbol{H}^{\omega} \| \boldsymbol{x}, \boldsymbol{y}$, Fig. 8.9g shows V_x^{DC} for $\boldsymbol{H}^{\omega} \| z$, and Fig. 8.9h shows V_y^{DC} for $\boldsymbol{H}^{\omega} \| z$. The results show that the DC component is enhanced significantly when the frequency of the microwave is tuned to an eigenfrequency of the spin-wave modes, which converts the microwave power to a DC voltage with high efficiency. The results also show that the sign of the DC voltage depends on the excited spin-wave mode and the microwave polarization, which indicates that the sign of the voltage can be switched by tuning the microwave frequency or the microwave polarization. Note that a large DC voltage is obtained for the counterclockwise rotation mode activated by $\boldsymbol{H}^{\omega} \| \boldsymbol{x}, \boldsymbol{y}$ and for the breathing mode activated by $\boldsymbol{H}^{\omega} \| z$,

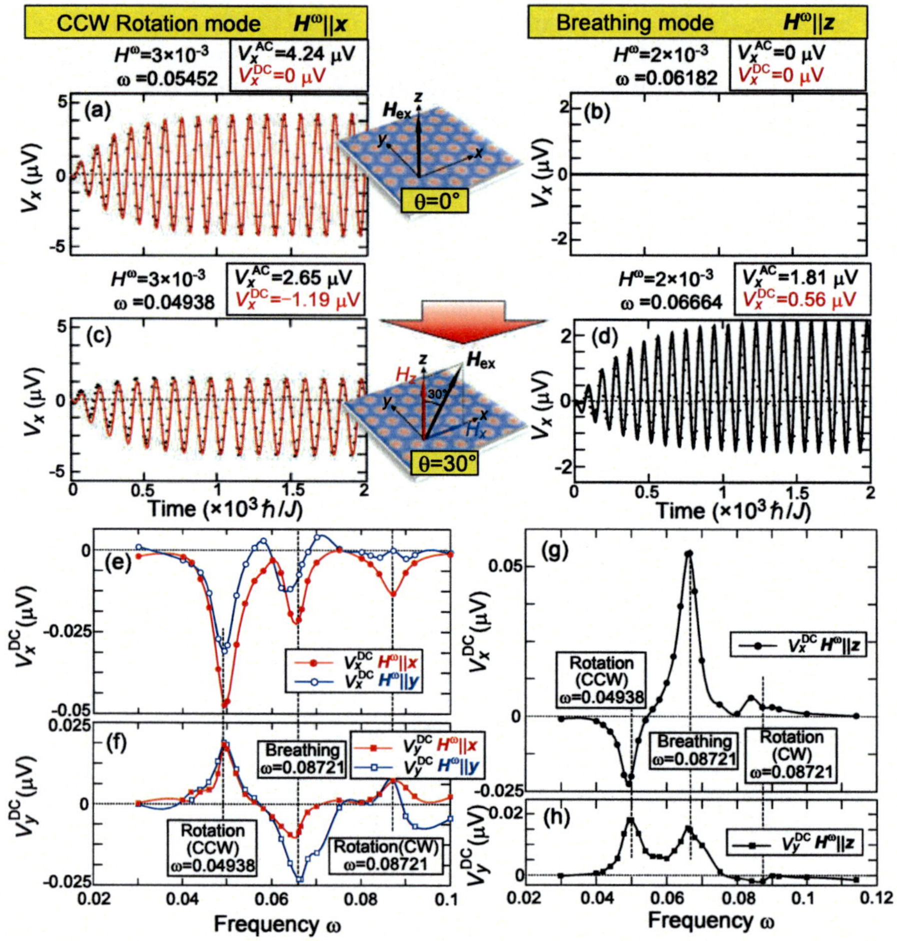

Fig. 8.9 **a**, **b** Calculated time profiles of spin voltages induced by **a** the counterclockwise (CCW) rotation mode activated by a microwave field $\boldsymbol{H}^\omega \| \boldsymbol{x}$ and **b** the breathing mode activated by a microwave field $\boldsymbol{H}^\omega \| z$ of a skyrmion crystal confined in a quasi-two-dimensional magnet under a perpendicular $\boldsymbol{H}_{\rm ex}$ field with $\theta = 0°$. **c**, **d** Those under a tilted $\boldsymbol{H}_{\rm ex}$ field with $\theta = 30°$. The amplitude H^ω and frequency ω of microwave are presented where $\omega(=2\pi f)$ is fixed at the eigenfrequency of the spin-wave mode. The DC component $V_x^{\rm DC}$ evaluated by fitting the simulated time profiles are shown as well. **e–h** Calculated microwave-frequency dependence of the DC component $V_\mu^{\rm DC}$ ($\mu = x, y$) of the spin voltage under a tilted $\boldsymbol{H}_{\rm ex}$ field with $\theta = 30°$, i.e., **e** $V_x^{\rm DC}$ for $\boldsymbol{H}^\omega \| \boldsymbol{x}, \boldsymbol{y}$, (f) $V_y^{\rm DC}$ for $\boldsymbol{H}^\omega \| \boldsymbol{x}, \boldsymbol{y}$, (g) $V_x^{\rm DC}$ for $\boldsymbol{H}^\omega \| z$, (h) $V_y^{\rm DC}$ for $\boldsymbol{H}^\omega \| z$. The microwave amplitude is fixed at $H^\omega = 0.6 \times 10^{-3}$ for **e–h**. The parameters are fixed at $J = 1$, $D = 0.27$, $H_z = 0.036$, and $\alpha = 0.04$, whereas the system of $N = 96 \times 111$ sites with a skyrmion crystal composed of twelve magnetic skyrmions is used for all the simulations (Reproduced from [31].)

which indicates that these sets of microwave polarization and the spin-wave mode are suitable for efficient conversion of microwaves to a DC electric voltage.

It should be also noted that an advantage of using a skyrmion crystal is that the arrays of numerous magnetic skyrmions in a skyrmion crystal behaves as batteries connected in series, which give a large electric voltage as a sum of each contribution. We expect several orders of magnitude larger electric voltage by using a larger-sized device, which offers a promising technique of an efficient conversion of microwaves to DC electric voltages.

8.8 Concluding Remarks

To summarize this chapter, we have discussed several microwave-related phenomena that magnetic skyrmions confined in a quasi-two-dimensional magnetic specimen are expected to show under a tilted external magnetic field by particularly taking the microwave-driven translational motion and the microwave-induced DC spin-motive forces as topics. The researches on the dynamical phenomena and device functions of magnetic skyrmions are now extended over a wide area. The researches on magnetic skyrmions which bring technical applications into a scope can be classified roughly into two categories. One category of the researches aims for application to magnetic memories and storage devices and seek the elementary techniques to write, erase, read, and drive magnetic skyrmions in controlled ways [3, 49]. Another category of the researches focuses on the microwave-device functions of magnetic skyrmions based on their peculiar spin-wave modes such as microwave generation, microwave detection, microwave diode, and magnonic-crystal devices [4, 6, 7]. In addition, magnetic skyrmions nowadays became to attract novel research interest as promising building blocks of logic-gate devices [50] and brain-inspired computing devices such as reservoir computing [51–53] and neuromorphic computing [54, 55]. It is yet to be clarified what kinds of technical applications can be expected for the phenomena discussed in this chapter. However, we naively anticipate that the magnetic skyrmions under a tilted $\boldsymbol{H}_{\mathrm{ext}}$ field are still hiding a lot of intriguing phenomena, useful device functions, and novel physics, which might be clarified in future studies.

References

1. S. Seki, M. Mochizuki, *Skyrmions in Magnetic Materials* (Springer, Berlin, 2015)
2. N. Nagaosa, Y. Tokura, Nat. Nanotechnol. **8**, 899 (2013)
3. A. Fert et al., Nat. Nanotechnol. **8**, 152 (2013)
4. G. Finocchio et al., J. Phys. D: Appl. Phys. **49**, 423001 (2016)
5. K. Everschor-Sitte et al., J. Appl. Phys. **124**, 240901 (2018)
6. M. Garst et al., J. Phys. D: Appl. Phys. **50**, 293002 (2017)
7. M. Mochizuki, S. Seki, J. Phys.: Condens. Matter **27**, 503001 (2015)
8. T.H.R. Skyrme, Proc. R. Soc. A **260**, 127 (1961)

9. T.H.R. Skyrme, Nucl. Phys. **31**, 556 (1962)
10. A.N. Bogdanov, D.A. Yablonskii, Sov. Phys. JETP **68**, 101 (1989)
11. A. Bogdanov, A. Hubert, J. Magn. Magn. Mat. **138**, 255 (1994)
12. S. Mühlbauer et al., Science **323**, 915 (2009)
13. X.Z. Yu et al., Nature **465**, 901 (2010)
14. X.Z. Yu et al., Nat. Mater. **10**, 106 (2011)
15. S. Seki et al., Science **336**, 198 (2012)
16. Y. Tokunaga et al., Nat. Commun. **6**, 7638 (2015)
17. I. Kezsmarki et al., Nat. Mater. **14**, 1116 (2012)
18. S. Bordács et al., Sci. Rep. **7**, 7584 (2017)
19. Y. Fujima et al., Phys. Rev. B **95**, 180410(R) (2017)
20. A. Nayak et al., Nature **548**, 561 (2017)
21. T. Kurumaji et al., Phys. Rev. Lett. **119**, 237201 (2017)
22. C. Moreau-Luchaire et al., Nat. Nanotechnol. **11**, 444 (2016)
23. G. Chen et al., Appl. Phys. Lett. **106**, 242404 (2015)
24. O. Boulle et al., Nat. Nanotechnol. **11**, 449 (2016)
25. I.E. Dzyaloshinskii, Sov. Phys. JETP **5**, 1259 (1957)
26. T. Moriya, Phys. Rev. **120**, 91 (1960)
27. M. Mochizuki, Phys. Rev. Lett. **108**, 017601 (2012)
28. W. Wang et al., Phys. Rev. B **92**, 020403(R) (2015)
29. M. Ikka et al., Phys. Rev. B **98**, 184428 (2018)
30. A. Takeuchi, M. Mochizuki, Appl. Phys. Lett. **113**, 072404 (2018)
31. T. Koide et al., Phys. Rev. B **100**, 014408 (2019)
32. P. Bak, M.H. Jensen, J. Phys. C **13**, L881 (1980)
33. S.D. Yi et al., Phys. Rev. B **80**, 054416 (2009)
34. A. Tonomura et al., Nano Lett. **12**, 1673 (2012)
35. A. Takeuchi et al., Sci. Rep. **9**, 9528 (2019)
36. K. Nawaoka et al., Appl. Phys. Exp. **8**, 063004 (2015)
37. T. Srivastava et al., Nano Lett. **18**, 4871 (2018)
38. I.E. Rashba, Sov. Phys. Solid State **2**, 1109 (1960)
39. A. Manchon et al., Nat. Mater. **14**, 871 (2015)
40. J. Nitta et al., Phys. Rev. Lett. **78**, 1335 (1997)
41. C.R. Ast et al., Phys. Rev. Lett. **98**, 186807 (2007)
42. T. Nakagawa et al., Phys. Rev. B **75**, 155409 (2007)
43. G.E. Volovik, J. Phys. C **20**, L83 (1987)
44. S.E. Barnes, S. Maekawa, Phys. Rev. Lett. **98**, 246601 (2007)
45. S.A. Yang et al., Phys. Rev. Lett. **102**, 067201 (2009)
46. K. Tanabe et al., Nat. Commun. **3**, 845 (2012)
47. J. Ohe, Y. Shimada, Appl. Phys. Lett. **103**, 242403 (2013)
48. Y. Shimada, J.I. Ohe, Phys. Rev. B **91**, 174437 (2015)
49. W. Koshibae et al., Jpn. J. Appl. Phys. **54**, 053001 (2015)
50. X. Zhang et al., Sci. Rep. **5**, 9400 (2015)
51. D. Pinna et al., Phys. Rev. Appl. **9**, 064018 (2018)
52. D. Prychynenko et al., Phys. Rev. Appl. **9**, 014034 (2018)
53. G. Bourianoff et al., AIP Adv. **8**, 055602 (2018)
54. Z. He, D. Fan, arXiv:1705.02995
55. X. Chen et al., Nanoscale **10**, 6139 (2018)

Chapter 9
Symmetry Approach to Chiral Optomagnonics in Antiferromagnetic Insulators

Igor Proskurin and Robert L. Stamps

Abstract We discuss several aspects of chiral optomagnonics in antiferromagnetic insulators by considering common symmetries between the electromagnetic field and spin excitations. This approach allows us to look at optical and magnetic materials from similar perspectives, and discuss useful analogies between them. We show that spin waves in collinear antiferromagnets and the electromagnetic field in vacuum are both invariant under the same eight-dimensional algebra of symmetry transformations. By such analogy, we can extend the concept of optical chirality to antiferromagnetic insulators, and demonstrate that the spin-wave dynamics in these materials in the presence of a spin current is similar to that of the light inside chiral metamaterials. Photo-excitation of magnonic spin currents is also discussed from the symmetry point of view. It is demonstrated that a direct magnonic spin photocurrent can be exited by circularly polarized light, which can be considered as a magnonic analogue of the photogalvanic effect. We also note that the *Zitterbewegung* process should appear and may play a role in photo-excitation processes.

9.1 Introduction

Modern spintronics is now a well-developed area that aims at bringing new functionality to conventional electronics by making use of the spin degrees of freedom [1], which may help to overcome looming saturation of Moore's Law [2]. There are a number of different trends in the development of the spintronics today. Among different materials, antiferromagnets play an important role, which brings us to the

I. Proskurin (✉) · R. L. Stamps
Department of Physics & Astronomy, University of Manitoba, Winnipeg R3T 2N2, Canada
e-mail: Igor.Proskurin@umanitoba.ca

R. L. Stamps
e-mail: Robert.Stamps@umanitoba.ca

I. Proskurin
Institute of Natural Sciences and Mathematics, Ural Federal University,
Ekaterinburg 620002, Russia

E. Kamenetskii (ed.), *Chirality, Magnetism and Magnetoelectricity*,
Topics in Applied Physics 138,
https://doi.org/10.1007/978-3-030-62844-4_9

field of antiferromagnetic spintronics [3, 4]. Their abundance in Nature and zero net magnetization make antiferromagnets potentially useful for applications, while the existence of two or more magnetic sublattices allows one to explore various topological effects [4]. The focus on optical manipulation of the spin states in magnetic insulators constitutes the scope of the optospintronics [5]. A prominent direction in optospintronics is related to the application of microwave cavity resonators [6], which has already seen a rapid development during the last several years [7].

Being interdisciplinary, spintronics in general, and optomagnonics in particular, can benefit by looking at the concept of chirality. Chirality or handedness, which according to the original definition given by Lord Kelvin in his Baltimore Lectures is related to the lack of symmetry between an object and its mirror image [8]. It is a universal phenomenon that has proved its significance in various scientific areas from high-energy physics to life sciences and soft matter [9]. Kelvin's definition, which is purely geometric, was generalized later to accommodate dynamical phenomena by Barron [10]. Thus, according to Barron's definition, one should distinguish between *true* and *false* chiralities. The former is to be found in the systems that break inversion symmetry, but at the same time are invariant under a time-reversal transformation combined with any proper rotation, while the latter is characterized by breaking time-reversal and inversion symmetries simultaneously [11].

How can the concept of chirality be useful for the development of optospintronics? A general observation is that the goal of the spintronics is manipulation and transformation of pure spin currents, and spin currents are chiral. Indeed, in agreement with the definition of true chirality, a flow of angular momentum reverses its sign under spatial inversion, while it remains invariant under the time reversal transformation, which reverses both velocities and spins. Thus, from the symmetry point of view, pure spin currents are in the same category as, for example, natural optical activity and circular dichroism in optics. This argument also suggests that materials with structural chirality may have unique properties for hosting and transferring spin currents that makes them interesting for applications, which is reflected in the rapid development of molecular spintronics [12, 13] and related topics such as chiral spin selectivity [14].

Another observation helpful to establish a link between optics and spintronics is that not only geometric structures but also physical fields can be characterized by chirality. Chirality density of the electromagnetic field, for example, has been known for a long time. Lipkin first noticed that the Maxwell's equations in vacuum have a hidden conservation law for a chiral density, which he dubbed *zilch* due to the lack of clear physical meaning of this quantity at that time [15]. Later, it was demonstrated that this conservation law is closely related to electromagnetic duality [16, 17]. This eventually led to the formulation of the *nongeometric* symmetries of the Maxwell's equations [18], i. e. the symmetries, which are not reduced to space-time transformations. For several decades, the formal properties of optical chirality, helicity, and dual symmetries were discussed [19–26] but it was not until Tang and Cohen showed how electromagnetic chirality density can be used to characterize dichroism in light interacting with a chiral metamaterial that this was understood for materials [27]. This revived interest in optical chirality [28–31], which has found a number of applications in optics and plasmonics [32–37].

The results of Tang and Cohen [27] can be understood as follows. In order to observe effects related to the chirality of light, we have to put the electromagnetic field in contact with a chiral environment. This principle suggests a way for finding similar effects in other systems. For example, spin-wave dynamics in collinear antiferromagnets can be represented in a form that closely resembles the Silberstein-Bateman formulation of the Maxwell's equations. Since collinear antiferromagnets have two magnetic sublattices, the concept of electromagnetic duality and nongeometric symmetries can be generalized to transformations between the antiferromagnetic sublattices [38]. This allows to establish a conservation law for a spin-wave analogue of the optical chirality. Injection of a spin current into the antiferromagnet in this case has an effect similar to a chiral environment for light-matter interactions inside a metamaterial [38].

It is also remarkable that both the Maxwell's equations [39] and the dynamics of antiferromagnetic spin waves [40] allow a formulation in the form of the Dirac equation for an ultra-relativistic particle. Such particles are characterized by conserving helicity—a projection of spin on the linear momentum [41], which also satisfies the definition of true chirality. Breaking the symmetry between right and left, in this case, corresponds to a Weyl material [42], wherein quasi-particles with different helicities are spatially separated. Symmetry considerations suggest that as far as single particle dynamics is concerned, there should be some analogy between optical metamaterials, Weyl semimetals, and chiral antiferromagnets. There has been several proposals in these directions. For example, one can emulate the chiral magnetic effect in metallic antiferromagnets [43].

These arguments have a direct impact on optospintronics. Since optical chirality and spin currents share the same symmetry properties, it is possible to use polarized light to excite magnon spin-photocurrents in antiferromagnetic insulators [44]. Circular polarized light in this case creates a direct *flow* of magnon angular momentum, whose direction is controlled by helicity of light. This effect resembles the circular photogalvanic effect in metals [45], which recently attracted attention in topological electron materials [46]. It has been demonstrated that for a separated Weyl node, the photocurrent excitation rate is determined by the product of the topological charge of the node and the helicity of light [46].

In this Chapter, we review chiral excitations in optics and antiferromagnetic insulators together with their applications in optomagnonics. Our discussion is organized as follows. In Sect. 9.2, we give a brief review of optical chirality and nongeometric symmetries, which is generalized to antiferromagnetic spin-waves in Sect. 9.3, where we discuss potential applications such as spin-current induced magnon dichroism. Section 9.4 is reserved for photo-excitation of magnon spin currents with polarized light. Summary and conclusions are in Sect. 9.5.

9.2 Optical Chirality and Nongeometric Symmetries of the Maxwell's Equations

Since the early developments of electrodynamics, it has been well established that the electromagnetic field in vacuum can be characterized by conserving energy, momentum, angular momentum, which reflects the invariance of the Maxwell's equations with respect to the translations and rotations in the four-dimensional space-time [18]. It was found almost by chance [15] that in addition to these conservation laws, the electromagnetic field has another invariant given by a combination of the electric, $\boldsymbol{E}$, and magnetic, $\boldsymbol{B}$, fields

$$\rho_\chi(t, \boldsymbol{r}) = \frac{\varepsilon_0}{2}\boldsymbol{E}\cdot(\nabla\times\boldsymbol{E}) + \frac{1}{2\mu_0}\boldsymbol{B}\cdot(\nabla\times\boldsymbol{B}), \tag{9.1}$$

which is odd under the spatial inversion and even under the time reversal transformations (ε_0 and μ_0 are the vacuum permittivity and permeability respectively). For this quantity, Lipkin coined a special term—optical *zilch* to emphasize the lack of a clear physical interpretation at that time [15]. According to its symmetry properties, ρ_χ is *truly chiral* [10], and can be considered as a chirality density of the electromagnetic field.

Using the Maxwell's equations, it is straightforward to demonstrate that in vacuum ρ_χ satisfies the continuity equation

$$\frac{\partial\rho_\chi}{\partial t} + \nabla\cdot\boldsymbol{J}_\chi = 0, \tag{9.2}$$

where

$$\boldsymbol{J}_\chi(t, \boldsymbol{r}) = \frac{\varepsilon_0}{2}\boldsymbol{E}\times\frac{\partial\boldsymbol{E}}{\partial t} + \frac{1}{2\mu_0}\boldsymbol{B}\times\frac{\partial\boldsymbol{B}}{\partial t}, \tag{9.3}$$

determines the corresponding zilch flow.

In this section, we will show that this conservation law belongs to the class of so-called "hidden" or *nongeometric* symmetries of the Maxwell's equations. One of these symmetries, which has been known since the time of Heaviside, Larmor, and Rainich, is the duality symmetry [47, 48]. If we consider Maxwell's equations in free space

$$\nabla\times\boldsymbol{E} = 0, \qquad \nabla\times\boldsymbol{B} = 0, \tag{9.4}$$

$$\nabla\cdot\boldsymbol{E} = 0, \qquad \nabla\cdot\boldsymbol{B} = 0, \tag{9.5}$$

(we set $c = 1$ throughout this section) the electromagnetic duality is a symmetry with respect to the rotation in the pseudo-space of the electric and magnetic fields, which leaves Maxwell's equations invariant

$$\boldsymbol{E} \to \boldsymbol{E}' = \boldsymbol{E}\cos\theta + \boldsymbol{B}\sin\theta, \tag{9.6}$$

$$\boldsymbol{B} \to \boldsymbol{B}' = -\boldsymbol{E}\sin\theta + \boldsymbol{B}\cos\theta, \tag{9.7}$$

where θ is a real parameter of the transformation. This symmetry is usually broken inside materials, unless we deal with a dual symmetric medium [49].

The existence of duality symmetry guarantees the conservation of optical helicity, i. e. the projection of spin angular momentum of the photon onto its linear momentum [16, 17, 23, 24]. It should be mentioned, however, that the formulation of helicity conservation law in classical electrodynamics is not straightforward, because the standard Lagrangian for the electromagnetic field is not dual symmetric [48]. Using the dual symmetric representation for the electromagnetic Lagrangian combined with the Noether's approach, it is possible to express the optical helicity density in the form similar to (9.1)

$$\rho_{\text{hel}}(t, \boldsymbol{r}) = \frac{1}{2}\left[\boldsymbol{A}\cdot(\nabla\times\boldsymbol{A}) + \boldsymbol{C}\cdot(\nabla\times\boldsymbol{C})\right], \tag{9.8}$$

where in addition to the magnetic vector potential $\boldsymbol{A}$, we also introduced the electric vector potential $\boldsymbol{C}$, which satisfies the following equations, $\boldsymbol{E} = -\nabla\times\boldsymbol{C} = -\partial_t\boldsymbol{A}$ and $\boldsymbol{B} = \nabla\times\boldsymbol{A} = \partial_t\boldsymbol{C}$. These are invariant under the transformations in (9.6) and (9.7) [47, 48].

The definition of electromagnetic helicity depends on a specific representation of the Lagrangian. It suggests that it would be useful to have a general formalism for deriving "hidden" symmetries and conservation laws directly from the equations of motion formulated exclusively in terms of the electromagnetic fields, and independent of any gauge choice. Such a formalism has been developed by Fushchich and Nikitin [18]. Below, we give a brief review of this formalism, which is necessary for further discussions.

9.2.1 Symmetry Analysis of the Maxwell's Equations

For the symmetry analysis, it is convenient to formulate Maxwell's equations in the form that resembles the Dirac equation for a massless relativistic particle. This representation is called the Silberstein-Bateman form [18]. In this form, the first pair of the Maxwell's equations in (9.4) is rewritten in terms of a Schrödinger-like equation for the six-component vector column composed of the components of the electric and magnetic fields $\phi = (\boldsymbol{E}, \boldsymbol{B})^T$

$$i\frac{\partial\phi(t, \boldsymbol{p})}{\partial t} = \mathcal{H}(\boldsymbol{p})\phi(t, \boldsymbol{p}), \tag{9.9}$$

where for convenience, we work in the momentum space, $\boldsymbol{p}$, defined by the following Fourier transformations

$$\boldsymbol{E}(t,\boldsymbol{r}) = \frac{1}{(2\pi)^{3/2}} \int d^3 p e^{i\boldsymbol{p}\cdot\boldsymbol{r}} \boldsymbol{E}(t,\boldsymbol{p}), \tag{9.10}$$

$$\boldsymbol{B}(t,\boldsymbol{r}) = \frac{1}{(2\pi)^{3/2}} \int d^3 p e^{i\boldsymbol{p}\cdot\boldsymbol{r}} \boldsymbol{B}(t,\boldsymbol{p}). \tag{9.11}$$

The matrix on the right-hand side of equation (9.9) has the following structure

$$\mathcal{H}(\boldsymbol{p}) = \begin{pmatrix} 0 & i(\hat{\boldsymbol{S}}\cdot\boldsymbol{p}) \\ -i(\hat{\boldsymbol{S}}\cdot\boldsymbol{p}) & 0 \end{pmatrix}, \tag{9.12}$$

which can be considered as a direct product of the Pauli matrix σ_2, which interchanges $\boldsymbol{E}$ and $\boldsymbol{B}$, and the "helicity" operator $(\hat{\boldsymbol{S}}\cdot\boldsymbol{p})$, where the matrices $\hat{S}_\alpha$ $(\alpha = x, y, z)$ form a representation of the three-dimensional rotation group, $(\hat{S}_\alpha)_{\beta\gamma} = i\epsilon_{\alpha\beta\gamma}$, with $\epsilon_{\alpha\beta\gamma}$ being the Levi-Civita symbol.

The second pair of the Maxwell's equations (9.5) in this formalism impose an additional constraint on the components of $\phi(t,\boldsymbol{p})$ [18]

$$(\hat{\boldsymbol{S}}\cdot\boldsymbol{p})^2\phi(t,\boldsymbol{p}) = p^2\phi(t,\boldsymbol{p}), \tag{9.13}$$

which acknowledges transversality of the electromagnetic field in vacuum.

9.2.1.1 Invariance Algebra of the Maxwell's Equations

Now, we can find the symmetry operations that transform a solution $\phi(t,\boldsymbol{p})$ of (9.9) into another solution $\tilde{\phi}(t,\boldsymbol{p}) = Q(\boldsymbol{p})\phi(t,\boldsymbol{p})$. We look for these transformations in the form of the six-dimensional matrices $Q(\boldsymbol{p})$, which may depend on the momentum $\boldsymbol{p}$. Formal resemblance of our representation with the quantum mechanics implies that these matrices should commute with $\mathcal{H}(\boldsymbol{p})$.

The problem of finding all such transformation becomes almost trivial if we transform to the helicity basis, where $\mathcal{H}(\boldsymbol{p})$ is diagonal. This transformation is reached by a combination of the rotation in the three-dimensional space

$$\hat{U}_\Lambda = \begin{pmatrix} -\dfrac{p_x p_z + i p_y p}{\sqrt{2} p p_\perp} & \dfrac{p_x p_z - i p_y p}{\sqrt{2} p p_\perp} & \dfrac{p_x}{p} \\ -\dfrac{p_y p_z - i p_x p}{\sqrt{2} p p_\perp} & \dfrac{p_y p_z + i p_x p}{\sqrt{2} p p_\perp} & \dfrac{p_y}{p} \\ \dfrac{p_\perp}{\sqrt{2} p} & -\dfrac{p_\perp}{\sqrt{2} p} & \dfrac{p_z}{p} \end{pmatrix}, \tag{9.14}$$

where $p_\perp = (p_x^2 + p_y^2)^{1/2}$, which diagonalizes the "helicity" operator, $\hat{U}_\Lambda^\dagger(\hat{\boldsymbol{S}}\cdot\boldsymbol{p})\hat{U}_\Lambda = \mathrm{diag}(-p, p, 0)$ (it transforms to the basis where the electric and magnetic fields are written in terms of circularly polarized components), with the $SU(2)$ transformation in the pseudo-space of electric and magnetic fields

$$U_2 = \frac{1}{\sqrt{2}} \begin{pmatrix} 1 & -i \\ -i & 1 \end{pmatrix}. \tag{9.15}$$

The resulting transformation $\mathcal{U} = U_2 \otimes \hat{U}_\Lambda$ diagonalizes $\mathcal{H}(\boldsymbol{p})$ so that in the transformed frame

$$\tilde{\mathcal{H}} = \mathcal{U}^\dagger \mathcal{H} \mathcal{U} = \text{diag}(-p, p, 0, p, -p, 0). \tag{9.16}$$

The eigenvalues of $\tilde{\mathcal{H}}$ correspond to the left and right polarized electromagnetic modes with the linear frequency dispersion cp (we have recovered the speed of light c here), which are degenerate in the absence of light-matter interactions.

Straightforward calculations show that in the diagonal frame, any matrix that commutes with $\tilde{\mathcal{H}}$, and at the same time leaves (9.13) invariant, is parameterized by eight parameters, $a, \ldots h$, and has the following structure

$$\tilde{Q} = \begin{pmatrix} a & 0 & 0 & 0 & e & 0 \\ 0 & b & 0 & f & 0 & 0 \\ 0 & 0 & 0 & 0 & 0 & 0 \\ 0 & g & 0 & c & 0 & 0 \\ h & 0 & 0 & 0 & d & 0 \\ 0 & 0 & 0 & 0 & 0 & 0 \end{pmatrix}. \tag{9.17}$$

The basis in the linear space of $\tilde{Q}$ can be chosen such as its basis elements, $\tilde{Q}_i$, ($i = 1, \ldots, 8$) form the algebra isomorphic to the Lie algebra of the group $U(2) \otimes U(2)$

$$\begin{aligned} &\tilde{Q}_1 = -\sigma_2 \otimes \hat{S}_y, \quad \tilde{Q}_2 = -i\sigma_3 \otimes \hat{I} \\ &\tilde{Q}_3 = -i\sigma_1 \otimes \hat{S}_y, \tilde{Q}_4 = \sigma_1 \otimes \hat{S}_x \\ &\tilde{Q}_5 = -\sigma_0 \otimes \hat{S}_z, \quad \tilde{Q}_6 = \sigma_2 \otimes \hat{S}_x \\ &\tilde{Q}_7 = \sigma_0 \otimes \hat{I}, \qquad \tilde{Q}_8 = i\sigma_3 \otimes \hat{S}_z, \end{aligned} \tag{9.18}$$

where σ_0 and $\hat{I}$ denote 2×2 and 3×3 unit matrices respectively.

Returning into original frame and taking into account that $\hat{U}_\Lambda \hat{S}_z \hat{U}_\Lambda^\dagger = -(\hat{\boldsymbol{S}} \cdot \boldsymbol{p})/p$, we obtain the generators of the symmetry transformations in the following form

$$\begin{aligned} &Q_1 = \sigma_3 \otimes (\hat{\boldsymbol{S}} \cdot \tilde{\boldsymbol{p}})\hat{D}, \quad Q_2 = i\sigma_2 \otimes \hat{I}, \\ &Q_3 = -\sigma_1 \otimes (\hat{\boldsymbol{S}} \cdot \tilde{\boldsymbol{p}})\hat{D}, Q_4 = -\sigma_1 \otimes \hat{D}, \\ &Q_5 = \sigma_0 \otimes (\hat{\boldsymbol{S}} \cdot \tilde{\boldsymbol{p}}), \qquad Q_6 = -\sigma_3 \otimes \hat{D}, \\ &Q_7 = \sigma_0 \otimes \hat{I}, \qquad\qquad Q_8 = i\sigma_2 \otimes (\hat{\boldsymbol{S}} \cdot \tilde{\boldsymbol{p}}), \end{aligned} \tag{9.19}$$

where $\tilde{\boldsymbol{p}} = \boldsymbol{p}/p$, and $\hat{D} = -p\hat{U}_\Lambda \hat{S}_x \hat{U}_\Lambda^\dagger$. These equations form the eight-dimensional invariance algebra of the Maxwell's equations in vacuum [18].

9.2.1.2 Nongeometric Symmetries

The basis elements in (9.19) generate continuous symmetries that Fushchich and Nikitin called the *nongeometric symmetries* of the Maxwell's equations [18]

$$\phi(t, \boldsymbol{p}) \to \phi'(t, \boldsymbol{p}) = \exp(Q_i\theta_i)\phi(t, \boldsymbol{p}), \tag{9.20}$$

where θ_i denotes the real parameter of the transformation.

Some symmetry generators have a clear physical meaning. For example, Q_7 is a unit element. Q_2 interchanges electric and magnetic fields in $\phi(t, \boldsymbol{p})$, so that the corresponding continuous transformation $\exp(i\sigma_2\theta)$ is the duality symmetry in (9.6) and (9.7). Q_5 has the form of the helicity operator. Q_8 is proportional to $\mathcal{H}$, which means that similar to Q_7 it commutes with every element of the algebra. It reflects the symmetry with respect to ∂_t (the time derivative of $\phi(t, \boldsymbol{p})$, which solves the Maxwell's equations, is again a solution for the same $\boldsymbol{p}$). The basis elements Q_2, Q_5, Q_7, and Q_8 form a trivial Abelian part of the algebra in (9.19). The existence of non-Abelian elements is related to the degeneracy between left and right polarized eigenvalues in (9.16).

The conservation laws that correspond to the symmetry transformations in (9.20) can be conveniently written in terms of the bilinear forms by analogy with the quantum-mechanics

$$\langle Q_i \rangle = \frac{1}{2}\int d^3p\phi^\dagger(t, \boldsymbol{p})Q_i\phi(t, \boldsymbol{p}). \tag{9.21}$$

It can be demonstrated that the electromagnetic field in vacuum can be characterized by an infinite number of invariants generated from the eight symmetry transformations [18]. For example, the unit element Q_7 in this formalism corresponds to the conservation of the electromagnetic energy

$$\langle Q_7 \rangle = \frac{1}{2}\int d^3p\phi^\dagger(t, \boldsymbol{p})\phi(t, \boldsymbol{p}) = \frac{1}{2}\int d^3p\left(E^2 + B^2\right). \tag{9.22}$$

9.2.1.3 Conservation Law for Optical Chirality

Using this formalism, optical zilch can be expressed as a conservation law for the helicity operator Q_5

$$C_\chi = \int d^3r\rho_\chi(t, \boldsymbol{r}) = \frac{1}{2}\int d^3p\phi^\dagger(t, \boldsymbol{p})(\hat{\boldsymbol{S}}\cdot\boldsymbol{p})\phi(t, \boldsymbol{p}). \tag{9.23}$$

Using the fact that the helicity operator, duality symmetry, and ∂_t are related to each other by the algebraic property, $pQ_5Q_2 = -i\mathcal{H} = \partial_t$, we establish a relation between zilch conservation and duality symmetry as it was originally discussed in [16, 17], which allows to write the conservation law above in the following equivalent form

$$C_\chi = -\frac{i}{2}\int d^3p\phi^\dagger(t,\boldsymbol{p})Q_2\partial_t\phi(t,\boldsymbol{p}) = \frac{1}{2}\int d^3r\left(\boldsymbol{B}\cdot\frac{\partial\boldsymbol{E}}{\partial t} - \boldsymbol{E}\cdot\frac{\partial\boldsymbol{B}}{\partial t}\right). \quad (9.24)$$

This expression can be easily generalized to accommodate higher order terms in space and time derivatives. By replacing $Q_2\partial_t$ with $-(ip)^{2n}Q_2(i\partial_t)^{2m+1}$, which is again a symmetry operation, we can find a hierarchy of conserving zilches

$$C_\chi^{(m,n)} = \frac{1}{2}\int d^3r\left(\boldsymbol{B}\cdot\nabla^{2n}\partial_t^{2m+1}\boldsymbol{E} - \boldsymbol{E}\cdot\nabla^{2n}\partial_t^{2m+1}\boldsymbol{B}\right), \quad (9.25)$$

where 00-zilch corresponds to the optical chirality [23, 24, 31].

It is possible to derive the conservation law for the optical zilch using the Noether's formalism by applying a specific "hidden" gauge transformation to the Lagrangian of the electromagnetic field [31], which leads to the same results as in (9.23) and (9.25). The advantage of the approach discussed in this section, based on the symmetry analysis of the Maxwell's equations, is that it does not depend on any specific gauge choice. This fact makes it easy to extend this formalism to other physical systems with similar form of the equations of motion.

9.2.2 *Optical Chirality in Gyrotropic Media*

Having now a complete picture of the nongeometric symmetries in vacuum, we discuss how this approach can be applied for the light-matter interactions. Electromagnetic field in dielectric medium is usually described by the material form of the Maxwell equations

$$\nabla\times\boldsymbol{E} = -\frac{\partial\boldsymbol{B}}{\partial t}, \qquad \nabla\cdot\boldsymbol{B} = 0, \quad (9.26)$$

$$\nabla\times\boldsymbol{H} = \frac{\partial\boldsymbol{D}}{\partial t}, \qquad \nabla\cdot\boldsymbol{D} = 0, \quad (9.27)$$

supplemented by the constituent relations between the fields $\boldsymbol{E}$, $\boldsymbol{H}$, $\boldsymbol{D}$, and $\boldsymbol{B}$. The constituent relations impose additional constraints on the form of the symmetry transformations for the electromagnetic field, which reflect the intrinsic symmetries of the medium. This often leads to the reduction of the invariance algebra in (9.19) to lesser number of elements [50]. In the case of common constituent relations, $\boldsymbol{D}(\boldsymbol{p}) = \hat{\varepsilon}(\boldsymbol{p})\boldsymbol{E}(\boldsymbol{p})$ and $\boldsymbol{B}(\boldsymbol{p}) = \hat{\mu}(\boldsymbol{p})\boldsymbol{H}(\boldsymbol{p})$, where $\hat{\varepsilon}(\boldsymbol{p})$ and $\hat{\mu}(\boldsymbol{p})$ denote the electric permittivity and magnetic permeability tensors in the Fourier space, Maxwell's equations in (9.9) are replaced by

$$i\frac{\partial}{\partial t}\begin{pmatrix} \boldsymbol{D}(\boldsymbol{p}) \\ \boldsymbol{B}(\boldsymbol{p}) \end{pmatrix} = \begin{pmatrix} 0 & i(\hat{\boldsymbol{S}}\cdot\boldsymbol{p})\hat{\mu}^{-1}(\boldsymbol{p}) \\ -i(\hat{\boldsymbol{S}}\cdot\boldsymbol{p})\hat{\varepsilon}^{-1}(\boldsymbol{p}) & \end{pmatrix}\begin{pmatrix} \boldsymbol{D}(\boldsymbol{p}) \\ \boldsymbol{B}(\boldsymbol{p}) \end{pmatrix}. \tag{9.28}$$

The symmetry analysis of the previous sections can be generalized for this case (see [50] for detailed discussion). In particular, for dual symmetric medium, the optical chirality is given by

$$C_\chi = \frac{1}{2}\int d^3p \left(\boldsymbol{D}^*(\boldsymbol{p})(\hat{\boldsymbol{S}}\cdot\boldsymbol{p})\hat{\varepsilon}^{-1}(\boldsymbol{p})\boldsymbol{D}(\boldsymbol{p}) + \boldsymbol{B}^*(\boldsymbol{p})(\hat{\boldsymbol{S}}\cdot\boldsymbol{p})\hat{\mu}^{-1}(\boldsymbol{p})\boldsymbol{B}(\boldsymbol{p})\right) \tag{9.29}$$

In the real space this expression becomes

$$C_\chi = \frac{1}{2}\int d^3r \left(\boldsymbol{B}\cdot\partial_t\boldsymbol{D} - \boldsymbol{D}\cdot\partial_t\boldsymbol{B}\right), \tag{9.30}$$

which also acknowledges spatial dispersion of the electromagnetic field.

As an important example, let us consider propagation of the electromagnetic field in chiral media where structural chirality of the material leads to the existence of such physical phenomena as natural optical activity and circular dichroism. There exists several approaches for the electrodynamics of chiral gyrotropic media [51–53]. One of these approaches, which is frequently adopted for characterizing metamaterials [54, 55], is based on the following constituent relations

$$\boldsymbol{D} = \varepsilon\varepsilon_0\boldsymbol{E} + i\varkappa\boldsymbol{H}, \tag{9.31}$$

$$\boldsymbol{B} = \mu\mu_0\boldsymbol{H} - i\varkappa\boldsymbol{E}, \tag{9.32}$$

where $\varkappa$ characterizes chirality of the material. This approach requires complex representation for the electromagnetic fields and can be derived from the relativistic covariance principle [51, 56].

By applying our general formalism to the Maxwell's equations (9.26) and (9.27) with the constituent relations (9.31) and (9.32), we obtain the same equation of motion as in (9.9), where ϕ is replaced by for the vector column $\boldsymbol{\phi}(t, \boldsymbol{p}) = (\boldsymbol{D}, \boldsymbol{B})^T$, and the matrix on the right-hand side is now given by (we use the units $\varepsilon\varepsilon_0 = \mu\mu_0 = 1$)

$$\mathcal{H}(\boldsymbol{p}) = -\frac{1}{1-\varkappa^2}\begin{pmatrix} \varkappa(\hat{\boldsymbol{S}}\cdot\boldsymbol{p}) & -i(\hat{\boldsymbol{S}}\cdot\boldsymbol{p}) \\ i(\hat{\boldsymbol{S}}\cdot\boldsymbol{p}) & \varkappa(\hat{\boldsymbol{S}}\cdot\boldsymbol{p}) \end{pmatrix}. \tag{9.33}$$

This matrix can be diagonalized by a combination of the same unitary transformations as in (9.14) and (9.15) that yields the following diagonal form

$$\tilde{\mathcal{H}} = \mathcal{U}^\dagger\mathcal{H}\mathcal{U} = \mathrm{diag}(-p_-, p_-, 0, p_+, -p_+, 0), \tag{9.34}$$

where $p_\pm = p/(1 \mp \varkappa)$.

Lifted degeneracy between left (p_-) and right (p_+) polarized eigenmodes in (9.34) leads to the reduction of the eight-dimensional invariance algebra to four basis elements, which commute with each other

$$\begin{array}{ll} Q_2 = i\sigma_2 \otimes \hat{I},\ Q_5 = \ \ \sigma_0 \otimes (\hat{\boldsymbol{S}} \cdot \tilde{\boldsymbol{p}}) \\ Q_7 = \ \ \sigma_0 \otimes \hat{I},\ Q_8 = i\sigma_2 \otimes (\hat{\boldsymbol{S}} \cdot \tilde{\boldsymbol{p}}). \end{array} \tag{9.35}$$

These symmetries, however, still contain the duality transformation Q_2, which means that the medium is dual-symmetric and supports the conservation of the electromagnetic helicity [49] and, as a consequence, optical zilches.

Definition of the optical chirality density in chiral media requires some attention. This situation is similar to the definition of the electromagnetic energy density where one should take care of the continuity of the energy flow at the boundary between two chiral media [51]. It can be demonstrated that chirality density in the medium with constituent relations (9.31) and (9.32) that provides continuity of chirality flow at the boundary of two chiral media with different $\varkappa$ can be introduced in the following way [50]

$$\rho_\chi = \frac{\varepsilon\varepsilon_0}{2}\boldsymbol{B}^* \cdot \frac{\partial \boldsymbol{E}}{\partial t} - \frac{\mu\mu_0}{2}\boldsymbol{D}^* \cdot \frac{\partial \boldsymbol{H}}{\partial t}, \tag{9.36}$$

This expression remains valid even if $\varepsilon(\boldsymbol{r})$, $\mu(\boldsymbol{r})$, and $\varkappa(\boldsymbol{r})$ become position dependent. In this case, it satisfies the continuity equation with a source term on the right-hand side

$$\partial_t \rho_\chi + \nabla \cdot \boldsymbol{J}_\chi = F(t, \boldsymbol{r}). \tag{9.37}$$

where

$$\boldsymbol{J}_\chi = \frac{\varepsilon_0\varepsilon}{2}\, \boldsymbol{E}^* \times \partial_t \boldsymbol{E} + \frac{\mu_0\mu}{2}\, \boldsymbol{H}^* \times \partial_t \boldsymbol{H}, \tag{9.38}$$

and the source term contains only gradients of ε and μ, but does not depend on the gradient of $\varkappa$

$$F(t, \boldsymbol{r}) = \frac{\varepsilon_0}{2}\, \nabla\varepsilon \cdot \boldsymbol{E}^* \times \partial_t \boldsymbol{E} + \frac{\mu_0}{2}\, \nabla\mu \cdot \boldsymbol{H}^* \times \partial_t \boldsymbol{H}. \tag{9.39}$$

In order to understand the physical meaning of ρ_χ, let us look at energy absorption in a dissipative gyrotropic medium with the constituent relations (9.31) and (9.32). As was demonstrated in [27], the electromagnetic energy absorption rate in this case has an asymmetric part, which has opposite signs for left and right polarized electromagnetic waves. This part is proportional the product between the chirality of the material, given by the imaginary part of $\varkappa$, and the chirality density of the electromagnetic field ρ_χ. The flow of optical chirality in (9.3), in this situation, can be associated with the asymmetric components of the electromagnetic forces in the medium, which can be used, for example, for optical separation of chiral molecules [37].

In the next section, we will show how these arguments can be generalized to spin excitations in antiferromagnetic materials. Similar to the results of this section, the symmetry analysis will play a principal role in our discussion.

9.3 Spin-Wave Chirality in Antiferromagnetic Insulators

The symmetry analysis developed in the previous section for Maxwell's equations can be generalized to other dynamical systems. Here, we develop such generalization for spin-wave excitations in an antiferromagnetic insulator. A key observation that helps us to draw the analogy between spin-wave dynamics and electrodynamics is that the antiferromagnetic spin waves can be also characterized by two polarization states. This stems from the fact that the magnetization dynamics in antiferromagnets involves two coupled magnetic sublattices. We, therefore, examine the symmetry transformation in the extended space that includes three-dimensional rotations and transformations between the sublattices, in order to find an algebra of nongeometric symmetries for spin waves equivalent to that of the electrodynamics.

9.3.1 Equations of Motion for Antiferromagnetic Spin Waves

We start our discussion with a simple case of a collinear antiferromagnet with two equivalent magnetic sublattices $\boldsymbol{M}_1(t, \boldsymbol{r})$ and $\boldsymbol{M}_2(t, \boldsymbol{r})$. The energy for such antiferromagnet can be written in the following form

$$W = \int d^3r \left[\frac{\alpha}{2} \left(\frac{\partial \boldsymbol{M}_1}{\partial x_\mu} \cdot \frac{\partial \boldsymbol{M}_1}{\partial x_\mu} + \frac{\partial \boldsymbol{M}_2}{\partial x_\mu} \cdot \frac{\partial \boldsymbol{M}_2}{\partial x_\mu} \right) + \alpha' \frac{\partial \boldsymbol{M}_1}{\partial x_\mu} \cdot \frac{\partial \boldsymbol{M}_2}{\partial x_\mu} + \frac{\delta}{2} \boldsymbol{M}_1 \cdot \boldsymbol{M}_2 - \frac{\beta}{2} \left((\boldsymbol{M}_1 \cdot \boldsymbol{n})^2 + (\boldsymbol{M}_2 \cdot \boldsymbol{n})^2 \right) \right], \quad (9.40)$$

where α, α', and δ are the antiferromagnetic exchange parameters and $\beta > 0$ describes the uniaxial magnetic anisotropy with $\boldsymbol{n}$ being the unit vector along the anisotropy axis [57]. In the ground state, the anisotropy stabilizes a uniform magnetic ordering along $\boldsymbol{n}$ where two sublattices compensate each other, $\boldsymbol{M}_1 = -\boldsymbol{M}_2$, so that the total magnetization vanishes.

In the semi-classical limit, magnetization dynamics are described by the Landau–Lifshitz–Gilbert equations of motion

$$\frac{\partial \boldsymbol{M}_i}{\partial t} = \gamma \boldsymbol{M}_i \times \boldsymbol{H}_i^{\text{eff}} - \eta \boldsymbol{M}_i \times \frac{\partial \boldsymbol{M}_i}{\partial t}, \quad (i = 1, 2), \quad (9.41)$$

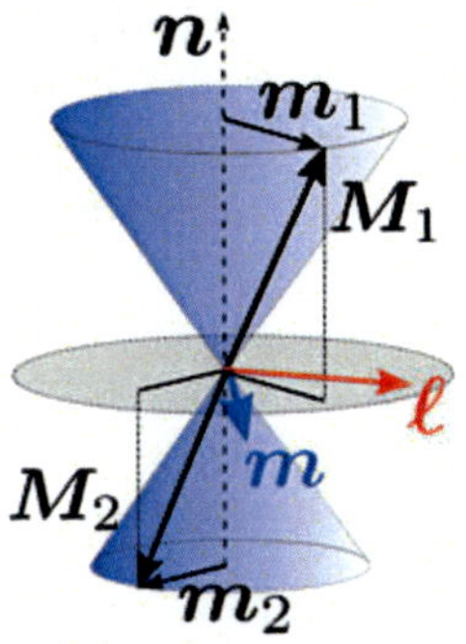

Fig. 9.1 Sublattice magnetizations $\boldsymbol{M}_1$ and $\boldsymbol{M}_2$ precessing along the anisotropy axis $\boldsymbol{n}$; $\boldsymbol{m} = \boldsymbol{m}_1 + \boldsymbol{m}_2$ is the resulting dynamic magnetization, and $\boldsymbol{l} = \boldsymbol{m}_1 - \boldsymbol{m}_2$ shows the dynamic part of the antiferromagnetic vector

where γ is the gyromagnetic ratio, $\boldsymbol{H}_i^{\text{eff}} = -\delta W/\delta \boldsymbol{M}_i$ is the effective field acting on the magnetization on the ith sublattice and η is the Gilbert damping that takes dissipation into account [57].

For small excitations around the ground state configuration a linear form of the Landau–Lifshitz–Gilbert equations of can be used. This is reached by breaking the sublattice magnetizations into static $M_s\boldsymbol{n}$ and dynamic $\boldsymbol{m}_i(t, \boldsymbol{t})$ parts, $\boldsymbol{M}_i = (-1)^{i+1} M_s \boldsymbol{n} + \boldsymbol{m}_i$, and keeping only the linear terms in $\boldsymbol{m}_i$ in the resulting equations of motion (M_s denotes the saturation magnetization). For convenience, we transform $\boldsymbol{m}_i(\boldsymbol{r})$ to momentum space, such that $\boldsymbol{m}_i(t, \boldsymbol{r}) = \int d^3p \exp(i\boldsymbol{p} \cdot \boldsymbol{r}) \boldsymbol{m}_{i\boldsymbol{p}}(t)$, and introduce the dynamic vectors of the magnetization, $\boldsymbol{m}_{\boldsymbol{p}} = \boldsymbol{m}_{1\boldsymbol{p}} + \boldsymbol{m}_{2\boldsymbol{p}}$, and antiferromagnetism, $\boldsymbol{l}_{\boldsymbol{p}} = \boldsymbol{m}_{1\boldsymbol{p}} - \boldsymbol{m}_{2\boldsymbol{p}}$, see Fig. 9.1. The resulting linear system of the equations of motions is given by

$$\frac{\partial \boldsymbol{m}_{\boldsymbol{p}}}{\partial t} = -\varepsilon_{\boldsymbol{p}}^{(l)} \boldsymbol{n} \times \boldsymbol{l}_{\boldsymbol{p}} + \eta \boldsymbol{n} \times \frac{\partial \boldsymbol{l}_{\boldsymbol{p}}}{\partial t}, \tag{9.42}$$

$$\frac{\partial \boldsymbol{l}_{\boldsymbol{p}}}{\partial t} = -\varepsilon_{\boldsymbol{p}}^{(m)} \boldsymbol{n} \times \boldsymbol{m}_{\boldsymbol{p}} + \eta \boldsymbol{n} \times \frac{\partial \boldsymbol{m}_{\boldsymbol{p}}}{\partial t}, \tag{9.43}$$

where $\varepsilon_{\boldsymbol{p}}^{(m)} = \gamma M_s(\delta + \beta + (\alpha + \alpha')p^2)$ and $\varepsilon_{\boldsymbol{p}}^{(l)} = \gamma M_s(\beta + (\alpha - \alpha')p^2)$.

For the equations of motion (9.42) and (9.43), it is possible to find a representation that is similar to the Silberstein-Bateman form of the Maxwell's equations [38]. For this purpose, we introduce a vector column $\psi(t, \boldsymbol{p}) = (\boldsymbol{m}_{\boldsymbol{p}}, \boldsymbol{l}_{\boldsymbol{p}})^T$, which allows us to rewrite the equations of motion for the spin waves in the form (9.9), where the matrix in the right-hand side is now given by

$$\mathcal{H}_{\text{m}} = \begin{pmatrix} 0 & -\varepsilon_{\boldsymbol{p}}^{(l)}(\hat{\boldsymbol{S}} \cdot \boldsymbol{n}) \\ -\varepsilon_{\boldsymbol{p}}^{(m)}(\hat{\boldsymbol{S}} \cdot \boldsymbol{n}) & 0 \end{pmatrix}. \tag{9.44}$$

Here, we omit damping terms, which we discuss later. In this form, the equations of motion for the spin waves resemble the Maxwell's equations in a dispersive medium where the roles of the electric permittivity and magnetic permeability is played by $\varepsilon_{\boldsymbol{p}}^{(m)}$ and $\varepsilon_{\boldsymbol{p}}^{(l)}$.

The matrix in (9.44) can be symmetrized by an appropriate choice of the units that can be expressed in the form of the transformation $\psi = \mathcal{N}\bar{\psi}$, where $\mathcal{N} = \mathrm{diag}([\varepsilon_{\boldsymbol{p}}^{(m)}]^{-1/2}, [\varepsilon_{\boldsymbol{p}}^{(l)}]^{-1/2})$. In the symmetric units, the equation of motion for the antiferromagnetic spin waves is written as

$$i\frac{\partial\bar{\psi}(t,\boldsymbol{p})}{\partial t} = \mathcal{H}_0(\boldsymbol{p})\bar{\psi}(t,\boldsymbol{p}), \tag{9.45}$$

where the matrix on the right-hand side becomes symmetric

$$\mathcal{H}_0(\boldsymbol{p}) = \begin{pmatrix} 0 & -\omega_{\boldsymbol{p}}(\hat{\boldsymbol{S}}\cdot\boldsymbol{n}) \\ -\omega_{\boldsymbol{p}}(\hat{\boldsymbol{S}}\cdot\boldsymbol{n}) & 0 \end{pmatrix} = -\omega_{\boldsymbol{p}}\sigma_1\otimes(\hat{\boldsymbol{S}}\cdot\boldsymbol{n}), \tag{9.46}$$

with $\omega_{\boldsymbol{p}} = \sqrt{\varepsilon_{\boldsymbol{p}}^{(m)}\varepsilon_{\boldsymbol{p}}^{(l)}}$.

This expression has a structure similar to $\mathcal{H}(\boldsymbol{p})$ in (9.12) for the Maxwell's equations. The important difference between $\mathcal{H}_0$ and $\mathcal{H}$ comes from their properties under spatial inversion ($\mathcal{P}$) and time-reversal ($\mathcal{T}$) transformations. For example, in the case of the time-reversal transformation, $\phi(t,\boldsymbol{p})$ in (9.9) transforms as $\mathcal{T}\phi(t,\boldsymbol{p}) \to \sigma_3\phi(-t,\boldsymbol{p})$. The Pauli matrix σ_3 appears on the right-hand side due to the different transformation properties of the electric and magnetic field with respect to $\mathcal{T}$. In contrast, both components of $\psi(t,\boldsymbol{p})$ are odd under $\mathcal{T}$, so that $\mathcal{T}\psi(t,\boldsymbol{p}) \to -\psi(-t,\boldsymbol{p})$. This means that if we want to transform from the spin wave dynamics to the electrodynamics, we should replace σ_1 in (9.46) with $\sigma_2 = i\sigma_1\sigma_3$ to ensure correct properties under the $\mathcal{PT}$ transformations.

9.3.2 Nongeometric Symmetries for Spin-Wave Dynamics

Formal analogy between the equations of motion for the antiferromagnetic spin waves and the Maxwell's equations enables us to generalize the concept of nongeometric symmetries. We may ask a question about all the transformations $\bar{\psi}(t,\boldsymbol{p}) \to \bar{\psi}'(t,\boldsymbol{p})$ that leave the equation of motion (9.45) invariant.

In order to find all such symmetries, one can repeat the steps of Sect. 9.2.1.1. First, we have to transform to the basis where $\mathcal{H}_0(\boldsymbol{p})$ is diagonal. For this purpose, we make a unitary transformation $\bar{\psi} = \mathcal{U}_{\mathrm{m}}\tilde{\psi}$, where the transformation matrix, $\mathcal{U}_{\mathrm{m}} = U_1\otimes\hat{U}_\Lambda$, is given by the rotation matrix to the helicity basis in (9.14) (where $\boldsymbol{p}$ is replaced by $\boldsymbol{n}$) combined with the $SU(2)$ rotation in the subspace of $\boldsymbol{m_p}$ and $\boldsymbol{l_p}$

$$U_1 = \frac{1}{\sqrt{2}}\begin{pmatrix} 1 & 1 \\ -1 & 1 \end{pmatrix}. \tag{9.47}$$

The resulting equation of motion for $\tilde{\psi}$ is given by (9.45) with the diagonal matrix on the right-hand side

$$\tilde{\mathcal{H}}_0 = \mathcal{U}_{\rm m}^{\dagger}\mathcal{H}_0\mathcal{U}_{\rm m} = \text{diag}(-\omega_{\boldsymbol{p}}, \omega_{\boldsymbol{p}}, 0, \omega_{\boldsymbol{p}}, -\omega_{\boldsymbol{p}}, 0). \tag{9.48}$$

This describes two antiferromagnetic spin waves with an energy dispersion $\omega_{\boldsymbol{p}}$ degenerate with respect to the two polarization states. In an antiferromagnet, magnetization precession is locked in the real space to the direction of $\boldsymbol{n}$, so that these polarization states correspond to left and right circular polarizations along the anisotropy axis. This is in contrast to electrodynamics, where we deal with real helicity—precession around the direction of wave vector $\boldsymbol{p}$.

Secondly, we have to find all the matrices Q that commute with $\tilde{\mathcal{H}}_0$, which can be done precisely in the same way as in (9.17). It should be mentioned that in the region $(\alpha - \alpha')p^2 \gg \beta$, antiferromagnetic spin waves have almost linear dispersion, $\omega_{\boldsymbol{p}} = c_s p$, where the velocity is given by $c_s = \gamma M_s\sqrt{\delta(\alpha - \alpha')}$. This fact gives them the appearance similar to the electromagnetic waves. However, we emphasize that the linear dispersion is not essential for our symmetry analysis.

What is important is that the eigenvalues of $\tilde{\mathcal{H}}_0$ are degenerate. This fact allows us find the eight-dimensional algebra of the symmetry transformations, which is isomorphic to invariance algebra of the Maxwell's equations. The generators of this algebra can be chosen as follows

$$\begin{aligned} &Q_1 = i\sigma_2 \otimes (\hat{\boldsymbol{S}}\cdot\boldsymbol{n})\hat{D}, && Q_2 = \sigma_1 \otimes \hat{I}, \\ &Q_3 = \sigma_3 \otimes (\hat{\boldsymbol{S}}\cdot\boldsymbol{n})\hat{D}, && Q_4 = i\sigma_2 \otimes \hat{D}, \\ &Q_5 = \sigma_0 \otimes (\hat{\boldsymbol{S}}\cdot\boldsymbol{n}), && Q_6 = \sigma_3 \otimes \hat{D}, \\ &Q_7 = \sigma_0 \otimes \hat{I}, && Q_8 = \sigma_1 \otimes (\hat{\boldsymbol{S}}\cdot\boldsymbol{n}), \end{aligned} \tag{9.49}$$

where $\hat{D} = 2[(\hat{\boldsymbol{S}}\cdot\boldsymbol{n}_\perp)^2 - \hat{I}_3\boldsymbol{n}_\perp^2]/n_\perp^2 - (\hat{\boldsymbol{S}}\cdot\boldsymbol{n})^2$, $\hat{I}_3 = \text{diag}(0, 0, 1)$, and $\boldsymbol{n}_\perp = (n_1, n_2, 0)$. The interpretation of these basis elements is similar to that in (9.19). We have the unit element Q_7, Q_8 up to the factor of $\omega_{\boldsymbol{p}}$ coincides with $\mathcal{H}_0(\boldsymbol{p})$ and, therefore, commutes with all the other basis elements, and Q_5 generates rotations along $\boldsymbol{n}$.

Remarkably, Q_2 plays a role of the duality transformation of the electromagnetic field. It generates a continuous symmetry transformation, the Bogolyubov's rotation, in the subspace of $\boldsymbol{m}_{\boldsymbol{p}}$ and $\boldsymbol{l}_{\boldsymbol{p}}$

$$\boldsymbol{m}_{\boldsymbol{p}} \to \boldsymbol{m}'_{\boldsymbol{p}} = \boldsymbol{m}_{\boldsymbol{p}}\cosh\theta + \sqrt{\frac{\varepsilon_{\boldsymbol{p}}^{(l)}}{\varepsilon_{\boldsymbol{p}}^{(m)}}}\boldsymbol{l}_{\boldsymbol{p}}\sinh\theta, \tag{9.50}$$

$$\boldsymbol{l}_{\boldsymbol{p}} \to \boldsymbol{l}'_{\boldsymbol{p}} = \boldsymbol{l}_{\boldsymbol{p}}\cosh\theta + \sqrt{\frac{\varepsilon_{\boldsymbol{p}}^{(m)}}{\varepsilon_{\boldsymbol{p}}^{(l)}}}\boldsymbol{m}_{\boldsymbol{p}}\sinh\theta, \tag{9.51}$$

which leaves (9.42) and (9.43) invariant for any real parameter θ. Similar to the electrodynamics, we have an algebraic property $Q_2 Q_2 = Q_8$, which establishes a relation between the duality, the rotation symmetry along $\boldsymbol{n}$, and ∂_t.

9.3.3 Conserving Chirality of Spin Waves

The existence of the symmetry transformations makes possible a formulation of the conservation laws that correspond to these symmetries. Conserving quantities can be expressed in terms of bilinear forms similar to (9.21)

$$C = \frac{1}{2} \int d^3 p \psi^\dagger(t, \boldsymbol{p}) \rho Q \psi(t, \boldsymbol{p}), \tag{9.52}$$

where Q is a symmetry transformation, which can be expressed as a linear combination of Q_i $(i = 1, \ldots, 8)$, and the measure $\rho = \text{diag}(\varepsilon_{\boldsymbol{p}}^{(m)}, \varepsilon_{\boldsymbol{p}}^{(l)})$ is necessary for transforming from the symmetric representation of the equations of motions in (9.45) and (9.46) to the original units.

The conservation law for spin-wave chirality can be formulated similar to the expression for the optical zilch in Sect. 9.2.1.3. Since the rotation symmetry is preserved only along the direction of $\boldsymbol{n}$, we take the component of the spin wave momentum along this direction $\boldsymbol{p}_n = (\boldsymbol{p} \cdot \boldsymbol{n})\boldsymbol{n}$, and apply the conservation law in (9.52) for the symmetry transformation $p_n Q_5 = (\hat{\boldsymbol{S}} \cdot \boldsymbol{p}_n)$. As a result, the expression for conserving spin-wave chirality is given by

$$C_\chi^{(\text{m})} = \frac{i}{2} \int d^3 p \left[\varepsilon_{\boldsymbol{p}}^{(m)} \boldsymbol{m}_{\boldsymbol{p}}^* \cdot (\boldsymbol{p}_n \times \boldsymbol{m}_{\boldsymbol{p}}) + \varepsilon_{\boldsymbol{p}}^{(l)} \boldsymbol{l}_{\boldsymbol{p}}^* \cdot (\boldsymbol{p}_n \times \boldsymbol{l}_{\boldsymbol{p}}) \right], \tag{9.53}$$

which is a direct analogue of the Lipkin's zilch for the electromagnetic field. In real space, the chirality density for spin waves can be written as

$$\rho_\chi^{(\text{m})}(t, \boldsymbol{r}) = \frac{1}{2} \left(\nabla_n \boldsymbol{m} \cdot \frac{\partial \boldsymbol{l}}{\partial t} + \nabla_n \boldsymbol{l} \cdot \frac{\partial \boldsymbol{m}}{\partial t} \right), \tag{9.54}$$

where $\nabla_n = \boldsymbol{\nabla} \cdot \boldsymbol{n}$.

Physical meaning of $C_\chi^{(\text{m})}$ becomes clear if we rewrite the expression (9.53) in terms of circularly polarized magnon operators. In this case, total spin wave chirality is determined by the difference between the number of left $(N_{\boldsymbol{p}}^{(R)})$ and right $(N_{\boldsymbol{p}}^{(R)})$ polarized magnons [38]

$$C_\chi^{(\text{m})} = 2 \sum_{\boldsymbol{p}} p_n \omega_{\boldsymbol{p}} \left(N_{\boldsymbol{p}}^{(L)} - N_{\boldsymbol{p}}^{(R)} \right). \tag{9.55}$$

Similar expression exists for the Lipkin's zilch written in terms of the polarized photon modes [30]. For a monochromatic spin wave, $C_\chi^{(\mathrm{m})}$ becomes proportional to the spin angular momentum component along $\boldsymbol{n}$, which in terms of magnon number operators is given by $S^{(n)} = \sum_p (N_p^{(L)} - N_p^{(R)})$ [30].

9.3.4 Spin-Wave Chirality in Dissipative Media

By now, we have established that spin waves in antiferromagnets can be characterized by the chiral invariant $C_\chi^{(\mathrm{m})}$, which is analogous to the Lipkin's zilch in optics. Similar to the optical case, we may ask a question: how can we make this chirality of the spin waves visible? To answer this question, we should look at the symmetries. Since $C_\chi^{(\mathrm{m})}$ is a pseudoscalar that is odd under $\mathcal{P}$ and even under $\mathcal{T}$, we have to break the same symmetries inside the antiferromagnet following the idea discussed in Sect. 9.2.2 for the light-matter interactions in chiral metamaterials.

Since our model in (9.40) is not chiral, we should provide some symmetry breaking mechanism. One interesting possibility of such mechanism that is relevant for spintronic applications is based on electron spin current [38]. The flow of spin angular momentum is odd under the spatial inversion and even under the time reversal transformation, therefore, its interaction with antiferromagnetic spin waves is able to provide the necessary symmetry breaking.

The microscopic mechanism beyond this symmetry breaking is as follows. Let us consider an electron spin current flowing along the magnetic ordering direction $\boldsymbol{n}$, which can be injected into an antiferromagnetic insulator film by a proximity effect or can be created in bulk metallic antiferromagnets. A pure spin current consists of a number of spin majority electrons ($\uparrow$) polarized along $\boldsymbol{n}$ flowing with the velocity $\boldsymbol{v}_s$ parallel to $\boldsymbol{n}$ balanced by the same amount of spin minority electrons ($\downarrow$) moving with the velocity $-\boldsymbol{v}_s$, so that the net electric charge transport is zero. Since the spin-wave dynamics is slow with respect to that of the electrons, the latter are able to exert a spin transfer torque on the magnetization dynamics via the Zhang-Li mechanism [58]. If the local s-d interactions between the electrons and sublattice magnetizations are in the exchange dominant regime [59], which means that we can neglect the intersublattice electron scattering, the spin majority (minority) electrons couple mostly to $\boldsymbol{M}_1$ ($\boldsymbol{M}_2$) sublattice magnetization. In this situation, the spin-$\uparrow$ electrons create the spin transfer torque acting mostly on the magnetization $\boldsymbol{M}_1$

$$\mathfrak{T}_1 = -\frac{1}{M_s^2}\boldsymbol{M}_1 \times (\boldsymbol{M}_1 \times (\boldsymbol{v}_s \cdot \nabla)\boldsymbol{M}_1) - \frac{\xi}{M_s}\boldsymbol{M}_1 \times (\boldsymbol{v}_s \cdot \nabla)\boldsymbol{M}_1, \tag{9.56}$$

where the first (second) term is the adiabatic (non-adiabatic) torque component, and $\xi \lesssim 1$ is the dimensionless parameter [58, 59]. At the same time, spin-$\downarrow$ electron flow produce the spin transfer torque $\mathfrak{T}_2 = -\mathfrak{T}_1$ applied to $\boldsymbol{M}_2$. Therefore, a pure spin current in the exchange dominant regime of the electron-spin interaction is able

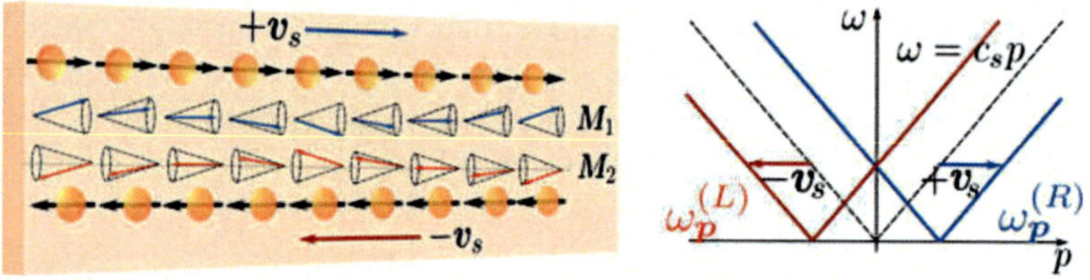

Fig. 9.2 Schematic picture of a pure spin current inside an antiferromagnet. Spin majority (minority) electrons moving with the velocity $+v_s$ $(-v_s)$ create adiabatic spin torque applied to $\boldsymbol{M}_1$ $(\boldsymbol{M}_2)$. These torques Doppler shift the energy dispersion of the left, $\omega_{\boldsymbol{p}}^{(L)}$, and right, $\omega_{\boldsymbol{p}}^{(R)}$, polarized modes in the opposite directions lifting the degeneracy between magnons of different polarizations

to create a pair of equal anti-parallel spin transfer torques $\mathfrak{T}_1$ and $\mathfrak{T}_2$ acting on magnetizations $\boldsymbol{M}_1$ and $\boldsymbol{M}_2$ respectively, as schematically shown in Fig. 9.2.

9.3.4.1 Doppler Shift from a Pure Spin Current

The Landau–Lifshitz–Gilbert equations of motion for the magnetizations in the presence of the spin-transfer torques are written as follows

$$\frac{\partial \boldsymbol{M}_1}{\partial t} = \gamma \boldsymbol{M}_1 \times \boldsymbol{H}_1^{\text{eff}} + \eta \boldsymbol{M}_1 \times \frac{\partial \boldsymbol{M}_1}{\partial t} - \frac{v_s}{M_s^2} \boldsymbol{M}_1 \times (\boldsymbol{M}_1 \times \nabla_n \boldsymbol{M}_1), \quad (9.57)$$

$$\frac{\partial \boldsymbol{M}_2}{\partial t} = \gamma \boldsymbol{M}_2 \times \boldsymbol{H}_2^{\text{eff}} + \eta \boldsymbol{M}_2 \times \frac{\partial \boldsymbol{M}_2}{\partial t} + \frac{v_s}{M_s^2} \boldsymbol{M}_2 \times (\boldsymbol{M}_2 \times \nabla_n \boldsymbol{M}_2), \quad (9.58)$$

where we neglect non-adiabatic contribution to the spin torque. Taking into account that $|\boldsymbol{M}_i| = M_s$ $(i = 1, 2)$, these expressions can be rewritten as follows

$$\left(\frac{\partial}{\partial t} \mp v_s \nabla_n\right) \boldsymbol{M}_i = \gamma \boldsymbol{M}_i \times \boldsymbol{H}_i^{\text{eff}} + \eta \boldsymbol{M}_i \times \frac{\partial \boldsymbol{M}_i}{\partial t}, \quad (9.59)$$

where the upper (lower) sign is for $i = 1$ $(i = 2)$. This expression shows that the role of the adiabatic spin transfer torque is to produce a Doppler shift of the spin waves by the velocity v_s. This effect is well-known for ferromagnetic and antiferromagnetic spin waves when the Doppler shift is caused by a spin polarized electric current [59–61]. In our case, the pure spin current produces two Doppler shifts in the opposite directions for the magnetization dynamics on each sublattice.

By solving the equations of motion (9.57) and (9.58), it is possible to show that in the presence of the spin current, the degeneracy between left and right polarizations in the dispersion relations for the spin waves propagating along $\boldsymbol{n}$ becomes lifted, and it can be approximated as follows [38]

$$\omega_p^{(R)} = c_s|p - p_s| + i\eta(\Delta_s - pv_s), \tag{9.60}$$

$$\omega_p^{(L)} = c_s|p + p_s| + i\eta(\Delta_s + pv_s), \tag{9.61}$$

where $p_s = \gamma M_s v_s \delta/(2c_s^2)$, $\Delta = \gamma M_s \delta/2$, and $p \gg p_s$ is the wave vector of the spin waves along $\boldsymbol{n}$, see Fig. 9.2.

This effect is in contrast to the Doppler shift from a spin polarized current where both modes are shifted in the same direction so that the degeneracy holds [59]. The imaginary parts of the frequencies $\omega_p^{(R)}$ and $\omega_p^{(L)}$ also have contributions from the spin current of the opposite signs for the waves with left and right polarizations. This can be considered as a spin-current-induced circular dichroims of spin waves, which occurs at the characteristic length scale $\ell_{\rm CD} = c_s/(\eta v_s p)$.

Interestingly, the effect of spin current on the spin waves in the linear approximation is analogous to the existence of the additional Dzyaloshinskii-Moriya interaction (DMI) term in the antiferromagnetic energy in (9.40)

$$W_{\rm DMI} = \frac{v_s}{2\gamma M_s} \int d^3r \left[\boldsymbol{m}_1 \cdot (\nabla_n \times \boldsymbol{m}_1) + \boldsymbol{m}_2 \cdot (\nabla_n \times \boldsymbol{m}_2)\right], \tag{9.62}$$

between the magnetizations on the same sublattices.

9.3.4.2 Asymmetric Energy Absorption

Let us now look at the spin-wave energy absorption. The dissipation rate for the magnetization dynamics can be expressed through the Rayleigh dissipation function

$$\frac{dW}{dt} = -\frac{\eta}{\gamma} \int d^3r \left[\left(\frac{\partial \boldsymbol{M}_1}{\partial t}\right)^2 + \left(\frac{\partial \boldsymbol{M}_2}{\partial t}\right)^2\right]. \tag{9.63}$$

According to the equations of motion (9.57) and (9.58), in the presence of the spin current we replace ∂_t with $\partial_t - v_s \nabla_n$ for $\boldsymbol{M}_1$ and with $\partial_t + v_s \nabla_n$ for $\boldsymbol{M}_2$. The energy dissipation rate in (9.63) in this case acquires the asymmetric contribution proportional to v_s that is written as

$$\left(\frac{dW}{dt}\right)_\chi = \frac{2\eta v_s}{\gamma} \int d^3r \left(\nabla_n \boldsymbol{m}_1 \cdot \frac{\partial \boldsymbol{m}_1}{\partial t} - \nabla_n \boldsymbol{m}_2 \cdot \frac{\partial \boldsymbol{m}_2}{\partial t}\right). \tag{9.64}$$

The expression in parentheses is nothing but the spin-wave chirality density $\rho_\chi^{(\rm m)}$ written in terms of $\boldsymbol{m}_1$ and $\boldsymbol{m}_2$.

As a result, when a pure spin current is injected into an antiferromagnet, the asymmetry in the spin-wave energy absorption rate becomes proportional to the spin-wave chirality, $(dW/dt)_\chi = 2\eta v_s \gamma^{-1} C_\chi^{\rm m}$. This result is a direct analogy with the result of Tang and Cohen [27] for the electromagnetic energy absorption rate in chiral metamaterials, see Sect. 9.2.2. In antiferromagnetic materials, the microscopic mechanism beyond this phenomenon can be based on the adiabatic spin transfer torque from a pure spin current, or on the DMI between the same sublattices, which breaks the inversion symmetry and lifts the degeneracy between the left and right polarized magnon modes. In contrast to optical metamaterials, where the asymmetry in light-matter interactions is related to structural chirality, the symmetry breaking mechanism, which is based on the spin current, induces chirality of the material in controllable way. For a spin current density $j_s \approx 10^{11}$ A/m^2 (in the electric units), we obtain $v_s = \mu_B j_s/(eM_s) \approx 30$ m/s for $M_s \approx 3.5 \times 10^5$ A/m. This parameter should be compared to the typical velocity of the spin waves in antiferromagnetic insulators $c_s \approx 10^{-4}$ m/s, which gives $v_s/c_s \approx 10^{-3}$. The characteristic length of the magnon circular dichroism, in this situation, $\ell_{\mathrm{CD}} \approx 5$ mm for the magnon frequencies about 1 THz and $\eta \approx 10^4$. Curiously, the effective strength of the DMI, $D_{\mathrm{eff}} = \hbar v_s/(k_B a_0)$ is about 0.5 K (a_0 is the lattice spacing), which is comparable to a typical DMI strength in magnetic materials.

9.4 Excitation of Magnon Spin Photocurrents with Polarized Fields

Among the major goals of spintronics are generation of spin currents, their transmission over large distances, and conversion from one form to another because the spin angular momentum can be carried by different types of carriers. Since magnons are able to carry spin angular momentum, spin excitations in low damping magnetic insulators are good candidates for being spin current mediators. The absence of the net magnetization and the existence of two polarization states per magnon make antiferromagnetic insulators particularly suitable for applications as spin current conductors. It was demonstrated that an introduction of a thin layer of the antiferromagnetic insulator can enhance the spin current transmission in interface systems [62, 63].

Magnon spin currents in antiferromagnetic insulators can be excited by several methods. For example, it can be done by pumping a magnon spin current from a neighboring ferromagnetic layer [62]. Thermal excitation of spin currents via the spin versions of the Seebeck and Nernst effects also has attracted considerable attention [64–68]. The latter is especially interesting in low-dimensional materials, where it is provided by topological terms in magnon dynamics [69–71].

Optical control of spin states in antiferromagnetic insulators [72, 73] is a feature in the emerging field of antiferromagnetic optospintronics [5]. In this respect, it is an intriguing problem to investigate whether it is possible to find some sort of magnon photo-effect [44]. Symmetry considerations suggest that this is indeed possible. As we have already mentioned, spin currents satisfy the definition of true chirality [11], which can be directly seen from the conservation law for the μth component of the spin density

$$\frac{\partial s^{\mu}(t, \boldsymbol{r})}{\partial t} + \nabla \cdot \boldsymbol{j}^{\mu}(t, \boldsymbol{r}) = 0. \tag{9.65}$$

Since $s^\mu(t, \boldsymbol{r})$ is $\mathcal{T}$ odd and $\mathcal{P}$ even, the spin current density $\boldsymbol{j}^\mu(t, \boldsymbol{r})$ has opposite transformation properties. As we have seen in Sect. 9.2, the electromagnetic field can be characterized by optical chirality $\rho_\chi(t, \boldsymbol{r})$ with the same transformations properties as $\boldsymbol{j}^\mu(t, \boldsymbol{r})$. Therefore, we may expect that by exposing an antiferromagnetic insulator to a circularly polarized electromagnetic field, we can excite a spin photocurrent, which direction should be determined by the helicity of light.

In this section, we will consider these arguments in detail, and show that this photo-excitation process requires the frequency of the electromagnetic field to be in the region of the antiferromagnetic resonance. We begin with a semiclassical theory. Nonlinear response and geometric effects in low dimensional materials are discussed at the end of this section. First we consider an interesting phenomenon analogous to the *Zitterbewegung* effect for magnons.

9.4.1 Magnon Spin Currents in Antiferromagnets

Equations (9.42) and (9.43) preserve rotation symmetry along the magnetic ordering direction that warrants conservation of the total angular momentum component along $\boldsymbol{n}$. From these equations, the time evolution of the nth component of the magnetization $M^{(n)} = \frac{1}{2M_s}(m_2^2 - m_1^2)$ is written in the following form

$$\frac{\partial M^{(n)}(t, \boldsymbol{r})}{\partial t} = \frac{1}{4M_s} \sum_{\boldsymbol{pq}} e^{-i\boldsymbol{q}\cdot\boldsymbol{r}} \boldsymbol{n} \cdot \Big\{ \left(\varepsilon^{(l)}_{\boldsymbol{p}-\boldsymbol{q}} - \varepsilon^{(l)}_{-\boldsymbol{p}} \right) \left[\boldsymbol{l}^*_{\boldsymbol{p}-\boldsymbol{q}} \times \boldsymbol{l}_{\boldsymbol{p}} \right] + \left(\varepsilon^{(m)}_{-\boldsymbol{p}+\boldsymbol{q}} - \varepsilon^{(m)}_{\boldsymbol{p}} \right) \left[\boldsymbol{m}^*_{\boldsymbol{p}-\boldsymbol{q}} \times \boldsymbol{m}_{\boldsymbol{p}} \right] \Big\} . \tag{9.66}$$

In the limit $\boldsymbol{q} \to 0$, this equation can be rewritten in the form of a continuity equation $\partial_t M^{(n)}_{\boldsymbol{q}} + i\boldsymbol{q} \cdot \boldsymbol{J}^{(n)}_s = 0$, where

$$\boldsymbol{J}^{(n)}_s = \frac{i}{4M_s} \sum_{\boldsymbol{p}} \left(\frac{\partial \varepsilon^{(m)}_{\boldsymbol{p}}}{\partial \boldsymbol{p}} \boldsymbol{m}^*_{\boldsymbol{p}} \cdot (\boldsymbol{n} \times \boldsymbol{m}_{\boldsymbol{p}}) + \frac{\partial \varepsilon^{(l)}_{\boldsymbol{p}}}{\partial \boldsymbol{p}} \boldsymbol{l}^*_{\boldsymbol{p}} \cdot (\boldsymbol{n} \times \boldsymbol{l}_{\boldsymbol{p}}) \right) \tag{9.67}$$

is the total magnon spin current. This expression looks similar to our definition of the spin-wave chirality in (9.53), especially if we consider the spin current flow along $\boldsymbol{n}$. However, as we shall see below, in contrast to magnon chirality, $\boldsymbol{J}^{(n)}_s$ does not obey any conservation law. It should be mentioned that the same expression for the spin current can be obtained directly from the antiferromagnetic Lagrangian using Noether's theorem (see Appendix).

It is interesting to discuss the analogy between antiferromagnetic magnon spin currents and charge currents in pseudo-relativistic Dirac materials. In the latter case, it was demonstrated that interband effects make a significant contribution near the Dirac point and can explain, for example, the universal conductivity of graphene [74]. In the relativistic language, interband effects in the dynamics of an electron wave packet correspond to the *Zitterbewegung*, or the trembling motion of an ultra-relativistic particle [74]. The *Zitterbewegung* effect has also been proposed for antiferromagnetic magnons [40]. It can be easily understood by looking at the time evolution of $\bar{\psi}_{\boldsymbol{p}}(t)$ calculated from (9.45) to (9.46)

$$\bar{\psi}_{\boldsymbol{p}}(t) = \frac{1}{2}\left\{\left[1 + \sigma_1 \otimes (\hat{\boldsymbol{S}} \cdot \boldsymbol{n})\right] e^{i\omega_{\boldsymbol{p}} t} + \left[1 - \sigma_1 \otimes (\hat{\boldsymbol{S}} \cdot \boldsymbol{n})\right] e^{-i\omega_{\boldsymbol{p}} t}\right\} \bar{\psi}_{\boldsymbol{p}}(0), \quad (9.68)$$

which is similar to the analogous equation for relativistic particles [74]. This expression contains the off-diagonal elements responsible for the mixing of $\boldsymbol{m}_{\boldsymbol{p}}$ and $\boldsymbol{l}_{\boldsymbol{p}}$ components of $\bar{\psi}_{\boldsymbol{p}}$ while evolving in time.

By applying (9.68) to the time evolution of the spin current in (9.67), we find that the spin current has two contributions, $\boldsymbol{J}_s^{(n)}(t) = \boldsymbol{J}_{s0}^{(n)} + \boldsymbol{J}_{s1}^{(n)}(t)$. The first contribution is conserved part of the spin current. It does not depend on time and is proportional to the group velocity of magnons $v_{\boldsymbol{p}} = \partial\omega_{\boldsymbol{p}}/\partial\boldsymbol{p}$. In our matrix notations, it can be written as

$$\boldsymbol{J}_s^{(n)} = \frac{1}{4M_s}\sum_{\boldsymbol{p}} v_{\boldsymbol{p}} \bar{\psi}^{\dagger}(0)(\hat{\boldsymbol{S}} \cdot \boldsymbol{n})\bar{\psi}_{\boldsymbol{p}}(0). \quad (9.69)$$

The second term in the spin current oscillates at the double frequency, and can be attributed to the *Zitterbewegung* of magnons

$$\boldsymbol{J}_{s1}^{(n)}(t) = \frac{1}{16M_s}\sum_{\boldsymbol{p}} e^{2i\omega_{\boldsymbol{p}} t} \boldsymbol{K}_{\boldsymbol{p}} \bar{\psi}_{\boldsymbol{p}}^{\dagger}(0) \begin{pmatrix} (\hat{\boldsymbol{S}} \cdot \boldsymbol{n}) & 1 \\ -1 & -(\hat{\boldsymbol{S}} \cdot \boldsymbol{n}) \end{pmatrix} \bar{\psi}_{\boldsymbol{p}}(0) + \text{H.c.}, \quad (9.70)$$

where

$$\boldsymbol{K}_{\boldsymbol{p}} = \frac{1}{\omega_{\boldsymbol{p}}}\left(\varepsilon_{\boldsymbol{p}}^{(l)} \frac{\partial \varepsilon_{\boldsymbol{p}}^{(m)}}{\partial \boldsymbol{p}} - \varepsilon_{\boldsymbol{p}}^{(m)} \frac{\partial \varepsilon_{\boldsymbol{p}}^{(l)}}{\partial \boldsymbol{p}}\right). \quad (9.71)$$

The physical meaning of these terms becomes clear if we transform to the helicity basis, $\tilde{\psi}_{\boldsymbol{p}} = (\tilde{\psi}_{\boldsymbol{p}}^{(R)}, \tilde{\psi}_{\boldsymbol{p}}^{(L)})^T$, where we have well-defined left and right polarized magnon modes, see (9.14), (9.47) and (9.48). In this basis, the first term is determined by the difference in numbers of magnons with opposite polarizations

$$\boldsymbol{J}_s^{(n)} = \frac{1}{4M_s}\sum_{\boldsymbol{p}} v_{\boldsymbol{p}} \left(\tilde{\psi}_{\boldsymbol{p}}^{*(R)}\tilde{\psi}_{\boldsymbol{p}}^{(R)} - \tilde{\psi}_{\boldsymbol{p}}^{*(L)}\tilde{\psi}_{\boldsymbol{p}}^{(L)}\right), \quad (9.72)$$

while the second term is purely off-diagonal and corresponds to the interband processes

$$\boldsymbol{J}_{s1}^{(n)}(t) = -\frac{1}{8M_s}\sum_{\boldsymbol{p}} \tilde{\psi}_{\boldsymbol{p}}^{\dagger}(0) \begin{pmatrix} 0 & \boldsymbol{K}_{\boldsymbol{p}}\hat{S}^z e^{-2i\omega_{\boldsymbol{p}}\hat{S}^z t} \\ \boldsymbol{K}_{\boldsymbol{p}}\hat{S}^z e^{2i\omega_{\boldsymbol{p}}\hat{S}^z t} & 0 \end{pmatrix} \tilde{\psi}_{\boldsymbol{p}}(0). \quad (9.73)$$

It should be mentioned that the contribution of the oscillating term in total spin current may seem insignificant. Indeed, in the theory the spin Seebeck effect only the term given by (9.72) was taken into account in the definition of the spin current [65, 66]. In this case, the second term, which mixes magnons of different helicities, has vanishing contribution. However, as we discuss below, such processes as the photo-excitation require both terms being considered with equal attention. Moreover, the contribution of the second term in (9.73) may become dominant in low-dimensional systems where it may contain geometric phase effects.

9.4.2 Photo-Excitation of Magnon Spin Currents

Let us now turn to a semi-classical theory of photo-excitation of magnon spin currents. For this purpose, we add a magneto-dipole interaction between the magnetic field component of the electromagnetic wave $\boldsymbol{h}(t, \boldsymbol{r})$ and the magnetization of the antiferromagnet, so that the total energy is written as

$$W_t = W - \int d^3r(\boldsymbol{M}_1 + \boldsymbol{M}_2) \cdot \boldsymbol{h}(t, \boldsymbol{r}), \tag{9.74}$$

where W is determined by (9.40). In this case, (9.43) acquires the additional term $-2\gamma M_s[\boldsymbol{n} \times \boldsymbol{h}_{\boldsymbol{p}}(t)]$, where $\boldsymbol{h}_{\boldsymbol{p}}(t)$ is the Fourier component of the magnetic field. The system of equations of motion (9.42) and (9.43) can be easily solved by transforming the ω-domain, which gives

$$\boldsymbol{m}_{\boldsymbol{p}}(\omega) = 2\gamma M_s \frac{\varepsilon_{\boldsymbol{p}}^{(l)} \boldsymbol{h}_{\boldsymbol{p}}(\omega)}{\omega_{\boldsymbol{p}}^2 - \omega^2}, \tag{9.75}$$

$$\boldsymbol{l}_{\boldsymbol{p}}(\omega) = 2i\gamma M_s \frac{\omega[\boldsymbol{n} \times \boldsymbol{h}_{\boldsymbol{p}}(\omega)]}{\omega_{\boldsymbol{p}}^2 - \omega^2}. \tag{9.76}$$

The Gilbert damping can be phenomenologically introduced in these equations by considering complex parameters $\varepsilon_{\boldsymbol{p}}^{(\alpha)} \to \varepsilon_{\boldsymbol{p}}^{(\alpha)} - i\eta\omega$ ($\alpha = m, l$). Using the definition of the spin current in (9.67), we find the current excited by the magnetic field vector (Fig. 9.3)

$$\boldsymbol{J}_s^{(n)} = i\gamma^2 M_s \sum_{\boldsymbol{p}\omega} \frac{\varepsilon_{\boldsymbol{p}}^{(l)2} \partial_{\boldsymbol{p}} \varepsilon_{\boldsymbol{p}}^{(m)} + \omega^2 \partial_{\boldsymbol{p}} \varepsilon_{\boldsymbol{p}}^{(l)}}{\left(\omega^2 - \omega_{\boldsymbol{p}}^2\right)^2} \boldsymbol{h}_{\boldsymbol{p}}^*(\omega) \cdot [\boldsymbol{n} \times \boldsymbol{h}_{\boldsymbol{p}}(\omega)]. \tag{9.77}$$

This expression shows that the direct spin current excited by the electromagnetic wave is the second order effect in $\boldsymbol{h}_{\boldsymbol{p}}(\omega)$, and is determined by the asymmetric combination $\boldsymbol{h}_{\boldsymbol{p}}^* \times \boldsymbol{h}_{\boldsymbol{p}}$, so that the direction of the current is determined by helicity of the electromagnetic wave. The effect is resonantly amplified near the antiferromagnetic resonance $\omega \approx \omega_{\boldsymbol{p}}$.

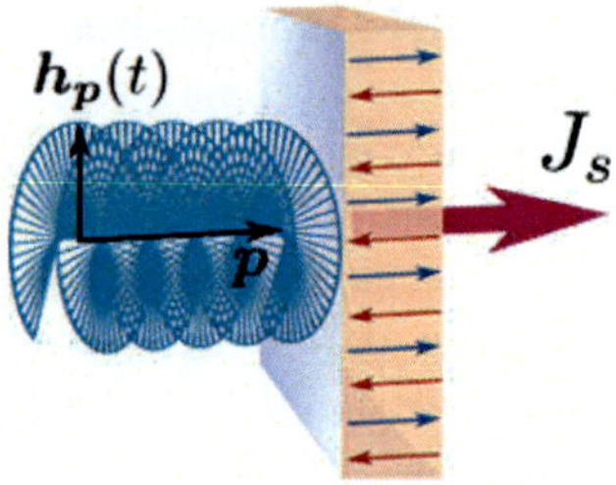

Fig. 9.3 Schematic picture of the magnon photocurrent $J_s^{(n)}$ induced inside an antiferromagnet by the circularly polarized electromagnetic wave propagating along the direction of magnetic ordering

Photo-excitation of magnon spin currents in antiferromagnetic insulators shows some similarity with the circular photogalvanic effect in noncentrosymmetric metals [45]. In the latter case, a direct electric photocurrent is generated by the helical combination the electric-field vector of the electromagnetic wave, $\boldsymbol{E}^*(\omega) \times \boldsymbol{E}(\omega)$, so that the direction of the current is reversed whenever circular polarization of light is switched to the opposite.

In order to have further insight into magnon spin photocurrents, let us consider a quantum variant of our theory.

9.4.3 Microscopic Theory of Magnon Spin Photocurrents

The spin Hamiltonian for an antiferromagnetic insulator with two magnetic sublattices A and B can be written in the following form

$$\hat{H} = \sum_{ij} \frac{1}{2}\left(J_{ij} S_i^{(+)} S_j^{(-)} + J_{ij}^* S_i^{(-)} S_j^{(+)}\right) + \sum_{ij} J'_{ij} S_i^z S_j^z - K \sum_i (S_i^z)^2, \tag{9.78}$$

where J_{ij} and J'_{ij} are the exchange interaction constants such as $\mathrm{Re}\, J_{ij} > 0$ and $J'_{ij} > 0$ for the nearest neighboring sites on A and B sublattices, and $K \sim \beta a_0^{-3}$ is the magnetic anisotropy that stabilizes the antiferromagnetic ordering along the z direction. We do not specify any lattice configuration at this stage. However, we note that J_{ij} may have a complex phase factors in the presence of DMI.

The spin-wave approximation for the Hamiltonian (9.78) is conveniently expressed by the Holstein–Primakoff transformation of the spin operators

$$\begin{aligned} S_{iA}^{(+)} &= \sqrt{2S} a_i, \quad & S_{iB}^{(+)} &= \sqrt{2S} b_i^\dagger, \\ S_{iA}^{(-)} &= \sqrt{2S} a_i^\dagger, \quad & S_{iB}^{(-)} &= \sqrt{2S} b_i, \\ S_{iA}^z &= S - a_i^\dagger a_i, \quad & S_{iB}^z &= -S + b^\dagger b_i^\dagger, \end{aligned} \tag{9.79}$$

where a_i and b_i are boson operators at the A and B sublattice respectively, which satisfy boson commutation relations. By transforming these operators to the reciprocal space, $a_i = \sum_k \exp(i\boldsymbol{k}\cdot\boldsymbol{r}_i)a_k$ and $b_i = \sum_k \exp(i\boldsymbol{k}\cdot\boldsymbol{r}_i)b_k$, we can rewrite (9.78) in the following form

$$\hat{H} = \sum_{\boldsymbol{k}} \left[A_k \left(a_k^\dagger a_k + b_{-k}^\dagger b_{-k} \right) + B_k a_k b_{-k} + B_k^* a_k^\dagger b_{-k}^\dagger \right], \tag{9.80}$$

where parameters A_k and B_k include microscopic details. For example, in the case when the exchange interactions are limited by the nearest neighboring sites so that $J_{ij} = J'_{ij} = J_1$, we obtain $A_k = 2KS + ZJ_1S$ and $B_k = J_1S\sum_{\boldsymbol{\delta}} \exp(-i\boldsymbol{k}\cdot\boldsymbol{\delta})$, where $\boldsymbol{\delta}$ connects a site on the A sublattice with its Z nearest neighboring sites on the B sublattice.

9.4.3.1 Magnon Spin Currents: Quantum Version

The expression for a magnon spin current can be derived following the same steps as in Sect. 9.4.1. Considering the equation of motion for the z component of the local spin density, $n(\boldsymbol{r}_i) = b_i^\dagger b_i - a_i^\dagger a_i$, we find the total magnon spin current

$$\hat{\boldsymbol{J}}_s = \sum_{\boldsymbol{k}} \left[\frac{\partial A_k}{\partial \boldsymbol{k}} \left(a_k^\dagger a_k + b_{-k}^\dagger b_{-k} \right) + \frac{\partial B_k}{\partial \boldsymbol{k}} a_k b_{-k} + \frac{\partial B_k^*}{\partial \boldsymbol{k}} a_k^\dagger b_{-k}^\dagger \right]. \tag{9.81}$$

This expression can be conveniently written in the matrix form

$$\hat{\boldsymbol{J}}_s = \sum_{\boldsymbol{k}} \chi_k^\dagger \frac{\partial \mathcal{H}_k}{\partial \boldsymbol{k}} \chi_k, \tag{9.82}$$

where we introduced $\chi_k = \begin{pmatrix} a_k \\ b_{-k}^\dagger \end{pmatrix}$ and $\mathcal{H}_k = \begin{pmatrix} A_k & B_k^* \\ B_k & A_k \end{pmatrix}$. Note that in this representation, χ_k does not satisfy boson communication relations; instead one has $[\chi_k, \chi_{k'}^\dagger] = \sigma_z \delta_{k,k'}$, which should be kept in mind.

Let us find how $\hat{\boldsymbol{J}}_s$ transforms under the Bogolyubov's transformation that preserves boson commutation relations of magnon operators. In the matrix form, this transformation is expressed as $\chi_k = U_k \tilde{\chi}_k$, where the transformation matrix is determined by two real parameters θ_k and ϕ_k:

$$U_k = \begin{pmatrix} \cosh\theta_k e^{i\phi_k} & -\sinh\theta_k \\ -\sinh\theta_k & \cosh\theta_k e^{-i\phi_k} \end{pmatrix}. \tag{9.83}$$

Since the definition of spin current involves $\partial_{\boldsymbol{k}}$, its transformation properties invoke covariant derivatives with respect to U_k. Explicit calculations show that in an arbitrary basis

$$\hat{\boldsymbol{J}}_s = \sum_{\boldsymbol{k}} \tilde{\chi}_k^\dagger \frac{\partial \tilde{\mathcal{H}}_k}{\partial \boldsymbol{k}} \tilde{\chi}_k - \frac{\partial \hat{\boldsymbol{A}}}{\partial t}, \tag{9.84}$$

where $\tilde{\mathcal{H}}_k = U_k^\dagger \mathcal{H}_k U_k$ is the Hamiltonian in the transformed basis, and $\hat{A} = \sum_k \tilde{\chi}_k^\dagger \mathcal{A}_k \tilde{\chi}_k$ with

$$\mathcal{A}_k = -i\sigma_z U_k^{-1} \frac{\partial U_k}{\partial k} \tag{9.85}$$

being the connection associated with the transformation U_k.

Among the various representations, there is one specific basis, where the Hamiltonian in (9.80) becomes diagonal. This basis is reached by choosing $\tanh 2\theta_k = |B_k|/A_k$ and $\phi_k = \arg B_k$, which gives

$$\hat{H} = \sum_k \varepsilon_k \left(\alpha_k^\dagger \alpha_k + \beta_{-k}^\dagger \beta_{-k} \right), \tag{9.86}$$

where $\varepsilon_k = \sqrt{A_k^2 - |B_k|^2}$ is the magnon energy dispersion. To find the expression for the spin current in this basis, we notice that in (9.84)

$$-\frac{\partial \hat{A}}{\partial t} = i[\hat{A}, \hat{H}] = \sum_k (\alpha_k^\dagger, \beta_{-k}) \begin{pmatrix} 0 & K_k^* \\ K_k & 0 \end{pmatrix} \begin{pmatrix} \alpha_k \\ \beta_{-k}^\dagger \end{pmatrix}, \tag{9.87}$$

is purely off-diagonal with $K_k = e^{i\phi_k} \left[\varepsilon_k^{-1} (A_k \partial_k |B_k| - |B_k| \partial_k A_k) - i |B_k| \partial_k \phi_k \right]$. Therefore, the total magnon spin current is written as

$$\hat{J}_s = \sum_k (\alpha_k^\dagger, \beta_{-k}) \begin{pmatrix} v_k & K_k^* \\ K_k & v_k \end{pmatrix} \begin{pmatrix} \alpha_k \\ \beta_{-k}^\dagger \end{pmatrix}, \tag{9.88}$$

where $v_k = \partial_k \varepsilon_k$ is the group velocity of magnons [44]. This expression generalizes two contributions to the spin current in (9.72) and (9.73) identified earlier in our semi-classical approach.

9.4.3.2 Nonlinear Response Theory for Magnon Spin Photocurrents

By using semi-classical equations of motion in Sect. 9.4.2, we have already demonstrated that magnon spin photocurrent is the second order effect in the magnetic field of the electromagnetic wave. Here, we show how the process of photo-excitation can be described via the nonlinear response theory.

Considering interaction of magnons with the electromagnetic wave as a perturbation, we can express the excited spin current using the second-order Kubo formula [75]

$$\langle \hat{\boldsymbol{J}}_s(t)\rangle = -\sum_{\omega_1\omega_2}\int_{-\infty}^{t} dt_1 \int_{-\infty}^{t_1} dt_2 e^{\epsilon(t_1+t_2-t)} e^{i\omega_1 t_1 + i\omega_2 t_2} \times \left\langle \left[\left[\hat{\boldsymbol{J}}_s(t), \hat{H}_I^{(\omega_1)}(t_1)\right], \hat{H}_I^{(\omega_2)}(t_2)\right]\right\rangle, \quad \epsilon \to 0^+, \tag{9.89}$$

where the interacting part of the Hamiltonian is taken in the form of dipole interaction between the magnetic field vector $\mathfrak{B}_{\boldsymbol{k}}(\omega)$ and the local magnetization of the antiferromagnet, $\hat{H}_I = -g\mu_B \sum_i \mathfrak{B}(t, \boldsymbol{r}_i)(\boldsymbol{S}_{iA} + \boldsymbol{S}_{iB})$, where g is the Landé factor. In terms of magnon operators, it is expressed as

$$\hat{H}_I^{(\omega)} = -g\mu_B\sqrt{\frac{S}{2}}\sum_{\boldsymbol{k}}\left[\mathfrak{B}_{\boldsymbol{k}}^{(-)}(\omega)\left(a_{\boldsymbol{k}} + b_{-\boldsymbol{k}}^{\dagger}\right) + \text{H.c.}\right]. \tag{9.90}$$

In (9.89), the operators are in the Heisenberg picture, e.g. $\hat{H}_I^{(\omega_1)}(t_1) = \exp(i\hat{H}t_1)\hat{H}_I^{(\omega_1)} \exp(-i\hat{H}t_1)$, and the statistical average is with the density matrix of the noninteracting system $\rho_0 = \exp(-\hat{H}/k_BT)$.

Straightforward algebra shows that the spin current is calculated from (9.89) as follows [44]

$$\langle \hat{\boldsymbol{J}}_s(t)\rangle = \frac{1}{4}\sum_{\omega\boldsymbol{k}}\left[\frac{\boldsymbol{v}_{\boldsymbol{k}}\mu_{\boldsymbol{k}}}{(\varepsilon_{\boldsymbol{k}}-\omega)^2+\Gamma^2} + \frac{\boldsymbol{v}_{\boldsymbol{k}}\mu_{\boldsymbol{k}}}{(\varepsilon_{\boldsymbol{k}}+\omega)^2+\Gamma^2} + \frac{\lambda_{\boldsymbol{k}}\boldsymbol{K}_{\boldsymbol{k}}}{(\varepsilon_{\boldsymbol{k}}-\omega-i\Gamma)(\varepsilon_{\boldsymbol{k}}+\omega-i\Gamma)} + \frac{\lambda_{\boldsymbol{k}}^*\boldsymbol{K}_{\boldsymbol{k}}^*}{(\varepsilon_{\boldsymbol{k}}-\omega+i\Gamma)(\varepsilon_{\boldsymbol{k}}+\omega+i\Gamma)}\right] h_{\boldsymbol{k}}^{(-)}(\omega) h_{-\boldsymbol{k}}^{(+)}(-\omega), \tag{9.91}$$

where $\boldsymbol{h} = -g\mu_B\sqrt{2S}\mathfrak{B}$, $h^{(\pm)} = h^x \pm h^y$, and the coefficient are given by

$$\mu_{\boldsymbol{k}} = \frac{A_{\boldsymbol{k}} - |B_{\boldsymbol{k}}|\cos\phi_{\boldsymbol{k}}}{\sqrt{A_{\boldsymbol{k}}^2 - |B_{\boldsymbol{k}}|^2}}, \tag{9.92}$$

$$\lambda_{\boldsymbol{k}} = e^{-i\phi_{\boldsymbol{k}}}\left(\frac{A_{\boldsymbol{k}}\cos\phi_{\boldsymbol{k}} - |B_{\boldsymbol{k}}|}{\sqrt{A_{\boldsymbol{k}}^2 - |B_{\boldsymbol{k}}|^2}} - i\sin\phi_{\boldsymbol{k}}\right). \tag{9.93}$$

This expression contains two kinds of terms. The first is proportional to the group velocity of magnons, and, therefore, can be associated with actual motion of magnon wave packets. The second, proportional to $\boldsymbol{K}_{\boldsymbol{k}}$, is related to intersublattice dynamics; it contains the phase gradient, $\partial_{\boldsymbol{k}}\phi_{\boldsymbol{k}}$. This phase can be interpreted as an offset in dynamics of the magnetizations on A and B sublattices given by $a_{\boldsymbol{k}}(t) \sim \exp(i\varepsilon_{\boldsymbol{k}}t)$ and $b_{-\boldsymbol{k}}^{\dagger}(t) \sim \exp(i\varepsilon_{\boldsymbol{k}}t - i\phi_{\boldsymbol{k}})$ respectively. It may be accumulated as a result of the DMI combined with a specific lattice configuration [76], or be generated by the external electric field via the Aharonov-Casher effect [77, 78].

In the case when both $\boldsymbol{v_k}$ and $\boldsymbol{K_k}$ are odd under the transformation $\boldsymbol{k} \to -\boldsymbol{k}$, the spin current is determined by the asymmetric part of $\boldsymbol{h}_{\boldsymbol{k}}^{(-)}(\omega)\boldsymbol{h}_{-\boldsymbol{k}}^{(+)}(-\omega)$, which is proportional to $i[\boldsymbol{h}_{\boldsymbol{k}}^{*}(\omega) \times \boldsymbol{h}_{\boldsymbol{k}}(\omega)]_z$. In the limiting case $\Gamma \to 0$ and $\phi_{\boldsymbol{k}} = 0$, we can combine both kinds of terms in (9.91), which eventually gives

$$\langle \hat{\boldsymbol{J}}_s(t) \rangle = \frac{i}{2} \sum_{\omega \boldsymbol{k}} \frac{q_{\boldsymbol{k}}^2 \partial_{\boldsymbol{k}} p_{\boldsymbol{k}} - \omega^2 \partial_{\boldsymbol{k}} q_{\boldsymbol{k}}}{(\varepsilon_{\boldsymbol{k}}^2 - \omega^2)^2} [\boldsymbol{h}_{\boldsymbol{k}}^{*}(\omega) \times \boldsymbol{h}_{\boldsymbol{k}}(\omega)]_z, \tag{9.94}$$

where $p_{\boldsymbol{k}} = A_{\boldsymbol{k}} + |B_{\boldsymbol{k}}|$ and $q_{\boldsymbol{k}} = A_{\boldsymbol{k}} - |B_{\boldsymbol{k}}|$, which coincides with (9.77) obtained from the semi-classical equations of motion [44].

9.4.4 *Magnon Spin Photocurrents in Antiferromagnetic Insulators and Low Dimensional Materials*

We have demonstrated that in antiferromagnetic materials magnon spin currents contain intraband terms, proportional to the group velocity of magnons, and interband terms, which by analogy to the relativistic mechanics can be associated with the *Zitterbewegung* effect of magnons. The latter is proportional to the fast-oscillating factors, which makes these terms irrelevant as far as response to a static perturbation is concerned. For the thermal excitation of spin currents, for example, the antiferromagnetic spin current can be taken in the form of (9.72) [65, 69, 70].

The response to a dynamic perturbation is different. Since spin photocurrent is the second-order effect, the interband terms that oscillate at the double frequency should be taken into account together with the intraband contributions, so that the resulting response current is given by (9.94).

For practical applications, the most interesting frequency region is near the antiferromagnetic resonance, $\omega \approx \varepsilon_{\boldsymbol{k}}$. In this area, the response current is resonantly amplified and determined by the damping of the material. In the case of ballistic magnon transport, when $\varepsilon_{\boldsymbol{k}} \gg \Gamma$, we can replace $\omega - \varepsilon_{\boldsymbol{k}} \pm i\Gamma \to \pm i\Gamma$ and $\omega + \varepsilon_{\boldsymbol{k}} \pm i\Gamma \to 2\omega_r$ near the resonance ω_r. In this limit, the dominant contribution in (9.91) comes from the first term proportional to $\boldsymbol{v_k}$

$$\langle \hat{J}_s \rangle_{\text{res}} \approx \frac{i q_{\boldsymbol{k}}}{4\hbar\omega_r} \frac{\boldsymbol{v_k}}{\Gamma^2} [\boldsymbol{h}^{*}(\omega_r) \times \boldsymbol{h}(\omega_r)]_z, \tag{9.95}$$

where we used monochromatic microwave field with $\boldsymbol{h_k}(\omega)$ [44]. This expression allows to estimate the order of magnitude for the spin photocurrent excited with circularly polarized light as $\langle \hat{J}_s \rangle_{\text{res}} \approx \chi g^2 \mu_B^2 J_1 S^2 c_s I_B / (2a_0 c^2 \hbar \eta^2 \omega_r)$, where we take $\Gamma = \hbar\eta\omega_r$, $\chi = \pm$ denotes helicity of the wave, $I_B = |\mathfrak{B}(\omega_r)|^2$ is intensity, and linear magnon energy disperison is implied, $|v_{\boldsymbol{k}}| = c_s$. For a typical material with $c_s = 3 \times 10^4$ m/s, $J_s = 200$ K, $\omega_r = 3 \times 10^{13}$ s^{-1}, $\eta = 10^{-4}$, and $a_0 = 0.5$ nm, we

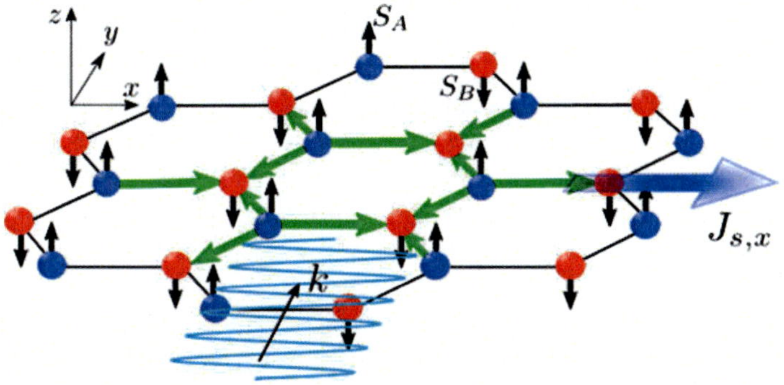

Fig. 9.4 Two-dimensional antiferromagnetic insulator with two magnetic sublattices $\boldsymbol{S}_A$ and $\boldsymbol{S}_B$ on the honeycomb lattice. Green arrows show the DMI configuration. The sign of D_{ij} is positive for $i \to j$ pointing from A to B

estimate $\langle \hat{J}_s \rangle_{\text{res}} \approx 1.5 \times 10^4$ A/m^2 (in electric units $e/\hbar$) for the microwave field strength $|\mathfrak{B}| \approx 10$ mT.

Relative contributions of different terms in (9.91) depend on the lattice configuration and on the details of microscopic interactions. We may expect that in low-dimensional antiferromagnets interband contribution determined by the phase gradient becomes more significant. We can separate this contribution from (9.91) as follows

$$\langle \hat{\boldsymbol{J}}_s \rangle_\phi = \frac{1}{2} \sum_{\omega \boldsymbol{k}} \frac{|B_{\boldsymbol{k}}| \sin \phi_{\boldsymbol{k}} \partial_{\boldsymbol{k}} \phi_{\boldsymbol{k}}}{\omega^2 - \varepsilon_{\boldsymbol{k}}^2} \boldsymbol{h}_{\boldsymbol{k}}^{(-)}(\omega) \boldsymbol{h}_{-\boldsymbol{k}}^{(+)}(-\omega). \tag{9.96}$$

Let us find a model system where this term in the spin current can be excited individually. For this purpose, we consider a two-dimensional antiferromagnet on the honeycomb lattice, as schematically shown in Fig. 9.4. This model is interesting because antiferromagnetic magnons on the honeycomb lattice have finite $\phi_{\boldsymbol{k}}$ even without DMI. Indeed, straightforward algebra shows that $B_{\boldsymbol{k}} = J_1 S C_{\boldsymbol{k}}$, where the structure factor is $C_{\boldsymbol{k}} = 2\cos(k_x/2)\cos(\sqrt{3}k_y/2) - 1 + 2i \sin(k_x/2)[\cos(k_x/2) - \cos(\sqrt{3}k_y/2)]$, which in the long-wavelength limit gives the phase $\phi_{\boldsymbol{k}} \approx k_x(3k_y^2 - k_x^2)/8$.

Note that $\phi_{\boldsymbol{k}}$ is odd under $\boldsymbol{k} \to -\boldsymbol{k}$. In order to break this symmetry, we add the specific DMI configuration $D_{ij}(\boldsymbol{S}_i \times \boldsymbol{S}_j)_z$ between the nearest neighboring sites i and j on the honeycomb lattice, such as $D_{ij} = D$ if $i \in A$ and $j \in B$, and $D_{ij} = -D$ otherwise. Adding such term does not modify the energy dispersion, but instead leads to the constant phase accumulation $B_{\boldsymbol{k}} = J_1 S C_{\boldsymbol{k}} \exp(i\phi_0)$ where $\tan\phi_0 = D/J_1$. In this case, $\sin(\phi_{\boldsymbol{k}} + \phi_0)\partial_{k_x}\phi_{\boldsymbol{k}}$ remains finite even in the $k_x \to 0$ limit. Therefore, by using (9.96), we are able to excite magnon spin current along x by the linearly polarized electromagnetic wave propagating along the y axis, see Fig. 9.4. The magnitude of the spin current is estimated as $\langle J_s^x \rangle_\phi \approx 3g^2\mu_B^2 J_1 S/(8\hbar^2 c^2) \sin\phi_0 I_B \omega^2/(\omega^2 - \varepsilon_{\boldsymbol{k}}^2)$, and its sign is proportional to the sign of ϕ_0.

9.5 Conclusions

We have discussed how symmetry analysis can help to bring new ideas from optics to antiferromagnetic spintronics. Our discussion started with an observation that a formal similarity between the electromagnetic field and spin waves in an antiferromagnetic insulator allows to find a generalization of optical chirality. This forms a background for establishing a link between optics of chiral metamaterials and magnonics. For example, spin wave absorption in chiral antiferromagnets can be described in the same terms as the electromagnetic energy dissipation in metamaterials. Moreover, in antiferromagnets a pure spin current can provide a chiral symmetry breaking in a controllable way through the spin torque mechanism.

Fundamentally, this follows from the fact that spin currents are truly chiral; they have the same $\mathcal{PT}$ transformation properties as e.g. optical chirality density. The latter suggests that chiral electromagnetic fields can be used for magnon spin current generation. We discussed that a direct magnon spin current appears as a second-order response to the circularly polarized microwave field, which frequency is near the antiferromagnetic resonance. The direction of the current is determined by helicity of light that makes it similar to the circular photogalvanic effect in metals.

Lastly, we discuss how magnon spin currents in antiferromagnets have an interesting dynamics that can come into play for photo-excitation. Besides the transport terms proportional to the group velocity of the spin waves, there is a contribution from the trembling motion of magnons, which can be identified by analogy with motion of ultra-relativistic particles. Although these fast oscillating terms can be safely omitted in some applications, they contribute to the photo-excitation process.

Acknowledgements R.L.S. acknowledges the support of the Natural Sciences and Engineering Research Council of Canada (NSERC) RGPIN 05011-18.

Appendix: Magnon Spin Current Definition from the Antiferromagnetic Lagrangian

Let us consider a classical spin model for an antiferromagnet with two sublattices $\boldsymbol{S}_A$ and $\boldsymbol{S}_B$ with the energy given by

$$\mathcal{H}_{\mathrm{AFM}} = J\sum_{\langle ij\rangle} \boldsymbol{S}_i \cdot \boldsymbol{S}_j - K\sum_i (S_i^z)^2, \tag{9.97}$$

where $J > 0$ is a nearest neighboring exchange interaction, K is the anisotropy constant along the z-axis, and summation is over the nearest neighboring sites on the A and B sublattices. For simplicity of notations, we consider one-dimensional arrangement of $\boldsymbol{S}_i$ along x. Semi-classical dynamics of this model can be captured from the following Lagrangian [79]

$$\mathcal{L} = \int dx \left[\rho \boldsymbol{M} \cdot \left(\boldsymbol{L} \times \frac{\partial \boldsymbol{L}}{\partial t} \right) - \frac{a}{2} |\boldsymbol{M}|^2 - A \frac{\partial}{\partial x} \left[\left(\frac{\partial \boldsymbol{L}}{\partial x} \right)^2 - \left(\frac{\partial \boldsymbol{M}}{\partial x} \right)^2 \right] - \ell \boldsymbol{M} \cdot \frac{\partial \boldsymbol{L}}{\partial x} + \frac{\tilde{\beta}}{2} (\boldsymbol{L}^z)^2 \right], \tag{9.98}$$

where $\boldsymbol{M} = \frac{1}{2S}(\boldsymbol{S}_A + \boldsymbol{S}_B)$ and $\boldsymbol{L} = \frac{1}{2S}(\boldsymbol{S}_A - \boldsymbol{S}_B)$, which satisfy the constraints $\boldsymbol{M} \cdot \boldsymbol{L} = 0$ and $\boldsymbol{M}^2 + \boldsymbol{L}^2 = 1$. The parameters of the Lagrangian are as follows: $\rho = 2\hbar S$, $a = 8JS^2$, $\ell = 2JS^2 a_0$, $A = JS^2 a_0^2$, and $\beta = 4KS^2$. Note that this expression contains so-called topological term proportional to ℓ, which breaks the inversion symmetry in the Lagrangian [79].

The expression for the spin current can be obtained applying the Noether's theorem to the Lagrangian transformation under the local infinitesimal rotation around z

$$\boldsymbol{M} \to \boldsymbol{M} + \delta\phi(\hat{z} \times \boldsymbol{M}), \tag{9.99}$$

$$\boldsymbol{L} \to \boldsymbol{L} + \delta\phi(\hat{z} \times \boldsymbol{L}), \tag{9.100}$$

where $\delta\phi(x)$ is the local rotation angle. The corresponding change in the Lagrangian is given by

$$\delta\mathcal{L} = -\int dx \delta\phi \left\{ \rho \frac{\partial}{\partial t} \left[M^z (1 - |\boldsymbol{M}|^2) \right] - A \frac{\partial}{\partial x} \left[(\hat{z} \times \boldsymbol{L}) \cdot \frac{\partial \boldsymbol{L}}{\partial x} - (\hat{z} \times \boldsymbol{M}) \cdot \frac{\partial \boldsymbol{M}}{\partial x} \right] - \ell \frac{\partial}{\partial x} \left[\boldsymbol{M} \cdot (\hat{z} \times \boldsymbol{L}) \right] \right\}, \tag{9.101}$$

which gives the following expression for the spin current density

$$j_s^z = -A\hat{z} \left[\left(\boldsymbol{L} \times \frac{\partial \boldsymbol{L}}{\partial x} \right) - \left(\boldsymbol{M} \times \frac{\partial \boldsymbol{M}}{\partial x} \right) \right] - \ell \hat{z} \cdot (\boldsymbol{L} \times \boldsymbol{M}). \tag{9.102}$$

The first term in this expression is consistent with the expression for the spin current obtained from the equations of motion. The second is the contribution from the topological terms, which has different symmetry. In particular, it changes the sign if we interchange $\boldsymbol{S}_A$ and $\boldsymbol{S}_B$.

References

1. I. Žutić, J. Fabian, S. Das Sarma, Rev. Mod. Phys. **76**, 323 (2004). https://doi.org/10.1103/RevModPhys.76.323
2. S.E. Thompson, S. Parthasarathy, Mater. Today **9**(6), 20 (2006). https://doi.org/10.1016/S1369-7021(06)71539-5

3. V. Baltz, A. Manchon, M. Tsoi, T. Moriyama, T. Ono, Y. Tserkovnyak, Rev. Mod. Phys. **90**, 015005 (2018). https://doi.org/10.1103/RevModPhys.90.015005
4. L. Šmejkal, Y. Mokrousov, B. Yan, A.H. MacDonald, Nat. Phys. **14**(3), 242 (2018). https://doi.org/10.1038/s41567-018-0064-5
5. P. Němec, M. Fiebig, T. Kampfrath, A.V. Kimel, Nat. Phys. **14**, 229 (2018)
6. M. Harder, C.M. Hu, Solid State Phys. **69**, 47 (2018). https://doi.org/10.1016/bs.ssp.2018.08.001
7. S.V. Kusminskiy (2019), p. 1911.11104
8. L. Kelvin, *Baltimore Lectures on Molecular Dynamics and the Wave Theory of Light* (CJ Clay & Sons, London, 1904)
9. L.D. Barron, Chirality **24**(11), 879 (2012). https://doi.org/10.1002/chir.22017
10. L. Barron, Chem. Phys. Lett. **123**(5), 423 (1986). https://doi.org/10.1016/0009-2614(86)80035-5
11. L.D. Barron, *Molecular Light Scattering and Optical Activity* (Cambridge University Press, 2004)
12. R. Naaman, D.H. Waldeck, Annu. Rev. Phys. Chem. **66**(1), 263 (2015). https://doi.org/10.1146/annurev-physchem-040214-121554
13. K. Michaeli, N. Kantor-Uriel, R. Naaman, D.H. Waldeck, Chem. Soc. Rev. **45**, 6478 (2016). https://doi.org/10.1039/C6CS00369A
14. R. Naaman, Y. Paltiel, D.H. Waldeck, Nat. Rev. Chem. **3**, 250 (2019). https://doi.org/10.1038/s41570-019-0087-1
15. D.M. Lipkin, J. Math. Phys. **5**(5), 696 (1964). https://doi.org/10.1063/1.1704165
16. M.G. Calkin, Am. J. Phys. **33**(11), 958 (1965). https://doi.org/10.1119/1.1971089
17. D. Zwanziger, Phys. Rev. **176**, 1489 (1968). https://doi.org/10.1103/PhysRev.176.1489
18. W.I. Fushchich, A.G. Nikitin, *Symmetries of Maxwell's Equations*. Mathematics and Its Applications (Springer, Netherlands, 1987)
19. I.Y. Krivskii, V.M. Simulik, Theor. Math. Phys. **80**(2), 864 (1989). https://doi.org/10.1007/BF01016113
20. I.Y. Krivskii, V.M. Simulik, Theor. Math. Phys. **80**(3), 912 (1989). https://doi.org/10.1007/BF01016183
21. G.N. Afanasiev, Y.P. Stepanovsky, Nuovo Cimento A **109**(3), 271 (1996). https://doi.org/10.1007/BF02731014
22. I. Bialynicki-Birula, *Coherence and Quantum Optics VII* (Springer, 1996), pp. 313–322
23. P.D. Drummond, Phys. Rev. A **60**(5), R3331 (1999)
24. P.D. Drummond, J. Phys. B: At. Mol. Opt. Phys. **39**(15), S573 (2006)
25. N.H. Ibragimov, Acta Appl. Math. **105**(2), 157 (2008). https://doi.org/10.1007/s10440-008-9270-y
26. M.V. Berry, J. Opt. A: Pure Appl. Opt. **11**(9), 094001 (2009), http://stacks.iop.org/1464-4258/11/i=9/a=094001
27. Y. Tang, A.E. Cohen, Phys. Rev. Lett. **104**, 163901 (2010). https://doi.org/10.1103/PhysRevLett.104.163901
28. K.Y. Bliokh, F. Nori, Phys. Rev. A **83**, 021803 (2011). https://doi.org/10.1103/PhysRevA.83.021803
29. S.M. Barnett, R.P. Cameron, A.M. Yao, Phys. Rev. A **86**, 013845 (2012). https://doi.org/10.1103/PhysRevA.86.013845
30. M.M. Coles, D.L. Andrews, Phys. Rev. A **85**, 063810 (2012). https://doi.org/10.1103/PhysRevA.85.063810
31. T.G. Philbin, Phys. Rev. A **87**, 043843 (2013). https://doi.org/10.1103/PhysRevA.87.043843
32. E. Hendry, T. Carpy, J. Johnston, M. Popland, R. Mikhaylovskiy, A. Lapthorn, S. Kelly, L. Barron, N. Gadegaard, M. Kadodwala, Nat. Nanotechn. **5**(11), 783 (2010)
33. Y. Tang, A.E. Cohen, Science **332**(6027), 333 (2011). https://doi.org/10.1126/science.1202817
34. M. Schäferling, D. Dregely, M. Hentschel, H. Giessen, Phys. Rev. X **2**, 031010 (2012). https://doi.org/10.1103/PhysRevX.2.031010

35. E. Hendry, R.V. Mikhaylovskiy, L.D. Barron, M. Kadodwala, T.J. Davis, Nano Lett. **12**(7), 3640 (2012). https://doi.org/10.1021/nl3012787
36. E.O. Kamenetskii, R. Joffe, R. Shavit, Phys. Rev. E **87**, 023201 (2013). https://doi.org/10.1103/PhysRevE.87.023201
37. A. Canaguier-Durand, J.A. Hutchison, C. Genet, T.W. Ebbesen, New J. Phys. **15**(12), 123037 (2013)
38. I. Proskurin, R.L. Stamps, A.S. Ovchinnikov, J.I. Kishine, Phys. Rev. Lett. **119**, 177202 (2017). https://doi.org/10.1103/PhysRevLett.119.177202
39. S.M. Barnett, New J. Phys. **16**(9), 093008 (2014), http://stacks.iop.org/1367-2630/16/i=9/a=093008
40. W. Wang, C. Gu, Y. Zhou, H. Fangohr, Phys. Rev. B **96**, 024430 (2017). https://doi.org/10.1103/PhysRevB.96.024430
41. L. Landau, E. Lifshitz, *Quantum Electrodynamics* (Pergamon Press, 1983)
42. B. Yan, C. Felser, Annu. Rev. Condens. Matter Phys. **8**(1), 337 (2017). https://doi.org/10.1146/annurev-conmatphys-031016-025458
43. A. Sekine, K. Nomura, Phys. Rev. Lett. **116**, 096401 (2016). https://doi.org/10.1103/PhysRevLett.116.096401
44. I. Proskurin, A.S. Ovchinnikov, J.I. Kishine, R.L. Stamps, Phys. Rev. B **98**, 134422 (2018). https://doi.org/10.1103/PhysRevB.98.134422
45. V. Belinicher, B.I. Sturman, Sov. Phys. Uspekhi **23**(3), 199 (1980)
46. F. de Juan, A.G. Grushin, T. Morimoto, J.E. Moore, Nat. Commun. **8**(1), 1 (2017)
47. R.P. Cameron, S.M. Barnett, New J. Phys. **14**(12), 123019 (2012), http://stacks.iop.org/1367-2630/14/i=12/a=123019
48. K.Y. Bliokh, A.Y. Bekshaev, F. Nori, New J. Phys. **15**(3), 033026 (2013), http://stacks.iop.org/1367-2630/15/i=3/a=033026
49. I. Fernandez-Corbaton, G. Molina-Terriza, Phys. Rev. B **88**, 085111 (2013). https://doi.org/10.1103/PhysRevB.88.085111
50. I. Proskurin, A.S. Ovchinnikov, P. Nosov, J.I. Kishine, New J. Phys. **19**(6), 063021 (2017). https://doi.org/10.1088/1367-2630/aa6acd
51. F.I. Fedorov, *Teorija girotropii* (Izd. Nauka i Technika, 1976)
52. J. Lekner, Pure Appl. Opt.: J. Eur. Opt. Soc. Part A **5**(4), 417 (1996)
53. K. Cho (2015)
54. D. Jaggard, A. Mickelson, C. Papas, Appl. Phys. **18**(2), 211 (1979)
55. S. Tomita, K. Sawada, A. Porokhnyuk, T. Ueda, Phys. Rev. Lett. **113**, 235501 (2014). https://doi.org/10.1103/PhysRevLett.113.235501
56. E.J. Post, *Formal Structure of Electromagnetics: General Covariance and Electromagnetics* (North-Holland Publishing Company, Amsterdam, 1962)
57. A.I. Akhiezer, V.G. Bar'yakhtar, S.V. Peletminskii, *Spin Waves* (North-Holland Publishing Company, Amsterdam, 1968)
58. S. Zhang, Z. Li, Phys. Rev. Lett. **93**, 127204 (2004). https://doi.org/10.1103/PhysRevLett.93.127204
59. Y. Yamane, J. Ieda, J. Sinova, Phys. Rev. B **94**, 054409 (2016). https://doi.org/10.1103/PhysRevB.94.054409
60. V. Vlaminck, M. Bailleul, Science **322**(5900), 410 (2008). https://doi.org/10.1126/science.1162843
61. A.C. Swaving, R.A. Duine, Phys. Rev. B **83**, 054428 (2011). https://doi.org/10.1103/PhysRevB.83.054428
62. H. Wang, C. Du, P.C. Hammel, F. Yang, Phys. Rev. Lett. **113**, 097202 (2014). https://doi.org/10.1103/PhysRevLett.113.097202
63. R. Khymyn, I. Lisenkov, V.S. Tiberkevich, A.N. Slavin, B.A. Ivanov, Phys. Rev. B **93**, 224421 (2016). https://doi.org/10.1103/PhysRevB.93.224421
64. S. Seki, T. Ideue, M. Kubota, Y. Kozuka, R. Takagi, M. Nakamura, Y. Kaneko, M. Kawasaki, Y. Tokura, Phys. Rev. Lett. **115**, 266601 (2015). https://doi.org/10.1103/PhysRevLett.115.266601

65. S.M. Rezende, R.L. Rodríguez-Suárez, A. Azevedo, Phys. Rev. B **93**, 014425 (2016). https://doi.org/10.1103/PhysRevB.93.014425
66. S.M. Rezende, R.L. Rodríguez-Suárez, A. Azevedo, Phys. Rev. B **93**, 054412 (2016). https://doi.org/10.1103/PhysRevB.93.054412
67. S.M. Wu, W. Zhang, A. KC, P. Borisov, J.E. Pearson, J.S. Jiang, D. Lederman, A. Hoffmann, A. Bhattacharya, Phys. Rev. Lett. **116**, 097204 (2016). https://doi.org/10.1103/PhysRevLett.116.097204
68. J. Holanda, D.S. Maior, O.A. Santos, L.H. Vilela-Leão, J.B.S. Mendes, A. Azevedo, R.L. Rodríguez-Suárez, S.M. Rezende, Appl. Phys. Lett. **111**(17), 172405 (2017). https://doi.org/10.1063/1.5001694
69. R. Cheng, S. Okamoto, D. Xiao, Phys. Rev. Lett. **117**, 217202 (2016). https://doi.org/10.1103/PhysRevLett.117.217202
70. V.A. Zyuzin, A.A. Kovalev, Phys. Rev. Lett. **117**, 217203 (2016). https://doi.org/10.1103/PhysRevLett.117.217203
71. Y. Shiomi, R. Takashima, E. Saitoh, Phys. Rev. B **96**, 134425 (2017). https://doi.org/10.1103/PhysRevB.96.134425
72. T. Satoh, S.J. Cho, R. Iida, T. Shimura, K. Kuroda, H. Ueda, Y. Ueda, B.A. Ivanov, F. Nori, M. Fiebig, Phys. Rev. Lett. **105**, 077402 (2010). https://doi.org/10.1103/PhysRevLett.105.077402
73. C. Tzschaschel, K. Otani, R. Iida, T. Shimura, H. Ueda, S. Günther, M. Fiebig, T. Satoh, Phys. Rev. B **95**, 174407 (2017). https://doi.org/10.1103/PhysRevB.95.174407
74. M.I. Katsnelson, Eur. Phys. J. B **51**(2), 157 (2006). https://doi.org/10.1140/epjb/e2006-00203-1
75. S.V. Tiablikov, *Methods in the Quantum Theory of Magnetism* (Springer, 2013)
76. M. Kawano, C. Hotta (2018), arXiv:1805.05872
77. K. Nakata, J. Klinovaja, D. Loss, Phys. Rev. B **95**, 125429 (2017). https://doi.org/10.1103/PhysRevB.95.125429
78. S. Owerre, J. Phys. Commun. **1**(2), 021002 (2017)
79. E.G. Tveten, T. Müller, J. Linder, A. Brataas, Phys. Rev. B **93**, 104408 (2016). https://doi.org/10.1103/PhysRevB.93.104408

Chapter 10
Realization of Artificial Chirality in Micro-/Nano-Scale Three-Dimensional Plasmonic Structures

Younghwan Yang, Yeseul Kim, Junho Gwak, Sunae So, Jungho Mun, Minkyung Kim, Heonyeong Jeong, Inki Kim, Trevon Badloe, and Junsuk Rho

Abstract Recent advancements in nano- and micro-fabrication technology have allowed the realization of artificial structured materials with strong electromagnetic chirality, far-exceeding natural chiral materials. This chapter categorizes the fabrication methods for realizing chiral structures based on the feature sizes, which closely relates to the operating wavelengths. Conventional top-down and bottom-up approaches are discussed along with their respective advantages and disadvantages, and the recently developed unconventional fabrication methods are also provided. Additionally, the chiral responses of the fabricated structures are briefly introduced. This chapter will contribute to the understanding of possible chiral structure designs and help to develop further fabrication methods for improving chiroptical activity.

10.1 Introduction

A chiral object cannot be superimposed onto its mirror image by only translations and rotations. In general, a chiral object and its mirror image (enantiomer) have identical physical characteristics such as mass, so they cannot be distinguished easily [1]. However, they may interact differently with other chiral objects; this feature has significant implications in pharmaceutics and synthetic chemistry, because a chiral molecule and its enantiomer have different chemical properties. Such chiral objects interact enantioselectively with circularly-polarized light, i.e., the chiral

Y.Yang and Y. Kim have contributed equally to this work.

Y. Yang · Y. Kim · J. Gwak · S. So · M. Kim · H. Jeong · I. Kim · T. Badloe · J. Rho (✉)
Department of Mechanical Engineering, Pohang University of Science and Technology (POSTECH), Pohang 37673, Republic of Korea
e-mail: jsrho@postech.ac.kr

J. Mun · J. Rho
Department of Chemical Engineering, Pohang University of Science and Technology (POSTECH), Pohang 37673, Republic of Korea

E. Kamenetskii (ed.), *Chirality, Magnetism and Magnetoelectricity*,
Topics in Applied Physics 138,
https://doi.org/10.1007/978-3-030-62844-4_10

objects interact differently with left-circularly-polarized (LCP) and right-circularly-polarized (RCP) light. Their absorption difference under LCP and RCP incidences is circular dichroism (CD), and the polarization rotation of light interacting with the chiral objects is optical rotatory dispersion. However, naturally occurring materials have intrinsically weak chiroptical responses, or circular birefringence, due to the large mismatch between the wavelength and the sizes of chiral molecules. This weak response hinders application of chiroptical properties.

Recent advances in artificial chirality in structured materials suggest a possibility of generating materials that have strong chiroptical properties that far exceed those found in natural materials, and chiral effects have been observed universally across different length scales. According to the antenna theory [2], the resonant wavelength of a single antenna is comparable to its geometric feature size. This principle also applies to chiral effects; the working wavelength of the chiral effects becomes generally similar to the geometric feature size of the chiral antenna. Therefore, chiral antennas that operate at wavelengths of micrometers or longer are relatively easy to fabricate. This principle applies because the optical properties of metals at long wavelengths are well represented as conductor (metallic) [3]. However, the optical properties of metals near the plasma frequency become strongly dispersive and deviate from those of a perfect conductor. Due to this feature, metallic antennas become lossy at wavelengths near the visible regime. With these points in mind, fabrication methods should be considered with regards to the desired working wavelength and the properties of the material.

Recently, subwavelength metallic particles have been actively investigated for surface-enhanced Raman scattering [4], photothermal effects [5], and optical antennas [6] due to their strong near-field enhancement and scattering effects, which originate from localized surface plasmon resonance (LSPR) coming the collective oscillations of electrons and photons trapped on the surface of the small particles [7]. When the particle size is much smaller than the wavelength, the LSPR can be predicted using the quasistatic theory. Therefore, the metallic antennas near the visible regime are distinguished from those at longer wavelengths as plasmonic antennas. The wavelength scaling principle above does not apply to plasmonic antennas [3], and subwavelength (~10 nm) chiral plasmonic nanoparticles (NPs) have strong chiral effects in the visible spectrum. In addition, the LSPR wavelength can be further redshifted without increasing the particle size by using high refractive index core and plasmonic shell [8, 9].

In this chapter, we discuss a few fabrication methods of chiral structures and their applications. Sections are divided according to the structure size, because it determines working wavelengths and the fabrication methods, and applications strongly depend on the working wavelengths.

10.2 Chirality at the Micrometer-Scale or Higher: Top-Down Approach

This chapter will introduce chiral plasmonic structures and their realization in micrometer-scale or beyond by using top-down fabrication. Artificial microscale structures generally interact with electromagnetic (EM) waves of 1 μm or more. The wavelength range ($\lambda > 1.0$ μm) includes infrared and microwave regions; they can be utilized for thermal detectors, light detection and ranging (LiDAR) devices, short-range wireless communication antenna, and spectroscopy. With these applications in mind, chiral plasmonic structures have been demonstrated from 1 μm size structures to millimeter-scale chiral ones. In this subchapter, their fabrication methods will be categorized by scales of artificial structures sizes; brief descriptions of chiroptical phenomena will be added.

10.2.1 Direct Laser Writing

A few micrometer-scale chiral structures (e.g. helical structures) have been fabricated using direct laser writing (DLW). This method can yield structures that have high aspect ratio, and that demonstrate a multi-pitch helix (Fig. 10.1a–c). These structures exhibit different chiroptical properties depending on how their helicities are designed. For example, two-pitch helical structures have been investigated as metamaterials to induce broadband, strong CD (Fig. 10.1a) [10]. In the wavelength range $3 \leq \lambda \leq 6.5$ μm, chiral structures interact only with EM waves that have the same handedness of polarization.

The corresponding structures are fabricated (Fig. 10.1d) by combining DLW and electroplating with a positive-tone photoresist (PR). First, a substrate coated with

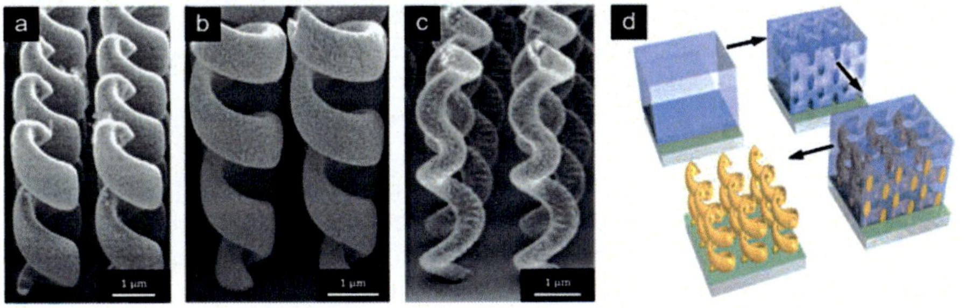

Fig. 10.1 SEM images and schematic of fabrication methods with direct laser writing. This method can fabricate **a** multi-pitch gold helixes, **b** tapered helixes and **c** handedness-converted helical structures. **d** The illustration shows fabrication process of gold helixes by combining direct laser writing and electroplating process. A positive photoresist is spun onto a substrate. And then, patterns are exposed by direct laser writing. Vacancies are filled with gold with electroplating. Gold structures are finally fabricated with removing the residual photoresist. **(a)** and **(d)** from [10], **(b)** from [11], **(c)** from [12]

thin indium-tin-oxide (ITO) is prepared, and then a positive-tone PR is spin-coated on it. The ITO has electrical conductivity, which is required to allow use of the electroplating processes. Second, the positive-tone PR is polymerized using focused photons, and the exposed spot is removed after the development process. Third, electrochemical deposition is conducted to grow gold structures in the voids in the polymer. Gold is a typical material for chiral plasmonics due to its high conductivity to enable electron oscillations, and to its resistance to corrosion. Finally, the remaining PR is removed using oxygen plasma etching to leave sophisticated gold structures.

The main strength of DLW is that three-dimensional printing induced by multi-photon polymerization yields the highest degree of design freedom [13, 14]. DLW can fabricate various helical structures, which have different chiroptical activities depending on the helical parameters and design. For example, tapered helical structures [11] increase extinction ratio and broaden bandwidth simultaneously. Handedness-converted helical structures achieve spin-conversion efficiency within the frequency range of 50–90 THz ($6 \geq \lambda \geq 3.33$ μm) (Fig. 1d) [12]. However, DLW has the disadvantage of limited resolution, so the wavelength band in which it works has a lower limit. The method is also slow, so it cannot be used practically to pattern large areas.

10.2.2 Buckling Process Using Focused Ion Beam

Thin-film buckling process enables curved surfaces with sizes of tens of micrometers; the curvature enables three-dimensional chiral structures. This thin-film-buckling method includes cutting and folding flat objects. Some three-dimensional chiral structures have gigantic chiral phenomena, compared to two-dimensional chiral structures [15]. Furthermore, this buckling process does not require stacking and aligning, which are necessary steps in multi-layer fabrications. The residual stress can be induced by various stimuli, including temperature, mechanical forces, and capillary forces [16–19]. In this subchapter, we will consider focused ion beam (FIB)-induced buckling which is a recently emerging method.

Residual induced compressive forces have been evaluated as methods to create chiral structures and strengthen their chiroptical responses [20]. One method to generate residual stress in a surface of thin film is to use FIB to implant ions. This technique requires only a single step, and has both high accuracy and high resolution, so the method can fabricate desirable chiral structures. For example [15], one chiral structure that had a strengthened chiroptical response (Fig. 10.2a); to generate plasmonic responses, the authors used a gold thin film in which residual stress was imposed by global ion beam irradiation (Fig. 10.2b). Curvatures of the structures can be modulated by controlling the dose intensity during irradiation, so chiral structures can be shaped with quite high accuracy (Fig. 10.2c).

Compared to two-dimensional structures, buckled chiral techniques enable versatile geometries. Chirality is derived from the difference in interaction depending

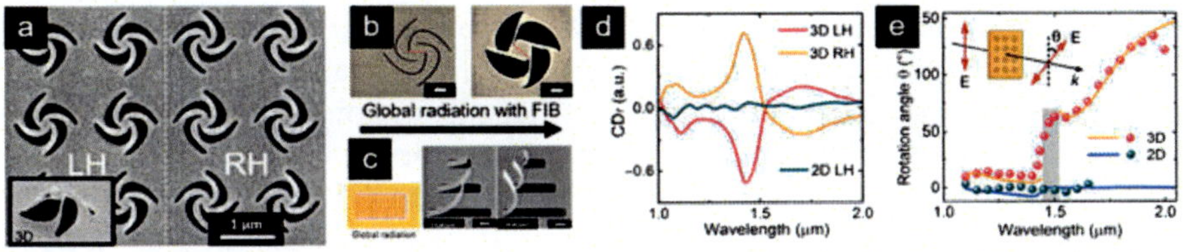

Fig. 10.2 Understanding of buckled structures fabricated by focused ion beam, **a** The rosette arrays are fabricated with buckling process with focused ion beam. Two different handedness can be fabricated, **b** A global radiation buckle a thin films. Right image shows a pre-buckled structure. the left-one shows post-bucked structures, **c** A film are buckled by global radiation, **d** measured chiral reponses of bucked structures, **e** The buckling process dramatically increase an optical rotation. (**a**–**e**) from [15]

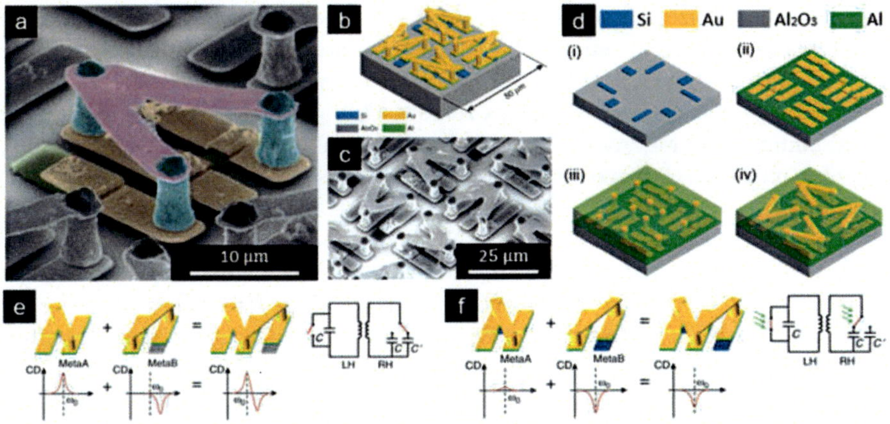

Fig. 10.3 Artificial chiral structures using photolithography and electroplating. **a** SEM image shows fabricated 3D chiral structures, **b** A unit-cell is consist of four-chiral structures and they are composed of four different materials, **c** SEM image shows a rotating arrangement of chiral structures by consisting unit-cell. The rotating arrangement induces chiral responses, **d** Schematic shows a fabrication process for chiral structures, **e**-**f** Handedness switch can be described by LC-circuit interpretation. (**a**–**f**) from [21]

on the spin-states of the circularly-polarized light and can be expanded to three-dimensional twisted structures. The different optical response is related to the electromagnetic coupling, which can be enhanced by three-dimensional structures [22]. Figure 10.2a, b show twisted chiral structures fabricated using a FIB, achieving a strong chiral response from three-dimensional structures using the residual stress of films. This method can be used to create versatile geometric structures with ion beam dose splitting (Fig. 10.2c). The fabricated buckled structures have CD of ~0.5 at 1.45 μm that can be used in telecommunications (Fig. 10.2d). Buckled structures have increased angles of optical rotation, compared to unbuckled structures in the telecommunications region (Fig. 10.2e). These results imply that three-dimensional fabrication methods can strengthen chiroptical phenomena.

Chiral structures with sizes of 10–100 μm have been realized by combining photolithography and electroplating methods. In the terahertz frequency range, silicon can be used as the photoactive materials, and artificial silicon meta-atoms may be described using inductor-capacitor (LC) circuits.

A device that had the above characteristics achieved active metasurfaces that could switch the their handedness [21] (Fig. 10.3a–c); consisted of meta-atoms composed of gold, silicon, aluminum oxide, and aluminum components that had been fabricated using a combination of photolithography and electroplating. To fabricate photoactive chiral metamaterials, firstly, a patterned silicon layer was deposited using photolithography and reactive ion etching. Then aluminum was deposited by sputtering. The aluminum was a conductive layer for electroplating, as the ITO layer, which used in DLW. Sputtering was used due to its high step-coverage. Gold layers were also deposited using photolithography with aligning processes. Gold columns were fabricated using photolithography to exposure photoresist (PR) and electroplating to infill vacancies. This process is similar to DLW with electroplating, but photolithography is more suitable than DLW to fabricate large-area metasurfaces. Finally, the residual resist was removed using oxygen plasma etching, and the aluminum also removed using aluminum etchant. Oxygen plasma etching is a common method to remove PR in complex structures and it can be applied to other processes (e.g. removing PR when helical structures are fabricated using DLW).

These chiral structures can switch the handedness of circularly-polarized light without geometrical reconfiguration. The handedness modulation is achieved by photoactive material and well-designed LC circuits (Fig. 10.3e–f). Gold acts as an ideal conductor, and loops of gold respond as inductors. The gaps between bottom gold plates function as capacitors, so the meta-atom has a resonance frequency like an LC (Fig. 10.3e, MetaA). The short length of MetaB causes a resonant shift, which leads to the chiral response of the meta-atom. The total atom can be interpreted as fusion of a right-handed meta-atom and frequency-shifted left-handed meta-atom (Fig. 10.3e). When photoexcitation of silicon is induced by near-infrared laser pulses, electrons jump from the silicon pad to the conduction bands, so the silicon becomes conductive. This change cancels the capacitance of MetaA, and including frequency shifting of MetaB. As a result, the handedness of chirality is switched by photoactive material property of Si and combination of fabrication methods (Fig. 10.3f).

10.3 Chirality at the Nanometer to Micrometer Scale

10.3.1 Electron Beam Lithography Overlay

The chiral behaviors in a chiral molecule can be explained by movements of electron clouds. Under illumination by circularly-polarized light, electron clouds of the molecules are displaced; this change induces a magnetic moment. It has components that are parallel to the electric dipole moment, and the interaction between these moments yields the chiral behaviors. This conceptual explanation of chiral behaviors can be transferred to plasmonic molecules. In plasmonic chiral systems, displacement currents are excited inside the plasmonic particles, where free electrons are displaced. Accordingly, a strong chiral response can occur, especially when the currents are accompanied by a resonant excitation of localized plasmons. One way to realize plasmonic chiral molecules is to arrange plasmonic structures in a 'handed' way [21]. One straightforward method to realize such handed structures is to use conventional electron beam lithography (EBL) techniques with a precise overlay process. An accurate EBL overlay method that has sub-20 nm alignment accuracy has been used to realize three-dimensional chiral structures on scales of a few nanometers by stacking plasmonic structures layer-by-layer.

To realize sub-20-nm alignment accuracy, a precise patterning of well-designed alignment mark is necessary. However, the EBL alignment process cannot be seen directly, even using scanning electron microscope. Instead, the relative position of a stage can be predicted by using a set of alignment marks. Therefore, precise alignment marks with sharp corners must be fabricated for use as reference points. Furthermore, by defining alignment marks first, then using them at the beginning of fabrication of the first layer, the alignment accuracy can be improved further. Stage movement is almost identical for a given substrate, so the relative position of first layer and second layer should be affected only by the alignment process. This EBL-overlay process can be used to fabricate many different configurations of 3D nanostructures, such as 3D suspended/connected, interlayered and hierarchical [23]. Artificial chiral structures composed of mirror-symmetric suspended and connected nanostructures show strong chirality at near-infrared (NIR) wavelengths of around 2.1–2.5 μm [23]. The structure has a chiral resonator, in which gold structures form an artificial LC circuit. A tilted metallic loop and the gap effectively function as EM circuit components of an inductor and a capacitor, respectively. Inside the LC circuit, the electric and magnetic dipoles are excited simultaneously because they share the same structural resonances. As a result, the structures can exhibit strong chiral response under the same structural resonances. The structures may exhibit negative refraction if the induced chirality is strong.

The similar idea an also be applied to design planarized chiral helical structures [24]. Metasurfaces are sequentially stacked with a tailored rotational twist that forms anisotropic arrays (Fig. 10.4). The structural anisotropy is effectively converted to strong magneto-electric coupling, which is responsible for artificial chirality. The twisted metasurface structure provides strong coupling between each pair of spaced surfaces over a broad range. If the number of stack sequential layers is increased, the bandwidth broadens, and the reflections of opposite waves increases.

Recently, the concept of chiral molecules to plasmonic structures has been transferred to realize effective chiral molecular structures using a precise EBL overlay. In [25], four gold nano-disks are closely arranged in a handed fashion to form a plasmonic oligomer (Fig. 10.5). The first three particles are placed in the first layer to make an L-shape, and the fourth particle is located in the second layer. The handedness is determined by the location of the last gold particle in the second layer, and the structure exhibits configurational chirality. In this construction rule, the ligands

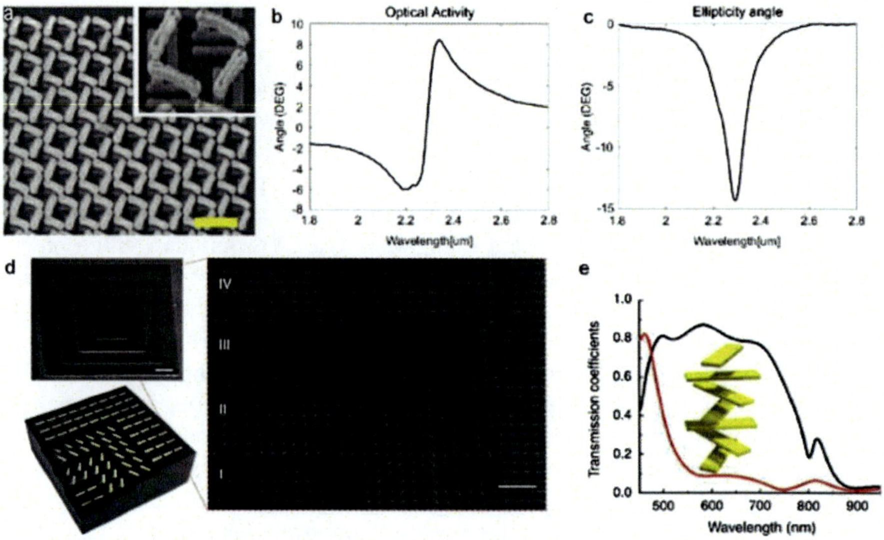

Fig. 10.4 **a** SEM image of the hierarchical, interlayered 3D nanostructure with misalignment kept below 20 nm. The scale bar indicates 500 nm. **b–c** Optical activity and ellipticity angle calculated from artificial chirality simulation at NIR frequencies. **d** SEM images of the twisted metamaterial multilayer, where each layer is revealed using FIB to show the corresponding layer of nanorods. **e** Transmission of left and right-handed circularly polarized light through seven layers of twisted nanoslab along the direction of propagation. (**a–c**) from [23], (**d–e**) from [24]

in chiral molecules can be represented by aligning the metal particles. However, this study showed that the structural handedness is not sufficient to generate a strong chiral response. Resonant plasmonic coupling is also an important prerequisite but can only be accompanied by the interactions of particles with similar resonance wavelengths. Therefore, due to weak coupling, the structures consisting of different size of particles did not show a strong chiral response, despite their geometrical handedness (Fig. 10.5).

The idea of transferring the concept of chiral molecules to plasmonic structures have been extended to the more-complex chiral structures of diastereomers [27, 28]. Diastereomers are two molecules are not mirror images to each other but have chiral centers that are not superimposable. The plasmonic analogue of diastereomers can also be constructed by stacking gold particles, where two chiral centers consisting of four gold NPs are added in a handed way (Fig. 10.5). The S-shaped first layer in the combined structures already shows two-dimensional chirality, which results in three-dimensional chiral structures regardless of the handedness. The additional gold NP placed on top of the first layer allows breaking of mirror symmetry and results in three-dimensional chirality. Interestingly, if diastereomers are weakly coupled to each other, their optical properties can be decoupled into those of two chiral centers, then added, i.e., the chiral response of the diastereomer is the sum of those of the

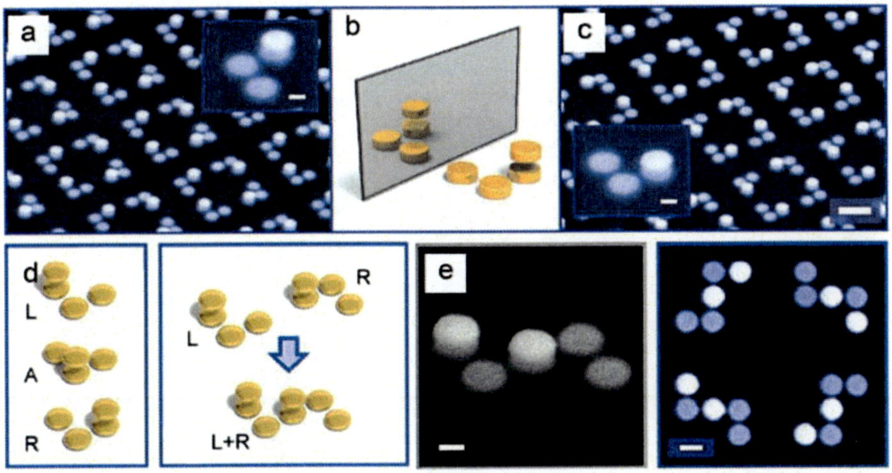

Fig. 10.5 Quadrumeric structures consisting of two layers of particles fabricated with EBL, to give chirality. The first layer consists of three particles arranged in an 'L' shape, and the second layer has one particle of which its position determines the handedness of the superstructure in a C_4 symmetric arrangement. **a, c** Tilted SEM images of the chiral plasmonic molecules. Scale bar: 500 nm; inset scale bar: 100 nm. **b** Schematic illustrating that the designed plasmonic molecules have chirality. **d** Process to fabricate plasmonic diastereomers. Each chiral center consists of four identical particles with three in the first and one in the second layer. As two chiral centers merge, the position of two particles in the second layer determine the chirality of the composite structure. **e** Close-up tilted and normal view of SEM images of (L+R) molecules, showing good alignment. (**a–c**) from [26], (**d–e**) from [27]

chiral centers (Fig. 10.5). However, this additivity of chiral responses does not occur if the two chiral centers are close together, and thus strongly coupled.

10.3.2 Glancing Angle Deposition

The glancing angle deposition (GLAD) method uses a bottom-up approach that is compatible with large-area fabrication, while retaining high-resolution capability down to 20 nm. The GLAD method utilizes a physical vapor deposition process with several parameters controlled, such as nanoseed pattern, substrate rotation speed and temperature. By changing these process parameters, many different shapes of 3D nanostructures can be fabricated on a large area.

Due to the nature of GLAD, the material grows as it is being rotated, and a helical nanostructure forms; this process allows easy fabrication of chiral nanostructures that operate in the visible spectrum. A typical GLAD fabrication starts with nanoseed fabrication from block copolymer micelles. The gold-loaded block copolymer is spin-coated on the wafer; subsequent plasma treatments remove the polymer to leave and array of gold dots with uniform size and spacing on the entire wafer surface. Then

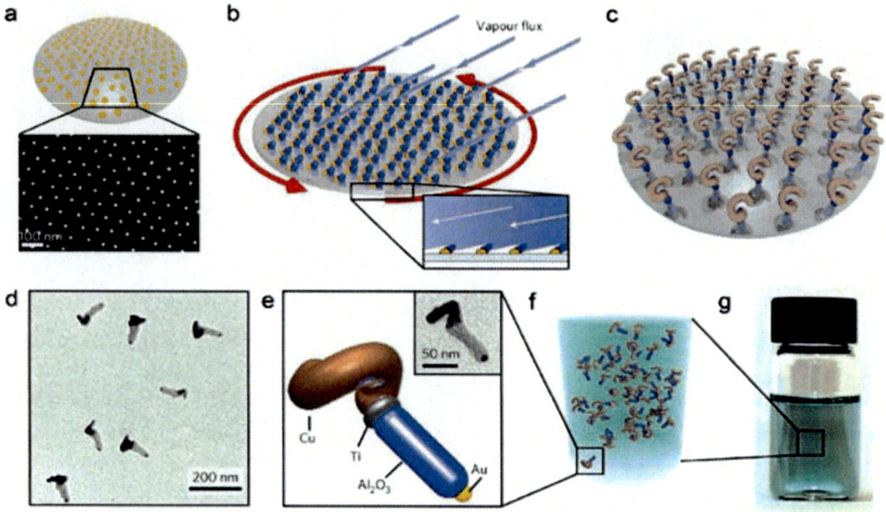

Fig. 10.6 **a** SEM image of gold nanodots regularly patterned on a wafer by micellar nanolithography. **b** Gold nanodots as nucleation sites during subsequent shadow growth. **c** Complex 3D structure can be designed on each nucleation site by manipulating the substrate angle and deposition material. **d** TEM image of hybrid insulator-metal nanohooks. **e** Schematic of the designed structure; inset: TEM image of a nanohook, suspended in a solution by using sonification. **f** Schematic and **g** photograph of the solution. (**a–g**) from [26]

the target material is deposited on the substrate by physical vapor deposition with grazing incidence. By manipulating the tilt angle and rotation speed of the substrate, many different kinds of nanostructures can be fabricated, including bars, zigzags and helices. Another advantage of GLAD technology is that the nanostructures made on the wafer can be removed from the substrate and immersed in a solution for use in a suspension, such suspensions in liquid may enable novel applications such as fluidic (chiral) molecular sensors and nano-robotic systems (Fig. 10.6).

Helical plasmonic NPs offer diverse set of optical response depending on thier geometrical parameters and their materials. Under irradiation, LSPR is generated along the helical axis; the strength is proportional to the total length of the helix. Therefore, nanohelices offer multiple variables for the manipulation of optical fields, including structural features such as pitch length, total number of turns, and material composition. However, fabrication of regularly shaped multi-turn helices in well-defined orientation is a difficult task.

Therefore, plasmonic helical NPs have recently been evaluated, along with advancement in fabrication techniques. Two-turn gold nanohelix structure shows a strong chiroptical response (g-factor ~0.01) in visible light [26]. The structure has 34-nm pitch and 100-nm height, which is only 1/40 as large as the similarly-shaped nanostructure fabricated by two-photon lithography, and is also smaller than the overlain structures fabricated by EBL (Fig. 10.7). To sculpt such complex helical structures and maximize their chiroptical response, requires a technique to cool them to

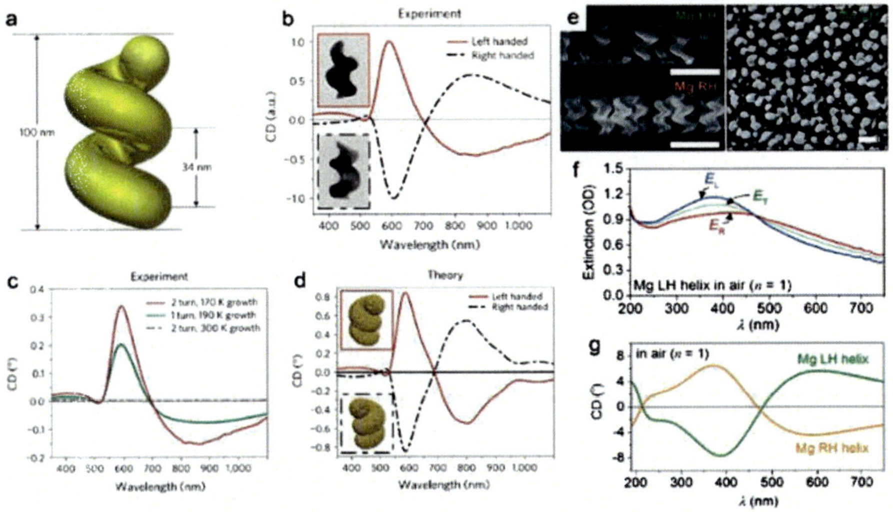

Fig. 10.7 The chiroptical response of two-turn gold nanohelix. **a** Geometrical features of the nanohelix structure. **b** Normalized circular dichroism spectra of left-handed and right-handed nanohelices; inset: TEM image of the nanohelices. **c** Circular dichroism spectra of one-turn and two-turn nanohelices grown under different cooling conditions. **d** Circular dichroism spectra simulated using a model of (**a**). The inset shows the discrete dipole model used in the simulation. **e** Side-view SEM images of left-handed (top left) and right-handed (bottom left) Mg nanohelices; scale bar: 200 nm. Right panel: top view of a left-handed array of nanohelices. **f** Extinction spectra in response to unpolarized (E_T) and circularly polarized light (E_L, E_R). **g** Circular dichroism spectra of 178-nm-thick film containing nanohelices of left or right handedness in air. (**a**–**d**) from [26], (**e**–**g**) from [29]

~170 K, particularly for materials that have high surface energy, such as noble metals. The fabricated wafer-scale helical nanostructure array (a 3 inch wafer can support roughly 10^{11} particles) can be released in the solution and can form uniform/isotropic chiral medium. The advantage of this approach is that the wavelength range in which chiral properties occur can be controlled by using an appropriate material for fabrication.

Magnesium (Mg)-based chiral NPs that work in the ultraviolet (UV) spectral range have been demonstrated [29]. Mg is widely used as a UV plasmonic material because in the UV region Mg has substantially high far-field absorption efficiency and strong near-field enhancement. Furthermore, Mg has high diffusion rate, so the Mg nanohelix structure should be sculpted in a far cooler condition ~100 K. However, Mg is highly reactive in air, so the nanohelix should be coated with a thin layer of HfO_2 or Al_2O_3 to prevent structural and chemical deformation (Fig. 10.7).

Multi-turn nickel and silver nanohelix structures show specular reflectance in the visible-and-NIR spectrum [30]. The associated optical reflectance spectrum showed minima at wavelengths that coincided with the total height of nanohelices; this result demonstrates the existence of LSPR along the axis of nanohelices. This further confirmed that in the visible-and-NIR spectrum, the optical response was mainly dominated by the scattering from individual nanohelies (Fig. 10.7).

10.3.3 Unconventional Approaches

Chiral planar or 3D structures with chirality has merits in a way that it can have chiroptical responses. Conventional spectroscopy cannot easily detect enantiomers, because they have the same chemical and physical properties. However, CD spectroscopy can distinguish enantiomers. When they interact with circularly polarized light, the responses depend on the handedness of the molecules. This effect can be exploited to distinguish enantiomers by their chiroptical responses. However, the EM coupling of chiral molecules is typically weak, so they are commonly accompanied by plasmonic NPs. When irradiated by an EM wave of frequency that coincides with the surface plasmon frequency of the plasmonic NPs, the CD response is generated [31]. The CD response can be further improved when the coalesced plasmonic particles are also chiral; this is the motivation for creating chiral plasmonic NPs.

Single-layered chiral plasmonic NPs were demonstrated first, but such planar structures show very weak chiral response and typically require oblique incidence. Chiral interaction between light and matter is naturally increased in proportion to the propagation length of light, so a 3D structure that can offer structural variation along the incident path shows a pronounced chiroptical response.

One possible fabrication method is EBL, which can create 3D geometries with high precision. Examples include stacked gammadions, twisted crosses, and twisted layers of nanorods. However, EBL can only generate plasmonic planar and 3D structures on areas of a few square micrometers, and the method is expensive. Commercialization of chiral plasmonic NPs coupled with enantiomers requires large-area fabrication at low cost.

Hole-mask lithography (HML) combined with tilted-angle rotation evaporation has been proposed as a fabrication method (Fig. 10.8) [32]. HML is a derivative of colloidal lithography [33] that patterns surfaces with nanostructures that are created by evaporation through holes between close-packed colloidal polystyrene beads that are self-aligned throughout a thin film mask. By rotating the tilt angle through which the evaporated gold was deposited on the substrate, 3D chiral gold nanostructures can be fabricated on an area of a few square centimeters. The structure had the average outer diameter of 260 nm and structure widths from 20 to 90 nm. Rotation of the tilt angle with gradually increasing angular velocity in either positive or negative directions yielded nanostructures that had right-handed and left-handed chirality. The fabricated chiral nanostructures were illuminated with linearly polarized light in x and y directions, and with RCP and LCP lights. The transmittance spectra showed frequency modes at 150, 250, and 320 THz, which all matched simulation results.

A bilayered twisted-arc photonic metamaterial has plasmonic chiral structure that exhibits CD in the NIR spectrum (Fig. 10.8) [34]. The unit cell is composed of a pair of silver twisted arcs in different azimuthal orientation, situated on two

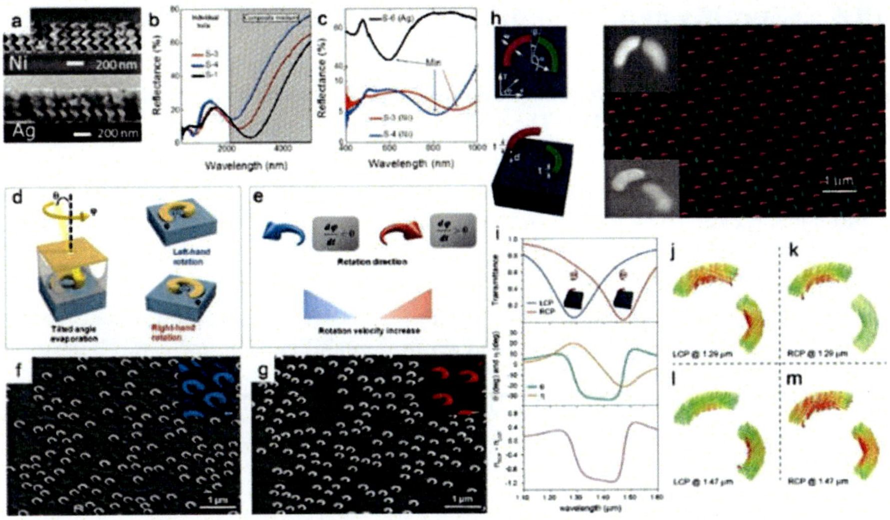

Fig. 10.8 **a** Side-view SEM images of Ni and Ag nanohelices with multiple turns. **b** Reflectance spectra of three different nanohelices at normal incidence. **c** Experimental reflectance spectra of different helices at normal incidence. **d** Hole-mask lithography combined with tilted-angle evaporation. **e** Schematic of two parameters, rotation direction and velocity, that result in 3D structure of different handedness. **f–g** SEM of 3D chiral structures fabricated on a large area of 1 cm^2; blue: left-handed sample; red: right-handed sample. **h** Structural geometry and SEM image of bilayered chiral metamaterial; inset: a unit cell. **i** Top: transmission spectrum. Middle: polarization rotation angle of a linearly polarized incident light and the resultant ellipticity. Bottom: difference in refractive indices for circularly polarized light of opposite handedness. **j–k** Induced current flow within the structure. **(a–c)** from [30], **(d–g)** from [32], **(h–m)** from [34]

distinct transparent layers that have given height. The unit cell is essentially a pair of planar structures, so it is fabricated using aligned EBL, in which three lithography steps are used to consecutively deposit two silver arcs separated by a transparent dielectric. The NIR transmission spectrum of LCP and RCP through the bilayered lattices agreed with full-wave simulation, with minima occurring at 1.29 μm (LCP) and 1.47 μm (RCP). The induced electric current in the dual-layered arcs generally rotated along the curved arc path and showed chiral interaction between the structure and circularly-polarized light. The plasmonic resonance modes of LCP at 1.29 μm and RCP at 1.47 μm arise by different phenomena: the induced currents collide with each other in LCP, but flow with coalescence in RCP. These features indicate that an antisymmetric mode is excited in LCP, whereas a symmetric mode is excited in RCP, and because antisymmetric resonance requires more energy than symmetric resonance, the transmission minimum is lower in LCP than in RCP.

10.4 Chirality at a Nanometer Scale: Bottom-Up Approach

Bottom-up techniques are versatile approaches to fabricate complicated nanostructures on a large scale [35–39]. These techniques have several advantages over top-down methods such as lithography and direct writing, which enable patterning of predefined structures, but can fabricate only 2D or stacked 2D structures, and are impractically expensive. On the contrary, bottom-up methods can fabricate truly-3D structures at reasonable cost. Bottom-up techniques can arrange NPs in a complex configuration with high precision and can fabricate complicated 3D geometry. They can also use diverse components including plasmonic NPs, organic materials and quantum dots. Dynamic tunability and programmability also make the bottom-up methods a powerful tool to realize 3D chiral nanostructures. In this section, we describe several such bottom-up techniques to make 3D chiral nanostructures and describe their optical characteristics.

10.4.1 Molecular Self-assembly

Molecular self-assembly is a bottom-up approach that exploits spontaneous assembly of inorganic materials such as metals, semiconductors, ceramics and biological materials including peptides, fibers and DNA [40, 41]. Synthesis from molecules in solution gives rise to plasmonic NPs that have chiral morphology. NPs that had broken mirror-symmetric geometry have been synthesized using thiolated chiral biomolecules [42]. Tellurium and selenium bind strongly with the thiol group and hence are transformed into chiral shape (Fig. 10.9a, b). Such NPs act as chiral resonators and show polarization-dependent extinction rate in the visible spectrum (Fig. 10.9c, d). The chiral tellurium nanostructures can be transformed into chiral gold and silver telluride nanostructures that have very large chiroptical activity, demonstrating a simple colloidal chemistry path to realize chiral plasmonic and semiconductor metamaterials. These materials are natural candidates for studies related to interactions of chiral biomolecules with chiral inorganic surfaces, with relevance to asymmetric catalysis, chiral crystallization and the evolution of homochirality in biomolecules.

Chiral gold NPs can be synthesized with the aid of amino acids and peptides [43–52]. During particle synthesis, amino acids and peptides interact with the particles and produce twisted high-Miller-index surfaces (Fig. 10.9e). The handedness of the input molecules determines the growth rates of chiral high-index planes that have opposite handedness. Thus, the chiral morphology of the molecules is transferred to the NPs (Fig. 10.9f). The chiral geometry of the NPs makes them interact differently with LCP and RCP. CD is significantly increased near the resonant wavelength of the gold NPs (Fig. 10.9g).

Chiroptical phenomena such as optical activity and CD do not necessarily require chiral NPs. Achiral NPs that are arranged in a chiral pattern can also exhibit an

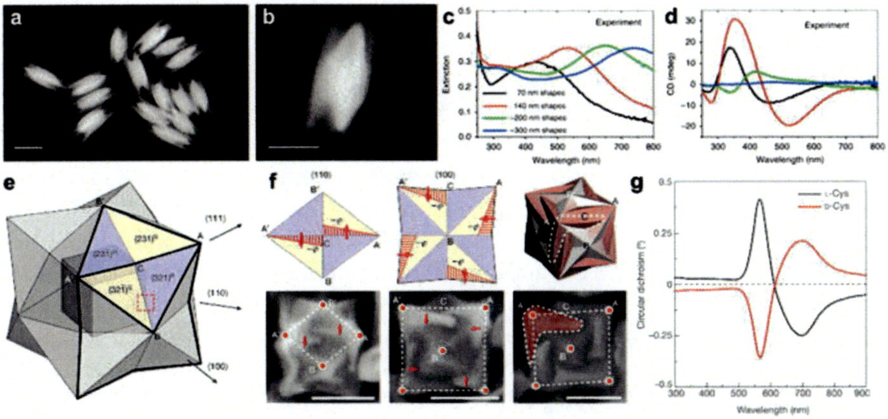

Fig. 10.9 Molecular self-assembly to produce 3D chiral nanostructures. **a**, **b** Dark-field STEM imaging of Te nanostructure with chiral shape. Experimentally-measured **c** extinction and **d** circular dichroism spectra. **e** Schematic of asymmetric growth of high-index surfaces. **f** Schematics and SEM images of chiral NPs from different view angles. **g** Circular dichroism spectra of particles synthesized using peptides that have opposite handedness. (**a**–**d**) from [42] and (**e**–**g**) from [43]

optical resonance effect as a result of inter-particle interaction [53]. Chiral assembly can be manufactured using self-organization of achiral molecules. Liquid crystals spontaneously assemble helical structures (Fig. 10.10a) that exhibit high sensitivity on plasmonic perturbation, in which handedness of the superstructure is determined by whether NIR irradiation is exposed or removed [54]. Achiral particles can also arranged in a mirror-symmetry broken pattern by using twisted fibers or layers called

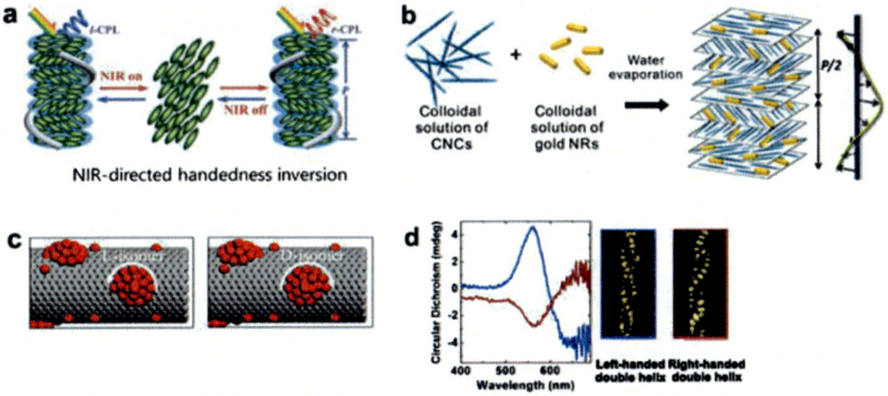

Fig. 10.10 Molecular chiral self-assembly with achiral NP. **a** Superstructures of chiral liquid crystal and their handedness differ when NIR irradiation is on or off. **b** Chiral plasmonic structures with layered CNC scaffold. **c** Assembly of gold NPs on peptide nanotube; arrows: opposite chirality. **d** Gold NP double helices based on peptide superstructures. (**a**) from [54], (**b**) from [56], (**c**) from [57], and (**d**) from [58]

scaffolds [55]. Gold nanorods load on layered twisted cellulose nanocrystal scaffolds (Fig. 10.10b) [56]. The nanoclusters that self-assemble on peptide nanotubes (Fig. 10.10c) assume either left-handed or right-handed chirality; this chiral arrangement yields CD at the surface plasmon frequency [57]. Gold NP double helices (Fig. 10.10d) are peptide-based superstructures that show plasmonic CD [58].

Molecular assembly provides a versatile route to synthesize complex 3D nanostructures, of which final morphology is determined by individual NPs and chiral symmetry groups [40]. As a result of complexity of shape, structures made by molecular assembly exhibit strong CD and chiroptical reactions. Also, the assembled plasmonic structures have no resolution limit. However, molecular assembly has some drawbacks. Control of inter-particle spacing is challenging in molecular assembly [59]. Furthermore, tuning the plasmonic coupling strength is only adjustable by particle concentration and average inter-particle range [60].

10.4.2 DNA Self-assembly

DNA, a biological molecule that is composed of double-stranded helices, can be used in self-assembly [61]. A nanotechnology that uses DNA as a building block to render 3D nanostructure is called DNA self-assembly and has unique molecular recognition capabilities. Long single-stranded DNA (ssDNA) can be folded to form a designated shape by using base pairing with short ssDNA. The long ssDNA is called the scaffold and the short ssDNA is called the staple. This fabrication technique yields 2D structure out of a 1D strand of DNA, and is often called DNA origami [62].

When NPs are functionalized with thiol-modified ssDNA, the base pairing provides 3D assembly of the NPs in a desired arrangement such as dimers [63, 64], trimers [64, 65], tetramers [66–68] and chains [69, 70]. A 3D chiral nanostructure composed of four gold NPs has been synthesized in a tetrahedral configuration by using double-stranded DNA as a scaffold to link the NPs [67]. Mirror symmetry of the tetrahedron was broken by assigning NPs with different sizes to each vertex (Fig. 10.11a). However, different sizes of the NPs and relatively large inter-particle spacing hinder efficient coupling of them, so CD was not observed.

Self-assembly of 3D chiral nanostructure can also be achieved by using a DNA template to arrange NPs in a predesignated manner. DNA origami has been used to fabricate 3D chiral nanostructures by positioning plasmonic NPs in a chiral arrangement [59]. A DNA origami bundle was used as a template to arrange gold nanospheres in a helical geometry; the nanospheres were attached to DNA strands that are complementary to the staples, which therefore attached at specified positions (Fig. 10.11b). The helically-arranged gold nanospheres yield strong CD in the visible spectrum (Fig. 10.11c). A similar structure was fabricated by rolling up a rectangular DNA sheet that had gold nanospheres attached [75]; the nanospheres were functionalized with ssDNA that was complementary with ssDNA attached to the DNA sheet. The handedness of the arrangement can be adjusted by changing the position of the

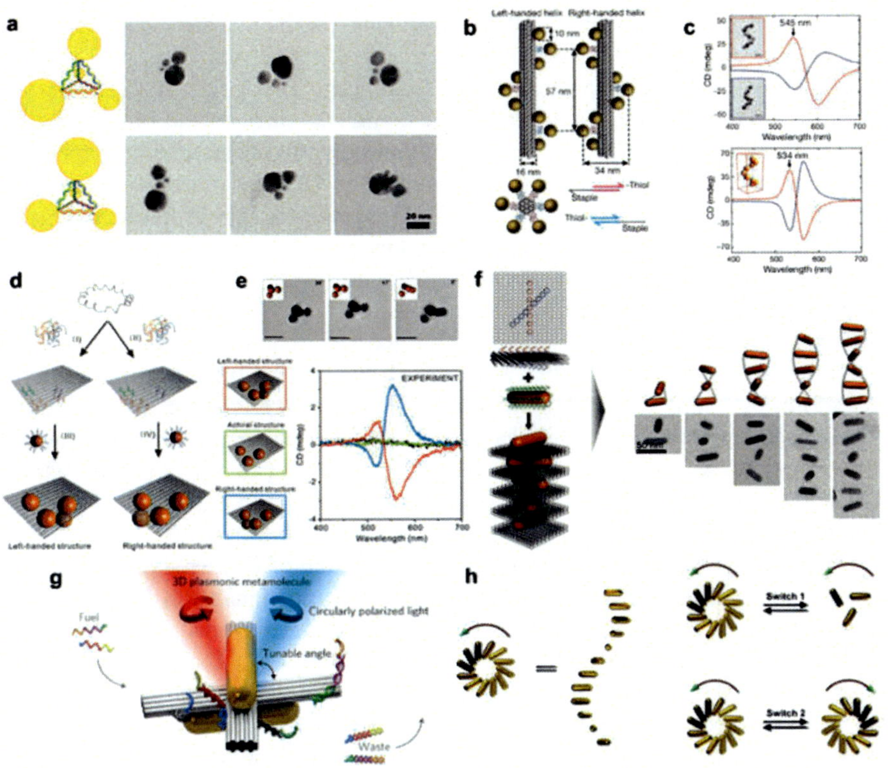

Fig. 10.11 3D chiral nanostructures synthesized by DNA self-assembly. **a** Schematics and TEM images of a tetramer of four gold NPs assembled by double-stranded DNA. **b** An assemble of gold NPs positioned by using a DNA origami bundle [59] and **c** its circular dichroism (top: measured, bottom: simulated) [59]. **d** Four gold NPs bound in a chiral arrangement by using a DNA origami sheet [71] and **e** measured circular dichroism [71]. **f** Twisted layer of gold nanorods stacked by using origami sheets [65]. **g** Schematics of a reconfigurable 3D chiral nanostructure [72]. **h** Schematics of dynamic reconfiguration. The nanorods can be switched between a tightly folded state and an extended state, and between two folded states with opposite handedness. (**a**) from [67], (**b**) and (**c**) from [59], (**d**) and (**e**) from [71], (**f**) from [73], (**g**) from [72], (**h**) from [74]

dressed nanospheres. 3D chiral nanostructures have been also made with a small number of nanospheres in a tetramer arrangement [71]; to realize a chiral geometry, four binding sites were defined on a rectangular DNA origami template, i.e., three on the top surface and one on the bottom surface (Fig. 10.11d). The position of the binding sites on the bottom surface breaks the mirror symmetry and determines the handedness. Four gold NPs were functionalized using complementary DNA strands and positioned, then one was positioned at each binding site by exploiting DNA hybridization. The fabricated structure showed CD in the visible spectrum (Fig. 10.11e).

The arrangement can be simplified by using plasmonic nanorods instead of nanospheres. A nanorod has an anisotropic geometry, which can be used as an additional degree of freedom to break mirror symmetry. Two gold nanorods oriented at 90° to each other, one on the top surface and the other on the bottom surface of the DNA origami template, were fabricated using the base-pairing mechanism [76]. The sample showed CD at $\lambda \sim 730$ nm. The CD was remarkably amplified by increasing the number of nanorods [73]. To synthesize this structure, both sides of the origami sheets were dressed with capture strands (Fig. 10.11f), which were defined in a twisted manner to achieve the chiral geometry. The twisted layer of gold nanorods had strong CD in the visible spectrum.

Tunability of chiroptical responses is an important goal. In most structures, these responses are determined by the geometrical chirality of the 3D nanostructures. Therefore, the chiroptical signal of a fabricated structure is static: its magnitude, peak wavelength and handedness cannot be changed once the structure is formed. However, DNA self-assembly facilitates programmable synthesis of particle assembles and thus provides a pathway to produce reconfigurable and multifunctional structures.

The first tunable 3D chiral nanostructure consisted of a twisted bilayer of gold nanorods [72] (Fig. 10.11g), which were attached individually to two connected origami bundles. The relative angle between two bundles and hence, the handedness of the chiral structure was dynamically controlled by two DNA locks. Here, DNA strands were used both as structural components, and as a tool to drive structural change. The fabricated sample showed CD near a $\lambda = 700$ nm, and the sign of the response could be flipped by applying external stimuli. Time-domain CD confirmed the tunable and reversible chiroptical responses. Alternatively, the distance between two nanorods, rather than the relative angle, can be adjusted to actively control chiroptical responses [77]. Two nanorods were positioned perpendicularly at opposite sides of a double-layer DNA origami. One of the nanorods walked on the surface of the sheet as a result of interacting fuel strands while the other was stationary.

Reconfigurable 3D chiral nanostructure has also been realized by folding a helical assembly of gold nanorods [74] (Fig. 10.11h), which had been positioned helically by using self-assembled DNA origami. DNA-toehold-mediated conformational changes in the DNA template enabled conversion between a tightly-folded state and an extended state, and between tightly-folded states with opposite handedness. The transformation was reversible, but the recovery efficiency is low because of the leakage of the strand-displacement reactions.

10.4.3 Block Copolymer Self-assembly

Block copolymer (BCP) self-assembly exploits phase separation and reconstruction of block copolymers to make light-matter interacting plasmonic nanostructures. A BCP consists more than one species of monomers, which are repeating units joined by covalent bonds to a organize a polymer chain. BCPs are classified according to the number of blocks and the shape of the polymer. Segments of BCP aggregate to reach thermal equilibrium, and this process leads to spontaneous assembly of BCPs. For example, a chiral gyroid structure can be synthesized by controlling the volume fraction of each block [78]. Even though the BCPs are composed of the same species of monomer units, the final morphology of the assembled BCP after order-to-disorder transition differs depending on the interaction between segments, the volume fraction of the blocks, the polymer structures and the solution-processing routes.

Pure BCPs do not readily yield a plasmonic surface effect, so inorganic materials must be combined with them. Two synthesis strategies have been used: templating and coassembly.

Templating transfers polymer morphology into an inorganic phase by deposition after BCP self-assembly. A non-centrosymmetric thin film plasmonic nanostructure could be synthesized using a templating method with linear ABC triblock copolymer, which has two end groups with single backbone chain and three distinct blocks (Fig. 10.12). A chiral alternating gyroid network (G^A), which has chiral spirals, could be generated by manipulating the volume fraction of the block copolymer (Fig. 10.12a) [79]. After an isoprene-block-styrene-block-ethylene oxide (ISO) polymer is assembled as alternating gyroid structures on the substrate (Fig. 10.12b), one of the gyroid networks is eliminated by selective UV and chemical etching (Fig. 10.12c). The resulting space is back-filled by gold electrodeposition, then the rest of the polymer is removed by plasma etching (Fig. 10.12d) to leave a continuous

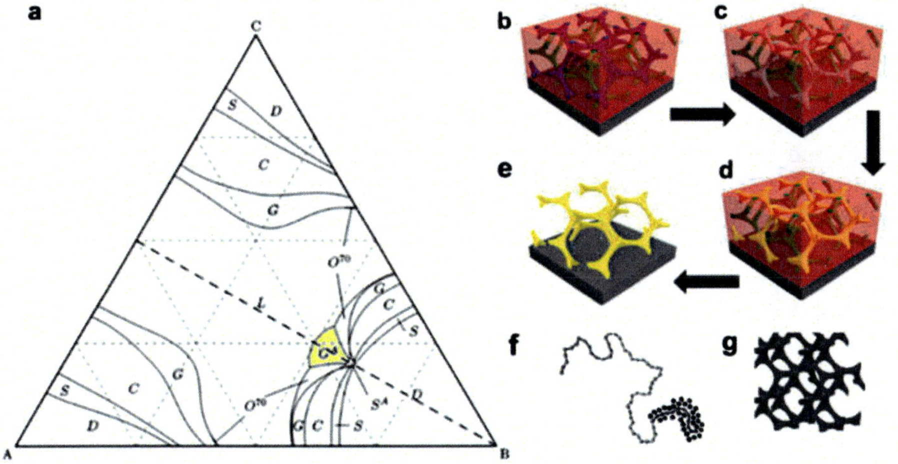

Fig. 10.12 **a** ABC triblock copolymer morphology diagram. The triblock copolymer corresponding volume fraction of G^A part self assembles as chiral alternating gyroid network. (**b**–**e**) BCP self-assembled chiral plasmonic structure by templating method. **b** ISO polymers assembled into alternating gyroid structures on the substrate. **c** Removing one gyroid by etching **d** Gold electrodeposition. **e** Final gyroid structure by templating. (**f**–**g**) BCP self-assembly chiral plasmonic structure by coassembly method. **f** Triblock poly(isoprene-b-styrene-b-ethylene oxide) and introduced sol(black particles) **g** Final gyroid structure by coassembly. (**a**) from [79], (**b**–**e**) from [78], and (**f**–**g**) from [81]

triple-periodic gold network (Fig. 10.12e) [80]. The templating approach has major disadvantages that deposited amorphous materials may crystallize, and that strain can accumulate.

Coassembly exploits intermolecular force to get an amphiphilic block copolymer to interact with inorganic materials in solution. This method has been used to fabricate an isoprene-block-styrene-block-ethylene oxide polymer with sol (black particles in Fig. 10.12f); the intermolecular forces drive the particles into the hydrophilic PEO blocks to yield an alternative gyroid structure [81] (Fig. 10.12g). The coassembly approach has disadvantages that the amorphous assemblies can shrink and crack during forming and crystallization, but it could be used to construct superlattices for novel photonic applications.

Fabrication using BCP self-assembly is a solution process and uses inexpensive constituent monomers, so it is less expensive than photolithography. Bottom up BCP self-assembly has a great advantage to fabricate sub-wavelength highly ordered nanostructures over large areas of a surface. These abilities are advantages over top-down approaches such as lithography, which have resolution limits. In addition, BCP self-assembly is much simpler sequences and requires fewer repeating units than biological molecular assembly that uses DNA or peptides [78]. BCP self-assembled plasmonic structure has been applied for organic solar cells, optoelectronics, and plasmonic nanoantennas, and can help to fabricate subminiature drug delivery systems and biosensing devices [82, 83].

10.5 Conclusion

In this chapter, fabrication methods realizing three-dimensional chiral structures were discussed and categorized by their manufacturing feature sizes since the working wavelength of enhanced artificial chirality is generally close to the feature sizes of the chiral structures. Top-down approaches can be used to fabricate micron-scale structures with high precision. Direct laser writing, focused ion beam, and photolithography can be used to realize three-dimensional structures with a large degree-of-freedom; however, they are time-consuming and expensive. Bottom-up approaches are more suitablefor realizing nano-scale structures. Solution processing and block copolymer self-assembly are cost-effective methods, but the achievable morphologies are limited.

Acknowledgements This work was financially supported by the National Research Foundation (NRF) grant (NRF-2019R1A2C3003129) funded by the Ministry of Science and ICT, Republic of Korea. Y.Y. and Y.K. acknowledge the fellowships from the Hyundai Motor *Chung Mong-Koo* Foundation. S.S., M.K. and I.K. acknowledge the NRF Global Ph.D. fellowships (NRF-2017H1A2A1043322, NRF-2017H1A2A1043204, NRF-2016H1A2A1906519), respectively, funded by the Ministry of Education, Republic of Korea.

Author Contributions J. R., Y. Y. and Y. K. conceived and initiated the project. Y. Y., Y. K. and J. M. mainly wrote the manuscript. J. G, S. S., M. K., H. J., I. K. and T. B. are partially involved in writing the manuscript. Y. Y and J. G organized the top-down fabrication parts with support from J. G, S. S., Y. K., H. J., I. K., Y. Y. and M. K. organized bottom-up fabrication parts with support from J. M.. All authors read and approved the final manuscript. J. R. guided the entire project.

Additional Information The authors declare no competing interests.

References

1. J. Mun, M. Kim, Y. Yang, T. Badloe, J. Ni, Y. Chen, C-W. Qiu, J. Rho, Light Sci. Appl. **9**, 139 (2020)
2. P. Bharadwaj, B. Deutsch, L. Novotny, Adv. Opt. Photonics **1**, 438 (2009)
3. L. Novotny, Phys. Rev. Lett. **98**, 266802 (2007)
4. A. Campion, P. Kambhampati, Chem. Soc. Rev. **27**, 241 (1998)
5. D. Jaque, L. Martínez Maestro, B. Del Rosal, P. Haro-Gonzalez, A. Benayas, J. L. Plaza, E. Martín Rodríguez, and J. García Solé, Nanoscale **6**, 9494 (2014).
6. L. Novotny, N. Van Hulst, Nat. Photonics **5**, 83 (2011)
7. X. Fan, W. Zheng, D.J. Singh, Light Sci. Appl. **3**, e179 (2014)
8. J. Mun, S. So, J. Rho, Phys. Rev. Appl. **10**, 1 (2019)
9. J. Mun, J. Rho, Opt. Lett. **43**, 2856 (2018)
10. J.K. Gansel, M. Thiel, M.S. Rill, M. Decker, K. Bade, V. Saile, G. Von Freymann, S. Linden, M. Wegener, Science **325**, 1513 (2009)
11. A. Phys, J.K. Gansel, M. Latzel, A. Frölich, J. Kaschke, M. Thiel, Opt. Express **101109**, 19936 (2017)
12. J. Kaschke, L. Blume, L. Wu, M. Thiel, K. Bade, Z. Yang, M. Wegener, Adv. Opt. Mater. **3**, 1411 (2015)
13. Z. Wang, F. Cheng, T. Winsor, Y. Liu, Nanotechnology **27**, 412001 (2016)
14. A. Selimis, V. Mironov, M. Farsari, Microelectron. Eng. **132**, 83 (2015)
15. Z. Liu, H. Du, J. Li, L. Lu, Z.Y. Li, N.X. Fang, Sci. Adv. **4**, 1 (2018)
16. A. Rafsanjani, K. Bertoldi, Phys. Rev. Lett. **118**, 084301 (2017)
17. Z. Chen, G. Huang, I. Trase, X. Han, Y. Mei, Phys. Rev. Appl. **5**, 017001 (2016)
18. E.A. Peraza-hernandez, D.J. Hartl, R.J.M. Jr, D.C. Lagoudas, Computer-aid Design **78**, 93 (2014)
19. L. Xu, T.C. Shyu, N.A. Kotov, ACS Nano **11**, 7587 (2017)
20. S. Xu, Z. Yan, K. Jang, W. Huang, H. Fu, J. Kim, Z. Wei, M. Flavin, J. Mccracken, R. Wang, A. Badea, Y. Liu, D. Xiao, G. Zhou, J. Lee, H.U. Chung, H. Cheng, W. Ren, A. Banks, X. Li, U. Paik, R.G. Nuzzo, Y. Huang, Science **347**, 154 (2015)
21. S. Zhang, J. Zhou, Y.S. Park, J. Rho, R. Singh, S. Nam, A.K. Azad, H.T. Chen, X. Yin, A.J. Taylor, X. Zhang, Nat. Commun. **3**, 942 (2012)
22. S. Zhang, Y.-S. Park, J. Li, X. Lu, W. Zhang, X. Zhang, Phys. Rev. Lett. **102**, 23901 (2009)
23. G. Yoon, I. Kim, S. So, J. Mun, M. Kim, J. Rho, Sci. Rep. **7**, 6668 (2017)
24. Y. Zhao, M.A. Belkin, A. Alù, Nat. Commun. **3**, 870 (2012)
25. M. Hentschel, M. Schäferling, T. Weiss, N. Liu, H. Giessen, Nano Lett. **12**, 2542 (2012)
26. A.G. Mark, J.G. Gibbs, T.C. Lee, P. Fischer, Nat. Mater. **12**, 802 (2013)
27. M. Hentschel, M. Schäferling, B. Metzger, H. Giessen, Nano Lett. **13**, 600 (2013)
28. Y. Yang, M. Kim, J. Mun, J. Rho, Adv. Theory Simul. **3**, 1900229 (2020)
29. H.-H. Jeong, A.G. Mark, P. Fischer, Chem. Commun. **52**, 12179 (2016)
30. J.M. Caridad, D. McCloskey, F. Rossella, V. Bellani, J. F. Donegan, V. Krstić, ACS Photonics **2**, 675 (2015)
31. M. Kim, J. Rho, Opt. Express **26**, 14051 (2018)
32. B. Frank, X. Yin, M. Schäferling, J. Zhao, S.M. Hein, P.V. Braun, H. Giessen, ACS Nano **7**, 6321 (2013)

33. H. Fredriksson, Y. Alaverdyan, A. Dmitriev, C. Langhammer, D.S. Sutherland, M. Zäch, B. Kasemo, Adv. Mater. **19**, 4297 (2007)
34. Y. Cui, L. Kang, S. Lan, S. Rodrigues, W. Cai, Nano Lett. **14**, 1021 (2014)
35. Y.Y. Lee, R.M. Kim, S.W. Im, M. Balamurugan, K.T. Nam, Nanoscale **12**, 58 (2020)
36. A spin-encoded all-dielectric metahologram for visible light. Laser Photon. Rev. **13**, 1900065 (2019)
37. Complex-amplitude metasurface-based orbital angular momentum holography in momentum space. Nature Nanotechnol. **15**, 948–955 (2020)
38. Planar achiral metasurfaces-induced anomalous chiroptical effect of optical spin isolaton. ACS Appl. Mater. Inter. **12**, 43 (2020)
39. Multipole decomposition for interactions between structured optical fields and meta-atoms. Optic. Express **28**, 36756–36770 (2020)
40. H.-E. Lee, H.-Y. Ahn, J. Lee, K.T. Nam, ChemNanoMat **3**, 685 (2017)
41. W. Ma, L. Xu, A.F. De Moura, X. Wu, H. Kuang, C. Xu, N.A. Kotov, Chem. Rev. **117**, 8041 (2017)
42. A. Ben-Moshe, S.G. Wolf, M.B. Sadan, L. Houben, Z. Fan, A.O. Govorov, G. Markovich, Nat. Commun. **5**, 4302 (2014)
43. H.-E. Lee, H.-Y. Ahn, J. Mun, Y.Y. Lee, M. Kim, N.H. Cho, K. Chang, W.S. Kim, J. Rho, K.T. Nam, Nature **556**, 360 (2018)
44. H.-E. Lee, R.M. Kim, H.-Y. Ahn, Y.Y. Lee, G.H. Byun, S.W. Im, J. Mun, J. Rho, K.T. Nam, Nat. Commun. **11**, 263 (2020)
45. L. Ohnoutek, N.H. Cho, A.W.A. Murphy, H. Kim, D.M. Rasadean, G.D. Pantoş, K.T. Nam, V.K. Valev, Nano Lett. **8**, 5792 (2020)
46. N.H. Cho, H. Lee, H. Ahn, Y.Y. Lee, S.W. Im, H. Kim, K.T. Nam, Part. Part. Syst. Charact. **36**, 1900062 (2019)
47. J. Karst, N.H. Cho, H. Kim, H.-E. Lee, K.T. Nam, H. Giessen, M. Hentschel, ACS Nano **13**, 8659 (2019)
48. H.-Y. Ahn, S. Yoo, N.H. Cho, R.M. Kim, H. Kim, J.-H. Huh, S. Lee, K.T. Nam, Acc. Chem. Res. **52**, 2768 (2019)
49. S.W. Im, H.Y. Ahn, R.M. Kim, N.H. Cho, H. Kim, Y.C. Lim, H.E. Lee, K.T. Nam, Adv. Mater. **31**, 1905758 (2019)
50. Y.Y. Lee, N.H. Cho, S.W. Im, H. Lee, H. Ahn, K.T. Nam, Chem. Nano Mat. **6**, 362 (2020)
51. N.H. Cho, G.H. Byun, Y.-C. Lim, S.W. Im, H. Kim, H.-E. Lee, H.-Y. Ahn, K.T. Nam, ACS Nano **14**, 3595 (2020)
52. H. Kim, S.W. Im, N.H. Cho, D. Hye Seo, R. M. Kim, Y. Lim, H. Lee, H. Ahn, K.T. Nam, Angew. Chemie. **59**, 12976 (2020)
53. J. Mun, J. Rho, Nanophotonics **8**, 941 (2019)
54. L. Wang, K.G. Gutierrez-Cuevas, A. Urbas, Q. Li, Adv. Opt. Mater. **4**, 247 (2016)
55. A. Guerrero-Martínez, B. Auguié, J.L. Alonso-Gómez, Z. Džolić, S. Gómez-Graña, M. Žinić, M.M. Cid, L.M. Liz-Marzán, Angew. Chemie Int. Ed. **50**, 5499 (2011)
56. M. Hentschel, M. Schäferling, X. Duan, H. Giessen, N. Liu, Sci. Adv. **3**, e1602735 (2017)
57. J. George, K.G. Thomas, J. Am. Chem. Soc. **132**, 2502 (2010)
58. C. Song, M.G. Blaber, G. Zhao, P. Zhang, H.C. Fry, G.C. Schatz, N.L. Rosi, Nano Lett. **13**, 3256 (2013)
59. A. Kuzyk, R. Schreiber, Z. Fan, G. Pardatscher, E.-M. Roller, A. Högele, F.C. Simmel, A.O. Govorov, T. Liedl, Nature **483**, 311 (2012)
60. W. Ma, H. Kuang, L. Wang, L. Xu, W.-S. Chang, H. Zhang, M. Sun, Y. Zhu, Y. Zhao, L. Liu, Sci. Rep. **3**, 1934 (2013)
61. N.C. Seeman, J. Theor. Biol. **99**, 237 (1982)
62. P.W.K. Rothemund, Nature **440**, 297 (2006)
63. A.P. Alivisatos, K.P. Johnsson, X. Peng, T.E. Wilson, C.J. Loweth, M.P. Bruchez, P.G. Schultz, Nature **382**, 609 (1996)
64. K.L. Wustholz, A.-I. Henry, J.M. McMahon, R.G. Freeman, N. Valley, M.E. Piotti, M.J. Natan, G.C. Schatz, R.P. Van Duyne, J. Am. Chem. Soc. **132**, 10903 (2010)

65. C.J. Loweth, W.B. Caldwell, X. Peng, A.P. Alivisatos, P.G. Schultz, Angew. Chemie Int. Ed. **38**, 1808 (1999)
66. A. Fu, C.M. Micheel, J. Cha, H. Chang, H. Yang, A.P. Alivisatos, J. Am. Chem. Soc. **126**, 10832 (2004)
67. A.J. Mastroianni, S.A. Claridge, A.P. Alivisatos, J. Am. Chem. Soc. **131**, 8455 (2009)
68. W. Chen, A. Bian, A. Agarwal, L. Liu, H. Shen, L. Wang, C. Xu, N.A. Kotov, Nano Lett. **9**, 2153 (2009)
69. Z. Deng, Y. Tian, S.-H. Lee, A.E. Ribbe, C. Mao, Angew. Chemie Int. Ed. **44**, 3582 (2005)
70. S. Beyer, P. Nickels, F.C. Simmel, Nano Lett. **5**, 719 (2005)
71. X. Shen, A. Asenjo-Garcia, Q. Liu, Q. Jiang, F.J. García de Abajo, N. Liu, B. Ding, Nano Lett. **13**, 2128 (2013)
72. A. Kuzyk, R. Schreiber, H. Zhang, A.O. Govorov, T. Liedl, N. Liu, Nat. Mater. **13**, 862 (2014)
73. X. Lan, X. Lu, C. Shen, Y. Ke, W. Ni, Q. Wang, J. Am. Chem. Soc. **137**, 457 (2014)
74. X. Lan, T. Liu, Z. Wang, A.O. Govorov, H. Yan, Y. Liu, J. Am. Chem. Soc. **140**, 11763 (2018)
75. X. Shen, C. Song, J. Wang, D. Shi, Z. Wang, N. Liu, B. Ding, J. Am. Chem. Soc. **134**, 146 (2012)
76. X. Shen, P. Zhan, A. Kuzyk, Q. Liu, A. Asenjo-Garcia, H. Zhang, F.J. García de Abajo, A. Govorov, B. Ding, N. Liu, Nanoscale **6**, 2077 (2014)
77. C. Zhou, X. Duan, N. Liu, Nat. Commun. **6**, 8102 (2015)
78. M. Stefik, S. Guldin, S. Vignolini, U. Wiesner, U. Steiner, Chem. Soc. Rev. **44**, 5076 (2015)
79. C.A. Tyler, J. Qin, F.S. Bates, D.C. Morse, Macromolecules **40**, 4654 (2007)
80. S. Vignolini, N. A. Yufa, P. S. Cunha, S. Guldin, I. Rushkin, M. Stefik, K. Hur, U. Wiesner, J. J. Baumberg, U. Steiner, Adv. Mater. **24**, OP23 (2012)
81. P. Docampo, M. Stefik, S. Guldin, R. Gunning, N.A. Yufa, N. Cai, P. Wang, U. Steiner, U. Wiesner, H.J. Snaith, Adv. Energy Mater. **2**, 676 (2012)
82. S.J. Go, D.-E. Lee, D.H. Lee, B.D. Chin, J. Korean Phys. Soc. **68**, 257 (2016)
83. I. Jung, M. Kim, M. Kwak, G. Kim, M. Jang, S.M. Kim, D.J. Park, S. Park, Nat. Commun. **9**, 1010 (2018)

Chapter 11
Floquet Theory and Ultrafast Control of Magnetism

Masahiro Sato

Abstract The development of laser science and technology has stimulated the study of condensed matter physics, especially, dynamical or non-equilibrium nature in solids. The laser technique in terahertz (THz) regime, whose photon energy is comparable to those of typical collective modes in solids such as magnetic excitations, phonons, etc., has remarkably proceeded in the last decade. Theoretical tools for non-equilibrium states have also progressed. Thanks to these backgrounds, magneto-optics, especially, the study of controlling magnetism with laser, now enters a new stage. For such controls, Floquet engineering is a key concept, which means the method of controlling static properties of targets with high-frequency external fields like laser. I review the theoretical foundation of Floquet engineering and its application to magnetic insulators. Basic magnetic quantities such as magnetization, spin chirality, and spin current are shown to be controlled with intense THz laser or wave.

11.1 Introduction

Laser science and technology have continuously developed in the last decades. The application of laser to solids is being one of the hottest topics in condensed-matter physics. If we apply intense laser to materials, their quantum states quickly change into a non-equilibrium one and we can observe non-equilibrium or relaxation dynamics, nonlinear responses to the intense AC field, ultrafast change of physical quantities, etc. In recent years, the significant development of terahertz (THz) laser science in the range of ∼0.1–10 THz [1–4] (THz = 10^{12} Hz) has accelerated the study of ultrafast control of magnetism with THz laser because its photon energy is comparable to the energy of magnetic excitations, especially, those of antiferromagnets [5]. The maximum intensity of currently available THz laser has attained the electric-field amplitude 1–10 [MV/cm] which corresponds to a few Tesla of the AC magnetic field amplitude. In addition to THz laser science, the magnetic resonance study with

M. Sato (✉)
Department of Physics, Ibaraki University, Mito, Ibaraki 310-8512, Japan
e-mail: masahiro.sato.phys@vc.ibaraki.ac.jp

E. Kamenetskii (ed.), *Chirality, Magnetism and Magnetoelectricity*,
Topics in Applied Physics 138,
https://doi.org/10.1007/978-3-030-62844-4_11

THz or gigahertz (GHz) waves (10 GHz–1.0 THz) have been long investigated [6, 7] (GHz = 10^9 Hz). The control of magnetism with laser or electromagnetic wave [5, 8] has also gathered much attention as a large branch of spintronics [9].

Recently, several theoretical methods for non-equilibrium systems have also developed, and they have gradually become widespread in broad fields of condensed-matter physics. Non-equilibrium Green's function method [10–14], approaches based on quantum master equation [15–19], and Floquet theory [20–23] are representative of them. These techniques have a high potential to capture different aspects of laser-driven non-equilibrium dynamics. In fact, with such methods, I and collaborators have theoretically explored/proposed several ways of controlling physical properties of materials, especially, focusing on magnetic systems [24–32].

Among studies for laser-driven phenomena, the concept "Floquet engineering" has been an important keyword and provided us various research directions. This terminology stands for controlling physical (especially static) properties of target systems by applying an AC field whose frequency (photon energy) is much higher than the energy scale of the systems. Excitations or quasi particles of focused systems cannot be directly coupled to such a high-frequency AC field, but it is known that static or low-frequency properties of the systems can be changed through the nonlinear effects of the AC field. This effect is theoretically formulated by Floquet theorem and related techniques developed in recent years [20–23].

In this chapter, I would like to review the theoretical basis of Floquet engineering and its application to simple, realistic magnetic systems. I will explain that basic magnetic quantities such as magnetization, spin chirality, and spin current can be controlled by application of intense THz laser or wave to magnets.

11.2 Floquet Engineering

This section is devoted to the explanation about the theoretical basis of Floquet engineering [20–23]. As I mentioned, Floquet engineering means creating non-equilibrium states with desirable (static) physical properties by periodically driving (i.e., by applying an external AC field to) a material. This concept stems from Floquet theorem, and so I start from the explanation about the theorem. Then I will derive the Floquet effective Hamiltonian through so-called high-frequency (Floquet-Magnus) expansion. The effective Hamiltonian is the most important instrument in Floquet engineering from its conceptual viewpoint, and it describes slow dynamics of the driven system. Finally, I will state some remarks on the physical meaning of Floquet Hamiltonian.

As one will see soon later, the Floquet theory based on Floquet theorem assumes that (i) the external AC field is treated as a classical number (not operator) in the Hamiltonian considered, and (ii) the driven system is decoupled to any environment, i.e., we consider "isolated" quantum systems. The assumption (i) might be sometimes justified. For instance, if the intensity of applied laser is large enough, its AC electric and magnetic fields can be approximated by classical external fields. The condition

(ii) is too ideal, especially, if we consider usual materials such as solids, liquids, and gases. However, as I will explain below, the ideal conditions (i) and (ii) enable us to reveal a clear, simple picture of Floquet engineering. The engineering in dissipative driven systems [14, 15, 19] is a front line of the non-equilibrium physics.

11.2.1 Floquet Theorem

Floquet theorem is an old mathematical result for a class of linear differential equations with a time periodic term, but in the field of physics, it has been recognized as a theorem for Schrødinger equation (i.e., equation of motion) for periodically-driven quantum systems. Following this convention, I will prove Floquet theorem for quantum systems in this subsection. Hereafter, I will often use the unit of $\hbar = 1$.

The statement of Floquet theorem is as follows. We start from the time-dependent Schrødinger equation for a periodically driven quantum system

$$i\frac{\partial}{\partial t}\Psi(t) = \hat{H}(t)\Psi(t). \tag{11.1}$$

Here, we assume that the Hamiltonian $\hat{H}(t)$ is time periodic, $\hat{H}(t+T) = \hat{H}(t)$, where $T = 2\pi/\omega$ is the period and ω is the (angular) frequency. If we focus on a system driven by laser or electromagnetic wave, ω is the laser frequency. For this driven quantum system, the theorem shows that the solution of (11.1) is given by

$$\Psi(t) = \exp(-i\epsilon t)\Phi(t), \tag{11.2}$$

where the "wave function" $\Phi(t)$ is a periodic one satisfying $\Phi(t+T) = \Phi(t)$ and the real number ϵ is called Floquet quasi energy. Namely, Floquet theorem states that the solution of Schrødinger equation for a periodically driven system is given by the product of a plane wave $e^{-i\epsilon t}$ (i.e., solution of vacuum) and a periodic function $\Phi(t)$. In this sense, Floquet theorem can be viewed as the time version of Bloch theorem (See, e.g., [33, 34]) for spatially-periodic quantum systems.

Let us prove the above statement. We define the one-cycle time-evolution operator as

$$\hat{U}(t+T,t) \equiv \mathcal{T}\Big[\exp\Big(-\frac{i}{\hbar}\int_t^{t+T} d\tau \hat{H}(\tau)\Big)\Big], \tag{11.3}$$

where the symbol $\mathcal{T}$ denotes time-ordered product. From the periodicity of the Hamiltonian, $\hat{U}(t+T,t) = \hat{U}(t+(n+1)T, t+nT)$ for $n \in Z$. If $\hat{U}(t+T,t)$ acts on $i\partial_t\Psi(t) = \hat{H}(t)\Psi(t)$ from the left in both sides, we obtain $i\partial_t\Psi(t+T) = \hat{H}(t+T)\Psi(t+T) = \hat{H}(t)\Psi(t+T)$. That is, $\Psi(t)$ and $\Psi(t+T)$ both satisfy the same Schrødinger equation. Therefore, for a solution $\Psi(t)$, there always exists another solution $\Psi(t+T)$ which is proportional to $\Psi(t)$: $\Psi(t+T) = c_T(t)\Psi(t)$.

The coefficient $c_T(t)$ is regarded as the eigenvalue of $\hat{U}(t+T,t)$. Namely, we have

$$\hat{U}(t+T,t)\Psi(t) = c_T(t)\Psi(t). \tag{11.4}$$

Since $\hat{U}(t+T,t)$ is unitary, its eigenvalue $c_T(t)$ is just a phase factor, $|c_T(t)| = 1$.

First, I show that $c_T(t)$ is t-independent ($c_T(t) = c_T$), i.e., a time-evolved state $\Psi(t') = \hat{U}(t',t)\Psi(t)$ with any t' is also the eigenstate of $\hat{U}(t'+T,t')$ with the same eigenvalue c_T. To this end, let us make $\hat{U}(t,t')\hat{U}(t',t) = \hat{1}$ act on (11.4) from the left side. The left hand side is calculated as $\hat{U}(t,t')\hat{U}(t',t)\hat{U}(t+T,t)\Psi(t) = \hat{U}(t,t')\hat{U}(t'+T,t+T)\hat{U}(t+T,t)\Psi(t) = \hat{U}(t,t')\hat{U}(t'+T,t)\Psi(t) = \hat{U}(t,t')\hat{U}$ $(t'+T,t')\Psi(t')$, where we have used the relation $\hat{U}(t',t) = \hat{U}(t'+T,t+T)$. The right hand side is done as $\hat{U}(t,t')\hat{U}(t',t)c_T(t)\Psi(t) = c_T(t)\hat{U}(t,t')\Psi(t')$. If we further multiply both the sides by $\hat{U}(t',t)$, then we obtain

$$\hat{U}(t'+T,t')\Psi(t') = c_T(t)\Psi(t'). \tag{11.5}$$

Equations (11.4) and (11.5) reveal that $c_T(t)$ is independent of t.

Next, let us turn to the remaining part, the proof of $c_T = e^{-i\epsilon T}$. Thanks to the periodicity of $\hat{H}(t+T) = \hat{H}(t)$, the Hamiltonian and the one-cycle time-evolution operator $\hat{U}(t+T,t)$ commute with each other: $[\hat{U}(t+T,t), \hat{H}(t)] = 0$, which means that $\hat{H}(t)$ and $\hat{U}(t+T,t)$ can be simultaneously diagonalized. Moreover, the operator $\hat{U}(t+T,t)$ follows the relation,

$$\begin{aligned}\hat{U}(t+(m+n)T,t+mT)\hat{U}(t+mT,t) &= \hat{U}(t+mT,t)\hat{U}(t+(m+n)T,t+mT)\\ &= \hat{U}(t+T,t)^{m+n}\\ &= \hat{U}(t+(m+n)T,t).\end{aligned} \tag{11.6}$$

This means that the eigenvalue c_T of $\hat{U}(t+T,t)$ satisfies $c_{nT}c_{mT} = c_{mT}c_{nT} = c_{(m+n)T} = c_T^{m+n}$. Therefore, to realize this equality, c_T has to be exponential, that is, $c_T = e^{-i\epsilon T}$. Combining this eigenvalue and the nature of simultaneous diagonalization, we can say that the solution $\Psi(t)$ follows

$$\Psi(t+T) = e^{-i\epsilon T}\Psi(t). \tag{11.7}$$

If we introduce $\Phi(t) = e^{i\epsilon t}\Psi(t)$, $\Phi(t)$ is shown to be a periodic function as follows:

$$\Phi(t+T) = e^{i\epsilon(t+T)}\Psi(t+T) = e^{i\epsilon(t+T)}e^{-i\epsilon T}\Psi(t) = e^{i\epsilon t}\Psi(t) = \Phi(t). \tag{11.8}$$

We thereby arrive at (11.2).

11.2.2 Discretized Fourier Transformation and Matrix Form of Schrødinger Equation

Bloch theorem for solid crystals leads to their electron band structure [33, 34]. Following a similar manner, I here show that a generalized eigenvalue problem appears by applying Floquet theorem to periodically-driven systems. It tells us a basic physical picture of the driven systems.

The periodicity of $\hat{H}(t+T) = \hat{H}(t)$ and $\Phi(t+T) = \Phi(t)$ indicate that $\hat{H}(t)$ and $\Phi(t)$ can both be Fourier-transformed along time direction in a discretized way. We define their Fourier transforms as

$$\hat{H}(t) = \sum_{m\in Z} e^{-im\omega t}\hat{H}_m, \qquad \Phi(t) = \sum_{m\in Z} e^{-im\omega t}\Phi_m. \tag{11.9}$$

The inverse transformation is given by

$$\hat{H}_m = T^{-1}\int_0^T dt\ e^{im\omega t}\hat{H}(t), \qquad \Phi_m = T^{-1}\int_0^T dt\ e^{im\omega t}\Phi(t). \tag{11.10}$$

Substituting these and (11.2) into the Schrødinger equation, we obtain the following generalized eigenvalue problem for $\{\hat{H}_m\}$ and $\{\Phi_m\}$:

$$\sum_{n\in Z}(\hat{H}_{m+n} - m\omega\delta_{m,n})\Phi_n = \epsilon\Phi_m. \tag{11.11}$$

Therefore, with the set of Hamiltonians $\{\hat{H}_m\}$, we can compute the wave functions $\{\Phi_m\}$ and the corresponding quasi energy ϵ (i.e., eigenvalue) in principle. Since the real number ϵ is introduced in the form of $e^{-i\epsilon T}$, its physical relevant range is ristricted in the "Brillouin" zone $-\frac{\pi\hbar}{T} = -\frac{\hbar\omega}{2} \le \epsilon < \frac{\hbar\omega}{2} = \frac{\pi\hbar}{T}$ like crystal momentum k in solids. Due to this periodicity, ϵ is called *quasi* energy. We emphasize that (11.11) does not explicitly depend on time t, and in that sense, it may be called a *static* eigenvalue equation.

In order to more deeply understand this static equation, let us re-write it in a matrix form. Equation (11.11) can be expressed as

$$\begin{pmatrix} \ddots & \ddots & \ddots & \ddots & & & \\ \cdots & \hat{H}_0 - 2\hbar\omega & \hat{H}_1 & \hat{H}_2 & \hat{H}_3 & \cdots & \\ \cdots & \hat{H}_{-1} & \hat{H}_0 - \hbar\omega & \hat{H}_1 & \hat{H}_2 & \hat{H}_3 & \cdots \\ \cdots & \hat{H}_{-2} & \hat{H}_{-1} & \hat{H}_0 & \hat{H}_1 & \hat{H}_2 & \cdots \\ \cdots & \hat{H}_{-3} & \hat{H}_{-2} & \hat{H}_{-1} & \hat{H}_0 + \hbar\omega & \hat{H}_1 & \cdots \\ & \cdots & \hat{H}_{-3} & \hat{H}_{-2} & \hat{H}_{-1} & \hat{H}_0 + 2\hbar\omega & \cdots \\ & & \ddots & \ddots & & \ddots & \ddots \end{pmatrix} \begin{pmatrix} \vdots \\ \Phi_2 \\ \Phi_1 \\ \Phi_0 \\ \Phi_{-1} \\ \Phi_{-2} \\ \vdots \end{pmatrix} = \epsilon \begin{pmatrix} \vdots \\ \Phi_2 \\ \Phi_1 \\ \Phi_0 \\ \Phi_{-1} \\ \Phi_{-2} \\ \vdots \end{pmatrix}. \tag{11.12}$$

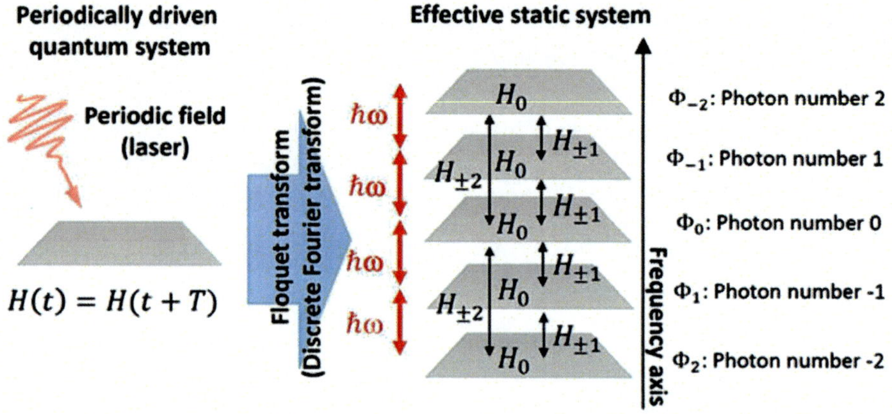

Fig. 11.1 Images of a periodically driven quantum system and the mapped static system through Floquet (Fourier) transform

Here we have explicitly restored the symbol $\hbar$. The wave function Φ_n on each line is a vector living in a Hilbert space whose width is the same as that of the original driven system. An image of this matrix representation is given in Fig. 11.1. We can obtain an intuitive picture of the eigenvalue equation (11.12), i.e., (11.11) if the external AC field is viewed as laser with photon energy $\hbar\omega$. From the diagonal components of (11.12), we find that the energy decreases (increases) by $\hbar\omega$ whenever one goes up (down) the lines one by one. Therefore, $\{\cdots, \Phi_{-2}, \Phi_{-1}, \Phi_0, \Phi_1, \Phi_2, \cdots\}$ may be viewed as wave functions in subspaces with different photon numbers $\{\cdots, 2, 1, 0, -1, -2, \cdots\}$, respectively. The diagonal part of the subspace with photon number n is given by $\hat{H}_0 + n\hbar\omega$, where $\hat{H}_0 = T^{-1}\int_0^T dt\,\hat{H}(t)$ is the time averaged Hamiltonian. On the other hand, from the off-diagonal part of (11.12), we see that $\hat{H}_{\pm n}$ $(n \neq 0)$ connects two subspaces whose difference of photon numbers is n.

In summary, Floquet theorem enables us to exactly map a non-equilibrium quantum system with periodic driving to a "static" eigenvalue problem (11.11) or (11.12). As well known, various theoretical tools have been developed to analyze static many-body (eigenvalue) problems thanks to the long history of equilibrium statistical and condensed-matter physics (See e.g.., [35, 36]). In this sense, Floquet theory makes a periodically-driven system transformed to an easier problem. However, we note that instead of the emergent static nature, a new index n (photon number) appears, and it means that the "spatial" dimension increases by unity if the direction of n is viewed as a new spatial axis.

11.2.3 Floquet-Magnus Expansion and Floquet Hamiltonian

It is known that one can compute an effective static Hamiltonian (called Floquet Hamiltonian) [37, 38] acting on the original space of the Schrødinger equation (11.1) from the eigenvalue equation (11.11) or (11.12). The approximation is based on the power series expansion of $1/\omega$. There are several sorts of the expansions and they are related with each other [39]. In general, they are called Floquet-Magnus (high-frequency) expansion.

Here, I explain one of the expansion methods which is based on degenerate perturbation theory, by making use of the matrix representation (11.12) and Fig. 11.1. Since the bra-ket notation is generally useful for the perturbation calculation, we re-write wave functions Φ_n and $\Psi(t)$ as $|\Phi_n\rangle$ and $|\Psi(t)\rangle$, respectively, in this subsection. We consider the case where the photon energy $\hbar\omega$ is sufficiently larger than other energy scales (i.e., all the eigenvalues of $\{\hat{H}_n\}$). In this condition, the "distance" $\hbar\omega$ between neighboring subspaces is large enough and therefore the averaged Hamiltonian $\hat{H}_0$ and Fourier components $\hat{H}_{\pm n}$ $(n \neq 0)$ connecting different subspaces may be viewed as the perturbation with respect to the the diagonal photon energies $\mathrm{diag}(\cdots, 2\hbar\omega, \hbar\omega, , 0, -\hbar\omega, -2\hbar\omega, \cdots)$.

Below, I will construct the effective Hamiltonian for a subspace with a fixed photon number. Hereafter, we call the subspace with photon number n as nth subspace, and ignore the unperturbed photon energies $n\hbar\omega$ since they are just constants. In fact, owing to periodicity of the quasi energy ϵ, we arrive at the same effective Hamiltonian in each subspace with photon number n, expect for the constant $n\hbar\omega$. We define the orthonormal basis $\{|\Phi_{-n}\rangle_j\}$ $(j = 1, 2, \cdots, d)$ for nth subspace, where d is the dimension of the subspace. Let us introduce the photon-number state $|n\rangle$ that is convenient for the perturbation calculation. With it, the projection operator $\hat{P}_{-n}$ to the nth subspace may be expressed as $\hat{P}_{-n} = \sum_{j=1}^{d} |\Phi_{-n}\rangle_{jj}\langle\Phi_{-n}| = |n\rangle\langle n| \times \sum_{j=1}^{d} |\Phi\rangle_{jj}\langle\Phi|$. The basis $|\Phi\rangle_j$ is common to all the subspaces and in that sense, two projection operators $\hat{P}_{-n}$ and $\hat{P}_{-m}$ $(n \neq m)$ are equal to each other, $\hat{P}_{-n} \sim \hat{P}_{-m}$, except for their photon states. Similarly, the averaged Hamiltonian $\hat{H}_0$ in the diagonal part of (11.12) should be represented as $\hat{H}_0|n\rangle\langle n|$, and $\hat{H}_m (m \neq 0)$ connecting nth and $(n+m)$th subspaces as $\hat{H}_m|n+m\rangle\langle n|$. One can perform the perturbation calculation utilizing these instruments.

The first-order Hamiltonian is equivalent to putting the perturbation term between the projection operators. Therefore, we obtain

$$\hat{H}^{(1)} \equiv \hat{P}_{-n}\hat{H}_0\hat{P}_{-n} = \hat{H}_0. \tag{11.13}$$

Fourier components $\hat{H}_{n\neq 0}$ all disappear due to the sandwich of $\hat{P}_{-n}$. We also note that $\sum_{j=1}^{d} |\Phi\rangle_{jj}\langle\Phi|$ is unity when we focus on a subspace with a fixed photon number. Following the standard formula of the second-order perturbation, we write down the second-order term as

$$\hat{H}^{(2)} = \hat{P}_{-n} \sum_{m=1}^{\infty} \Big(\sum_j \frac{\hat{H}_{+m} |\Phi_{-(n+m)}\rangle_{j\,j} \langle \Phi_{-(n+m)}| \hat{H}_{-m}}{n\hbar\omega - (n+m)\hbar\omega} + \sum_j \frac{\hat{H}_{-m} |\Phi_{-(n-m)}\rangle_{j\,j} \langle \Phi_{-(n-m)}| \hat{H}_{+m}}{n\hbar\omega - (n-m)\hbar\omega} \Big) \hat{P}_{-n}. \tag{11.14}$$

The projection operator $\hat{P}_{-(n\pm m)}$ in the intermediate process is also viewed as unity except for the photon bra-ket $|n \pm m\rangle\langle n \pm m|$. Therefore, we arrive at

$$\hat{H}^{(2)} = \hat{P}_{-n} \sum_{m=1}^{\infty} \left(-\frac{\hat{H}_{+m}\hat{H}_{-m}}{m\hbar\omega} + \frac{\hat{H}_{-m}\hat{H}_{+m}}{m\hbar\omega} \right) \hat{P}_{-n} = -\sum_{m=1}^{\infty} \frac{[\hat{H}_{+m},\ \hat{H}_{-m}]}{m\hbar\omega}. \tag{11.15}$$

Namely, the second-order Hamiltonian is given by the sum of commutators between two off-diagonal terms $\hat{H}_{\pm m}$.

Up to the second-order terms, the effective Hamiltonian (Floquet Hamiltonian) is

$$\hat{H}_{\text{eff}} = \hat{H}_0 - \sum_{m=1}^{\infty} \frac{[\hat{H}_{+m},\ \hat{H}_{-m}]}{m\hbar\omega} + \mathcal{O}\left((\hbar\omega)^{-2}\right). \tag{11.16}$$

The first term is the time averaged Hamiltonian and often identified with that before applying the external AC field. The second and higher-order terms with power of $1/\omega$ emerge only when the AC field is applied. The formula (11.16) indicates that these AC-field-driven terms can be controlled by tuning kinds, wave forms, and frequency ω of the AC field. That is, the Hamiltonian can be desirably changed by applying a high-frequency AC field in a clever way. This statement gives contrast to a stereotype idea that the Hamiltonian is fixed for each material. Equation (11.16) clearly shows the basic idea of Floquet engineering. In principle, we can compute higher-order terms by continuing the degenerate perturbation theory. For instance, the third-order term [39] is

$$\hat{H}^{(3)} = \sum_{\substack{m=-\infty \\ (m\neq 0)}}^{\infty} \frac{[[\hat{H}_{-m}, \hat{H}_0], \hat{H}_m]}{2m^2(\hbar\omega)^2} + \sum_{\substack{m=-\infty \\ (m\neq 0)}}^{\infty} \sum_{\substack{n=-\infty \\ (n\neq 0,m)}}^{\infty} \frac{[[\hat{H}_{-m}, \hat{H}_{m-n}], \hat{H}_n]}{3mn(\hbar\omega)^2}. \tag{11.17}$$

Before ending this subsection, we shortly comment on another expansion method [20–23, 39]. The eigenvalue of (11.12) is the quasi energy ϵ, and (as we already mentioned) it is defined in the eigenvalue of the one-cycle time evolution operator $\hat{U}(t+T,t)$ as $e^{-i\epsilon T}$. From these facts, the static Floquet Hamiltonian $\hat{H}_{\text{eff}}$ may seem to be defined as

$$\exp(-i\hat{H}_{\text{eff}}T) = \mathcal{T}\Big[\exp\Big(-\frac{i}{\hbar}\int_0^T d\tau\,\hat{H}(\tau)\Big)\Big] = \hat{U}(T,0). \tag{11.18}$$

If the phase factor of the right hand side can be expanded with respect to power of $1/\omega$, we obtain the series expansion formula of $\hat{H}_{\rm eff}$. It is known that if the time evolution operator is assumed to be decomposed in the following form, one can compute the expanded form of the effective Hamiltonian:

$$\hat{U}(t_2, t_1) = \exp(-i\hat{G}(t_2)) \exp\left[-i\hat{\mathcal{H}}\ (t_2 - t_1)\right] \exp(i\hat{G}(t_1)). \tag{11.19}$$

Here, $\hat{G}(t)$ is a periodic operator $\hat{G}(t) = \hat{G}(t+T)$, and the exponential $\exp(\pm i\hat{G}(t))$ describes the high-frequency fluctuation in one cycle T. $\hat{G}(t)$ is called kick operator or micro-motion operator. On the other hand, $\hat{\mathcal{H}}$ is the time-independent operator describing the slow dynamics longer than the period T. Generally, $\hat{\mathcal{H}}$ depends on the initial time t_1, but the t_1 dependence is eliminated under the condition that $\hat{G}(t)$ satisfies $\int_0^T dt\hat{G}(t) = 0$. In this case, the high-frequency expansion of $\hat{\mathcal{H}}$ is shown to be equal to that of $\hat{H}_{\rm eff}$ in (11.16). Similarly, the kick operator can be expanded as $\hat{G}(t) = \sum_{n=1} G^{(n)}$. The lower-order terms are given by

$$i\hat{G}^{(1)}(t) = -\sum_{\substack{m=-\infty \\ (m\neq 0)}}^{\infty} \frac{\hat{H}_m}{m\hbar\omega} e^{-im\omega t}, \tag{11.20}$$

$$i\hat{G}^{(2)}(t) = \sum_{\substack{m=-\infty \\ (m\neq 0)}}^{\infty} \frac{[\hat{H}_m, \hat{H}_0]}{m^2(\hbar\omega)^2} e^{-im\omega t} + \sum_{\substack{m=-\infty \\ (m\neq 0)}}^{\infty} \sum_{\substack{n=-\infty \\ (n\neq 0,m)}}^{\infty} \frac{[\hat{H}_n, \hat{H}_{m-n}]}{2mn(\hbar\omega)^2} e^{-im\omega t}. \tag{11.21}$$

11.2.4 Physical Meaning of Floquet Hamiltonian

In the previous subsection, I explained that the static Floquet Hamiltonian $\hat{H}_{\rm eff}$ of (11.16) is computed through the Floquet-Magnum high-frequency expansion. What can we understand from $\hat{H}_{\rm eff}$? One might expect that thermal equilibrium or ground states of the "Hamiltonian" $\hat{H}_{\rm eff}$ is realized by continuously applying an AC field. However, such a naive expectation is not correct. Below, I will comment on a few important results related with the Floquet Hamiltonian $\hat{H}_{\rm eff}$, focusing mainly on many-body systems.

[Short time behavior] When one applies the expansion formulas of $\hat{H}_{\rm eff}$ and $\hat{G}(t)$ to a periodically-driven system, the Floquet-Magnus expansion should be terminated at a certain order in a practical sense. If we focus on a small finite-size system, the expansion is often well-defined. On the other hand, it is known that if we consider a wide class of locally-interacting many-body systems, their Floquet-Magnus expansion is usually of an asymptotic-expansion type like the perturbation expansion of quantum field theories. Let us define a truncated Floquet Hamiltonian as $\hat{H}_q \equiv \sum_{k=0}^{q} \hat{H}^{(k)}$. The following statements about $\hat{H}_q$ have been theoretically shown for locally interacting many-body systems [40, 41]. (i) There is an optimal number $q = q_0 \propto \hbar\omega/G$ which is the best truncation order of the Floquet-Magnus expansion. Here, G is the typical local energy scale of the system including the coupling between the system and AC field. (ii) $\hat{H}_{q<q_0}$ behaves as an almost conserved quantity during a short time $\tau_1 \sim \exp[\mathcal{O}(\hbar\omega/G)]\hbar/G$ from the beginning of application of AC field. (iii) During a short time $\tau_2 \ll \tau_1$, the time evolution operator can be approximated as

$$\hat{U}(t_2, t_1) \approx \exp\left[-i\hat{H}_{q<q_0}\,(t_2 - t_1)\right]. \tag{11.22}$$

as long as we focus on the slow dynamics longer than the period T. These statements indicate that $\hat{H}_{q<q_0}$ is considered as the Hamiltonian for describing the time evolution at least within a short time. If we attach a truncated kick operator $G_p(t) = \sum_{n=1}^{p} G^{(n)}$ to (11.22), a more accurate short-time evolution operator is given by

$$\hat{U}(t_2, t_1) \approx \exp(i\hat{G}_p(t_2)) \exp\left[-i\hat{H}_{q<q_0}\,(t_2 - t_1)\right] \exp(i\hat{G}_p(t_1)). \tag{11.23}$$

From these arguments, the truncated Floquet Hamiltonian is physically relevant in a short time.

[Heating effect] In the present Floquet-theory formalism, we consider isolated (closed) quantum systems decoupled to any environment. It is believed that if we apply an intense AC field to such a closed system for a long time, the system is eventually heated up, except for a class of toy models including integrable systems. In fact, this feature is confirmed by theoretical studies for some concrete models driven by AC fields [42, 43]. This result seems to be very natural because if a crystal is irradiated by laser, it is generally heated and sometimes evaporates. Namely, everyone knows that the heating effect of applied laser is usually unavoidable. We also note that the heating effect of AC fields is consistent with the above theoretical results (i)–(iii), because the results indicate that a long-time Floquet engineering is generally impossible if we consider an isolated system.

[Efficient engineering] The form of the Floquet Hamiltonian (11.16) tells us that the dimensionless expansion parameter is roughly given by $A/(\hbar\omega)$, where A is the typical energy scale of the local interaction between the system and the AC field. In order to enhance the accuracy of predictions from the Floquet Hamiltonian truncated at a low order, one should increase AC-field frequency ω or decrease the coupling strength A (i.e., tuning the value of $A/(\hbar\omega)$). On the other hand, for a small value of $A/(\hbar\omega)$, AC-field driven low-order terms such as $\hat{H}^{(0)}$, $\hat{H}^{(1)}$, and $\hat{H}^{(2)}$ are also weak. Namely, the change of physical quantities via Floquet engineering is quite small for a small $A/(\hbar\omega)$. Therefore, one should take a moderate value of $A/(\hbar\omega)$ to perform Floquet engineering in an efficient manner (although the Floquet Hamiltonian gradually becomes invalid with increase of $A/(\hbar\omega)$). The requirement of a large A is quite natural because the Floquet engineering is a typical nonlinear phenomenon driven by a strong AC field. As I will mention in the next section, it is not easy to increase A if one use laser or electromagnetic wave as the driving field of Floquet engineering.

From these statements, we see that the present Floquet theory can correctly predict the short-time behavior when we consider isolated periodically driven many-body systems. For a long-time driving, the systems are generally heated and the Floquet theory does not work well. However, I emphasize that the forms of low-order terms such as $\hat{H}^{(0)}$, $\hat{H}^{(1)}$, and $\hat{H}^{(2)}$ are very important to qualitatively understand which kinds of quantities can be controlled by an AC field. A more quantitative computation beyond Floquet Hamiltonian (e.g.., numerical analysis) is often necessary to more accurately predict the values and time evolution of AC-field driven quantities, but finding set of possible engineering observables is the most fundamental process to propose a novel Floquet engineering.

11.3 Laser and Typical Excitations in Solids

Some parts of theoretical proposals for Floquet engineering have been experimentally realized by using cold-atom systems [21, 22] in which one can prepare nearly isolated (closed) quantum systems in a few (sub-)seconds. It means that Floquet engineering in cold-atom systems has already grown to a certain degree, while that in materials still offers various open issues. Therefore, now is the time to develop theories and experiments for Floquet engineering in usual materials such as solids, liquids, and gases. Use of usual materials has the advantage compared with cold atoms. For instance, the lifetime of materials may be considered as infinity, and thereby one can repeatedly perform Floquet-engineering process with single material. Floquet engineering in materials can be done in a table-top manner by applying a suitable AC field while that in cold atoms requires application of many sorts of tuned lasers together with atom-trapping techniques.

As I mentioned in Introduction, I will review a few examples of the Floquet engineering in solid crystals in the next section. A typical high-frequency field for solids is electromagnetic wave or laser beams. Therefore, the information about currently-available laser and excitations in solids is important. Below I discuss it from a quantitative viewpoint.

Figure 11.2 depicts typical excitations (quasi particles) of solids in a wide range of laser frequency (photon energy). Magnetic (electron spin) excitations in solids are usually distributed from 1 GHz to 10 THz. Nuclear spin dynamics is much slower than electron spin motion and nuclear magnetic resonance (NMR) [6] is usually done in a megahertz (MHz $=10^6$Hz) range. Phonons (lattice vibration) and molecular oscillations are located from 1 THz to infrared wave ($\sim$ 100 THz). Electron charge excitations are around the visible-light regime (0.1 $-$ 10 petahertz (PHz $= 10^{15}$ Hz)) in both band and Mott insulators.

Electromagnetic waves and quasi-particle dynamics in MHz and GHz (low-frequency) range have been widely used in electronics and spintronics [9]. In this low-frequency regime, it is difficult to create intense coherent waves like laser beams (although there exists maser technique), and instead various methods of generating electromagnetic waves have been established. For instance, Gunn diode has been

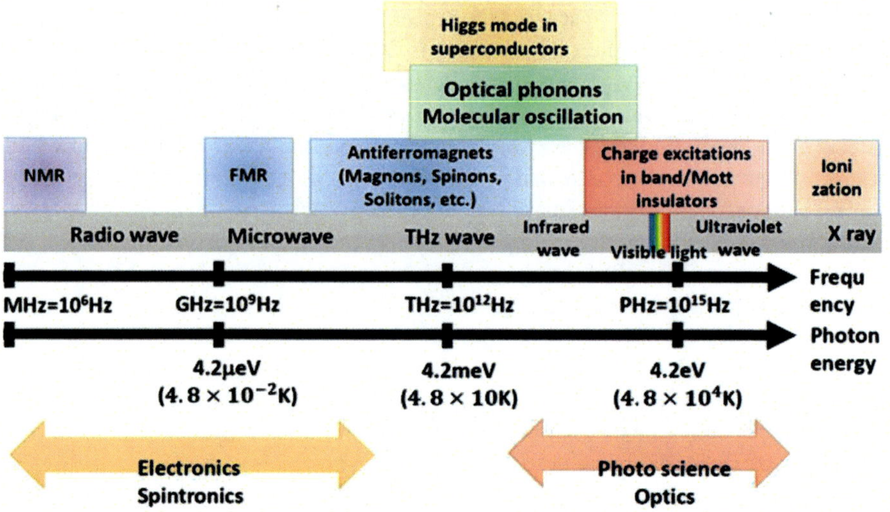

Fig. 11.2 Typical excitations in solids and their typical frequency (energy scales) in a wide frequency range of electromagnetic wave

often used to create micro waves for electron spin resonance (ESR) experiments. Techniques based on electromagnetic induction have been often used in NMR. On the other hand, the high-frequency waves including infrared, visible and ultraviolet light and the corresponding high-energy excitations in solids (various charge excitations) have been central instruments in the research of photo science, optics, and magneto-optics [8]. In the high-frequency regime, several methods of creating laser beams have been established.

The laser science and technology in THz range (0.1–10 THz), which is located between the frequencies used in electronics (spintronics) and optics (photo science), have massively proceeded in the last decades [1–5]. As shown in Fig. 11.2, their photon energy is comparable to the collective modes in solids like magnetic excitations, phonons, molecular oscillations, Higgs modes in superconductors, etc. Therefore we can now directly control these modes in ultrafast ways with intense laser. The intensity of THz laser is still weaker than those of higher-frequency laser, but methods of controlling both intensity and shape of THz waves have been continuously developed. In fact, non-equilibrium magnetic phenomena induced by THz laser or wave have been actively investigated in various experimental groups [2, 44–49]. As every one well knows, the strongest light-matter coupling is the interaction between electric charge and electric field. Thus so far charge dynamics in metals and semiconductors has been the central target in photo physics and optics. However, thanks to the development of THz laser science, non-equilibrium magnetic phenomena induced by direct spin-light couplings have joined in photo science [5, 8].

Usually, laser beams stand for the strong, coherent electromagnetic wave propagating for a long time. Such a beam is called continuous wave (CW). In addition

Table 11.1 Properties of electromagnetic wave (or laser) with 1[MV/cm]. c is speed of light

Electric field of electromagnetic wave	E_0 =1 MV/cm
Magnetic flux density, $B_0 = E_0/c$	0.33 T
Energy flux, I	1.3×10^9 W/cm^2 = 1.3 GW/cm^2

to CW, laser pulses with a short-time Gaussian envelop curve have been intensively studied, especially, in condensed-matter physics. A few cycle, one-cycle and half cycle pulses have been often utilized in the study of photo science. Particularly, as I said, since it is difficult to make the intensity of THz laser strong compared to the other lasers, the pulse techniques have been often used to create intense THz laser pulse. This is very important to generate nonlinear laser-driven phenomena including Floquet engineering in the THz range (0.1–10 THz). On the other hand, the Floquet theory discussed in Sect. 11.2 is reliable for systems driven by CW. One thereby carefully apply the Floquet approach to theoretically study Floquet engineering in pulse-driven systems.

The laser intensity of 1 [MV/cm] may be viewed as a reference value for observing nonlinear photo-induced phenomena, and the corresponding amplitude of the AC magnetic field is $\sim$ 0.3 T. See Table 11.1. Let us here reminder that 1 [MV/cm] = 0.1 [V/nm], Bohr radius $a_B = 0.0529$ nm, and the energy levels of a hydrogen atom are given by $-13.6/n^2$[eV] ($n = 1, 2, 3, \cdots$ is the quantum number). Therefore, 1 [MV/cm] of electric field in an atomic size or lattice spacing of crystals ($\sim$1nm) is the same as about 10 % of a typical energy gap between neighboring atomic levels. In addition, since typical strength of exchange interactions in magnetic insulators is 10–100 Tesla, the magnetic field of 0.3 T is the order of 1–10% of typical exchange interactions. I note that intensities of external electromagnetic waves used in usual condensed-matter experiments are quite smaller than 1 [MV/cm] and 1 T. Such a weak field is sufficient to observe linear responses of materials, but is not enough to do Floquet engineering. In the range of infrared and visible light, it is relatively easy to generate strong laser with intensity 1.0–10 [MV/cm], while such a strong laser in THz range is created only in the pulse form [1–4]. One should also note that (i) if we apply a strong CW beam with more than 10 [MV/cm] to crystals, most of them burn or evaporate, and (ii) the strength of external electric field usually become weaken in materials due to the relative permittivity.

I summarize frequencies of electromagnetic wave and related physical quantities (electric and magnetic fields) in Tables 11.1 and 11.2. We have to carefully consider these values of electromagnetic waves, when we propose a realistic set up of Floquet engineering with a moderate value of $A/(\hbar\omega)$ (See Sect. 11.3). For example, if magnetic excitations are located in the range of 0.1–1 THz in a magnet considered, visible light is not suitable to perform the Floquet engineering through the spin-light coupling since $A/(\hbar\omega)$ (See Sect. 11.2.4) is too small. Instead, an intense laser with frequency of 2–3 THz would be better for the engineering.

Table 11.2 Electromagnetic waves with frequencies 1GHz and 1THz and related physical quantities. Symbols of k_B, $g = 2$, μ_B, c, e, and a_B are respectively Boltzmann constant, electron g factor in vacuum, Bohr magneton, speed of light, elementary charge, and Bohr radius

Electromagnetic wave	1 THz	1 GHz
Frequency, $\omega/(2\pi)$	10^{12} Hz	10^9 Hz
Photon energy, $\hbar\omega$	4.1 meV	4.1 μeV
Temperature, $T = \hbar\omega/k_B$	48 K	48 mK
Magnetic flux density, $B_0 = \hbar\omega/g\mu_B$	36 T	36 mT
Electric field (vol. 1), $E_1 = cB_0$	107 MV/cm	107 kV/cm
Electric field (vol. 2), $E_2 = \hbar\omega/(ea_B)$	0.8 MV/cm	0.8 kV/cm

Finally, I comment on wave length and spatial distribution of laser beams. Usually, the diffraction limit of electromagnetic wave is the order of its wave length. It is generally difficult to spatially focus laser beams beyond their diffraction length. For instance, 1 THz wave has the wave length of $\sim$300 [μm] which is much larger than lattice spacing of crystals. Therefore, AC electric or magnetic fields of low-frequency coherent waves can be viewed as spatially uniform AC fields for electrons in solids. In other words, it is hard to introduce a microscopic spatial modulation in the THz range. However, in recent years, techniques based on meta-material and plasmonics [50, 51] have made possible to tightly focus AC fields beyond their diffraction limit. With these methods, it gradually becomes possible to create spatially modulated THz waves in micro- to nano-scales [27, 51, 52].

11.4 Floquet Engineering in Magnets

In this section, I review two proposals of Floquet engineering in magnetic insulators (quantum spin systems). Inverse Faraday effect in spin-orbit (SO) coupled electron systems [53], dynamical localization [54–57], and Floquet topological insulators [58–61] are named as representative phenomena of Floquet engineering in solids. Particularly, the prediction of Floquet topological insulators have triggered the popularization of Floquet engineering in broad condensed-matter fields. Since this prediction, plenty of Floquet theories for electron or quasi-particle systems have been proposed.

On the other hand, since the spin-light coupling is generally much weaker than the charge-light one, Floquet engineering in magnets had not been developed well. However, as I discussed in Sect. 11.3, THz laser science has strikingly grown and we can use intense THz laser pulses, which can be used to perform the Floquet engineering with spin-light couplings. I will explain theories for the inverse Faraday effect (ultrafast control of magnetization) with THz laser and ultrafast control of Dzyaloshinskii-Moriya (DM) interaction [62–64] in a class of multiferroic systems [65–67], respectively, in Sects. 11.4.1 and 11.4.2.

11.4.1 Inverse Faraday Effect by THz Laser

In this subsection, I consider a wide class of standard magnetic insulators (quantum spin systems) [24, 68]. As I discussed in Sect. 11.3, magnetic excitations are distributed from 1 GHz to 10 THz, and thereby THz laser or wave are suitable for their Floquet engineering. I concentrate on a generic quantum spin model under the application of a static magnetic field and a circularly polarized wave, whose Hamiltonian (See Fig. 11.3) is given by

$$\hat{\mathcal{H}}_{\rm mag}(t) = \hat{H}_{\rm mag} - \boldsymbol{B}\cdot\hat{\boldsymbol{S}}_{\rm tot} - \boldsymbol{B}(t)\cdot\hat{\boldsymbol{S}}_{\rm tot}. \tag{11.24}$$

Here, $\boldsymbol{B} = (0, 0, B)$ and $\boldsymbol{B}(t) = B_0(\sin(\omega t), -\cos(\omega t), 0)$ respectively denote the Zeeman coupling constants of the static field and circularly polarized AC one with frequency ω. They are defined as $B = g\mu_{\rm B}H_{\rm dc}$ and $B_0 = g\mu_{\rm B}H_{\rm ac}$ ($g, \mu_{\rm B}, H_{\rm dc}$ and $H_{\rm ac}$ are respectively electron g factor, Bohr magneton, and the strength of the static and AC magnetic fields), and $\boldsymbol{S}_{\rm tot} = \sum_{\boldsymbol{r}} \boldsymbol{S}_{\boldsymbol{r}}$ is the total spin of the system. The AC field is in the x-y plane and it means that the laser is irradiated from the z direction. The first term $\hat{H}_{\rm mag}$ represents the static multiple-spin interaction and usually a Heisenberg-type exchange interaction is dominant there.

For this driven system, only three Fourier components $\hat{H}_{0,+1,-1}$ are finite: $\hat{H}_0 = \hat{H}_{\rm mag} - B\hat{S}^z_{\rm tot}$ and $\hat{H}_{\pm 1} = \mp\frac{i}{2}B_0\hat{S}^{\pm}_{\rm tot}$. Here, we define operators $\hat{S}^{\pm}_{\rm tot} = \hat{S}^x_{\rm tot} \pm i\hat{S}^y_{\rm tot}$. The truncated Floquet Hamiltonian is thereby estimated as

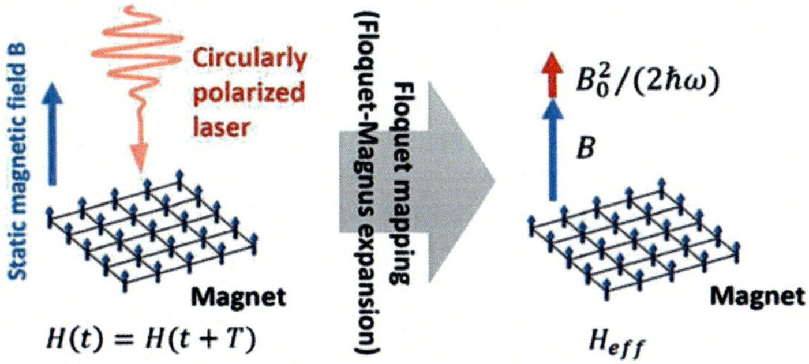

Fig. 11.3 Set up of our model (11.24) of a quantum spin system under a circularly polarized THz laser, and an image of Floquet mapping for the model

$$\hat{H}_{\text{eff}} = \hat{H}_{\text{mag}} - \left(B + \frac{B_0^2}{2\hbar\omega}\right)\hat{S}_{\text{tot}}^z + \mathcal{O}\left((\hbar\omega)^{-2}\right). \tag{11.25}$$

One finds that an effective Zeeman interaction $-(B_0^2/(2\hbar\omega))\hat{S}_{\text{tot}}^z$ emerges owing to the applied laser. The Floquet Hamiltonian indicates that circularly polarized THz laser can change the value of magnetization $\langle\hat{S}_{\text{tot}}^z\rangle$. An image of this Floquet theory is depicted in Fig. 11.3. It is well known that if we apply a circularly polarized high-frequency wave (such as visible light) to metallic systems with SO coupling, an effective Zeeman interaction also appears through the combination between charge-light and SO couplings. This Floquet engineering is called inverse Faraday effect [53] and it has been utilized in various studies of condensed-matter and applied physics [8]. Equation (11.25) shows that an inverse Faraday effect can also be generated even by applying circularly polarized THz (low-frequency) laser through the direct spin-light coupling.

However, we should note that if the spin Hamiltonian $\hat{H}_{\text{mag}}$ is SU(2)-symmetric (i.e., spin-rotation symmetric) like the Heisenberg model $\sum_{\langle \boldsymbol{r},\boldsymbol{r}'\rangle} \hat{\boldsymbol{S}}_{\boldsymbol{r}} \cdot \hat{\boldsymbol{S}}_{\boldsymbol{r}'}$, the static (time averaged) magnetization $\langle\hat{S}_{\text{tot}}^z\rangle$ does not change even under the existence of the laser-driven Zeeman term [24]. This is understood by mapping the driven system to that on a rotating frame via the Unitary transform

$$\hat{\mathcal{U}}(t) = \exp(i\hat{S}_{\text{tot}}^z\omega t). \tag{11.26}$$

This mapping has been often used in the study of magnetic resonance [6] and it eliminates the time-periodic Zeeman coupling since the Unitary rotation has the same frequency ω as the laser. After acting $\hat{\mathcal{U}}(t)$ to the Schrødinger equation from the left side, the transformed Hamiltonian $\hat{H}_{\text{u}} = \hat{\mathcal{U}}(t)\hat{H}(t)\hat{\mathcal{U}}^\dagger(t) - i\hat{\mathcal{U}}(t)(\partial_t\hat{\mathcal{U}}^\dagger(t))$ is given by

$$\hat{H}_{\text{u}} = \hat{H}_{\text{mag}} - B\hat{S}^z + B_0\hat{S}^y - \hbar\omega\hat{S}^z = \hat{H}_{\text{mag}} - \boldsymbol{B}_{\text{u}} \cdot \hat{\boldsymbol{S}}, \tag{11.27}$$

where we have defined a new field $\boldsymbol{B}_{\text{u}} = (0, -B_0, B + \hbar\omega)$. We note that the spin Hamiltonian $\hat{H}_{\text{mag}}$ is invariant through the mapping due to the SU(2) symmetry. As expected, the in-plane AC field is mapped to a static field B_0 along the S^y axis, while an additional Zeeman term $-\hbar\omega\hat{S}^z$ emerges. From this static model (11.27), we see that the spin along the new field $\boldsymbol{B}_{\text{u}}$ is conserved and therefore the static magnetization cannot grow even by applying any circularly polarized laser. For instance, a magnetic anisotropy such as single-ion terms $D_z\sum_{\boldsymbol{r}}(\hat{S}_{\boldsymbol{r}}^z)^2$ and Ising ones $\Delta_z\sum_{\langle \boldsymbol{r},\boldsymbol{r}'\rangle}\hat{S}_{\boldsymbol{r}}^z\hat{S}_{\boldsymbol{r}'}^z$ is necessary to break this conservation. Such anisotropies stem from SO coupling and after all it means that even the THz inverse Faraday effect requires an interaction connecting real and spin spaces like the usual high-frequency inverse Faraday effect. In [24, 68], the detailed condition for generating a large $\langle\hat{S}_{\text{tot}}^z\rangle$ is discussed and its value is numerically computed.

11.4.2 Ultrafast Control of Spin Chirality and Spin Current in Multiferroic Magnets

In this subsection, we consider another type of magnetic insulators, a class of multiferroic magnets [66, 67]. Multiferroics stands for the system with cross-correlated multiple ferro-type orders in the broad sense. Particularly, in the last decade, researchers have actively studied a class of multiferroic magnetic insulators with a strong coupling (called magneto-electric (ME) coupling) between magnetic moments and electric polarization. Therefore, recently the word "multiferroics" have been used as this sort of magnets in a limited sense. In this subsection, I will use terminology "multiferroics" according to this convention. In such multiferroic magnets, there are several types of ME couplings and the electric polarization may be almost always approximated by a function of electron spins.

When we theoretically analyze the essential aspects of the Floquet engineering for multiferroics [25], it is enough to consider a simple two-spin multiferroic magnetic model described by Fig. 11.4. In this model, two spins resides on the x-y plane and circularly polarized THz laser propagates parallel to the z axis. The Hamiltonian is given by

$$\hat{\mathcal{H}}_{2\text{spin}}(t) = \hat{H}_{\text{mag}} - \boldsymbol{B}(t) \cdot \hat{\boldsymbol{S}} - \boldsymbol{E}(t) \cdot \hat{\boldsymbol{P}}. \tag{11.28}$$

The first term $\hat{H}_{\text{mag}}$ is the two-spin interaction like (11.24). The second and third terms are driven by the circularly polarized laser: The second is the Zeeman interaction of the AC field $\boldsymbol{B}(t) = B_0(\sin(\omega t), -\cos(\omega t), 0)$ with $B_0 = g\mu_B H_{\text{ac}}$ and the third is the coupling between the AC electric field $\boldsymbol{E}(t) = E_0(\cos(\omega t), \sin(\omega t), 0)$ and the electric polarization $\hat{\boldsymbol{P}}$. The symbol $\hat{\boldsymbol{S}} = \hat{\boldsymbol{S}}_1 + \hat{\boldsymbol{S}}_2$ is the sum of two spins $\hat{\boldsymbol{S}}_{1,2}$, and the strength of magnetic field H_{ac} satisfies $H_{\text{ac}} = E_0/c$ with c being speed of light.

Let us compute the truncated Floquet Hamiltonian for the model (11.28). The Fourier components are given by $\hat{H}_0 = \hat{H}_{\text{mag}}$ and $\hat{H}_{\pm 1} = -\frac{1}{2}(E_0\hat{P}^{\pm} \pm iB_0\hat{S}^{\pm})$, where $\hat{S}^{\pm} = \hat{S}^x \pm i\hat{S}^y$ and $\hat{P}^{\pm} = \hat{P}^x \pm i\hat{P}^y$. Therefore, the effective Hamiltonian up to the $1/\omega$ order is

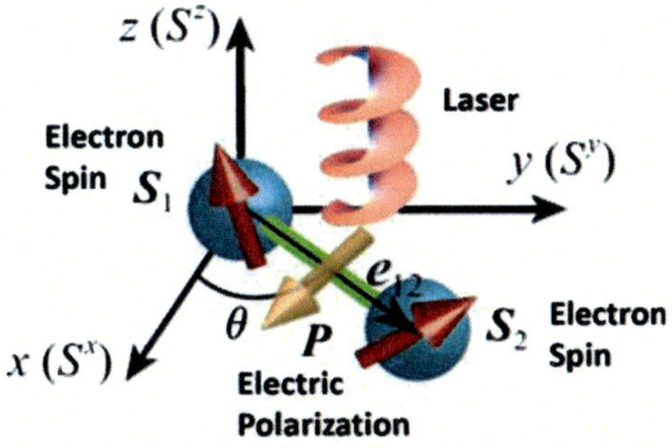

Fig. 11.4 Our two-spin multiferroic magnet under a circularly polarized THz laser

$$\hat{H}_{\text{eff}} = \hat{H}_{\text{mag}} - \frac{1}{2\hbar\omega}\Big[B_0^2\hat{S}^z - iE_0^2[\hat{P}^x, \hat{P}^y] - iE_0B_0\Big([\hat{P}^x, \hat{S}^x] + [\hat{P}^y, \hat{S}^y]\Big)\Big]. \tag{11.29}$$

This may be viewed as the generic formula for the Floquet Hamiltonian of multiferroic magnets under a circularly polarized laser [25]. The terms proportional to $1/\omega$ are all the laser-driven interactions. The first term with $\hat{S}^z$ corresponds to the inverse Faraday effect discussed in the previous subsection. The other terms proportional to E_0B_0 appear only in multiferroics with ME coupling, and they disappear in usual magnetic insulators (See Sect. 11.4.1).

With this formula of (11.29), let us consider a multiferroic model with a concrete ME coupling. Among several ME couplings, I here focus on so-called inverse DM type coupling, in which the electric polarization is given by the outer product of two neighboring spins $\hat{\boldsymbol{V}} = \hat{\boldsymbol{S}}_1 \times \hat{\boldsymbol{S}}_2$ (vector spin chirality) [66]:

$$\hat{\boldsymbol{P}} = g_{\text{me}}\boldsymbol{e}_{12} \times (\hat{\boldsymbol{S}}_1 \times \hat{\boldsymbol{S}}_2), \tag{11.30}$$

where g_{me} is the ME coupling constant and generally depends on the frequency ω. The vector $\boldsymbol{e}_{12} = (\cos\theta, \sin\theta, 0)$ is the unit vector connecting two spins as shown in Fig. 11.4 and the symbol $\times$ denotes outer product. This inverse DM type coupling is originated from SO coupling and is known to often emerge in multiferroics with super-exchanges between a transition metal ion and an oxygen ion (such as Mn oxides and Cu oxides) [65–67, 69, 70]. Substituting the polarization $\hat{\boldsymbol{P}} = g_{\text{me}}(\sin\theta\hat{V}^z, -\cos\theta\hat{V}^z, \cos\theta\hat{V}^y - \sin\theta\hat{V}^x)$ to the Floquet Hamiltonian (11.29), we obtain

$$\hat{H}_{\text{eff}} = \hat{H}_{\text{mag}} - \frac{B_0^2}{2\hbar\omega}\hat{S}^z - \frac{g_{\text{me}}E_0B_0}{2\hbar\omega}\boldsymbol{e}_{12}\cdot\hat{\boldsymbol{V}}. \tag{11.31}$$

The final term is generated by the cross-correlation between E_0 and B_0, and may be called a laser-driven DM interaction. DM interactions generally make two neighboring spins perpendicularly oriented with each other, while standard exchange interactions prefer a collinear spin structure (parallel or anti-parallel). Therefore, the co-existence of exchange and DM interactions usually creates a non-collinear spin structure, in which neighboring spins take a certain angle $\phi \neq 0, \pi$. The effective model (11.31) thus indicates that if we apply a circularly polarized laser to a multiferroic magnet with inverse DM coupling, a non-collinear magnetic structure can be created or annihilated.

It is straightforward to extend the result of (11.31) to many-spin multiferroic models. For instance, an one-dimensional (1D) multiferroic model along the x direction under a circularly polarized laser is described by the following Hamiltonian

$$\hat{\mathcal{H}}_{\text{1D}} = \sum_j J\hat{\boldsymbol{S}}_j\cdot\hat{\boldsymbol{S}}_{j+1} - \boldsymbol{B}\cdot\hat{\boldsymbol{S}}_{\text{tot}} - \boldsymbol{B}(t)\cdot\hat{\boldsymbol{S}}_{\text{tot}} - \boldsymbol{E}(t)\cdot\hat{\boldsymbol{P}}_{\text{tot}}, \tag{11.32}$$

where S_j is the electron spin on jth site, $S_{\rm tot} = \sum_j S_j$ is the total spin, and the polarization is given by $\hat{P}_{\rm tot} = g_{\rm em} \sum_j e \times (\hat{S}_j \times \hat{S}_{j+1})$ with $e = (1, 0, 0)$ being the unit vector along the chain (x) direction. The first term is the 1D exchange interaction between neighboring spins and the second is the static Zeeman interaction. The third and fourth terms are generated by applied laser fields $B(t)$ and $E(t)$. After some algebra, we arrive at the Floquet Hamiltonian

$$\hat{H}_{\rm eff} = \sum_j J\hat{S}_j \cdot \hat{S}_{j+1} - B \cdot \hat{S}_{\rm tot} - \frac{B_0^2}{2\hbar\omega}\hat{S}^z_{\rm tot} - \frac{g_{\rm me}E_0B_0}{2\hbar\omega}\sum_j (\hat{S}_j \times \hat{S}_{j+1})^x. \tag{11.33}$$

The final term is the laser-driven uniform DM interaction which prefers a spiral spin structure. A finite spin chirality $\langle V^x_{\rm tot}\rangle = \sum_j \langle(\hat{S}_j \times \hat{S}_{j+1})^x\rangle$ is shown [25] to be created when we numerically solve the Schrødinger equation for the driven multiferroic model (11.32). This is owing to the competition between exchange and laser-driven DM interactions in (11.33). An image of this Floquet engineering is depicted in Fig. 11.5.

A local spin chirality $\langle V^x_j\rangle = \langle(\hat{S}_j \times \hat{S}_{j+1})^x\rangle$ can be viewed as a local spin current [9, 66]. From the equation of continuity of spin, one sees that a local spin flow can appear if neighboring spin chiralities have different expectation values: $\langle V^x_j\rangle - \langle V^x_{j+1}\rangle \neq 0$. Such a spin current is numerically shown [25] to be created if we apply a "spatially modulated" laser to the 1D multiferroic model. As I mentioned in Sect. 11.3, spatially modulated THz laser could be generated with meta-material techniques [50–52].

In addition to inverse DM interaction, the magneto-striction mechanism is also famous as a representative of ways of generating ME couplings [67]. This mechanism usually stems from spin-phonon coupling and leads to a coupling between a local electric polarization and a local exchange energy $\hat{S}_r \cdot \hat{S}_{r'}$. Of course, it is generally possible to propose a Floquet engineering with such a magneto-striction

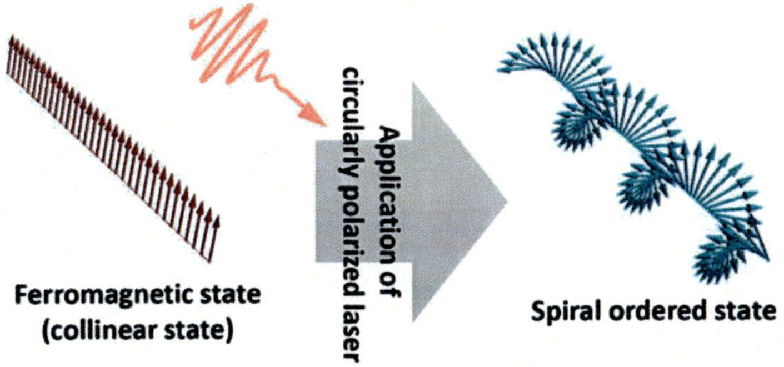

Fig. 11.5 Floquet engineering of a multiferroic chain with inverse DM coupling under a circularly polarized THz laser

ME coupling. In fact, it has been predicted [26] that a topological spin liquid can be created if we apply a circularly polarized laser to a honeycomb Kitaev model [71–73] with a striction type ME coupling.

11.5 Summary and Outlook

In this chapter, I have reviewed the basic theoretical part of Floquet engineering [20–23], focusing on isolated periodically-driven quantum systems. Floquet theorem, Floquet-Magnus expansion for deriving the effective Floquet Hamiltonian, and its physical meanings have been explained in Sect. 11.2. The truncated Floquet Hamiltonian can be used to correctly describe the short-time behavior of driven systems. Then, in Sect. 11.3, I have discussed some information about currently-available laser and electromagnetic waves which is important to perform Floquet engineering in materials, especially, solids. The intensity of 1 [MV/cm] gives a reference amplitude of AC electric field for effective engineering of physical quantities. Finally, I have explained two examples of Floquet engineering in magnetic insulators: inverse Faraday effect with THz laser in generic magnetic insulators [24, 68] and ultrafast control of spin chirality in multiferroics [25].

Further development of Floquet theory is necessary to more accurately predict Floquet engineered quantities in materials. An important research direction is the development of sophisticated theories treating effect of environment (i.e., dissipation) because dissipation effect cannot be ignored in real materials driven by laser. Approaches based on non-equilibrium Green's function [12, 14] and quantum master equation [15, 18, 19] have high potential to provide a deep insight to dissipative periodically driven systems. These studies would also contribute to the development of the fundamental of non-equilibrium physics.

Acknowledgements I would like to thank all the collaborators of our recent works for laser-driven phenomena, especially, Shintaro Takayoshi, Takashi Oka, Tatsuhiko N. Ikeda, Hiroaki Ishizuka, Horoyuki Fujita, and Sho Higashikawa. I was supported by JSPS KAKENHI (Grant No. 17K05513 and No. 20H01830) and a Grant-in-Aid for Scientific Research on Innovative Areas "Quantum Liquid Crystals" (Grant No. JP19H05825).

References

1. H. Hirori, A. Doi, F. Blanchard, K. Tanaka, Appl. Phys. Lett. **98**, 91106 (2011)
2. Y. Mukai, H. Hirori, T. Yamamoto, H. Kageyama, K. Tanaka, New J. Phys. **18**, 013045 (2016)
3. B. Liu, H. Bromberger, A. Cartella, T. Gebert, M. Forst, A. Cavalleri, Opt. Lett. **42**, 129 (2017)
4. M. Sato, T. Higuchi, N. Kanda, K. Konishi, K. Yoshioka, T. Suzuki, K. Misawa, M. Kuwato-Gonokami, Nature Photo. **7**, 724 (2013)
5. P. Nemec, M. Fiebig, T. Kampfrath, A.V. Kimel, Nature Phys. **14**, 229 (2018)
6. C.P. Slichter, *Principles of Magnetic Resonance* (Springer, 1989)
7. H. Nojiri, Z.W. Ouyang, Terahertz Sci. Tech. **5**, 1 (2012)

8. A. Kirilyuk, A.V. Kimel, T. Rasing, Rev. Mod. Phys. **82**, 2731 (2010)
9. *Spin Current*, edited by S. Maekawa, S. O. Valenzuela, E. Saitoh, and T. Kimura (Oxford Univ. Press, 2012)
10. A.M. Zagoskin, *Quantum Theory of Many-Body Systems: Techniques and Applications*, 2nd edn. (Springer, New York, 2014)
11. G. Stefanucci and R. v. Leeuwen, *Nonequilibrium Many-Body Theory of Quantum Systems: A Modern Introduction* (Cambridge University Press, Cambridge, England, 2013)
12. H.J.W. Haug, A.-P. Jauho, *Quantum Kinetics in Transport and Optics of Semiconductors* (Springer, 2007)
13. A. Kamenev, *Field Theory of Non-Equilibrium Systems*, (Cambridge Univ. 2011)
14. H. Aoki, N. Tsuji, M. Eckstein, M. Kollar, T. Oka, P. Werner, Rev. Mod. Phys. **86**, 779 (2014)
15. H.-P. Breuer, F. Petruccione, *The Theory of Open Quantum Systems* (Oxford University Press, Oxford, 2007)
16. R. Alicki, K. Lendi, *Quantum Dynamical Semigroups and Applications* (Springer, 2007)
17. U. Weiss, *Quantum Dissipative Systems* (World Scientific, 1999)
18. H.-P. Breuer, W. Huber, F. Petruccione, Phys. Rev. E **61**, 4883 (2000)
19. T. N. Ikeda and M. Sato, Sci. Adv. **6**, eabb4019 (2020)
20. M. Bukov, L. D'Alessio, A. Polkovnikov, Adv. in Phys. **64**, 139 (2015)
21. A. Eckardt, E. Anisimovas, New J. Phys. **17**, 93039 (2015)
22. A. Eckardt, Rev. Mod. Phys. **89**, 011004 (2017)
23. T. Oka, S. Kitamura, Ann. Rev. Cond. Mat. Phys. **10**, 387 (2019)
24. S. Takayoshi, M. Sato, T. Oka, Phys. Rev. B **90**, 214413 (2014)
25. M. Sato, S. Takayoshi, T. Oka, Phys. Rev. Lett. **117**, 147202 (2016)
26. M. Sato, Y. Sasaki, T. Oka,. arXiv:1404.2010
27. H. Fujita, M. Sato, Phys. Rev. B **95**, 054421 (2017); Phys. Rev. B **96**, 060407(R) (2017)
28. H. Fujita, M. Sato, Sci. Rep. **8**, 15738 (2018)
29. H. Fujita, Y. Tada, M. Sato, New. J. Phys. **21**, 073010 (2019)
30. S. Higashikawa, H. Fujita, M. Sato,. arXiv:1810.01103
31. H. Ishizuka, M. Sato, Phys. Rev. Lett. **122**, 197702 (2019); Phys. Rev. B **100**, 224411 (2019)
32. T.N. Ikeda, M. Sato, Phys. Rev. B **100**, 214424 (2019)
33. N. W. Ashcroft and N. D. Mermin, *Solid State Physics*, (Thomson Learning, 1976)
34. G. Grosso, G.P. Parravicini, *Solid State Physics*, 2nd edn. (Academic Press, 2013)
35. A.A. Abrikosov, L.P. Gorkov, I.E. Dzyaloshinski, *Methods of Quantum Field Theory in Statistical Physics* (Dover Publications, 1975)
36. G.D. Mahan, *Many-Particle Physics*, 3rd edn. (Springer, 2000)
37. J.H. Shirley, Phys. Rev. **138**, B979 (1965)
38. H. Sambe, Phys. Rev. A **7**, 2203 (1973)
39. T. Mikami, S. Kitamura, K. Yasuda, N. Tsuji, T. Oka, H. Aoki, Phys. Rev. B **93**, 144307 (2016)
40. T. Kuwahara, T. Mori, K. Saito, Ann. Phys. **367**, 96 (2016)
41. T. Mori, T. Kuwahara, K. Saito, Phys. Rev. Lett. **116**, 120401 (2016)
42. A. Lazarides, A. Das, R. Moessner, Phys. Rev. E **90**, 012110 (2014)
43. L. D'Alessio, M. Rigol, Phys. Rev. X **4**, 041048 (2014)
44. S. Baierl, J.H. Mentink, M. Hohenleutner, L. Braun, T.-M. Do, C. Lange, A. Sell, M. Fiebig, G. Woltersdorf, T. Kampfrath, R. Huber, Phys. Rev. Lett. **117**, 197201 (2016)
45. J. Lu, X. Li, H.Y. Hwang, B.K. Ofori-Okai, T. Kurihara, T. Suemoto, K.A. Nelson, Phys. Rev. Lett. **118**, 207204 (2017)
46. A. Pimenov, A.A. Mukhin, V.Y. Ivanov, V.D. Travkin, A.M. Balbashov, A. Loidl, Nat. Phys. **2**, 97 (2006)
47. Y. Takahashi, R. Shimano, Y. Kaneko, H. Murakawa, Y. Tokura, Nat. Phys. **8**, 121 (2011)
48. T. Kubacka, J.A. Johnson, M.C. Hoffmann, C. Vicario, S. de Jong, P. Beaud, S. Grubel, S.-W. Huang, L. Huber, L. Patthey, Y.-D. Chuang, J.J. Turner, G.L. Dakovski, W.-S. Lee, M.P. Minitti, W. Schlotter, R.G. Moore, C.P. Hauri, S.M. Koohpayeh, V. Scagnoli, G. Ingold, S.L. Johnson, U. Staub, Science **343**, 1333 (2014)

49. A.A. Sirenko, P. Marsik, C. Bernhard, T.N. Stanislavchuk, V. Kiryukhin, S.-W. Cheong, Phys. Rev. Lett. **122**, 237401 (2019)
50. A. Pors, E. Moreno, L. Martin-Moreno, J.B. Pendry, F.J. Garcia-Vidal, Phys. Rev. Lett. **108**, 223905 (2012)
51. R.W. Heeres, V. Zwiller, Nano Lett. **14**, 4598 (2014)
52. T. Arikawa, S. Morimoto, K. Tanaka, Optics Express **25**, 13728 (2017)
53. P.S. Pershan, J.P. Van Der Ziei, L.D. Malmstrom, Phys. Rev. **143**, 574 (1966)
54. D.H. Dunlap, V.M. Kenkre, Phys. Rev. B **34**, 3625 (1986)
55. F. Grossmann, T. Dittrich, P. Jung, P. Hanggi, Phys. Rev. Lett. **67**, 516 (1991)
56. T. Ishikawa, Y. Sagae, Y. Naitoh, Y. Kawakami, H. Itoh, K. Yamamoto, K. Yakushi, H. Kishida, T. Sasaki, S. Ishihara, Y. Tanaka, K. Yonemitsu, S. Iwai, Nature Comm. **5**, 5528 (2014)
57. H. Lignier, C. Sias, D. Ciampini, Y. Singh, A. Zenesini, O. Morsch, E. Arimondo, Phys. Rev. Lett. **99**, 220403 (2007)
58. T. Oka, H. Aoki, Phys. Rev. B **79**, 081406(R) (2009)
59. T. Kitagawa, T. Oka, A. Brataas, L. Fu, E. Demler, Phys. Rev. B **84**, 235108 (2011)
60. N.H. Lindner, G. Refael, V. Galitski, Nature Phys. **7**, 490 (2011)
61. Y.H. Wang, H. Steinberg, P. Jarillo-Herrero, N. Gedik, Science **342**, 453 (2013)
62. I. Dzyaloshinskii, J. Phys. Chem. Solids **4**, 241 (1958)
63. T. Moriya, Phys. Rev. **120**, 91 (1960)
64. K. Yoshida, *Theory of Magnetism* (Springer, 1996)
65. T. Kimura, T. Goto, H. Shintani, K. Ishizaka, T. Arima, Y. Tokura, Nature **426**, 55 (2003)
66. H. Katsura, N. Nagaosa, A.V. Balatsky, Phys. Rev. Lett. **95**, 057205 (2005)
67. Y. Tokura, S. Seki, N. Nagaosa, Rep. Prog. Phys. **77**, 076501 (2014)
68. S. Takayoshi, H. Aoki, T. Oka, Phys. Rev. B **90**, 085150 (2014)
69. D. Huvonen, U. Nagel, T. Room, Y.J. Choi, C.L. Zhang, S. Park, S.-W. Cheong, Phys. Rev. B **80**, 100402(R) (2009)
70. S. Furukawa, M. Sato, S. Onoda, Phys. Rev. Lett. **105**, 257205 (2010)
71. Y.A. Kitaev, Ann. Phys. **321**, 2 (2006)
72. G. Jackeli, G. Khaliullin, Phys. Rev. Lett. **102**, 017205 (2009)
73. J. Chaloupka, G. Jackeli, G. Khaliullin, Phys. Rev. Lett. **105**, 027204 (2010)

Chapter 12
Magnetoelastic Waves in Thin Films

Frederic Vanderveken, Florin Ciubotaru, and Christoph Adelmann

Abstract This chapter discusses the physics of magnetoelasticity and magnetoelastic waves in thin films as well as their mathematical description. Magnetoelastic waves occur as a result of strong coupling between spin waves and elastic waves in magnetostrictive ferromagnetic media. In a first part, the basic behavior of spin waves is reviewed in both bulk ferromagnets as well as in thin films. Next, elastic waves are discussed with a focus on thin films. Then, the interactions between the elastic and magnetic domains are described and it is shown how this results in the formation of magnetoelastic waves. The description and the mathematical formalism of magnetoelastic waves in infinitesimally thin films is extended to magnetoelastic waves in thin films with finite thickness. The dispersion relations and eigenstates are derived and graphically visualised for such magnetoelastic waves. It is shown that the behavior strongly depends on the geometry of the system, especially on the polarization of the spin and elastic waves and the direction of the magnetization of the magnetostrictive ferromagnetic medium.

12.1 Introduction

The coupling between different physical properties of a system is of great interest for transducer elements. In recent years, the field of spintronics, which includes applications of magnetism and magnetic materials in electronics, has gained enor-

F. Vanderveken
Departement Materiaalkunde, KU Leuven, 3001 Leuven, Belgium

F. Vanderveken (✉) · F. Ciubotaru · C. Adelmann
Imec, 3001 Leuven, Belgium
e-mail: Frederic.Vanderveken@imec.be

F. Ciubotaru
e-mail: Florin.Ciubotaru@imec.be

C. Adelmann
e-mail: Christoph.Adelmann@imec.be

E. Kamenetskii (ed.), *Chirality, Magnetism and Magnetoelectricity*,
Topics in Applied Physics 138,
https://doi.org/10.1007/978-3-030-62844-4_12

mous attention, which has led to the introduction of commercial magnetic memory technologies. However, the efficient energy conversion from the electric to the magnetic domain and vice versa still remains challenging. Current magnetic memories are based on spin-transfer or spin-orbit torques to switch the magnetization [1, 2]; however, these mechanisms depend on the current density in the device and require typically energies on the order of (10 s of) femtojoules to reverse the magnetization of a nanomagnet, despite a much lower intrinsic energy barrier of the order of attojoules. While femtojoule switching energies are promising for nonvolatile memories, they are not competitive in other spintronic applications, such as spintronic logic circuits. Therefore, much research has been devoted to developing more efficient transducers between electric and magnetic subsystems of spintronic devices.

One of the most promising devices to efficiently couple electric and magnetic properties in a spintronic system is the magnetoelectric transducer. Magnetoelectric transducers consist of composite materials, which comprise piezoelectric and magnetostrictive layers [3–7]. Applying a voltage to the piezoelectric layer(s) leads to the formation of strain in the compound. This strain introduces an effective magnetic anisotropy field in the magnetostrictive ferromagnetic component, leading to an coupling between voltage (electric field) and magnetization. The coupling is bidirectional since rotating the magnetization of a magnetostrictive layer also induces strain in the compound and consequently a polarization in the piezoelectric. Hence, such magnetoelectric schemes provide indirect coupling between electricity and magnetism mediated by elastodynamics. Since generating large electric fields in the piezoelectric layers can be energy efficiency, capacitive magnetoelectric transducers promise a much higher energy efficiency than their current-based counterparts.

The coupling scheme of a magnetoelectric transducer can be split into two parts, (i) piezoelectric coupling between electric and elastic domains, and (ii) magnetostrictive coupling between elastic and magnetic domains. Here, we investigate the second part and focus especially on the behavior at GHz frequencies that are relevant for fast electronic devices. In this frequency range, elastic waves (hypersound) interact with magnetic waves (spin waves), forming hybrid magnetoelastic waves under resonant conditions. The physics of the magnetoelastic resonance and the resulting magnetoelastic waves have been described in bulk and infinitesimally thin films decades ago [8–17]. However, modern applications of magnetoelastic waves in magnetoelectric and spintronic devices are based on nm-thick films. The finite thickness of these magnetic films alters the dynamic dipolar field with respect to infinitesimally thin films. Consequently, also the magnetoelastic coupling and the behavior of the magnetoelastic waves change due to the finite film thickness. In this chapter, the magnetoelastic theory and equations are extended to describe magnetoelastic waves in thin films of finite thickness.

The chapter begins by introducing basic magnetic interactions and by reviewing the properties of spin waves in bulk and thin film ferromagnets. The waves are described by a general formalism to calculate the eigensystem. The effect of the finite film thickness is then incorporated in this formalism. In the second part, linear elasticity and elastic waves in thin films are discussed. In the third part, the magnetoelastic interactions together with the combination of the magnetodynamic and elastodynamic equations of motion are described. Finally, the fundamental properties of magnetoelastic waves in thin films with finite thickness are derived and illustrated both analytically and graphically for different magnetization orientations.

12.2 Spin Waves

Spin waves are collective excitations of the magnetization in magnetic materials. The properties of spin waves are strongly affected by the geometry and the dominant interactions inside the material. Hence, the relevant magnetic interactions will be shortly introduced, followed by the derivation of the properties of spin waves in bulk and thin film ferromagnets using the plane wave method.

12.2.1 Magnetic Interactions and Magnetization Dynamics

The magnetization dynamics in a ferromagnet can be described by the Landau–Lifshitz–Gilbert (LLG) equation [18, 19]

$$\frac{d\mathbf{M}}{dt} = -\gamma\mu_0(\mathbf{M} \times \mathbf{H}_{\text{eff}}) + \frac{\alpha}{M_s}\left(\mathbf{M} \times \frac{d\mathbf{M}}{dt}\right) \tag{12.1}$$

with γ the absolute value of the gyromagnetic ratio ($s^{-1}T^{-1}$), μ_0 the vacuum permeability (TmA^{-1}), α the Gilbert damping constant, M_s the saturation magnetization (Am^{-1}), and H_{eff} the effective magnetic field (Am^{-1}). The first term in the LLG equation describes the precession of the magnetization around the effective magnetic field. The second term in the LLG equation leads to the damping of the magnetization precession towards the direction of the effective magnetic field.

Multiple magnetic interactions and effects exist that influence the magnetization dynamics such as the exchange interaction, dipolar interaction, magnetocrystalline effect, magnetoelastic effect, etc.. It is possible to derive a magnetic field that corresponds to every interaction via

$$\mathbf{H} = -\frac{1}{\mu_0}\frac{\delta U(\mathbf{M})}{\delta \mathbf{M}} \quad \text{and} \quad U(\mathbf{M}) = \int_V \mathcal{E}(\mathbf{M})dV \tag{12.2}$$

with $\mathcal{E}(\mathbf{M})$ the corresponding energy density of that interaction. The total effective magnetic field H_{eff}, which is governing the magnetization dynamics in the LLG equation, is given by the sum of all individual magnetic fields, including externally applied fields. Below, the dipolar and exchange interaction are explained in more detail since these lead to spin waves. Fully elaborated discussions of spin waves and their properties can be found in [16, 20, 21].

The dipolar interaction describes the direct interaction between magnetic dipoles. Following (12.2), the interaction can be represented by a dipolar magnetic field. This field is found by solving Maxwell's equations. For spin waves at GHz frequencies, the magnetostatic approximation is valid since the wavelengths of such spin waves are several orders of magnitude shorter than those of electromagnetic waves in vacuum at the same frequency, i.e. $k_0 \ll k_{sw}$ with k_0 the wavenumber of an electromagnetic wave in vacuum and k_{sw} the wavenumber of a spin wave. This approximation implies that the change in electric field over time, $\partial \mathbf{E}/\partial t$, has a negligible effect on the generation of the magnetic field. Assuming further that no free charges and no electrical currents are present inside the material, Maxwell's equations become

$$\nabla \cdot \mathbf{E} = 0 \tag{12.3}$$

$$\nabla \cdot \mathbf{B} = 0 \tag{12.4}$$

$$\nabla \times \mathbf{E} = -\frac{\partial \mathbf{B}}{\partial t} \tag{12.5}$$

$$\nabla \times \mathbf{H} = 0 \tag{12.6}$$

with $\mathbf{B} = \mu_0(\mathbf{H} + \mathbf{M})$ the magnetic induction (T). Hence, in the magnetostatic limit (as in the electrostatic limit), electric and magnetic fields are decoupled from each other. Equation (12.6) indicates that the curl of the magnetic field equals zero. This allows for the definition of a magnetic scalar potential ϕ as

$$\mathbf{H}_{\text{dip}} = -\nabla \phi\,. \tag{12.7}$$

Using (12.4), the definition of the magnetic scalar potential and the magnetic induction $\mathbf{B}$, one finds the magnetic Poisson relation

$$\nabla^2 \phi = \nabla \cdot \mathbf{M}\,. \tag{12.8}$$

This relation indicates that the divergence of the magnetization $\nabla \cdot \mathbf{M}$, also called the magnetic charge, acts as a source of the magnetic scalar potential and hence as a source of the dipolar field. Two types of magnetic charges can be identified: first, a surface charge, originating from surfaces between two materials with different magnetization magnitude or direction. Secondly, a magnetic volume charge, originating from the change of the magnetization in the bulk of a ferromagnetic material. Both surface and volume magnetic charges generate dipolar fields. The field outside the magnetic material is called the stray field and the field inside the material is called the demagnetization field.

By solving the magnetic Poisson equation (12.8) and using (12.7), it is possible to derive a general expression for the demagnetization field given by [22, 23]

$$\mathbf{H}_{\text{demag}} = \frac{1}{4\pi} \int_{V'} \bar{D}(\mathbf{r} - \mathbf{r}')\mathbf{M}(\mathbf{r}')dV' \tag{12.9}$$

with V' the volume of the magnetic material and $\bar{D}(\mathbf{r}-\mathbf{r}')$ the tensorial magnetostatic Green's function given by [16]

$$\bar{D}(\mathbf{r}-\mathbf{r}') = -\nabla_{\mathbf{r}}\nabla_{\mathbf{r}'}\frac{1}{|\mathbf{r}-\mathbf{r}'|}\,. \tag{12.10}$$

For uniform magnetization, the demagnetizing field is only generated by surface charges, and (12.9) reduces to

$$\mathbf{H}_{\text{demag}} = \frac{1}{4\pi}\mathbf{M}\int_{V'}\bar{D}(\mathbf{r}-\mathbf{r}')dV' = -\bar{N}(\mathbf{r})\mathbf{M} \tag{12.11}$$

with $\bar{N}(\mathbf{r})$ the demagnetization tensor, which only depends on the shape of the magnetic volume. The anisotropy introduced by the demagnetization field is thus often called the shape anisotropy.

The second magnetic interaction necessary to describe spin waves is the exchange interaction between individual magnetic dipoles. This interaction gives rise to ferromagnetic coupling below the Curie temperature [24]. The exchange energy density is given by

$$\mathcal{E}_{\text{ex}} = \frac{A_{\text{ex}}}{M_{\text{s}}^2}\left[(\nabla M_{\text{x}})^2 + (\nabla M_{\text{y}})^2 + (\nabla M_{\text{z}})^2\right] \tag{12.12}$$

with A_{ex} the exchange stiffness constant (J/m). Following (12.2), the exchange field is

$$\mathbf{H}_{\text{ex}} = \frac{2A_{\text{ex}}}{\mu_0 M_{\text{s}}^2}\Delta\mathbf{M} = l_{\text{ex}}^2\Delta\mathbf{M} \equiv \lambda_{\text{ex}}\Delta\mathbf{M} \tag{12.13}$$

with Δ the Laplace operator and l_{ex} the exchange length (m). In ferromagnets, the exchange interaction tries to keep the individual magnetic moments parallel. The exchange length l_{ex} characterizes the competition between the dipolar and the exchange interaction [21, 25]. At length scales below the exchange length l_{ex}, the exchange interaction is dominant and magnetic moments align parallel with each other. At length scales above the exchange length, the dipolar interaction is dominant, and it becomes possible for domains to form. Analogously, the properties of spin waves with short wavelengths are dominated by the exchange interaction, whereas the dipolar interaction strongly affects the properties of spin waves with large wavelengths.

12.2.2 Spin Waves in the Bulk Ferromagnets

Consider a ferromagnetic material with static magnetic field $\mathbf{H}_{\text{ext}}$ applied in the z-direction. In absence of any anisotropy, the external field forces the equilibrium magnetization along the z-direction. In such a system, stable wave-like excitations exist, which can be described by weak perturbations of the equilibrium magnetization. For a plane wave, the magnetization at a specific point in space and time can be written as

$$\mathbf{M}(\mathbf{r},t) = \mathbf{M}_0 + \mathbf{m}(\mathbf{r},t) = \begin{bmatrix} 0 \\ 0 \\ M_0 \end{bmatrix} + \begin{bmatrix} m_x \\ m_y \\ 0 \end{bmatrix} e^{i(\omega t + \mathbf{k}\cdot\mathbf{r})} \tag{12.14}$$

with ω the angular frequency of the wave (rad s^{-1}), and $\mathbf{k}$ the wavevector with norm $||\mathbf{k}|| = k = 2\pi/\lambda$ (rad m^{-1}) and direction perpendicular to the phase front. For weak perturbations, i.e. $||\mathbf{m}|| \ll M_0$, $\mathbf{m}(\mathbf{r},t)$ describes a wave-like perturbation which is called a spin wave.

In (12.14), the z-component of the dynamic magnetization, m_z, is neglected. This approximation is only valid if the perturbations are weak. Since the angular momentum, *i.e.* the norm of the magnetization vector, is conserved, the z-component is given by $m_z^2 = M_0^2 - m_x^2 - m_y^2$. Therefore, the m_z component can be considered as a second order perturbation and is neglected in the remainder of this chapter.

For a uniform bulk material, the dipolar and exchange fields that correspond to the perturbed magnetization state can be found via (12.9) and (12.13), respectively, and are given by [26]

$$\mathbf{h}_{\text{dip}}(\mathbf{r},t) = -\frac{\mathbf{k}\cdot\mathbf{m}(\mathbf{r},t)}{||\mathbf{k}||^2}\mathbf{k} = -\frac{1}{k^2}\begin{bmatrix} k_x^2 & k_x k_y & 0 \\ k_x k_y & k_y^2 & 0 \\ 0 & 0 & 0 \end{bmatrix} \mathbf{m}(\mathbf{r},t) \tag{12.15}$$

and

$$\mathbf{h}_{\text{ex}}(\mathbf{r},t) = -\lambda_{\text{ex}} k^2 \mathbf{m}(\mathbf{r},t)\,. \tag{12.16}$$

The wavevector $\mathbf{k} = [k_x, k_y, k_z]$ is determined by a single parameter θ because of the axial symmetry around the magnetization vector. Hence, the wavevector can be written as $\mathbf{k} = k[\sin(\theta), 0, \cos(\theta)]$ with θ the angle between the magnetization and the propagation direction of the wave. With this substitution, the dipolar field can be simplified to

$$\mathbf{h}_{\text{dip}}(\mathbf{r},t) = -\begin{bmatrix} \sin^2(\theta) & 0 & 0 \\ 0 & 0 & 0 \\ 0 & 0 & 0 \end{bmatrix} \mathbf{m}(\mathbf{r},t)\,. \tag{12.17}$$

The magnetization dynamics corresponding to the spin wave is found by solving the LLG equation (12.1) including the perturbation $\mathbf{m}(\mathbf{r},t)$. Neglecting the damping term, the LLG equation becomes

$$\frac{d[\mathbf{M}_0 + \mathbf{m}(\mathbf{r}, t)]}{dt} = -\gamma_0[(\mathbf{M}_0 + \mathbf{m}(\mathbf{r}, t)) \times (\mathbf{H}_{\text{ext}} + \mathbf{h}_{\text{dip}}(\mathbf{r}, t) + \mathbf{h}_{\text{ex}}(\mathbf{r}, t))] \tag{12.18}$$

with $\gamma_0 = \gamma\mu_0$. Terms quadratic in $\mathbf{m}$ can be neglected because the perturbation is assumed to be weak, which results in the linearized LLG equation given by

$$i\omega\mathbf{m}(\mathbf{r}, \omega) = -\gamma_0[\mathbf{M}_0 \times (\mathbf{h}_{\text{ex}}(\mathbf{r}, \omega) + \mathbf{h}_{\text{dip}}(\mathbf{r}, \omega)) + \mathbf{m}(\mathbf{r}, \omega) \times \mathbf{H}_{\text{ext}}]\,. \tag{12.19}$$

Rearranging the terms and rewriting the system in matrix notation leads to

$$\begin{bmatrix} \omega_{\text{bx}} & -i\omega \\ i\omega & \omega_{\text{by}} \end{bmatrix} \begin{bmatrix} m_{\text{x}} \\ m_{\text{y}} \end{bmatrix} = 0 \tag{12.20}$$

with

$$\omega_{\text{bx}} = \omega_0 + \omega_{\text{M}}(\lambda_{\text{ex}}k^2 + \sin^2(\theta)) \tag{12.21}$$

$$\omega_{\text{by}} = \omega_0 + \omega_{\text{M}}\lambda_{\text{ex}}k^2 \tag{12.22}$$

$\omega_0 = \gamma_0 H_{\text{ext}}$, and $\omega_{\text{M}} = \gamma_0 M_{\text{s}}$. The parameters ω_{bx} and ω_{by} are related to the effective magnetic fields that interact with the x- and y-components of the dynamic magnetization, respectively.

The properties of the stable perturbations of the magnetization, i.e. the spin waves, can be extracted by analyzing the eigenvalues and corresponding eigenstates of (12.20). Equation (12.20) has nontrivial solutions only if its determinant is zero. This condition can be utilized to obtain the dispersion relations of the spin waves. Considering only positive frequencies, the spin wave angular frequency is given by

$$\omega = \sqrt{\omega_{\text{bx}}\omega_{\text{by}}} = \sqrt{(\omega_0 + \omega_{\text{M}}\lambda_{\text{ex}}k^2)[\omega_0 + \omega_{\text{M}}(\lambda_{\text{ex}}k^2 + \sin^2(\theta))]}\,. \tag{12.23}$$

This equation is the dispersion relation for spin waves in bulk ferromagnets. It is also called the Herring–Kittel equation.

Equation (12.23) indicates that there is a nonzero minimum frequency, above which resonant magnetization dynamics are obtained. Exciting a ferromagnet at frequencies below the spin wave resonance generates evanescent waves. If the excitation source is removed, these waves disappear after a certain time (their lifetime) even in absence of intrinsic damping. Moreover, they do not propagate and thus do not contribute to steady state wave patterns at distances from the excitation source that are much larger than their wavelength. However, they are important to satisfy boundary conditions and in transient regimes.

The eigenstates corresponding to the eigenvalues of the linearized LLG equation are

$$\mathbf{m}(k) = \frac{N}{\sqrt{\omega_{\rm bx}\omega_{\rm by}}}\begin{bmatrix} i\omega_{\rm by} \\ \sqrt{\omega_{\rm bx}\omega_{\rm by}} \end{bmatrix} \tag{12.24}$$

with N a normalization constant. Note that $\omega_{\rm bx}$ and $\omega_{\rm by}$ both depend on k and thus, via the dispersion relation, also on the frequency. The eigenstate indicates that the precession of the magnetization (the polarization of the wave) is always clockwise in the direction of propagation. Furthermore, the precession of the magnetization is generally elliptical with an ellipticity equal to

$$\epsilon_{\rm b} = \frac{|m_{\rm x}|}{|m_{\rm y}|} = \frac{\omega_{\rm by}}{\sqrt{\omega_{\rm bx}\omega_{\rm by}}}\,. \tag{12.25}$$

In the limit of small k, the exchange interaction can be neglected since $\lambda_{\rm ex}k^2 \ll 1$, and the dispersion relation becomes $\omega = \sqrt{\omega_0(\omega_0 + \omega_{\rm M}\sin^2(\theta))}$. This dispersion relation characterizes dipolar spin waves that are degenerate. Hence, in this limit, multiple spin waves with different wavelengths exist at the same frequency. For $\theta = 0$, the dispersion relation becomes $\omega = \omega_0$ and the effect of the dipolar self-interaction disappears. In this case, the dynamic magnetization components only interact with the external field $H_{\rm ext}$. For $\theta = \pi/2$, the interaction between the dynamic dipolar field and the spin wave is strongest. In this case, the dispersion relation is $\omega = \sqrt{\omega_0(\omega_0 + \omega_M)}$. Therefore, the spin wave frequencies in the dipolar regime are limited to a specific interval

$$\omega_0 \le \omega \le \sqrt{\omega_0(\omega_0 + \omega_M)}\,. \tag{12.26}$$

On the other hand, in the limit of large k, when $\lambda_{\rm ex}k^2 \gg 1$, a quadratic dispersion relation is obtained

$$\omega = \omega_{\rm M}\lambda_{\rm ex}k^2\,. \tag{12.27}$$

This dispersion characterizes spin waves for which the exchange interaction is dominant. It is worth noting that these exchange spin waves are isotropic with respect to the propagation direction. By contrast, dipolar spin waves are anisotropic because they depend on the propagation direction via the parameter θ.

12.2.3 Spin Waves in Ferromagnetic Thin Films

In the previous section, the properties of spin waves in an infinite bulk medium were discussed. In this section, we introduce boundaries in the ferromagnetic medium and derive the properties of spin waves in ferromagnetic thin films of finite thickness.

Consider an infinite magnetic thin film of thickness d with its normal parallel to the y-direction. In the previous section, electrical currents were neglected in Maxwell's equations, which is only a good approximation for ferromagnetic insulators. However, for thin films, this approximation is even valid for conductors as long as the thickness of the film is sufficiently small with respect to the skin depth of the ferromagnet [27]. It should also be noted that in the derivations below, the dynamic magnetization and the fields are averaged over the thickness of the film and are thus uniform in the y-direction. This is a valid approximation when the wavelength is much larger than the thickness of the film, i.e. $kd \ll 1$. If this is not the case, it is possible for thickness modes to arise which have varying amplitude along the thickness [16]. These thickness modes will however not be considered here.

The magnetization is again defined as in (12.14) with the magnetization saturated in-plane along an external field $\mathbf{H}_{\text{ext}}$ in the z-direction. The components of the dynamic magnetization are in the x- and y-direction and form a plane wave, the spin wave. The exchange field is not affected by the thin film boundaries and is given by (12.13). As indicated by (12.8), the boundaries generate magnetic surface charges and therefore act as a source of the dipolar field. Therefore, in contrast with the exchange field, the dipolar field is affected by the boundaries.

For a thin film of finite thickness, the dipolar field can be approximated by [28–30]

$$\mathbf{h}_{\text{dip}}(\mathbf{r}, t) = -[P\frac{\mathbf{k}\cdot\mathbf{m}}{||\mathbf{k}||^2}\mathbf{k} + (1-P)(\mathbf{n}\cdot\mathbf{m})\mathbf{n}] \tag{12.28}$$

$$= -\begin{bmatrix} P\sin^2(\theta) & 0 & P\sin(\theta)\cos(\theta) \\ 0 & 1-P & 0 \\ P\sin(\theta)\cos(\theta) & 0 & P\cos^2(\theta) \end{bmatrix}\mathbf{m}(\mathbf{r}, t) \tag{12.29}$$

with

$$P = 1 - \frac{1-e^{-kd}}{kd}, \tag{12.30}$$

$k^2 = k_x^2 + k_z^2$, as well as θ the angle between the static magnetization $\mathbf{M}_0$ and wavevector $\mathbf{k}$. In the limit of an infinitesimally thin film, this simplifies to

$$\lim_{d\to 0}\mathbf{h}_{\text{dip}}(\mathbf{r}, t) = -\begin{bmatrix} 0 \\ 1 \\ 0 \end{bmatrix}\mathbf{m}(\mathbf{r}, t) \tag{12.31}$$

and thus only the out-of-plane magnetization component contributes to the spin wave dipolar field. Hence, for a thin film of finite thickness, the spin wave behavior is markedly different as compared to a thin film of infinitesimal thickness.

The linearized LLG equation (12.19) with the modified dipolar field in (12.28) can then be written as

$$\begin{bmatrix} \omega_{\text{fx}} & -i\omega \\ i\omega & \omega_{\text{fy}} \end{bmatrix}\begin{bmatrix} m_x \\ m_y \end{bmatrix} = 0 \tag{12.32}$$

with

$$\omega_{\mathrm{fx}} = \omega_0 + \omega_{\mathrm{M}}(\lambda_{\mathrm{ex}}k^2 + P\sin^2(\theta)) \tag{12.33}$$

$$\omega_{\mathrm{fy}} = \omega_0 + \omega_{\mathrm{M}}(\lambda_{\mathrm{ex}}k^2 + 1 - P)\,. \tag{12.34}$$

Again, nontrivial solutions of this equation only exist when the determinant of the matrix is zero. This condition leads to the dispersion relation of spin waves in thin ferromagnetic films, given by [30, 31]

$$\omega = \sqrt{\omega_{\mathrm{fx}}\omega_{\mathrm{fy}}} = \sqrt{(\omega_0 + \omega_{\mathrm{M}}\lambda_{\mathrm{ex}}k^2)(\omega_0 + \omega_{\mathrm{M}}\lambda_{\mathrm{ex}}k^2 + \omega_{\mathrm{M}}F_{\mathrm{m}})} \tag{12.35}$$

with

$$F_{\mathrm{m}} = 1 - P\cos^2(\theta) + \frac{\omega_{\mathrm{M}}P(1-P)\sin^2(\theta)}{\omega_0 + \omega_{\mathrm{M}}\lambda_{\mathrm{ex}}k^2}\,. \tag{12.36}$$

The corresponding eigenstate has the same form and properties as the eigenstate of spin waves in bulk ferromagnets and is given by

$$\mathbf{m}(k) = \frac{N}{\sqrt{\omega_{\mathrm{fx}}\omega_{\mathrm{fy}}}}\begin{bmatrix} i\omega_{\mathrm{fy}} \\ \sqrt{\omega_{\mathrm{fx}}\omega_{\mathrm{fy}}} \end{bmatrix} \tag{12.37}$$

with N a dimensionless normalization constant.

In the limit of large wavevectors, i.e. the exchange limit $\lambda_{\mathrm{ex}}k^2 \gg 1$, the dispersion relation reduces to (12.27) that was derived for bulk magnetic media. However, in the dipolar limit of small k-values, $\lambda_{\mathrm{ex}}k^2 \ll 1$, the dispersion relation differs from that in bulk ferromagnetic media. Again, two limiting cases are found for $\theta = 0$ and $\theta = \pi/2$.

For $\lambda_{\mathrm{ex}}k^2 \ll 1$ and $\theta = 0$, the dispersion relation becomes

$$\omega_{\mathrm{BVW}}^2 = \omega_0\left(\omega_0 + \omega_{\mathrm{M}}\frac{1-e^{-kd}}{kd}\right)\,. \tag{12.38}$$

The propagation direction of these waves is parallel to the direction of the static equilibrium magnetization. Their dispersion relation is plotted in Fig. 12.1 for a 30 nm thick Ni film with $M_{\mathrm{s}} = 480$ kA/m [32], $A_{\mathrm{ex}} = 8$ pJ/m [33], and an external magnetic field of $\mu_0 H_{\mathrm{ext}} = 50$ mT. According to the dispersion relation, the frequency decreases with increasing wavenumber, and thus the group velocity, which is defined as $\mathbf{v}_{\mathrm{g}} = \partial\omega/\partial\mathbf{k}$, is negative. On the other hand, the phase velocity $\mathbf{v}_{\mathrm{p}} = \mathbf{k}\,\omega/k^2$, which describes the velocity and direction of the phase front, is positive. The energy flow of a wave is always parallel to the group velocity, and thus in this geometry, the energy flow and the group velocity are antiparallel to the wavevector and the phase velocity. For this reason, such waves are called backward volume waves (BVWs). As shown in Fig. 12.1, when the exchange interaction becomes non-negligible at larger wavevectors, the dispersion relation shifts to higher frequencies. This effect increases for higher k-values, finally reaching the limiting case of exchange-dominated spin waves.

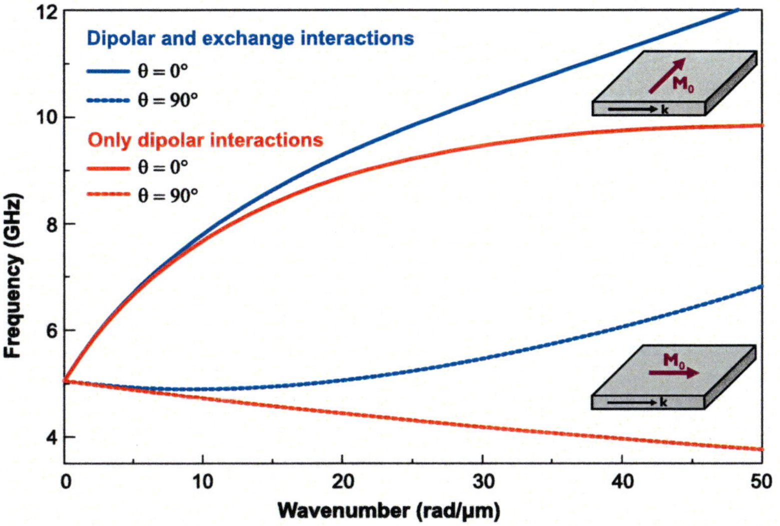

Fig. 12.1 Spin wave dispersion according to (12.35) for a 30 nm thick Ni film. Material parameters are $M_s = 480\,\text{kA/m}$ and $A_{ex} = 8\,\text{pJ/m}$, whereas the external magnetic field is $\mu_0 H_{ext} = 50\,\text{mT}$. The solid red and blue lines correspond to dispersion relations of dipolar and dipolar–exchange surface spin waves, respectively. The dashed red and blue lines correspond to dispersion relations of dipolar and dipolar–exchange backward volume spin waves, respectively

In the dipolar limit ($\lambda_{ex}k^2 \ll 1$), the dispersion relation for $\theta = \pi/2$ becomes

$$\omega_{SW}^2 = \omega_0(\omega_0 + \omega_M) + \omega_M^2 \left(1 - \frac{1 - e^{-kd}}{kd}\right) \frac{1 - e^{-kd}}{kd}. \tag{12.39}$$

These waves are called surface waves since their amplitude decays exponentially away from the surface. However, if the film is sufficiently thin, the magnetization can be considered uniform over the film thickness as mentioned earlier. The dispersion relations of spin waves both in the dipolar approximation and when the dipolar and exchange interaction are simultaneously present are plotted in Fig. 12.1. The group velocity of these waves is positive and thus points in the same direction as the phase velocity.

It should also be mentioned that spin waves are accompanied by a dynamic electric field. This electric field $\mathbf{e}$ is obtained from Maxwell's equations (12.3) and (12.5) which can be rewritten as

$$\nabla \cdot \mathbf{e} = 0 \tag{12.40}$$

$$\nabla \times \mathbf{e} = -i\mu_0\omega(\mathbf{h}_{dip} + \mathbf{m}) = -i\mu_0\omega\left(\bar{N}_{dip} + \bar{I}\right)\mathbf{m} \tag{12.41}$$

with $\bar{I}$ the identity matrix. Equation (12.41) indicates that both the dynamic dipolar field and the dynamic magnetization contribute to the generation of the dynamic electric field. However, in the magnetostatic limit when $k_0 \ll k$, the effect of the dipolar field is much smaller than that of the dynamic magnetization. Taking this into account and solving Maxwell's equations for plane waves results in

$$\mathbf{e} = -\frac{\mu_0 \omega}{k^2} \mathbf{k} \times \mathbf{m} \tag{12.42}$$

for the dynamic electric field [20].

For spin waves at GHz frequencies in ferromagnetic media, the energy stored in the electric field is much smaller than the energy stored in the magnetic system [20]. Therefore, the magnetostatic waves can be considered as "magnetization waves". Note that this applies to spin waves in both the dipolar and exchange regime. In both cases, at GHz frequencies, the wavelength of a spin wave is much shorter than the wavelength of an electromagnetic wave in vacuum and the magnetostatic approximation is thus valid.

All calculations in this section are only valid at GHz frequencies in the magnetostatic limit. At higher frequencies near the THz regime, the spin wave wavelength becomes comparable to the wavelength of the electromagnetic wave in vacuum, $k_{sw} \approx k_0$, and the magnetostatic approximation does not longer hold. At these higher frequencies, the influence of the time-varying electric field alters the wave behavior. This can be seen by considering both generation mechanisms of the magnetic field. As mentioned earlier, the dipolar magnetic field can be generated by both varying electric fields over time and by varying magnetization over space. The generation mechanism via the time varying electric field is proportional to the regular electromagnetic wave wavenumber k_0, whereas the generation mechanism via the magnetization is proportional to the magnetization wavenumber k_{sw}. Hence, the mechanism which governs the highest wavenumber dominates the generation of the magnetic dipolar field.

In the GHz regime and magnetostatic limit $k_0 \ll k_{sw}$, the dipolar field generation is thus dominated by the variation of the magnetization over space. However, at much higher frequencies near the THz regime, both wavenumbers are of the same order and thus both generation mechanisms are of similar magnitude. This means that the generation of the magnetic dipolar field by the time varying electric field cannot be neglected anymore.

The frequency, for which the magnetostatic approximation breaks down, ω_{crit}, can be found by relating the wavenumbers to the frequency via the dispersion relations. The crossing point of the spin-wave dispersion relation, (12.27), with the linear electromagnetic dispersion relation, $\omega_0 = ck_0$, determines ω_{crit} and is given by

$$\omega_{crit} = \frac{c^2}{\omega_M \lambda_{ex}} \tag{12.43}$$

with c the speed of light in vacuum. For frequencies $\omega \ll \omega_{crit}$, the magnetostatic limit is valid and regular spin waves are obtained. For frequencies above ω_{crit}, spin waves behave similarly to classical electromagnetic waves with a considerable fraction of energy stored in the dynamic electric field. As a result, besides the dipolar and exchange regime, there also exist a regular electromagnetic regime at higher frequencies, which corresponds to electromagnetic waves with a linear dispersion relation.

12.3 Elastic Waves

In the previous section, the properties of spin waves in ferromagnetic media, both in bulk materials and in thin films, have been discussed. In this section, we turn to the properties of wave-like oscillations of the displacement, i.e. elastic waves. We start with a short derivation of the fundamental equations of linear elasticity. Then, the different types of elastic waves and their characteristics are described.

12.3.1 Elastodynamic Equations of Motion

The equation of motion for the displacement $\mathbf{u}$ is given by

$$\rho \frac{d^2\mathbf{u}}{dt^2} = \nabla \cdot \bar{\sigma} + \mathbf{f}_\mathrm{b} \tag{12.44}$$

with ρ the mass density (kgm^{-3}), $\bar{\sigma}$ the two dimensional stress tensor with components σ_ij (Nm^{-2}), and $\mathbf{f}_\mathrm{b}$ the body forces acting on the material (Nm^{-3}). For linear elastic materials, the stress tensor is related to the strain tensor via Hooke's law

$$\bar{\sigma} = \bar{\bar{C}} : \bar{\varepsilon} \qquad \text{or} \qquad \sigma_\mathrm{ij} = \sum_{k=1}^{3}\sum_{l=1}^{3} C_\mathrm{ijkl}\varepsilon_\mathrm{kl}\,. \tag{12.45}$$

Here, $\bar{\bar{C}}$ is the fourth-order stiffness tensor and $\bar{\varepsilon}$ is the second-order strain tensor. The symmetries of the stiffness tensor allows to rewrite Hooke's law in reduced dimensionality [34, 35] as

$$\begin{bmatrix}\sigma_{11}\\ \sigma_{22}\\ \sigma_{33}\\ \sigma_{12}\\ \sigma_{13}\\ \sigma_{23}\end{bmatrix} = \begin{bmatrix} C_{11} & C_{12} & C_{13} & C_{14} & C_{15} & C_{16}\\ & C_{22} & C_{23} & C_{24} & C_{25} & C_{26}\\ & & C_{33} & C_{34} & C_{35} & C_{36}\\ & & & C_{44} & C_{45} & C_{46}\\ & \text{symm.} & & & C_{55} & C_{56}\\ & & & & & C_{66}\end{bmatrix} \begin{bmatrix}\varepsilon_{11}\\ \varepsilon_{22}\\ \varepsilon_{33}\\ 2\varepsilon_{12}\\ 2\varepsilon_{13}\\ 2\varepsilon_{23}\end{bmatrix}. \tag{12.46}$$

This is called the Voigt notation for Hooke's law. Important to note is the notation for the shear strain elements. In some works, this is given by the engineering strains $\gamma_{ij} = 2\varepsilon_{ij}$ which gives a factor of 2 difference with the real shear strains.

Equation (12.46) indicates that a material with a nonsymmetric (e.g. triclinic) crystal structure is described by 21 independent stiffness coefficients [36, 37]. In a crystal system with a certain symmetry, the number of independent stiffness constants can be greatly reduced. For example, only three independent stiffness constants are required to describe cubic crystal systems. The stiffness tensor then becomes

$$\bar{C}_{\text{cubic}} = \begin{bmatrix} C_{11} & C_{12} & C_{12} & 0 & 0 & 0 \\ & C_{11} & C_{12} & 0 & 0 & 0 \\ & & C_{11} & 0 & 0 & 0 \\ & & & C_{44} & 0 & 0 \\ & \text{symm.} & & & C_{44} & 0 \\ & & & & & C_{44} \end{bmatrix} . \tag{12.47}$$

In this case, the residual anisotropy can be quantified by the Zener factor A, which is given by

$$A = \frac{2C_{44}}{C_{11} - C_{12}} . \tag{12.48}$$

A Zener factor of 1 indicates fully isotropic elastic properties. In this isotropic limit, only two independent constants are necessary to describe the stiffness tensor. Note that different combinations of parameters can be used to represent the isotropic case, such as Young's modulus and the Poisson ratio, Young's modulus and the shear modulus, or the Lamé moduli. All descriptions are fully equivalent [37, 38].

For small displacements, the relation between the strain and displacement is given by [34, 36]

$$\bar{\varepsilon} = \frac{1}{2}\left(\nabla\mathbf{u} + (\nabla\mathbf{u})^T - \nabla\mathbf{u}\,(\nabla\mathbf{u})^T\right) \approx \frac{1}{2}\left(\nabla\mathbf{u} + (\nabla\mathbf{u})^T\right) . \tag{12.49}$$

Combining (12.44), (12.47), and (12.49) results in the elastodynamic equations of motion with the displacement as the only variable. For a material with cubic symmetry, the equations are given by

$$\begin{aligned} \rho\frac{\partial^2 u_x}{\partial t^2} &= C_{11}\frac{\partial^2 u_x}{\partial x^2} + C_{44}\left(\frac{\partial^2 u_x}{\partial y^2} + \frac{\partial^2 u_x}{\partial z^2}\right) + (C_{12} + C_{44})\left(\frac{\partial^2 u_y}{\partial x \partial y} + \frac{\partial^2 u_z}{\partial x \partial z}\right) + f_x \\ \rho\frac{\partial^2 u_y}{\partial t^2} &= C_{11}\frac{\partial^2 u_y}{\partial y^2} + C_{44}\left(\frac{\partial^2 u_y}{\partial x^2} + \frac{\partial^2 u_y}{\partial z^2}\right) + (C_{12} + C_{44})\left(\frac{\partial^2 u_x}{\partial x \partial y} + \frac{\partial^2 u_z}{\partial y \partial z}\right) + f_y \\ \rho\frac{\partial^2 u_z}{\partial t^2} &= C_{11}\frac{\partial^2 u_z}{\partial z^2} + C_{44}\left(\frac{\partial^2 u_z}{\partial x^2} + \frac{\partial^2 u_z}{\partial y^2}\right) + (C_{12} + C_{44})\left(\frac{\partial^2 u_x}{\partial x \partial z} + \frac{\partial^2 u_y}{\partial y \partial z}\right) + f_z . \end{aligned} \tag{12.50}$$

Note that the above equations do not contain any damping terms and the system is assumed to be lossless. In practice, materials always possess some degree of viscoelasticity. In this case, the energy in the elastic wave is lost by different mechanisms such as phonon–phonon scattering due to the anharmonicity of the vibrational potential or the scattering of phonons by impurities. This can be taken into account by considering complex stiffness coefficients [37, 38]. However, in the following, perfect elasticity without loss is assumed for simplicity.

12.3.2 Elastic Waves in Thin Films

In this section, we introduce the properties of elastic waves in an idealized thin film with free surfaces. This corresponds to an isolated thin film in vacuum in which the elastodynamics is perfectly confined inside the film. The perfect confinement is achieved by large acoustical impedance mismatch between the film and vacuum. Therefore, the model also approximately represents a thin film surrounded by materials with strongly different acoustic impedances, e.g. a film with a free top surface and a large acoustic impedance mismatch with the supporting substrate. For more realistic approaches, appropriate stress and velocity boundary conditions need to be applied at the interfaces. In the next section, when the magnetoelastic interaction is included, it is demonstrated that the magnetization dynamics also generate elastic stresses, which further complicates the description at the boundaries. In such cases, an analytical treatment of the system is difficult and accurate studies require numerical simulations, e.g. by finite element methods. Nonetheless, the treatment of an idealized system presented here provides analytical insights in the basic elastic (and magnetoelastic) behavior. This insight will help in the understanding of the magnetoelastic waves in the next section and can be used in the future to interpret numerical simulations of more realistic systems.

For the case of thin films with free surface boundary conditions, the variation of the displacement along the thickness of the film is much smaller than the in-plane variation. Hence, the derivative of the displacement along the film surface normal can be neglected with respect to the derivatives in the in-plane directions, i.e. $\partial u/\partial y \ll \partial u/\partial x$, $\partial u/\partial z$. The elastodynamic equations of motion for a thin film with a surface normal in the y-direction are then given by

$$\begin{aligned}
\rho\frac{\partial^2 u_x}{\partial t^2} &= C_{11}\frac{\partial^2 u_x}{\partial x^2} + C_{44}\frac{\partial^2 u_x}{\partial z^2} + (C_{12}+C_{44})\frac{\partial^2 u_z}{\partial x \partial z} \\
\rho\frac{\partial^2 u_y}{\partial t^2} &= C_{44}\left(\frac{\partial^2 u_y}{\partial x^2} + \frac{\partial^2 u_y}{\partial z^2}\right) \\
\rho\frac{\partial^2 u_z}{\partial t^2} &= C_{11}\frac{\partial^2 u_z}{\partial z^2} + C_{44}\frac{\partial^2 u_z}{\partial x^2} + (C_{12}+C_{44})\frac{\partial^2 u_x}{\partial x \partial z}\,.
\end{aligned} \tag{12.51}$$

The above set of linear differential equations has wave-like solutions of the form [15, 37]

$$\mathbf{u}(\mathbf{r}, t) = \begin{bmatrix} u_x \\ u_y \\ u_z \end{bmatrix} e^{i(\omega t + \mathbf{k}\cdot\mathbf{r})} . \tag{12.52}$$

To determine the dispersion relation of elastic waves in thin films, (12.52) is substituted into the wave equations (12.51). Rewriting the system in matrix notation and considering that the wavevector **k** points along the x-direction, results in

$$\begin{bmatrix} \omega^2 - v_l^2 k^2 & 0 & 0 \\ 0 & \omega^2 - v_t^2 k^2 & 0 \\ 0 & 0 & \omega^2 - v_t^2 k^2 \end{bmatrix} \begin{bmatrix} u_x \\ u_y \\ u_z \end{bmatrix} = 0 \tag{12.53}$$

with $v_l = \sqrt{C_{11}/\rho}$ the velocity of the longitudinal wave, $v_t = \sqrt{C_{44}/\rho}$ the velocity of the transversal wave, and ω the angular frequency of the elastic wave. As a result, three independent elastic waves are found, which correspond to the three components of the displacement vector.

When only the u_x component is nonzero, longitudinal waves are formed since the displacement oscillation is in the same direction as the wavevector. This wave is also called a compressional or dilational wave. The dispersion relation, $\omega_l(k)$, of this wave is easily found from (12.53) to be $\omega_l = v_l k$ [15, 36, 37]. The dispersion relation is linear, and thus the group velocity v_l equals the phase velocity, independently of frequency.

Waves with nonzero displacement components u_y and u_z oscillate perpendicular to the propagation direction. Therefore, these waves are transversal waves, also called shear or rotational waves. Their dispersion relation is also linear and equals $\omega_t = v_t k$ [15, 36, 37]. The phase and group velocities are thus both equal to v_t. It is further possible to classify these waves based on their polarization with respect to the film surface. The u_y component corresponds to shear vertical (SV) waves and the u_z component corresponds to shear horizontal (SH) waves. It is important to note that the velocity of the longitudinal wave is always larger than the velocity of the shear waves because $C_{11} > C_{44}$ [36, 37].

The energy of elastic waves oscillates between the elastic potential energy and the kinetic energy. The elastic energy density is given by [15, 36, 37]

$$\mathcal{E}_{el} = \frac{1}{2}\bar{\sigma} : \bar{\varepsilon} = \frac{1}{2} C_{ijkl}\varepsilon_{ij}\varepsilon_{kl} = \frac{1}{2}\sum_{i=1}^{3}\sum_{j=1}^{3}\sum_{k=1}^{3}\sum_{l=1}^{3} C_{ijkl}\varepsilon_{ij}\varepsilon_{kl} \tag{12.54}$$

or in Voigt notation

$$\mathcal{E}_{el} = \frac{1}{2}\bar{\sigma} : \bar{\varepsilon} = \frac{1}{2} C_{ij}\varepsilon_j\varepsilon_i , \tag{12.55}$$

whereas the kinetic energy density is given by

$$\mathcal{E}_{\text{kin}} = \frac{\rho||\mathbf{v}||^2}{2} \quad \text{with} \quad \mathbf{v} = \frac{\partial \mathbf{u}}{\partial t}. \tag{12.56}$$

Hence, for an elastic wave, the total energy is $\mathcal{E}_{\text{tot}} = \mathcal{E}_{\text{el}} + \mathcal{E}_{\text{kin}}$ and $\mathcal{E}_{\text{el}} = \mathcal{E}_{\text{kin}}$.

12.4 Magnetoelastic Waves

In the two previous sections, magnetic and elastic waves in thin films were studied. This section connects the two previous sections by introducing magnetoelastic interactions. In the first part of this section, the magnetoelastic interaction terms are described, which couple magnetic and elastic waves. In the second part, the properties of these magnetoelastic waves are derived and explained. Magnetoelastic waves have been studied in detail in bulk materials [8–10], at free surfaces [39–41], and in infinitesimally thin films [11, 12]. This section reviews the most important aspects of these magnetoelastic waves together with the corresponding equations. Beyond this review, we subsequently derive the influence of finite film thickness on the properties of the magnetoelastic waves by taking into account the appropriate dipolar and exchange fields.

12.4.1 Magnetoelastic Interactions

Magnetoelastic interactions can be separated in two different effects: firstly, the influence of the direction of the magnetization on the internal strain in a ferromagnet, called the magnetostrictive effect; and secondly, the effect of strain on the magnetization state, called the Villari effect. If both effects are considered simultaneously, one speaks about magnetoelasticity.

12.4.1.1 Magnetostriction

Magnetostriction describes how the magnetization affects the elastic behavior of a material. Therefore, in a magnetostrictive material, different magnetization states result in different strain states. For a material with cubic symmetry, the magnetoelastic energy density is given by [8]

$$\begin{aligned}\mathcal{E}_{\text{mel}} =& \frac{B_1}{M_s^2}\left(\varepsilon_{xx}\left(M_x^2 - \frac{1}{3}\right) + \varepsilon_{yy}\left(M_y^2 - \frac{1}{3}\right) + \varepsilon_{zz}\left(M_z^2 - \frac{1}{3}\right)\right)\\ &+ \frac{2B_2}{M_s^2}\left(\varepsilon_{xy}M_xM_y + \varepsilon_{yz}M_yM_z + \varepsilon_{zx}M_xM_z\right)\end{aligned} \tag{12.57}$$

with B_1 and B_2 the linear isotropic and anisotropic magnetoelastic coupling constants, respectively (Jm^{-3}). It is worth noting that the magnitude of the saturation magnetization has no influence on the magnetoelastic energy or strain state, which are rather determined by the orientation of the magnetization vector. The magnetization orientation is defined by the vector

$$\zeta = \begin{bmatrix} \zeta_x \\ \zeta_y \\ \zeta_z \end{bmatrix} = \frac{1}{M_s} \begin{bmatrix} M_x \\ M_y \\ M_z \end{bmatrix} . \tag{12.58}$$

Substituting this into (12.57) leads to

$$\begin{aligned} \mathcal{E}_{\mathrm{mel}} =& B_1 \left(\varepsilon_{\mathrm{xx}} \left(\zeta_x^2 - \frac{1}{3} \right) + \varepsilon_{\mathrm{yy}} \left(\zeta_y^2 - \frac{1}{3} \right) + \varepsilon_{\mathrm{zz}} \left(\zeta_z^2 - \frac{1}{3} \right) \right) \\ &+ 2B_2 \left(\varepsilon_{\mathrm{xy}} \zeta_x \zeta_y + \varepsilon_{\mathrm{yz}} \zeta_y \zeta_z + \varepsilon_{\mathrm{zx}} \zeta_x \zeta_z \right) . \end{aligned} \tag{12.59}$$

Based on this expression for the magnetoelastic energy density, it is possible to calculate the magnetostrictive body force, which is given by

$$\mathbf{f}_{\mathrm{mel}} = \nabla \cdot \bar{\sigma}_{\mathrm{mel}} = \nabla \cdot \left(\frac{d\mathcal{E}_{\mathrm{mel}}}{d\varepsilon_{\mathrm{ij}}} \right) , \tag{12.60}$$

leading to

$$\mathbf{f}_{\mathrm{mel}} = 2B_1 \begin{bmatrix} \zeta_x \frac{\partial \zeta_x}{\partial x} \\ \zeta_y \frac{\partial \zeta_y}{\partial y} \\ \zeta_z \frac{\partial \zeta_z}{\partial z} \end{bmatrix} + B_2 \begin{bmatrix} \zeta_x \left(\frac{\partial \zeta_y}{\partial y} + \frac{\partial \zeta_z}{\partial z} \right) + \zeta_y \frac{\partial \zeta_x}{\partial y} + \zeta_z \frac{\partial \zeta_x}{\partial z} \\ \zeta_y \left(\frac{\partial \zeta_x}{\partial x} + \frac{\partial \zeta_z}{\partial z} \right) + \zeta_x \frac{\partial \zeta_y}{\partial x} + \zeta_z \frac{\partial \zeta_y}{\partial z} \\ \zeta_z \left(\frac{\partial \zeta_x}{\partial x} + \frac{\partial \zeta_y}{\partial y} \right) + \zeta_x \frac{\partial \zeta_z}{\partial x} + \zeta_y \frac{\partial \zeta_z}{\partial y} \end{bmatrix} . \tag{12.61}$$

There are three important parameters that determine the strength of the magnetostrictive body force: the magnetoelastic coupling constants, the magnetization orientation, and the gradient of the magnetization orientation.

The magnetostrictive body force affects the elastodynamics and thus needs to be added to the elastodynamic equation (12.44) in magnetostrictive media. This allows for the analytical description of the influence of the magnetization direction on the elastodynamics and the properties of (magneto-)elastic waves.

Another important quantity is the magnetostrictive strain, which is the additional strain originating from the magnetostrictive effect. For a material with cubic symmetry, the magnetostrictive strain is given by [24, 42, 43]

$$\bar{\varepsilon}_{\mathrm{mel}} = \frac{3}{2} \begin{bmatrix} \lambda_{100} \left(\zeta_x^2 - \frac{1}{3} \right) & \lambda_{111} \zeta_x \zeta_y & \lambda_{111} \zeta_x \zeta_z \\ \lambda_{111} \zeta_y \zeta_x & \lambda_{100} \left(\zeta_y^2 - \frac{1}{3} \right) & \lambda_{111} \zeta_y \zeta_z \\ \lambda_{111} \zeta_z \zeta_x & \lambda_{111} \zeta_z \zeta_y & \lambda_{100} \left(\zeta_z^2 - \frac{1}{3} \right) \end{bmatrix} . \tag{12.62}$$

with λ_{100} and λ_{111} the magnetostriction coefficients representing the maximum magnetostrictive strain for fully-saturated magnetization along the $\langle 100\rangle$ or $\langle 111\rangle$ crystallographic directions, respectively.

The magnetostriction coefficients are related to the magnetoelastic coupling constants by

$$\lambda_{100} = \frac{2}{3}\frac{B_1}{C_{12}-C_{11}}, \qquad \lambda_{111} = -\frac{B_2}{3C_{44}}. \tag{12.63}$$

Hence, it is also possible to express the magnetostrictive body force as a function of the magnetostrictive strain via

$$\mathbf{f}_{\mathrm{mel}} = \nabla\cdot\bar{\sigma}_{\mathrm{mel}} = \nabla\cdot\left(\bar{\bar{C}} : \bar{\varepsilon}_{\mathrm{mel}}\right). \tag{12.64}$$

Note that for an isotropic material, $\lambda_{100} = \lambda_{111} = \lambda_{\mathrm{eq}}$ and $B_1 = B_2 = B$.

12.4.1.2 The Villari Effect

The Villari effect describes how elastic strain affects the magnetization state and is also called the inverse magnetostrictive effect. Strain in a magnetostrictive material results in an effective magnetoelastic field, which can be derived from (12.2) and (12.57). For a material with cubic symmetry, the magnetoelastic effective field is

$$\mathbf{H}_{\mathrm{mel}} = -\frac{1}{\mu_0}\frac{d\mathcal{E}_{\mathrm{mel}}}{d\mathbf{M}} = -\frac{2}{\mu_0 M_s}\begin{bmatrix} B_1\varepsilon_{\mathrm{xx}}\zeta_{\mathrm{x}} + B_2(\varepsilon_{\mathrm{xy}}\zeta_{\mathrm{y}} + \varepsilon_{zx}\zeta_{\mathrm{z}}) \\ B_1\varepsilon_{\mathrm{yy}}\zeta_{\mathrm{y}} + B_2(\varepsilon_{\mathrm{xy}}\zeta_{\mathrm{x}} + \varepsilon_{yz}\zeta_{\mathrm{z}}) \\ B_1\varepsilon_{\mathrm{zz}}\zeta_{\mathrm{z}} + B_2(\varepsilon_{\mathrm{zx}}\zeta_{\mathrm{x}} + \varepsilon_{yz}\zeta_{\mathrm{y}}) \end{bmatrix} \tag{12.65}$$

with ζ_i the normalized magnetization components, as defined by (12.58). The resulting magnetization dynamics are described by the LLG equation (12.1), including the above magnetoelastic field as a contribution to $\mathbf{H}_{\mathrm{eff}}$.

12.4.2 *Magnetoelastic Waves in Thin Films*

When the magnetoelastic interaction terms $\mathbf{f}_{\mathrm{mel}}$ and $\mathbf{H}_{\mathrm{mel}}$ are combined with the magnetodynamic equation (12.1) and the elastodynamic equation (12.51), a set of coupled differential equations is obtained. Formally, these differential equations are nonlinear because the terms originating from the magnetoelastic interaction show a quadratic dependence on the magnetization and the displacement. Therefore, the magnetoelastic effect formally results in a nonlinear interaction. However, when the dynamic components are assumed to be weak, the differential equations can be linearized. In this case, wave-like solutions for the magnetization and the displacement, given by (12.14) and (12.52), exist for the coupled set of equations. These solutions correspond to magnetoelastic waves. However, it is important to keep in mind that for large dynamic components, a system of nonlinear differential equations has to be solved, including nonlinear magnetoelastic interaction effects.

To reduce the complexity of the calculations, a homogeneous and isotropic material is assumed. The geometry of the structure remains the same as in the previous sections with the film in the xz-plane and the y-direction normal to the film surface. The static magnetization and the static external field are chosen along the z-direction, as in Sect. 12.2.3. Then, substituting the wave-like ansatz into the equations of motion and neglecting terms quadratic in $\mathbf{m}$ or $\mathbf{u}$, leads to the following linearized system of equations:

$$
\begin{aligned}
-\rho\omega^2 u_x &= -C_{11}k_x^2 u_x - C_{44}k_z^2 u_x - (C_{12}+C_{44})k_x k_z u_z + \frac{B_2}{M_s} i k_z m_x \\
-\rho\omega^2 u_y &= -C_{44}\left(k_x^2 u_y + k_z^2 u_y\right) + \frac{B_2}{M_s} i k_z m_y \\
-\rho\omega^2 u_z &= -C_{11}k_z^2 u_z - C_{44}k_x^2 u_z - (C_{12}+C_{44})k_x k_z u_x + \frac{B_2}{M_s} i k_x m_x \\
i\omega m_x &= -\omega_{fy} m_y - \gamma B_2 i k_z u_y \\
i\omega m_y &= \omega_{fx} m_x + \gamma B_2 i\left(k_z u_x + k_x u_z\right) .
\end{aligned}
\tag{12.66}
$$

Assuming that the elastic properties of the thin film are isotropic, the C_{12} stiffness constant can be replaced by $C_{12} = C_{11} - 2C_{44}$. Moreover, as discussed above, two types of elastic waves exist in in an isotropic material, i.e. longitudinal and transversal waves. Therefore, it is convenient to define new displacement variables parallel (u_l) and perpendicular (u_t) to the propagation direction such that

$$
u_x = u_l \sin(\theta) + u_t \cos(\theta) , \qquad u_z = u_l \cos(\theta) - u_t \sin(\theta) . \tag{12.67}
$$

Here, θ is the angle between the static magnetization $\mathbf{M}_0$ and the propagation direction $\mathbf{k}$. Substituting these redefined displacement components into the dynamic equation of motion together with $k_x = k\sin(\theta)$ and $k_z = k\cos(\theta)$ results in

$$
\begin{aligned}
(\omega^2-\omega_l^2)\sin(\theta)u_l + (\omega^2-\omega_H)\cos(\theta)u_t + \frac{iBk\cos(\theta)}{\rho M_s}m_x &= 0 \\
(\omega^2-\omega_V^2)u_y + \frac{iBk\cos(\theta)}{\rho M_s}m_y &= 0 \\
(\omega^2-\omega_l^2)\cos(\theta)u_l - (\omega^2-\omega_H^2)\sin(\theta)u_t + \frac{iBk\sin(\theta)}{\rho M_s}m_x &= 0 \\
i\gamma Bk\cos(\theta)u_y + i\omega m_x + \omega_{fy} m_y &= 0 \\
i\gamma Bk\sin(2\theta)u_l + i\gamma Bk\cos(2\theta)u_t + \omega_{fx} m_x - i\omega m_y &= 0
\end{aligned}
\tag{12.68}
$$

with ω the angular frequency of the magnetoelastic wave, $\omega_l = v_l k = \sqrt{\frac{C_{11}}{\rho}} k$ the dispersion relation of longitudinal elastic waves, $\omega_{\mathrm{H}} = v_{\mathrm{t}} k = \sqrt{\frac{C_{44}}{\rho}} k$ the dispersion relation of horizontally-polarized (in-plane) transversal elastic waves, and $\omega_{\mathrm{V}} = \omega_{\mathrm{H}}$ the dispersion relation of vertically-polarized (out-of-plane) transversal elastic waves. Here, the distinction between ω_{V} and ω_{H} is made to keep track of the origin of different terms in the equations of motion.

Note that this set of equations describes magnetoelastic waves in thin films with finite thickness. The finite thickness of the film changes the dipolar field according to (12.30) and consequently also the properties of the magnetoelastic waves. The thickness influence is captured by the parameters ω_{fx} and ω_{fy}. In the following, different cases and geometries of magnetoelastic wave solutions of the coupled equations of motion are discussed.

12.4.2.1 Wave Propagation Perpendicular to the Magnetization

We first consider the case in which the wave propagation direction is perpendicular to the static equilibrium magnetization, i.e. $\theta = \pi/2$. In this geometry, (12.68) indicates that the magnetoelastic body force $\mathbf{f}_{\mathrm{mel}}$ only acts on the u_{t} component of the displacement. Conversely, only the displacement component u_{t} generates a magnetoelastic field that interacts with the magnetic system. Hence, only the in-plane transversal elastic wave couples to surface spin waves and vice versa. This means that the longitudinal and out-of-plane transversal elastic waves are independent of the magnetic system in a first-order approximation. As a consequence, their dispersion relations remain unchanged, i.e. $\omega_l = v_l k$ and $\omega_{\mathrm{V}} = v_t k$, respectively, as described in Sect. 12.3.

Eliminating all uncoupled equations and using $\theta = \pi/2$ in (12.68), the system becomes

$$\begin{bmatrix} \omega^2 - \omega_{\mathrm{H}}^2 & \frac{iBk}{\rho M_s} & 0 \\ -i\gamma Bk & \omega_{\mathrm{fx}} & -i\omega \\ 0 & i\omega & \omega_{\mathrm{fy}} \end{bmatrix} \begin{bmatrix} u_{\mathrm{t}} \\ m_{\mathrm{x}} \\ m_{\mathrm{y}} \end{bmatrix} = 0 \tag{12.69}$$

with ω the angular frequency of the magnetoelastic wave and $\omega_{\mathrm{H}} = v_{\mathrm{t}} k$ the resonance frequency of the uncoupled horizontally-polarized transversal elastic wave. Note that in this geometry, the in-plane transversal displacement component is fully aligned in the z-direction, i.e. $u_{\mathrm{t}} = u_{\mathrm{z}}$. To obtain nontrivial solutions, the determinant of the linear system must vanish, which leads to the condition

$$(\omega^2 - \omega_{\mathrm{H}}^2)(\omega^2 - \omega_{\mathrm{fm}}^2) - Jk^2\omega_{\mathrm{fy}} = 0 \tag{12.70}$$

with

$$J = \frac{\gamma B^2}{\rho M_{\mathrm{s}}} \tag{12.71}$$

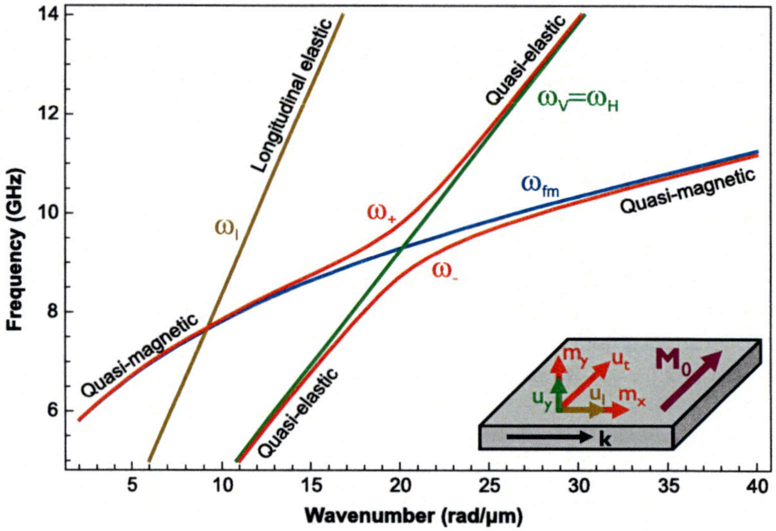

Fig. 12.2 Magnetoelastic wave dispersion relations according to (12.70) for a 30 nm thick Ni film and propagation direction perpendicular to the magnetization (red lines). The external magnetic field is $\mu_0 H_{\text{ext}} = 50\,\text{mT}$. For comparison, the dispersion relations of longitudinal and transversal elastic waves (brown and green lines, respectively) as well as uncoupled spin waves (blue line) are also shown

and $\omega_{\text{fm}} = \sqrt{\omega_{\text{fx}}\omega_{\text{fy}}}$ the uncoupled spin wave resonance frequency. Equation (12.70) has the general form of a dispersion relation of two interacting waves. Here, the first wave is a transversal elastic wave characterized by $\omega^2 - \omega_{\text{H}}^2 = 0$ and the second wave is a spin wave characterized by $\omega^2 - \omega_{\text{fm}}^2 = 0$. The interaction between these two waves is quantified by $Jk^2\omega_{\text{fy}}$. As expected, setting the magnetoelastic coupling constant B to zero leads to the original dispersion relations of uncoupled elastic and magnetic waves.

Equation (12.70) has two physically-meaningful solutions for ω, which are given by

$$\omega_{\pm}^2 = \frac{\omega_{\text{H}}^2 + \omega_{\text{fm}}^2}{2} \pm \sqrt{\left(\frac{\omega_{\text{fm}}^2 - \omega_{\text{H}}^2}{2}\right)^2 + Jk^2\omega_{\text{fy}}}\,. \tag{12.72}$$

These two solutions represent the dispersion relations of the resulting magnetoelastic waves. These dispersion relations together with the dispersion relations of the uncoupled elastic waves are plotted in Fig. 12.2 for a 30 nm thick Ni film. The magnetic parameters are the same as used in Fig. 12.1. The magnetoelastic coupling constant is $B = 10$ MJ/m^3 [44, 45], the stiffness constants are $C_{11} = 245$ GPa and $C_{44} = 75$ GPa [46], and the mass density is $\rho = 8900$ kg/m^3 [47]. The two linear dispersion relations correspond to the uncoupled elastic waves whereas the two red curves represent the dispersion relations of the magnetoelastic waves.

Figure 12.2 clearly shows that the two branches of the magnetoelastic wave dispersion relations do not cross each other. If the transversal elastic waves and the spin waves were not interacting, their dispersion relations would intersect. However, due to the magnetoelastic interaction, this crossing is avoided, leading to a gap between the two curves. This so-called anticrossing behavior of the dispersion relations is a typical characteristic of interacting waves [15, 16].

The gap formation is also visible in the equation of the dispersion relations, i.e. (12.72). At the point in reciprocal space where the dispersion relations of the uncoupled waves would intersect, i.e. $(\omega_{\text{cross}}, k_{\text{cross}})$, the term $\omega_{\text{fm}}^2 - \omega_{\text{H}}^2$ vanishes. At this condition, the interaction coefficient $Jk^2\omega_{\text{fy}}$ has a strong influence on the dispersion relation. When $Jk^2\omega_{\text{fy}} \gtrsim \omega_{\text{fm}}^2 - \omega_{\text{H}}^2$, the interaction between the magnetic and elastic system is strong, leading to the formation of coupled magnetoelastic waves. As a result, the anticrossing is formed, with a frequency gap that quantifies the strength of the interaction. This frequency gap is, to first order, given by

$$\Delta\omega(k_{\text{cross}}) \approx \frac{\sqrt{Jk_{\text{cross}}^2\omega_{\text{fy}}}}{\omega_{\text{cross}}} = \sqrt{\frac{\gamma B^2 \omega_{\text{fy}}}{C_{44} M_0}} \tag{12.73}$$

where the relation $\omega_{\text{cross}} = \sqrt{\frac{C_{44}}{\rho}} k_{\text{cross}}$ was used. Note that ω_{fy} also depends on k_{cross} and that this approximation is only valid when $\Delta\omega(k_{\text{cross}}) < \omega_{\text{cross}}$. On the other hand, when $Jk^2\omega_{\text{fy}} \ll \omega_{\text{fm}}^2 - \omega_{\text{H}}^2$, the interaction term can be neglected, leading to nearly uncoupled elastic and magnetic waves. In this regime, the waves are called quasi-elastic or quasi-magnetic [15, 16]. Hence, the interaction between the elastic and magnetic waves is strongest when they are (nearly) degenerate, resulting in coupled magnetoelastic waves. By contrast, quasi-noninteracting waves are obtained when their frequencies and/or their wavelengths differ strongly.

The wavenumber at the crossing, k_{cross}, can be found by equalizing the dispersion relations of the noninteracting systems, i.e. $\omega_{\text{H}}(k_{\text{cross}}) = \omega_{\text{fm}}(k_{\text{cross}})$, and solving for k_{cross}. For the geometry considered here, the noninteracting dispersion relations are equal when

$$v_t k_{\text{cross}} = (\omega_0 + \omega_{\text{M}}\lambda_{\text{ex}}k_{\text{cross}}^2)^2 + \omega_{\text{M}}\left(\omega_0 + \omega_{\text{M}}\lambda_{\text{ex}}k_{\text{cross}}^2 + \omega_{\text{M}}(1-P)P\right), \tag{12.74}$$

which needs to be solved iteratively. Note that P is also a function of k_{cross} according to (12.30). Once k_{cross} is determined, the interaction coefficient $Jk^2\omega_{\text{fy}}$ and the gap amplitude can be calculated. In general, the coupling increases strongly for higher wavenumbers k_{cross}. This originates from the behavior of the magnetostriction and the Villari effect: a shorter wavelength leads to larger gradients of both displacement and magnetization. As a result, the magnetoelastic body force in (12.60) and the magnetoelastic field in (12.65) increase, leading to stronger interactions for higher k_{cross} values. This behavior opens possibilities to control the interaction strength by external parameters. For example, increasing an external magnetic field shifts the spin wave dispersion relation to higher frequencies, leading to a larger value of k_{cross} and thus a stronger magnetoelastic coupling.

According to the dispersion relation in (12.72), two different wave-like solutions exist that correspond to two different magnetoelastic waves. To describe the characteristics of these waves, the corresponding eigenstates need to be calculated. They are given by

$$\begin{bmatrix} u_{\text{t}} \\ m_{\text{x}} \\ m_{\text{y}} \end{bmatrix} = N \begin{bmatrix} 1 \\ i\frac{\rho M_{\text{s}}}{Bk}(\omega_{\pm}^2 - \omega_{\text{H}}^2) \\ \frac{\rho M_{\text{s}}\omega_{\pm}}{Bk\omega_{\text{fy}}}(\omega_{\pm}^2 - \omega_{\text{H}}^2) \end{bmatrix} = N \begin{bmatrix} 1 \\ i\frac{\gamma Bk\omega_{\text{fy}}}{\omega_{\pm}^2 - \omega_{\text{fm}}^2} \\ \frac{\gamma Bk\omega_{\pm}}{\omega_{\pm}^2 - \omega_{\text{fm}}^2} \end{bmatrix} \tag{12.75}$$

with N a dimensionless normalization factor. Note that the polarization of the two magnetization components, for both cases ω_+ and ω_-, is clockwise (right-hand) elliptically polarized with ellipticity $\epsilon = |m_{\text{x}}|/|m_{\text{y}}| = \omega_{\text{fy}}/\omega_{\pm}$. The precession described by the u_{t} displacement and the m_{x} magnetization components is clockwise or counterclockwise (right-hand or left-hand) polarized, depending on the eigenstate ω_+ or ω_-.

Based on the eigenstate, it is possible to determine the variation of the energy associated with the different wave components during propagation. There is always a phase difference of $\pi/2$ between m_{x} and m_{y} as well as m_{x} and u_{t}. This indicates that, during propagation, the energy in the m_{x} component is transferred partially to the m_{y} and partially to the u_{t} component. Hence, for magnetoelastic waves, there is resonant energy transfer between the elastic and magnetic domains.

The three different regimes described by the dispersion relation in (12.72), i.e. the quasi-elastic, quasi-magnetic, and magnetoelastic regimes, are also seen from the eigenstates. In the quasi-elastic regime, the dispersion relation approaches the linear dispersion of the elastic waves, i.e. $\omega_{\pm}^2 - \omega_{\text{H}}^2 \approx 0$ and thus $m_{\text{x}}, m_{\text{y}} \approx 0$ according to (12.75). In other words, in the quasi-elastic regime, the total energy is almost completely dominated by the elastic energy [15, 16] and the energy transfer to the magnetic system during propagation can be neglected. On the other hand, in the quasi-magnetic regime, the dynamic displacement component u_{t} is very small and thus the total energy is dominated by the magnetic energy. In the magnetoelastic regime near the anticrossing, the total energy of the wave is distributed between the magnetic and elastic systems. Hence, a large part of the total wave energy resonantly oscillates between the magnetic and elastic domains [15, 16]. This is also seen in Fig. 12.3, which shows the magnetization components for the two branches of the dispersion relation, ω_+ and ω_-, as a function of the frequency. In keeping with the above discussion, the magnetization components have strong amplitudes in the quasi-magnetic and weak amplitudes in the quasi-elastic regime.

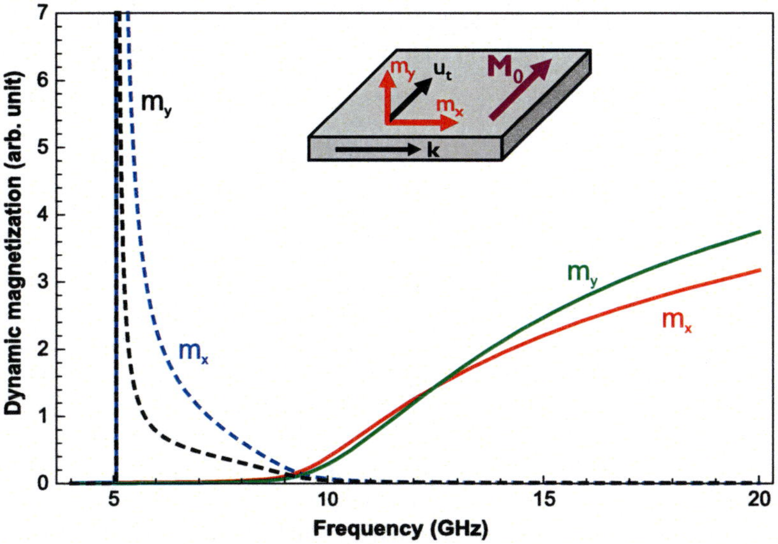

Fig. 12.3 Frequency dependence of the dynamic magnetization components of magnetoelastic waves in a 30 nm thick Ni film. The propagation direction is perpendicular to the magnetization, as shown in the inset. The dashed lines represent the m_x and m_y components of the ω_+ state, whereas the solid lines represent the m_x and m_y components of the ω_- state. The external magnetic field is $\mu_0 H_{ext} = 50\,\mathrm{mT}$

12.4.2.2 Wave Propagation Parallel to the Magnetization

When the propagation direction of the magnetoelastic wave is parallel to the equilibrium magnetization direction, i.e. $\theta = 0$ in (12.72), the magnetic body force $\mathbf{f}_{mel}$ acts on both the in-plane u_t and out-of-plane u_y transversal displacement components. Note that in this geometry, the transversal in-plane displacement component is fully aligned along the x-direction, i.e. $u_t = u_x$. Analogously, both the in-plane and the out-of-plane transversal elastic waves generate magnetoelastic fields that interact with the dynamic magnetization. Hence, both transversal displacement components couple to backward volume spin waves and only the longitudinal elastic wave is decoupled from the magnetic system. Neglecting longitudinal elastic waves, the system of equations in matrix notation becomes

$$\begin{bmatrix} \omega^2 - \omega_H^2 & 0 & \frac{iBk}{\rho M_s} & 0 \\ 0 & \omega^2 - \omega_V^2 & 0 & \frac{iBk}{\rho M_s} \\ i\gamma Bk & 0 & \omega_{fx} & -i\omega \\ 0 & i\gamma Bk & i\omega & \omega_{fy} \end{bmatrix} \begin{bmatrix} u_t \\ u_y \\ m_x \\ m_y \end{bmatrix} = 0\,. \qquad (12.76)$$

It is worth noting that both transversal elastic waves have the same dispersion relation and thus $\omega_H = \omega_V = v_t k$, as discussed earlier.

Again, the homogeneous linear system has only nontrivial solutions when its determinant is zero. This condition leads to the dispersion relation of the resulting magnetoelastic waves, given by

$$(\omega^2 - \omega_{\mathrm{fm}}^2)(\omega^2 - \omega_{\mathrm{H}}^2)(\omega^2 - \omega_{\mathrm{V}}^2) - Jk^2[\omega_{\mathrm{fx}}(\omega^2 - \omega_{\mathrm{H}}^2) + \omega_{\mathrm{fy}}(\omega^2 - \omega_{\mathrm{V}}^2) + Jk^2] = 0\,. \tag{12.77}$$

Three different interaction terms can be identified in this equation. The first interaction term $Jk^2\omega_{\mathrm{fx}}(\omega^2 - \omega_{\mathrm{H}}^2)$ represents the interaction between the out-of-plane u_{y} transversal elastic wave and the backward volume spin wave. The second term $Jk^2\omega_{\mathrm{fx}}(\omega^2 - \omega_{\mathrm{V}}^2)$ characterizes the interaction between the in-plane u_{t} transversal elastic wave and the backward volume spin wave. As a result, these two terms induce an anticrossing near the point where the dispersion relations of the noninteracting elastic and magnetic waves would intersect each other. The third interaction term J^2k^4 couples all three different components with each other and thus also generates an interaction between the two transversal elastic waves.

Figure 12.4 shows the different dispersion relations for material parameters corresponding to Ni, as mentioned above. To better understand their behavior, the corresponding eigenstates of the different magnetoelastic waves are calculated. The eigenstates are given as a function of the angular frequency of the magnetoelastic wave, ω, by

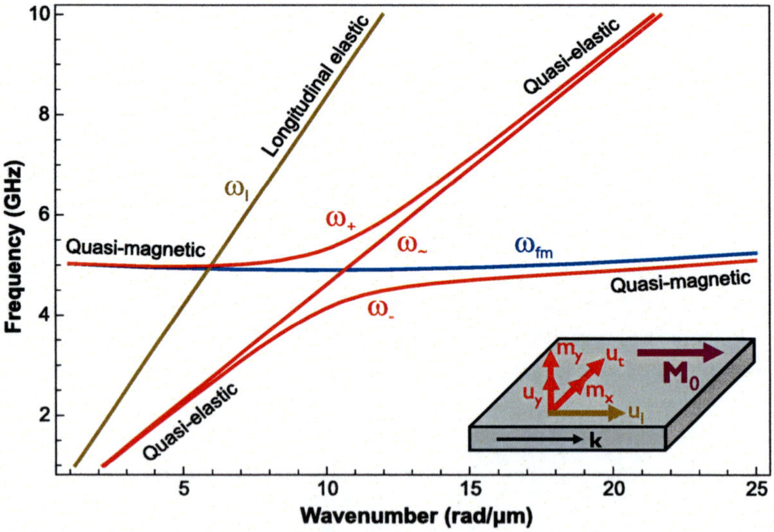

Fig. 12.4 Magnetoelastic wave dispersion relations (red lines) according to (12.77) for a 30 nm thick Ni film and propagation directions parallel with the magnetization, as shown in the inset. The external magnetic field is $\mu_0 H_{\mathrm{ext}} = 50\,\mathrm{mT}$. For comparison, the dispersion relations of longitudinal elastic waves (brown line) and uncoupled spin waves (blue line) are also shown

$$\begin{bmatrix} u_t \\ u_y \\ m_x \\ m_y \end{bmatrix} = N \begin{bmatrix} \frac{-i}{\omega(\omega^2-\omega_H^2)}(\omega_{fy}(\omega^2-\omega_V^2)+Jk^2) \\ 1 \\ -\frac{\rho M_s}{Bk\omega}(\omega_{fy}(\omega^2-\omega_V^2)+Jk^2) \\ i\frac{\rho M_s}{Bk}(\omega^2-\omega_V^2) \end{bmatrix} \quad (12.78)$$

with N a dimensionless normalization constant. The different displacement components of such magnetoelastic waves are plotted in Fig. 12.5 as a function of frequency, whereas Fig. 12.6 shows the magnetization components. As above, Ni material parameters were assumed, and the external magnetic field was $\mu_0 H_{ext} = 50\,\text{mT}$.

In the following, the different eigenstates and their properties are discussed. The upper ω_+ and lower ω_- branches of the dispersion relation both correspond to clockwise (right-hand) elliptically polarized waves for the magnetization and displacement, i.e. $m_y/m_x = i|m_y|/|m_x|$ and $u_y/u_x = i|u_y|/|u_x|$ [15, 16]. In both cases, the in-plane magnetization component is always larger than the out-of-plane component, i.e. $m_x > m_y$, since the demagnetization field is strongest in the out-of-plane direction. Concerning the displacement components, the two branches behave differently. For the ω_+ eigenstate, the in-plane and out-of-plane displacement components have the same order of magnitude at GHz frequencies. On the other hand, for the ω_- state,

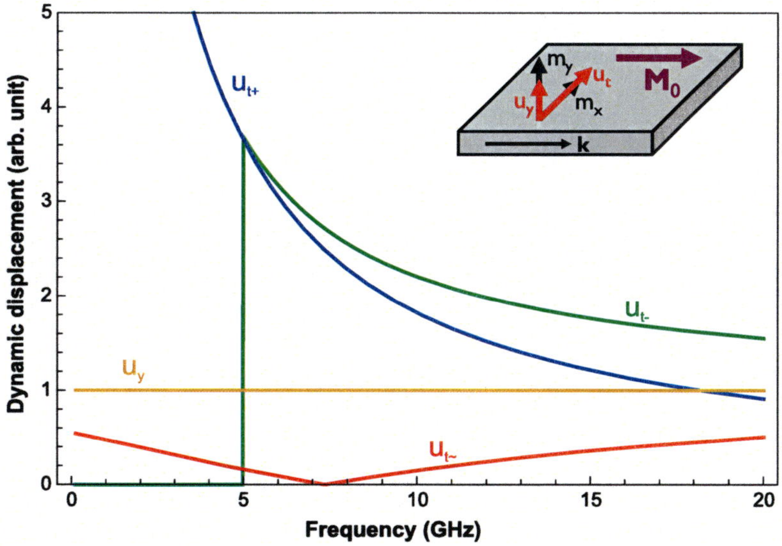

Fig. 12.5 Frequency dependence of the dynamic displacement components for the different magnetoelastic waves in a 30 nm thick Ni film and an external magnetic field of $\mu_0 H_{ext} = 50\,\text{mT}$. The propagation direction is parallel to the magnetization, as shown in the inset. All displacement values are normalized to the out-of-plane component of the displacement u_y (yellow line). The blue, green and red lines correspond to the in-plane displacement components of the ω_+, ω_- and $\omega_\sim$ modes, respectively

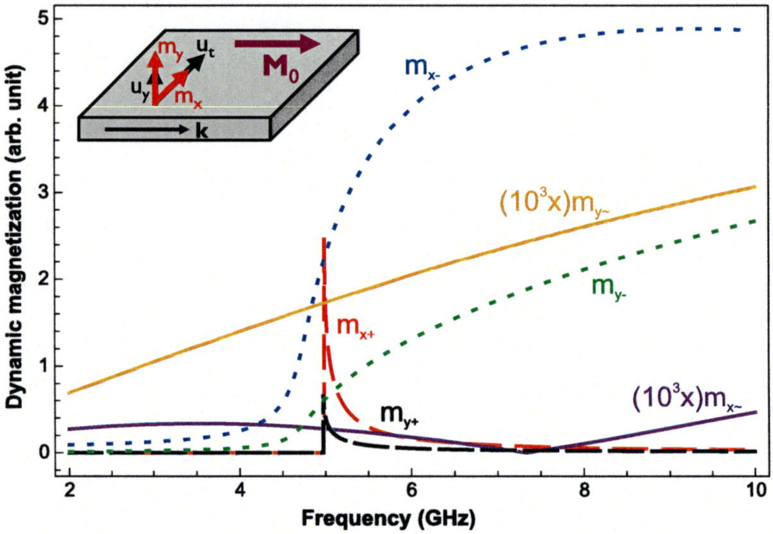

Fig. 12.6 Frequency dependence of the dynamic magnetization components of magnetoelastic waves in a 30 nm thick Ni film for an external magnetic field of $\mu_0 H_{ext} = 50\,\text{mT}$. The propagation direction is parallel to the magnetization, as shown in the inset. The dashed blue and green lines correspond to the ω_- mode, whereas the dashed red and black line correspond to the ω_+ modes, and the solid lines correspond to the $\omega_\sim$ mode. Note that the magnetization components corresponding to the $\omega_\sim$ mode are multiplied by a factor of 10^3

u_t is dominant at low frequencies, whereas u_y becomes dominant at high frequencies. This behavior is also visualized in Fig. 12.5.

The dispersion relation corresponding to the third magnetoelastic eigenstate is also shown in Fig. 12.4, labelled $\omega_\sim$. This dispersion relation is nearly linear and falls slightly below the dispersion relation for uncoupled transversal elastic waves, which was discussed in Sect. 12.3 [15, 16]. The magnetization and the displacement components corresponding to this state are both counterclockwise (left-hand) elliptically polarized. For uncoupled backward volume spin waves, counterclockwise polarization corresponds to evanescent spin waves. However, such evanescent spin waves can still couple to left-hand polarized displacement waves, resulting in left-hand polarized *propagating* magnetoelastic waves. Nonetheless, the magnetization components for this magnetoelastic mode remain very weak. This is also seen in Fig. 12.6, where the magnetization components corresponding to the $\omega_\sim$ branch have three orders of magnitude lower amplitude than the magnetization components of the ω_+ and ω_- branches. In terms of displacement, the u_y displacement component is dominant for the $\omega_\sim$ mode for a wide frequency range. This is also illustrated in Figs. 12.5 and 12.6.

Because of the coupling to the spin wave system, magnetoelastic waves show some peculiarities in the quasi-elastic regime, where the wave energy is largely dominated by the elastic energy. As shown above, the J^2k^4 interaction term couples all waves with each other. Consequently, the two transversal elastic waves become also coupled. For interacting waves, it is impossible to share the same frequency–wavenumber couple, i.e. it is impossible to have degenerate points in the dispersion relations. As a result, the two quasi-elastic branches do not overlap anymore which is in contrast to their original behavior without magnetoelastic interactions (see Sect. 12.3). Therefore, at all frequencies, a small wavenumber shift remains present between the two quasi-elastic branches, even in the quasi-elastic regime where the displacement components are large and magnetization components are weak. Hence, even though almost all the wave energy is in the elastic system, the interaction between the two transversal displacement components is mediated by the magnetic system, leading to an indirect coupling of the two elastic waves via the magnetic system. This interaction is proportional to k^4 and thus strongly depends on the wavelength.

Moreover, the polarization of the displacement in the quasi-elastic regime also shows a peculiar behavior. One of the two waves in the quasi-elastic regime corresponds to a clockwise (right-hand) polarized wave and the other to a counterclockwise (left-hand) polarized wave, as discussed above. Hence, excitation at a single angular frequency ω in the quasi-elastic regime leads to two different magnetoelastic waves with different wavelengths and opposite polarization. Their amplitudes as a function of time and space can be written as

$$\mathbf{u}_{\pm} = \begin{bmatrix} |u_{\mathrm{t}\pm}| \\ i\,|u_{\mathrm{y}\pm}| \end{bmatrix} e^{i\omega t + k_{\pm} z} \quad \text{and} \quad \mathbf{u}_{\sim} = \begin{bmatrix} |u_{\mathrm{t}\sim}| \\ -i\,|u_{\mathrm{y}\sim}| \end{bmatrix} e^{i\omega t + k_{\sim} z} \tag{12.79}$$

with $u_{\mathrm{t}\pm} = u_{\mathrm{t}}(k_{\pm})$ and $u_{\mathrm{t}\sim} = u_{\mathrm{t}}(k_{\sim})$ given by (12.78). The total wave is the sum of both individual waves:

$$\mathbf{u}_{\mathrm{tot}} = \begin{bmatrix} |u_{\mathrm{t}\pm}| e^{ik_{\pm} z} + |u_{\mathrm{t}\sim}| e^{ik_{\sim} z} \\ i\left(|u_{\mathrm{y}\pm}| e^{ik_{\pm} z} - |u_{\mathrm{y}\sim}| e^{ik_{\sim} z}\right) \end{bmatrix} e^{i\omega t}\,. \tag{12.80}$$

The difference in amplitude between the clockwise and counterclockwise polarized components results in an elliptical polarization of the total displacement. The different wavenumbers of the two individual waves ($k_{\pm}$ and $k_{\sim}$) results in the rotation of the major and minor axes of the ellipsoid described by the tip of the displacement vectors during wave propagation [48, 49]. This is similar to the Faraday effect for electromagnetic waves and also called acoustic wave rotation.

The dispersion relation of backward volume spin waves is rather flat in the dipolar–exchange regime, leading to an interesting property of magnetoelastic waves in this geometry. As shown in Fig. 12.4 at frequencies around 4–5 GHz, the magnetoelastic coupling leads to the formation of a pseudobandgap for clockwise (right-hand) polarized elastic waves at the anticrossing. On the other hand, due to the flatness of the dispersion relation, counterclockwise (left-hand) polarized magnetoelastic waves can still exist in this pseudobandgap. Hence, in this frequency range, only pure magnetoelastic waves or quasi-magnetic waves with weak displacement components can be excited. This pseudobandgap formation is a general result when waves with a rather flat dispersion relation interact with waves with a steep dispersion relation near the crossing point.

12.4.2.3 Arbitrary Propagation Direction

We now consider an arbitrary propagation direction of the magnetoelastic wave with respect to the equilibrium magnetization. In this case, the magnetoelastic body force interacts with all displacement components. Conversely, all displacement components generate magnetic fields that interact with the magnetization. Hence, all magnetization and displacement components become coupled with each other. Again, nontrivial wave-like solutions only exist when the determinant of the linear system in (12.68) is zero, which leads to the dispersion relation

$$\begin{aligned}(\omega^2-\omega_l^2)[(\omega^2-\omega_{\mathrm{t}}^2)^2(\omega^2-\omega_{\mathrm{fm}}^2)\\ -(\omega^2-\omega_{\mathrm{t}}^2)Jk^2(\omega_{\mathrm{fx}}\cos^2(\theta)+\omega_{\mathrm{fy}}\cos^2(2\theta))-J^2k^4\cos^2(2\theta)\cos^2(\theta)]\\ -(\omega^2-\omega_{\mathrm{t}}^2)Jk^2[\omega_{\mathrm{fy}}(\omega^2-\omega_{\mathrm{t}}^2)\sin^2(2\theta)+Jk^2\sin^2(2\theta)\cos^2(\theta)]=0\,.\end{aligned} \quad (12.81)$$

Note that for $\theta = \pi/4$, the coupling between the magnetic and the longitudinal elastic wave reaches a maximum, whereas for $\theta = 0$ and $\theta = \pi/2$, this coupling is zero.

The dispersion relations of the resulting magnetoelastic waves are plotted in Fig. 12.7 for material parameters of Ni and $\theta = \pi/6$. For each frequency, multiple magnetoelastic waves exist with different wavelengths. Since the system of equations is reduced to a set of linear differential equations by assuming weak dynamic components, every linear combination of these different magnetoelastic waves is also a solution of the system. The waves can be excited by dynamic magnetic fields and/or mechanical forces. Therefore, it is possible to generate elastodynamics via the magnetization or, vice versa, magnetization dynamics via the displacement in magnetostrictive materials.

It can also be seen from the dispersion relations that the group velocity of the magnetoelastic waves is different from the group velocity of the magnetic and elastic waves. As mentioned earlier, the group velocity is defined as $\mathbf{v}_{\mathrm{g}} = \partial\omega/\partial\mathbf{k}$ and thus proportional to the slope of the dispersion relation. Hence, near the anticrossing, this change in group velocity is most pronounced. On the other hand, the group velocity of quasi-elastic and quasi-magnetic waves is nearly the same as their purely elastic and magnetic counterparts, respectively.

The total energy of a magnetoelastic wave consists of several contributions. The magnetic energy contribution is determined by the dynamic components m_{x} and m_{y}. In this chapter, only dipolar, and exchange energy interactions were considered, although other magnetic interactions, such as the magnetocrystalline [40, 41, 50] or the Dzyaloshinskii–Moriya interaction [51] may also contribute to the total energy. The magnetic energy is complemented by the energy of the elastic waves, which consists of both elastic and kinetic energy contributions and is fully determined by the

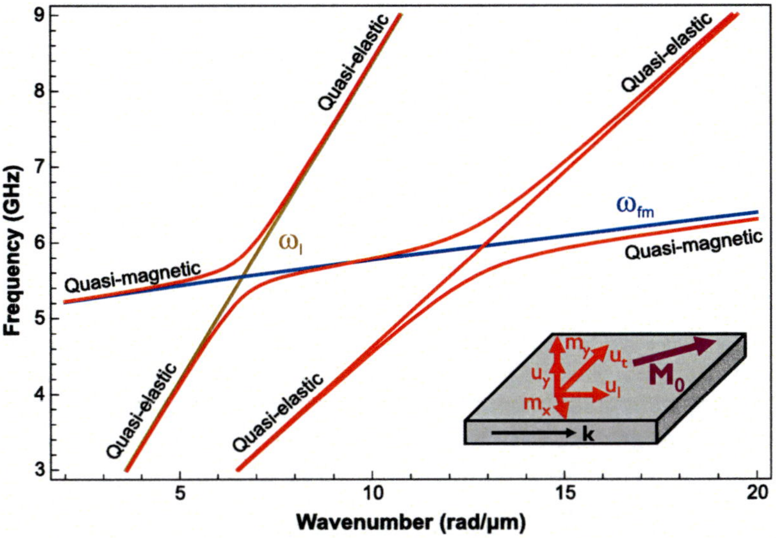

Fig. 12.7 Magnetoelastic wave dispersion relations (red lines) according to (12.81) for a 30 nm thick Ni film nm and an angle of 30° between the propagation direction and the magnetization. The external field is $\mu_0 H_{\text{ext}} = 50\,\text{mT}$. For comparison, the dispersion relations of longitudinal elastic waves (brown line) and uncoupled spin waves (blue line) are also shown

displacement components and their time and space derivatives, as given by (12.54) and (12.56), respectively. A third energy contribution stems from the magnetoelastic interaction, as described by (12.57). Just as spin waves (cf. (12.41)), magnetoelastic waves also comprise an electric field component. However, at GHz frequencies, the magnetostatic approximation is typically valid and therefore the energy contribution of the electric field is small and can be neglected. Nonetheless, this ceases to be accurate when frequencies approach the THz range where the magnetostatic approximation no longer holds.

During the propagation of a magnetoelastic wave, the energy oscillates between the different energy contributions. For strongly interacting waves near the anticrossing point, a large part of the energy resonantly oscillates between the elastic and magnetic domains. This energy transfer is characterized by a specific energy transfer length L_t which describes the distance necessary to transfer all energy from the elastic to the magnetic system and vice versa [52]. On the other hand, the time necessary for a complete magnetoelastic energy oscillation between the elastic and magnetic domain is given by $\Gamma = 2/\Delta f$ with Δf the frequency gap between the two dispersion relations (cf. (12.73)) [53]. By contrast, in the quasi-elastic regime, most of the energy remains in the elastic system during propagation, whereas in the quasi-magnetic regime most energy remains in the magnetic system [15, 16].

In this chapter, it was assumed that the wavelength of the magnetoelastic wave is much larger than the thickness of the film. In this case, the dynamic magnetization and displacement are approximately uniform over the film thickness. However, if the thickness becomes comparable to the wavelength, different thickness modes can arise. In the magnetic domain, these are called perpendicularly standing spin waves and in the elasticity domain, these are called Lamb waves. The magnetoelastic coupling of such waves is beyond the scope of this chapter and will generally require numerical calculations.

12.4.3 Damping of Magnetoelastic Waves

So far, all waves have been considered to be lossless and their intrinsic damping was neglected. However, in real systems, magnetoelastic waves are expected to decay during propagation. Since their decay length is of great practical interest, we will present in this last part a brief introduction on the damping of magnetoelastic waves. More detailed discussions can be found in [54–58].

Several different energy loss mechanisms exist, which dampen the magnetization and displacement dynamics. In the semi-classical continuum theory used in this chapter, it is common to subsume all different loss mechanisms in a single phenomenological damping term, which is then included in the equation of motion. The damping of the magnetization dynamics is captured by the damping term in the LLG equation, characterized by the phenomenological Gilbert damping parameter α. Analogously, for elastic waves, damping can be introduced into the equations of motion via phenomenological complex stiffness constants.

The addition of the damping terms to the equations of motion results in energy dissipation of the dynamic system. As a consequence, the amplitude of the plane waves considered above decays in time and space. Therefore, the plane wave ansatz to solve the equations of motion needs to be modified by adding an exponential decay factor. This damping factor can be seen as originating from the complex frequency, i.e.

$$\mathbf{w}(\mathbf{r},t)e^{i((\omega_r+i\omega_i)t+\mathbf{k}\cdot\mathbf{r})} = \mathbf{w}(\mathbf{r},t)e^{-t/\tau}e^{i(\omega_r t+\mathbf{k}\cdot\mathbf{r})} = \mathbf{w}(\mathbf{r},t)e^{-x/\delta}e^{i(\omega_r t+\mathbf{k}\cdot\mathbf{r})} \quad (12.82)$$

with $\tau = 1/\omega_i$ the lifetime, $\delta = v_g\tau$ the mean free path or attenuation length, and $\mathbf{w}(\mathbf{r},t) = [u_\mathrm{x}, u_\mathrm{y}, u_\mathrm{z}, m_\mathrm{x}, m_\mathrm{y}]^T$ the dynamic components of the wave. Note that the lifetime characterizes the decay of the wave in time and the mean free path characterizes the attenuation of the wave in space.

To determine the decay characteristics of a wave, the imaginary part of its frequency needs to be assessed. This can be achieved within the above approach, which is based on finding nontrivial solutions of homogeneous linear systems by calculating the roots of their determinants. The real part of the resulting frequency still represents the dispersion relation, whereas the imaginary part originates from the additional damping terms and represents the inverse of the lifetime.

For spin waves in a ferromagnetic medium, the lifetime can be found by solving the LLG equation and is given by

$$\tau_{\mathrm{fm}} = \frac{2}{\alpha(\omega_{\mathrm{fx}} + \omega_{\mathrm{fy}})}. \tag{12.83}$$

The lifetime at GHz frequencies is typically of the order of ns in metallic ferromagnets, such as Ni considered in this chapter, and of the order of μs for low-damping magnetic insulators such as Yttrium Iron Garnet (YIG). On the other hand, much less is know for elastic waves at GHz frequencies although estimates suggest that the lifetime is similar to that of spin waves. Experimentally, it is typically found that the mean free path of (surface) elastic waves at these frequencies is somewhat larger than these of spin waves [40, 51, 52, 59], however, the topic still requires further research.

In the case of magnetoelastic waves, analytical derivations of the lifetimes and decay lengths are rather complex. In the quasi-elastic regime, it is clear that the lifetime is strongly determined by the lifetime of the elastic wave. The energy of quasi-elastic waves is almost completely stored in the elastic system, with only a negligible part in the magnetic system. Hence, the dissipation due to the magnetic loss has negligible influence on the overall dissipation. An analogous argument can be made for the quasi-magnetic regime, where magnetic properties and lifetimes should determine the decay of the magnetoelastic waves.

In the strongly coupled magnetoelastic regime, i.e. near the anticrossing, no simple conclusion can be drawn. In this regime, the energy is distributed between magnetic and elastic domains and is transferred forth and back during propagation. Therefore both magnetic and elastic losses contribute to the total energy dissipation. One may expect in such a case that the lifetime of a magnetoelastic wave is given by a suitable weighted average of the lifetimes of magnetic and elastic waves. In general, the lifetime depends on multiple parameters, such as the orientation of the static magnetization, the interaction coefficient, the wavenumber, etc.. Further work is required to fully understand in particular the effect of the magnetoelastic interactions on the lifetime of strongly coupled magnetoelastic waves. By contrast, the group velocities of magnetoelastic waves are well understood and can be calculated from the dispersion relations, so the assessment of mean free paths is straighforward once the lifetime is known.

12.5 Conclusion

The first part of this chapter presented a review of magnetic and elastic interactions as well as the formation of magnetic (spin) and elastic waves. It has been shown that the dynamic behavior of the magnetization and displacement can be seen as an eigensystem with eigenvalues corresponding to the dispersion relations and eigenstates describing the polarization and ellipticity of the resulting waves. Based on this formalism of eigensystems, both magnetic and elastic waves have been studied in bulk and thin film materials. For magnetic (spin) waves, different regimes have been identified and their correlation with Maxwell's equations has been explained.

In the second part of this chapter, the coupling between magnetic and elastic waves, due to the direct and inverse magnetoelastic interactions, was described. By combining the magnetoelastic interaction terms with the magnetodynamic and the elastodynamic equations, the magnetoelastic eigensystem has been derived for an arbitrary in-plane magnetization orientation. Within this framework, both the exchange and dipolar interaction have been taken into account. Previous descriptions of magnetoelastic waves in infinitesimally thin films have been extended to thin films of finite thickness by considering the appropriate dipolar field based on the magnetostatic Green's function. Two limiting cases, i.e. static magnetization perpendicular or parallel to the propagation direction, have been studied in more detail and their dispersion relations and eigenstates have been mathematically and graphically described. In addition, several properties of magnetoelastic waves, such as the energy transfer length and the magnetoelastic bandgap, and concepts, such as wave anticrossings and polarization rotations, have been discussed in detail. The fundamental framework of magnetoelastic phenomena and waves described in this chapter can be utilised for the theoretical description and modeling of the next generation of magnetoelectric transducers. These transducers need to operate at GHz frequencies and should be miniaturized to the nanometer scale. Despite the technical challenges, such transducers show high potential for efficient energy transfer between the electric and magnetic domains.

Acknowledgements This work has been supported by imec's industrial affiliate program on beyond-CMOS logic and by the European Union's Horizon 2020 research and innovation program within the FET-OPEN project CHIRON under grant agreement No. 801055. F.V. acknowledges financial support from the Research Foundation – Flanders (FWO) through grant No. 1S05719N.

References

1. Z. Diao, Z. Li, S. Wang, Y. Ding, A. Panchula, E. Chen, L. Wang, Y. Huai, Spin-transfer torque switching in magnetic tunnel junctions and spin-transfer torque random access memory. J. Phys.-Condens. Matter **19**, 165209 (2007)
2. M. Bapna, B. Parks, S.D. Oberdick, H. Almasi, W. Wang, S.A. Majetich, Spin-orbit-torque switching in 20-nm perpendicular magnetic tunnel junctions. Phys. Rev. Appl. **10**, 024013 (2018)
3. N.I. Polzikovaa, S.G. Alekseev, I.I. Pyataikin, I.M. Kotelyanskii, V.A. Luzanov, A.P. Orlov, Acoustic spin pumping in magnetoelectric bulk acoustic wave resonator. AIP Adv. **6**, 056306 (2016)
4. S. Cherepov, P.K. Amiri, J.G. Alzate, K. Wong, M. Lewis, P. Upadhyaya, J. Nath, M. Bao, A. Bur, T. Wu, G.P. Carman, A. Khitun, K.L. Wang, Electric-field-induced spin wave generation using multiferroic magnetoelectric cells. Appl. Phys. Lett. **104**, 082403 (2014)
5. N.I. Polzikova, S.G. Alekseev, V.A. Luzanov, A.O. Raevskiy, Electroacoustic excitation of spin waves and their detection due to the inverse spin hall effect. Phys. Solid State **60**, 2211–2217 (2018)
6. M. Balinskiy, A.C. Chavez, A. Barra, H. Chiang, G.P. Carman, A. Khitun, Magnetoelectric spin wave modulator based on synthetic multiferroic structure. Sci. Rep. **8**, 10867 (2018)
7. H. Zhou, A. Talbi, N. Tiercelin, O. Bou Matar, Multilayer magnetostrictive structure based surface acoustic wave devices. Appl. Phys. Lett. **104**, 114101 (2014)

8. C. Kittel, Interaction of spin waves and ultrasonic waves in ferromagnetic crystals. Phys. Rev. **110**, 836 (1958)
9. A.I. Akhiezer, V.G. Bar'iakhtar, S.V. Peletminskii, Coupled magnetoelastic waves in ferromagnetic media and ferroacoustic resonance. JETP **8**, 157 (1959)
10. P.A. Fedders, Theory of acoustic resonance and dispersion in bulk ferromagnets. Phys. Rev. B **9**, 3835 (1974)
11. T. Kobayashi, R.C. Barker, J.L. Bleustein, A. Yelon, Ferromagnetoelastic resonance in thin films I. Formal treatment. Phys. Rev. B **7**, 3273 (1973)
12. T. Kobayashi, R.C. Barker, A. Yelon, Ferromagnetoelastic resonance in thin films. II. Application to nickel. Phys. Rev. B **7**, 3286 (1973)
13. A. Kamra, G.E.W. Bauer, Actuation, propagation, and detection of transverse magnetoelastic waves in ferromagnets. Solid State Commun. **198**, 35–39 (2014)
14. S.M. Keller, A. Sepulveda, G.P. Carman, Effective magnetoelectric properties of magnetoelectroelastic (multiferroic) materials and effects on plane wave dynamics. Prog. Electromagn. Res. **154**, 115–126 (2015)
15. V.W. Tucker, J.W. Rampton, *Microwave Ultrasonics in Solid State Physics* (North Holland, 1972)
16. A.G. Gurevich, G.A. Melkov, *Magnetization Oscillations and Waves* (CRC Press, 1996)
17. E. Schlomann, Generation of phonons in high power ferromagnetic resonance experiments. J. Appl. Phys. **31**, 1647 (1960)
18. L. Landau, E. Lifshitz, On the theory of the dispersion of magnetic permeability in ferromagnetic bodies. Phys. Z. Sowjetunion **8**, 153–164 (1935)
19. T.L. Gilbert, A phenomenological theory of damping in ferromagnetic materials. IEEE Trans. Magn. **40**, 3443 (2004)
20. D.D. Stancil, A. Prabhakar, *Spin Waves Theory and Applications* (Springer, 2009)
21. M.G. Cottam, *Linear and Nonlinear Spin Waves in Magnetic Films and Superlattices* (World Scientific Publishing Company, 1994)
22. R.I. Joseph, E. Schlömann, Demagnetizing field in nonellipsoidal bodies. J. Appl. Phys. **36**, 1579 (1965)
23. A. Smith, K.K. Nielsen, D.V. Christensen, C.R.H. Bahl, R. Bjork, J. Hattel, The demagnetizing field of a nonuniform rectangular prism. J. Appl. Phys. **107**, 103910 (2010)
24. S. Chikazumi, *Physics of Magnetism* (Krieger Pub Co, 1986)
25. B. Hillebrands, K. Ounadjela, *Spin Dynamics in Confined Magnetic Structures I* (Springer, 2002)
26. C. Herring, C. Kittel, On the theory of spin waves in ferromagnetic media. Phys. Rev. **81**, 869 (1951)
27. S. Tamaru, J.A. Bain, M.H. Kryder, D.S. Ricketts, Green's function for magnetostatic surface waves and its application to the study of diffraction patterns. Phys. Rev. B **84**, 064437 (2011)
28. R.W. Damon, J.R. Eshbach, Magnetostatic modes of a ferromagnet slab. J. Phys. Chem. Solids **19**, 308–320 (1961)
29. K.J. Harte, Theory of magnetization ripple in ferromagnetic films. J. Appl. Phys. **39**, 1503 (1968)
30. B.A. Kalinikos, A.N. Slavin, Theory of dipole-exchange spin wave spectrum for ferromagnetic films with mixed exchange boundary conditions. J. Phys. C: Solid State Phys. **19**, 7013–7033 (1986)
31. B.A. Kalinikos, Spectrum and linear excitation of spin waves in ferromagnetic films. Sov. Phys. J. **24**, 718–731 (1981)
32. B.D. Cullity, C.D. Graham, *Introduction to Magnetic Materials* (Wiley, 2011)
33. C. Wilts, S. Lai, Spin wave measurements of exchange constant in Ni-Fe alloy films. IEEE Trans. Magn. **8**, 280 (1972)
34. M. Ewing, *Elastic Waves in Layered Media*. McGraw-Hill Series in the Geological Sciences (1957)
35. J.R. Barber, *Elasticity* (Kluwer Academic Publishers, 2004)
36. K.F. Graff, *Wave Motion in Elastic Solids* (Dover Publications, 1975)

37. J. D. Achenbach, *Wave Propagating in Elastic Solids* (North-Holland Publishing, 1973)
38. A.H. Nayfeh, *Wave Propagation in Layered Anisotropic Media* (North-Holland Publishing, 1995)
39. A.K. Ganguly, K.L. Davis, D.C. Webb, C. Vittoria, Magnetoelastic surface waves in a magnetic film-piezoelectric substrate configuration. J. Appl. Phys. **47**, 2696 (1976)
40. L. Dreher, M. Weiler, M. Pernpeintner, H. Huebl, R. Gross, M.S. Brandt, S.T.B. Goennenwein, Surface acoustic wave driven ferromagnetic resonance in nickel thin films: theory and experiment. Phys. Rev. B **86**, 134415 (2012)
41. L. Thevenard, C. Gourdon, J.Y. Prieur, H.J. von Bardeleben, S. Vincent, L. Becerra, L. Largeau, J.-Y. Duquesne, Surface-acoustic-wave-driven ferromagnetic resonance in (Ga, Mn)(As, P) epilayers. Phys. Rev. B **90**, 094401 (2014)
42. E. du Tremolet, de Lacheisseirie, *Magnetostriction—Theory and Applications of Magnetoelasticity* (CRC Press, Boca Raton, 1993)
43. R.C. O'Handley, *Modern Magnetic Materials: Principles and Applications* (John Wiley & Sons, Inc., 2000)
44. J. Walowski, M. Djordjevic Kaufmann, B. Lenk, C. Hamann, J. McCord, M. Münzenberg, Intrinsic and non-local Gilbert damping in polycrystalline nickel studied by Ti: sapphire laser fs spectroscopy. J. Phys. D **41**, 164016 (2008)
45. D. Sander, The correlation between mechanical stress and magnetic anisotropy in ultrathin films. Rep. Prog. Phys. **62**, 809 (1999)
46. M. Yamamoto, On elastic constants of nickel crystals. Phys. Rev. **77**, 566 (1950)
47. A.F. Mills, *Basic Heat and Mass Transfer*, 2nd edn. (Prentice Hall, New Jersey, 1999)
48. H. Matthews, R.C. Le Craw, Acoustic wave rotation by magnon-phonon interaction. Phys. Rev. Lett. **8**, 397 (1962)
49. K.B. Vlasov, B.K. Ishmukhametov, Rotation of the plane of polarization of elastic waves in magnetically polarized magneto-elastic media. Sov. Phys. JETP **10**, 531 (1960)
50. P.G. Gowtham, T. Moriyama, D.C. Ralph, R.A. Buhrman, Travelling surface spin-wave resonance spectroscopy using surface acoustic waves. J. Appl. Phys. **118**, 233910 (2015)
51. R. Verba, I. Lisenkov, I. Krivorotov, V. Tiberkevich, A. Slavin, Nonreciprocal surface acoustic waves in multilayers with magnetoelastic and interfacial Dzyaloshinskii-Moriya interactions. Phys. Rev. Appl. **9**, 064014 (2018)
52. P. Graczyk, M. Krawczyk, Coupled-mode theory for the interaction between acoustic waves and spin waves in magnonic-phononic crystals: Propagating magnetoelastic waves. Phys. Rev. B **96**, 024407 (2017)
53. C. Berk, M. Jaris, W. Yang, S. Dhuey, S. Cabrini, H. Schmidt, Strongly coupled magnon-phonon dynamics in a single nanomagnet. Nat. Commun. **10**, 2652 (2019)
54. S.M. Rezende, F.R. Morgenthaler, Magnetoelastic waves in time-varying magnetic fields. I. Theory. J. Appl. Phys. **40**, 524 (1969)
55. A. Widom, C. Vittoria, S.D. Yoon, Gilbert ferromagnetic damping theory and the fluctuation-dissipation theorem. J. Appl. Phys. **108**, 073924 (2010)
56. V.L. Gurevich, Dissipation function in a magnetic field (Review). Phys. Solid State **57**, 1271–1276 (2015)
57. E. Rossi, O.G. Heinonen, A.H. MacDonald, Dynamics of magnetization coupled to a thermal bath of elastic modes. Phys. Rev. B **72**, 174412 (2005)
58. S. Streib, H. Keshtgar, G. Bauer, Damping of magnetization dynamics by phonon pumping. Phys. Rev. Lett. **121**, 027202 (2018)
59. X. Li, D. Labanowski, S. Salahuddin, C.S. Lynch, Spin wave generation by surface acoustic waves. J. Appl. Phys. **122**, 043904 (2017)

Chapter 13
Theoretical Generalization of the Optical Chirality to Arbitrary Optical Media

J. Enrique Vázquez-Lozano and Alejandro Martínez

Abstract Chiroptical light-matter interaction is largely boosted in the surroundings of complex-shaped metallic nanostructures. Multiple enhancement schemes have been proposed, from twisted structures, such as spirals or helices, arrays of chiral and even achiral plasmonic nanostructures, to stacked planar metasurfaces. Furthermore, there is a steadily growing trend in using assemblies of high-index dielectric nanostructures, which are actually revealing promising results in terms of the enhancement of chiroptical effects. At any rate, whatever the type of material is, the effects of dispersion and absorption need to be accounted for, with the only exception of the vacuum. These considerations are often neglected, presuming media with an ideal lossless and dispersionless behavior. However, when matter is nanostructured to achieve more complex behaviors, as for the case of metamaterials or plasmonic nanostructures, the effects of dispersion and losses in chiroptical interactions cannot be disregarded at all. Hence, as with the energy, the optical chirality should also be generalized to the case of arbitrary dispersive and lossy optical media. This is the matter of the present chapter; namely, a thorough derivation of the optical chirality, extending it so as to include both dispersive and dissipative effects. For simplicity, as well as for constructiveness, we shall elaborate this theoretical analysis upon the basis of the most complete form of the conservation law for the optical chirality.

13.1 Introduction

Just by taking a glance at the current literature on optics and nanophotonics, one can realize that optical chirality is an active research topic that often goes hand in hand with plasmonics and metamaterials [1, 2]. These kind of systems are actually being regarded as the best-suited platform for strengthening, and thus for investigating, chiral light-matter interactions [3–9]. More recently, high-index dielectric nanopar-

J. E. Vázquez-Lozano (✉) · A. Martínez
Nanophotonics Technology Center, Universitat Politècnica de València,
Camino de Vera s/n, 46022 Valencia, Spain
e-mail: juavazlo@ntc.upv.es

E. Kamenetskii (ed.), *Chirality, Magnetism and Magnetoelectricity*,
Topics in Applied Physics 138,
https://doi.org/10.1007/978-3-030-62844-4_13

ticles, in very distinct arrangements, have also been suggested as suitable systems to boost near-field chirality [10, 11]. Besides displaying lower absorption losses than its plasmonic counterparts, dielectric resonators are proving to give promising results with a number of important practical advantages for chiral applications such as sensing, spectroscopy, and enantioselectivity [12]. Specifically, they can provide large areas of high and uniform-sign chirality [13, 14], spectral accessibility and tunability of chiral interactions [15], and switchability upon reversing the input polarization [16]. Ultimately, all these approaches, either based on metallic or dielectric nanostructures, have been strongly fostered on account of the early proposal of superchiral fields [17], as they can be used to get high levels of resolution and sensitivity in order to check and characterize the chirality of matter or nanostructures.[1] This necessity of enhancing chiroptical responses arises from the fact that interaction between chiral light and chiral analytes (typically biomolecules such as sugars or proteins [20], or artificial metamolecules at the micro or even nanoscale [21]) is in general extremely weak. Indeed, leaving aside the complexity that the experimental arrangements might entail by themselves, this smallness is mainly due to the large scale difference (and the subsequent mismatch) between the operational wavelength of the input light and the typical size of the chiral objects [22]. It is here where plasmonic nanostructures and metamaterials come into play, as they ease the full spatiotemporal control of light-matter interactions [23], thereby leading to sculpted and highly contorted three-dimensional electromagnetic (EM) fields [24]. In fact, it should be noted that the main requirement for the occurrence of strong optical chirality, and thus for enhancing the chiroptical interactions, actually lies on the complexity in the shapes traced out by the EM field distributions [24–26].

It is very well known that plasmonic nanostructures and metamaterials are inherently dispersive systems [27]. In fact, all the media and materials, except the vacuum, are dispersive, and, consequently, feature absorption losses [28–30]. These characteristics are to be accounted for when addressing dynamical properties such as, for example, EM energy. Hence, as regards the matter we are concerned, the same should be done for the optical chirality [31, 32]. However, it is certainly surprising that, in most of the previous studies on optical chirality and its interaction with matter, contributions of material dispersion as well as dissipation have mostly been ignored. Rather, it is a common practice to apply the earliest definition originally derived for monochromatic optical fields in free space [17, 31–33]:

$$C_{\text{vacuum}}(\boldsymbol{r}, t) \equiv \frac{1}{2}\left[\varepsilon_0 \boldsymbol{\mathcal{E}} \cdot (\nabla \times \boldsymbol{\mathcal{E}}) + \mu_0 \boldsymbol{\mathcal{H}} \cdot (\nabla \times \boldsymbol{\mathcal{H}})\right], \tag{13.1}$$

[1] It is worth pointing out that, *superchirality* (or superchiral light), is well defined only in the case of plane-wave propagation in free space, because it is actually defined with respect to the chirality of circularly polarized light. Notice that, for example, in the case of waveguiding systems, the term of superchirality may be misunderstood, as it would depend on the specific structure [18]. So, in lieu of superchirality, henceforth we shall refer to it simply as the enhanced chirality. In any case, it is noteworthy to mention that there exists a subtle controversy regarding superchiral fields and its apparent unlimited enhancement factor (for further details on this issue see, e.g., [19]).

where ε_0 and μ_0 are, respectively, the permittivity and permeability of the vacuum, and $\boldsymbol{\mathcal{E}} = \boldsymbol{\mathcal{E}}(\boldsymbol{r}, t)$ and $\boldsymbol{\mathcal{H}} = \boldsymbol{\mathcal{H}}(\boldsymbol{r}, t)$ are the local, time-dependent electric and magnetic fields. This definition for the optical chirality as given here, in the time-dependent representation, has been successfully used in enhanced circular dichroism spectroscopic measurements for the experimental detection and characterization of chiral biomolecules [20], thus confirming its physical significance, and highlighting its feasibility for practical applications. Notwithstanding, its applicability can be highly questionable beyond the simplest case of the vacuum, even further when used for looking into chiral effects either in metallic nanostructures or metamaterials, often regarded as the paradigmatic examples of dispersive media. That in itself is reason enough for wanting to address a generalization of such a quantity. Furthermore, recent advances in nanofabrication are opening up new possibilities for the experimental measurement and investigation of dynamical properties such as the EM energy-momentum, the optical orbital and spin angular momentum, and the EM helicity, so far only accessible theoretically.[2] This fact has likely led to reexamine both the theoretical treatment and the formulation of these dynamical properties taking into account their dispersive features [39–42]. As for the optical chirality, it should be noted that there are also few works attempting to extend its original definition to arbitrarily defined linear [19], gyrotropic [43], or dispersive media [44]. However, as will be shown throughout this chapter, at the anomalous dispersion region, i.e., the spectral range where the real part of the permittivity decreases abruptly, or, to put it simply, near the resonance frequencies, the optical chirality can be strongly enhanced. This evidences the important role played by the absorption losses of a given medium in the analysis of the optical response by a dispersive material,[3] in particular, as far as chiral light-matter interactions are concerned. Hence, a full description of optical chirality in dispersive media, extending it to account for the medium's dissipative effects as well would be highly valuable and enlightening.

There are many different ways in which one might want to conduct such a generalization; either from phenomenological aspects, just by fitting experimental outcomes or via arguments stemming from the chiroptical responses; or directly through a thorough theoretical analysis. The latter will be the approach followed up herein, specifically, we shall take advantage of the underlying mathematical structure of the corresponding continuity equation (or conservation law) for the optical chirality

[2] In this regard, it is noteworthy to mention the so-called *Abraham-Minkowski dilemma*, a long-standing problem concerning with an ambiguity that arises from the real definition of the linear and angular momentum for optical radiation in media [34]. Even though there are a number of influential papers claiming to have solved it (see, e.g., [35, 36]), this challenging problem still remains as a subject of current interest and debate [37, 38].

[3] Notice that, strictly speaking, dispersion is necessarily tied to dissipation. This connection is well established by the so-called *Kramers-Kronig relations* [28], according to which the real and imaginary parts of the material parameters, i.e., the electric permittivity and the magnetic permeability ($\varepsilon(\omega) = \varepsilon' + i\varepsilon''$ and $\mu(\omega) = \mu' + i\mu''$), appear to be coupled together. In addition, it has been demonstrated that Kramers-Kronig relations underpin the fundamental principle of causality [45], and hence, initial assumptions regarding dispersion and dissipation have to be carefully considered, otherwise they may lead to misleading outcomes. Still, one can find many cases where is assumed a dispersive medium neglecting the losses.

[46]. Notice that just as much the energy, linear momentum, angular momentum, or helicity, optical chirality is also a conserved quantity for free-space EM fields [31, 32, 47]. Upon this basis, in Sect. 13.2 we will recall some general aspects on conservation laws, particularizing to the simplest case of EM energy. In Sect. 13.3 we will elaborate on the most complete form of the continuity equation for the optical chirality, without any restrictions on the nature of the medium. From this, and building on previous approaches addressing the EM energy density considering dispersion as well as dissipation, in Sect. 13.4, we will put forward an alternative derivation for the optical chirality density in dispersive media, distinguishing between the lossless and lossy cases [46]. Remarkably, our description will be completely general, i.e., it will be valid for arbitrarily varying radiation fields, and will be applicable to any kind of medium, including dielectrics, semiconductors, as well as highly dispersive material systems such as metals (or plasmonic structures) and metamaterials [48]. Finally, in Sect. 13.5, we will summarize the main results of this chapter and conclude with a general outlook towards possible lines of future research.

13.2 Electromagnetic Energy Density in Dispersive and Lossy Media: A General Approach from the Continuity Equation

Aiming to provide a comprehensive and easy-to-follow guideline for generalizing the optical chirality (and any other dynamical property), we will start by revisiting basic aspects on the conservation laws focusing upon the most well-known dynamical property: the EM energy. As will be seen below, this will enable a straightforward procedure for obtaining a general expression of the EM energy density valid in any kind of medium, either non-dispersive or dispersive, and, concerning this latter case, extending it to both the lossless and the lossy approaches.

13.2.1 Poynting's Theorem and Energy Density in Non-Dispersive Media

Like Maxwell's equations and the conservation of charge, the EM energy conservation law, often referred to as the Poynting's theorem, is a fundamental piece of classical electrodynamics. It is in fact a recurring matter all over the main textbooks on electromagnetism [28, 30], optics [49], nano-optics [29], and photonics [50]. For our purposes, it will serve as a guideline for mathematics and also for the interpretation when addressing the optical chirality density, specially those regarding the source-like contributions, since its physical meaning in such a case may not be so obvious [51, 52]. Notice that, even though it could seem quite burdensome, for the sake of completeness, as well as for teaching purposes, we start from the basic foundations. Hence, a proficient reader might skip the following part of the section.

In time domain, Poynting's theorem can be directly obtained from the vector identity $\nabla \cdot (\mathcal{A} \times \mathcal{B}) = (\nabla \times \mathcal{A}) \cdot \mathcal{B} - \mathcal{A} \cdot (\nabla \times \mathcal{B})$, along with the structural (or curl-like) Maxwell's equations $\nabla \times \mathcal{E} = -\partial_t \mathcal{B}$ and $\nabla \times \mathcal{H} = \partial_t \mathcal{D} + \mathcal{J}$:

$$\nabla \cdot \mathcal{S} = -\mathcal{E} \cdot \partial_t \mathcal{D} - \mathcal{H} \cdot \partial_t \mathcal{B} - \mathcal{J} \cdot \mathcal{E}, \tag{13.2}$$

where $\mathcal{S} = \mathcal{S}(\boldsymbol{r}, t) \equiv \mathcal{E} \times \mathcal{H}$ is the Poynting vector, representing the energy flux density [53], and $\mathcal{D} = \mathcal{D}(\boldsymbol{r}, t)$, $\mathcal{B} = \mathcal{B}(\boldsymbol{r}, t)$, and $\mathcal{J} = \mathcal{J}(\boldsymbol{r}, t)$ are the time-dependent electric displacement, magnetic induction, and electric current density, respectively. This statement, typically dubbed as the *differential form of Poynting's theorem*, is an absolutely general result, i.e., it holds for any EM field that propagates through any arbitrary medium, either lossy or lossless [28–30]. It establishes a local energy balance over the whole system, thereby relating the rate of change of electric and magnetic energy density stored in the EM fields, the EM energy flux density, and the energy absorption losses, or gain, of the medium.

Notwithstanding, what is most usual to find in basic literature, is the expression of the simplest case considering linear and non-dispersive systems. In such a case it is implicitly presumed that $\mathcal{D} = \varepsilon_0 \tilde{\varepsilon} \mathcal{E}$ and $\mathcal{B} = \mu_0 \tilde{\mu} \mathcal{H}$, with $\tilde{\varepsilon}$ and $\tilde{\mu}$ being real numbers. Therefore, instead of equation (13.2), one usually finds that

$$\nabla \cdot \mathcal{S} + \partial_t \mathcal{W} = -\mathcal{J} \cdot \mathcal{E}, \tag{13.3}$$

where

$$\mathcal{W} = \mathcal{W}(\boldsymbol{r}, t) \equiv \frac{1}{2} \left[\mathcal{E} \cdot \mathcal{D} + \mathcal{H} \cdot \mathcal{B} \right], \tag{13.4}$$

is the local time-dependent EM field energy density, and $\mathcal{J} \cdot \mathcal{E}$ stands for the power loss (or gain) dissipated (or pumped) to (from) the medium owing to external currents.

We can further simplify the above expressions by considering EM fields with a harmonic time dependence of the form $e^{-i\omega t}$, i.e., those that can be written as

$$\mathcal{F} = \mathrm{Re}\left[\boldsymbol{F} e^{-i\omega t} \right] = \frac{1}{2} \left[\boldsymbol{F}^{-i\omega t} + \boldsymbol{F}^* e^{i\omega t} \right], \tag{13.5}$$

where ω is the angular frequency and the asterisk denotes complex conjugation. It should be noted that this consideration does not undermine the generality of the treatment at all, since one can always express any arbitrary time-dependent EM field from its spectral representation by means of the Fourier transform [28–30]:

$$\boldsymbol{F}(\boldsymbol{r}, \omega) = \frac{1}{2\pi} \int_{-\infty}^{+\infty} \mathcal{F}(\boldsymbol{r}, t) e^{i\omega t} dt, \tag{13.6}$$

where $\mathcal{F}(\boldsymbol{r}, t)$ and $\boldsymbol{F}(\boldsymbol{r}, \omega)$ stand, respectively, for the real-valued EM fields and their corresponding complex-like counterparts. This way, we could consider independently every spectral component as if each of them were a single monochromatic wave. Of course, one can always turn back to the time-dependent representation via the inverse Fourier transform:

$$\mathcal{F}(\boldsymbol{r}, t) = \int_{-\infty}^{+\infty} \boldsymbol{F}(\boldsymbol{r}, \omega) e^{-i\omega t} d\omega. \tag{13.7}$$

Notice that, for the sake of clarity, throughout this chapter we will use the above distinct notation either for the real-valued EM fields in the time domain, or for the complex fields in the frequency domain, thus allowing us to omit, hereinafter, the arguments in the field expressions so as to avoid cumbersome notation. After outlining these basics, it is easy to see from (13.4) that the EM energy density of each spectral component in non-dispersive media reads as follows

$$\mathcal{W} = \mathcal{W}^{\mathrm{e}} + \mathcal{W}^{\mathrm{m}} = \frac{1}{4}\left[\varepsilon_0 \varepsilon \left(|\boldsymbol{E}|^2 + \boldsymbol{E}^2 e^{-2i\omega t}\right) + \mu_0 \mu \left(|\boldsymbol{H}|^2 + \boldsymbol{H}^2 e^{-2i\omega t}\right)\right]. \tag{13.8}$$

The last terms in both the electric and magnetic energy contributions have a relative phase that is rapidly oscillating. Thus, for simplicity, one is usually interested in dealing with dynamical properties, in this case, the EM energy density, averaged over an optical cycle:

$$W \equiv \langle \mathcal{W} \rangle_{\mathrm{T}} = \frac{1}{T}\int_0^{2\pi/\omega} \left[\mathcal{W}^{\mathrm{e}} + \mathcal{W}^{\mathrm{m}}\right] dt = \frac{1}{4}\left[\varepsilon_0 \varepsilon \,|\boldsymbol{E}|^2 + \mu_0 \mu \,|\boldsymbol{H}|^2\right]. \tag{13.9}$$

This is the time-averaged form of the EM energy density in a non-dispersive medium (i.e., strictly speaking, it only applies to the vacuum). It is actually the most familiar expression for representing this dynamical property, likely, because it resembles its time-dependent counterpart given in (13.4) (but taking care not to confuse or mix up the real-valued EM fields in the time domain, with the complex field amplitudes in the frequency domain). It should be remarked that this resemblance is because we are assuming EM fields in linear and non-dispersive media, so ε and μ ought to be real numbers independent of frequency, and hence, the corresponding constitutive relations, accounting for the influence of the EM radiation on matter, have the same form in both the time- and the frequency-dependent representations:

$$\mathcal{D} = \varepsilon_0 \tilde{\varepsilon} \mathcal{E}, \quad \Longleftrightarrow \quad \boldsymbol{D} = \varepsilon_0 \varepsilon \boldsymbol{E}; \tag{13.10a}$$

$$\mathcal{B} = \mu_0 \tilde{\mu} \mathcal{H}, \quad \Longleftrightarrow \quad \boldsymbol{B} = \mu_0 \mu \boldsymbol{H}. \tag{13.10b}$$

Notice the deliberate difference between the material parameters with and without tilde, which are intended to distinguish between time- or frequency dependence.

13.2.2 Electromagnetic Energy Density in Dispersive Media: Lossless (Brillouin's Approach) and Lossy (Loudon's Approach) Cases

The situation may become slightly trickier when the optical field is thought to propagate through a dispersive medium. In the time domain field representation this means that the material response is not instantaneous, but it relies on all the past history. According to the basic properties of Fourier transform, this should be described in terms of a convolution in the time domain of EM fields characterizing the material influence [28, 29]. Specifically, the dynamic response of a time-dependent EM field passing through a dispersive medium is to be expressed as [30]:

$$\mathcal{D}(\mathbf{r},t) = \int_{-\infty}^{t} \tilde{\varepsilon}(t-t')\mathcal{E}(\mathbf{r},t')dt' = \int_{-\infty}^{+\infty} \tilde{\varepsilon}(t-t')\mathcal{E}(\mathbf{r},t')dt'; \tag{13.11a}$$

$$\mathcal{B}(\mathbf{r},t) = \int_{-\infty}^{t} \tilde{\mu}(t-t')\mathcal{H}(\mathbf{r},t')dt' = \int_{-\infty}^{+\infty} \tilde{\mu}(t-t')\mathcal{H}(\mathbf{r},t')dt'. \tag{13.11b}$$

In the frequency domain, this behavior translates into a description in which the electric permittivity ε, and the magnetic permeability, μ, both depend on the frequency ω, and, thus, **D** and **B** are simply described by the corresponding constitutive relations (i.e., equations (13.10a) and (13.10b) in the case of linear media), which account for the influence of the EM radiation on matter.

The above considerations are essential aiming to derive a general and closed expression for the EM energy density in dispersive and lossy media. Indeed, it is worth observing that equation (13.2), representing the most general form for the conservation law of EM energy, involves time derivatives of the fields $\mathcal{D}$ and $\mathcal{B}$, which, in turn, ought to be written as the convolution integrals given in (13.11a) and (13.11b). This fact entails the main difficulty in conducting the sought generalization of the EM energy density (and, consequently, that for the optical chirality density). Nevertheless, there are several routes to deal with this [54–65], each of them subject to well distinct prescriptions concerning both the characteristics of the medium as well as the time dependence of the EM fields. At this respect, it should be emphasized that, even though we are searching for an approach as general as possible, expressions accounting for both the EM energy density stored and the dissipation, will crucially depend upon the specific features of the model characterizing the medium. For practical purposes, we will focus on the treatment provided by Loudon [54], considering an absorbing classical dielectric with a single resonance frequency, i.e., a Lorentz-like medium. This approach has been further extended by many other authors to account for dispersive magnetic permeabilities [55–59], as well as the possibility of multiple resonance frequencies describing interband transition effects [60, 61]. For completeness, and for convenience in subsequent analysis, in the rest of this section, we shall take into account these considerations. Furthermore, for comparison, we will also outline the classical procedure for determining the EM energy density in dispersive and lossless media, which leads to the so-called Brillouin formula [30, 65].

13.2.2.1 Characterization of Linear Dispersive Media: Drude-Lorentz Model

A physically realistic description of dispersive media would require a careful analysis of dissipative effects. In classical theory there are many models to characterize the optical properties of a medium, i.e., the electric permittivity ε and the magnetic permeability μ. In the simplest case of a linear medium, the most commonly used is the Drude-Lorentz (D-L) model, which is generally described as a collection of N oscillators coupled all together [27]:

$$\varepsilon_{\mathrm{D-L}}(\omega) = 1 - \frac{f_0\omega_{\mathrm{p}}^2}{\omega^2 + i\omega\gamma_0} - \sum_{n=1}^{N-1} \frac{f_n\omega_{\mathrm{p}}^2}{\omega^2 - \omega_n^2 + i\omega\gamma_n} = 1 - \sum_{n=0}^{N-1} \frac{f_n\omega_{\mathrm{p}}^2}{\omega^2 - \omega_n^2 + i\omega\gamma_n}, \tag{13.12}$$

where f_n, ω_{p}, ω_n, ω and γ_n are, respectively, the relative strength of the oscillators, the plasma frequency, the nth resonance (or restoring) frequency, the excitation frequency, and the nth damping constant (or characteristic collision frequency). Notice that the Drude model only holds for intraband effects, i.e., it only accounts for the response owing to free electrons moving within the conduction band, so there is no resonant behavior for the first term of (13.12), and thus $\omega_0 \equiv 0$. On the other side, the Lorentz model allows us to complete this description by including bound carriers giving rise to interband transitions.

Each oscillator, described as a pole in the dielectric function (13.12), actually comes from the motion equation of a bound electron with undamped resonance frequency ω_n, experiencing a damping force, characterized by γ_n, and subjected to a time-varying external electric field $\boldsymbol{\mathcal{E}}_{\mathrm{loc}}$:

$$\frac{\partial^2 \mathbf{r}_n}{\partial t^2} + \gamma_n \frac{\partial \mathbf{r}_n}{\partial t} + \omega_n^2 \mathbf{r}_n = -\frac{q_{\mathrm{e}}}{m_{\mathrm{e}}} \boldsymbol{\mathcal{E}}_{\mathrm{loc}}, \tag{13.13}$$

where m_{e}, q_{e} and $\mathbf{r}_n$ are the effective mass, charge, and the displacement of the electrons, respectively. By treating each mode n as an electron gas of uniform density ρ_n, the collective effect emerging from all individual displacement leads to a polarization field $\boldsymbol{\mathcal{P}}_n = \rho_n \mathbf{p}_n = (-q_{\mathrm{e}}\rho_n)\,\mathbf{r}_n$, where $\mathbf{p}_n$ is the electric dipole moment associated to the nth mode. Hence, (13.13) can be rewritten as

$$\frac{\partial^2 \boldsymbol{\mathcal{P}}_n}{\partial t^2} + \gamma_n \frac{\partial \boldsymbol{\mathcal{P}}_n}{\partial t} + \omega_n^2 \boldsymbol{\mathcal{P}}_n = \varepsilon_0 f_n \omega_{\mathrm{p}}^2 \boldsymbol{\mathcal{E}}_{\mathrm{loc}}, \tag{13.14}$$

where $\omega_{\mathrm{p}} \equiv \sqrt{q_{\mathrm{e}}^2 \rho_{\mathrm{e}}/(m_{\mathrm{e}}\varepsilon_0)}$, and $f_n \equiv \rho_n/\rho_{\mathrm{e}}$, with ρ_{e} being the total density of D-L oscillators. For time-harmonic, monochromatic, EM fields, the individual contribution of each oscillator is given by:

$$\mathbf{P}_n = \rho_n \alpha_n^{(\mathrm{e})}(\omega)\mathbf{E}_{\mathrm{loc}} = \varepsilon_0 \left[\frac{f_n \omega_{\mathrm{p}}^2}{\omega_n^2 - \omega^2 - i\omega\gamma_n} \right] \mathbf{E}_{\mathrm{loc}}, \tag{13.15}$$

where $\alpha_n^{(\mathrm{e})}(\omega)$ is the electric dipole polarizability. Therefore, every D-L pole contributes to the electric permittivity given in (13.12) in such a way that

$$\mathbf{D} = \varepsilon_0 \left[1 - \sum_{n=0}^{N-1} \frac{f_n \omega_{\mathrm{p}}^2}{\omega^2 - \omega_n^2 + i\omega\gamma_n} \right] \mathbf{E}, \tag{13.16}$$

where the total polarization field is defined as $\mathbf{P} \equiv \sum_n \langle \mathbf{P}_n \rangle$, and the macroscopic electric field is $\mathbf{E} \equiv \langle \mathbf{E}_{\mathrm{loc}} \rangle$, with the angular brackets indicating an average over space. Notice that the external electric field, $\boldsymbol{\mathcal{E}}_{\mathrm{loc}}$, is actually the local microscopic field acting as a driving force [66, 67]. Thus, the corresponding macroscopic counterpart arises when averaging over sufficiently large spatial distances [27–29].

The above procedure is specific for getting the electric permittivity in a linear medium that is modeled as a combination of D-L oscillators [27]. As long as the magnetic permeability can be properly tailored by Lorentzian line shapes (e.g., when describing negative-index metamaterials such as split-ring resonator or fishnet-like structures [68–71]), it might also be extended to magnetically dispersive media [55–59]. For such a case, the corresponding dynamic equation for the magnetization field $\boldsymbol{\mathcal{M}}_n$ would read as:

$$\frac{\partial^2 \boldsymbol{\mathcal{M}}_n}{\partial t^2} + \tilde{\gamma}_n \frac{\partial \boldsymbol{\mathcal{M}}_n}{\partial t} + \tilde{\omega}_n^2 \boldsymbol{\mathcal{M}}_n = \tilde{f}_n \tilde{\omega}_n^2 \boldsymbol{\mathcal{H}}_{\mathrm{loc}}, \tag{13.17}$$

where, analogously to the previous case, $\tilde{\omega}_n$, $\tilde{\gamma}_n$, and $\tilde{f}_n$ are, respectively, the nth resonance frequency of the magnetic dipole oscillators, the nth magnetic damping constant, and the magnetic-like oscillator strength. It should be noted that in earlier works addressing magnetic dispersion, only a single Lorentzian resonance was there. Nonetheless, for completeness, we shall express the magnetic permeability by considering the possible presence of N magnetic oscillators. Therefore, each the individual contribution to the magnetization field is of the form

$$\mathbf{M}_n = \left[\frac{\tilde{f}_n \tilde{\omega}_n^2}{\tilde{\omega}_n^2 - \omega^2 - i\omega\tilde{\gamma}_n} \right] \mathbf{H}_{\mathrm{loc}}. \tag{13.18}$$

Hence, if we define the total magnetization field as $\mathbf{M} \equiv \sum_n \langle \mathbf{M}_n \rangle$, and the macroscopic magnetic field $\mathbf{H} \equiv \langle \mathbf{H}_{\mathrm{loc}} \rangle$, the magnetic permeability is given by

$$\mu_{\mathrm{D-L}}(\omega) = 1 - \sum_{n=0}^{N-1} \frac{\tilde{f}_n \tilde{\omega}_n^2}{\omega^2 - \tilde{\omega}_n^2 + i\omega\tilde{\gamma}_n}. \tag{13.19}$$

Notice that, alike to the electric case, the external magnetic field, $\mathcal{H}_{\mathrm{loc}}$, stands for the local microscopic field, and thus, the corresponding macroscopic counterpart turns up when performing the spatial average [27–29].

This multi-resonant model for characterizing linear dispersive media by Lorentzian line shapes, has been proved to fit very well with experiments [72–74]. Since it is applicable to any frequency, bandwidth, and material system, including dielectrics, semiconductors, metals, as well as metamaterials, it may be regarded as completely general, thereby providing a description as accurate and reliable as needed. As shown below, these basics, along with the mathematical structure of the continuity equation, are going to be the key points to derive the most general expression of the EM energy density in dispersive and lossy media, and, similarly, for carrying out the corresponding generalization of the optical chirality.

13.2.2.2 Energy Density in Dispersive and Lossy Media: Loundon's Approach

Whatever the specific EM properties of the medium, it can always be represented in terms of the polarization and the magnetization vector fields as follows:

$$\mathcal{D} = \varepsilon_0 \mathcal{E} + \mathcal{P}; \tag{13.20a}$$

$$\mathcal{B} = \mu_0 \mathcal{H} + \mu_0 \mathcal{M}. \tag{13.20b}$$

From these general relationships, the first two terms in the right-hand side of (13.2) can be straightforwardly evaluated to give:

$$\mathcal{E} \cdot \partial_t \mathcal{D} = \varepsilon_0 \mathcal{E} \cdot \partial_t \mathcal{E} + \mathcal{E} \cdot \partial_t \mathcal{P}; \tag{13.21a}$$

$$\mathcal{H} \cdot \partial_t \mathcal{B} = \mu_0 \mathcal{H} \cdot \partial_t \mathcal{H} + \mu_0 \mathcal{H} \cdot \partial_t \mathcal{M}. \tag{13.21b}$$

By comparing the mathematical structure of the continuity equation, i.e.,

$$\nabla \cdot [\text{FLUX}] + \partial_t[\text{CONSERVED QUANTITY}] = [\text{SOURCE or SINK}], \tag{13.22}$$

with the general form of the Poynting's theorem as given in (13.2), one can infer that the above expressions should encapsulate both the energy density and the source-like contributions. To determine which one is either accounting for the conserved quantity or the source- or sink-like term, one should observe the time derivative operator, which, according to (13.22), ought to precede the conserved quantity, namely, in this case, the EM energy density. In the lossy approach, this identification should be performed by means of the dynamic equations of the polarization and the magnetization fields [i.e., (13.14) and (13.17)]. Then, attempting to find a total time derivative for the electric and the magnetic contributions, it can be demonstrated that (13.21a) and (13.21b) can be recast as

$$\boldsymbol{\mathcal{E}} \cdot \partial_t \boldsymbol{\mathcal{D}} = \partial_t \left\{ \frac{\varepsilon_0}{2} \boldsymbol{\mathcal{E}}^2 + \sum_{n=0}^{N-1} \frac{1}{2\varepsilon_0 f_n \omega_\mathrm{p}^2} \left[(\partial_t \langle \boldsymbol{\mathcal{P}}_n \rangle)^2 + \omega_n^2 \langle \boldsymbol{\mathcal{P}}_n \rangle^2 \right] \right\}$$
$$+ \sum_{n=0}^{N-1} \frac{\gamma_n}{\varepsilon_0 f_n \omega_\mathrm{p}^2} (\partial_t \langle \boldsymbol{\mathcal{P}}_n \rangle)^2; \tag{13.23a}$$

$$\boldsymbol{\mathcal{H}} \cdot \partial_t \boldsymbol{\mathcal{B}} = \partial_t \left\{ \frac{\mu_0}{2} \boldsymbol{\mathcal{H}}^2 + \sum_{n=0}^{N-1} \frac{\mu_0}{2 \tilde{f}_n \tilde{\omega}_n^2} \left[(\partial_t \langle \boldsymbol{\mathcal{M}}_n \rangle)^2 + \tilde{\omega}_n^2 \left(\langle \boldsymbol{\mathcal{M}}_n \rangle \right)^2 \right] \right\}$$
$$+ \sum_{n=0}^{N-1} \frac{\mu_0 \tilde{\gamma}_n}{\tilde{f}_n \tilde{\omega}_n^2} (\partial_t \langle \boldsymbol{\mathcal{M}}_n \rangle)^2. \tag{13.23b}$$

Equations (13.23a) and (13.23b) result in a total time derivative plus a residual term, i.e.,

$$\boldsymbol{\mathcal{E}} \cdot \partial_t \boldsymbol{\mathcal{D}} = \partial_t \left[\mathcal{W}_\mathrm{vac}^\mathrm{e} + \mathcal{W}_\mathrm{med}^\mathrm{e} \right] + \mathcal{L}^\mathrm{e}; \tag{13.24a}$$
$$\boldsymbol{\mathcal{H}} \cdot \partial_t \boldsymbol{\mathcal{B}} = \partial_t \left[\mathcal{W}_\mathrm{vac}^\mathrm{m} + \mathcal{W}_\mathrm{med}^\mathrm{m} \right] + \mathcal{L}^\mathrm{m}. \tag{13.24b}$$

Hence, from (13.23a) and (13.23b) it is easy to see that the electric and magnetic contributions to the energy density are given by

$$\mathcal{W}^\mathrm{e} = \mathcal{W}_\mathrm{vac}^\mathrm{e} + \mathcal{W}_\mathrm{med}^\mathrm{e} = \frac{\varepsilon_0}{2} \boldsymbol{\mathcal{E}}^2 + \sum_{n=0}^{N-1} \frac{1}{2\varepsilon_0 f_n \omega_\mathrm{p}^2} \left[(\partial_t \langle \boldsymbol{\mathcal{P}}_n \rangle)^2 + \omega_n^2 \langle \boldsymbol{\mathcal{P}}_n \rangle^2 \right]; \tag{13.25a}$$

$$\mathcal{W}^\mathrm{m} = \mathcal{W}_\mathrm{vac}^\mathrm{m} + \mathcal{W}_\mathrm{med}^\mathrm{m} = \frac{\mu_0}{2} \boldsymbol{\mathcal{H}}^2 + \sum_{n=0}^{N-1} \frac{\mu_0}{2 \tilde{f}_n \tilde{\omega}_n^2} \left[(\partial_t \langle \boldsymbol{\mathcal{M}}_n \rangle)^2 + \tilde{\omega}_n^2 \left(\langle \boldsymbol{\mathcal{M}}_n \rangle \right)^2 \right], \tag{13.25b}$$

where the contributions related to the vacuum and the medium, are, in turn, separated from each other. Furthermore, the terms accounting for the power loss densities are

$$\mathcal{L}^\mathrm{e} = \sum_{n=0}^{N-1} \frac{\gamma_n}{\varepsilon_0 f_n \omega_\mathrm{p}^2} (\partial_t \langle \boldsymbol{\mathcal{P}}_n \rangle)^2; \tag{13.26a}$$
$$\mathcal{L}^\mathrm{m} = \sum_{n=0}^{N-1} \frac{\mu_0 \tilde{\gamma}_n}{\tilde{f}_n \tilde{\omega}_n^2} (\partial_t \langle \boldsymbol{\mathcal{M}}_n \rangle)^2; \tag{13.26b}$$
$$\mathcal{L}^\mathrm{c} = -\boldsymbol{\mathcal{J}} \cdot \boldsymbol{\mathcal{E}}. \tag{13.26c}$$

Putting it all together, the EM energy conservation law finally reads as

$$\nabla \cdot \mathcal{S} + \partial_t \left[\mathcal{W}^{\mathrm{e}} + \mathcal{W}^{\mathrm{m}} \right] = - \left[\mathcal{L}^{\mathrm{e}} + \mathcal{L}^{\mathrm{m}} + \mathcal{L}^{\mathrm{c}} \right]. \tag{13.27}$$

To complete this analysis, it only remains to obtain the corresponding time-averaged form of the EM energy density. Under the assumption of time-harmonic fields, the direct substitution of (13.15) into (13.25a), and (13.18) into (13.25b) gives

$$W^{\mathrm{e}} \equiv \langle \mathcal{W}^{\mathrm{e}} \rangle_{\mathrm{T}} = \frac{\varepsilon_0}{4} \left[1 + \sum_{n=0}^{N-1} \frac{f_n \left(\omega^2 + \omega_n^2 \right) \omega_{\mathrm{p}}^2}{\left(\omega_n^2 - \omega^2 \right)^2 + \omega^2 \gamma_n^2} \right] |\mathbf{E}|^2 = \frac{\varepsilon_0 \varepsilon_{\mathrm{eff}}(\omega)}{4} |\mathbf{E}|^2 ; \tag{13.28a}$$

$$W^{\mathrm{m}} \equiv \langle \mathcal{W}^{\mathrm{m}} \rangle_{\mathrm{T}} = \frac{\mu_0}{4} \left[1 + \sum_{n=0}^{N-1} \frac{\tilde{f}_n \left(\omega^2 + \tilde{\omega}_n^2 \right) \tilde{\omega}_n^2}{\left(\tilde{\omega}_n^2 - \omega^2 \right)^2 + \omega^2 \tilde{\gamma}_n^2} \right] |\mathbf{H}|^2 = \frac{\mu_0 \mu_{\mathrm{eff}}(\omega)}{4} |\mathbf{H}|^2 , \tag{13.28b}$$

where the angle brackets indicate time averaging over one period of oscillation [see (13.9)], and $\varepsilon_{\mathrm{eff}}$ and μ_{eff} are the real-valued effective material parameters (i.e., the electric permittivity and magnetic permeability), which are defined as

$$\varepsilon_{\mathrm{eff}}(\omega) \equiv 1 + \sum_{n=0}^{N-1} \left(\chi_n' + \frac{2\omega \chi_n''}{\gamma_n} \right); \tag{13.29a}$$

$$\mu_{\mathrm{eff}}(\omega) \equiv 1 + \sum_{n=0}^{N-1} \left(\xi_n' + \frac{2\omega \xi_n''}{\tilde{\gamma}_n} \right), \tag{13.29b}$$

with $\chi = \sum_n \chi_n = \chi' + i\chi'' \equiv \varepsilon - 1$ and $\xi = \sum_n \xi_n = \xi' + i\xi'' \equiv \mu - 1$ being the electric and magnetic susceptibilities:

$$\chi_n = \frac{f_n \omega_{\mathrm{p}}^2 \left(\omega_n^2 - \omega^2 \right)}{\left(\omega_n^2 - \omega^2 \right)^2 + \omega^2 \gamma_n^2} + i \left(\frac{f_n \omega_{\mathrm{p}}^2 \omega \gamma_n}{\left(\omega_n^2 - \omega^2 \right)^2 + \omega^2 \gamma_n^2} \right); \tag{13.30a}$$

$$\xi_n = \frac{\tilde{f}_n \tilde{\omega}_n^2 \left(\tilde{\omega}_n^2 - \omega^2 \right)}{\left(\tilde{\omega}_n^2 - \omega^2 \right)^2 + \omega^2 \tilde{\gamma}_n^2} + i \left(\frac{\tilde{f}_n \tilde{\omega}_n^2 \omega \tilde{\gamma}_n}{\left(\tilde{\omega}_n^2 - \omega^2 \right)^2 + \omega^2 \tilde{\gamma}_n^2} \right). \tag{13.30b}$$

13.2.2.3 Energy Density in Dispersive and Lossless Media: Brillouin's Approach

Unlike in the lossy case, the lossless approach does not require any specific characterization about the medium under consideration. Still, for the sake of simplicity, the only assumptions made hereinafter are that the medium is linear, homogeneous, and isotropic. This derivation simply involves the direct evaluation of the Fourier integrals. So, starting from the general form of the Poynting's theorem as given in (13.2), let us expand the right-hand side in the following form:

$$\begin{aligned}\boldsymbol{\mathcal{E}}\cdot\partial_t\boldsymbol{\mathcal{D}} &= \int_{-\infty}^{+\infty}\mathbf{E}(\omega')e^{-i\omega' t}d\omega'\cdot\partial_t\left\{\int_{-\infty}^{+\infty}\mathbf{D}(\omega)e^{-i\omega t}d\omega\right\}\\ &= \int_{-\infty}^{+\infty}\int_{-\infty}^{+\infty}(-i\omega)\mathbf{E}(\omega')\cdot\mathbf{D}(\omega)e^{-i(\omega'+\omega)t}d\omega' d\omega.\end{aligned}\tag{13.31}$$

According to (13.24a), we now know that this is actually the time rate of change of energy. The instantaneous distribution of the EM energy density can then be obtained by integrating the latter expression over time:

$$\begin{aligned}\mathcal{W}^{e}(t) &= \int_{-\infty}^{t}\int_{-\infty}^{+\infty}\int_{-\infty}^{+\infty}(-i\omega)\mathbf{E}(\omega')\cdot\mathbf{D}(\omega)e^{-i(\omega'+\omega)t'}d\omega' d\omega dt'\\ &= \int_{-\infty}^{+\infty}\int_{-\infty}^{+\infty}\left[\frac{\omega}{\omega'+\omega}\right]\mathbf{E}(\omega')\cdot\mathbf{D}(\omega)e^{-i(\omega'+\omega)t}d\omega' d\omega,\end{aligned}\tag{13.32}$$

where it has been implicitly assumed EM fields tending sufficiently rapidly to zero as $t' \to -\infty$. This is indeed a crucial assumption that necessarily restricts the applicability of this procedure to the so-called *slowly varying amplitude approximation* [29]. In linear, homogeneous and isotropic media, $\mathbf{D} = \varepsilon_0\varepsilon\mathbf{E}$, so that

$$\mathcal{W}^{e}(t) = \varepsilon_0\int_{-\infty}^{+\infty}\int_{-\infty}^{+\infty}\left[\frac{\omega\varepsilon(\omega)}{\omega'+\omega}\right]\mathbf{E}(\omega')\cdot\mathbf{E}(\omega)e^{-i(\omega'+\omega)t}d\omega' d\omega.\tag{13.33}$$

By regarding monochromatic optical fields, i.e., such that their complex amplitudes are of the form $\mathbf{E}(\omega) = \left[\mathbf{E}_0\delta(\omega-\omega_0) + \mathbf{E}_0^*\delta(\omega+\omega_0)\right]/2$, it can be shown that

$$\begin{aligned}\mathbf{E}(\omega')\cdot\mathbf{E}(\omega) &= \frac{1}{4}\left[\mathbf{E}_0^2\delta(\omega'-\omega_0)\delta(\omega-\omega_0) + (\mathbf{E}_0^*)^2\delta(\omega'+\omega_0)\delta(\omega+\omega_0)\right]\\ &\quad + \frac{1}{4}\left|\mathbf{E}_0\right|^2\left[\delta(\omega'-\omega_0)\delta(\omega+\omega_0) + \delta(\omega'+\omega_0)\delta(\omega-\omega_0)\right].\end{aligned}\tag{13.34}$$

Substituting this into (13.33) the time-averaged electric energy density is given by

$$W^{e} \equiv \langle\mathcal{W}^{e}\rangle_{T} = \frac{\varepsilon_0}{2}\left[\frac{-\omega_0\mathrm{Re}[\varepsilon(\omega_0)]}{\omega_0'-\omega_0}\right]\left|\mathbf{E}_0\right|^2,\tag{13.35}$$

where only the terms in the second line of (13.34) effectively contribute to the time average. Here it is noteworthy to observe that the expression (13.35) turns out to be quite oddly written, as the numerator solely depends on ω_0, and the denominator, for its part, have a double dependency on both ω_0 and ω_0'. This distinction between ω_0' and ω_0 is actually artificial, and it is intended to simplify the calculations. So, in order to deal with this inconsistency, we make the following algebraic manipulation:

$$W^{\mathrm{e}} = \frac{\varepsilon_0}{4}\left[\frac{\omega_0'\mathrm{Re}\big[\varepsilon(\omega_0')\big] - \omega_0\mathrm{Re}[\varepsilon(\omega_0)]}{\omega_0' - \omega_0}\right]|\mathbf{E}_0|^2. \tag{13.36}$$

In this way, ω_0' and ω_0 are treated on equal footing, thus showing much more clearly a singular behavior when $\omega_0' = \omega_0$. To overcome this issue one should take the limit $\omega_0' \to \omega_0$, transforming the above expression into a derivative with respect to ω. By performing the same calculation for the magnetic contribution one obtains that

$$W^{\mathrm{e}} \equiv \langle \mathcal{W}^{\mathrm{e}} \rangle_{\mathrm{T}} = \frac{\varepsilon_0 \tilde{\varepsilon}_{\mathrm{eff}}(\omega)}{4}|\mathbf{E}_0|^2\bigg|_{\omega\to\omega_0}; \tag{13.37a}$$

$$W^{\mathrm{m}} \equiv \langle \mathcal{W}^{\mathrm{m}} \rangle_{\mathrm{T}} = \frac{\mu_0 \tilde{\mu}_{\mathrm{eff}}(\omega)}{4}|\mathbf{H}_0|^2\bigg|_{\omega\to\omega_0}, \tag{13.37b}$$

where $\tilde{\varepsilon}_{\mathrm{eff}}$ and $\tilde{\mu}_{\mathrm{eff}}$ are the real-valued effective material parameters (i.e., the electric permittivity and magnetic permeability), which are defined as

$$\tilde{\varepsilon}_{\mathrm{eff}}(\omega) \equiv \frac{d\left[\omega\varepsilon'\right]}{d\omega}, \qquad \tilde{\mu}_{\mathrm{eff}}(\omega) \equiv \frac{d\left[\omega\mu'\right]}{d\omega}, \tag{13.38}$$

with $\varepsilon' \equiv \mathrm{Re}[\varepsilon(\omega)]$ and $\mu' \equiv \mathrm{Re}[\mu(\omega)]$.

By comparing (13.37a) and (13.37b) with (13.28a) and (13.28b), respectively, one can observe that the differences among them are found in the dispersion-modified material parameters. After some lengthy but straightforward calculations it can be shown that $\tilde{\varepsilon}_{\mathrm{eff}}(\omega) = \varepsilon_{\mathrm{eff}}(\omega)$ and $\tilde{\mu}_{\mathrm{eff}}(\omega) = \mu_{\mathrm{eff}}(\omega)$ if and only if $(\omega^2 + \omega_n^2)\gamma_n^2 = 0$, and $(\omega^2 + \tilde{\omega}_n^2)\tilde{\gamma}_n^2 = 0$. Namely, if $\gamma_n = 0$ and/or $\omega = \pm i\omega_n$, and $\tilde{\gamma}_n = 0$ and/or $\omega = \pm i\tilde{\omega}_n$, for all n. Since ω must be real (otherwise one would deal with, a priori, unphysical imaginary frequencies [64]), the only possible solution is that $\gamma_n = \tilde{\gamma}_n = 0$, i.e., a lossless media. Furthermore, (13.37a) and (13.37b) equal (13.9) if and only if $\varepsilon = \varepsilon'$, $\mu = \mu'$, and they do not depend on frequency (which is certainly redundant on account of the Kramers-Kronig relations, since non-dispersive necessarily implies lossless, and vice versa). Hence, we can state that the sum of (13.28a) and (13.28b), provide ultimately the most general definition of the EM energy density in dispersive and lossy media, since it allows us to recover the more particular expressions successively, just by relaxing further assumptions.

13.3 Generalizing the Conservation Law for the Optical Chirality

As we have just seen, conservation laws, in particular, that for the EM energy, can be straightforwardly treated from its corresponding continuity equation. From this viewpoint, we have shown an insightful manner of identifying the corresponding conserved quantity, the EM energy density. Yet, this task of digging out conserved quantities typically relies upon the search of the underlying symmetry properties of the physical system [75]. Indeed, as established by the Noether's theorem, conserved quantities and symmetries can be regarded as equivalent features [76]. For example, the well-known conservation laws of energy, linear momentum and angular momentum, are actually associated with continuous symmetries of the system, namely, they follows from the invariance under the universal space-time transformations [77]. These theoretical concepts can be mathematically described by means of symmetry groups, which are in turn related to the corresponding physical transformations [78]. Furthermore, drawing on a formalism resembling that often used in quantum mechanics [79], one can deal with the conserved quantities through differential operators representing the generators of the corresponding infinitesimal symmetry transformations. For the above dynamical properties, i.e., the energy, the linear momentum, and the orbital angular momentum, these generators simply involve first derivatives with respect to the space-time coordinates acting on the EM fields, and are given by $\{i\partial_t, i\nabla\}$, for the space-time translations [39], and $i\,(\mathbf{r} \times \nabla)$, for the spatial rotations [40]. As for the optical chirality, it was demonstrated that its conservation is underpinned by $i\,(\partial_t \nabla \times)$ [44], but noticing that it must be applied to the vector potentials [51]. Importantly, these generators allow us to find the corresponding eigenstates associated to the conserved quantities. In this regard, one can find that the plane waves are the eigenstates of the energy-momentum differential operator. Similarly, for the optical chirality, it is found that the associated eigenstates are the circularly polarized plane waves [44].

Noether's theorem, therefore, constitutes a powerful tool for identifying and analyzing conserved quantities from the symmetries of the physical system. However, this approach only holds in the absence of external sources. In the presence of sources (i.e., external charges and/or currents), though, conservation laws and the subsequent identification and analysis of the conserved quantities ought to be addressed through the continuity equations [51, 52]. This would be the case if we wanted to analyze a given dynamical property in a dispersive medium, wherein we also include the presence of absorption losses. Hence, by building upon the sound and self-consistent framework described above for the EM energy conservation law, we are now ready to proceed with the generalization of the conservation law for the optical chirality.

Akin to the EM energy continuity equation, the procedure to derive the most general form of the optical chirality conservation law starts from the definition of the corresponding optical chirality flux density. In this regard, it is worth recalling that the literature concerning optical chirality and its interaction with matter has mostly dealt with EM fields in free space [31–33]. That is why one can often find several expressions for the chirality flux density wherein the fields $\mathcal{B}$ and $\mathcal{H}$ are used interchangeably, yet leading to same results. However, this does not occur in dispersive media, and special care should be taken in dealing with EM fields either in free space ($\mathcal{E}$ and $\mathcal{H}$), or within a medium ($\mathcal{D}$ and $\mathcal{B}$). Be that as it may, for symmetry reasons, one may heuristically assume that the chirality flux density actually reads as

$$\mathcal{F} \equiv \frac{1}{2}\left[\mathcal{E} \times (\nabla \times \mathcal{H}) - \mathcal{H} \times (\nabla \times \mathcal{E})\right]. \tag{13.39}$$

This authorative choice of the definition is in fact that used in [80], and coincides with that originally introduced by Tang and Cohen [31] for EM fields in free space. Following a similar procedure as for the EM energy conservation law, one can readily calculate the divergence of the chirality flux density with the aid of the Maxwell's equations:

$$\begin{aligned} \nabla \cdot \mathcal{F} &= \frac{1}{2}\left[\mathcal{H} \cdot \nabla \times (\nabla \times \mathcal{E}) - \mathcal{E} \cdot \nabla \times (\nabla \times \mathcal{H})\right] \\ &= -\frac{1}{2}\left[\mathcal{H} \cdot \partial_t (\nabla \times \mathcal{B}) + \mathcal{E} \cdot \partial_t (\nabla \times \mathcal{D}) + \mathcal{E} \cdot (\nabla \times \mathcal{J})\right]. \end{aligned} \tag{13.40}$$

To find out the conserved quantity and its accompanying source-like term, one should compare the latter expression with the mathematical structure of the continuity equation as given in (13.22), and identify the total time derivative operator. To this aim, it follows that

$$\mathcal{E} \cdot \partial_t (\nabla \times \mathcal{D}) = \partial_t \left[\mathcal{E} \cdot (\nabla \times \mathcal{D})\right] - \partial_t \mathcal{E} \cdot (\nabla \times \mathcal{D}); \tag{13.41a}$$

$$\mathcal{H} \cdot \partial_t (\nabla \times \mathcal{B}) = \partial_t \left[\mathcal{H} \cdot (\nabla \times \mathcal{B})\right] - \partial_t \mathcal{H} \cdot (\nabla \times \mathcal{B}). \tag{13.41b}$$

In this manner, (13.40) can be recast as a true continuity equation, i.e.,

$$\nabla \cdot \mathcal{F} + \partial_t C = \mathcal{S}, \tag{13.42}$$

where the optical chirality density and the source-like terms are defined as

$$C \equiv \frac{1}{2}\left[\mathcal{E} \cdot (\nabla \times \mathcal{D}) + \mathcal{H} \cdot (\nabla \times \mathcal{B})\right], \tag{13.43}$$

$$\mathcal{S} \equiv \frac{1}{2}\left[\partial_t \mathcal{E} \cdot (\nabla \times \mathcal{D}) + \partial_t \mathcal{H} \cdot (\nabla \times \mathcal{B}) - \mathcal{E} \cdot (\nabla \times \mathcal{J})\right]. \tag{13.44}$$

It is worth pointing out that the above expressions represent the most general result for the optical chirality conservation law, without any restrictions on the nature of the medium. Nonetheless, they differ considerably from the previously established for EM fields in free space [17, 31–33, 44, 77, 80]:

$$\nabla \cdot \mathcal{F}_{\text{vacuum}} + \partial_t C_{\text{vacuum}} = \mathcal{S}_{\text{vacuum}}, \tag{13.45}$$

where $\mathcal{F}_{\text{vacuum}} \equiv \mathcal{F}$, but $C_{\text{vacuum}} \equiv \left[\varepsilon_0 \mathcal{E} \cdot (\nabla \times \mathcal{E}) + \mu_0 \mathcal{H} \cdot (\nabla \times \mathcal{H})\right]/2$, i.e., that given in (13.1), and $\mathcal{S}_{\text{vacuum}} \equiv -\left[\mathcal{J} \cdot (\nabla \times \mathcal{E}) + \mathcal{E} \cdot (\nabla \times \mathcal{J})\right]/2$ being, respectively, the chirality flux density, the optical chirality density, and the source-like term in free space. Curiously, albeit (13.45) was initially posed for optical fields in the vacuum, it has been widely used for investigating chiroptical effects occurring in material systems, including metals (or plasmonic structures) as well as metamaterials. However, it can be demonstrated that the general result given in (13.42) reduces to (13.45) only for linear and lossless media, i.e., when $\mathcal{D} = \varepsilon_0 \mathcal{E}$ and $\mathcal{B} = \mu_0 \mathcal{H}$, and assuming $\varepsilon = \mu = 1$:

$$\begin{aligned} C &= \frac{1}{2}\left[\mathcal{E} \cdot (\nabla \times \mathcal{D}) + \mathcal{H} \cdot (\nabla \times \mathcal{B})\right] \\ &= \frac{1}{2}\left[\varepsilon_0 \mathcal{E} \cdot (\nabla \times \mathcal{E}) + \mu_0 \mathcal{H} \cdot (\nabla \times \mathcal{H})\right] = C_{\text{vacuum}}, \end{aligned} \tag{13.46}$$

and

$$\begin{aligned} \mathcal{S} &= \frac{1}{2}\left[\partial_t \mathcal{E} \cdot (\nabla \times \mathcal{D}) + \partial_t \mathcal{H} \cdot (\nabla \times \mathcal{B}) - \mathcal{E} \cdot (\nabla \times \mathcal{J})\right] \\ &= \frac{1}{2}\left[\varepsilon_0 \partial_t \mathcal{E} \cdot (-\mu_0 \partial_t \mathcal{H}) + \mu_0 \partial_t \mathcal{H} \cdot (\varepsilon_0 \partial_t \mathcal{E} + \mathcal{J}) - \mathcal{E} \cdot (\nabla \times \mathcal{J})\right] \\ &= \frac{1}{2}\left[\mu_0 \partial_t \mathcal{H} \cdot \mathcal{J} - \mathcal{E} \cdot (\nabla \times \mathcal{J})\right] \\ &= -\frac{1}{2}\left[\mathcal{J} \cdot (\nabla \times \mathcal{E}) + \mathcal{E} \cdot (\nabla \times \mathcal{J})\right] = \mathcal{S}_{\text{vacuum}}, \end{aligned} \tag{13.47}$$

where $\mu_0 \partial_t \mathcal{H} = -(\nabla \times \mathcal{E})$. As one can see at a glance, the essential discrepancy arises on account of the dispersion-related terms. This becomes much more evident by rewriting the above expressions in terms of those for vacuum. Indeed, since

$$\mathcal{E} \cdot (\nabla \times \mathcal{D}) = \mathcal{D} \cdot (\nabla \times \mathcal{E}) + \nabla \cdot (\mathcal{P} \times \mathcal{E}); \tag{13.48a}$$

$$\mathcal{H} \cdot (\nabla \times \mathcal{B}) = \mathcal{B} \cdot (\nabla \times \mathcal{H}) + \mu_0 \nabla \cdot (\mathcal{M} \times \mathcal{H}), \tag{13.48b}$$

the optical chirality density as given in (13.43) can be generally expressed as

$$\begin{aligned} C &= \frac{1}{2}\left\{\mathcal{D} \cdot (\nabla \times \mathcal{E}) + \mathcal{B} \cdot (\nabla \times \mathcal{H}) + \nabla \cdot \left[\mathcal{P} \times \mathcal{E} + \mu_0 (\mathcal{M} \times \mathcal{H})\right]\right\} \\ &= \frac{1}{2}\left\{\varepsilon_0 \mathcal{E} \cdot (\nabla \times \mathcal{E}) + \mathcal{E} \cdot (\nabla \times \mathcal{P}) + \mu_0 \mathcal{H} \cdot (\nabla \times \mathcal{H}) + \mu_0 \mathcal{H} \cdot (\nabla \times \mathcal{M})\right\} \\ &= C_{\text{vacuum}} + C_{\text{medium}} = C_{\text{vacuum}} + C^{\text{e}}_{\text{medium}} + C^{\text{m}}_{\text{medium}}, \end{aligned} \tag{13.49}$$

where the free-space contribution C_{vacuum}, i.e., the original expression as given in (13.1), has been separated from the one accounting for the medium contribution, $C_{\text{medium}} \equiv \left[\mathcal{E} \cdot (\nabla \times \mathcal{P}) + \mu_0 \mathcal{H} \cdot (\nabla \times \mathcal{M})\right]/2$. Similarly, the source-like term can also be recast as follows

$$\mathcal{S} = \mathcal{S}_{\text{vacuum}} + \mathcal{S}_{\text{medium}} = \mathcal{S}_{\text{vacuum}} + \mathcal{S}^{\text{e}}_{\text{medium}} + \mathcal{S}^{\text{m}}_{\text{medium}}, \tag{13.50}$$

where $\mathcal{S}_{\text{vacuum}}$ is the free-space contribution as given in (13.47), and $\mathcal{S}^{\text{e/m}}_{\text{medium}}$ stand for the electric and magnetic components due to the medium contribution:

$$\mathcal{S}^{\text{e}}_{\text{medium}} = \frac{1}{2}\left\{\partial_t \mathcal{E} \cdot (\nabla \times \mathcal{P}) - \partial_t \mathcal{P} \cdot (\nabla \times \mathcal{E})\right\}; \tag{13.51a}$$

$$\mathcal{S}^{\text{m}}_{\text{medium}} = \frac{\mu_0}{2}\left\{\partial_t \mathcal{H} \cdot (\nabla \times \mathcal{M}) - \partial_t \mathcal{M} \cdot (\nabla \times \mathcal{H})\right\}. \tag{13.51b}$$

Therefore, the material systems brings about important corrections into the optical chirality conservation law [compare (13.42) with (13.45)], and consequently into the original expressions for the optical chirality density [compare (13.43) with (13.46)] and its associated source-like contribution [compare (13.44) with (13.47)]. Of course, these considerations should not be disregarded, and will be the subject under which we focus on the rest of this chapter.

13.4 Optical Chirality Density in Linear Dispersive Media

So far, we have shown that the conservation law for the optical chirality established up to now, i.e., that expressed in (13.45) (see, e.g., [17, 31–33, 44, 77, 80]), is only valid for EM waves in free space. Starting from the definition of the optical chirality flux density as given in (13.39), we have provided a detailed derivation showing that there are also additional terms accounting for the presence of a material system, which have often been neglected. Noteworthily, their dispersion characteristics lead to important corrections into the original expressions of the optical chirality density as well as the source-like terms appearing in the continuity equation, that should be carefully considered. These kinds of theoretical implications have already been demonstrated in other works generalizing the EM energy, the linear momentum, the orbital and the spin angular momenta [39–41], and the EM helicity [42], to dispersive media. Likewise, in this section, we shall perform an alternative derivation for the optical chirality density in dispersive media, considering both the lossless and the lossy approaches. Specifically, as for the lossless case, the derivation will be analogous to the classical procedure leading to the Brillouin formula for the EM energy density in dispersive media [30, 65], i.e., just involving Fourier integrals [29]. On the other side, the corresponding expression for the optical chirality density in dispersive media including the medium's dissipation, will be tackled by following a similar approach as already provided by Loudon [54] (and further extended by Ruppin [55]) for the EM energy. Herein, it is worth reminding that, just like for the EM energy case outlined above, these results will be valid as long as the material parameters (i.e., the electric permittivity and the magnetic permeability) can be properly fitted by Lorentzian line shapes. In this regard, the multi-resonant models for both the permittivity, (13.12), and for the permeability, (13.19), have proven to fit very well with experiments [72–74], and thus, can be regarded as absolutely general for characterizing the optical response of whatever material system, for any frequency and bandwidth.

13.4.1 Optical Chirality Density in Dispersive and Lossless Media: Brillouin's Approach

By regarding the original definition of the optical chirality density in the time-dependent representation, it is easy to show that, for monochromatic electric and magnetic fields in free space, $\mathcal{E}(\mathbf{r}, t) = \mathrm{Re}[\mathbf{E}(\mathbf{r})e^{-i\omega t}]$ and $\mathcal{H}(\mathbf{r}, t) = \mathrm{Re}[\mathbf{H}(\mathbf{r})e^{-i\omega t}]$, the time-averaged optical chirality density is given by [31, 32]:

$$C_{\text{vacuum}} \equiv \langle C_{\text{vacuum}} \rangle_{\mathrm{T}} = \frac{\omega}{2c^2}\mathrm{Im}\big[\mathbf{E} \cdot \mathbf{H}^*\big]. \tag{13.52}$$

From this, it is straightforward to verify that the maximum value of C is achieved when assuming freely propagating EM plane waves with circular polarization (CP), i.e., corresponding to the eigenstates associated to the optical chirality [44]:

$$C^{(\pm)\mathrm{CP}}_{\text{vacuum}} = \pm\frac{\omega}{2c^2}\frac{1}{Z_0}|\mathbf{E}|^2, \tag{13.53}$$

where $Z_0 \equiv \sqrt{\mu_0/\varepsilon_0}$ is the vacuum impedance, and the signs + and − correspond to the left- and right-handed circular polarizations.

Now, the aim of this section is finding out a closed expression generalizing the above to dispersive media in the lossless case. For simplicity, hereinafter it will be assumed a linear, homogeneous and isotropic medium such that $\mathbf{D} = \varepsilon_0\varepsilon(\omega)\mathbf{E}$ and $\mathbf{B} = \mu_0\mu(\omega)\mathbf{H}$. Then, starting from the continuity equation as given in (13.40), and taking into account the above discussion, let us expand the right-hand side by using the Fourier integrals:

$$\begin{aligned}\frac{1}{2}\mathcal{E} \cdot \partial_t\left(\nabla \times \mathcal{D}\right) &= \frac{1}{2}\int_{-\infty}^{+\infty} \mathbf{E}(\omega')e^{-i\omega' t}d\omega' \cdot \partial_t\left\{\nabla \times \left[\int_{-\infty}^{+\infty} \mathbf{D}(\omega)e^{-i\omega t}d\omega\right]\right\}\\ &= \frac{-i}{2}\int_{-\infty}^{+\infty}\int_{-\infty}^{+\infty} \omega\mathbf{E}(\omega') \cdot \left[\nabla \times \mathbf{D}(\omega)\right] e^{-i(\omega'+\omega)t}d\omega' d\omega;\end{aligned} \tag{13.54a}$$

$$\begin{aligned}\frac{1}{2}\mathcal{H} \cdot \partial_t\left(\nabla \times \mathcal{B}\right) &= \frac{1}{2}\int_{-\infty}^{+\infty} \mathbf{H}(\omega')e^{-i\omega' t}d\omega' \cdot \partial_t\left\{\nabla \times \left[\int_{-\infty}^{+\infty} \mathbf{B}(\omega)e^{-i\omega t}d\omega\right]\right\}\\ &= \frac{-i}{2}\int_{-\infty}^{+\infty}\int_{-\infty}^{+\infty} \omega\mathbf{H}(\omega') \cdot \left[\nabla \times \mathbf{B}(\omega)\right] e^{-i(\omega'+\omega)t}d\omega' d\omega.\end{aligned} \tag{13.54b}$$

According to the continuity equation (13.42), this is actually the time rate of change of optical chirality. Still, one might guess that the electric and magnetic contributions to the optical chirality density are indeed encoded in such a way that $\mathcal{E} \cdot \partial_t (\nabla \times \mathcal{D}) \to \partial_t C^{\mathrm{e}}$ and $\mathcal{H} \cdot \partial_t (\nabla \times \mathcal{B}) \to \partial_t C^{\mathrm{m}}$, respectively. Therefore, the instantaneous distribution of the electric and magnetic contributions to the optical chirality density can then be obtained by integrating the latter expression over time:

$$\begin{aligned} C^{\mathrm{e}}(t) &= \frac{-i}{2} \int_{-\infty}^{t} \int_{-\infty}^{+\infty} \int_{-\infty}^{+\infty} \omega \mathbf{E}(\omega') \cdot [\nabla \times \mathbf{D}(\omega)]\, e^{-i(\omega'+\omega)t'} d\omega' d\omega dt' \\ &= \frac{1}{2} \int_{-\infty}^{+\infty} \int_{-\infty}^{+\infty} \left[\frac{\omega}{\omega' + \omega}\right] \mathbf{E}(\omega') \cdot [\nabla \times \mathbf{D}(\omega)]\, e^{-i(\omega'+\omega)t} d\omega' d\omega; \end{aligned} \tag{13.55a}$$

$$\begin{aligned} C^{\mathrm{m}}(t) &= \frac{-i}{2} \int_{-\infty}^{t} \int_{-\infty}^{+\infty} \int_{-\infty}^{+\infty} \omega \mathbf{H}(\omega') \cdot [\nabla \times \mathbf{B}(\omega)]\, e^{-i(\omega'+\omega)t'} d\omega' d\omega dt' \\ &= \frac{1}{2} \int_{-\infty}^{+\infty} \int_{-\infty}^{+\infty} \left[\frac{\omega}{\omega' + \omega}\right] \mathbf{H}(\omega') \cdot [\nabla \times \mathbf{B}(\omega)]\, e^{-i(\omega'+\omega)t} d\omega' d\omega. \end{aligned} \tag{13.55b}$$

Notice that the integral convergence is constrained by the *slowly varying amplitude approximation* [29], i.e., assuming that the EM fields tend sufficiently rapidly to zero as $t' \to -\infty$. In linear, homogeneous and isotropic media, $\nabla \times \mathbf{D} = i\omega\varepsilon_0\mu_0\varepsilon\mu\mathbf{H}$, and $\nabla \times \mathbf{B} = -i\omega\varepsilon_0\mu_0\varepsilon\mu\mathbf{E} + \mu_0\mu\mathbf{J}$, so that

$$C^{\mathrm{e}}(t) = \frac{i}{2c^2} \int_{-\infty}^{+\infty} \int_{-\infty}^{+\infty} \left[\frac{\omega^2 \varepsilon(\omega)\mu(\omega)}{\omega' + \omega}\right] \mathbf{E}(\omega') \cdot \mathbf{H}(\omega) e^{-i(\omega'+\omega)t} d\omega' d\omega; \tag{13.56a}$$

$$\begin{aligned} C^{\mathrm{m}}(t) &= \frac{-i}{2c^2} \int_{-\infty}^{+\infty} \int_{-\infty}^{+\infty} \left[\frac{\omega^2 \varepsilon(\omega)\mu(\omega)}{\omega' + \omega}\right] \mathbf{E}(\omega) \cdot \mathbf{H}(\omega') e^{-i(\omega'+\omega)t} d\omega' d\omega \\ &+ \frac{\mu_0}{2} \int_{-\infty}^{+\infty} \int_{-\infty}^{+\infty} \left[\frac{\omega\sigma(\omega)\mu(\omega)}{\omega' + \omega}\right] \mathbf{E}(\omega) \cdot \mathbf{H}(\omega') e^{-i(\omega'+\omega)t} d\omega' d\omega, \end{aligned} \tag{13.56b}$$

where $\sigma(\omega) = i\varepsilon_0\omega\,[1 - \varepsilon(\omega)]$ is the complex-valued electric conductivity. Therefore, by summing both the electric and magnetic contributions it follows that

$$C(t) \equiv C^{\mathrm{e}} + C^{\mathrm{m}} = \frac{i}{2c^2} \int_{-\infty}^{+\infty} \int_{-\infty}^{+\infty} \left[\frac{\omega^2 \varepsilon(\omega)\mu(\omega)\Xi(\omega', \omega) e^{-i(\omega'+\omega)t}}{\omega' + \omega}\right] d\omega' d\omega, \tag{13.57}$$

where $\Xi(\omega', \omega) = \mathbf{E}(\omega') \cdot \mathbf{H}(\omega) - \mathbf{E}(\omega) \cdot \mathbf{H}(\omega')$. It should be noted that the above expression only accounts for the first term of the right-hand side of (13.56b); the second one involves a current-like contribution, and is skipped for the moment. In this way, just by regarding monochromatic optical fields, it can be shown that

$$\Xi_{\rm av}(\omega', \omega) = \frac{i}{2}{\rm Im}\big[\mathbf{E}^* \cdot \mathbf{H}\big]\big[\delta(\omega - \omega_0)\delta(\omega' + \omega_0) - \delta(\omega + \omega_0)\delta(\omega' - \omega_0)\big], \tag{13.58}$$

where there are only outlined the terms contributing to the time average. Hence, substituting this into (13.57) one find that the time-averaged optical chirality density is given by

$$C \equiv \langle C^{\rm e} + C^{\rm m} \rangle_{\rm T} = \frac{1}{2c^2}{\rm Re}\left[\frac{\omega_0^2 \varepsilon(\omega_0)\mu(\omega_0)}{\omega_0' - \omega_0}\right]{\rm Im}\big[\mathbf{E}^* \cdot \mathbf{H}\big], \tag{13.59}$$

where it has been used that $\varepsilon(-\omega) = \varepsilon^*(\omega)$ and $\mu(-\omega) = \mu^*(\omega)$. Similarly to that already pointed out in (13.35), this expression is oddly written, since the numerator only involves ω_0, whereas the denominator also brings in the corresponding primed frequencies. This drawback can be readily overcome by means of an algebraic symmetrization procedure, in which the expression is split into two parts interchanging $\omega_0 \leftrightarrow \omega_0'$, to finally add them up together:

$$C = \frac{1}{4c^2}{\rm Re}\left[\frac{\omega_0'^2 \varepsilon(\omega_0')\mu(\omega_0') - \omega_0^2 \varepsilon(\omega_0)\mu(\omega_0)}{\omega_0' - \omega_0}\right]{\rm Im}\big[\mathbf{E} \cdot \mathbf{H}^*\big]. \tag{13.60}$$

Akin to the energy density in (13.36), this expression exhibits a singularity at $\omega_0' = \omega_0$. Therefore, taking the limit $\omega_0' \to \omega_0$, it can be nicely expressed in a compact form as a derivative with respect to ω:

$$C_{\rm lossless} = \frac{1}{4c^2}{\rm Re}\left[\frac{d\left[\omega^2 \varepsilon(\omega)\mu(\omega)\right]}{d\omega}\right]{\rm Im}\big[\mathbf{E} \cdot \mathbf{H}^*\big]. \tag{13.61}$$

A more elegant form for expressing the above result may be made as follows

$$C_{\rm lossless} = {\rm Re}\big[n(\omega)\tilde{n}(\omega)\big]C_{\rm vacuum} = \frac{\omega}{2}\frac{{\rm Im}[\mathbf{E} \cdot \mathbf{H}^*]}{v_p(\omega)v_g(\omega)}, \tag{13.62}$$

where $v_p(\omega) \equiv c/{\rm Re}[n(\omega)]$ and $v_g(\omega) \equiv c/{\rm Re}[\tilde{n}(\omega)]$, are the phase and group velocities [81, 82], respectively, which can be in turn expressed in terms of the phase refractive index, $n(\omega) = \sqrt{\varepsilon\mu}$, and the corresponding dispersion-modified group refractive index, $\tilde{n}(\omega) \equiv n(\omega) + \omega\,[\partial n(\omega)/\partial\omega]$ [41, 42]. It should be noted that the same expression for the optical chirality density in dispersive media has also been obtained but using a more complicated formalism (see equation (33) in [44]). At any rate, it is easy to prove that this definition, as given in (13.61) or (13.62), reduces to the original one for freely propagating optical fields, (13.52), when $n = 1$, as expected. In addition, this simple relation provides important insights on account of the dependence on the dispersion-related phase and group velocities. Indeed, it is easy to realize that the stored optical chirality density may be naturally enhanced in artificially engineered materials just by lowering both velocities. This is specifically accomplished in the vicinity of the resonance frequency, i.e., in the anomalous dispersion region, wherein the real part of the permittivity decreases abruptly. Notwithstanding, in a dispersive and lossy media, there are, incidentally, some spectral ranges where the precise physical meaning of the group velocity turns out to be somewhat unclear [30, 65]. Let's see how to proceed on this matter.

13.4.2 Optical Chirality Density in Dispersive and Lossy Media: Loudon's Approach

According to the Kramers-Kronig relations [28], a physically realistic description of dynamical properties in dispersive media would require careful considerations of dissipative effects. As outlined above, this is very well known for the EM field energy in metals, for which a general treatment has been developed [27, 29]. Herein we shall do the same for the optical chirality density, thereby incorporating properly these aspects to finally get a general expression. Starting from the continuity equation as given in (13.40), we now expand both the electric and magnetic contributions as follows:

$$\frac{1}{2}\mathcal{E}\cdot\partial_t\left(\nabla\times\mathcal{D}\right)=\frac{1}{2}\left[\varepsilon_0\mathcal{E}\cdot\partial_t\left(\nabla\times\mathcal{E}\right)+\mathcal{E}\cdot\partial_t\left(\nabla\times\mathcal{P}\right)\right]; \tag{13.63a}$$

$$\frac{1}{2}\mathcal{H}\cdot\partial_t\left(\nabla\times\mathcal{B}\right)=\frac{\mu_0}{2}\left[\mathcal{H}\cdot\partial_t\left(\nabla\times\mathcal{H}\right)+\mathcal{H}\cdot\partial_t\left(\nabla\times\mathcal{M}\right)\right]. \tag{13.63b}$$

In these expressions each of the first term of the right-hand side can be recast as

$$\frac{\varepsilon_0}{2}\mathcal{E}\cdot\partial_t\left(\nabla\times\mathcal{E}\right)=\frac{\varepsilon_0}{2}\left\{\partial_t\left[\mathcal{E}\cdot\left(\nabla\times\mathcal{E}\right)\right]-\partial_t\mathcal{E}\cdot\left(\nabla\times\mathcal{E}\right)\right\}; \tag{13.64a}$$

$$\frac{\mu_0}{2}\mathcal{H}\cdot\partial_t\left(\nabla\times\mathcal{H}\right)=\frac{\mu_0}{2}\left\{\partial_t\left[\mathcal{H}\cdot\left(\nabla\times\mathcal{H}\right)\right]-\partial_t\mathcal{H}\cdot\left(\nabla\times\mathcal{H}\right)\right\}, \tag{13.64b}$$

leading to a total time derivative plus a residual term for each case. As shown below, these residual terms exactly cancel to each other in the vacuum, thus recovering the original expression for the optical chirality in free space. The second term in the right-hand side of (13.63a) and (13.63b) are to be addressed from a material standpoint by means of the dynamic equations for the polarization and the magnetization fields:

$$\frac{1}{2}\mathcal{E}\cdot\partial_t\left(\nabla\times\mathcal{P}\right)=\partial_t\left\{\sum_{n=0}^{N-1}\frac{\partial_t\langle\mathcal{P}_n\rangle\cdot(\nabla\times\partial_t\langle\mathcal{P}_n\rangle)+\omega_n^2\langle\mathcal{P}_n\rangle\cdot(\nabla\times\langle\mathcal{P}_n\rangle)}{2\varepsilon_0 f_n\omega_{\mathrm{p}}^2}\right\}$$
$$-\sum_{n=0}^{N-1}\frac{\omega_n^2\partial_t\langle\mathcal{P}_n\rangle\cdot(\nabla\times\langle\mathcal{P}_n\rangle)+\partial_t\langle\mathcal{P}_n\rangle\cdot\left(\nabla\times\partial_t^2\langle\mathcal{P}_n\rangle\right)}{2\varepsilon_0 f_n\omega_{\mathrm{p}}^2}$$
$$+\sum_{n=0}^{N-1}\frac{\gamma_n\partial_t\langle\mathcal{P}_n\rangle\cdot(\nabla\times\partial_t\langle\mathcal{P}_n\rangle)}{2\varepsilon_0 f_n\omega_{\mathrm{p}}^2}; \tag{13.65a}$$

$$\frac{1}{2}\mathcal{H}\cdot\partial_t\left(\nabla\times\mathcal{M}\right)=\partial_t\left\{\sum_{n=0}^{N-1}\frac{\partial_t\langle\mathcal{M}_n\rangle\cdot(\nabla\times\partial_t\langle\mathcal{M}_n\rangle)+\tilde{\omega}_n^2\langle\mathcal{M}_n\rangle\cdot(\nabla\times\langle\mathcal{M}_n\rangle)}{2\tilde{f}_n\tilde{\omega}_n^2}\right\}$$
$$-\sum_{n=0}^{N-1}\frac{\tilde{\omega}_n^2\partial_t\langle\mathcal{M}_n\rangle\cdot(\nabla\times\langle\mathcal{M}_n\rangle)+\partial_t\langle\mathcal{M}_n\rangle\cdot\left(\nabla\times\partial_t^2\langle\mathcal{M}_n\rangle\right)}{2\tilde{f}_n\tilde{\omega}_n^2}$$
$$+\sum_{n=0}^{N-1}\frac{\tilde{\gamma}_n\partial_t\langle\mathcal{M}_n\rangle\cdot(\nabla\times\partial_t\langle\mathcal{M}_n\rangle)}{\tilde{f}_n\tilde{\omega}_n^2}. \tag{13.65b}$$

Then, taking into account the structure of the continuity equation, (13.22), one can readily identify the electric and magnetic contributions of the optical chirality density stored either by the fields or the medium, as well as the source-like terms accounting for the loss (or gain) rate of the chirality [51, 52]:

$$\frac{1}{2}\mathcal{E}\cdot\partial_t\left(\nabla\times\mathcal{D}\right)=\partial_t\left[C^{\mathrm{e}}_{\mathrm{vacuum}}+C^{\mathrm{e}}_{\mathrm{medium}}\right]+\mathcal{L}^{\mathrm{e}}; \tag{13.66a}$$
$$\frac{1}{2}\mathcal{H}\cdot\partial_t\left(\nabla\times\mathcal{B}\right)=\partial_t\left[C^{\mathrm{m}}_{\mathrm{vacuum}}+C^{\mathrm{m}}_{\mathrm{medium}}\right]+\mathcal{L}^{\mathrm{m}}. \tag{13.66b}$$

In order to compare these results with those obtained via the Fourier transform, we should calculate the corresponding time average of the optical chirality density by considering time harmonic fields in a linear medium. Therefore, from (13.64a), (13.65a), (13.64b) and (13.65b), and with the aid of (13.15) and (13.18), it can be demonstrated that

$$C^{\mathrm{e}}\equiv\langle C^{\mathrm{e}}_{\mathrm{vacuum}}+C^{\mathrm{e}}_{\mathrm{medium}}\rangle_{\mathrm{T}}=\frac{\omega}{4c^2}\varepsilon_{\mathrm{eff}}(\omega)\mathrm{Im}\left[\mu^*(\omega)\mathbf{E}\cdot\mathbf{H}^*\right]; \tag{13.67a}$$
$$C^{\mathrm{m}}\equiv\langle C^{\mathrm{m}}_{\mathrm{vacuum}}+C^{\mathrm{m}}_{\mathrm{medium}}\rangle_{\mathrm{T}}=\frac{\omega}{4c^2}\mu_{\mathrm{eff}}(\omega)\mathrm{Im}\left[\varepsilon(\omega)\mathbf{E}\cdot\mathbf{H}^*\right], \tag{13.67b}$$

where $\varepsilon_{\mathrm{eff}}$ and μ_{eff} are the real-valued effective material parameters defined in (13.29a) and (13.29b), respectively. Hence, by summing both the electric and magnetic contributions shown above, one finally gets the generalization of the optical chirality density to any arbitrary optical media [46]:

$$C_{\mathrm{lossy}}=\frac{\omega}{4c^2}\mathrm{Im}\left[\left(\varepsilon(\omega)\mu_{\mathrm{eff}}(\omega)+\varepsilon_{\mathrm{eff}}(\omega)\mu^*(\omega)\right)\mathbf{E}\cdot\mathbf{H}^*\right]. \tag{13.68}$$

It should be noticed that, in the magnetic contribution, there would also be an additional term associated with the current density **J**:

$$C^{c}_{\text{lossy}} \equiv \langle C^{c}_{\text{lossy}-\text{medium}} \rangle_{\text{T}} = \frac{\omega}{4c^2} \mu_{\text{eff}}(\omega) \text{Im}\big[(\varepsilon(\omega) - 1)\mathbf{E} \cdot \mathbf{H}^*\big]. \tag{13.69}$$

In the lossless case, the corresponding term accounting for this current-related contribution is included in the second line of (13.56b), that would lead to the following

$$C^{c}_{\text{lossless}} \equiv \langle C^{c}_{\text{lossless}-\text{medium}} \rangle_{\text{T}} = \frac{1}{8c^2} \text{Im}\left[\frac{d\left[\omega^2 \left(\varepsilon(\omega) - 1\right) \mu(\omega)\right]}{d\omega} \mathbf{E} \cdot \mathbf{H}^*\right]. \tag{13.70}$$

This way, these terms allow one to complete the optical chirality conservation law:

$$\nabla \cdot \mathcal{F} + \partial_t \left[C^{e}_{\text{vacuum+medium}} + C^{m}_{\text{vacuum+medium}} + C^{c}_{\text{medium}}\right] = -\left[\mathcal{L}^{e} + \mathcal{L}^{m} + \mathcal{L}^{c}\right], \tag{13.71}$$

where $\mathcal{L}^{c} = \mathcal{E} \cdot (\nabla \times \mathcal{J})/2$.

It is worth remarking that both developments for generalizing the optical chirality density to dispersive lossless and lossy media, actually relies on the optical chirality conservation law. Accordingly, the efforts were specially undertaken in order to express the right-hand side of (13.40) in terms of total time derivatives. For simplicity, we followed a similar treatment as the one already employed for the derivation of the energy density in dispersive and lossless media [30, 65], and for lossy media [54, 55]. However, in comparison with the energy case, the optical chirality exhibits many curl-like terms, thus hindering the mathematics. Therefore, special care must be taken with the residual terms so as to avoid misleading outcomes. In this regard, it should be noted that, when combining the electric and magnetic contributions, given in (13.63a) and (13.63b), respectively, one directly gets (13.64a) and (13.64b), provided that the EM fields are assumed to propagate in free space, i.e., so that $\mathcal{P} = \mathcal{M} = 0$. In such a case, it is straightforward to check that the residual terms, $\partial_t \mathcal{E} \cdot (\nabla \times \mathcal{E})$ and $\partial_t \mathcal{H} \cdot (\nabla \times \mathcal{H})$, cancel each other, so that

$$\frac{1}{2}\left[\mathcal{E} \cdot \partial_t (\nabla \times \mathcal{D}) + \mathcal{H} \cdot \partial_t (\nabla \times \mathcal{B})\right] = \partial_t C_{\text{vacuum}},$$

thus allowing us to recover the original expression for the optical chirality in free space. Yet, looking at the electric and magnetic contributions to the optical chirality density, (13.66a) and (13.66b), these residual terms, along with the corresponding ones accounting for the material contribution (arising from the second and third summation in (13.65a) and (13.65b)), should be generally interpreted as the loss rate.

Equation (13.68) is the main result of this chapter. It provides the most general definition for the optical chirality density in dispersive and lossy media, i.e., it is valid for EM fields with arbitrary time dependence, and it is applicable to any material system, including dielectrics, semiconductors, as well as highly dispersive and lossy media, such as plasmonic nanostructures, and metamaterials [1]. It is important to realize that this expression differs significantly from the standard formula for fields in free space [31, 32], and even from other derivations claiming to work out the optical chirality in dispersive media [44, 80]. The main difference essentially raises on account of considering properly the dynamic response of the time-dependent EM fields within a dispersive medium. In this regard, it should be noted that the time derivative of the fields $\mathcal{D}$ and $\mathcal{B}$, must be expressed as convolution integrals in the time domain [28–30], as pointed out in (13.11a) and (13.11b). Notwithstanding the foregoing, departing from this general expression (13.68), it is, of course, possible to particularize it to the lossless case (13.62), as well as to the original definition for free-space EM fields (13.52), just by relaxing successively the corresponding conditions, i.e., by imposing $\gamma_n = \tilde{\gamma}_n = 0$, and $\varepsilon(\omega) = \varepsilon$ and $\mu(\omega) = \mu$, respectively. Hence, it can be concluded that this approach, relying upon the underlying mathematical structure of the continuity equation, leads to a sound and self-consistent definition for the optical chirality density in dispersive and lossy media [46, 48].

13.4.3 Brillouin's Approach Vs Loudon's Approach

The classical approaches put forward by Brillouin and Landau enable one to obtain a closed expression for the EM energy density, (13.37a) and (13.37b), and likewise for the optical chirality density, (13.62), which are only valid under the *slowly varying amplitude approximation* [29], i.e., in a relatively narrow frequency range where the effects of material absorption can be considered as negligible. Therefore, those expressions are suitable for describing such dynamical properties solely in lossless dispersive media. In order to further include the dissipation, one should perform a careful analysis from a material standpoint. This can be carried out by means of the corresponding dynamic equations characterizing the polarization and the magnetization fields, thereby introducing additional terms that must be explicitly accounted for. By doing so, one finally is able to find the most general definition for the EM energy density, (13.28a) and (13.28b), and the optical chirality density (13.68), thus capturing analytically the effects due to dispersion, as well as the absorption, for a complete and proper description of chiroptical interaction between light and matter.

Naturally, both approaches yield different behaviors, see Fig. 13.1. Indeed, assuming a nonmagnetic medium (i.e., such that $\mu = 1$) whose permittivity is described by a single Lorentz pole with $\omega_p = \omega_0$, one can observe that, whereas C_{lossless} displays both positive and negative values, the general expression C_{lossy} remains always positive, with a minimum value of C_{vacuum}, that is reached in the high-frequency limit. Importantly, the largest discrepancies are occurring close to the resonance frequency, as expected. Still, the peaks for both approaches are almost equal in absolute value. These signatures are actually better appreciated in Fig. 13.2, where we compare the dispersion-dependent features of the optical chirality density in both approaches, the lossless and lossy cases, for silver and silicon. Both materials have been modeled using (13.12) with parameters taken from [72] and [73] (see the specific values listed

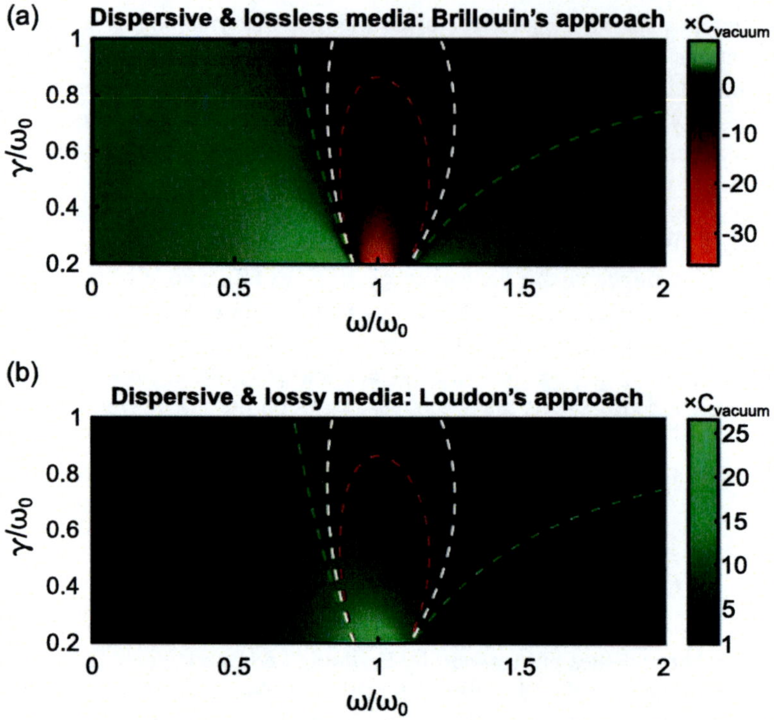

Fig. 13.1 Optical chirality density in **a** lossless and **b** lossy dispersive media. Material parameters correspond to a nonmagnetic medium ($\mu = 1$), with ε being described by a single Lorentz pole with $\omega_p = \omega_0$. Red, white and green dashed lines indicate the curves where the optical chirality in the lossless case is $-C_{\text{vacuum}}$, 0, and C_{vacuum}, respectively

in Table 13.1). From these results it is shown that C_{lossless} overlaps almost exactly with C_{lossy} for all frequencies, except in the vicinity of the region of anomalous dispersion, i.e., where $d\varepsilon'/d\omega < 0$. Therein, the curves drastically separate from each other, so highlighting the importance of considering dissipative effects. This fact should therefore be carefully accounted for and reexamined in experiments considering chiroptical interaction between light and dispersive media such as metamaterials or plasmonic systems.

To complete this analysis, let us check the condition under which the optical chirality density for both the lossless and the lossy approaches exactly coincide with each other. Indeed, it is easy to see that C_{lossless} is equal to C_{lossy} if and only if

$$2\text{Re}\left[\varepsilon(\omega)\mu(\omega) + \frac{\omega\mu(\omega)}{2}\frac{\partial\varepsilon(\omega)}{\partial\omega} + \frac{\omega\varepsilon(\omega)}{2}\frac{\partial\mu(\omega)}{\partial\omega}\right] = \varepsilon(\omega)\mu_{\text{eff}} + \varepsilon_{\text{eff}}\mu^*(\omega). \tag{13.72}$$

Therefore, from the latter equation it is straightforward to observe that the imaginary part of the right-hand side must be zero. This translates into the following condition,

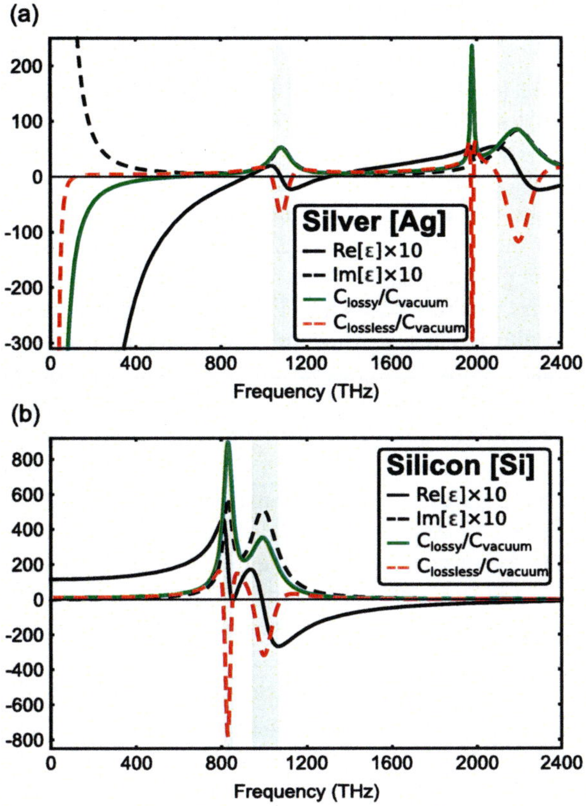

Fig. 13.2 Optical chirality density for **a** silver, and **b** silicon. Material parameters describing ε are directly taken from [72] and [73], respectively (see Table 13.1). For comparison, the curves for lossy (green solid lines) and lossless (red dashed lines) dispersive media are represented in terms of C_{vacuum}. The gray shaded areas indicate the spectral ranges showing anomalous dispersion

Table 13.1 Material parameters characterizing the permittivity of silver [72] and silicon [73]

Ag [A. D. Rakić et al., Appl. Opt. **37**, 5271 (1998)]

$$\varepsilon_{\text{D-L}}^{(\text{Ag})}(\omega) = 1 - \frac{f_0\omega_p^2}{\omega^2 + i\omega\gamma_0} - \frac{f_1\omega_p^2}{\omega^2 - \omega_1^2 + i\omega\gamma_1} - \frac{f_2\omega_p^2}{\omega^2 - \omega_2^2 + i\omega\gamma_2} - \frac{f_3\omega_p^2}{\omega^2 - \omega_3^2 + i\omega\gamma_3}$$

ω_p	f_0	γ_0	f_1	γ_1	ω_1	f_2	γ_2	ω_2	f_3	γ_3	ω_3
2178.61	0.845	11.61	0.065	939.63	197.31	0.124	109.29	1083.50	0.011	15.72	1979.12

Si [E. D. Palik (Academic Press, New York, 1985)]

$$\varepsilon_{\text{D-L}}^{(\text{Si})}(\omega) = 1 - \frac{f_0\omega_0^2}{\omega^2 - \omega_0^2 + i\omega\gamma_0} - \frac{f_1\omega_1^2}{\omega^2 - \omega_1^2 + i\omega\gamma_1}$$

f_0	γ_0	ω_0	f_1	γ_1	ω_1
7.5	150	1000	3	50	830

$$\chi_n''\left(1+\xi_n'\right) - \xi_n''\left(1+\chi_n'\right) + 2\omega\chi_n''\xi_n''\left(\frac{1}{\tilde{\gamma}_n} - \frac{1}{\gamma_n}\right) = 0, \tag{13.73}$$

which is met when $\gamma_n = \tilde{\gamma}_n = 0$ for all n. At the same time, one can verify that this solution is indeed the only one, just by substituting it into the real part of (13.72).

13.5 Conclusions and Outlook

In this chapter, we have put forward a step-by-step theoretical generalization of the optical chirality density in arbitrary dispersive and lossy media, as well as a thorough analysis addressing the corresponding conservation law in its most complete form. All along the whole exposure, we have always kept in mind the classical development for the EM energy, whose conservation is dictated by the very well-known Poynting's theorem [28–30, 53]. Even though these matters may seem somewhat trivial at a first glance, simply the fact of considering EM waves propagating through a dispersive medium poses a challenge, making the mathematical treatment much more complicated, but, at the same time, enriching the physics. Good evidence of this fact is the number of papers on this issue that still continue being published nowadays in renowned journals [54–64], wherein there is discussed and reexamined both the fundamentals and the interpretation of the EM energy density in dispersive media. This renewed interest in such seemingly basic aspects has likely been fostered by recent progresses on left-handed materials [83, 84], and metamaterial photonics in general [85–88].

In the same vein, as for the optical chirality, the latest efforts have been focused on the design and fabrication of metamaterials and plasmonic nanostructures [1], aiming to boost the chiral light-matter interactions for the development of advanced chiroptical applications [10], such as sensing, chiral spectroscopy, or enhanced enantioselectivity [5, 7, 8]. Surprisingly, however, from a theoretical point of view, contributions of material dispersion, as well as dissipation, have mostly been ignored; using instead the original definition for the vacuum [31, 32], even when its applicability can be questionable. Now, a decade later, we have a more comprehensive understanding and, a general and closed formulation for the optical chirality density [46, 48], which will not only enable us to account for such features (dispersion and dissipation) from now on, but it also would allow a thorough reexamination of prior results in order to check their validity as well as their accuracy, e.g., by reconsidering the widespread presumption that surface plasmons would have a null density of optical chirality everywhere [89]. Furthermore, by including the presence of a material system, the analysis of the continuity equation reveals the appearance of additional source-like terms describing the loss (or gain) rate of optical chirality. Remarkably, whatever the dynamical property is concerned, these contributions are important because enable one to get deeper insights into fundamental aspects of light-matter interaction [51, 52]. For example, in the familiar case of EM energy, $\mathcal{J} \cdot \mathcal{E}$, is directly related to the power lost (or the work exerted) by the EM fields on the sources. However, in the optical chirality case, the physical significance of its associated source-like terms is not so obvious, thereby limiting to some extent the concept of source (or sink) of optical chirality and thus hindering a proper interpretation. An in-depth understanding of the meaning and the physical implications of these contributions remain as yet unclear and would deserve further efforts [90, 91]. Finally, it is also worth to be mentioned the controversial debate currently existing around the meaningfulness and the differences between the EM helicity and the optical chirality [91–93]. In this regard, it is quite possible that a complete approach relying upon the role played by the material absorption losses may shed some light on the solution of this and many other intriguing questions.

References

1. M. Schäferling, *Chiral Nanophotonics: Chiral Optical Properties of Plasmonic Systems* (Springer, Berlin, 2017)
2. S. Boriskina, N.I. Zheludev, *Singular and Chiral Nanoplasmonics* (CRC Press, Boca Raton, FL, 2014)
3. M. Schäferling, D. Dregely, M. Hentschel, H. Giessen, Tailoring enhanced optical chirality: Design principles for chiral plasmonic nanostructures. Phys. Rev. X **2**, 031010 (2012). https://doi.org/10.1103/PhysRevX.2.031010
4. N. Meinzer, E. Hendry, W.L. Barnes, Probing the chiral nature of electromagnetic fields surrounding plasmonic nanostructures. Phys. Rev. B **88**, 041407 (2013). https://doi.org/10.1103/PhysRevB.88.041407
5. V.K. Valev, J.J. Baumberg, C. Sibilia, T. Verbiest, Chirality and chiroptical effects in plasmonic nanostructures: Fundamentals, recent progress, and outlook. Adv. Mater. **25**, 2517 (2013). https://doi.org/10.1002/adma.201205178
6. M.L. Nesterov, X. Yin, M. Schäferling, H. Giessen, T. Weiss, The role of plasmon-generated near fields for enhanced circular dichroism spectroscopy. ACS Photon. **3**, 578 (2016). https://doi.org/10.1021/acsphotonics.5b00637
7. J.T. Collins, C. Kuppe, D.C. Hooper, C. Sibilia, M. Centini, V.K. Valev, Chirality and chiroptical effects in metal nanostructures: Fundamentals and current trends. Adv. Opt. Mater. **5**, 1700182 (2017). https://doi.org/10.1002/adom.201700182
8. Y. Luo, C. Chi, M. Jiang, R. Li, S. Zu, Y. Li, Z. Fang, Plasmonic chiral nanostructures: Chiroptical effects and applications. Adv. Opt. Mater. **5**, 1700040 (2017). https://doi.org/10.1002/adom.201700040
9. M. Hentschel, M. Schäferling, X. Duan, H. Giessen, N. Liu, Chiral Plasmonics. Sci. Adv. **3**, e1602735 (2017). https://doi.org/10.1126/sciadv.1602735
10. A. García-Etxarri, J.A. Dionne, Surface-enhanced circular dichroism spectroscopy mediated by nonchiral nanoantennas. Phys. Rev. B **87**, 235409 (2013). https://doi.org/10.1103/PhysRevB.87.235409
11. E. Mohammadi, K.L. Tsakmakidis, A.N. Askarpour, P. Dehkhoda, A. Tavakoli, H. Altug, Nanophotonic platforms for enhanced chiral sensing. ACS Photonics **5**, 2669 (2018). https://doi.org/10.1021/acsphotonics.8b00270
12. G. Pellegrini, M. Finazzi, M. Celebrano, L. Duò, P. Biagioni, Chiral surface waves for enhanced circular dichroism. Phys. Rev. B **95**, 241402 (2017). https://doi.org/10.1103/PhysRevB.95.241402
13. M.L. Solomon, J. Hu, M. Lawrence, A. García-Etxarri, J.A. Dionne, Enantiospecific optical enhancement of chiral sensing and separation with dielectric metasurfaces. ACS Photonics **6**, 43 (2019). https://doi.org/10.1021/acsphotonics.8b01365
14. F. Graf, J. Feis, X. Garcia-Santiago, M. Wegener, C. Rockstuhl, I. Fernandez-Corbaton, Achiral, helicity preserving, and resonant structures for enhanced sensing of chiral molecules. ACS Photonics **6**, 482 (2019). https://doi.org/10.1021/acsphotonics.8b01454
15. E. Mohammadi, A. Tavakoli, P. Dehkhoda, Y. Jahani, K.L. Tsakmakidis, A. Tittl, H. Altug, Accessible superchiral near-fields driven by tailored electric and magnetic resonances in all-dielectric nanostructures. ACS Photonics **6**, 1939 (2019). https://doi.org/10.1021/acsphotonics.8b01767
16. X. Zhao, B.M. Reinhard, Switchable chiroptical hot-spots in silicon nanodisk dimers. ACS Photonics **6**, 1981 (2019). https://doi.org/10.1021/acsphotonics.9b00388

17. Y. Tang, A.E. Cohen, Enhanced enantioselectivity in excitation of chiral molecules by superchiral light. Science **332**, 333 (2011). https://doi.org/10.1126/science.1202817
18. J.E. Vázquez-Lozano, A. Martínez, Toward chiral sensing and spectroscopy enabled by all-dielectric integrated photonic waveguides. Laser Photonics Rev. **14**, 1900422 (2020). https://doi.org/10.1002/lpor.201900422
19. J.S. Choi, M. Cho, Limitations of a superchiral field. Phys. Rev. A **86**, 063834 (2012). https://doi.org/10.1103/PhysRevA.86.063834
20. E. Hendry, T. Carpy, J. Johnston, M. Popland, R.V. Mikhaylovskiy, A.J. Lapthorn, S.M. Kelly, L.D. Barron, N. Gadegaard, M. Kadodwala, Ultrasensitive detection and characterization of biomolecules using superchiral fields. Nat. Nanotechnol. **5**, 783 (2010). https://doi.org/10.1038/nnano.2010.209
21. A. Kuzyk, R. Schreiber, H. Zhang, A.O. Govorov, T. Liedl, N. Liu, Reconfigurable 3D plasmonic metamolecules. Nat. Mater. **13**, 862 (2014). https://doi.org/10.1038/nmat4031
22. L.D. Barron, *Molecular Light Scattering and Optical Activity* (Cambridge University Press, Cambridge, 2004)
23. T. Brixner, F.J. García de Abajo, J. Schneider, W. Pfeiffer, Nanoscopic ultrafast space-time-resolved spectroscopy. Phys. Rev. Lett. **95**, 093901 (2005). https://doi.org/10.1103/PhysRevLett.95.093901
24. N. Yang, Y. Tang, A.E. Cohen, Spectroscopy in sculpted fields. Nano Today **4**, 269 (2009). https://doi.org/10.1016/j.nantod.2009.05.001
25. C. Kramer, M. Schäferling, T. Weiss, H. Giessen, T. Brixner, Analytic optimization of near-field optical chirality enhancement. ACS Photonics **4**, 396 (2017). https://doi.org/10.1021/acsphotonics.6b00887
26. L.E. Barr, S.A.R. Horsley, I.R. Hooper, J.K. Eager, C.P. Gallagher, S.M. Hornett, A.P. Hibbins, E. Hendry, Investigating the nature of chiral near-field interactions. Phys. Rev. B **97**, 155418 (2018). https://doi.org/10.1103/PhysRevB.97.155418
27. S.A. Maier, *Plasmonics: Fundamentals and Applications* (Springer, New York, 2007)
28. J.D. Jackson, *Classical Electrodynamics* (Wiley, New York, 1999)
29. L. Novotny, B. Hecht, *Principles of Nano-Optics* (Cambridge University Press, Cambridge, 2012)
30. L.D. Landau, E.M. Lifshitz, L.P. Pitaevskii, *Electrodynamics of Continuous Media* (Pergamon, New York, 1984)
31. Y. Tang, A.E. Cohen, Optical chirality and its interaction with matter. Phys. Rev. Lett. **104**, 163901 (2010). https://doi.org/10.1103/PhysRevLett.104.163901
32. K.Y. Bliokh, F. Nori, Characterizing optical chirality. Phys. Rev. A **83**, 021803 (2011). https://doi.org/10.1103/PhysRevA.83.021803
33. M.M. Coles, D.L. Andrews, Chirality and angular momentum in optical radiation. Phys. Rev. A **85**, 063810 (2012). https://doi.org/10.1103/PhysRevA.85.063810
34. S.M. Barnett, R. Loudon, The enigma of optical momentum in a medium. Phil. Trans. R. Soc. A **368**, 927 (2010). https://doi.org/10.1098/rsta.2009.0207
35. D.F. Nelson, Momentum, pseudomomentum, and wave momentum: Toward resolving the Minkowski-Abraham controversy. Phys. Rev. A **44**, 3985 (1991). https://doi.org/10.1103/PhysRevA.44.3985
36. S.M. Barnett, Resolution of the Abraham-Minkowski dilemma. Phys. Rev. Lett. **104**, 070401 (2010). https://doi.org/10.1103/PhysRevLett.104.070401
37. M.G. Silveirinha, Reexamination of the Abraham-Minkowski dilemma. Phys. Rev. A **96**, 033831 (2017). https://doi.org/10.1103/PhysRevA.96.033831
38. K.Y. Bliokh, A.Y. Bekshaev, F. Nori, Optical momentum and angular momentum in complex media: from the Abraham-Minkowski debate to unusual properties of surface plasmon-polaritons. New J. Phys. **19**, 123014 (2017). https://doi.org/10.1088/1367-2630/aa8913
39. T.G. Philbin, Electromagnetic energy momentum in dispersive media. Phys. Rev. A **83**, 013823 (2011). https://doi.org/10.1103/PhysRevA.83.013823; Erratum, Phys. Rev. A 85, 059902 (2012) https://doi.org/10.1103/PhysRevA.85.059902

40. T.G. Philbin, O. Allanson, Optical angular momentum in dispersive media. Phys. Rev. A **86**, 055802 (2012). https://doi.org/10.1103/PhysRevA.86.055802
41. K.Y. Bliokh, A.Y. Bekshaev, F. Nori, Optical momentum, spin, and angular momentum in dispersive media. Phys. Rev. Lett. **119**, 073901 (2017). https://doi.org/10.1103/PhysRevLett.119.073901
42. F. Alpeggiani, K.Y. Bliokh, F. Nori, L. Kuipers, Electromagnetic helicity in complex media. Phys. Rev. Lett. **120**, 243605 (2018). https://doi.org/10.1103/PhysRevLett.120.243605
43. I. Proskurin, A.S. Ovchinnikov, P. Nosov, J. Kishine, Optical chirality in gyrotropic media: symmetry approach. New J. Phys. **19**, 063021 (2017). https://doi.org/10.1088/1367-2630/aa6acd
44. T.G. Philbin, Lipkin's conservation law, Noether's theorem, and the relation to optical helicity. Phys. Rev. A **87**, 043843 (2013). https://doi.org/10.1103/PhysRevA.87.043843
45. M.V. Gorkunov, V.E. Dmitrienko, A.A. Ezhov, V.V. Artemov, O.Y. Rogov, Implications of the causality principle for ultra chiral metamaterials. Sci. Rep. **5**, 9273 (2015). https://doi.org/10.1038/srep09273
46. J.E. Vázquez-Lozano, A. Martínez, Optical chirality in dispersive and lossy media. Phys. Rev. Lett. **121**, 043901 (2018). https://doi.org/10.1103/PhysRevLett.121.043901
47. D.M. Lipkin, Existence of a new conservation law in electromagnetic theory. J. Math. Phys. **5**, 696 (1964). https://doi.org/10.1063/1.1704165
48. J.E. Vázquez-Lozano, A. Martínez, Optics in 2018: Generalizing optical chirality to an arbitrary medium. Opt. Photon. News **29**, 39 (2018). https://www.osa-opn.org/home/articles/volume_29/december_2018/extras/generalizing_optical_chirality_to_an_arbitrary_med/
49. E. Hecht, *Optics*, 4th edn. (Pearson Education, Harlow, 2013)
50. B.E.A. Saleh, M.C. Teich, *Fundamentals of Photonics* (Wiley-Interscience, New York, 2007)
51. G. Nienhuis, Conservation laws and symmetry transformations of the electromagnetic field with sources. Phys. Rev. A **93**, 023840 (2016). https://doi.org/10.1103/PhysRevA.93.023840
52. I. Fernandez-Corbaton, C. Rockstuhl, Unified theory to describe and engineer conservation laws in light-matter interactions. Phys. Rev. A **95**, 053829 (2017). https://doi.org/10.1103/PhysRevA.95.053829
53. J.H. Poynting, On the transfer of energy in the electromagnetic field. Phil. Trans. R. Soc. London **175**, 343 (1884). https://doi.org/10.1098/rstl.1884.0016
54. R. Loudon, The propagation of electromagnetic energy through an absorbing dielectric. J. Phys. A: Gen. Phys. **3**, 233 (1970). https://doi.org/10.1088/0305-4470/3/3/008
55. R. Ruppin, Electromagnetic energy density in a dispersive and absorptive material. Phys. Lett. A **299**, 309 (2002). https://doi.org/10.1016/S0375-9601(01)00838-6
56. T.J. Cui, J.A. Kong, Time-domain electromagnetic energy in a frequency-dispersive left-handed medium. Phys. Rev. B **70**, 205106 (2004). https://doi.org/10.1103/PhysRevB.70.205106
57. S.A. Tretyakov, Electromagnetic field energy density in artificial microwave materials with strong dispersion and loss. Phys. Lett. A **343**, 231 (2005). https://doi.org/10.1016/j.physleta.2005.06.023
58. A.D. Boardman, K. Marinov, Electromagnetic energy in a dispersive metamaterial. Phys. Rev. B **73**, 165110 (2006). https://doi.org/10.1103/PhysRevB.73.165110
59. P.-G. Luan, Power loss and electromagnetic energy density in a dispersive metamaterial medium. Phys. Rev. E **80**, 046601 (2009). https://doi.org/10.1103/PhysRevE.80.046601
60. A. Raman, S. Fan, Photonic band structure of dispersive metamaterials formulated as a Hermitian eigenvalue problem. Phys. Rev. Lett. **104**, 087401 (2010). https://doi.org/10.1103/PhysRevLett.104.087401
61. W. Shin, A. Raman, S. Fan, Instantaneous electric energy and electric power dissipation in dispersive media. J. Opt. Soc. Am. B **29**, 1048 (2012). https://doi.org/10.1364/JOSAB.29.001048
62. F.S.S. Rosa, D.A.R. Dalvit, P.W. Milonni, Electromagnetic energy, absorption, and Casimir forces: Uniform dielectric media in thermal equilibrium. Phys. Rev. A **81**, 033812 (2010). https://doi.org/10.1103/PhysRevA.81.033812
63. K.J. Webb and Shivanand, Electromagnetic field energy in dispersive materials. J. Opt. Soc. Am. B **27**, 1215 (2010) https://doi.org/10.1364/JOSAB.27.001215

64. F.D. Nunes, T.C. Vasconcelos, M. Bezerra, J. Weiner, Electromagnetic energy density in dispersive and dissipative media. J. Opt. Soc. Am. B **28**, 1544 (2011). https://doi.org/10.1364/JOSAB.28.001544
65. L. Brillouin, *Wave Propagation and Group Velocity* (Academic, New York, 1960)
66. K.E. Oughstun, S. Shen, Velocity of energy transport for a time-harmonic field in a multiple-resonance Lorentz medium. J. Opt. Soc. Am. B **5**, 2395 (1988). https://doi.org/10.1364/JOSAB.5.002395
67. F.D. Nunes, B.-H.V. Borges, J. Weiner, Analysis of dispersive and dissipative media with optical resonances. Opt. Express **20**, 15679 (2012). https://doi.org/10.1364/OE.20.015679
68. D.R. Smith, W.J. Padilla, D.C. Vier, S.C. Nemat-Nasser, S. Schultz, Composite medium with simultaneously negative permeability and permittivity. Phys. Rev. Lett. **84**, 4184 (2000). https://doi.org/10.1103/PhysRevLett.84.4184
69. C. García-Meca, J. Hurtado, J. Martí, A. Martínez, W. Dickson, A.V. Zayats, Low-Loss Multilayered Metamaterial Exhibiting a Negative Index of Refraction at Visible Wavelengths. Phys. Rev. Lett. **106**, 067402 (2011). https://doi.org/10.1103/PhysRevLett.106.067402
70. S. Yoo, M. Cho, Q.-H. Park, Globally enhanced chiral field generation by negative-index metamaterials. Phys. Rev. B **89**, 161405 (2014). https://doi.org/10.1103/PhysRevB.89.161405
71. S. Yoo, Q.-H. Park, Chiral light-matter interaction in optical resonators. Phys. Rev. Lett. **114**, 203003 (2015). https://doi.org/10.1103/PhysRevLett.114.203003
72. A.D. Rakić, A.B. Djurišić, J.M. Elazar, M.L. Majewski, Optical properties of metallic films for vertical-cavity optoelectronic devices. Appl. Opt. **37**, 5271 (1998). https://doi.org/10.1364/AO.37.005271
73. E.D. Palik, *Handbook of Optical Constants of Solids* (Academic Press, New York, 1985)
74. H.S. Sehmi, W. Langbein, E.A. Muljarov, Optimizing the Drude-Lorentz model for material permittivity: Method, program, and examples for gold, silver, and copper. Phys. Rev. B **95**, 115444 (2017). https://doi.org/10.1103/PhysRevB.95.115444
75. W.I. Fushchich, A.G. Nikitin, *Symmetries of Maxwell's Equations* (Mathematics and its Applications (Springer, Amsterdam, 1987)
76. E. Noether, Gott. Nachr. **1918**, 235 (1918) [Transp. Theory Stat. Phys. **1**, 186 (1971) https://doi.org/10.1080/00411457108231446]
77. R.P. Cameron, J.B. Götte, S.M. Barnett, A.M. Yao, Chirality and the angular momentum of light. Phil. Trans. R. Soc. A **375**, 20150433 (2017). https://doi.org/10.1098/rsta.2015.0433
78. W.-K. Tung, *Group Theory in Physics* (World Scientific, Singapore, 1985)
79. S. Weinberg, 1st ed., vol. 1 *The Quantum Theory of Fields* (Cambridge University Press, Cambridge, 1995)
80. L.V. Poulikakos, P. Gutsche, K.M. McPeak, S. Burger, J. Niegemann, C. Hafner, D.J. Norris, Optical chirality flux as a useful far-field probe of chiral near fields. ACS Photonics **3**, 1619 (2016). https://doi.org/10.1021/acsphotonics.6b00201
81. R.W. Boyd, D.J. Gauthier, Controlling the velocity of light pulses. Science **326**, 1074 (2009). https://doi.org/10.1126/science.1170885
82. V. Gerasik, M. Stastna, Complex group velocity and energy transport in absorbing media. Phys. Rev. E **81**, 056602 (2010). https://doi.org/10.1103/PhysRevE.81.056602
83. J.B. Pendry, Negative refraction makes a perfect lens. Phys. Rev. Lett. **85**, 3966 (2000). https://doi.org/10.1103/PhysRevLett.85.3966
84. H.J. Lezec, J.A. Dionne, H.A. Atwater, Negative refraction at visible frequencies. Science **316**, 430 (2007). https://doi.org/10.1126/science.1139266
85. N. Engheta, Circuits with light at nanoscales: Optical nanocircuits inspired by metamaterials. Science **317**, 1698 (2007). https://doi.org/10.1126/science.1133268
86. J. Valentine, S. Zhang, T. Zentgraf, E. Ulin-Avila, D. A. Genov, G. Bartal, X. Zhang, Three-dimensional optical metamaterial with a negative refractive index. Nature (London) **455**, 376 (2008) https://doi.org/10.1038/nature07247
87. C.-G. Huang, Y.-Z. Zhang, Poynting vector, energy density, and energy velocity in an anomalous dispersion medium. Phys. Rev. A **65**, 015802 (2001). https://doi.org/10.1103/PhysRevA.65.015802

88. G. Dolling, C. Enkrich, M. Wegener, C.M. Soukoulis, S. Linden, Simultaneous negative phase and group velocity of light in a metamaterial. Science **312**, 892 (2006). https://doi.org/10.1126/science.1126021
89. K.Y. Bliokh, F. Nori, Transverse spin of a surface polariton. Phys. Rev. A **85**, 061801 (2012). https://doi.org/10.1103/PhysRevA.85.061801
90. P. Gutsche, L.V. Poulikakos, M. Hammerschmidt, S. Burger, F. Schmidt, Time-harmonic optical chirality in inhomogeneous space. Proc. SPIE **9756**, 97560X (2016). https://doi.org/10.1117/12.2209551
91. F. Crimin, N. Mackinnon, J.B. Götte, S.M. Barnett, Optical helicity and chirality: Conservation and sources. Appl. Sci. **9**, 828 (2019). https://doi.org/10.3390/app9050828
92. N. Mackinnon, On the differences between helicity and chirality. J. Opt. **21**, 125402 (2019). https://doi.org/10.1088/2040-8986/ab4db9
93. L.V. Poulikakos, J.A. Dionne, A. García-Etxarri, Optical helicity and optical chirality in free space and in the presence of matter. Symmetry **11**, 1113 (2019). https://doi.org/10.3390/sym11091113

Chapter 14
Topology in Magnetism

X. S. Wang and X. R. Wang

Abstract In this chapter, we review recent developments of two usages of topology in magnetism. One is to classify spin structures with different topological numbers (topology in real space). The other usage is to distinguish normal magnetic materials from those magnetic materials supporting topologically protected unidirectional surface spin waves inside spin wave band gaps (topology in reciprocal space).

14.1 Introduction

Our knowledge about magnetic phenomena has a long history that can be traced back to the 6th century B.C. when Thales of Miletus, one of the sages in ancient Greece, told people that lodestones can attract iron. In the 4th century B.C., two Chinese books, *Master Lü's Spring and Autumn Annual* and *Guiguzi* reported the same phenomenon [1]. Around 1000 A.D., Chinese people already knew how to magnetize a normal iron needle with a lodestone and make magnetic compass (vol. 24 of *Mengxi Bitan* by Kuo Shen). After the industrial revolution, our understanding of magnetism had greatly advanced. Based on the studies of Coulomb, Ampère, Faraday, and many others, James Clerk Maxwell wrote down the famous Maxwell's equations in 1861 [2],

X. S. Wang
School of Physics and Electronics, Hunan University, Changsha 410082, China

X. R. Wang (✉)
Physics Department, The Hong Kong University of Science and Technology, Clear Water Bay, Kowloon, Hong Kong
e-mail: phxwan@ust.hk

HKUST Shenzhen Research Institute, Shenzhen 518057, China

E. Kamenetskii (ed.), *Chirality, Magnetism and Magnetoelectricity*,
Topics in Applied Physics 138,
https://doi.org/10.1007/978-3-030-62844-4_14

$$\nabla \cdot \mathbf{E} = \frac{\rho}{\varepsilon_0}, \tag{14.1}$$

$$\nabla \cdot \mathbf{B} = 0, \tag{14.2}$$

$$\nabla \times \mathbf{E} = -\frac{\partial \mathbf{B}}{\partial t}, \tag{14.3}$$

$$\nabla \times \mathbf{B} = \mu_0 \left(\mathbf{J} + \varepsilon_0 \frac{\partial \mathbf{E}}{\partial t} \right), \tag{14.4}$$

which unify the optical, magnetic and electric phenomena. After entering the 20th century, people realized that, besides magnetism from the motion of electric charges in real space, spin, the intrinsic angular momentum of particles, can also generate magnetic moments [3]. In 1888, Oberlin Smith proposed the idea of recording audio on a magnetic wire [4], and in 1928, Fritz Pleumer developed the first magnetic tape recorder capable of recording analog audio signal. The basic principle is to convert a sound wave into a spatially dependent magnetization directions. After electrical computers were invented, there was a great demand of high-density data storage. The digital version of magnetic recording uses two distinct domains of different magnetization directions to represent binary bits "0" or "1". However, the recent revival of nano-magnetism, or spintronics, is largely due to the discovery of giant magnetoresistance (GMR) by Peter Grünberg and Albert Fert in 1988 [5, 6]. Peter and Albert received the Nobel Prize in Physics in 2007 for this important discovery. With GMR and much improved material engineering, the storage density of magnetic devices, hard disks, increases very fast in the 1990s and 2000s. However, the design of hard disks requires mechanical components to locate the patterned areas, which are susceptible to a tiny vibration. Furthermore, the superparamagnetic limit [7] indicates that to achieve long-time and higher density storage, higher crystalline anisotropy is required, which is challenging in material science [8]. Thus, people are looking for other paradigms of magnetic storage that include (1) three-dimensional (3D) racetrack memory, (2) to manipulate the data in a non-mechanical way and (3) to find smaller data storage elements.

Traditional electronics uses electron charge to store as well as to process information. The continuous demand of high computational capacity and miniaturization require to integrate more and more transistors in a unit area. As a result, the Joule heating becomes a bottleneck and electronics is reaching its limit [9]. Thus, it is important to find other information carriers to replace electron charges, and one promising candidate is the spin that can be non-volatile, and thus has much less energy consumption. This is a main topic of spintronics [10]. The other possible candidates include the charges of Cooper pairs (superconductivity) [11] and spins of magnons (magnonics) [12]. Magnons are quanta of spin waves, the elemental excitations in strongly-correlated spin systems such as ferromagnets, antiferromagnets, and ferrimagnets. A magnon carries a spin of $\hbar$, and the power consumption of magnon current is much lower than that of electric current [13]. Thus, the manipulation of magnons also becomes a hot topic in condensed matter physics.

In the 18th century, Leonhard Euler studied *Seven Bridges of Königsberg* problem and the properties of polyhedrons, which is considered as the birth of the topology [14]. Topology became a well-accepted branch of mathematics in the early part of the 20th century. The basic motivation of topology is to study preserved quantities and their possible relationships under continuous deformations, such as stretching/compressing and twisting, but not tearing or gluing. In 1940s, Shiing-Shen Chern formulated the concept of Chern class and Chern numbers. The vector bundles on a smooth manifold are classified into Chern classes with specific Chern numbers, and within the same Chern class, the vector bundles are *topologically equivalent*. Almost at the same time, physicists started to study gauge transformation and the gauge-field origin of the fundamental forces in nature. For example, it was realized that nature must have electromagnetic fields whose dynamics follows the Maxwell's equation if nature respects $U(1)$ gauge symmetry. We know now that the gauge charges are the Chern numbers of these fields [15]. In the 1970s and the 1980s, the interest on topology proliferates from particle physics to condensed matter physics. David J. Thouless, J. Michael Kosterlitz, F. Duncan M. Haldane and many others discovered topological phase transitions and possible topological phases of matter. David, Michael, and Duncan were awarded the Nobel Prize in Physics in 2016 for these pioneering work. Simply speaking, topology provides a way to certain invariants in nature. Within each class, some common properties exist and these properties are robust against continuous deformations. This robustness is called topological protection [16]. In many circumstances, an extra cost (energy, time, force, defects, etc.) has to be paid to break the topological protection. Thus, the topologically protected excitations are extraordinarily useful in practice.

In summary, magnetism and topology are academically interesting and practically useful. The hybridization of the two subjects is the central theme of this chapter. Here we review recent developments of two usages of topology in magnetism based on our own results. We focus on ferromagnets although there are increasing interest in anti-ferromagnetic systems. In Sect. 14.2, we introduce spin/magnetic textures with nontrivial topology in real space, including domain walls in one-dimensional (1D) wires to 3D bulk magnets, vortices and skyrmions in two-dimensional (2D) films and hopfions in 3D magnets. It should be emphasized that topological structures with various non-zero topological number does not necessarily imply stability although they cannot be mapped to a single domain with zero topological number under a continuous deformation. Nature respects energy, but not the topological deformations. In Sect. 14.3, we introduce topological spin waves (or magnons, in this chapter we use this two terms interchangeably). Spin waves are the propagation of small spin fluctuation, and we concentrate on unidirectional surfaces waves topologically protected by the band gap in the reciprocal space. These topological spin waves very often imply robustness because they can only be destroyed when the bulk topological charges are removed or annihilated through the external forces that consume a finite amount of energy.

14.2 Topological Spin Textures

The order parameter of ferromagnets is magnetization, **M**. The magnetization at different locations are strongly correlated. Generally speaking, a system is an object occupied a real space of $\mathbb{R}$ with real-space dimension d. The order parameter of the system has a dimension d'. d and d' are not related. d' of **M** is 2 when a ferromagnet is far below the Curie temperature, but it can be 3 near the Curie temperature or smaller than 2 when **M** is restricted. For ferrimagnets or anti-ferromagnets, the order parameters can be more than one [17]. In the ambient conditions, thermodynamic quantity **M** can be treated as a classical vector. When **M** at each point can take only two discrete values of "up" and "down", the system is usually called Ising-like with dimension $d' = 0$. When **M** has a fixed magnitude and can only move in a plane or on a surface, the magnet is XY-like with $d' = 1$ because **M** is on a 1D circle. For most ferromagnets, **M** of fixed magnitude can point to any direction in real space so their order parameters are unit vector direction **m** of **M**. $d' = 2$ and the system can be modelled by a Heisenberg model or its generalizations consisting of various anisotropies and all kinds of exchange interactions: symmetric or asymmetric, long- or short-range interactions. The simplest spin texture is the one with all spins pointing to the same direction. Since the spins in an Ising model are discrete, there is no continuous deformation and magnetization **m** acts as an index to distinguish the two states. For different d and $d' \geq 1$, one can classify various structures according to their topology, as shown in Fig. 14.1. Different topological indices are defined for different dimensionality.

In this section, we discuss various spin structures for domain walls (Sect. 14.2.1), vortices, skyrmions (Sect. 14.2.2), and hopfions (Sect. 14.2.3) in $d = 1$ to 3. We consider the continuous limit where the length scale of excitations is much larger than atom-atom distance. The order parameter, $\mathbf{m}(\mathbf{r}, t)$, is governed by the Landau-Lifshitz-Gilbert (LLG) equation [18],

$$\frac{\partial \mathbf{m}}{\partial t} = -\gamma \mathbf{m} \times \mathbf{B}_{\text{eff}} + \alpha \mathbf{m} \times \frac{\partial \mathbf{m}}{\partial t}, \tag{14.5}$$

where γ is the gyromagnetic ratio, $\mathbf{B}_{\text{eff}}$ is the effective field due to all kinds of anisotropic energies and exchange interactions, and α is the Gilbert damping. The LLG equation describes generally the magnetization dynamics for an arbitrary d and $d' = 2$.

14.2.1 Domain Walls

Domain walls (DWs) are transition regions between different domains. The minimal model for a DW is on $\mathbb{R}^1 \mapsto \mathbf{M}^0$, as shown in Fig. 14.1. In this case, the Ising DW is a topological defect [19] of trivial structure without a well-accepted dynamics. We

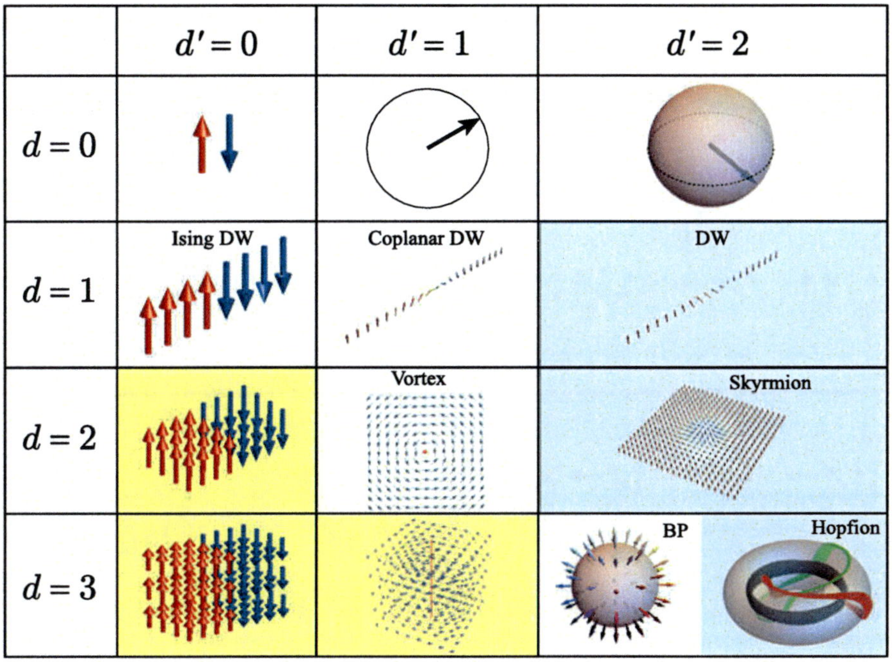

Fig. 14.1 Spin textures in different dimensionalities (DW = Domain Wall, BP = Bloch Point). The structures with yellow background are the trivial repetitions of their $(d-1)$-counterparty in the additional dimension. The structures with blue background are continuous everywhere without any singularity

start with 1D model on $\mathbb{R}^1 \mapsto \mathbf{M}^1$, which describes a coplanar DW on a wire. Let spin $\mathbf{m}_-$ at the left end of a wire point to the $+\hat{z}$-direction and spin $\mathbf{m}_+$ at the right end point to the $-\hat{z}$-direction, denoted respectively by the red and green dots on the unit circle in Fig. 14.2. Any DW can be mapped to a path connecting the two points. Within $\mathbf{M}^1$, there is an infinite number of topological distinct DWs. For instance, the DWs shown in Fig. 14.2a–c are topologically different, because they cannot be transformed to each other under continuous deformations. However, if $\mathbf{m}$ is allowed to tilt out of plane, i.e. consider the DWs in $\mathbb{R}^1 \mapsto \mathbf{M}^2$ as shown in Fig. 14.2d–f, each of the DW paths can continuously deform to another within the unit sphere surface of $\mathbf{m}$, which can be intuitively visualized in the figure. Thus all the DWs are topologically equivalent as long as the boundary values $\mathbf{m}_\pm$ are the same.

How does the topology affect the dynamics of DWs? We consider a 180° DW that separates two adjacent oppositely-oriented domains ($\mathbf{m}_+ = -\mathbf{m}_-$). Figure 14.3a shows three types of 180° DWs. Let us use the head-to-head/tail-to-tail (HH/TT) DWs as an example. Use polar angle θ (with respect to z axis), we can define a winding number or a charge as,

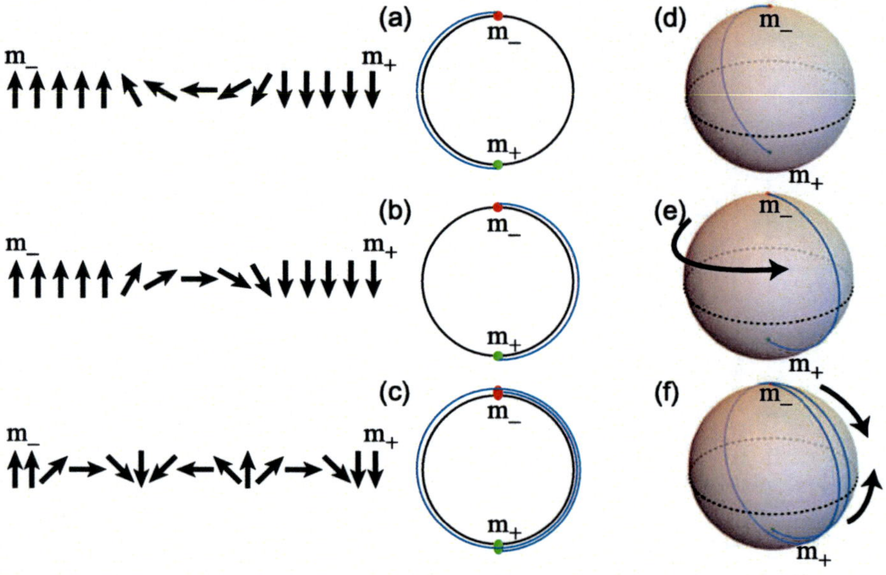

Fig. 14.2 Schematic diagrams of DW topology. The top panel illustrates three DWs. **a–c** are representative diagrams for three DWs in $\mathbb{R}^1 \mapsto \mathbf{M}^1$ model, and **d–f** are in $\mathbb{R}^1 \mapsto \mathbf{M}^2$ model

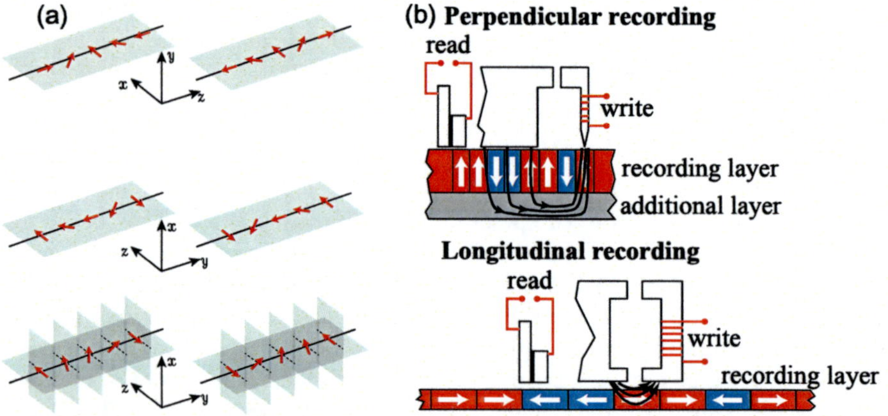

Fig. 14.3 **a** Three types of 180° DWs. From the top to the bottom: heat-to-head/tail-to-tail, up-down Néel/down-up Néel, up-down Bloch/down-up Bloch. **b** Schematic diagrams of DW racetrack memory: in-plane storage and perpendicular storage

Fig. 14.4 Comparison between a DW in a magnetic field and a charge in an electric field

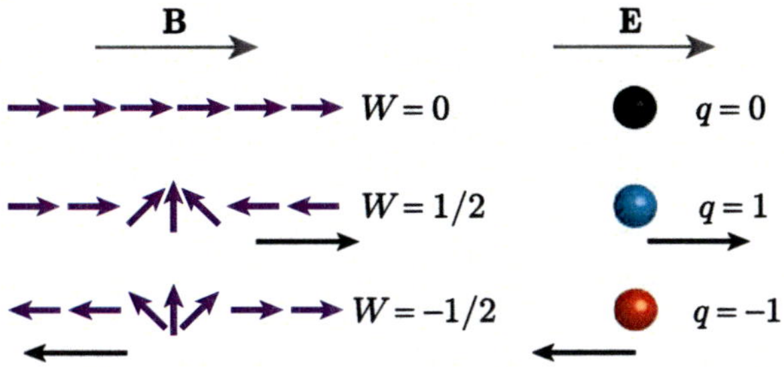

$$W = \frac{1}{2\pi}\int_{-\infty}^{+\infty}\frac{d\theta}{dz}dz = \frac{\theta(+\infty)-\theta(-\infty)}{2\pi}. \tag{14.6}$$

W describes how many times $\mathbf{m}$ wraps the unit circle when the spatial coordinate sweeps the whole magnetic wire. For the HH (TT) wall, $W = 1/2$ ($W = -1/2$). In the presence of an external field along the wire, a DW of $W = 1/2$ ($W = -1/2$) moves along (against) the field to lower the Zeeman energy. This is similar to a positive (negative) charge in an electric field as illustrated in Fig. 14.4. In $\mathbf{M}^1$, θ can be any real number. For fixed boundary values of $\mathbf{m}_{\pm}$, W can still differ by an integer (θ can differ by an integer multiple of 2π). Thus, there is an infinite number of topological distinct DWs. In $\mathbf{M}^2$, the range of θ is in $[0, \pi]$, so all the DWs with the same boundary values have the same W. This example also shows how the dimension of order parameter affect the kinetic paths from one structure to another. Therefore, W is a topological index, which is consistent with the intuition. As a comparison, the winding number of a single domain is 0, which is defined as topologically trivial.

The 180° DWs are basic building blocks of racetrack memory [20]. In a racetrack memory, magnetization directions of domains separated by a series of DWs encode binary data in a patterned wire. All DWs are required to move simultaneously along the same direction at the same pace under a current to address the data, as schematically illustrated in Fig. 14.3b. Thus efficient manipulations of DWs is a technologically important topic. There are a lot of theoretical and experimental studies of DW motion, such as [21–28], to name a few. A magnetic field is traditionally used to drive a DW to move. Our basic understanding of field-driven DW motion comes from, one way or another, the exact Walker solution of a 1D LLG equation [21]: Energy density functional of the model consists of an exchange energy $A|\nabla\mathbf{m}|^2$ with exchange stiffness A, a hard anisotropy $K_x m_x^2$ and an easy anisotropy $-K_z m_z^2$ along x and z directions and the Zeeman energy $-M_s\mathbf{B}\cdot\mathbf{m}$ due to external field $\mathbf{B}$. Here K_x and K_z are hard and easy anisotropy strengths, including the shape and crystalline anisotropies, and M_s is the saturation magnetization. The temperature is considered to be far below the Curie temperature. In Walker solution [21], a threshold field of $B_w = \alpha K_x/M_s$ (called Walker breakdown field) exists. For $B < B_w$, a DW undergoes a rigid-body translational motion. Beyond B_w, the translational motion is no longer allowed and the DW starts to rotate. DW speed v in this region oscillates with time [29], and average v dramatically decreases. From energy consideration, one can rigorously prove that no static DW is allowed in a homogeneous nanowire in the presence of an external magnetic field. A moving DW must dissipate energy because of various damping mechanisms. Thus, energy conservation requires that the dissipated energy comes from the energy decrease of the wire that should be supplied by the Zeeman energy released from the DW propagation [30]. This is the origin of DW propagation. The energy dissipation can come not only from the Gilbert damping as in aforementioned Walker solution, but also from spin wave emission [31, 32]. Somehow similar to the Walker solution, depending on K_x and B, breathing or more complicated periodic transformations such as a DW drilling emit spin waves. During the DW propagation, the energy dissipation rate due to the SW emission is balanced by the Zeeman energy release. Spin wave emission persists in the presence of Gilbert damping and make the usual Walker rigid-body propagation solution unstable in a region below the Walker breakdown field [33]. In strong field and low K_x, a broad DW undergoes a breathing motion. Both bow and stern SWs are emitted by the periodical breathing of the DW width. The main frequencies of the SWs are respectively of $3\gamma B$ and γB in the domains along and opposite to the field. When the sum of longitudinal and transverse anisotropy energies is of the same order of the exchange energy, the DW goes into drilling motion when the DW plane is greatly distorted. The DW propagates at a constant high speed, and stern waves are mainly emitted.

As mentioned before in Fig. 14.4, a magnetic field drives DWs of $W = 1/2$ and $W = -1/2$ to move in opposite directions so that they can annihilate with each other. Under a magnetic field, the data stored in the domain between the two annihilated DWs would be destroyed. In a racetrack memory, all domain walls are required to move in a synchronized way, regardless its topological charge. This requirement can be realized by spin-polarized electric current through spin-transfer torque [23–27]. Different from the field-driven motion which originates from energy conservation, the current-driven DW motion comes from the angular momentum transfer [34]. In most FM materials, the electrons are spin-polarized along the magnetization direction of the domain (with angular momentum $\hbar/2$). After passing through a DW, electrons can follow the local magnetization to reverse its spin (adiabatic process) to $-\hbar/2$ if the DW is wide enough. Thus there is a change in angular momentum of $-\hbar$. Due to angular momentum conservation, this change can be converted into the motion of DW towards the direction of electron flow. The torque on the local spin is called the spin-transfer torque (STT). No matter what the type of a DW is, STT always drives the DW to propagate along the same direction. STT-driven DW propagation soon attracts intensive theoretical and experimental studies due to the nice property, including but not limited to [20, 23–27, 35, 36]. STT has both adiabatic and nonadiabatic components[25, 37, 38] whose expressions are

$$\boldsymbol{\tau} = -(\mathbf{u}\cdot\nabla)\,\mathbf{m} + \beta\mathbf{m}\times(\mathbf{u}\cdot\nabla)\,\mathbf{m}, \tag{14.7}$$

where the first term is the adiabatic torque and the second term is the non-adiabatic torque with non-adiabaticity coefficient β. $\mathbf{u} = \mu_B p g_e \mathbf{J}/[2eM_s(1+\beta^2)]$ has the dimension of velocity that is proportional to electric current $\mathbf{J}$. μ_B, p, g_e, and e are respectively the Bohr magneto, the degree of current spin polarization, the electron g-factor and the electron charge.

To achieve a high DW speed, a large electric current has to be used. Thus, a large amount of Joule heat is produced [35] that is harmful to device performance. As explained before, spin waves can take the role of an electric current in the DW motion. Indeed, both electrons and magnons carry spins. A magnon carrying an angular momentum of $\hbar$. Similar to the STT from electrons to local magnetic moments, a STT from magnons to a DW also exists. Since the spins carried by magnons are always opposite to spins of the domain, magnon STT should drag a DW towards the opposite direction of magnon flows. The magnonic STT has been predicted [39] and experimentally verified [40, 41]. The magnons can not only be coherent ones generated by oscillating electromagnetic fields but also be incoherent ones generated by nonlinear effects or thermal gradients. The STT from thermal magnons also has been predicted [42, 43] and experimentally observed [44]. Under a thermal gradient, the hotter side has a larger magnon population than the colder side. So the magnons diffuse from the hotter side to the colder side, forming a magnon flow. This flow drags the DW towards the hotter side.

The $d = 1$ model is a good approximation for small diameter (much less than exchange length and DW width) nanowires or very large bulky magnets whose magnetization varies only along one direction. For DWs in nanostrips which are usually used in experiments and devices, a $d = 2$ model is necessary. We now discuss a $\mathbb{R}^2 \mapsto \mathbf{M}^1$ model, i.e. the 2D classical XY model, which is applicable to in-plane magnetized strips [45]. The pioneering work of McMichael and Donahue [46] show numerically that a HH DW can be a transverse wall or a vortex wall, depending on the width and the thickness of the strip, as shown in Fig. 14.5. To discuss the topology, similar to the winding number defined in (14.6), we can define a winding number along a path C in $\mathbb{R}^2$,

$$W_C = \frac{1}{2\pi}\int_C \nabla\theta \cdot d\mathbf{r}, \tag{14.8}$$

where C is a closed loop inside a strip denoted as Ω, on which θ is continuum and smooth. W_C in this case is a bulk winding number that must be an integer. If $W_C = 0$, there is no singularity inside C. If $W_C = +1$ (-1), there is a singularity inside C, called a vortex (antivortex), as shown in Fig. 14.5 [45]. At the edge of the strip denoted as $\partial\Omega$, C can be a segment of $\partial\Omega$, and the definition of W_C is modified as

$$W_C = \frac{1}{2\pi}\int_C \nabla(\theta - \theta_C) \cdot d\mathbf{r}, \tag{14.9}$$

where θ_C is the angle of the sample edge. W_C of edge defects takes the values $\pm 1/2$, as shown in Fig. 14.5. The sum of all the winding numbers over $\Omega + \partial\Omega$ is conserved. Total winding number of $\sum W_C$ is a topological invariant.

A transverse wall has a pair of edge defects of $W_C = \pm 1/2$, and a vortex wall with one vortex of $W_C = 1$ has two edge defects of $W_C = -1/2$. Both walls have the same total winding numbers of 0. Thus the transverse wall and the vortex wall

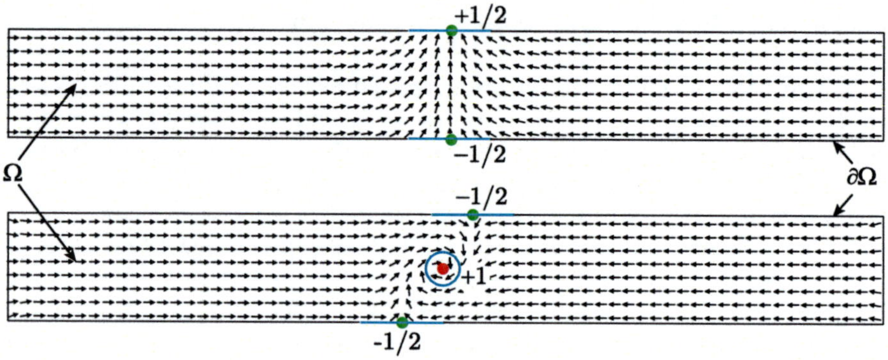

Fig. 14.5 Schematic diagrams of transverse wall and vortex wall, showing the bulk (red dot) and edge (green dots) winding numbers. The blue lines are the paths of integral C

are topologically equivalent. The field-driven motion of transverse walls has a very interesting dynamics [47]. Due to the dipolar interactions, the transverse wall has a strawberry shape such that the width of the DW is not uniform along the width of the strip. When the field is larger than a certain value but substantially below the usual Walker breakdown field, a series of DW structure transformations, including periodic birth and death of antivortices, are observed and explained. Two types of antivortex evolution and their causes are identified for system with large damping constant while only one type of evolution is found for system with small damping constant.

For large damping and intermediate field, antivortices generated at the edge defect near the tail of DW center line move along its DW center line to the other edge and die there. At the same time the polarity of the transverse wall is reversed. Figure 14.6 shows this DW transformations. As shown in the left plot of Fig. 14.6a, the initial DW of polarity $+1$ is strawberry-like and it is located at the center of the strip. It composes two edge defects whose winding numbers are $1/2$ (the filled circle) and $-1/2$ (the open circle) on the top and bottom edges, respectively. The DW center line is parallel to y axis as illustrated by the black vertical line in the right plot of Fig. 14.6a. After the field is applied along the $+x$-direction, the center line tilts its direction away from y axis and elongates at the same time (Fig. 14.6b–e). As the center line tilts far enough from the y axis as shown in the left plot of Fig. 14.6b, a small antivortex (the blue dot) is born near the $-1/2$ edge defect, the tail of DW center line. Since the defects are topological objects with well-defined winding numbers, edge defect of $-1/2$ winding number can only give birth to an antivortex of winding number -1 and change itself to a $+1/2$ edge defect. This is illustrated in the right plot of Fig. 14.6b with the blue dot representing the antivortex core and two filled circles for the edge defects of $+1/2$ winding number. Then, this antivortex moves toward the other side of the strip (Fig. 14.6c, d) along the DW center line. Later, the antivortex reaches the top edge defect and dies there as shown in Fig. 14.6e. At the same time, top edge defect changes its winding number from $+1/2$ to $-1/2$, and

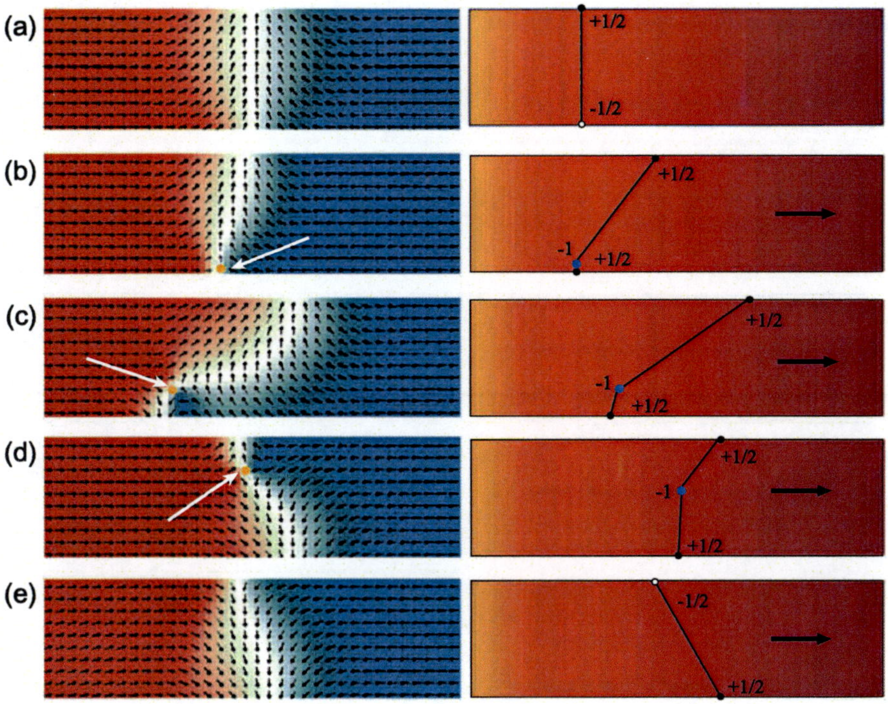

Fig. 14.6 **a–e** Evolution of the DW structure during propagation for large damping and intermediate field. The vortex core is indicated by a dot and the corresponding winding numbers are labelled. Figure redrawn using the results in [47]

transverse wall reverses its polarity from $+1$ to -1. After a transient period, the above mentioned process will repeat again, starting from the top edge of the strip this time. The birth-death process of antivortices repeats periodically while the transverse DW propagates along the wire.

For larger field or smaller damping, the antivortex is not able to reach the other edge of the strip because distortion of the DW center line is too large so that the corner space is not large enough to accommodate the antivortex. Figure 14.7 shows such DW transformations. The left plot of Fig. 14.7a shows the initial spin configuration of the strawberry-like transverse wall. The edge defects with corresponding winding numbers and the DW center line are illustrated in the right plot of Fig. 14.7a. The internal DW structure right before the generation of an antivortex at the edge defect on the bottom edge, is shown in the left plot of Fig. 14.7b. The right plot of Fig. 14.7b shows the corresponding DW center line and antivortex core (the blue dot) and two edge defects of winding number $+1/2$. Different from the previous case, this newly born antivortex cannot reach the other edge. As shown by the the DW configuration (left) and the corresponding location of the antivortex, edge defects, as well as the DW center line (right) in Fig. 14.7c, the antivortex makes a U-turn back to bottom

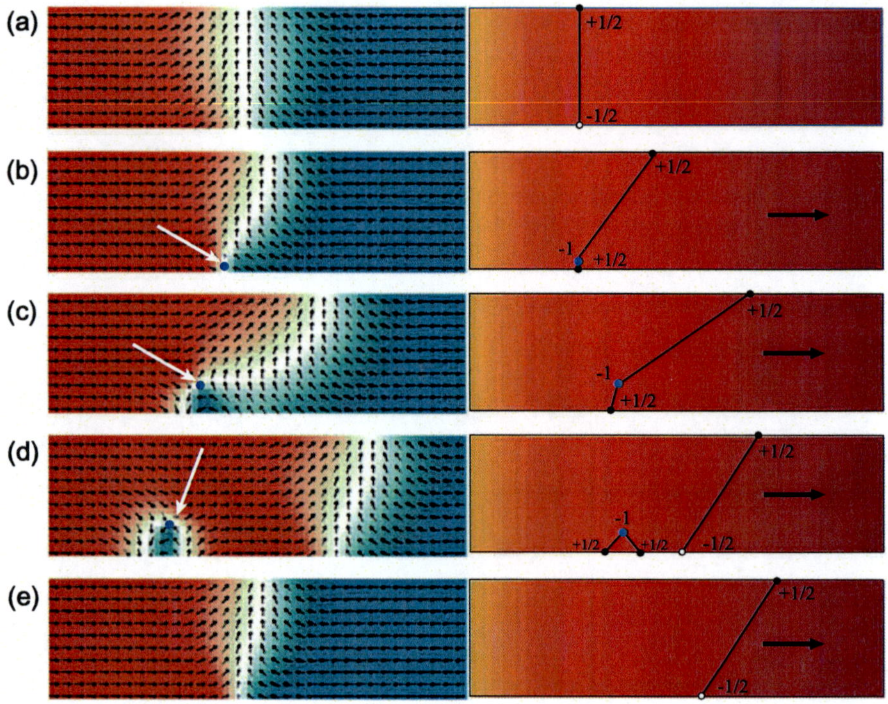

Fig. 14.7 **a–e** Evolution of the DW structure during the propagation for smaller damping and larger field. Figure redrawn using the results in [47]

edge. Together with two edge defects of winding number $+1/2$ at the bottom edge, the antivortex detaches itself from the DW by creating an isolated defect region, keeping the total winding number unchanged as shown in Fig. 14.7d. The isolated defect eventually vanishes after dissipating its extra energy. As shown in in Fig. 14.7e, a pure transverse DW appears again. This birth-death process of an antivortex repeats itself periodically.

To manipulate DWs, one needs not only DW propagation but also DW pinning and depinning [36]. The birth-death processes of (anti)vortices during the DW transformation are also closely related to the depinning of DWs from notches [48, 49]. A transverse DW can be depined from a notch assisted by the birth of a vortex [48]. For a strip with periodic notches, the DW motion can be even faster than that in a perfect strip due to the vortex birth-death processes [49].

We have also studied DW models in higher dimensions. Consider $\mathbb{R}^3 \mapsto \mathbf{M}^2$ model where the DW is a plane in 3D space. Due to the well-known bound states inside the DW and the spin wave phase shift after crossing a DW [39, 50], the density of states of spin waves in a DW and in a domain is different. In the presence of a DW, there is an increase in number of low-energy states and a depletion in number of high-energy states, keeping the total number of states equals the total degree of freedom. For bosons with the Bose-Einstein distribution, the low-energy states are always easier to be excited. Thus, the DW has a larger entropy S. The free energy of the system is $F = U - TS$ where U is the internal energy and T is the temperature. Therefore, when the DW has higher temperature, the total free energy of the system is lower. This theory also explains why the DW moves to the hotter side in the presence of a thermal gradient, which is consistent with the magnon STT theory [51].

14.2.2 Vortices and Skyrmions

In previous section, we discussed vortices in the context of vortex walls and winding numbers in 2D XY model ($d = 2, d' = 1$). In this section, we discuss a more general model on $\mathbb{R}^2 \mapsto \mathbf{M}^2$. Since $\mathbf{m}$ is allowed to tilt out-of plane, the spin at vortex (antivortex) center points perpendicularly out-of-plane rather than be singular as discussed before. A new topological charge/number called skyrmion number on $\mathbb{R}^2 \mapsto \mathbf{M}^2$ can be defined [52–54], as a generalization of (14.6),

$$Q = -\frac{1}{4\pi}\int \mathbf{m}\cdot\left(\frac{\partial\mathbf{m}}{\partial x}\times\frac{\partial\mathbf{m}}{\partial y}\right)d^2\mathbf{r}. \tag{14.10}$$

Geometrically, the integrand is the solid angle spanned by $\mathbf{m}$ when one moves all spins to the Bloch sphere center in the spin space. Thus, Q measures how many times the vector $\mathbf{m}$ wraps the unit Bloch sphere.

For vortices whose $\mathbf{m}$ are out-of-plane at the centers and in-plane at the periphery, the skyrmion numbers are $\pm 1/2$ since all $\mathbf{m}$ can only wrap a hemisphere. Different types of vortices are shown in Fig. 14.8, with ansatz profiles

$$\Phi(\phi) = c\phi + \Delta\Phi, \quad \Theta(r) = \frac{\pi}{2}(1 - pe^{-r/R}), \tag{14.11}$$

where r, ϕ are radial and angular coordinates of a point in $\mathbb{R}^2$ centered at the vortex center, and Θ, Φ are polar angle and azimuthal angle of $\mathbf{m}$ in $\mathbf{M}^2$. c takes integer values, measuring how many turns $\mathbf{m}$ goes around the vortex center. $p = \pm 1$ denotes the spin direction at the vortex center ($p = 1$ for $m_z = 1$ and $p = -1$ for $m_z = -1$). c and p are usually called chirality and polarity of a vortex, respectively [55]. Equation (14.10) can be written in polar coordinate with the ansatz (14.11) as $Q = \frac{1}{2}cp$.

The topological charge Q can not only characterize the topology of a vortex, but also directly enters the equation of motion. Thiele derived an equation of motion for a spin texture, known as Thiele equation [54], under the assumption of a rigid-body motion. After the discovery of STT [25], the Thiele equation including STT becomes [37, 53, 56],

$$\frac{\gamma}{M_s}\mathbf{T} + \mathbf{G}\times(\mathbf{v}-\mathbf{u}) - \overleftrightarrow{\mathcal{D}}\cdot(\alpha\mathbf{v} - \beta\mathbf{u}) = 0, \tag{14.12}$$

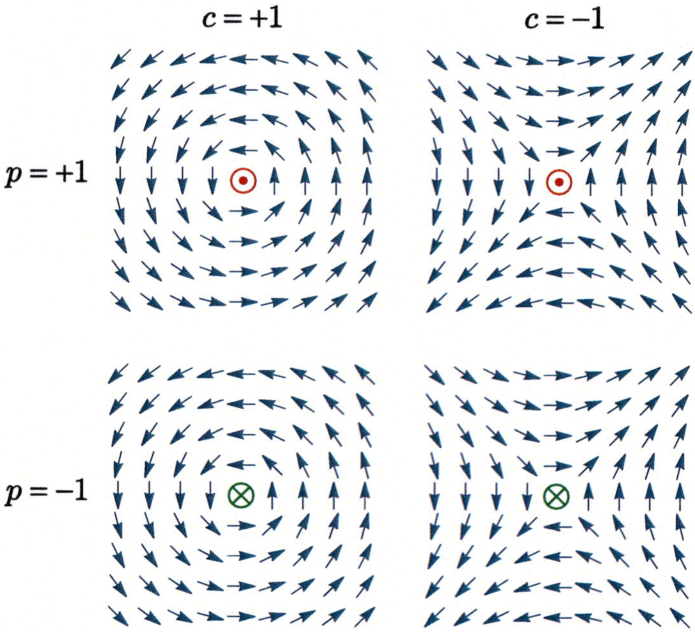

Fig. 14.8 Schematic diagrams for vortices of different chiralities and polarities

where $\mathbf{v}$ is the velocity of the texture; $\mathbf{u} = -\mu_B p\mathbf{J}/[eM_s(1+\beta^2)]$ is a vector with the dimension of a velocity that is proportional to the current density $\mathbf{J}$, in which p is the spin polarization and e is the electron charge; $\mathbf{G}$ is the gyrovector; $\overleftrightarrow{\mathcal{D}}$ is the dissipation tensor defined as $\mathcal{D}_{ij} = \int \partial_i\mathbf{m}\cdot\partial_j\mathbf{m}dV$; and $\mathbf{T}$ is the force on the texture and its i'th component is $T_i = -\frac{\partial\int\mathcal{F}dV}{\partial X_i} - \int\frac{\partial\mathbf{m}}{\partial x_i}\cdot(\mathbf{m}\times\boldsymbol{\tau})dV$, where $\mathcal{F}$ is the free-energy functional and $\mathbf{X}$ is the position of the texture center. The components of gyrovector $\mathbf{G}$ is defined as

$$G_i = \int\frac{1}{2}\varepsilon_{ijk}\mathbf{m}_i\cdot(\partial_j\mathbf{m}\times\partial_k\mathbf{m})\,d^2\mathbf{r}, \tag{14.13}$$

where ε_{ijk} is the Levi-Civita symbol and $i, j, k = 1, 2, 3$ stand for x, y, z. For a 2D film in the xy plane, only G_z is present, and it is easy to see $G_z = 2\pi Q$. The term $\mathbf{G}\times\mathbf{v}$ says that a finite Q induces a transverse motion to the velocity, similar to the Lorentz force or Magnus force.

Due to the non-zero winding number, an isolated vortex cannot be a localized object in an infinite film. However, they can exist locally in confined geometries such as nanodisks [57–63]. The vortices in vortex domain walls (VDWs) could also be used as information carriers in a racetrack memory designs [46]. In the presence of STT only [$\mathbf{T} = 0$ in (14.12)], if $\alpha \neq \beta$, the vortices not only move along the electron flow, but also move in the transverse direction, which hinders the practical

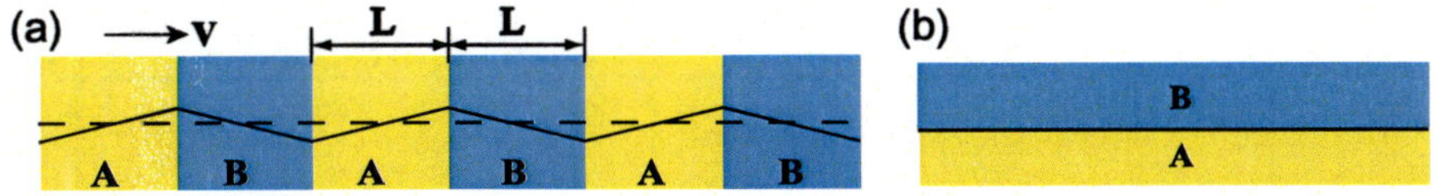

Fig. 14.9 **a** Schematic diagram of superlattice VDW guide, and **b** bilayer VDW guide. $\alpha > \beta$ (A) and $\alpha < \beta$ (B). Figure reproduced from [53]

application. Since Q is a topological invariant that is fixed, the direction of the transverse motion depends on the relative magnitude of α and β. Although $\alpha = \beta$ is unrealistic in practice, heterostructures made of $\alpha > \beta$ and $\alpha < \beta$ materials can be used to build a VDW guide [53]. Figure 14.9 shows two types of VDW guides. The superlattice VDW guide (Fig. 14.9a) is made of superlattice of two materials with $\alpha > \beta$ and $\alpha < \beta$. It requires precise match between α, β and the superlattice period. The vortex moves in a wavy trajectory. A more robust and simple design is the bilayer VDW guide (Fig. 14.9b). The vortex is almost confined near the interface of the $\alpha > \beta$ (A) and $\alpha < \beta$ (B) materials.

As mentioned above, vortices are not particle-like localized textures. Magnetic skyrmions are localized topological spin textures that have attracted a lot of attentions in recent years. Skyrmion is named after Tony Skyrme, who proposed a model to explain why particles are stable in field theory [64, 65]. In the model, particles are topologically protected field configurations (skyrmions), so that they cannot be deformed into trivial configurations topologically equivalent to the vacuum. A topological charge called "Bayon number", which is a winding number defined on $\mathbb{R}^3 \mapsto \mathbf{M}^3$, is defined as

$$B = \frac{1}{24\pi^2} \int \varepsilon_{ijk} \mathrm{Tr}\left(U^\dagger \partial_i U U^\dagger \partial_j U U^\dagger \partial_k U\right) d^3\mathbf{r}, \tag{14.14}$$

where U is the SU(2) matrix order parameter (a 3D version of $\mathbf{m}$). In magnetic systems, because of the 2D nature of $\mathbf{m}$, magnetic skyrmion are 2D topological textures with the topological charge (skyrmion number) Q defined in (14.10) [66]. To distinguish from the original skyrmion, the 2D version is sometimes called "baby skyrmion" [67].

We consider a 2D ferromagnetic film of thickness d with perpendicular magnetic anisotropy (PMA) and DMI. The total energy functional consists of the exchange energy E_{ex}, the DMI energy E_{DM}, the anisotropy energy E_{an}, and the Zeeman energy E_{Ze},

$$E = E_{\mathrm{ex}} + E_{\mathrm{DM}} + E_{\mathrm{an}} + E_{\mathrm{Ze}}, \tag{14.15}$$

where $E_{\mathrm{ex}} = Ad \iint |\nabla \mathbf{m}|^2 \mathrm{d}S$, $E_{\mathrm{an}} = Kd \iint (1 - m_z^2) \mathrm{d}S$, and $E_{\mathrm{Ze}} = M_s B d \iint (1 - m_z) \mathrm{d}S$. The ground state is $m_z = \pm 1$ single domain state. $K = K_u - \mu_0 M_s^2/2$, where K_u is the perpendicular magnetocrystalline anisotropy, and $-\mu_0 M_s^2/2$ is the thin-film shape anisotropy, which is a good approximation when the film thickness d is much smaller than the length scale of magnetization variation. In

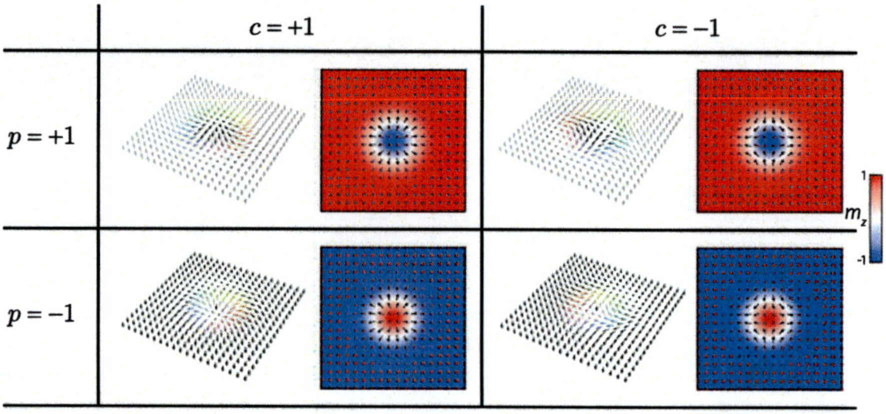

Fig. 14.10 Schematic diagrams for different types of skyrmions

bulk noncentrosymmetric materials such as FeGe and MnSi, the DMI is bulk-like $E_{\mathrm{DM}} = \int D_b\mathbf{m} \cdot (\nabla \times \mathbf{m})$, where D_b is the bulk DMI strength in units of J/m^2 [52]. In inversion-symmetry-broken films such as Pt/Co/AlO$_x$, the DMI is interfacial-like $E_{\mathrm{DM}} = D_i\left[(\hat{\mathbf{z}} \cdot \mathbf{m})\nabla \cdot \mathbf{m} - (\mathbf{m} \cdot \nabla)(\hat{\mathbf{z}} \cdot \mathbf{m})\right]$, where $\hat{\mathbf{z}}$ is the direction normal to the film and D_i is the interfacial DMI strength in units of J/m^2 [68, 69].

Similar to the approximate vortex profile, skyrmion profile in isotropic films is well described by

$$\Theta(r,\phi) = \Theta(r), \quad \Phi(r,\phi) = c\phi + \Delta\phi. \tag{14.16}$$

The skyrmion number (14.10) can be simplified as [52],

$$Q = -\frac{1}{4\pi}\int_0^\infty \int_0^{2\pi} \frac{d\Theta}{dr}\frac{d\Phi}{d\phi} \sin\Theta d\phi dr = \left(\frac{1}{2}\cos\Theta\Big|_0^\infty\right)\left(\frac{1}{2\pi}\Phi\Big|_0^{2\pi}\right) = cp, \tag{14.17}$$

where c is the chirality, and p is the polarity, similar to the vortices. However, different from the vortices, the magnetization of skyrmions can wrap the whole unit sphere so that the skyrmion number takes integer values. Skyrmions of negative c such as $c = -1$ are often called antiskyrmions. The highly twisted skyrmions of $c \leq 2$ usually have higher energies. Figure 14.10 shows different types of skyrmions. The value of $\Delta\phi$ depends on the form of DMI.

We now focus on $c = 1$ skyrmions which are observed in most experiments [52, 68, 70, 71]. The skyrmions can form lattices [72, 73] or be isolated [52, 68, 70, 71, 74, 75]. From the application point of view, the isolated skyrmions are more useful since they can be manipulated individually so that reading and writing of data [76, 77] are possible. It is very interesting to know the size of skyrmions in a given material [78–83]. We present a theory for skyrmion size below.

As an example, we consider the interfacial DMI generally existed in heterostructures. Four energy terms in terms of Θ for $c = 1$ are

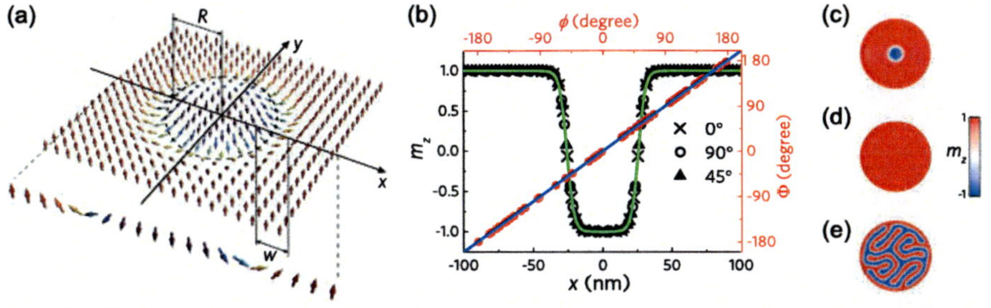

Fig. 14.11 **a** Schematic diagram of a Néel skyrmion of radius R and wall (between the skyrmion core and the outer domain) width w in a 2D film. The arrows denote the spin direction. Spin orientations of the skyrmion along x axis are sketched below the main figure. **b** (Right axis) ϕ-dependence of Φ (red circles) and (Left axis) radial (r) distribution of $m_z = \cos\Theta$ along the diameters of $\phi = 0$ (crosses), 45° (triangles), and 90° (circles) for $A = 15\,\text{pJm}^{-1}$, $D = 3.7\,\text{mJm}^{-2}$, $K_u = 0.8\,\text{MJm}^{-3}$, and $B = 0$. The green solid line is the fit to $\Theta_{\text{dw}}(r) = 2\arctan\left[\frac{\sinh(R/w)}{\sinh(r/w)}\right]$ with fitting parameters of $R = 25.77$ nm and $w = 4.94$ nm. The blue solid line is $\Phi = \phi$. **c–e** Three typical equilibrium states obtained in the numerical simulations for different parameters. **c** Isolated skyrmion. **d** Single-domain state of $m_z = 1$ (or $m_z = -1$). **e** Stripe domains. The pixel color encodes the m_z component with the color bar shown in the figure. Figure reproduced from [84]

$$E_{\text{ex}} = 2\pi Ad \int_0^\infty \left[\left(\frac{\text{d}\Theta}{\text{d}r}\right)^2 + \frac{\sin^2\Theta}{r^2}\right] r\text{d}r, \tag{14.18}$$

$$E_{\text{DM}} = 2\pi Dd\cos\Delta\phi \int_0^\infty \left(\frac{\text{d}\Theta}{\text{d}r} + \frac{\sin 2\Theta}{2r}\right) r\text{d}r, \tag{14.19}$$

$$E_{\text{an}} = 2\pi Kd \int_0^\infty \sin^2\Theta r\text{d}r, \tag{14.20}$$

$$E_{\text{Ze}} = 2\pi M_s Bd \int_0^\infty (1-\cos\Theta) r\text{d}r. \tag{14.21}$$

$\Delta\phi = 0$ or π minimizes the DMI term. This kind of skyrmions are called Néel skyrmions or hedgehog skyrmions. Along a radial direction, the skyrmion profile looks like a 360° Néel domain wall as illustrated in Fig. 14.11b. This leads us to model a skyrmion profile by a 360° domain wall originally proposed by Braun [75, 85, 86],

$$\Theta_{\text{dw}}(r) = 2\arctan\left[\frac{\sinh(R/w)}{\sinh(r/w)}\right]. \tag{14.22}$$

Whether this is a good approximation subjects to its comparison with micromagnetic simulations. To test how good ansatz (14.22) is for a skyrmion, we use MuMax3 [87] to simulate various stable spin structures. $A = 15$ pJm^{-1}, $M_s = 580$ kAm^{-1}, and perpendicular easy-axis anisotropy $K_u = 0.8$ MJm^{-3} [68] are used to mimic Co layer in Pt/Co/MgO system. The initial state is $m_z = 1$ for $r > 10$ nm and $m_z = -1$ for $r \leq 10$ nm. The final stable structures depend on the values of D and B. The lower left inset of Fig. 14.11 is three typical stable structures. Figure 14.11c is a skyrmion for $D = 3.7$ mJm^{-2} and $B = 0$. Figure 14.11d is a single-domain state of $m_z = 1$ (or $m_z = -1$) for $D = 0$ and $B = 0$. Figure 14.11e is a stripe domains state for $D = 5$ mJm^{-2} and $B = 0$. Figure 14.11b shows the spatial distribution of m_z of the skyrmion in Fig. 14.11c along three radial directions, $\phi = 0$ (the crosses), 45° (the triangles), and 90° (the circles). All three sets of data are on the same smooth curve, showing that m_z is a function of r, but not ϕ. The curve can fit perfectly to (14.22) with $R = 25.77$ nm, $w = 4.94$ nm. We plotted also $\Phi(\phi)$ at randomly picked spins from the simulated skyrmion. All numerical data (red circles) are perfectly on the curve of $\Phi = \phi$. These results not only confirm the validity of (14.22), but also suggest that $m_z(r) \equiv \cos[\Theta_{\mathrm{dw}}(r)]$ follows the 360° DW profile (14.22) although it is not exact since (14.22) does not satisfy nonlinear partial differential equations for skyrmions [83].

The energy of a skyrmion can then be obtained from (14.18) by using the 360° domain wall profile $\Theta_{\mathrm{dw}}(r)$. Substituting $\Theta_{\mathrm{dw}}(r)$ into (14.15) and (14.18), the total energy is, in general, a function of R and w [instead of a functional of $\Theta(r)$ and $\Phi(\phi)$] as

$$\begin{aligned} E(R,w) = 4\pi d \Bigg\{ & A\left[f_1\left(\frac{R}{w}\right) + f_2\left(\frac{R}{w}\right)\right] \\ & + Dw\left[f_3\left(\frac{R}{w}\right) + f_4\left(\frac{R}{w}\right)\right] \\ & + Kw^2 f_5\left(\frac{R}{w}\right) + \mu_0 M_s B w^2 f_6\left(\frac{R}{w}\right)\Bigg\}, \end{aligned} \tag{14.23}$$

where $f_i(x)$ ($i = 1 \sim 6$) are non-elementary functions defined as,

$$\begin{aligned} f_1(x) &= \int_0^\infty \frac{2\sinh^2(x)\cosh^2(t)}{\left[\sinh^2(x) + \sinh^2(t)\right]^2} t\mathrm{d}t, \\ f_2(x) &= \int_0^\infty \frac{2\sinh^2(x)\sinh^2(t)}{t\left[\sinh^2(x) + \sinh^2(t)\right]^2} \mathrm{d}t, \\ f_3(x) &= -\int_0^\infty \frac{t\sinh x\cosh t}{\sinh^2 x + \sinh^2 t}\mathrm{d}t, \\ f_4(x) &= -\int_0^\infty \frac{\sinh x\sinh t\left(\sinh^2 x - \sinh^2 t\right)}{\left(\sinh^2 x + \sinh^2 t\right)^2}\mathrm{d}t, \\ f_5(x) &= \int_0^\infty \frac{2t\sinh^2 x\sinh^2 t}{\left[\sinh^2 x + \sinh^2 t\right]^2}\mathrm{d}t, \\ f_6(x) &= \int_0^\infty \frac{\sinh^2 x}{\sinh^2 t + \sinh^2 x} t\mathrm{d}t. \end{aligned}$$

The skyrmion size R and wall width w are the values that minimize $E(R, w)$. In [78], w was assumed to be a fixed value such as $w = \sqrt{A/K}$ that eventually leads to a wrong skyrmion size. Figure 14.12 is $D-$ (a), $A-$ (b), $K-$ (c) and $B-$ (d)

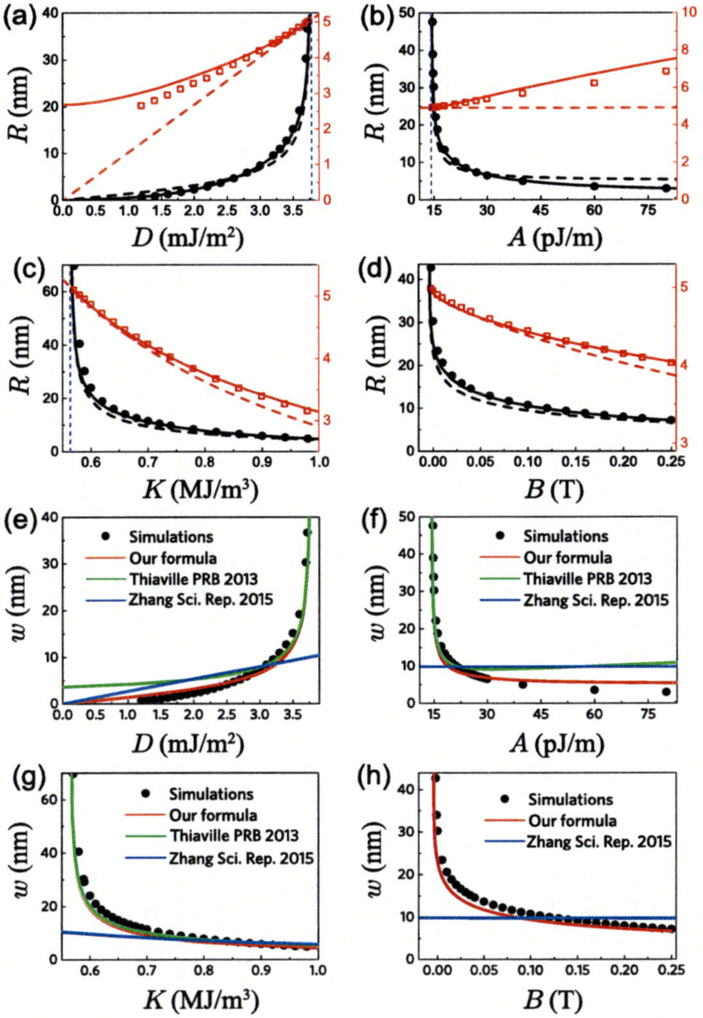

Fig. 14.12 **a–d** The D (a), A (b), K (c) and B (d) dependence of the skyrmion size R (the left axis) and wall width w (the right axis). Model parameters are $A = 15$ pJ/m, $K = K_u - \mu_0 M_s^2/2 = 0.588$ MJ/m^3, $D = 3.7$ mJm^{-2}, and $B = 0$. In each subfigure, one of these four parameters is treated as a tuning parameter, and the other three parameters are fixed to above values. The symbols are the micromagnetic simulation data. The solid lines are exact analytic results obtained from minimising E of (14.23). The dashed lines are approximate results of $R = \pi D\sqrt{\frac{A}{16AK^2-\pi^2 D^2 K}}$, $w = \frac{\pi D}{4K}$ [subfigures **a–c**] and solution of (14.25)(14.26) [subfigure **d**]. Vertical dashed lines are the upper (lower) bound of parameters above (below) which a stable skyrmion cannot exist. **e–h** The comparison of the skyrmion size R for the micromagnetic simulation (the black dots), our formula (14.27) [or solution of (14.25) (14.26) for non-zero B] (the red lines), and previous studies [78] (the green lines), [80] (the blue lines). Figure redrawn using the results in [84]

dependences of the skyrmion size R (left y-axis) and the skyrmion wall width w (right y-axis), with other parameters fixed to the values for Co mentioned earlier. The symbols are the micromagnetic simulation data [R is the size of $m_z = 0$ contour and w is the fit of the skyrmion profile to $\Theta_{\rm dw}(r)$]. The sample size used in the simulations is large enough to avoid the boundary effects. Solid lines are numerical results from minimising E of (14.23). The simulation results agree almost perfectly (except a slight deviation in the D-dependence of w for smaller D) with our analytic results of (14.23). Both micromagnetic simulations and analytical results clearly show that the skyrmion can exist for $D < 3.8$ mJ/m^2 in the current case. Above the upper limit, the stable state is not a skyrmion, but stripe domains as shown in Fig. 14.11e for $D = 5$ mJ/m^2. $E(R, w)$ of (14.23) has a minimum as long as $|D| < 3.8$ mJ/m^2 that indicates existence of the skyrmion. However, micromagnetic simulations show that the skyrmion can only exist in the window of 1.2 mJm^{-2} $< D <$ 3.8 mJ/m^2 when the skyrmion size is larger than 1 nm in the current case. Below 1.2 mJ/m^2, the stable state is a single domain with all spins pointing up or down as shown in Fig. 14.12d. This discrepancy is due to the discretization of continuous LLG equation in micromagnetic simulations. There exists also a minimal A of around 14 pJ/m as shown in Fig. 14.12b and a minimal K of around 0.56 MJm^{-3} as shown in Fig. 14.12c, below which the skyrmion does not exist, and the stable state is stripe domains as shown in Fig. 14.11e. The skyrmion size decreases with B, which is consistent with the experimental observations [70, 75, 88].

It is still unclear how R and w depend on A, D and K although (14.23) agrees almost perfectly with the simulation results. Thus it is highly desirable to have simple approximate expressions for R and w in terms of material parameters. The exchange and DMI energies come from the spatial magnetization variation rate. For a skyrmion, the magnetization variation rates in the radial and tangent directions scale respectively as $1/w$ and $1/R$. The exchange energy is then proportional to the skyrmion wall area of $\pi R w$ multiplying the square of the magnetization variation rates $1/R^2 + 1/w^2$, i.e. $E_{\rm ex} \propto (R/w + w/R)$. The integrand of the second term (tangential term) of $E_{\rm DM}$ is proportional to $\sin 2\Theta$ due to the triple product of the interfacial DMI vector (along the radial direction), the tangential variation of $\mathbf{m}$ (along the tangential direction), and $\mathbf{m}$ itself. Because $\sin 2\Theta$ is almost asymmetric about $r = R$ (positive when $r > R$ and negative when $r < R$) and nonzero only near $r = R$, the second term in $E_{\rm DM}$ is vanishingly small after integral over r. The DMI energy is then mainly contributed by the first term (radial term), which is proportional to wall area (Rw) multiplying the magnetization variation rate along radial direction ($1/w$), i.e. $E_{\rm DM} \propto R$. The isotropic energy is mainly from the skyrmion wall area. Thus, $E_{\rm an} \propto Rw$. The Zeeman energy of the skyrmion comes from the skyrmion core proportional to its area of $\pi(R - cw)^2$, where c is a coefficient depending on the magnetization profile, and from the wall area proportional to its area of $\pi R w$. To obtain the proportional coefficients, one needs to find approximate expressions for $f_i(R/w)$ ($i = 1, \ldots, 6$) in (14.23). In the case of $R \gg w$ (or $x \equiv R/w \gg 1$), $\sinh(x) \approx \cosh(x) \approx e^x$. Thus, function $g(t, x) = [2\sinh^2(x)\cosh^2(t)]/[\sinh^2(x) + \sinh^2(t)]^2 \approx 2e^{2(x-t)}/[e^{2(x-t)} + 1]^2$ is positive and significantly non-zero only near $t = x$, reflecting the fact that $E_{\rm ex}$, $E_{\rm DM}$, and $E_{\rm an}$ are mainly from the skyrmion wall region that is assumed to be very thin. Furthermore, the area bounded by $g(t, x)$-curve and t-axis is 1 so that $g(t, x) \approx \delta(t - x)$ resembles the properties of a delta function. f_i under this approximation are,

$$f_1(x) \approx x, \quad f_2(x) \approx \frac{1}{x},$$
$$f_3(x) \approx \frac{\pi}{2}x, \quad f_4(x) \approx 0,$$
$$f_5(x) \approx x \quad f_6(x) \approx \frac{x^2}{2} + \frac{\pi^2}{24}.$$

The total energy is then

$$E(R, w) = 4\pi d\left[A\left(\frac{R}{w} + \frac{w}{R}\right) - \frac{\pi}{2}DR + KwR + M_s B\left(\frac{R^2}{2} + \frac{\pi^2}{24}w^2\right)\right]. \quad (14.24)$$

Due to the specific form of the magnetization profile of $\Theta_{dw}(r)$, Rw-term in E_{Ze} vanishes and $E_{Ze} \approx 4\pi M_s B\left(\frac{R^2}{2} + \frac{\pi^2}{24}w^2\right)$. The skyrmion size and wall width are then the values that make $E(R, w)$ minimal, or

$$A\left(\frac{1}{w} - \frac{w}{R^2}\right) - \frac{\pi}{2}D + Kw + M_s BR = 0, \quad (14.25)$$

$$A\left(-\frac{R}{w^2} + \frac{1}{R}\right) + KR + \frac{\pi^2}{12}M_s Bw = 0. \quad (14.26)$$

For $B = 0$, (14.25) and (14.26) can be analytically solved. The results are

$$R = \pi D\sqrt{\frac{A}{16AK^2 - \pi^2 D^2 K}}, \quad w = \frac{\pi D}{4K}. \quad (14.27)$$

The dashed lines in Figs. 14.12a–c are the approximate formulas that compare quite well with simulation results too. When $B \neq 0$, the equation for w is a sextic equation,

$$432\pi^2 A^2 D^2 - 3456\pi A^2 DKw + [-(1728 + 288\pi^2 + 12\pi^4)A^2(M_s B)^2 + 6912\,A^2K^2]w^2$$
$$+ [3456\pi ADK^2 - (72 - 6\pi^2)\pi^3 AD(M_s B)^2]w^3 + [-6912K^2 + 576\pi^2(M_s B)^2]AKw^4$$
$$- 6\pi^5(M_s B)^2 DKw^5 + \pi^4(M_s B)^2[12K^2 - \pi^2(M_s B)^2]w^6 = 0, \quad (14.28)$$

and R, in terms of w, is

$$R = \frac{\pi^2 M_s Bw^3 \pm w\sqrt{576A^2 - 576AKw^2 + \pi^4(M_s B)^2 w^4}}{24(A - Kw^2)}. \quad (14.29)$$

The sextic equation does not have a closed-form solution, but their numerical solutions are easily obtained as plotted in dashed lines in Fig. 14.12d. In summary, our approximate formula agrees very well with the simulations for $R \gg w$ as expected from our approximation. For smaller skyrmions, the approximation is still not bad, and qualitatively gives correct parameter dependence. We can also determine the upper limit of D and lower limits of A, K, and B from the approximate formula. Since R must be real and finite, we have

$$16AK > \pi^2 D^2, \tag{14.30}$$

for $B = 0$, so that the upper limit of D and lower limits of A and K are obtained as $D < \frac{4}{\pi}\sqrt{AK}$, $A > \frac{\pi^2 D^2}{16K}$, and $K > \frac{\pi^2 D^2}{16A}$. The limit is also the critical value separating the uniform state from helical state [89]. These critical values are plotted in Figs. 14.12a–c as vertical dashed lines that agrees also well with simulations.

It is natural to extend our approach to Bloch skyrmions in the systems with bulk inversion symmetry broken. The bulk DMI energy $E_{\mathrm{DM}} = D \iint \mathbf{m} \cdot (\nabla \times \mathbf{m}) dS$ can be rewritten as

$$E_{\mathrm{DM}} = 2\pi D d \sin\gamma \int_0^\infty \left(\frac{\mathrm{d}\Theta}{\mathrm{d}r} + \frac{\sin 2\Theta}{2r} \right) r \mathrm{d}r, \tag{14.31}$$

where $\gamma = \pm\pi/2$ gives minimal energy. Since all other discussions are the same as those for Néel skyrmions, the formulas of R and w are applicable to the Bloch skyrmions.

It is worth mentioning that the topological protection does not mean an absolute stability. The energy barrier between a skyrmion state and the single domain state is still finite. The skyrmions can be generated from the edges without breaking the conservation of topology [90], and from the bulk with the conservation of topology broken [70, 71].

The non-trivial topology of the skyrmions is a double-edged sword. Despite of the better stability, similar to the vortices, the non-trivial topological number of skyrmions leads to a finite gyrovector [54] in the Thiele equation (14.12), so there is a Skyrmion Hall effect which means the existence of a transverse motion in a longitudinal driving force [52, 68, 79, 91–93]. Thus, there exists a threshold current density above which skyrmions can annihilate at the film edge [94]. This edge effect strongly limits the speed of skyrmion propagation which is of vital importance for real applications.

14.2.3 Hopfions

The 1D and 2D magnetic solitons have been discovered and studied for a long time. A natural question is whether there are 3D solitons and how they behave. Of course, there are transient states that can be regarded as 3D solitons, and some of them can be topologically non-trivial [95]. However, the existence of *stable* 3D solitons in magnetic systems is a highly non-trivial problem. The Derrick theorem [96] indicates that there is no stable solitons in three or higher dimension for Heisenberg ferromagnets. However, the theorem does not forbid the existence of stable 3D solitons in the presence of DMI or higher-order exchange interactions. In addition, the Derrick theorem is only applicable for infinite systems. In finite systems with boundary confinement, stable 3D solitons are also possible.

The existence of 3D string-like topological solitons has been proposed by Ludvig D. Faddeev [97] as a limit of the Skyrme model [64]. These 3D topological solitons are known as Faddeev-Hopf knots [98] or hopfions, which are classified by a topological charge called the Hopf index [99]. Hopfions have been discussed in many physical systems, such as gauge theories [97, 100], cosmic strings [101], ferromagnets [95] (as a special case of dynamical vortex rings), low-temperature bosonic systems [102–104], fluids [105], and liquid crystals [106–108]. Recently, stable magnetic hopfions were numerically predicted in finite-size noncentrosymmetric FM systems with Dzyaloshinskii-Moriya interaction (DMI) [109, 110] and interfacial perpendicular magnetic anisotropy (PMA) [111–113] or higher-order exchange interaction [114].

To study the statics and dynamics hopfions, we consider a film of thickness d_t, and start with the energy functional [56],

$$\mathcal{F} = \int_V \left[A_{ex} |\nabla \mathbf{m}|^2 + E_{\mathrm{DM}} \left(\mathbf{m}, \frac{\partial \mathbf{m}}{\partial x_i} \right) + K_b(1 - m_z^2) + BM_s(1 - m_z) \right] \mathrm{d}V + \int_{z=\pm d/2} K_s(1 - m_z^2) \mathrm{d}S + E_d, \quad (14.32)$$

where A_{ex} is the exchange constant; E_{DM} is the aforementioned DMI energy density functional, which depends on the symmetry of the system; K_b and K_s are the bulk PMA and the interfacial PMA, respectively; B is a perpendicular magnetic field; M_s is the saturation magnetization; and E_d is the demagnetization energy.

Figures 14.13a, b show the typical magnetization profiles of Bloch-type (corresponding to Bulk DMI) and Néel-type hopfions (corresponding to interfacial DMI), respectively, obtained by numerical relaxation of the total energy using conjugate gradient method [115]. The simulations are performed using MuMax3 [87] at zero temperature with mesh size 0.5 nm×0.5 nm×0.5 nm. We consider a 128 nm×128 nm×16 nm film with $A_{ex} = 0.16\,\mathrm{pJ\,m^{-1}}$ and $M_s = 1.51 \times 10^5\,\mathrm{A\,m^{-1}}$, representing MnSi parameters [111]. No external field is applied. The Bloch-type (Néel-type) hopfions are favorable in bulk (interfacial) DMI systems. In Fig. 14.13a, we use $K_s = 0.5$ mJ m^{-2}, $K_b = 41$ kJ m^{-3}, and $D_b = 0.115$ mJ m^{-2}, while in Fig. 14.13b, we use $K_s = 0.5$ mJ m^{-2}, $K_b = 20$ kJ m^{-3}, and $D_i = 0.115$ mJ m^{-2}. The surface PMA is modelled by imposing a very strong PMA $K = 10^6$ J m^{-3} on two additional layers attached to the top and bottom surfaces. This corresponds to a surface anisotropy $K_s = 0.5$ mJ m^{-2} by multiplying the mesh size.

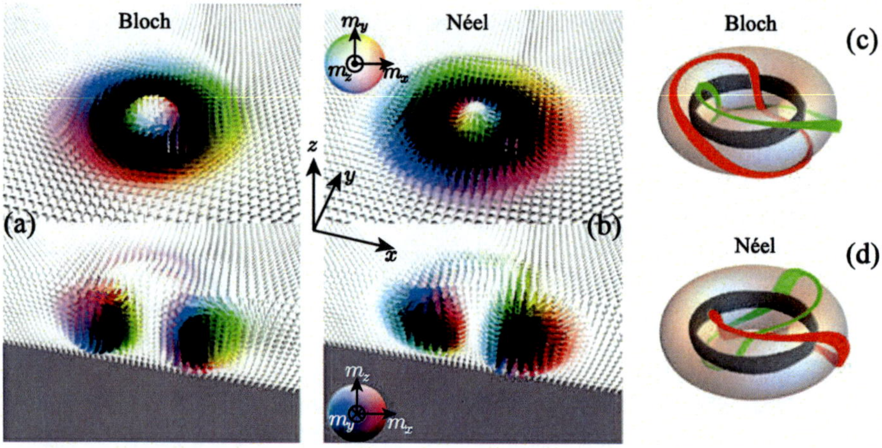

Fig. 14.13 **a, b** Midplane cross-sections in the xy-plane (the upper panel) and the xz-plane (the lower panel) of **a** a Bloch-type hopfion and **b** a Néel-type hopfion. **c, d** The preimages of $\mathbf{m} = (0, 0, -1)$, $(1, 0, 0)$ and $(0, 1, 0)$ for **c** a Bloch-type hopfion and **d** a Néel-type hopfion. The tori are the isosurfaces of $m_z = 0$. The colors of the arrows in **a, b** and the preimages in **c, d** depict the full orientation of the corresponding **m**. The color sphere and the coordinate system are shown in the insets. Figure redrawn using the results in [56]

Because the hopfions are non-isomorphic maps from $\mathbb{R}^3 \cup \{\infty\}$ to $\mathbb{S}^2$, the topological invariant of hopfions, known as the Hopf index H, differs from the winding number. This index is defined as

$$H = \frac{1}{(4\pi)^2} \int_V \mathbf{F} \cdot \mathbf{A} \mathrm{d}V, \tag{14.33}$$

where $F_i = \varepsilon_{ijk}\mathbf{m} \cdot \left(\partial_j \mathbf{m} \times \partial_k \mathbf{m}\right)/2$, in which $i, j, k = \{x, y, z\}$ and ε_{ijk} is the Levi-Civita symbol, and $\mathbf{A}$ is a vector potential, which satisfies $\nabla \times \mathbf{A} = \mathbf{F}$ [116]. The components of $\mathbf{F}$ are solid angle densities in different coordinate planes. $\mathbf{F}$ can be understood as the gyrovector density [54], emergent magnetic field [117], or topological charge [52]. The Hopf index takes integer value for localized textures (i.e. the order parameter is homogeneous at the infinity) [111, 118]. Geometrically, the absolute value of Hopf index means how many times the close-loop preimages (the constant-**m** lines) link with each other [119].

We first demonstrate that the Hopf index is well-defined for an infinite system. Straightforward derivation shows that $\mathbf{F}$ is divergenceless ($\nabla \cdot \mathbf{F} = 0$) when $|\mathbf{m}|$ =constant such that the vector potential $\mathbf{A}$ exists. However, obviously, $\mathbf{A}$ is not unique. For any continuous function $\varphi(\mathbf{r})$, $\mathbf{A}' = \mathbf{A} + \nabla\varphi$ is also a vector potential. The corresponding Hopf index is

$$H' = \frac{1}{(4\pi)^2} \int \mathbf{F} \cdot \mathbf{A}' \mathrm{d}V = H + \frac{1}{(4\pi)^2} \int \mathbf{F} \cdot \nabla\varphi \mathrm{d}V. \tag{14.34}$$

The integral in the extra term can be rewritten as

$$\int \mathbf{F} \cdot \nabla \varphi \mathrm{d}V = \int \nabla \cdot (\varphi \mathbf{F}) \mathrm{d}V - \int \varphi \nabla \cdot \mathbf{F} \mathrm{d}V = \oint \varphi \mathbf{F} \cdot d\mathbf{S} - 0 = \oint \varphi \mathbf{F} \cdot d\mathbf{S}, \tag{14.35}$$

where Gauss's theorem has been used and $\oint$ means the integration over the surface of the volume. In an infinite system, the surface is infinitely far away, and the **m** field should be homogenous such that, on the surface, **F** is 0, and the integral $\oint \varphi \mathbf{F} \cdot d\mathbf{S}$ vanishes. Thus, we have $H' = H$, meaning that the Hopf index is well-defined independent of the choice of **A**.

The hopfion profile centered at a certain location can be expressed via $\Theta(r, \phi, z)$, $\Phi(r, \phi, z)$, where (r, ϕ, z) are cylindrical spatial coordinates, and Θ, Φ are the polar and azimuthal angles of the magnetization. For a rotationally symmetric system, it is natural to assume that the hopfion profile following Θ is independent of ϕ, and $\Phi(r, \phi, z) = \Delta\Phi(r, z) + n\phi$, where n is an integer. This form means that the polar angle (or z component) of **m** is independent of ϕ, and when transversing a whole circle centered at the origin in real space (ϕ changes from 0 to 2π), the azimuthal angle Φ of **m** uniformly rotates by $2n\pi$. With this assumption, we can write the **F** field in cylindrical coordinates as

$$F_r = \mathbf{m} \cdot \left(\frac{\partial \mathbf{m}}{r \partial \phi} \times \frac{\partial \mathbf{m}}{\partial z} \right) = -n \frac{\sin \Theta}{r} \frac{\partial \Theta}{\partial z}, \tag{14.36}$$

$$F_\phi = \mathbf{m} \cdot \left(\frac{\partial \mathbf{m}}{\partial z} \times \frac{\partial \mathbf{m}}{\partial r} \right) = \sin \Theta \left(\frac{\partial \Theta}{\partial z} \frac{\partial \Delta\Phi}{\partial r} - \frac{\partial \Theta}{\partial r} \frac{\partial \Delta\Phi}{\partial z} \right), \tag{14.37}$$

$$F_z = \mathbf{m} \cdot \left(\frac{\partial \mathbf{m}}{\partial z} \times \frac{\partial \mathbf{m}}{r \partial \phi} \right) = n \frac{\sin \Theta}{r} \frac{\partial \Theta}{\partial r}. \tag{14.38}$$

The vector potential **A** is

$$A_r = -(1 + \cos \Theta) \frac{\partial \Delta\Phi}{\partial r}, \tag{14.39}$$

$$A_\phi = \frac{n}{r}(1 - \cos \Theta), \tag{14.40}$$

$$A_z = -(1 + \cos \Theta) \frac{\partial \Delta\Phi}{\partial z}. \tag{14.41}$$

Then, the Hopf index is

$$H = \frac{1}{(4\pi)^2} \int \mathbf{F} \cdot \mathbf{A} \mathrm{d}V = \frac{n}{4\pi} \int_{-\infty}^{+\infty} \int_0^{+\infty} \sin \Theta \left(\frac{\partial \Theta}{\partial z} \frac{\partial \Delta\Phi}{\partial r} - \frac{\partial \Theta}{\partial r} \frac{\partial \Delta\Phi}{\partial z} \right) \mathrm{d}r \mathrm{d}z. \tag{14.42}$$

Thus, the Hopf index equals the whirling number n along the ϕ direction multiplied by the skyrmion number at the rz half plane [118].

We compute that the Hopf indices are 0.96 (Bloch) and 0.95 (Néel) for Fig. 14.13 [120]. The two types of hopfions are topologically equivalent. The upper and lower panels are the midplane cross-sections in the xy-plane and xz-plane. The magnetization profile in each xy-midplane cross-sections is Bloch-type (a) or Néel-type (b) skyrmionium or the target skyrmion [121, 122], while the xz-midplane cross-section shows a pair of vortices with opposite chirality. The right ($x > 0$) xz-midplane contains a vortex (antivortex) with chirality $+1$ (-1) for an $H = +1$ ($H = -1$) hopfion. Outside the hopfions and at the center of the hopfions, the magnetization is along the z direction, and the donut-shape transition region is chiral (for Bloch-type hopfions) or hedgehog-like (for Néel-type hopfions). Figure 14.13c, d show the corresponding preimages (constant-$\mathbf{m}$ curves in real space) of Fig. 14.13a, b. The preimages link with each other once, which is consistent with the Hopf index calculation, justifying the hopfion nature of the textures in (a) and (b).

Unlike skyrmions, although the topology of a hopfion is nontrivial, the gyrovector $\mathbf{G} = \int \mathbf{F} \mathrm{d}V$ of a hopfion vanishes. Consequently, the main drawback of a FM skyrmion racetrack memory, the skyrmion Hall effect, is absent in the hopfion racetrack memory. In addition to the numerical verification, the vanishing gyrovector of a hopfion can be understood as follows. Consider a film that is isotropic in the xy plane. In cylindrical coordinates, $F_z = n \frac{\sin\Theta}{r} \frac{\mathrm{d}\Theta}{\mathrm{d}r}$. We can rewrite $G_z = \int F_z \mathrm{d}V$ as

$$G_z = \int_V F_z r \mathrm{d}r \mathrm{d}\phi \mathrm{d}z = -2n\pi \int_{-d/2}^{d/2} \left(\cos\Theta\Big|_{r=0}^{r=\infty}\right) \mathrm{d}z. \tag{14.43}$$

Since in a hopfion the magnetization directions are the same at both the periphery ($r = \infty$) and the center ($r = 0$), G_z vanishes. Since the two vortices in any xz (or yz) midplane cross-section have opposite chirality, as shown in the lower panels of Fig. 14.13a, b, the integration of F_x (or F_y) over the volume gives a vanishing contribution to G_x (or G_y). The components of $\mathbf{G}$ are invariant under continuous deformation [52]; therefore, $\mathbf{G} = 0$ applies to all the hopfions.

According to the Thiele equation (14.12), the direct consequence of the vanishing gyrovector is the absent of Hall effect. Under spin transfer torque, the hopfions move straightforward along the current. Under spin Hall torque which emerges in the heavy metal/ferromagnet heterostructures, Néel-type hopfions move along the current while the Bloch-type hopfions move transverse to the current.

14.3 Topological Spin Waves

Matter that possesses excitations with nontrivial band topology has attracted enormous attention in recent years because of their interesting and exotic properties [123–127]. One of such properties is the existence of unidirectional and topologically protected surface/edge states that are robust against internal and external perturbations. The study was initially exclusive for electron systems and was believed to be a quantum phenomenon. It is now known that the topological states can exist in classical mechanics [128] and photonics [129]. There are also intensive researches on topological magnetic states in recent years [130–146]. Magnons are quanta of spin waves, which are the propagation of spin fluctuation. Magnons are useful particles/excitations due to their low energy consumption and long coherence distance, as well as a control knob of magnetization dynamics [31, 33, 39]. Generation, detection, and manipulation of magnons is an emerging subfield called magnonics in condensed matter physics [147–150], and topologically protected unidirectional surface spin waves should be very useful in magnonics because they are highly non-susceptible to external perturbations.

In this section, we consider a magnetic thin film whose bulk spin waves are gapped, but its edge spin waves are chiral and gapless. This is a magnonic counterparty of quantum Hall systems or topological insulators [126, 127]. The film consists of a set of perpendicularly magnetized ferromagnetic spins on a honeycomb lattice with a nearest-neighbour (NN) pseudodipolar interaction and/or next-nearest-neighbour (NNN) Dzyaloshinskii-Moriya interaction (DMI) as well as the usual NN Heisenberg ferromagnetic interaction. Although the magnons are bosons, the bulk spin wave bands carry nontrivial Chern numbers so that one knows for sure the existence of topologically-protected unidirectional (chiral) gapless spin waves at edges inside the bulk gap according to the bulk-boundary correspondence [151, 152]. A staggered sublattice anisotropy can change the system from a topologically nontrivial phase to a trivial phse, and vice versa. It is also possible to tune the topological phase by changing DMI. In 14.3.1, we show how to calculate the Chern number of the bosonic spin wave bands, nontrivial spin wave bands of a finite sample and topologically protected edge spin waves [153–155]. In 14.3.2, we go a step further to consider 3D magnets that are either magnonic weak topological insulators or magnonic Weyl semimetals [156, 157].

14.3.1 Topologically Protected Edge Spin Waves

Various magnetic systems has been proposed for realizing nontrivial topology of spin wave bands, such as kagome ferromagnet with NN DMI [132, 135, 137], dipolar magnonic crystals [144, 145], honeycomb ferromaget with NNN DMI [142, 143], and so on. We call them topological magnonic systems. Here, we provide the basic ingredients and a theoretical framework for topological magnonic materials.

As a representative, we consider ferromagnetic spins on a honeycomb lattice as shown in Fig. 14.14a. $\mathbf{a}_{1(2,3)}$ are three neighbouring site-vectors of length a. The blue and red arrows represent the magnetic moments $\boldsymbol{\mu}_i = \mu \mathbf{m}_i$ of magnitude μ and direction $\mathbf{m}_i$ at site i of sublattices A and B, respectively. The system is described by a Hamiltonian,

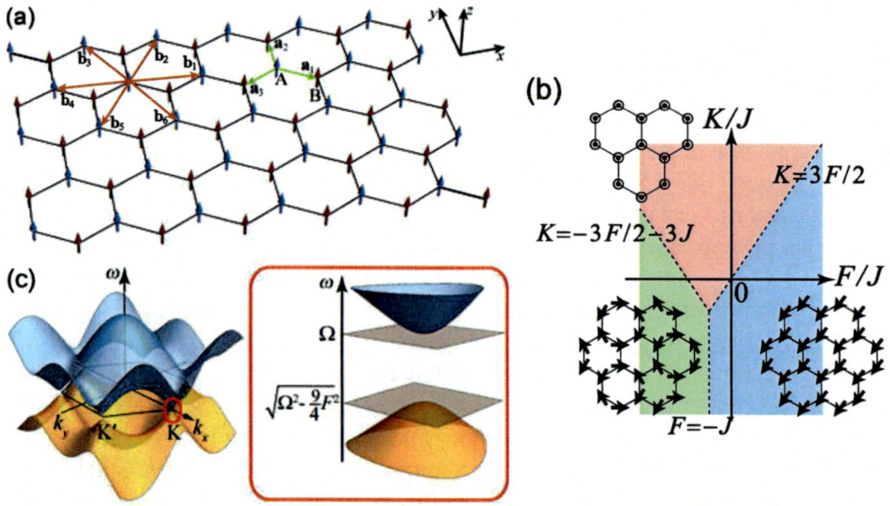

Fig. 14.14 **a** Schematic diagram of ferromagnetic spins on a honeycomb lattice with perpendicular anisotropy (along the z direction). The blue and red arrows denote magnetic moment vectors on A and B sublattices. The green arrows are three neighbouring site-vectors $\mathbf{a}_{1,2,3}$, and the brown arrows are next-nearest-neighbour vectors $\mathbf{b}_{1\sim 6}$. **b** Phase diagram of ground state in the $K/J-F/J$ plane when $D=\Delta=0$. The spin arrangements in these phases are shown in the insets, and the equations of phase boundaries are indicated. **c** The left panel is the spin wave spectrum of an infinite system in the first Brillouin zone (hexagon shape) for $J=0.1$, $F=0.5$ and $M=1.3$ (in units of $\mu_0\mu^2/a^3$). The right panel is a close-up view of the gap at K point (indicated by the red circle in the left panel)

$$\mathcal{H}=-\frac{J}{2}\sum_{\langle i,j\rangle}\mathbf{m}_i\cdot\mathbf{m}_j-\frac{F}{2}\sum_{\langle i,j\rangle}(\mathbf{m}_i\cdot\mathbf{e}_{ij})(\mathbf{m}_j\cdot\mathbf{e}_{ij})$$
$$-D\sum_{\langle\langle i,j\rangle\rangle}\nu_{ij}\hat{\mathbf{z}}\cdot(\mathbf{m}_i\times\mathbf{m}_j)-\sum_i\frac{K_i}{2}m_{iz}^2,\quad(14.44)$$

where the first term is the nearest-neighbour ferromagnetic Heisenberg exchange interaction with $J>0$. The second and third terms, arising from the spin-orbit coupling (SOC) [109, 110, 158], are respectively the nearest-neighbour pseudodipolar interaction of strength F and the next-nearest-neighbour DMI of strength D. $\mathbf{e}_{ij}$ is the unit vector pointing from site i to j, and $\nu_{ij}=\frac{2}{\sqrt{3}}\hat{\mathbf{z}}\cdot(\mathbf{e}_{li}\times\mathbf{e}_{lj})=\pm 1$, where l is the nearest neighbour site of i and j. the pseudodipolar interaction is the second-order effect of the SOC that is important here because the nearest-neighbour Dzyaloshinskii-Moriya interaction (DMI), the first-order effect, vanishes due to the inversion symmetry with respect to the center of the A-B bond. The last term is the sublattice-dependent anisotropy whose easy-axis is along the z direction with anisotropy coefficients of $K_i=K+\Delta$ for $i\in$ A and $K-\Delta$ for $i\in$ B.

Hamiltonian (14.44) can be treated either classically or quantum-mechanically. In the classical case, $\mathbf{m}_i$ are classical unit vectors whose dynamics is governed by the atomistic LLG equation,

$$\frac{\partial \mathbf{m}_i}{\partial t} = -\mathbf{m}_i \times \left[J \sum_j \mathbf{m}_j + F \sum_j (\mathbf{m}_j \cdot \mathbf{e}_{ij})\mathbf{e}_{ij} + K m_{iz}\mathbf{e}_z + B\mathbf{e}_z \right], \tag{14.45}$$

where $\mathbf{e}_z$ is the unit vector along the z direction. The length, time, magnetic field and energy are in units of a, $a^3(\gamma\mu)^{-1}$, $\mu_0\mu/a^3$ and $\mu_0\mu^2/a^3$, where γ and μ_0 are respectively the gyromagnetic ratio and the vacuum permeability. Out of the five model parameters in Hamiltonian (14.44), J can be the natural energy unit. The natural units of the time, length and magnetic field are $\frac{\mu_0\mu}{\gamma J}$, a and $\frac{J}{\mu_0\mu}$. Using the LLG equation, we first determine the ground state of the system by numerically relaxing the spins to their stable state starting from an initial configuration in which the spins are randomly and uniformly distributed in a cone with polar angle $\theta < 15°$. For $D = 0$ and $\Delta = 0$, the phase diagram of the model in the $K/J - F/J$ plane is shown in Fig. 14.14b. The system is in an out-of-plane ferromagnetic phase with spins aligning along the z direction (the pink) when $3J + K > -\frac{3}{2}F$ and $K > \frac{3}{2}F$. When $3J + K < -\frac{3}{2}F$ and $F < -J$ (the green), the ground state has a chiral spin structure in which spins lie in the xy plane with zero net magnetic moment on each hexagon. As shown in the lower left inset, six spins on each hexagon form three ferromagnetic pairs. The spins of each pair are perpendicular to the bond of the pair, and the three pairs are in all-in or all-out spin structure (120° with each other). For $K < \frac{3}{2}F$ and $F > -J$ (the cyan), the system prefers an in-plane ferromagnetic state (the lower right inset).

Now we focus on the out-of-plane ferromagnetic phase. To obtain spin wave spectrum, we can either expand $\mathbf{m}$ around its equilibrium state and keep only linear terms in the LLG equation (14.45) in the absence of damping, or treat the Hamiltonian (14.44) quantum-mechanically and use the Holstein-Primakoff transformation. Both methods lead to the same eigenvalue problem. To have an intuitive picture of spin waves, we use the classical picture in the following discussions. We consider a small deviation of $\mathbf{m}_i$ from $\mathbf{m}_0 = \mathbf{e}_z$, $\mathbf{m}_i = (\delta m_{ix}, \delta m_{iy}, 1)$, $(\sqrt{\delta m_{ix}^2 + \delta m_{iy}^2} \ll 1)$. The eigen-solutions of linearized (14.45) have, according to the Bloch theorem, the forms of $\delta m_{ix} = X_A e^{i(\mathbf{k}\cdot\mathbf{R}_A - \omega t)}$, $\delta m_{iy} = Y_A e^{i(\mathbf{k}\cdot\mathbf{R}_A - \omega t)}$ and $\delta m_{ix} = X_B e^{i(\mathbf{k}\cdot\mathbf{R}_B - \omega t)}$, $\delta m_{iy} = Y_B e^{i(\mathbf{k}\cdot\mathbf{R}_B - \omega t)}$ for sublattices A and B. We define $\psi_A^\pm = (X_A \pm iY_A)/\sqrt{2}$, $\psi_B^\pm = (X_B \pm iY_B)/\sqrt{2}$. From (14.45), the equation for the column vector $\Psi(\mathbf{k}) = (\psi_A^+, \psi_A^-, \psi_B^+, \psi_B^-)^T$ is

$$H(\mathbf{k})\Psi(\mathbf{k}) = \omega(\mathbf{k})\Psi(\mathbf{k}), \tag{14.46}$$

$H(\mathbf{k})$ is a 4×4 matrix of following expression

$$H = \begin{pmatrix} M_A + d(\mathbf{k}) & 0 & -f(\mathbf{k}) & g_1(\mathbf{k}) \\ 0 & -M_A + d(\mathbf{k}) & -g_2(\mathbf{k}) & f(\mathbf{k}) \\ -f^*(\mathbf{k}) & g_2^*(\mathbf{k}) & M_B - d(\mathbf{k}) & 0 \\ -g_1^*(\mathbf{k}) & f^*(\mathbf{k}) & 0 & -M_B - d(\mathbf{k}) \end{pmatrix}, \tag{14.47}$$

where $M_\alpha = K_\alpha + B + 3J$, $d(\mathbf{k}) = iD\sum_i(-1)^i e^{i\mathbf{k}\cdot\mathbf{b}_i}$, $f(\mathbf{k}) = \left(J + \frac{F}{2}\right)\sum_j e^{i\mathbf{k}\cdot\mathbf{a}_j}$, $g_1(\mathbf{k}) = \frac{F}{2}\sum_j e^{2i\theta_j}e^{i\mathbf{k}\cdot\mathbf{a}_j}$, $g_2(\mathbf{k}) = \frac{F}{2}\sum_j e^{-2i\theta_j}e^{i\mathbf{k}\cdot\mathbf{a}_j}$ ($\alpha = \mathrm{A, B}$; $j = 1, 2, 3$; $i = 1 \sim 6$). θ_j is the angle between $\mathbf{a}_j$ and $\mathbf{e}_x$ and η_i is the angle between $\mathbf{b}_i$ and $\mathbf{e}_x$. The solutions come in pairs $\pm\lambda$, obeying a particle-hole symmetry, so it is enough to consider only the positive bands. Although H is not Hermitian, H can be expressed as SH_1 where H_1 is Hermitian and $S = \sigma_0 \bigotimes \sigma_3$ (with σ_0 being the 2×2 identity matrix and σ_3 the Pauli matrix), and all the eigenvalues of H are real as long as the ferromagnetic state is a stable equilibrium state. H is pseudo-Hermitian with metric S, expressed as [159],

$$H^\dagger = SHS^{-1}. \tag{14.48}$$

The metric S comes from the bosonic nature of system in the presence of the pseudodipolar term [160]. Thus, for finite F, unlike previous studies [142, 143], this model is exclusive in bosonic systems and has no counterparty in electronic systems.

The spin wave dispersion relation $\omega(\mathbf{k})$ is shown in Fig. 14.14c for $J = 0.1$, $F = 0.5$, $\Delta = D = 0$, and $M \equiv 3J + K + B = 1.3$ (in units of $\mu_0\mu^2/a^3$). The band gap at K and K′ points is $\Delta_g = M - \sqrt{M^2 - 9F^2/4}$, as shown in the right panel of Fig. 14.14b. When $F = 0$, the bands linearly cross each other at K and K′ to form Dirac cones, similar to electron Dirac cones in graphene [161]. Thus, the pseudodipolar NN exchange interaction plays a crucial role in the band gap opening.

We consider now a long strip of width $100a$ with zigzag edges along x direction as shown in Fig. 14.14a. The density plot of the spectral function on the top edge is shown in Fig. 14.15a, for the same parameters as those in Fig. 14.14b. The negative slope of the dispersion curve (the bright line) of the top edge states in the band gap proves that these states propagate to the left. Similarly, the states on the bottom edge propagate unidirectionally to the right. The edge channels connect the upper and lower bands and linearly cross each other in the momentum space. Figure 14.15b shows spatial distribution of the edge spin wave eigenstate of $\omega = 1.2$ ($\gamma\mu/a^3$). The symbol shapes and sizes are the spin precession trajectory and the precession radius at each site, respectively. The azimuthal angle ϕ of $\mathbf{m}$ at $t = 0$ is encoded by the colors with the color ring shown in the inset. It is evident that the spin wave is mainly localized on the outermost sites. Here we use a zigzag edge as an example, and the existence of the topologically protected unidirectional edge states does not depend on edge types.

To understand the Berry phase of a spin wave, we consider the time evolution of a state of $\varphi_\mathbf{k}(t)$ with momentum $\mathbf{k}$ in an infinite sample. Without breaking the translational symmetry, $\mathbf{k}$ is a good quantum number that can be used to label eigenstates. $\varphi_\mathbf{k}(t)$ must be the Bloch state in the nth band $\Psi_n(\mathbf{k})$ multiplying a time dependent phase, and satisfies following time-dependent linearized LLG equation, similar to a Schrödinger equation,

$$i\frac{\mu_0\mu}{\gamma}\frac{\partial\varphi_\mathbf{k}}{\partial t} = H\varphi_\mathbf{k}. \tag{14.49}$$

Note that the same expression can be derived from quantum-mechanical model by using Bogoliubov transformation, given that $\frac{\mu_0\mu}{\gamma} = s\hbar$ where s is the total spin of each

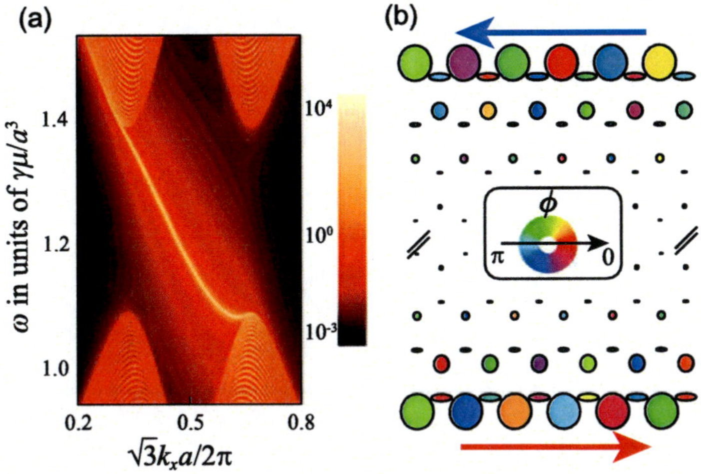

Fig. 14.15 **a** Density plot of the spectral function on the top edge. Gapless edge spin wave states are clearly shown in the band gap. The colors (from dark to bright) encode the value of the spectral function (from small to large) in logarithmic scale shown by the color bar. **b** Spatial distribution of the spin wave edge eigenstate of $\omega = 1.2\ (\gamma\mu/a^3)$. The symbol shape traces the spin precession trajectories, and the size of symbols denotes the amplitude of the spin wave at each site. The azimuthal angles of spins on the lattice at $t = 0$ are encoded by the symbol colors with the color ring shown in the inset. Figure reproduced from [153]

site [160]. Different from the Schrödinger equation, **k**-dependent 4×4 matrix H is not Hermitian, but pseudo-Hermitian. The orthogonality and normalization relations is given by $\Psi_i^\dagger(\mathbf{k}) S \Psi_j(\mathbf{k}) = S$ [144, 145]. Following exactly the same procedure as that in electron systems, the Berry connection of the nth band of our system is defined as

$$\mathbf{A}_n(\mathbf{k}) = i\frac{\Psi_n^\dagger(\mathbf{k}) S \partial_{\mathbf{k}} \Psi_n(\mathbf{k})}{\Psi_n^\dagger(\mathbf{k}) S \Psi_n(\mathbf{k})}, \tag{14.50}$$

and the Berry curvature $\Omega_n(\mathbf{k}) = \nabla_{\mathbf{k}} \times \mathbf{A}_n(\mathbf{k})$. The Chern number of nth band is defined as the integration over the Brillouin zone, similar to the electronic case. To obtain a gauge-invariant formula, we use a bosonic version of projector, written as [144, 145],

$$P_n = \Psi T_n S \Psi^\dagger S, \tag{14.51}$$

where Ψ is the 4×4 eigenmatrix whose nth column is Ψ_n, and T_n is the 4×4 matrix with only (n, n) element 1 and other elements 0, selecting the nth column from Ψ. Note that for an Hermitian system, $S = \mathbb{I}$ and (14.51) goes back to the conventional definition of projectors.

The Chern number is expressed in terms of P as [162],

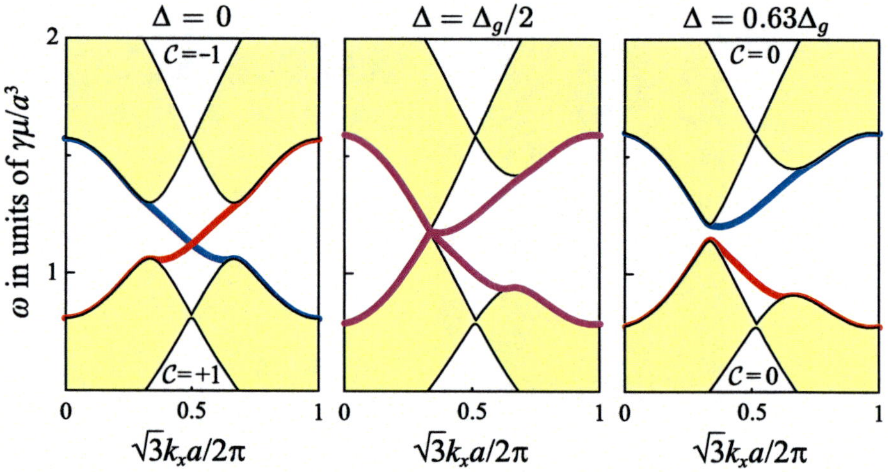

Fig. 14.16 Spin wave band structures of a zigzag strip for different Δ ($\Delta = 0$, $\Delta_g/2$, and $0.63\Delta_g$ from left to right, respectively). The light yellow areas are the bulk states and the color thick lines are the edge states. The Chern numbers of the corresponding bulk bands are given for $\Delta = 0$ and $\Delta = 0.63\Delta_g$. Figure redrawn using the results of [153]

$$\mathcal{C}_n = \frac{i}{2\pi}\iint dk_x dk_y \mathrm{Tr}\left[P\left(\frac{\partial P}{\partial k_x}\frac{\partial P}{\partial k_y} - \frac{\partial P}{\partial k_y}\frac{\partial P}{\partial k_x}\right)\right], \tag{14.52}$$

As long as $F \neq 0$, the Chern number of the lower band is $+1$ and that of the upper band is -1, consistent with the theorem that the sum of the Chern numbers of all bands must be zero [163]. The unidirectional spin wave edge states are similar to the chiral electron edge states in the quantum Hall and quantum anomalous Hall systems, and are robust against disorders. This nontrivial topological order can only be destroyed by closing and reopening the band gap. To see this, we let sublattices A and B have different anisotropy constants of $K_\mathrm{A} = K - \Delta$ and $K_\mathrm{B} = K + \Delta$, respectively. The band gap at the K point is in the range of $\sqrt{M^2 - 9F^2/4} + \Delta < \omega < M - \Delta$, while the band gap at the K′ point is in the range of $\sqrt{M^2 - 9F^2/4} - \Delta < \omega < M + \Delta$. If we gradually increase (decrease) Δ from 0, the band gap at the K (K′) point shrinks and the gap closes at $\Delta = \Delta_g/2$ ($-\Delta_g/2$), where $\Delta_g = M - \sqrt{M^2 - 9F^2/4}$. With further increase (decrease) of Δ, the gap reopens and the whole spectrum becomes fully gapped without topologically protected edge states. Figure 14.16 shows the band structures of zigzag strips for $\Delta = 0$, $\Delta = \Delta_g/2$ and $\Delta = 0.63\Delta_g$, with other parameters being the same as those in Figs. 14.14 and 14.15. The Chern numbers of the corresponding bulk bands are also given. For $|\Delta| < \Delta_g/2$, the edge states cross each other inside the gap in the momentum space, and the topological order is nontrivial, with Chern numbers $\mathcal{C} = \pm 1$ for the lower and upper bands. For $|\Delta| > \Delta_g/2$, the system is topologically trivial with Chern numbers $\mathcal{C} = 0$ for both lower and upper bands. In the topologically trivial case, the edge states could still exist, but do not cross each other, and are not topologically protected.

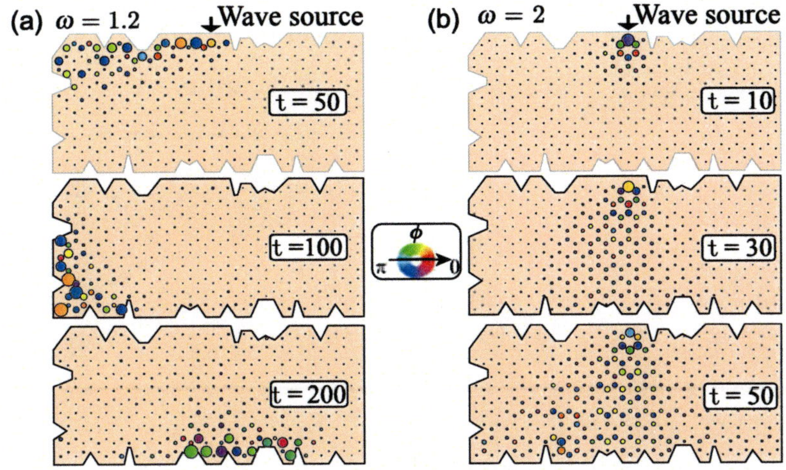

Fig. 14.17 **a, b** Snapshots of spatial distributions of the spin waves generated by a microwave pulse of $\Delta t = 50$ long in a $20\sqrt{3}a \times 14a$ strip with irregular edges, for $\omega = 1.2$ at $t = 50, 100, 200$ **a** and $\omega = 2$ at $t = 10, 30, 50$ **b** (from the top to the bottom). The radius of circles is proportional to the spin wave amplitude of $\sqrt{m_x^2 + m_y^2}$, and the colors encode the azimuthal angle ϕ shown by the color ring in the inset. Figure redrawn using the results of [153]

To see the fundamental differences between the usual bulk spin waves and topologically protected edge spin waves, we numerically solve (14.45) on a strip of $20\sqrt{3}a$ long and $14a$ wide. In order to see whether the edge states are robust against defects, we introduce irregular edge to the sample as shown in Fig. 14.17a, b. To excite spin waves, we apply a circularly polarized microwave field pulse of $\mathbf{h} = h[\cos(\omega t)\mathbf{e}_x + \sin(\omega t)\mathbf{e}_y]$ in a time duration of $0 \leq t \leq 50$, at the middle site of the top edge indicated by the black arrows in Fig. 14.17a, b. For $h = 0.01$ (μ/a^3) and $\omega = 1.2$ (inside the bulk band gap), Fig. 14.17a shows the snapshots of spin wave distributions at various times of t =50, 100, 200. The size of the circles is proportional to the local spin wave amplitude of $\sqrt{m_x^2 + m_y^2}$, and the color encodes the azimuthal angle ϕ. It is apparent that an edge mode is excited and is propagating counterclockwise along the sample edges. Backward scattering and leakage into the bulk can hardly be observed. For a comparison, we also excite spin waves of $\omega = 2$ that is inside the upper bulk band. The snapshots of spatial distributions of the excited spin waves at various times of $t = 10, 30, 50$ are shown in Fig. 14.17b. Clearly, the spin wave propagates into the whole sample.

There exist topologically protected edge spin waves (TPESWs) in the band gap that propagate unidirectionally for $F \neq 0$ and $D = 0$ when a bulk spin wave band is topologically non-trivial. The TPESWs are protected by the band gap, and always propagate counter-clockwise with respect to the magnetization direction. Let us further include non-zero DMI in the model. The topological phase diagram in D/J-Δ/J-F/J space is shown in Fig. 14.18. The topological phases are classified by

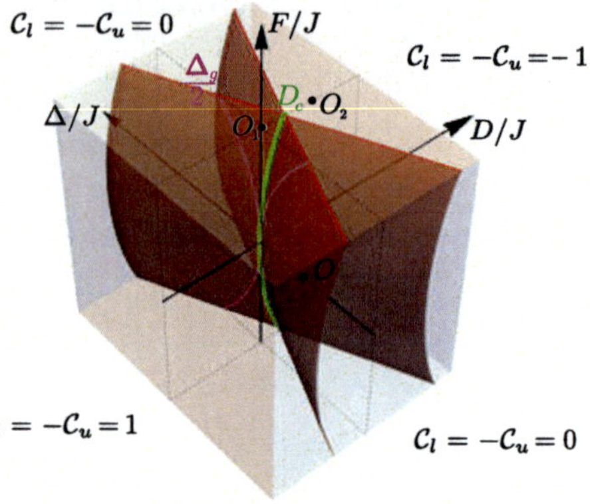

Fig. 14.18 Topological phase diagram in D/J-Δ/J-F/J space. Figure redrawn using the results of [155]

Chern numbers $\mathcal{C}_l$ and $\mathcal{C}_u$ of lower and upper magnon bands. $\mathcal{C}_l + \mathcal{C}_u = 0$ satisfies the zero sum rule [162, 163]. The magnon band Chern numbers change their values when the magnon band gap closes and reopens at valley K or K′. Thus, the band gap closing at K or K′ defines two phase boundary surfaces, which are derived to be $\Delta = \pm\Delta_c$, where

$$\Delta_c = \frac{1}{2}\left[(M - 3\sqrt{3}D) - \sqrt{(M + 3\sqrt{3}D)^2 - \frac{9}{4}F^2}\right]. \quad (14.53)$$

When $D = 0$, $\Delta_c = \frac{1}{2}\Delta_g$ as discussed before. The two phase boundaries intersect each other at the line of $D = D_c \equiv \frac{\sqrt{3}F^2}{16M}$ in the plane of $\Delta = 0$. In the non-trivial phase denoted by O_1 (O_2), the TPESWs propagate counter-clockwise (clockwise) with respect to the magnetization direction.

Utilizing the unidirectional property of TPESWs, one can design magnonic devices such as diodes and beam splitters. A segment of sample edge can be used as a spin wave diode and a DW of a strip can be used as a beam splitter. Since the TPESWs propagate in opposite directions in the two domains, a TPESW beam propagating towards the domain wall can neither penetrate it nor be reflected by it. It must move along the domain wall. When the spin wave beam reaches the other edge, it will split into two beams propagating in opposite directions, which is the functionality of a beam splitter. The idea can be generalized to 1-to-n spin wave splitting. A Mach-Zehnder-type spin wave interferometer can also be designed. A spin wave beam is first split into two and recombined later to form an interference pattern that varies periodically with the relative phase change of the two beams. The details of these proposed devices can be found in [153, 154]. The TPESWs provide a new paradigm for the design of magnonic devices, and open a new subfield of topological magnonics.

The non-trivial distribution of Berry curvature also has an impact on magnon transport. The Hall and the Nernst effects commonly refer to the generation of a transverse electric voltage or current by a longitudinal electric field or thermal gradient in an electronic system. It is natural to ask whether there is a similar effect for magnons. Magnons are charge neutral quasiparticles and do not have the usual Lorentz force. However, moving magnons experience gyroscopic forces because of nonzero Berry curvature of a magnetic system [130–134]. When the transverse components of the currents due to opposite Berry curvatures do not cancel each other, a net transverse magnon current is generated under a longitudinal force such as a thermal gradient in the absence of a magnetic field, leading to the anomalous magnon Nernst effect (AMNE) [155].

14.3.2 3D Topological Spin Waves

In the previous subsection, we have shown that a honeycomb ferromagnetic layer supports topologically protected edge states. Since pure 2D system is difficult to realize, it is useful to see whether a stacked 3D system can preserve the gapped bulk band and TPESWs. We discuss now a 3D ferromagnet made from a directly AA-stacking of honeycomb layers, as shown in Fig. 14.19a.

For simplicity and without loss of generality, we consider intralayer NN distance a and interlayer distance $d = a$. The generic Hamiltonian of the system is

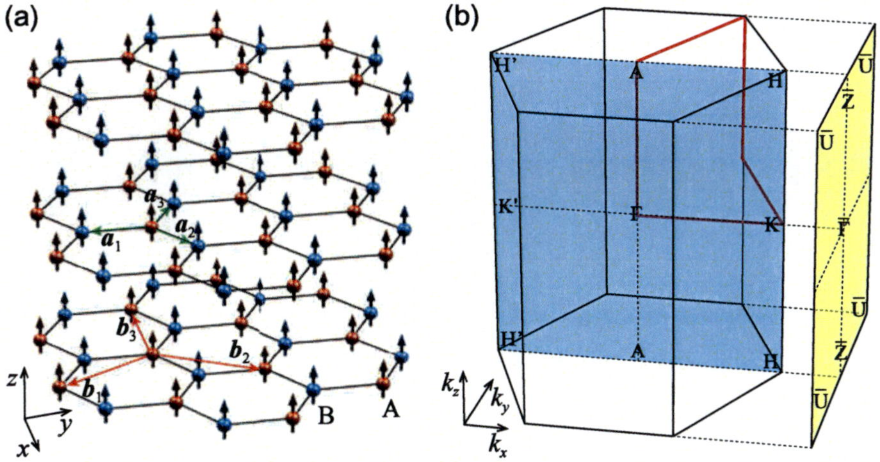

Fig. 14.19 **a** AA-stacked honeycomb ferromagnet. **b** The corresponding Brillouin Zone

$$\begin{aligned}\mathcal{H} = -J \sum_{\langle i,j\rangle,l} \mathbf{m}_{i,l} \cdot \mathbf{m}_{j,l} - F \sum_{\langle i,j\rangle,l} (\mathbf{m}_{i,l} \cdot \mathbf{e}_{ij})(\mathbf{m}_{j,l} \cdot \mathbf{e}_{ij}) \\ -D \sum_{\langle\langle i,j\rangle\rangle,l} \hat{\mathbf{d}}_{ij} \cdot (\mathbf{m}_{i,l} \times \mathbf{m}_{j,l}) - J_\sigma \sum_{i,\langle l,l'\rangle} \mathbf{m}_{i,l} \cdot \mathbf{m}_{i,l'} \\ -\frac{K_\sigma}{2} \sum_{i,l} (\mathbf{m}_{i,l} \cdot \hat{\mathbf{z}})^2 - \mu B \sum_{i,l} (\mathbf{m}_{i,l} \cdot \hat{\mathbf{z}}),\end{aligned} \tag{14.54}$$

where i and j label lattice sites in the same honeycomb layers, and l and l' are layer indices. $\mathbf{e}_{ij}$ is the unit vector from site i to j, and $\hat{\mathbf{z}}$ is the unit vector along z direction. As in the 2D case, J, F, D are intralayer NN exchange interaction, NN pseudodipolar interaction, NNN DMI, respectively. J_σ is the NN interlayer exchange interaction, and K_σ is the anisotropy. We define $J' = (J_A + J_B)/2$ and $\delta J = J_A - J_B$; $K = (K_A + K_B)/2$ and $\delta K = K_A - K_B$. Similar to the 2D case, the Hamiltonian $\mathcal{H}$ can be treated either classically [153] or quantum mechanically [157]. The linearized Landau-Lifshitz-Gilbert (LLG) equation [18] from the classical Hamiltonian and the Shrödinger equation from the quantum Hamiltonian give identical eigenvalue problem for magnon bands. The Brillouin zone is illustrated in Fig. 14.19b.

The band structure of an infinite system for $D = 0, K = 10J, J' = 0.4J, F = 6J$, $B = 0$, and $\delta J = \delta K = 0$ is shown in Fig. 14.20a. The main figure is the magnon frequency along a path from $(k_x, k_y, k_z) = (0, 0, 0)$ (Γ) to $(\frac{4\pi}{3\sqrt{3}}, 0, 0)$ (K); from K to $(\frac{2\pi}{3\sqrt{3}}, \frac{2\pi}{3}, 0)$ (K'); from K' to $(\frac{2\pi}{3\sqrt{3}}, \frac{2\pi}{3}, \pi)$ (H'); from H' to $(0, 0, \pi)$ (A); and then from A back to Γ (the red line in Fig. 14.19b). The frequency is in the

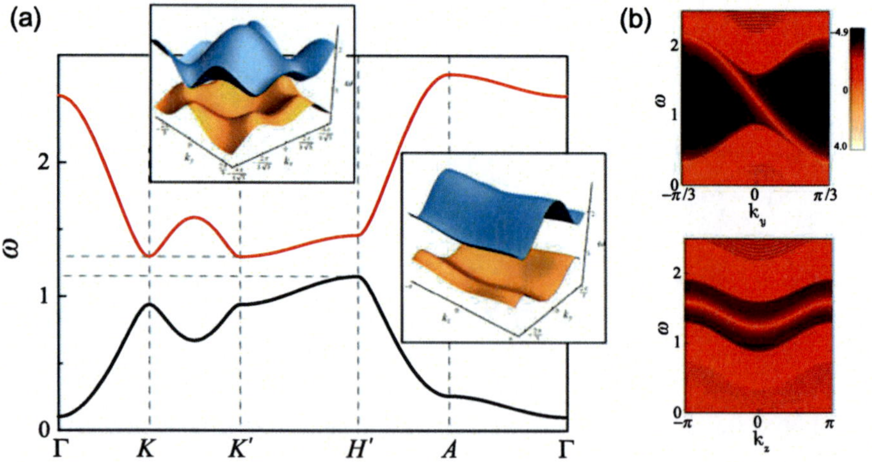

Fig. 14.20 **a** Bulk magnon band structure along Γ-K-K'-H'-A-Γ. The left (right) inset: Band structure in k_x-k_y (k_y-k_z) plane for $k_z = 0$ ($k_x = 0$). **b** Spectral function on the left ($-x$ side) surface of a slab infinite in yz plane and finite in x direction. The upper (lower) panel is the spectral function along k_y (k_z) direction with $k_z = 0$ ($k_y = 0$). The color bar is shown in the middle

units of $\gamma J/\mu$ where μ is the magnitude of magnetic moment of each spin. The top-left (the low-right) inset is the band structure in k_x-k_y (k_y-k_z) plane for $k_z = 0$ ($k_x = 0$). The magnon band is globally gapped with an indirect gap between K and H points. Figure 14.20b shows the spectral function on the left ($-x$ side) surface of a slab infinite in yz plane and 15 unit cells in x direction). The brighter color means larger local density of states. The upper (lower) panel is the spectral function along k_y (k_z) direction with $k_z = 0$ ($k_y = 0$). A surface magnon band is clearly shown inside the bulk band gap with a negative slope in ω-k_y plane, which means a negative group velocity and a unidirectional propagation along $-y$ direction. Similarly, on the rear surface, the surface modes propagate along $+y$ direction (not shown). In other words, the surface states propagate counterclockwise with respect to the ground state magnetization direction. The Chern number of the system can be calculated by the standard formula [145, 154, 157] to show that the system is a magnonic weak topological insulator, so that the side surface states along xz or yz surfaces are topologically protected.

To have a better picture about the surface spin waves, we numerically solve the LLG equation with a small damping $\alpha = 0.0001$. The sample contains 15 unit cells in x direction, 10 unit cells in y direction, and 5 layers in z direction. A localized circularly polarized microwave field $\mathbf{h} = h_0[\cos(\omega t)\hat{\mathbf{x}} + \sin(\omega t)\hat{\mathbf{y}}]$ is applied on the spin marked by the black arrow. The frequency is 1.25 which is in the bulk band gap so that only surface modes are excited. Figure 14.21a shows the snapshot of spin waves after $t = 100(\gamma J/\mu)^{-1}$. The radius of the cone denotes the oscillation amplitude, and the color of the cone encodes the azimuthal angle of each spin. One can seen from the figure that a beam of spin wave is emitted from the source, and propagate along the xz and yz surfaces unidirectionally.

We then study whether the stacked honeycomb ferromagnet can support Weyl magnons. The Weyl magnon is the bosonic counterparty of a Weyl fermion [164–

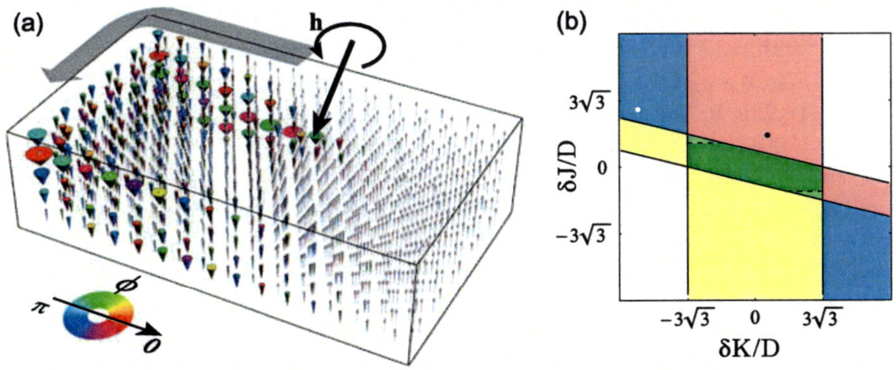

Fig. 14.21 **a** Snapshot of spin waves at $t = 100(\gamma J/\mu)^{-1}$ in a finite sample. The radius of the cone denotes the oscillation amplitude of each spin, and the color of the cone encodes the azimuthal angle of each spin with the color ring shown at the corner. **b** Topological phase diagrams for $D \neq 0$, $F = 0$

167] in electronic systems, similar to Weyl points in photonic systems [168]. Weyl magnons have also been identified in pyrochlore magnets [156, 169, 170]. For $F = 0$ and a finite D, the spin wave spectrum can be analytically solved. Two positive bands are $\omega(\mathbf{k}) = \omega_0(\mathbf{k}) \pm \sqrt{\sum_{\beta=x,y,z} h_\beta^2(\mathbf{k})}$, where $\omega_0(\mathbf{k}) = 3J + K + 2J'(1 - \cos kz)$, $h_x(\mathbf{k}) = -J\sum_{i=1}^3 \cos(\mathbf{k}\cdot\mathbf{a}_i), h_y(\mathbf{k}) = -J\sum_{i=1}^3 \sin(\mathbf{k}\cdot\mathbf{a}_i)$, and $h_z(\mathbf{k}) = 2D\sum_{i=1}^3 \sin(\mathbf{k}\cdot\mathbf{b}_i) + \delta K + 2\delta J(1 - \cos k_z)$ (the frequency is in units of $\gamma J/\mu$ as before). The negative bands are related to positive ones by $\omega(-\mathbf{k}) = -\omega(\mathbf{k})$ and they actually represent the same modes as the positive ones.

The two bands cross each other only when $h_\beta(\mathbf{k}) = 0$. The crossing points locate at $\mathbf{k}_\eta^\pm = (-\frac{4\pi}{3\sqrt{3}}, 0, \pm\cos^{-1} f_\eta)$, where $\eta = 1, 2$ and $f_\eta = \frac{\delta K}{2\delta J} + 1 + (-1)^\eta \frac{3\sqrt{3}D}{2\delta J}$. The requirement of $|f_\eta| \leq 1$ yields the phase boundaries of $\delta J = \pm 3\sqrt{3}D$ and $4\delta J + \delta K = \pm 3\sqrt{3}D$. Figure 14.21b shows the phase diagram in $\delta K/D - \delta J/D$ plane. In the green region, two bands do not cross each other, and each band has a nontrivial Chern number. This is the weak topological insulator phase and topologically protected surface modes exist on the side surfaces. The band is gapped (the dark green) when $|\delta J| < 3\sqrt{3}D - J'$ (the dashed lines). This constraint plays a role only when $J' > 3\sqrt{3}D/2$. Otherwise the whole green region is gapped. The white regions are topologically trivial. The rest of the regions belong to three different phases. In the pink (yellow) regions there is one pair of band crossing points at $\mathbf{k}_1^\pm$ ($\mathbf{k}_2^\pm$). In the blue regions, there are two pairs of band crossing points at $\mathbf{k}_{1,2}^\pm$.

To see whether the band crossing points are Weyl nodes (WNs), we calculate the Berry curvatures of the magnon bands. We fix the model parameters to $D = J' = 0.2J$, $K = 60J$, and choose $(\delta K, \delta J) = (0.2J, 0.5J)$ and $(-1.8J, 0.9J)$ (the black and the white dots in Fig. 14.21b). Energy bands in the $k_x - k_z$ plane for fixed $k_y = 0$ (the blue plane in Fig. 14.19b) of the WMs are shown in Fig. 14.22a, b in which one and two pairs of WNs appear. The red and blue dots denote chirality ± 1 of the WNs. The corresponding Berry curvatures are shown in Figs. 14.22c, d, respectively, in which the black arrows encode the direction of Berry curvatures projected onto the $k_x - k_z$ plane and the background color represents the divergence of Berry curvature $\nabla_\mathbf{k}\cdot\mathbf{\Omega}_\mathbf{k}$ with red for positive and blue for negative. Thus, the red and blue spots in Figs. 14.22c, d are indeed monopoles of Berry curvature and correspond to the WNs in Figs. 14.22a, b. The spectral functions on the front (100) surface along $\bar{Z}\bar{\Gamma}\bar{Z}$ of the first (100) surface BZ (see Fig. 14.19b) are shown in Figs. 14.22e, f, respectively. The surface states with high density (red color) on the front surface between WNs can be clearly seen. Near the energy of WNs, these surface states form magnon arcs (an analog to the Fermi arcs) on sample surfaces.

The Weyl magnons demonstrated above are type-I, in analogy to the type-I Weyl electrons [171]. Two spin wave bands cross at WNs. When $J' > \delta J$, the system can also be type-II Weyl semimetals. Two bands of type-II semimetals intersect with each other with a line called nodal-lines [171] so the constant-energy surface passing through WN is a line. In an inhomogeneous electric field, the magnons can also form Landau levels [157, 172]. Chiral anomaly for magnons results in the linear

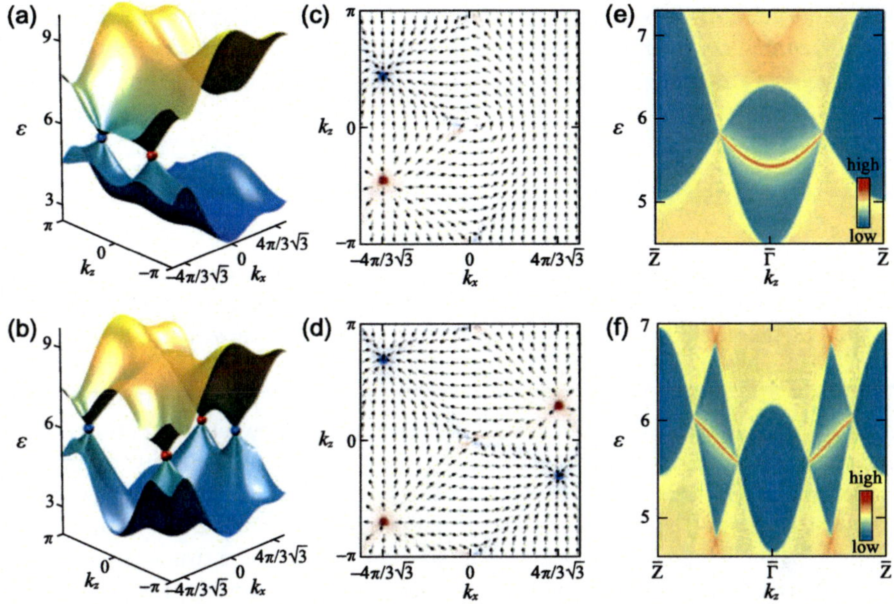

Fig. 14.22 **a** and **b** Band structures of WMs marked by the black and white dots in Fig. 14.19c, respectively. The WNs of chirality ±1 are marked by red and blue dots. **c** and **d** Corresponding Berry curvatures of the lower magnon bands in panels (**a**) and (**b**). The arrows represent the direction of Berry curvature vectors in the $k_x - k_z$ plane with $k_y = 0$. The background color denotes the divergence of Berry curvature, where red and blue represent positive and negative values. **e** and **f** Density plots of the front (100) surface spectral functions along $\bar{Z}\bar{\Gamma}\bar{Z}$ for the energy bands in panels (**a**) and (**b**). Figure reproduced from [157]

dependence of spin and heat conductance on the electric field gradient in mutually perpendicular inhomogeneous electric and magnetic fields.

The studies above show that the honeycomb magnetic systems are good playground with rich physics for topological spin waves. To specifically realize the model, the system needs not be a perfect honeycomb. A deformed honeycomb lattices with two spins in one unit cell should be enough. Such a structure is quite common in nature. The pseudodipolar interaction and the Dzyaloshinskii-Moriya interaction exist in most magnetic materials in principle. Heavy-metal magnetic materials are preferred because larger spin-orbit interaction. For example, in A_2BO_3 materials (where A is an alkali metal and B is a transition metal), the B atoms form a honeycomb structure normally with a strong spin-orbit interaction. Ferromagnetic order is also possible for certain special elements combinations or some special lattice directions [173]. For example, $\beta - Li_2IrO_3$ is ferromagnetic and has a hyperhoneycomb lattice structure [174] so it be a candidate for realizing the model studied here. AB_3 (where A is a transition metal and B is a halogen) is another family of candidate materials. A atoms form a honeycomb lattice in these materials. There are already first-principle predictions of ferromagnetism in $NiCl_3$ and $OsCl_3$ monolayers [175,

176], and the A-B-A bond is very promising for inducing a strong pseudodipolar exchange interaction. Our model may also be realized in magnonic crystals in which nanomagnets are arranged into a honeycomb lattice [177].

14.4 Conclusion

In conclusion [178], magnetism is an old and an important subject in physics. Our long lasting interest on the subject in the past hundred years is mainly because the studies of the subject keep providing us with new ideas, new concepts and new principles, as well as because of its applications in modern information technologies. Topology, an old mathematical concept, is becoming a popular physics concept since the recent discoveries of topologically protected surface states in both quantum and classical systems. The concept of topology has been used in two ways in physics. One is to classify structures of order parameters such as the magnetic domain walls, vortices, skyrmions, etc., in magnetism. Namely, spin structures cannot be mapped to a single magnetic domain under a continuous deformation of the structures. It should be emphasized that nature does not respect topology, but the energy. Thus, a topologically non-trivial structure does not imply stability or a potential barrier between the structure and other distinct ones. The second usage of topology refers to a class of materials that support chiral surface states. These states can only be destroyed when some topological charges existing in the bulk can be removed by external forces. This second usage of topology leads to a new emerging field called topological materials that include topological insulators and topological semimetals. The concept of topological protection provides a paradigm for stabilizing structures and excitations because the kinetic paths to topologically distinct structures are greatly reduced even in the case of only a finite potential barrier between different structures. Last but not least, concepts and principles reviewed here can also be applied to other fields of the physics.

References

1. S. Li, Origine de la boussole II. aimant et boussole. Isis **45**(2), 175–196 (1954)
2. J.D. Jackson, *Classical Electrodynamics*, 3rd ed. edn. (Wiley, New York, NY, 1999)
3. J.J. Sakurai, *Modern Quantum mechanics*, rev edn. (Addison-Wesley, Reading, MA, 1994)
4. W. Ley, The galactic giants. *Galaxy Science Fiction* (1965), pp. 130–142
5. M.N. Baibich, J.M. Broto, A. Fert, F. Nguyen Van Dau, F. Petroff, P. Etienne, G. Creuzet, A. Friederich, J. Chazelas, Giant magnetoresistance of (001)Fe/(001)Cr magnetic superlattices. Phys. Rev. Lett. **61**, 2472–2475 (1988)
6. G. Binasch, P. Grünberg, F. Saurenbach, W. Zinn, Enhanced magnetoresistance in layered magnetic structures with antiferromagnetic interlayer exchange. Phys. Rev. B **39**, 4828–4830 (1989)
7. N. Kurtl, *Selected Works of Louis Néel*, 1st edition (CRC Press, 1988)

8. E. Chen, D. Apalkov, Z. Diao, A. Driskill-Smith, D. Druist, D. Lottis, V. Nikitin, X. Tang, S. Watts, S. Wang, S.A. Wolf, A.W. Ghosh, J.W. Lu, S.J. Poon, M. Stan, W.H. Butler, S. Gupta, C.K.A. Mewes, T. Mewes, P.B. Visscher, Advances and future prospects of spin-transfer torque random access memory. IEEE Trans. Magn. **46**(6), 1873–1878 (2010)
9. G. Moore, *Progress in Digital Integrated Electronics* (IEEE, IEDM Tech Digest, 1975), pp. 11–13
10. Fundamentals and applications, Igor Žutić, Jaroslav Fabian, and S. Das Sarma. Spintronics. Rev. Mod. Phys. **76**, 323–410 (2004)
11. K. Fossheim, A. Sudbø, *Superconductivity: Physics and Applications* (John Wiley & Sons, Ltd, 2005)
12. V.V. Kruglyak, S.O. Demokritov, D. Grundler, Magnonics. J. Phys. D Appl. Phys. **43**(26), 264001 (2010)
13. A.A. Serga, A.V. Chumak, B. Hillebrands, YIG magnonics. J. Phys. D Appl. Phys. **43**(26), 264002 (2010)
14. D.S. Richeson, *Euler's Gem: The Polyhedron Formula and the Birth of Topology* (Princeton University Press, 2008)
15. C.N. Yang, Mo-Lin Ge, and Yang-Hui He (Topology and Physics, WORLD SCIENTIFIC, 2019)
16. A. Blanco-Redondo, B. Bell, D. Oren, B.J. Eggleton, M. Segev, Topological protection of biphoton states. Science **362**(6414), 568–571 (2018)
17. H.Y. Yuan, Q. Liu, K. Xia, Z. Yuan, X.R. Wang, Proper dissipative torques in antiferromagnetic dynamics. *EPL (Europhysics Letters)***126**(6), 67006 (2019)
18. T.L. Gilbert, A phenomenological theory of damping in ferromagnetic materials. IEEE Trans. Magn. **40**(6), 3443–3449 (2004)
19. A. Thiaville, J. Miltat. *Topology and Magnetic Domain Walls* (Springer International Publishing, Cham, 2018), pp. 41–73
20. S.P. Stuart, Parkin, Masamitsu Hayashi, and Luc Thomas. Magnetic domain-wall racetrack memory. Science **320**(5873), 190–194 (2008)
21. N.L. Schryer, L.R. Walker, The motion of 180° domain walls in uniform dc magnetic fields. J. Appl. Phys. **45**(12), 5406–5421 (1974)
22. A.P. Malozemoff, J.C. Slonczewski, *Magnetic Domain Walls in Bubble Materials* (Academic Press, 1979)
23. J.C. Slonczewski, Current-driven excitation of magnetic multilayers. J. Magn. Magn. Mater. **159**(1), L1–L7 (1996)
24. L. Berger, Emission of spin waves by a magnetic multilayer traversed by a current. Phys. Rev. B **54**, 9353–9358 (1996)
25. S. Zhang, Z. Li, Roles of nonequilibrium conduction electrons on the magnetization dynamics of ferromagnets. Phys. Rev. Lett. **93**, 127204 (2004)
26. A. Yamaguchi, T. Ono, S. Nasu, K. Miyake, K. Mibu, T. Shinjo, Real-space observation of current-driven domain wall motion in submicron magnetic wires. Phys. Rev. Lett. **92**, 077205 (2004)
27. M. Hayashi, L. Thomas, Y.B. Bazaliy, C. Rettner, R. Moriya, X. Jiang, S.S.P. Parkin, Influence of current on field-driven domain wall motion in permalloy nanowires from time resolved measurements of anisotropic magnetoresistance. Phys. Rev. Lett. **96**, 197207 (2006)
28. K. Kondou, N. Ohshima, S. Kasai, Y. Nakatani, T. Ono, Single shot detection of the magnetic domain wall motion by using tunnel magnetoresistance effect. Appl. Phys. Expr. **1**, 061302 (2008)
29. X.R. Wang, P. Yan, J. Lu, High-field domain wall propagation velocity in magnetic nanowires. EPL (Europhysics Letters) **86**(6), 67001 (2009)
30. X.R. Wang, P. Yan, J. Lu, C. He, Magnetic field driven domain-wall propagation in magnetic nanowires. Ann. Phys. **324**(8), 1815–1820 (2009)
31. X.S. Wang, P. Yan, Y.H. Shen, G.E.W. Bauer, X.R. Wang, Domain wall propagation through spin wave emission. Phys. Rev. Lett. **109**, 167209 (2012)

32. X.S. Wang, X.R. Wang, Spin wave emission in field-driven domain wall motion. Phys. Rev. B **90**, 184415 (2014)
33. B. Hu, X.R. Wang, Instability of walker propagating domain wall in magnetic nanowires. Phys. Rev. Lett. **111**, 027205 (2013)
34. G. Tatara, H. Kohno, Theory of current-driven domain wall motion: Spin transfer versus momentum transfer. Phys. Rev. Lett. **92**, 086601 (2004)
35. A. Yamaguchi, S. Nasu, H. Tanigawa, T. Ono, K. Miyake, K. Mibu, T. Shinjo, Effect of joule heating in current-driven domain wall motion. Appl. Phys. Lett. **86**(1), 012511 (2005)
36. M. Hayashi, L. Thomas, C. Rettner, R. Moriya, X. Jiang, S.S.P. Parkin, Dependence of current and field driven depinning of domain walls on their structure and chirality in permalloy nanowires. Phys. Rev. Lett. **97**, 207205 (2006)
37. A. Thiaville, Y. Nakatani, J. Miltat, Y. Suzuki, Micromagnetic understanding of current-driven domain wall motion in patterned nanowires. Europhysics Letters (EPL) **69**(6), 990–996 (2005)
38. Ya.B. Bazaliy, B.A. Jones, S-C. Zhang, Modification of the Landau-Lifshitz equation in the presence of a spin-polarized current in colossal- and giant-magnetoresistive materials. Phys. Rev. B **57**, R3213–R3216 (1998)
39. P. Yan, X.S. Wang, X.R. Wang, All-magnonic spin-transfer torque and domain wall propagation. Phys. Rev. Lett. **107**, 177207 (2011)
40. J. Han, P. Zhang, J.T. Hou, S.A. Siddiqui, L. Liu, Mutual control of coherent spin waves and magnetic domain walls in a magnonic device. Science **366**(6469), 1121–1125 (2019)
41. Y. Wang, D. Zhu, Y. Yang, K. Lee, R. Mishra, G. Go, S.-H. Oh, D.-H. Kim, K. Cai, E. Liu, S.D. Pollard, S. Shi, J. Lee, K. Leong Teo, Y. Wu, K.-J. Lee, H. Yang, Magnetization switching by magnon-mediated spin torque through an antiferromagnetic insulator Science **366**(6469),1125–1128 (2019)
42. D. Hinzke, U. Nowak, Domain wall motion by the magnonic spin seebeck effect. Phys. Rev. Lett. **107**, 027205 (2011)
43. A.A. Kovalev, Y. Tserkovnyak, Thermomagnonic spin transfer and Peltier effects in insulating magnets. EPL (Europhysics Letters) **97**(6), 67002 (2012)
44. W. Jiang, P. Upadhyaya, Y. Fan, J. Zhao, M. Wang, L.-T. Chang, M. Lang, K.L. Wong, M. Lewis, Y.-T. Lin, J. Tang, S. Cherepov, X. Zhou, Y. Tserkovnyak, R.N. Schwartz, K.L. Wang, Direct imaging of thermally driven domain wall motion in magnetic insulators. Phys. Rev. Lett. **110**, 177202 (2013)
45. O. Tchernyshyov, G.-W. Chern, Fractional vortices and composite domain walls in flat nanomagnets. Phys. Rev. Lett. **95**, 197204 (2005)
46. R.D. McMichael, M.J. Donahue, Head to head domain wall structures in thin magnetic strips. IEEE Trans. Magn. **33**(5), 4167–4169 (1997)
47. H.Y. Yuan, X.R. Wang, Birth, growth and death of an antivortex during the propagation of a transverse domain wall in magnetic nanostrips. J. Magn. Magn. Mater. **368**, 70–74 (2014)
48. H. Yuan, X. Wang, Vortex-assisted domain wall depinning and propagation in notched nanowires. Eur. Phys. J. B **88**(9), 214 (2015)
49. H.Y. Yuan, X.R. Wang, Boosting domain wall propagation by notches. Phys. Rev. B **92**, 054419 (2015)
50. R. Hertel, W. Wulfhekel, J. Kirschner, Domain-wall induced phase shifts in spin waves. Phys. Rev. Lett. **93**, 257202 (2004)
51. X.S. Wang, X.R. Wang, Thermodynamic theory for thermal-gradient-driven domain-wall motion. Phys. Rev. B **90**, 014414 (2014)
52. N. Nagaosa, Y. Tokura, Topological properties and dynamics of magnetic skyrmions. Nat. Nanotechnol. **8**(12), 899–911 (2013)
53. H.Y. Yuan, X.R. Wang, Nano magnetic vortex wall guide. AIP Adv. **5**(11), 117104 (2015)
54. A.A. Thiele, Steady-state motion of magnetic domains. Phys. Rev. Lett. **30**, 230–233 (1973)
55. S. Agramunt-Puig, N. Del-Valle, C. Navau, A. Sanchez, Controlling vortex chirality and polarity by geometry in magnetic nanodots. Appl. Phys. Lett. **104**(1), 012407 (2014)
56. X.S. Wang, A. Qaiumzadeh, A. Brataas, Current-driven dynamics of magnetic hopfions. Phys. Rev. Lett. **123**, 147203 (2019)

57. T. Shinjo, T. Okuno, R. Hassdorf, K. Shigeto, T. Ono, Magnetic vortex core observation in circular dots of permalloy. Science **289**(5481), 930–932 (2000)
58. A. Wachowiak, J. Wiebe, M. Bode, O. Pietzsch, M. Morgenstern, R. Wiesendanger, Direct observation of internal spin structure of magnetic vortex cores. Science **298**(5593), 577–580 (2002)
59. S.-B. Choe, Y. Acremann, A. Scholl, A. Bauer, A. Doran, J. Stöhr, H.A. Padmore, Vortex core-driven magnetization dynamics. Science **304**(5669), 420–422 (2004)
60. B. Van Waeyenberge, A. Puzic, H. Stoll, K.W. Chou, T. Tyliszczak, R. Hertel, M. Fähnle, H. Brückl, K. Rott, G. Reiss, I. Neudecker, D. Weiss, C.H. Back, G. Schütz, Magnetic vortex core reversal by excitation with short bursts of an alternating field. Nature **444**(7118), 461–464 (2006)
61. K.Yu. Guslienko, K-S. Lee, S-K. Kim, Dynamic origin of vortex core switching in soft magnetic nanodots. Phys. Rev. Lett. **100**, 027203 (2008)
62. D.-H. Kim, E.A. Rozhkova, I.V. Ulasov, S.D. Bader, T. Rajh, M.S. Lesniak, V. Novosad, Biofunctionalized magnetic-vortex microdiscs for targeted cancer-cell destruction. Nat. Mater. **9**(2), 165–171 (2010)
63. R. Streubel, D. Makarov, F. Kronast, V. Kravchuk, M. Albrecht, O.G. Schmidt, Magnetic vortices on closely packed spherically curved surfaces. Phys. Rev. B **85**, 174429 (2012)
64. T.H.R. Skyrme, B.F.J. Schonland, A non-linear field theory. Proc. Roy. Soc. Lond. Ser A. Math. Phys. Sci. **260**(1300), 127–138 (1961)
65. T.H.R. Skyrme, A unified field theory of mesons and baryons. Nuc. Phys. **31**, 556–569 (1962)
66. A. Bogdanov, New localized solutions of the nonlinear field equations. JETP Lett. **62**, 247 (1995)
67. B.M.A.G. Piette, B.J. Schroers, W.J. Zakrzewski, Dynamics of baby skyrmions. Nucl. Phys. B **439**(1), 205–235 (1995)
68. J. Sampaio, V. Cros, S. Rohart, A. Thiaville, A. Fert, Nucleation, stability and current-induced motion of isolated magnetic skyrmions in nanostructures. Nat. Nanotechnol. **8**(11), 839–844 (2013)
69. A. Qaiumzadeh, I.A. Ado, R.A. Duine, M. Titov, A. Brataas, Theory of the interfacial Dzyaloshinskii-Moriya interaction in Rashba antiferromagnets. Phys. Rev. Lett. **120**, 197202 (2018)
70. N. Romming, C. Hanneken, M. Menzel, J.E. Bickel, B. Wolter, K. von Bergmann, A. Kubetzka, R. Wiesendanger, Writing and deleting single magnetic skyrmions. Science **341**(6146), 636–639 (2013)
71. W. Jiang, P. Upadhyaya, W. Zhang, G. Yu, M. Benjamin Jungfleisch, F.Y. Fradin, J.E. Pearson, Y. Tserkovnyak, K.L. Wang, O. Heinonen, S.G.E. te Velthuis, A. Hoffmann, Blowing magnetic skyrmion bubbles. Science. **349**(6245), 283–286 (2015)
72. J. Zang, M. Mostovoy, J.H. Han, N. Nagaosa, Dynamics of skyrmion crystals in metallic thin films. Phys. Rev. Lett. **107**, 136804 (2011)
73. S. Mühlbauer, B. Binz, F. Jonietz, C. Pfleiderer, A. Rosch, A. Neubauer, R. Georgii, P. Böni, Skyrmion lattice in a chiral magnet. Science **323**(5916), 915–919 (2009)
74. X.Z. Yu, N. Kanazawa, W.Z. Zhang, T. Nagai, T. Hara, K. Kimoto, Y. Matsui, Y. Onose, Y. Tokura, Skyrmion flow near room temperature in an ultralow current density. Nat. Commun. **3**(1), 988 (2012)
75. N. Romming, A. Kubetzka, C. Hanneken, K. von Bergmann, R. Wiesendanger, Field-dependent size and shape of single magnetic skyrmions. Phys. Rev. Lett. **114**, 177203 (2015)
76. R. Tomasello, E. Martinez, R. Zivieri, L. Torres, M. Carpentieri, G. Finocchio, A strategy for the design of skyrmion racetrack memories. Sci. Rep. **4**(1), 6784 (2014)
77. S. Krause, R. Wiesendanger, Skyrmionics gets hot. Nat. Mater. **15**(5), 493–494 (2016)
78. S. Rohart, A. Thiaville, Skyrmion confinement in ultrathin film nanostructures in the presence of Dzyaloshinskii-Moriya interaction. Phys. Rev. B **88**, 184422 (2013)
79. J. Iwasaki, M. Mochizuki, N. Nagaosa, Universal current-velocity relation of skyrmion motion in chiral magnets. Nat. Commun. **4**(1), 1463 (2013)

80. X. Zhang, Y. Zhou, M. Ezawa, G.P. Zhao, W. Zhao, Magnetic skyrmion transistor: skyrmion motion in a voltage-gated nanotrack. Sci. Rep. **5**(1), 11369 (2015)
81. M.A. Castro, S. Allende, Skyrmion core size dependence as a function of the perpendicular anisotropy and radius in magnetic nanodots. J. Magn. Magn. Mater. **417**, 344–348 (2016)
82. N. Vidal-Silva, A. Riveros, J. Escrig, Stability of Néel skyrmions in ultra-thin nanodots considering Dzyaloshinskii-Moriya and dipolar interactions. J. Magn. Magn. Mater. **443**, 116–123 (2017)
83. A.O. Leonov, T.L. Monchesky, N. Romming, A. Kubetzka, A.N. Bogdanov, R. Wiesendanger, The properties of isolated chiral skyrmions in thin magnetic films. New J. Phys. **18**(6), 065003 (2016)
84. X.S. Wang, H.Y. Yuan, X.R. Wang, A theory on skyrmion size. Commun. Phys. **1**(1), 31 (2018)
85. H.-B. Braun, Fluctuations and instabilities of ferromagnetic domain-wall pairs in an external magnetic field. Phys. Rev. B **50**, 16485–16500 (1994)
86. A. Siemens, Y. Zhang, J. Hagemeister, E.Y. Vedmedenko, R. Wiesendanger, Minimal radius of magnetic skyrmions: statics and dynamics. New J. Phys. **18**(4), 045021 (2016)
87. A. Vansteenkiste, J. Leliaert, M. Dvornik, M. Helsen, F. Garcia-Sanchez, B. Van Waeyenberge, The design and verification of MuMax3. AIP Adv. **4**(10), 107133 (2014)
88. D. Haifeng, R. Che, L. Kong, X. Zhao, C. Jin, C. Wang, J. Yang, W. Ning, R. Li, C. Jin, X. Chen, J. Zang, Y. Zhang, M. Tian, Edge-mediated skyrmion chain and its collective dynamics in a confined geometry. Nat. Communications. **6**(1), 8504 (2015)
89. I.E. Dzyaloshinskii, The theory of helicoidal structures in antiferromagnets. II. metals. J. Exp. Heoretic. Phys. **20**(1), 223 (1965)
90. H.Y. Yuan, X.R. Wang, Skyrmion creation and manipulation by nano-second current pulses. Sci. Rep. **6**(1), 22638 (2016)
91. M. Stone, Magnus force on skyrmions in ferromagnets and quantum hall systems. Phys. Rev. B **53**, 16573–16578 (1996)
92. J. Iwasaki, M. Mochizuki, N. Nagaosa, Current-induced skyrmion dynamics in constricted geometries. Nat. Nanotechnol. **8**(10), 742–747 (2013)
93. W. Jiang, X. Zhang, G. Yu, W. Zhang, X. Wang, M. Benjamin Jungfleisch, J.E. Pearson, X. Cheng, O. Heinonen, K.L. Wang, Y. Zhou, A. Hoffmann, S.G.E. te Velthuis, Direct observation of the skyrmion Hall effect. Nat. Phys. **13**(2), 162–169 (2017)
94. M.-W. Yoo, V. Cros, J.-V. Kim, Current-driven skyrmion expulsion from magnetic nanostrips. Phys. Rev. B **95**, 184423 (2017)
95. N.R. Cooper, Propagating magnetic vortex rings in ferromagnets. Phys. Rev. Lett. **82**, 1554–1557 (1999)
96. G.H. Derrick, Comments on nonlinear wave equations as models for elementary particles. J. Math. Phys. **5**(9), 1252–1254 (1964)
97. L.D. Faddeev, Some comments on the many-dimensional solitons. Lett. Math. Phys. **1**(4), 289–293 (1976)
98. J. Hietarinta, P. Salo, Faddeev-Hopf knots: dynamics of linked un-knots. Phys. Lett. B **451**(1), 60–67 (1999)
99. H. Hopf, Über die Abbildungen der dreidimensionalen Sphäre auf die Kugelfläche. Math. Ann. **104**(1), 637–665 (1931)
100. L. Faddeev, A.J. Niemi, Partially dual variables in SU(2) Yang-Mills theory. Phys. Rev. Lett. **82**, 1624–1627 (1999)
101. L. Faddeev, A.J. Niemi, Stable knot-like structures in classical field theory. Nature. **387**(6628), 58–61 (1997)
102. E. Babaev, Dual neutral variables and knot solitons in triplet superconductors. Phys. Rev. Lett. **88**, 177002 (2002)
103. E. Babaev, L.D. Faddeev, A.J. Niemi, Hidden symmetry and knot solitons in a charged two-condensate bose system. Phys. Rev. B **65**, 100512 (2002)
104. Y. Kawaguchi, M. Nitta, M. Ueda, Knots in a spinor Bose-Einstein condensate. Phys. Rev. Lett. **100**, 180403 (2008)

105. D. Kleckner, W.T.M. Irvine, Creation and dynamics of knotted vortices. Nat. Phys. **9**(4), 253–258
106. P.J. Ackerman, I.I. Smalyukh, Diversity of knot solitons in liquid crystals manifested by linking of preimages in torons and hopfions. Phys. Rev. X **7**, 011006 (2017)
107. P.J. Ackerman, I.I. Smalyukh, Static three-dimensional topological solitons in fluid chiral ferromagnets and colloids. Nat. Mater. **16**(4), 426–432 (2017)
108. I. Georgescu, Out of the shadows. Nat. Phys. **13**(3), 208 (2017)
109. I. Dzyaloshinsky, A thermodynamic theory of weak ferromagnetism of antiferromagnetics. J. Phys. Chem. Solids **4**(4), 241–255 (1958)
110. T. Moriya, Anisotropic superexchange interaction and weak ferromagnetism. Phys. Rev. **120**, 91–98 (1960)
111. J.-S.B. Tai, I.I. Smalyukh, Static Hopf solitons and knotted emergent fields in solid-state noncentrosymmetric magnetic nanostructures. Phys. Rev. Lett. **121**, 187201 (2018)
112. Y. Liu, R.K. Lake, J. Zang, Binding a hopfion in a chiral magnet nanodisk. Phys. Rev. B **98**, 174437 (2018)
113. P. Sutcliffe, Hopfions in chiral magnets. J. Phys. A: Math. Theor. **51**(37), 375401 (2018)
114. F.N. Rybakov, N.S. Kiselev, A.B. Borisov, L. Döring, C. Melcher, S. Blügel, Magnetic hopfions in solids. *Preprint*, page arxiv:1904.00250 (2019)
115. M.R. Hestenes, E. Stiefel, Methods of conjugate gradients for solving linear systems. J. Res. Nat. Bureau Stand. **49**, 409–436 (1952)
116. J.H.C. Whitehead, An expression of Hopf's invariant as an integral. Proc. Natl. Acad. Sci. **33**(5), 117–123 (1947)
117. S. Zhang, S.S.-L. Zhang, Generalization of the Landau-Lifshitz-Gilbert equation for conducting ferromagnets. Phys. Rev. Lett. **102**, 086601 (2009)
118. J. Gladikowski, M. Hellmund, Static solitons with nonzero Hopf number. Phys. Rev. D **56**, 5194–5199 (1997)
119. R. Bott, L.W. Tu, *Differential Forms in Algebraic Topology* (Graduate Texts in Mathematics. Springer, New York, 1995)
120. C.P. McNally, Divergence-free interpolation of vector fields from point values – exact $\nabla \cdot \mathbf{B} = 0$ in numerical simulations. Month. Not. Roy. Astronomic. Soc.: Lett. **413**(1), L76–L80 (2011)
121. X. Zhang, J. Xia, Y. Zhou, D. Wang, X. Liu, W. Zhao, M. Ezawa, Control and manipulation of a magnetic skyrmionium in nanostructures. Phys. Rev. B **94**, 094420 (2016)
122. F. Zheng, H. Li, S. Wang, D. Song, C. Jin, W. Wei, A. Kovács, J. Zang, M. Tian, Y. Zhang, D. Haifeng, R.E. Dunin-Borkowski, Direct imaging of a zero-field target skyrmion and its polarity switch in a chiral magnetic nanodisk. Phys. Rev. Lett. **119**, 197205 (2017)
123. C.L. Kane, E.J. Mele, Z_2 topological order and the quantum spin Hall effect. Phys. Rev. Lett. **95**, 146802 (2005)
124. B. Andrei Bernevig, T.L. Hughes, S-C. Zhang, Quantum spin Hall effect and topological phase transition in HgTe quantum wells. Science **314**(5806), 1757–1761 (2006)
125. M. König, S. Wiedmann, C. Brüne, A. Roth, H. Buhmann, L.W. Molenkamp, X.-L. Qi, S.-C. Zhang, Quantum spin hall insulator state in HgTe quantum wells. Science **318**(5851), 766–770 (2007)
126. M.Z. Hasan, C.L. Kane, Colloquium: Topological insulators. Rev. Mod. Phys. **82**, 3045–3067 (2010)
127. X.-L. Qi, S.-C. Zhang, Topological insulators and superconductors. Rev. Mod. Phys. **83**, 1057–1110 (2011)
128. C.L. Kane, T.C. Lubensky, Topological boundary modes in isostatic lattices. Nat. Phys. **10**(1), 39–45 (2014)
129. F.D.M. Haldane, S. Raghu, Possible realization of directional optical waveguides in photonic crystals with broken time-reversal symmetry. Phys. Rev. Lett. **100**, 013904 (2008)
130. S. Fujimoto, Hall effect of spin waves in frustrated magnets. Phys. Rev. Lett. **103**, 047203 (2009)
131. H. Katsura, N. Nagaosa, P.A. Lee, Theory of the thermal Hall effect in quantum magnets. Phys. Rev. Lett. **104**, 066403 (2010)

132. Y. Onose, T. Ideue, H. Katsura, Y. Shiomi, N. Nagaosa, Y. Tokura, Observation of the magnon Hall effect. Science **329**(5989), 297–299 (2010)
133. R. Matsumoto, S. Murakami, Theoretical prediction of a rotating magnon wave packet in ferromagnets. Phys. Rev. Lett. **106**, 197202 (2011)
134. R. Matsumoto, R. Shindou, S. Murakami, Thermal Hall effect of magnons in magnets with dipolar interaction. Phys. Rev. B **89**, 054420 (2014)
135. L. Zhang, J. Ren, J.-S. Wang, B. Li, Topological magnon insulator in insulating ferromagnet. Phys. Rev. B **87**, 144101 (2013)
136. M. Mochizuki, X.Z. Yu, S. Seki, N. Kanazawa, W. Koshibae, J. Zang, M. Mostovoy, Y. Tokura, N. Nagaosa, Thermally driven ratchet motion of a skyrmion microcrystal and topological magnon Hall effect. Nat. Mater. **13**(3), 241–246 (2014)
137. A. Mook, J. Henk, I. Mertig, Edge states in topological magnon insulators. Phys. Rev. B **90**, 024412 (2014)
138. M. Mena, R.S. Perry, T.G. Perring, M.D. Le, S. Guerrero, M. Storni, D.T. Adroja, Ch. Rüegg, D.F. McMorrow, Spin-wave spectrum of the quantum ferromagnet on the pyrochlore lattice $Lu_2V_2O_7$. Phys. Rev. Lett. **113**, 047202 (2014)
139. H. Lee, g H. Han, P.A. Lee, Thermal Hall effect of spins in a paramagnet. Phys. Rev. B **91**, 125413 (2015)
140. R. Chisnell, J.S. Helton, D.E. Freedman, D.K. Singh, R.I. Bewley, D.G. Nocera, Y.S. Lee, Topological magnon bands in a kagome lattice ferromagnet. Phys. Rev. Lett. **115**, 147201 (2015)
141. C-E. Bardyn, T. Karzig, G. Refael, T.C. H. Liew. Chiral Bogoliubov excitations in nonlinear bosonic systems. Phys. Rev. B **93**, 020502 (2016)
142. S.A. Owerre, A first theoretical realization of honeycomb topological magnon insulator. J. Phys.: Condens. Matter **28**(38), 386001 (2016)
143. Se K. Kim, H. Ochoa, R. Zarzuela, Y. Tserkovnyak. Realization of the Haldane-Kane-Mele model in a system of localized spins. Phys. Rev. Lett. **117**, 227201 (2016)
144. R. Shindou, -ichiro Ohe, R. Matsumoto, S. Murakami, E. Saitoh, Chiral spin-wave edge modes in dipolar magnetic thin films. Phys. Rev. B **87**, 174402 (2013)
145. R. Shindou, R. Matsumoto, S. Murakami, J. Ohe, Topological chiral magnonic edge mode in a magnonic crystal. Phys. Rev. B **87**, 174427 (2013)
146. I. Lisenkov, V. Tyberkevych, A. Slavin, P. Bondarenko, B.A. Ivanov, E. Bankowski, T. Meitzler, S. Nikitov, Spin-wave edge modes in finite arrays of dipolarly coupled magnetic nanopillars. Phys. Rev. B **90**, 104417 (2014)
147. S.O. Demokritov, A.N. Slavin, *Magnonics: From Fundamentals to Applications* (Topics in Applied Physics. Springer, Berlin Heidelberg, 2012)
148. D. Grundler, Reconfigurable magnonics heats up. Nat. Phys. **11**(6), 438–441 (2015)
149. B. Lenk, H. Ulrichs, F. Garbs, M. Münzenberg, The building blocks of magnonics. Phys. Rep. **507**(4), 107–136 (2011)
150. A.V. Chumak, V.I. Vasyuchka, A.A. Serga, B. Hillebrands, Magnon spintronics. Nat. Phys. **11**(6), 453–461 (2015)
151. B.I. Halperin, Quantized hall conductance, current-carrying edge states, and the existence of extended states in a two-dimensional disordered potential. Phys. Rev. B **25**, 2185–2190 (1982)
152. X. Chen, A. Tiwari, S. Ryu, Bulk-boundary correspondence in (3+1)-dimensional topological phases. Phys. Rev. B **94**, 045113 (2016)
153. X.S. Wang, Y. Su, and X.R. Wang, Topologically protected unidirectional edge spin waves and beam splitter. Phys. Rev. B **95**, 014435 (2017)
154. X.S. Wang, H.W. Zhang, X.R. Wang, Topological magnonics: A paradigm for spin-wave manipulation and device design. Phys. Rev. Applied **9**, 024029 (2018)
155. X.S. Wang, X.R. Wang, Anomalous magnon Nernst effect of topological magnonic materials. J. Phys. D Appl. Phys. **51**(19), 194001 (2018)
156. Y. Su, X.S. Wang, X.R. Wang, Magnonic Weyl semimetal and chiral anomaly in pyrochlore ferromagnets. Phys. Rev. B **95**, 224403 (2017)

157. Y. Su, X.R. Wang, Chiral anomaly of Weyl magnons in stacked honeycomb ferromagnets. Phys Rev B **96**, 104437 (2017)
158. G. Jackeli, G. Khaliullin, Mott insulators in the strong spin-orbit coupling limit: From Heisenberg to a quantum compass and Kitaev models. Phys. Rev. Lett. **102**, 017205 (2009)
159. M. Sato, K. Hasebe, K. Esaki, M. Kohmoto, Time-reversal symmetry in non-hermitian systems. Progress Theor. Phys. **127**(6), 937–974 (2012)
160. R.M. White, M. Sparks, I. Ortenburger, Diagonalization of the antiferromagnetic magnon-phonon interaction. Phys. Rev. **139**, A450–A454 (1965)
161. A.H. Castro Neto, F. Guinea, N.M.R. Peres, K.S. Novoselov, A.K. Geim, The electronic properties of graphene. Rev. Mod. Phys. **81**, 109–162 (2009)
162. J.E. Avron, R. Seiler, B. Simon, Homotopy and quantization in condensed matter physics. Phys. Rev. Lett. **51**, 51–53 (1983)
163. A. Bohm, A. Mostafazadeh, H. Koizumi, Q. Niu, J. Zwanziger, *The Geometric Phase in Quantum Systems: Foundations, Mathematical Concepts, and Applications in Molecular and Condensed Matter Physics* (Theoretical and Mathematical Physics. Springer, Berlin Heidelberg, 2013)
164. X. Wan, A.M. Turner, A. Vishwanath, S.Y. Savrasov, Topological semimetal and Fermi-arc surface states in the electronic structure of pyrochlore iridates. Phys. Rev. B **83**, 205101 (2011)
165. A.A. Burkov, L. Balents, Weyl semimetal in a topological insulator multilayer. Phys. Rev. Lett. **107**, 127205 (2011)
166. S-Y. Xu, I. Belopolski, N. Alidoust, M. Neupane, G. Bian, C. Zhang, R. Sankar, G. Chang, Z. Yuan, C-C. Lee, S-M. Huang, H. Zheng, J. Ma, D.S. Sanchez, B. Wang, A. Bansil, F. Chou, P.P. Shibayev, H. Lin, S. Jia, M. Zahid Hasan, Discovery of a Weyl fermion semimetal and topological Fermi arcs. Science **349**(6248), 613–617 (2015)
167. B.Q. Lv, H.M. Weng, B.B. Fu, X.P. Wang, H. Miao, J. Ma, P. Richard, X.C. Huang, L.X. Zhao, G.F. Chen, Z. Fang, X. Dai, T. Qian, H. Ding, Experimental discovery of Weyl Semimetal TaAs. Phys. Rev. X **5**, 031013 (2015)
168. L. Ling, Z. Wang, D. Ye, L. Ran, F. Liang, J.D. Joannopoulos, M. Soljačić, Experimental observation of Weyl points. Science **349**(6248), 622–624 (2015)
169. F-Y. Li, Y-D. Li, Y. Baek Kim, L. Balents, Y. Yu, G. Chen, Weyl magnons in breathing pyrochlore antiferromagnets. Nat. Commun. **7**(1), 12691 (2016)
170. A. Mook, J. Henk, I. Mertig, Tunable magnon Weyl points in ferromagnetic pyrochlores. Phys. Rev. Lett. **117**, 157204 (2016)
171. A. Alexey, Soluyanov, Dominik Gresch, Zhijun Wang, QuanSheng Wu, Matthias Troyer, Xi Dai, and B. Andrei Bernevig. Type-II Weyl semimetals. Nature **527**(7579), 495–498 (2015)
172. F. Meier, D. Loss, Magnetization transport and quantized spin conductance. Phys. Rev. Lett. **90**, 167204 (2003)
173. J. Sugiyama, K. Mukai, H. Nozaki, M. Harada, M. Månsson, K. Kamazawa, D. Andreica, A. Amato, A.D. Hillier, Antiferromagnetic spin structure and lithium ion diffusion in Li_2MnO_3 probed by μ^+SR. Phys. Rev. B **87**, 024409 (2013)
174. T. Takayama, A. Kato, R. Dinnebier, J. Nuss, H. Kono, L.S.I. Veiga, G. Fabbris, D. Haskel, H. Takagi, Hyperhoneycomb iridate β-Li_2IrO_3 as a platform for Kitaev magnetism. Phys. Rev. Lett. **114**, 077202 (2015)
175. X.-L. Sheng, B.K. Nikolić, Monolayer of the 5*d* transition metal trichloride $OsCl_3$: A playground for two-dimensional magnetism, room-temperature quantum anomalous hall effect, and topological phase transitions. Phys. Rev. B **95**, 201402 (2017)
176. J. He, X. Li, P. Lyu, P. Nachtigall, Near-room-temperature Chern insulator and Dirac spin-gapless semiconductor: nickel chloride monolayer. Nanoscale **9**, 2246–2252 (2017)
177. M-H. Pan, H. Liu, Z. Wang, J-F. Jia, Q-K. Xue, J-L. Li, S. Qin, U.M. Mirsaidov, X.-R. Wang, J.T. Markert, Z. Zhang, C.-K. Shih, Quantum growth of magnetic nanoplatelets of Co on Si with high blocking temperature. Nano Lett. **5**(1), 87–90 (2005) PMID: 15792418

178. This work is supported by the NSFC Grants (No. 11974296, 11774296 and 11804045) and Hong Kong RGC Grants (No. 16301518 and 16301619)

Chapter 15
Topological Dynamics of Spin Texture Based Metamaterials

Zhixiong Li, Yunshan Cao, and Peng Yan

Abstract Pursuing topological phase and matter in a variety of systems is one central issue in current physical sciences and engineering. Similar to other (quasi-)particles, the collective gyration motion of magnetic spin textures (vortex, bubble, and skyrmion etc.) can also exhibit the behavior of waves. In this chapter, we review our recent work on topological dynamics of spin-texture based metamaterials. We first briefly introduce the topological structures, properties, and applications of magnetic solitons. Then we focus on the topological dynamics of spin texture lattice, uncovering the first-order topological insulator in two-dimensional honeycomb lattice of massive magnetic skyrmions, and the second-order topological insulator in breathing kagome and honeycomb lattice of vortices. Conclusion and outlook are drawn finally.

15.1 Introduction

In recent years, topological insulators (TIs) [1–3] are receiving considerable attention for their exotic physical properties. The most peculiar character of TIs is that they can support chiral edge states which are absent in conventional insulators. Topological edge states are modes confined at the boundary/surface of a system and generally have a certain chirality which enables them to be immune from small disturbances such as disorders and/or defects. Ever since its discovery in electronic systems [4, 5], the topological edge state has been readily predicted and observed in optics [6–10], mechanics [11–14], acoustics [15–18], and very recently in magnetics [19–22].

A conventional n-dimensional topological insulator only has $(n-1)$-dimensional (first-order) topological edge/surface modes according to the bulk-boundary correspondence [1, 2]. Very recently, the concept of higher-order topological insulators (HOTIs) was proposed [23–29] and confirmed by various experiments in

Z. Li · Y. Cao · P. Yan (✉)
School of Electronic Science and Engineering, University of Electronic Science and Technology of China, Chengdu 610054, China
e-mail: yan@uestc.edu.cn

E. Kamenetskii (ed.), *Chirality, Magnetism and Magnetoelectricity*, Topics in Applied Physics 138,
https://doi.org/10.1007/978-3-030-62844-4_15

photonic [30–35], acoustic [36–40], and electric-circuit [41–45] systems. Different from first-order topological insulators (FOTIs), a HOTI has $(n-2)$ or even $(n-3)$-dimensional (second-order or third-order) topological boundary states, which goes beyond the conventional bulk-boundary correspondence and is characterized by several new topological invariants, such as the nested Wilson loop [46], Green's function zeros [47], the quantized bulk polarization (Wannier center) [25, 36, 37], and the $\mathbb{Z}_N$ Berry phase (quantized to $2\pi/N$) [48–52]. The HOTIs thus have broadened our understanding on topological insulating phases of matter.

There are two important excitations in magnetic systems. One is the spin wave (or magnon) [53, 54], i.e., the collective motion of magnetic moments. The topological phase of magnons in magnetic materials is of great current interest in magnetism because of its fundamental significance as well as in spintronics because of its practical utility for robust information processing [20, 22, 55–60]. The other one is the collective oscillation of magnetic solitons (such as the magnetic vortex [61, 62], bubble [63–65], skyrmion [66, 67], and domain wall [68–70]), which are long-term topic in condensed matter physics for their interesting dynamics and promising application. These magnetic solitons generally have the characteristics of small size, easy manipulation and high stability. The spintronic devices based on magnetic solitons thus have advantages over other electronic devices. For example, the racetrack memory made of magnetic domain walls can greatly improve the data storage density and the reading speed [71, 72]; the critical current density required for encoding information can be significantly reduced by using the skyrmion as the carrier of information [73, 74]; the magnetic oscillators based on vortices or skyrmions are very robust and flexible [75–77]. On the other hand, it has been shown that the collective gyration motion of magnetic solitons can exhibits the behavior of waves [78–82]. Furthermore, the first-order topological chiral edge states based on two-dimensional honeycomb lattices of magnetic solitons (vortices and bubbles) have been predicted by Kim and Tserkovnyak [83]. In a word, the metamaterials based on topological spin texture are attracting more and more attention for both fundamental interest and the potential applications in spintronics and quantum computing [84–89].

In this chapter, we report the realization of TIs in magnetic soliton lattices. The exposition is organized as follows: Sect. 15.2 introduces the topological structures, properties, and applications of magnetic solitons; the first-order topological edge states of skyrmion lattice are discussed in Sects. 15.3, 15.4 and 15.5 focus on the second-order TIs in the breathing kagome and honeycomb lattices of magnetic vortices, respectively. We summarise the results in Sect. 15.6.

15.2 Topological Structures, Properties, and Applications of Magnetic Solitons

Topology is a study of geometry or space which can keep some properties invariant under a continuous transformation. Topological magnetic soliton is an application of topology in condensed matter physics. More precisely, magnetic solitons are the spin textures characterized by a topological charge [83],

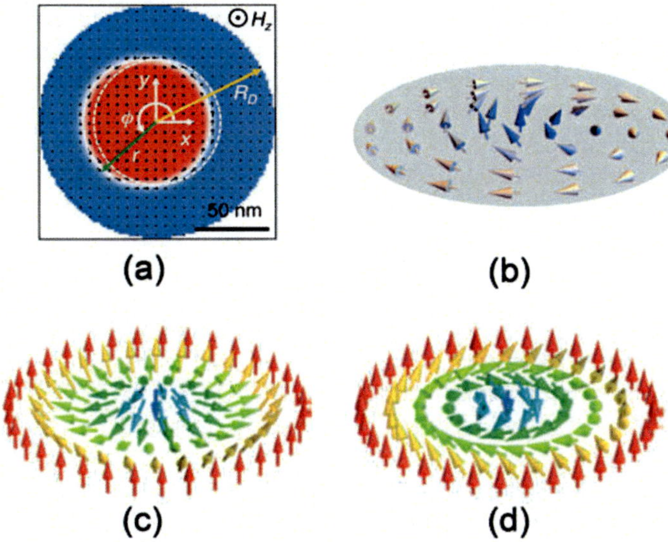

Fig. 15.1 The micromagnetic structures of magnetic bubble (**a**), vortex (**b**), Néel-type skyrmion (**c**), and Bloch-type skyrmion (**d**). Images are taken from [64, 83, 90]

$$Q = \frac{1}{4\pi} \iint dxdy\mathbf{m} \cdot (\frac{\partial \mathbf{m}}{\partial x} \times \frac{\partial \mathbf{m}}{\partial y}), \tag{15.1}$$

which counts how many times the local magnetization $\mathbf{m}$ wraps the unit sphere. Typical magnetic solitons include magnetic bubble, vortex, and skyrmion, with the micromagnetic structures shown in Fig. 15.1, respectively. The topological charge of magnetic bubble and skyrmion are ± 1, while this value changes to $\pm 1/2$ for vortex configuration. Topological charge is a topological invariant which indicates the trivial structure (for example, ferromagnetic state) can not continuously deformed into a topological spin texture because of the topological protection and the magnetic solitons with the same topological charge are homotopic.

The low-energy dynamics of the magnetic vortex can be described by the massless Thiele's equation [83, 91] whithin the approximation of the rigid model:

$$G\hat{z} \times \frac{d\mathbf{U}_j}{dt} - \alpha D \frac{d\mathbf{U}_j}{dt} + \mathbf{F}_j = 0, \tag{15.2}$$

where $\mathbf{U}_j \equiv \mathbf{R}_j - \mathbf{R}_j^0$ is the displacement of the vortex core from its equilibrium position $\mathbf{R}_j^0$; $G = -4\pi Q d M_s/\gamma$ is the gyroscopic constant with Q is the topological charge; d is the thickness of ferromagnetic layer; M_s is the saturation magnetization; γ is the gyromagnetic ratio. αD is the viscous coefficient with α being the Gilbert damping constant. The conservative force $\mathbf{F}_j = -\partial W/\partial \mathbf{U}_j$ where W is the potential energy of the system. For a single vortex, the potential energy have the parabolic type: $W = W_0 + K\mathbf{U}_j^2/2$, where W_0 is the energy of system when vortex core located at the center of the nanodisk and K is the spring constant. By neglecting the viscous force term, we can derive the gyration frequency of an isolated vortex with $\omega_0 = K/|G|$.

However, it is well known that magnetic bubbles and skyrmions in particular manifest an inertia in their gyration motion [63, 92]. The mass effect thus should be taken into account for describing the skyrmion (or bubble) oscillation. Therefore, the Thiele's equation can be generalized as:

$$-M\frac{d^2\mathbf{U}_j}{dt^2} + G\hat{z} \times \frac{d\mathbf{U}_j}{dt} - \alpha D\frac{d\mathbf{U}_j}{dt} + \mathbf{F}_j = 0, \tag{15.3}$$

with M being the inertial mass of magnetic soliton. Similarly, we can calculate the gyration frequency of skyrmion (or bubble) with:

$$\omega_{\pm} = -G/2M \pm \sqrt{(G/2M)^2 + K/M}. \tag{15.4}$$

The positive and negative values of the ω in (15.4) indicate that there are two kinds of gyration modes with clockwise and counterclockwise direction, respectively.

Magnetic skyrmions are typical magnetic solitons stabilized by Dzyaloshinskii Moriya interactions (DMIs) [66, 93, 94]. They can be manipulated by spin-polarized electrical current with extremely low density [73, 74]. Skyrmion also can be drived by other external force, such as spin wave [95, 96], microwave field [97, 98], and temperature gradient [99–101]. Interestingly, very recently, it has been proposed that twisted photons [102] and magnons [103] carrying orbital angular momentum (OAM) can act as "optical tweezers" and "magnetic tweezers" to drive the rotation of skyrmion, as shown in Fig. 15.2 and Fig. 15.3, respectively. Theoretical calculation and numerical simulation show that the topological charge of twisted photons (or magnons) can determine both the magnitude and the handedness of the rotation velocity of skyrmions.

Skyrmions are the ideal information carrier in spintronic devices. However, a skyrmion can not move in a straight line along the driving current direction because of the Magnus force which leads to the shift of its motion trajectory, such behavior is called skyrmion Hall effect (SHE) [104, 105]. In practical applications, the skyrmion may be destroyed when touching the device boundaries. To overcome this issue, a lot of methods have been proposed. For example, X. Zhang et al. [106] showed that the antiferromagnetically exchange-coupled bilayer system containing two skyrmions with different polarities can suppress the SHE, leading to a perfectly straight trajectory for skyrmion driven by a spin-polarized current. Moreover, the antiferromagnetic (AFM) skyrmion can also avoid SHE [107, 108]. Interestingly, recent research shows that the AFM skyrmion can emerge in a ferromagnet with gain (negative α) [109]. Figure 15.4 shows the dynamical process of the formation of AFM skyrmion in ferromagnet (here α is set to be -0.01), with the initial magnetization profile being random [see Fig. 15.4a]. At $t = 0.035$ ns, local magnetic moments quickly evolve to an antiparallelly aligned state, as shown in Fig. 15.4b. Then, all spins inside a circle of radius 5 nm in the film center are randomized. At $t = 0.14$ ns, an AFM skyrmion is stabilized (see Fig. 15.4d). When an in-plane spin-polarized electric current was

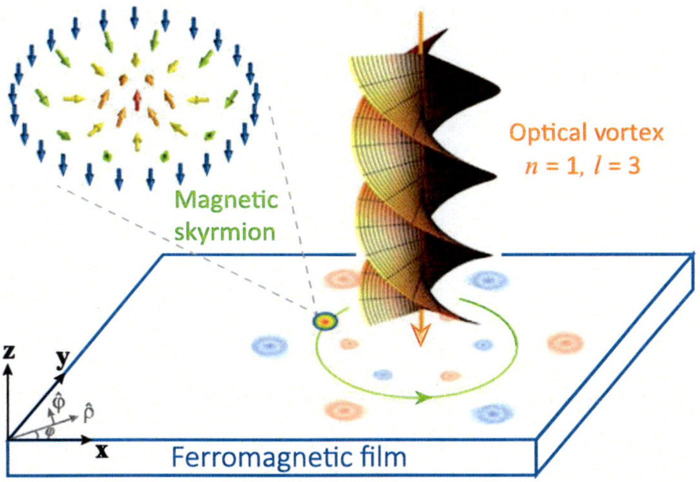

Fig. 15.2 Schematic diagram of the rotational motion of a Néel-type skyrmion in a thin ferromagnetic film driven by an optical vortex with radial index $n = 1$ and OAM quantum number $l = 3$. The solid circle with a red core represents the skyrmion. The flower-like pattern (pink and blue spots) sketches the induced magnetization profile by the optical vortex field shinning on the magnetic film. Images are taken from [102]

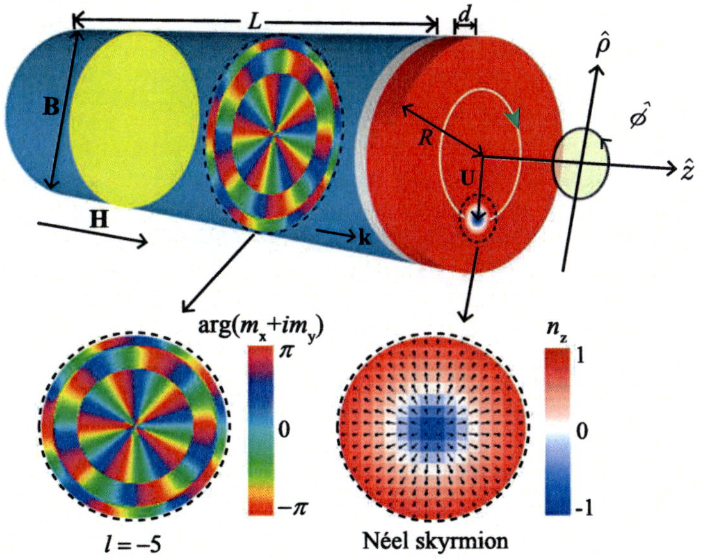

Fig. 15.3 Schematic illustration of a heterostructured nanocylinder exchange-coupled to a chiral magnetic nanodisk hosting a Néel-type skyrmion. An external static field **H** is applied along the z-direction. A spin-wave beam with the wavevector **k** and OAM quantum number $l = -5$ is excited by a localized microwave field **B**, leading to a steady skyrmion gyration around the disk center. Images are taken from [103]

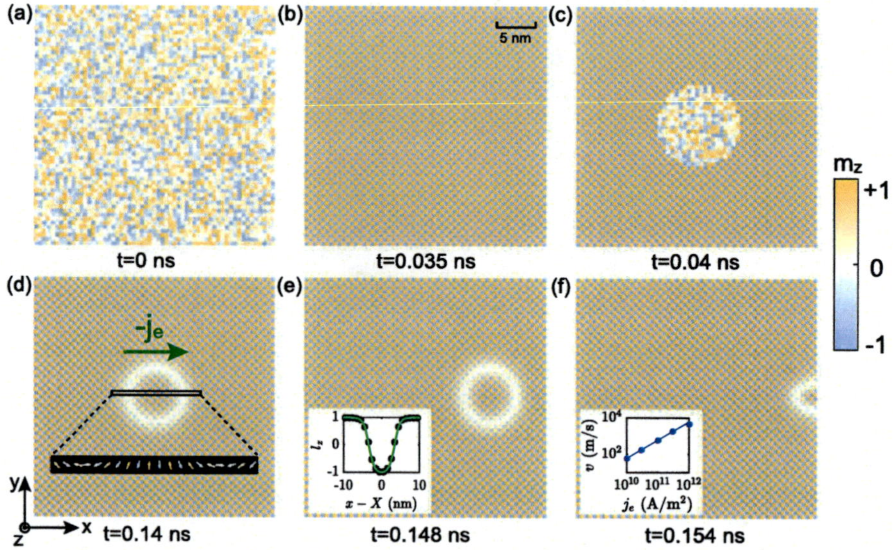

Fig. 15.4 **a** Random spin configuration at $t = 0$ ns. **b** AFM state evolved at $t = 0.035$ ns. **c** Randomizing spins inside the circle at $t = 0.04$ ns. **d** AFM skyrmion stabilized at $t = 0.14$ ns. (Inset) Magnetization profile of the cross section of the AFM skyrmion. **e** Current-driven AFM skyrmion motion. (Inset) Spatial distribution of the z component of the Néel vector $\mathbf{l}_i$. **f** AFM skyrmion annihilation at the film boundary. (Inset) Current dependence of the skyrmion velocity when it is far from the edge. Images are taken from [109]

applied to drive the AFM skyrmion, the skyrmion trajectory is exactly along the flowing direction of electrons, without the SHE. Besides, the method for the current-driven skyrmion motion on magnetic nanotubes [110] can also make the skyrmion away from the edge of system (avoid touching the edges of system), since the tube is edgeless for the tangential skyrmion motion. A stable skyrmion propagation can survive in the presence of a very large current density without any annihilation or accumulation. The nanotube can be viewed as a seamless, hollow tubular structure rolled from a planar strip, as illustrated in Fig. 15.5.

On the other hand, the magnetic skyrmions can be used to realize logical operation and thus have great potential application in logic devices [111–113]. Figure 15.6 shows typical double-track logic AND gate and OR gate based on twisted skyrmions [114]. The logical AND operation can be realized through the following processes (Fig. 15.7a1–a5): two $p = -1$ skyrmions (encoding 1) are placed at the left and right sides of the logic AND gate at $t = 0$. Here p represents the polarity of the skyrmion core. Then, driven by an electric current, the two skyrmions begin to move close to each other and become two twisted skyrmions, as shown in Fig. 15.7a2. Next, the two twisted skyrmions move along the AFM boundaries and merge into one twisted skyrmion at the intersection of the two boundaries, see Fig. 15.7a3, a4. Finally, the twisted skyrmion is pushed out of the boundary.

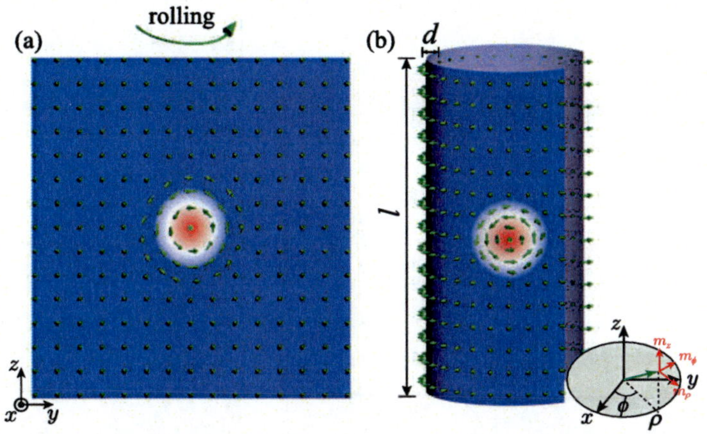

Fig. 15.5 **a** Schematic illustration of a Bloch-type skyrmion in a planar film. Green arrows refer to the local spin directions. **b** Skyrmion on a nanotube by rolling up (**a**). Colors refer to the ρ-component of the magnetization. The coordinate system is defined in the inset. Images are taken from [110]

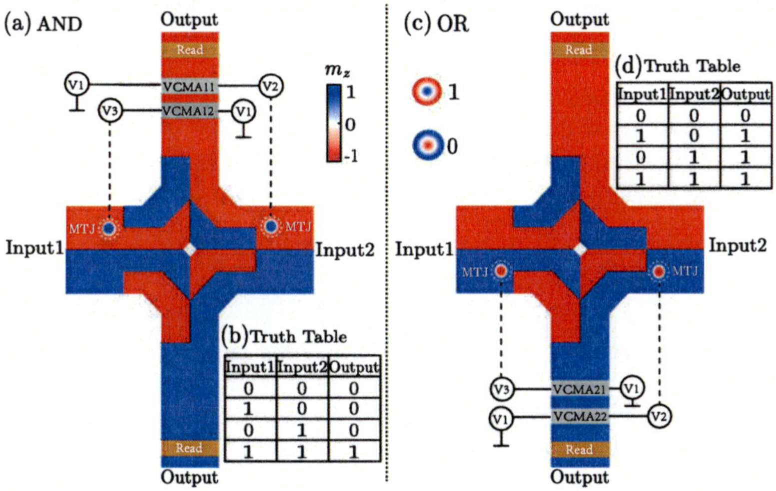

Fig. 15.6 Schematic plot of a logic AND (**a**) and OR (**c**) gate with two input skyrmions (top view). Truth table for the logic AND (**b**) and OR (**d**) gate. Images are taken from [114]

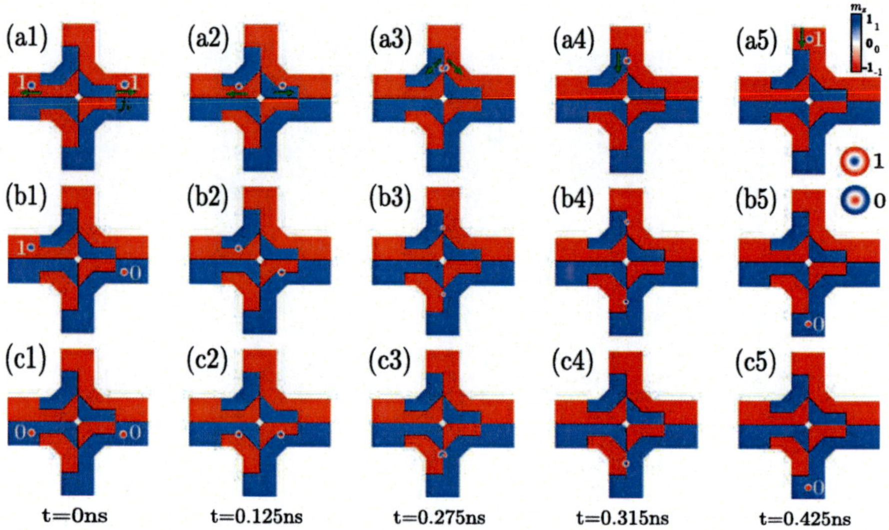

Fig. 15.7 Snapshots of the working process of skyrmions logic AND gate (top view). **a1–a5** Logic operation "1 + 1 = 1". **b1–b5** "1 + 0 = 0". **c1–c5** "0 + 0 = 0". Images are taken from [114]

Consequently, the final state is a skyrmion with $p = -1$ (encoding 1) such that "1 + 1 = 1", as shown in Fig. 15.7a5. Likewise, the other logical AND operations ("0 + 1 = 0" and "0 + 0 = 0") are plotted in Fig. 15.7b1–b5 and c1–c5, respectively.

15.3 The Topological Properties of Skyrmion Lattice

As mentioned in Sect. 15.1, Kim and Tserkovnyak have predicted theoretically the chiral edge modes in two-dimensional honeycomb lattice of vortices and bubbles by solving massless Thiele's equation [83]. However, the higher-order terms are important for describing the skyrmions (or bubbles) oscillation. In this section, we will discuss the edge states of honeycomb lattice of massive magnetic skyrmions for considering both a second-order inertial term of skyrmion mass and a third-order non-Newtonian gyroscopic term.

15.3.1 Large-Scale Micromagnetic Simulations

A large two-dimensional honeycomb lattice consisting of 984 identical magnetic nanodisks is considered to show the chiral edge states of magnetic soliton system. Figure 15.8a shows the sketch. Each disk contains a single magnetic skyrmion made of MnSi [115] which supports the Bloch-type skyrmion (depicted in Fig. 15.8b) due

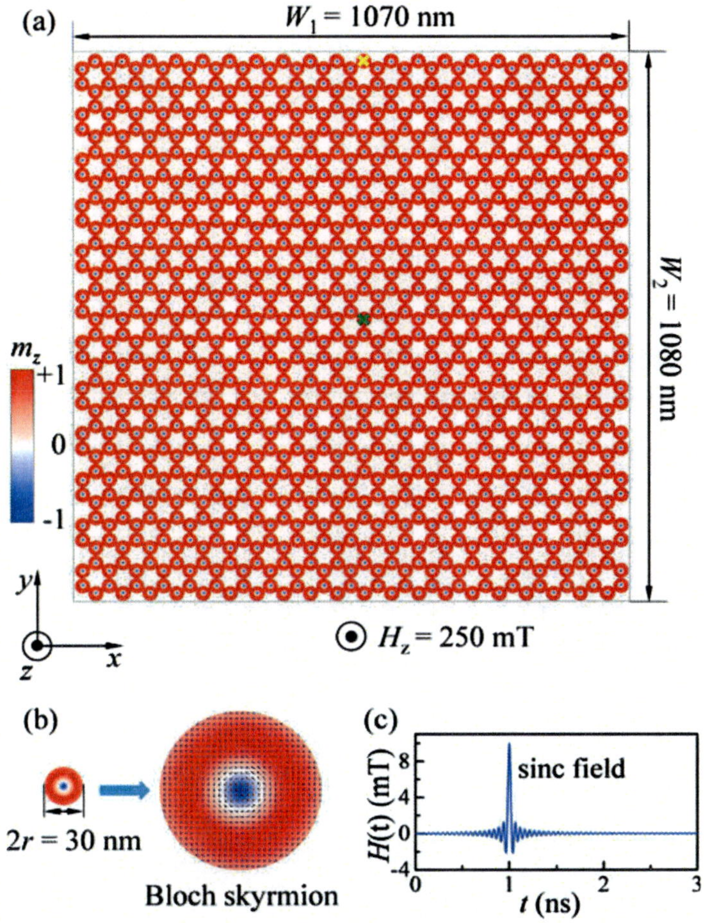

Fig. 15.8 **a** Illustration of the honeycomb lattice with size 1070×1080 nm^2, including 984 Bloch-type skyrmions. A uniform magnetic field is applied along the z axis to stabilize the skyrmions. Green and yellow crosses denote the positions of the driving fields in the center and at the edge of the lattice, respectively. **b** Zoomed in details of a nanodisk containing a Bloch-type skyrmion. **c** Time-dependence of the sinc-function field $H(t)$. Images are taken from [125]

to the bulk Dzyakoshinskii-Moriya interaction (DMI) [93, 116]. Here, the distance between nearest-neighbor disks is chosen to be equal to the disk diameter, indicating that nearest-neighbor skyrmions can strongly interact with each other mediated by the exchange spin-wave. It is worth noting that the dipolar interaction can not efficiently couple skyrmions when a physical gap between nearest-neighboring nanodisks is left. The micromagnetic simulations are performed with MUMAX3 [117].

The dispersion relation of skyrmion gyrations can be obtained by computing the spatiotemporal Fourier spectrum of the skyrmion positions over the lattice. Figure 15.9a shows the simulated band structure of skyrmion oscillation below a

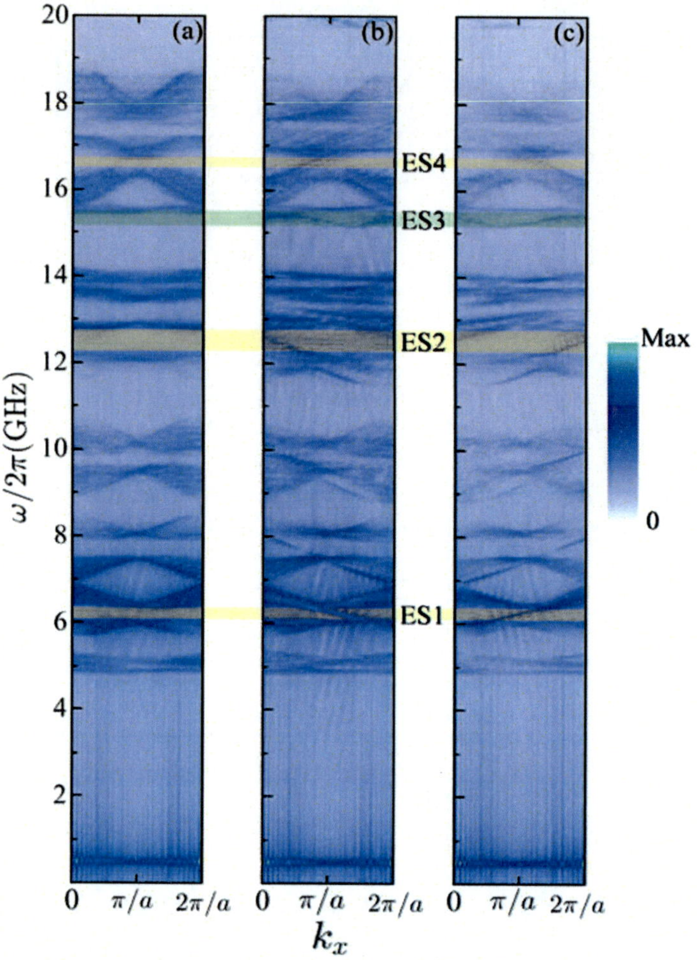

Fig. 15.9 The band structure of skyrmion gyrations when the exciting field is in the film center **a** and at the film edge by evaluating the Fourier spectrum over the upper **b** and the lower **c** parts of the honeycomb lattice. The constant $a = 2\sqrt{3}r$ represents the distance between the second-nearest neighboring nanodisks. Images are taken from [125]

cutoff frequency of 20 GHz when the exciting field (sinc-function magnetic field) locates in the lattice center. It can be seen that there is no bulk state in the gaps (areas shaded in both yellow and green). However, when the driving field is located at the edge of lattice, the situations are totally different. By implementing the Fourier analysis over the upper ($W_2/2 < y < W_2$) and the lower parts ($0 < y < W_2/2$) of the lattice, with results plotted in Fig. 15.9b, c, respectively, one can find that four edge states emerge in the spectrum gaps, labeled as ES1–ES4. Moreover, By evaluating the group velocity $d\omega/dk_x$ of each mode with ω the frequency and k_x the wave vector

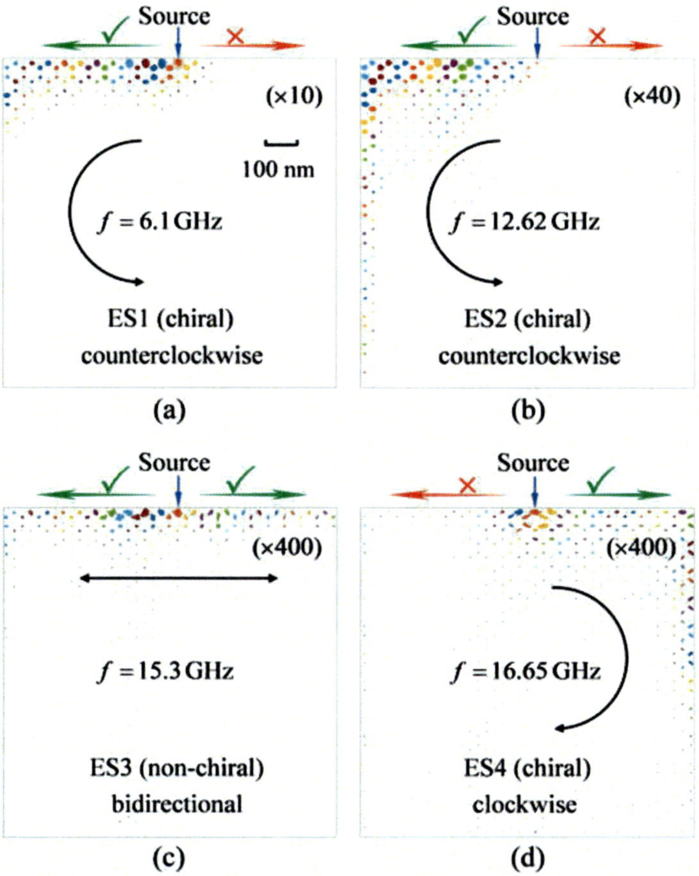

Fig. 15.10 Snapshot of the propagation of edge states with frequency $f = 6.1$ GHz (**a**), 12.62 GHz (**b**), 15.3 GHz (**c**), and 16.65 GHz (**d**) at $t = 40$ ns. Since the oscillation amplitudes of the skyrmion guiding centers are too small, we have magnified them by 10, 40 or 400 times labeled in each figure, correspondingly. Images are taken from [125]

along x direction, the chirality of these edge states can be identified: three edge states ES1, ES2 and ES4 (shaded in yellow) are unidirectional and chiral, in which ES1 and ES2 counterclockwise propagate, while ES4 behaves oppositely; ES3 (shaded in green) is bidirectional and thus non-chiral.

We choose four representative frequencies to visualize the propagation of edge wave by stimulating the dynamics of lattice under a sinusoidal field $\mathbf{h}(t) = h_0 \sin(2\pi ft)\hat{x}$ on one disk at the top edge, indicated by the blue arrows in Fig. 15.10. Figures 15.10a, b, d show the propagation of chiral edge states. One can clearly observe unidirectional wave propagation of these modes with either a counterclockwise manner (ES1 and ES2 shown in Fig. 15.10a, b respectively) or a clockwise one (ES4 plotted in Fig. 15.10d). It's very interesting and unique that multiband edge states with opposite chiralities coexist in a given soliton lattice. There is no analogue in condensed matter system, to the best of our knowledge. In contrast, the propagation of ES3 is bidirectional, as shown in Fig. 15.10c. This non-chiral mode can be simply explained in terms of the Tamm-Shockley mechanism [118, 119] which predicts that the periodicity breaking of the crystal potential at the boundary can lead to the formation of a conducting surface/edge state. Furthermore, it is rather straightforward to show the propagation of the chiral modes is immune from the defects and robust against the type of boundary, while the Tamm-Shockley mode is not.

15.3.2 Theoretical Model

We adopt the generalized Thiele's equation to theoretically understand the multiband chiral skyrmionic edge states carrying opposite chiralities. First of all, we assume that the steady-state magnetization $\mathbf{m}$ depends on not only the position of the guiding center $\mathbf{R}(t)$ but also its velocity $\dot{\mathbf{R}}(t)$ and acceleration $\ddot{\mathbf{R}}(t)$, and write $\mathbf{m} = \mathbf{m}(\mathbf{r} - \mathbf{R}(t), \dot{\mathbf{R}}(t), \ddot{\mathbf{R}}(t))$. After some algebra and by neglecting the damping term, one can obtain the generalized Thiele's equation including both a second-order inertial term of skyrmion mass M [63, 92, 102] and a non-Newtonian third-order gyroscopic term G_3 [120–122]:

$$G_3\hat{z} \times \frac{d^3\mathbf{U}_j}{dt^3} - M\frac{d^2\mathbf{U}_j}{dt^2} + G\hat{z} \times \frac{d\mathbf{U}_j}{dt} + \mathbf{F}_j = 0, \tag{15.5}$$

where $\mathbf{U}_j$ and G share the same definition as (15.2). The conservative force can be expressed as $\mathbf{F}_j = -\partial W/\partial \mathbf{U}_j$. Here W is the total potential energy for the sum of single disk and the interaction energy between nearest-neighbor disks: $W = \sum_j K\mathbf{U}_j^2/2 + \sum_{j\neq k} U_{jk}/2$ with $U_{jk} = I_\parallel U_j^\parallel U_k^\parallel - I_\perp U_j^\perp U_k^\perp$ [83, 123, 124]. $I_\parallel$ and $I_\perp$ are the longitudinal and the transverse coupling constants, respectively. By imposing $\mathbf{U}_j = (u_j, v_j)$ and defining $\psi_j = u_j + iv_j$, we have:

$$\hat{\mathcal{D}}\psi_j = \omega_K\psi_j + \sum_{k\in\langle j\rangle}(\zeta\psi_k + \xi e^{i2\theta_{jk}}\psi_k^*), \tag{15.6}$$

where the differential operator $\hat{\mathcal{D}} = i\omega_3\frac{d^3}{dt^3} - \omega_M\frac{d^2}{dt^2} + i\frac{d}{dt}$, $\omega_3 = G_3/|G|$, $\omega_M = M/|G|, \omega_K = K/|G|, \zeta = (I_\parallel - I_\perp)/2|G|$, and $\xi = (I_\parallel + I_\perp)/2|G|, \theta_{jk}$ is the angle of the direction $\hat{e}_{jk}$ from x axis with $\hat{e}_{jk} = (\mathbf{R}_k^0 - \mathbf{R}_j^0)/|\mathbf{R}_k^0 - \mathbf{R}_j^0|$, and $\langle j\rangle$ is the set of the nearest neighbors of j (here $Q = -1$). We then expand the complex variable to

$$\psi_j = \chi_j(t)\exp(-i\omega_0 t) + \eta_j(t)\exp(i\omega_0 t). \tag{15.7}$$

For counterclockwise (clockwise) skyrmion gyrations, one can justify $|\chi_j| \gg |\eta_j|$ ($|\chi_j| \ll |\eta_j|$). Substituting (15.7) into (15.6), one can obtain the following eigenvalue equation:

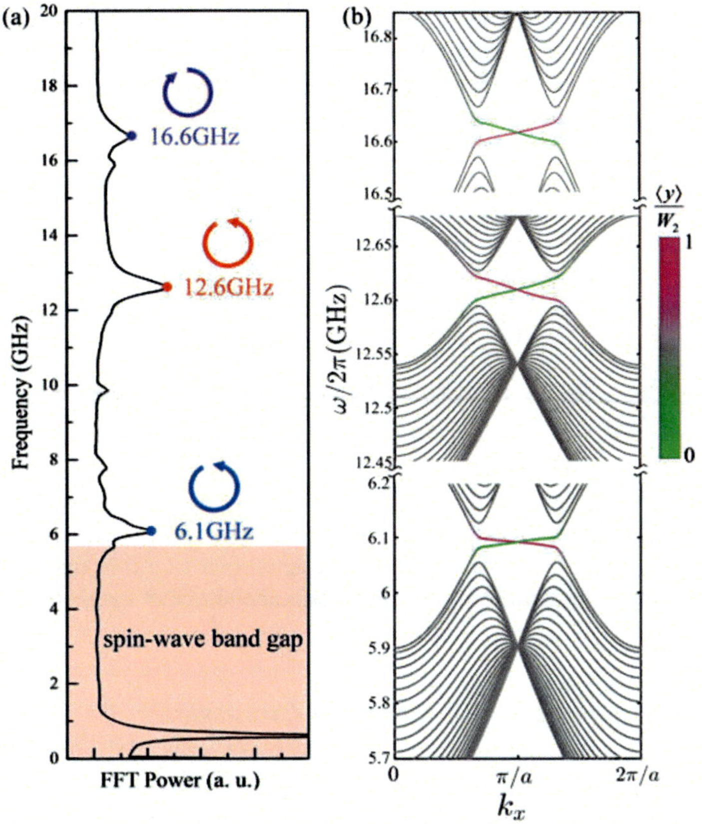

Fig. 15.11 **a** Resonant spectrum of skyrmion gyrations when the exciting field is applied over the whole system. Inset shows the chirality/handedness of the skyrmion guiding center of each mode. **b** Band structure by solving (15.8). Images are taken from [125]

$$\hat{\mathcal{D}}\psi_j = (\omega_K - 3\xi^2/2\bar{\omega}_K)\psi_j + \zeta \sum_{k\in\langle j\rangle} \psi_k - (\xi^2/2\bar{\omega}_K) \sum_{l\in\langle\langle j\rangle\rangle} e^{i2\bar{\theta}_{jl}}\psi_l, \tag{15.8}$$

with $\bar{\omega}_K = \omega_K - \omega_0^2\omega_M$, $\bar{\theta}_{jl} = \theta_{jk} - \theta_{kl}$ the relative angle from the bond $k \to l$ to the bond $j \to k$ with k between j and l, and $\langle\langle j\rangle\rangle$ the set of the second-nearest neighbors of j.

The key parameters G_3, M, K, $I_\parallel$ and $I_\perp$ can be determined from micromagnetic simulations in a self-consistent manner [125]. Figure 15.11a shows the spectrum of collective skyrmion oscillations with three strong resonance peaks at $\omega_{0,1}/2\pi = 6.1$ GHz, $\omega_{0,2}/2\pi = 12.6$ GHz, and $\omega_{0,3}/2\pi = 16.6$ GHz above the spin-wave band gap. Furthermore, Fig. 15.11b shows the computed band structure of the skyrmion gyrations near the resonance frequencies $\omega_0 = \omega_{0,1}, \omega_{0,2}$, and $\omega_{0,3}$ by solving (15.8) with the periodic boundary condition along x direction and the zigzag termination at $y = 0$ and $y = W_2$. The average vertical position of the modes $\langle y \rangle \equiv \sum_j R^0_{j,y}|\mathbf{U}_j|^2/\sum_j |\mathbf{U}_j|^2$ are also shown in Fig. 15.11, where $R^0_{j,y}$ is the equilibrium position of the skyrmion projected onto the y axis, represented by different colors: closer to magenta indicating more localized at the upper edge. It is interesting to note that the chirality of ES4 is opposite to those of ES1 and ES2. An interpretation is as following: The direction reversal of the skyrmion gyration generates a π phase accumulation in the next-nearest-neighbor hopping term of (15.8). The Chern numbers of two neighbouring bulk bands then switch their signs, so that the chirality of the edge state in between reverses.

15.4 Corner States in a Breathing Kagome Lattice of Vortices

In previous sections, we have discussed the topological insulating phases in magnetic system. All these phases, however, are first order by nature. In this section, we will discuss the higher-order topological edge states (corner states) in magnetic system, and demonstrate their features in breathing kagome lattice of magnetic vortices.

15.4.1 The Theoretical Results and Discussions

A breathing kagome lattice of magnetic nanodisks with vortex states is considered. Figure 15.12a plots the lattice structure with alternate distance parameters d_1 and d_2. We start with the generalized Thiele's equation (15.6) to describe the collective dynamics of vortex lattice:

$$\hat{\mathcal{D}}\psi_j = \omega_K \psi_j + \sum_{k \in \langle j \rangle, l} (\zeta_l \psi_k + \xi_l e^{i2\theta_{jk}} \psi_k^*), \tag{15.9}$$

where the differential operator $\hat{\mathcal{D}} = i\omega_3 \frac{d^3}{dt^3} - \omega_M \frac{d^2}{dt^2} - i\frac{d}{dt}$, $\zeta_l = (I_{\parallel,l} - I_{\perp,l})/2|G|$, and $\xi_l = (I_{\parallel,l} + I_{\perp,l})/2|G|$, with $l = 1$ (or $l = 2$) representing the distance d_1 (or d_2) between the nearest neighbor vortices, here the topological charge $Q = 1/2$.

The coupling strengths $I_\parallel$ and $I_\perp$ are strongly depend on the parameter d ($d = d'/r$ with d' the distance between two vortices and r being the radius of nanodisk) [127–129]. The analytical expression of $I_\parallel(d)$ and $I_\perp(d)$ are essential for evaluating the spectrum and the phase diagram of vortex gyrations. With the help of micromagnetic simulations for two-vortex system with different combinations of vortex polarities, one can obtain the best fit of the numerical data: $I_\parallel = \mu_0 M_s^2 r(-1.72064 \times 10^{-4} + 4.13166 \times 10^{-2}/d^3 - 0.24639/d^5 + 1.21066/d^7 - 1.81836/d^9)$ and $I_\perp = \mu_0 M_s^2 r\,(5.43158 \times 10^{-4} - 4.34685 \times 10^{-2}/d^3 + 1.23778/d^5 - 6.48907/d^7 +$

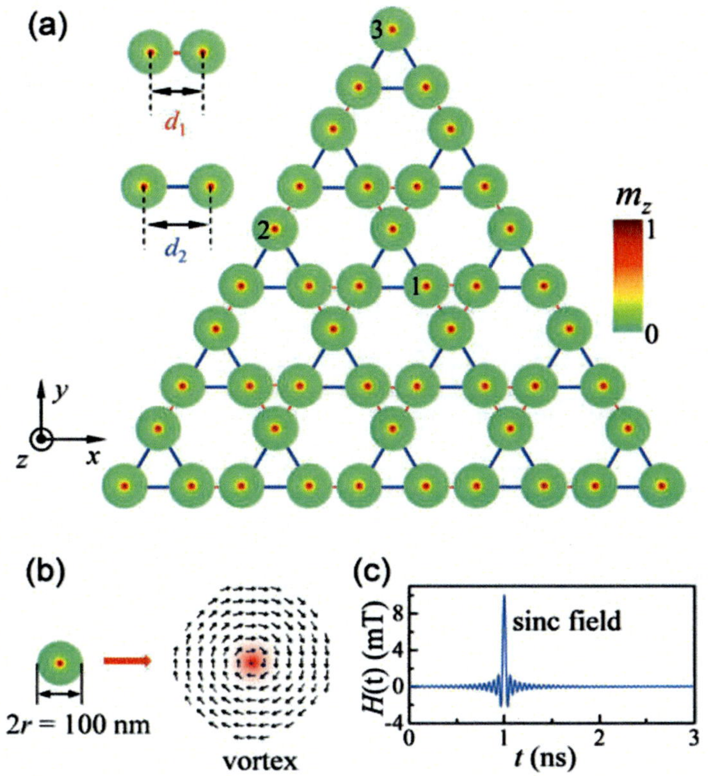

Fig. 15.12 **a** Illustration of the triangle-shape breathing kagome lattice including 45 nanodisks of the vortex state. d_1 and d_2 are the distance between two nearest-neighbor vortices. Arabic numbers 1, 2 and 3 denote the positions of spectrum analysis for bulk, edge and corner states, respectively. **b** Zoomed in details of a nanodisk with the radius $r = 50$ nm and the thickness $w = 10$ nm. **c** Time dependence of the sinc-function field $H(t)$ applied to the whole system. Images are taken from [126]

$13.6422/d^9$), as shown in Fig. 15.13a with symbols and curves representing simulation results and analytical formulas, respectively. In the calculations, the material parameters of Permalloy (Py: $Ni_{80}Fe_{20}$) [88, 130] was adopted, and $G = -3.0725 \times 10^{-13}$ J s rad^{-1}m^{-2}. The spring constant K, mass M, and non-Newtonian gyration G_3 can be obtained by analyzing the dynamics of a single vortex confined in the nanodisk [126]: $K = 1.8128 \times 10^{-3}$ J m^{-2}, $M = 9.1224 \times 10^{-25}$ kg, and $G_3 = -4.5571 \times 10^{-35}$J s^{3}rad^{-3}m^{-2}. Then, the eigenfrequencies of vortex gyrations in the breathing kagome lattice can be obtained by solving (15.9) numerically. Figure 15.13b shows the eigenfrequencies of the triangle-shape system for different values d_2/d_1 with a fixed $d_1 = 2.2r$. The results of the spatial distribution of the corresponding eigenfunctions show that corner states can exist only if $d_2/d_1 > 1.2$, as indicated by the red line segment. Different choices of d_1 gives almost the same conclusion. Furthermore, the complete phase diagram can be calculated by sys-

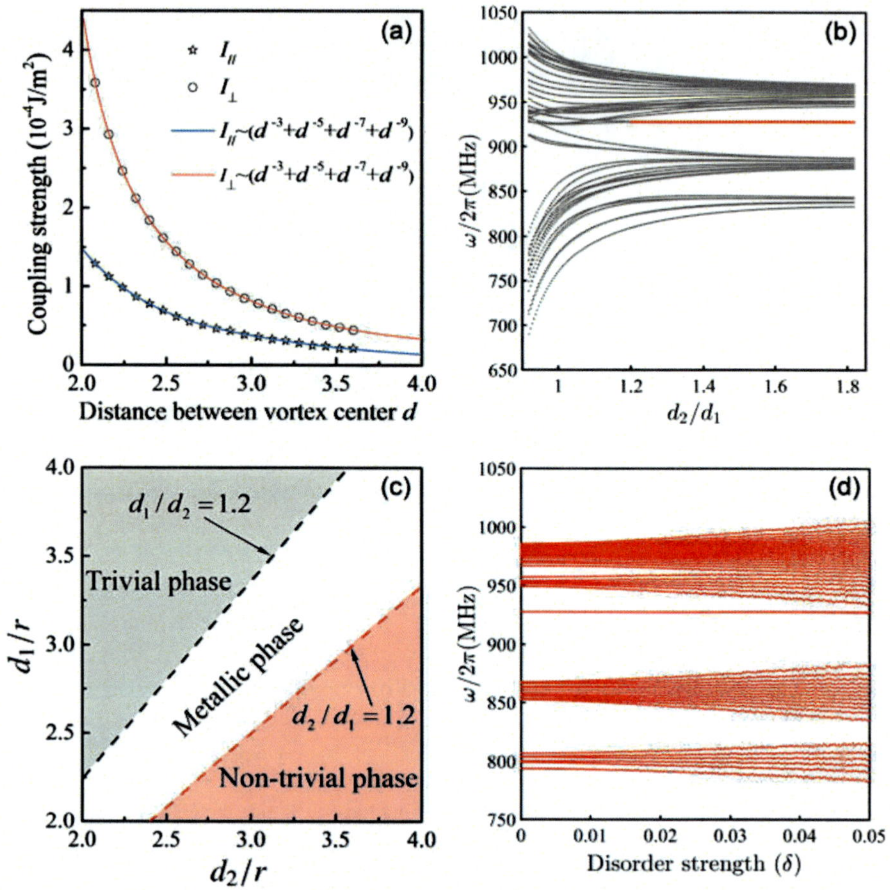

Fig. 15.13 **a** Dependence of the coupling strength $I_{\parallel}$ and $I_{\perp}$ on the vortex-vortex distance d (normalized by the disk radius r). Pentagrams and circles denote simulation results and solid curves represent the analytical fitting. **b** Eigenfrequencies of collective vortex gyration under different ratio d_2/d_1 with the red segment labeling the corner state phase. **c** The phase diagram. **d** Eigenfrequencies of the breathing kagome lattice of vortices under different disorder strength. Images are taken from [126]

tematically changing d_1 and d_2. It can be clearly seen that the boundary separating topologically non-trivial and metallic phases lies in $d_2/d_1 = 1.2$, while topologically trivial and metallic phases are separated by $d_1/d_2 = 1.2$, as shown in Fig. 15.13c. When $d_2/d_1 > 1.2$, the system is topologically non-trivial and can support second-order topological corner states. The system is trivial without any topological edge modes if $d_1/d_2 > 1.2$. Here, the trivial phase is the gapped (insulator) state, the metallic/conducting phase represents the gapless state such that vortices oscillations can propagate in the bulk lattice, and the non-trivial phase means the second-order corner state surviving in a gapped bulk. Besides, it is worth mentioning that the critical

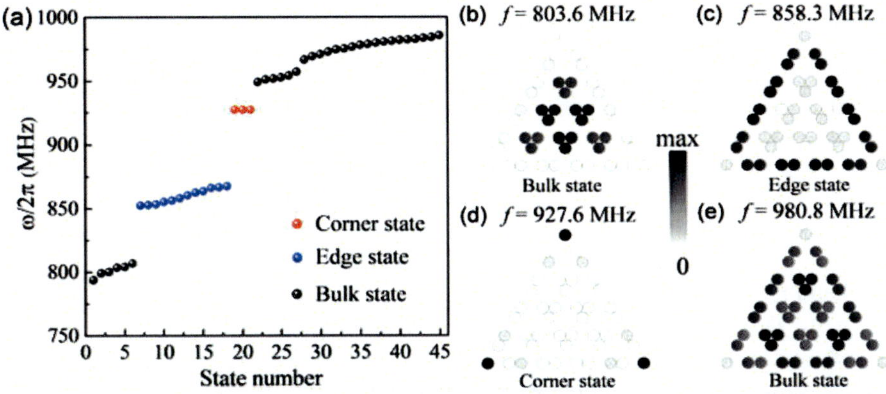

Fig. 15.14 **a** Eigenfrequencies of triangle-shape kagome vortex lattice with $d_1 = 2.08r$ and $d_2 = 3.60r$. The spatial distribution of vortex gyrations for the bulk [(**b**) and (**e**)], edge (**c**), and corner (**d**) states. Images are taken from [126]

condition ($d_2/d_1 > 1.2$) for HOTIs may vary with respect to materials parameters. For example, the critical value will slightly increase (decrease) if the radius of the nanodisk increases (decreases).

Topological corner states should be robust against disorders in the bulk. Figure 15.13d shows the eigenfrequencies of the triangle-shape breathing kagome lattice of vortices under bulk disorders of different strengths, where $d_1 = 2.08r$ and $d_2 = 3.60r$ ($d_2/d_1 = 1.73 > 1.2$). Here, disorders are introduced by assuming the resonant frequency ω_0 with a random shift, i.e., $\omega_0 \rightarrow \omega_0 + \delta Z \omega_0$, where δ indicates the strength of the disorder and Z is a uniformly distributed random number between -1 to 1. It can be seen from Fig. 15.13d that with the increasing of the disorder strength, the spectrum of both edge and bulk states is significantly modified, while the corner states are quite robust. What's more, it can be further confirmed that these corner states are also robust against defects.

The same geometric parameters as Fig. 15.13d are chosen to explicitly visualize the corner states and other modes in the phase diagram. The computed eigenfrequencies and eigenmodes are plotted in Fig. 15.14a–e, respectively. It is found that there are three degenerate modes with the frequency equal to 927.6 MHz, represented by red balls. These modes are indeed second-order topological states (corner states) with oscillations being highly localized at the three corners, see Fig. 15.14d. The edge states are also identified, denoted by blue balls in Fig. 15.14a. The spatial distribution of edge oscillations are confined on three edges, as shown in Fig. 15.14c. However, these edge modes are Tamm-Shockley type [118, 119], not chiral, which can be confirmed by micromagnetic simulations [126]. Bulk modes are plotted in Fig. 15.14b, e, where corners do not participate in the oscillations.

The higher-order topological properties can be interpreted in terms of the bulk topological index, i.e., the polarization [131, 132]:

$$P_j = \frac{1}{S}\iint_{\text{BZ}} A_j d^2k, \tag{15.10}$$

where S is the area of the first Brillouin zone, $A_j = -i\langle\psi|\partial k_j|\psi\rangle$ is Berry connection with $j = x, y$, and ψ is the wave function for the lowest band. It is shown that $(P_x, P_y) = (0.499, 0.288)$ for $d_1 = 2.08r$ and $d_2 = 3.60r$ and $(P_x, P_y) = (0.032, 0.047)$ for $d_1 = 3r$ and $d_2 = 2.1r$. The former corresponds to the topological insulating phase while the latter is for the trivial phase. Theoretically, for breathing kagome lattice, the polarization (P_x, P_y) is identical to Wannier center, which is restricted to two positions for insulating phases. If Wannier center coincides with (0, 0), the system is in trivial insulating phase and no topological edge states exist. Higher-order topological corner states emerge when the Wannier center lies at (1/2, $1/2\sqrt{3}$) [25, 36]. The distribution of bulk topological index is consistent with the computed phase diagram Fig. 15.13c.

The other type of breathing kagome lattice of vortices (parallelogram-shape) also supports the corner states, with the sketch plotted in Fig. 15.15a. Here, the same parameters as those in the triangle-shape lattice are adopted. Figure 15.15b shows the eigenfrequencies of system. Interestingly, it can be seen that there is only one corner state, represented by the red ball. Edge and bulk states are also observed, denoted by blue and black balls, respectively. The spatial distribution of vortices oscillation for different modes are shown in Fig. 15.15c–f. From Fig. 15.15e, one can clearly see that the oscillations for corner state are confined to one acute angle and the vortex at the position of two obtuse angles hardly oscillates. The spatial distribution of vortex gyration for edge and bulk states are plotted in Fig. 15.15c, d, f. Further, the robustness of the corner states are also confirmed [126].

It is interesting to note that the results of triangle-shape and parallelogram-shape lattices are closely related. Two opposite acuted-angle corners in the parallelogram are actually not equivalent: one via d_1 bonding while the other one via d_2 bonding; see Fig. 15.15a. Only the d_2 bonding (bottom-right) corner in the parallelogram-shape lattice is identical to three corners in the triangle-shape lattice. Therefore, for parallelogram-shape lattice, we can observe only one corner state.

15.4.2 Micromagnetic Simulations

Micromagnetic simulations are implemented to verify the theoretical predictions of corner states above. The breathing kagome lattice consisting of massive identical magnetic nanodisks in vortex states are considered, as shown in Figs. 15.12a and 15.15a, with the same geometric parameters as those in Figs. 15.14a and 15.15b, respectively. Micromagnetic software MUMAX3 [117] is used to simulate the dynamics of vortices.

Figure 15.16a shows the temporal Fourier spectra of the vortex oscillations at different positions, with black, blue, and red curves denoting the positions of bulk, edge, and corner bands, respectively. One can immediately see that, near the frequency of

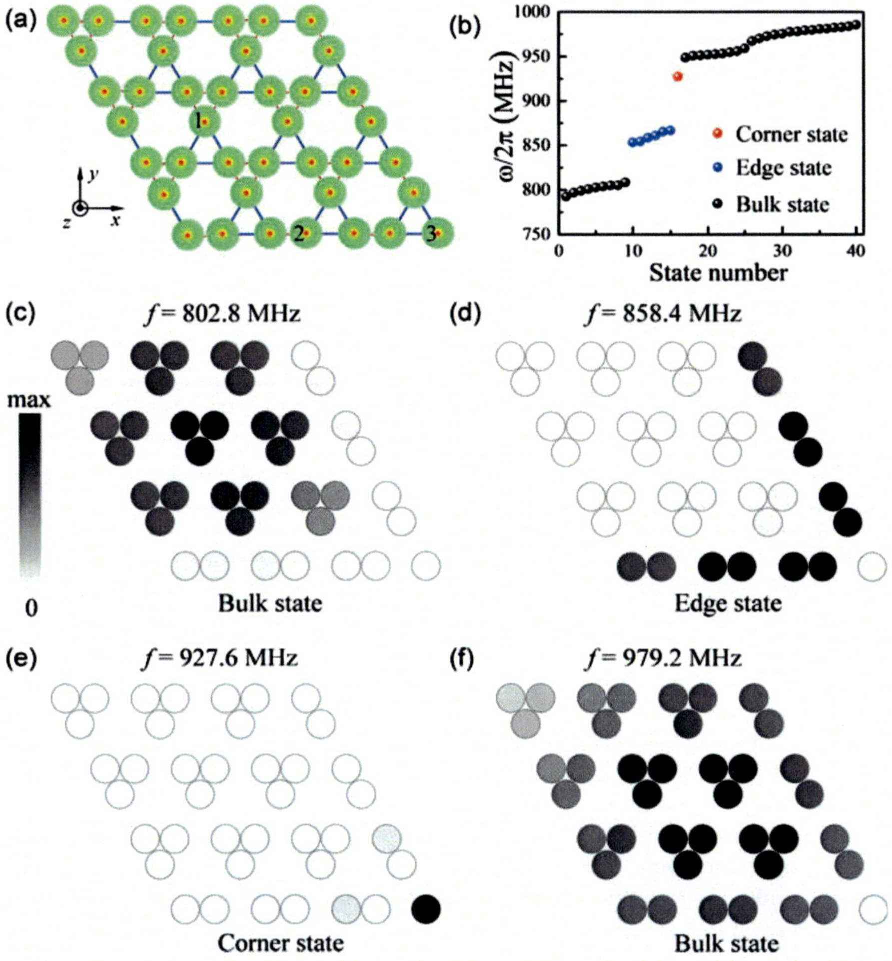

Fig. 15.15 **a** The sketch for parallelogram-shaped breathing kagome lattice of vortices. Arabic numbers 1, 2 and 3 denote the position of spectrum analysis for bulk, edge, and corner states, respectively. **b** Numerically computed eigenfrequencies for parallelogram-shaped system. The spatial distribution of vortices oscillation for the bulk [(**c**) and (**f**)], edge (**d**), and corner (**e**) states. Images are taken from [126]

940 MHz, the spectrum for the corner has a very strong peak, which does not happen for the edge and bulk. It can be inferred that this is the corner-state band with oscillations localized only at three corners. Similarly, one can identify the frequency range which allows the bulk and edge states, as shown by shaded area with different colors in Fig. 15.16a. Four representative frequencies are chosen to visualize the spatial distribution of vortex oscillations for different modes: 940 MHz for the corner state, 842 MHz for the edge state, and both 769 MHz and 959 MHz for bulk states, and then stimulate their dynamics by a sinusoidal magnetic field $\mathbf{h}(t) = h_0 \sin(2\pi f t)\hat{x}$ with

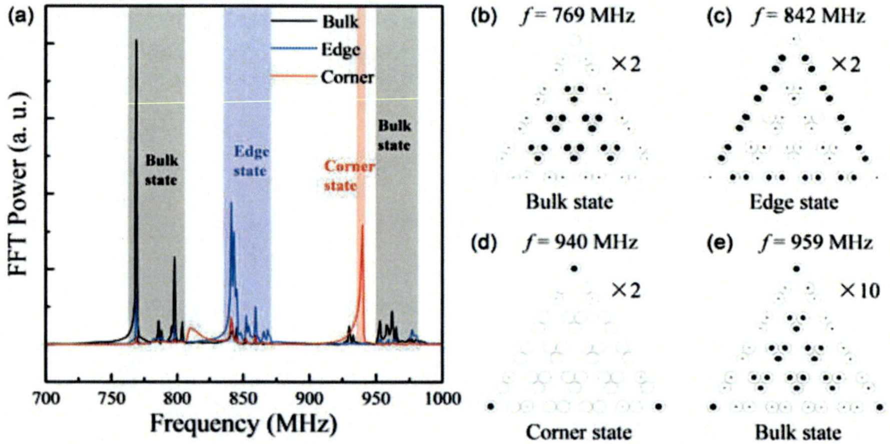

Fig. 15.16 Micromagnetic simulation of excitations in triangle-shape structure. **a** The temporal Fourier spectrum of the vortex oscillations at different positions. The spatial distribution of oscillation amplitude under the exciting field of various frequencies, 769 MHz (**b**), 842 MHz (**c**), 940 MHz (**d**), and 959 MHz (**e**). Since the oscillation amplitudes of the vortex centers are too small, we have magnified them by 2 or 10 times labeled in each figure. Images are taken from [126]

$h_0 = 0.1$ mT to the whole system for 100 ns. Figures 15.16b–e plot the spatial distribution of oscillation amplitude. One can clearly see the corner state in Fig. 15.16d, which is in a good agreement with theoretical results shown in Fig. 15.14d. Spatial distribution of vortices motion for bulk and edge states are shown in Fig. 15.16b, c, respectively. It is worth noting that vortices at three corners in Fig. 15.16e also oscillate with a sizable amplitude, which is somewhat quite unexpected for bulk states. This inconsistency may come from the strong coupling (or hybridization) between the bulk and corner modes, since their frequencies are very close to each other, as shown in Figs. 15.14a and 15.16a.

The simulations of parallelogram-shaped lattice show similar results to triangle-shaped lattice. The spectra are shown in Fig. 15.17a with the black, blue and red curves indicating the positions of bulk (Number 1), edge (Number 2) and corner (Number 3) bands, respectively. Shaded area with different colors denote different modes. The spatial distribution of oscillation amplitude is plotted in Fig. 15.17b–e. Figure 15.17d shows only one corner state at only one (bottom-right) acute angle, which is in a good agreement with theoretical results shown in Fig. 15.15e. Spatial distribution of vortices motion for bulk and edge states are shown in Fig. 15.17b, c, respectively. Interestingly, the hybridization between bulk mode and corner mode occurs as well in parallelogram-shaped lattice, see Fig. 15.17e.

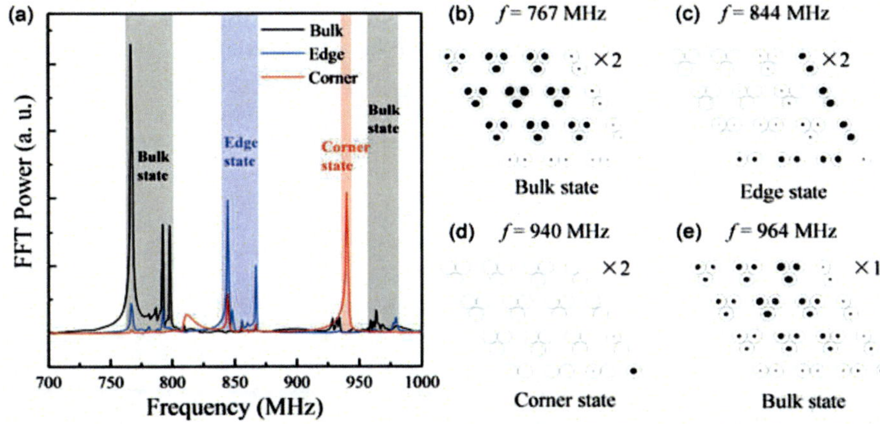

Fig. 15.17 Micromagnetic simulation of excitations in parallelogram-shape structure. **a** The temporal Fourier spectrum of the vortex oscillations at different positions. The spatial distribution of oscillation amplitude under the exciting field with different frequencies, 767 MHz (**b**), 844 MHz (**c**), 940 MHz (**d**), and 964 MHz (**e**). The simulation time is 100 ns. Since the oscillation amplitudes of the vortices centers are too small, we have magnified them by 2 or 10 times labeled in each figure. Images are taken from [126]

15.5 Corner States in a Breathing Honeycomb Lattice of Vortices

It is known that the perfect graphene lattice has a gapless band structure with Dirac cones in momentum space [133]. When spatially periodic magnetic flux [134] or spin-orbit coupling [135] are introduced, a gap opens at the Dirac point, leading to a FOTI. Interestingly, the realization of gap opening and closing by tuning the intercellular and intracellular bond distances has been demonstrated in photonic [31] and elastic [136] honeycomb lattices, in which HOTI emerges. In this section, we show that the higher-order topological insulating phase do exist in a breathing honeycomb lattice of vortices.

15.5.1 Theoretical Model

Figure 15.18 shows a breathing honeycomb lattice of magnetic nanodisks with vortex states. We use (15.9) to describe the collective dynamics of the breathing honeycomb lattice of vortices. For vortex (topological charge $Q = +1/2$) gyrations, one can justify $|\chi_j| \ll |\eta_j|$. By substituting (15.7) into (15.9), we obtain the eigenvalue equation of the system,

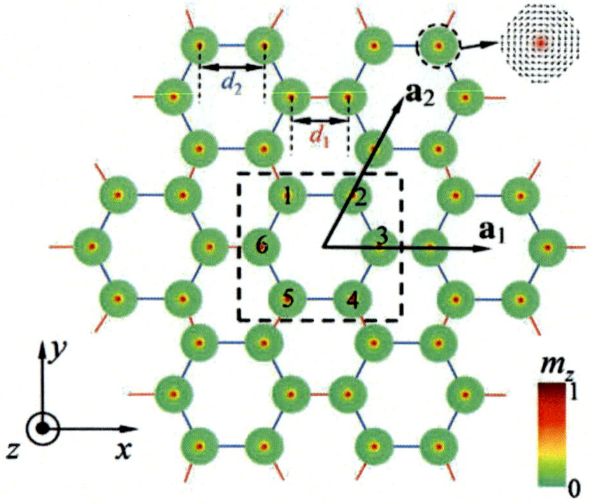

Fig. 15.18 Illustration (top view) of the breathing honeycomb lattice of magnetic vortices, with d_1 and d_2 denoting the alternating lengths of intercellular and intracellular bonds, respectively. The radius of each nanodisk is $r = 50$ nm, and the thickness is $w = 10$ nm. The dashed black rectangle is the unit cell used for calculating the band structure, with $\mathbf{a}_1$ and $\mathbf{a}_2$ denoting the basis vectors. Images are taken from [137]

$$\begin{aligned}\hat{\mathcal{D}}\psi_j = (\omega_0 - \frac{\xi_1^2 + 2\xi_2^2}{2\omega_K})\psi_j + \zeta_1 \sum_{k\in\langle j_1\rangle} \psi_k + \zeta_2 \sum_{k\in\langle j_2\rangle} \psi_k \\ - \frac{\xi_1\xi_2}{2\omega_K} \sum_{s\in\langle\langle j_1\rangle\rangle} e^{i2\bar{\theta}_{js}}\psi_s - \frac{\xi_2^2}{2\omega_K} \sum_{s\in\langle\langle j_2\rangle\rangle} e^{i2\bar{\theta}_{js}}\psi_s, \end{aligned} \tag{15.11}$$

with $\omega_K = \omega_0 - \omega_0^2\omega_M$, $\bar{\theta}_{js} = \theta_{jk} - \theta_{ks}$ is the relative angle from the bond $k \to s$ to the bond $j \to k$ with k between j and s, $\langle j_1\rangle$ and $\langle j_2\rangle$ ($\langle\langle j_1\rangle\rangle$ and $\langle\langle j_2\rangle\rangle$) are the set of nearest (next-nearest) intercellular and intracellular neighbors of j, respectively.

For an infinite lattice, with the dashed black rectangle indicating the unit cell, as shown in Fig. 15.18, $\mathbf{a}_1 = a\hat{x}$ and $\mathbf{a}_2 = \frac{1}{2}a\hat{x} + \frac{\sqrt{3}}{2}a\hat{y}$ are two basis vectors of the crystal, with $a = d_1 + 2d_2$. The band structure of system can be calculated by diagonalizing the Hamiltonian,

$$\mathcal{H} = \begin{pmatrix} Q_0 & \zeta_2 & Q_1 & Q_2 & Q_3 & \zeta_2 \\ \zeta_2 & Q_0 & \zeta_2 & Q_4 & \zeta_1\exp(i\mathbf{k}\cdot\mathbf{a}_2) & Q_5 \\ Q_1^* & \zeta_2 & Q_0 & \zeta_2 & Q_6 & \zeta_1\exp(i\mathbf{k}\cdot\mathbf{a}_1) \\ Q_2^* & Q_4^* & \zeta_2 & Q_0 & \zeta_2 & Q_7 \\ Q_3^* & \zeta_1\exp(-i\mathbf{k}\cdot\mathbf{a}_2) & Q_6^* & \zeta_2 & Q_0 & \zeta_2 \\ \zeta_2 & Q_5^* & \zeta_1\exp(-i\mathbf{k}\cdot\mathbf{a}_1) & Q_7^* & \zeta_2 & Q_0 \end{pmatrix}, \tag{15.12}$$

with elements explicitly expressed as

$$
\begin{aligned}
Q_0 &= \omega_0 - \frac{\xi_1^2 + 2\xi_2^2}{2\omega_K}, \\
Q_1 &= -\frac{\xi_1\xi_2}{2\omega_K}\exp(i\frac{2\pi}{3})\Big\{\exp[i\mathbf{k}\cdot(\mathbf{a}_2-\mathbf{a}_1)] + \exp(-i\mathbf{k}\cdot\mathbf{a}_1)\Big\} - \frac{\xi_2^2}{2\omega_K}\exp(i\frac{2\pi}{3}), \\
Q_2 &= \zeta_1\exp[i\mathbf{k}\cdot(\mathbf{a}_2-\mathbf{a}_1)], \\
Q_3 &= -\frac{\xi_1\xi_2}{2\omega_K}\exp(-i\frac{2\pi}{3})\Big\{\exp[i\mathbf{k}\cdot(\mathbf{a}_2-\mathbf{a}_1)] + \exp(i\mathbf{k}\cdot\mathbf{a}_2)\Big\} - \frac{\xi_2^2}{2\omega_K}\exp(-i\frac{2\pi}{3}), \\
Q_4 &= -\frac{\xi_1\xi_2}{2\omega_K}\exp(i\frac{2\pi}{3})\Big\{\exp[i\mathbf{k}\cdot(\mathbf{a}_2-\mathbf{a}_1)] + \exp(i\mathbf{k}\cdot\mathbf{a}_2)\Big\} - \frac{\xi_2^2}{2\omega_K}\exp(i\frac{2\pi}{3}), \\
Q_5 &= -\frac{\xi_1\xi_2}{2\omega_K}\exp(-i\frac{2\pi}{3})[\exp(i\mathbf{k}\cdot\mathbf{a}_2) + \exp(i\mathbf{k}\cdot\mathbf{a}_1)] - \frac{\xi_2^2}{2\omega_K}\exp(-i\frac{2\pi}{3}), \\
Q_6 &= -\frac{\xi_1\xi_2}{2\omega_K}\exp(i\frac{2\pi}{3})[\exp(i\mathbf{k}\cdot\mathbf{a}_2) + \exp(i\mathbf{k}\cdot\mathbf{a}_1)] - \frac{\xi_2^2}{2\omega_K}\exp(i\frac{2\pi}{3}), \\
Q_7 &= -\frac{\xi_1\xi_2}{2\omega_K}\exp(i\frac{2\pi}{3})\Big\{\exp[i\mathbf{k}\cdot(\mathbf{a}_1-\mathbf{a}_2)] + \exp(i\mathbf{k}\cdot\mathbf{a}_1)\Big\} - \frac{\xi_2^2}{2\omega_K}\exp(i\frac{2\pi}{3}).
\end{aligned}
\tag{15.13}
$$

Topological invariants can be used to distinguish different phases. For any insulator with translational symmetry, the gauge-invariant Chern number of bulk bands [58, 138]

$$
C = \frac{i}{2\pi}\iint_{\mathrm{BZ}} dk_x dk_y \mathrm{Tr}\big[P(\frac{\partial P}{\partial k_x}\frac{\partial P}{\partial k_y} - \frac{\partial P}{\partial k_y}\frac{\partial P}{\partial k_x})\big] \tag{15.14}
$$

is often adopted for determining the FOTI phase, where P is projection matrix $P(\mathbf{k}) = \phi(\mathbf{k})\phi(\mathbf{k})^\dagger$, with $\phi(\mathbf{k})$ being the normalized eigenstate (column vector) of Hamiltonian, and the integral is over the first Brillouin zone. However, to determine whether the system allows the HOTI phase, another different topological invariant should be considered.

In the presence of six-fold rotational (C_6) symmetry, a proper topological invariant is the $\mathbb{Z}_6$ Berry phase [48–52]:

$$
\theta = \int_{L_1} \mathrm{Tr}[\mathbf{A}(\mathbf{k})]\cdot d\mathbf{k} \ \ (\mathrm{mod}\ 2\pi), \tag{15.15}
$$

where $\mathbf{A}(\mathbf{k})$ is the Berry connection:

$$
\mathbf{A}(\mathbf{k}) = i\Psi^\dagger(\mathbf{k})\frac{\partial}{\partial\mathbf{k}}\Psi(\mathbf{k}). \tag{15.16}
$$

Here, $\Psi(\mathbf{k}) = [\phi_1(\mathbf{k}),\phi_2(\mathbf{k}),\phi_3(\mathbf{k})]$ is the 6×3 matrix composed of the eigenvectors of (15.12) for the lowest three bands. L_1 is an integral path in momentum space $G' \to \Gamma \to K'$; see the green line segment in Fig. 15.19a. The Wilson-loop approach is adopted for evaluating the Berry phase θ to avoid the difficulty of the gauge choice [23, 24]. It is worth mentioning that the six high-symmetry points G, K, G', K', G'',

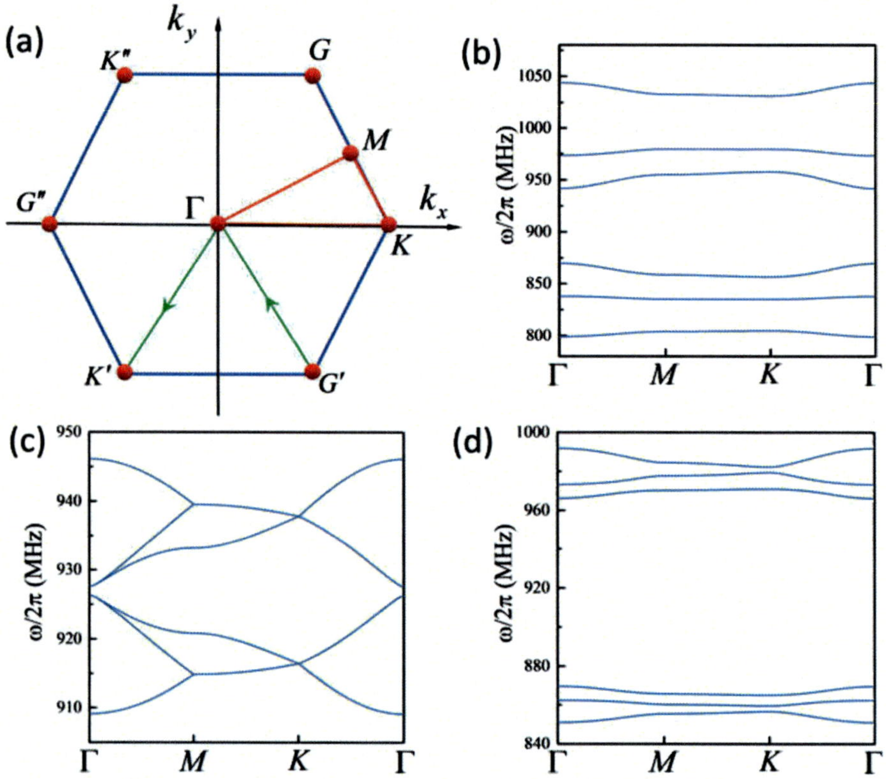

Fig. 15.19 **a** The first Brillouin zone of the breathing honeycomb lattice, with high-symmetry points Γ, G, M, and K located at $(k_x, k_y) = (0, 0)$, $(\frac{2\pi}{3a}, \frac{2\sqrt{3}\pi}{3a})$, $(\frac{\pi}{a}, \frac{\sqrt{3}\pi}{3a})$, and $(\frac{4\pi}{3a}, 0)$, respectively. Band structures along the loop Γ-M-K-Γ for different lattice parameters: $d_1 = 3.6r$, $d_2 = 2.08r$ (**b**), $d_1 = d_2 = 3.6r$ (**c**), and $d_1 = 2.08r$, $d_2 = 3.6r$ (**d**). Images are taken from [137]

and K'' in the first Brillouin zone are equivalent (see Fig. 15.19a), because of the C_6 symmetry. Therefore, there are other five equivalent integral paths ($L_2 : K' \to \Gamma \to G''$, $L_3 : G'' \to \Gamma \to K''$, $L_4 : K'' \to \Gamma \to G$, $L_5 : G \to \Gamma \to K$, and $L_6 : K \to \Gamma \to G'$) leading to the identical θ. It is also straightforward to see that the integral along the path $L_1 + L_2 + L_3 + L_4 + L_5 + L_6$ vanishes. Thus, the $\mathbb{Z}_6$ Berry phase must be quantized as $\theta = \frac{2n\pi}{6}$ $(n = 0, 1, 2, 3, 4, 5)$. By simultaneously quantifying the Chern number C and the $\mathbb{Z}_6$ Berry phase θ, the topological phases and their transition can be determined accurately.

Corner states are of particular interest and are deeply related to the symmetry of Hamiltonian (15.12). Below, we prove that the emergence of topological zero modes is protected by the generalized chiral symmetry. First of all, because $(\xi_1^2 + 2\xi_2^2)/2\omega_K \ll \omega_0$, the diagonal element of $\mathcal{H}$ can be regarded as a constant independent of d, i.e., $Q_0 = \omega_0$, which is the "zero-energy" of the original Hamiltonian. $Q_{1,2,3,4,5,6}$ are the next-nearest hopping terms. At first glance, the system does not possess any chiral symmetry to protect the "zero-energy" modes because the breathing honeycomb lattice is not a bipartite lattice. Inspired by the explanation of generalized chiral symmetry in the breathing kagome lattice [37, 139], the chiral symmetry for a unit cell containing six sites can be generalized by defining

$$\begin{aligned}\Gamma_6^{-1}\mathcal{H}_1\Gamma_6 &= \mathcal{H}_2,\\ \Gamma_6^{-1}\mathcal{H}_2\Gamma_6 &= \mathcal{H}_3,\\ \Gamma_6^{-1}\mathcal{H}_3\Gamma_6 &= \mathcal{H}_4,\\ \Gamma_6^{-1}\mathcal{H}_4\Gamma_6 &= \mathcal{H}_5,\\ \Gamma_6^{-1}\mathcal{H}_5\Gamma_6 &= \mathcal{H}_6,\\ \mathcal{H}_1+\mathcal{H}_2+\mathcal{H}_3+\mathcal{H}_4+\mathcal{H}_5+\mathcal{H}_6 &= 0,\end{aligned} \tag{15.17}$$

where the chiral operator Γ_6 is a diagonal matrix to be determined, and $\mathcal{H}_1 = \mathcal{H} - Q_0\mathrm{I}$. Here, to prove the system has generalized chiral symmetry, we divide the system into six subgroups with the components of matrix Hamiltonian being nonzero only between different subgroups, such a property is essential for chiral symmetry and indicates no interaction within sublattices. Therefore, the matrices (H_2, H_3, etc.) are the subgroups used for explaining the chiral symmetry of system. Upon combining the last equation with the previous five in (15.17), we have $\Gamma_6^{-1}\mathcal{H}_6\Gamma_6 = \mathcal{H}_1$, implying that $[\mathcal{H}_1, \Gamma_6^6] = 0$; thus, $\Gamma_6^6 = \mathrm{I}$, via the reasoning completely analogous to the Su-Schrieffer-Heeger model [140]. Hamiltonians $\mathcal{H}_{1,2,3,4,5,6}$ each have the same set of eigenvalues $\lambda_{1,2,3,4,5,6}$. The eigenvalues of Γ_6 are given by 1, $\exp(2\pi i/6)$, $\exp(4\pi i/6)$, $\exp(\pi i)$, $\exp(8\pi i/6)$, and $\exp(10\pi i/6)$. Therefore, the matrix Γ_6 can written as

$$\Gamma_6 = \begin{pmatrix} 1 & 0 & 0 & 0 & 0 & 0\\ 0 & e^{\frac{2\pi i}{6}} & 0 & 0 & 0 & 0\\ 0 & 0 & e^{\frac{4\pi i}{6}} & 0 & 0 & 0\\ 0 & 0 & 0 & e^{\pi i} & 0 & 0\\ 0 & 0 & 0 & 0 & e^{\frac{8\pi i}{6}} & 0\\ 0 & 0 & 0 & 0 & 0 & e^{\frac{10\pi i}{6}} \end{pmatrix}, \tag{15.18}$$

in the same bases as expressing the Hamiltonian (15.12). By taking the trace of the sixth line from (15.17), we find $\sum_{i=1}^{6} \mathrm{Tr}(\mathcal{H}_i) = 6\mathrm{Tr}(\mathcal{H}_1) = 0$, which means that the sum of the six eigenvalues vanishes $\sum_{i=1}^{6} \lambda_i = 0$. Given an eigenstate ϕ_j that has support in only sublattice j, it will satisfy $\mathcal{H}_1\phi_j = \lambda\phi_j$ and $\Gamma_6\phi_j = \exp[2\pi i(j-1)/6]\phi_j$ with $j = 1, 2, 3, 4, 5, 6$. From these formulas and (15.17), one can find that $\sum_{i=1}^{6} \mathcal{H}_i\phi_j = \sum_{i=1}^{6} \Gamma_6^{-(i-1)}\mathcal{H}_1\Gamma_6^{i-1}\phi_j = 6\lambda\phi_j = 0$, indicating $\lambda = 0$ for any mode that has support in only one sublattice, i.e., zero-energy corner state.

15.5.2 Corner States and Phase Diagram

Figure 15.19b–d shows the bulk band structures under a variety of lattice parameters. For $d_1 = d_2 = 3.6r$ (see Fig. 15.19c), it can be found that the highest three bands and the lowest three bands merged separately, leaving a next-nearest hopping-induced gap centered at 927 MHz. In this case, the FOTI phase was anticipated [83, 125]. However, the six bands are separated from each other when considering the other two kinds of parameters (see Fig. 15.19b, d), indicating that the system is in the insulating state. These insulating phases and the phase transition point can be further distinguished by calculating Chern number and $\mathbb{Z}_6$ Berry phase.

Figure 15.20a plots the dependence of the Chern number (C) and the $\mathbb{Z}_6$ Berry phase (θ) on the parameter d_2/d_1. Here the material parameters of Py ($Ni_{80}Fe_{20}$) [88, 130] are used and fix $d_1 = 2.5r$. In addition, the eigenfrequencies of collective vortex gyration under different ratios d_2/d_1 for a parallelogram-shaped (see Fig. 15.20b) structure are also shown in Fig. 15.20c. By considering the topological invariants and spectrum simultaneously, one can infer that the system is in the trivial phase when $d_2/d_1 < 0.9$ and $1.08 < d_2/d_1 < 1.49$, in the FOTI phase when $0.9 < d_2/d_1 < 1.08$, and in the HOTI phase when $d_2/d_1 > 1.49$. Haldane model is a well-known example for breaking the time-reversal symmetry [134]. If we consider the limit case of honeycomb lattice, i.e., $d_1 = d_2$, (15.9) can be exactly mapped to the Haldane model, as shown in [83] and [125]. The very existence of the chiral edge state is thus naturally expected. For a breathing honeycomb lattice, our (15.11) represents the generalized form of the mapping, where the last two terms in the right-hand side are the next-to-nearest hopping that breaks the time-reversal symmetry. The complete phase diagram of system can be obtained by systematically changing d_1 and d_2, with the results plotted in Fig. 15.20d: The regions labeled gray, white, and red represent the trivial, FOTI, and HOTI phases, respectively. Importantly, we find that the boundary for the phase transition between trivial and FOTI phases depends only weakly on the choice of the absolute values of d_1 and d_2 but is (almost) solely determined by their ratio, as indicated by dashed black lines ($l_1 : d_2/d_1 = 0.94$ and $l_2 : d_2/d_1 = 1.05$) in the figure. However, the boundary for the phase transition between trivial and HOTI phases is a linear function $l_3 : d_2 = 2.24d_1 - 1.88$ (see the dashed red line in Fig. 15.20d). It is worth noting that the topological charge of the vortex has no influence on higher-order topology for the reason that the sign of topological charge just determines the direction (clockwise or anti-clockwise) of gyration, see (15.5). However, it indeed can affect the chiral edge state (first-order topology). Namely the chirality of edge state will be reversed if the topological charge changes.

The existence of symmetry-protected states on boundaries is the hallmark of a topological insulating phase. Figure 15.21b–d shows the energy spectrum of the ribbon configuration with armchair edges (see Fig. 15.21a) for different choices of d_1 and d_2. For $d_1 = 3.6r$ and $d_2 = 2.08r$ (black star in Fig. 15.20d), the system is in the trivial phase without any topological edge mode (see Fig. 15.21b). For $d_1 = d_2 = 3.6r$ (blue star in Fig. 15.20d), the lattice considered is identical to a magnetic texture version of graphene (the perfect honeycomb lattice). In contrast to

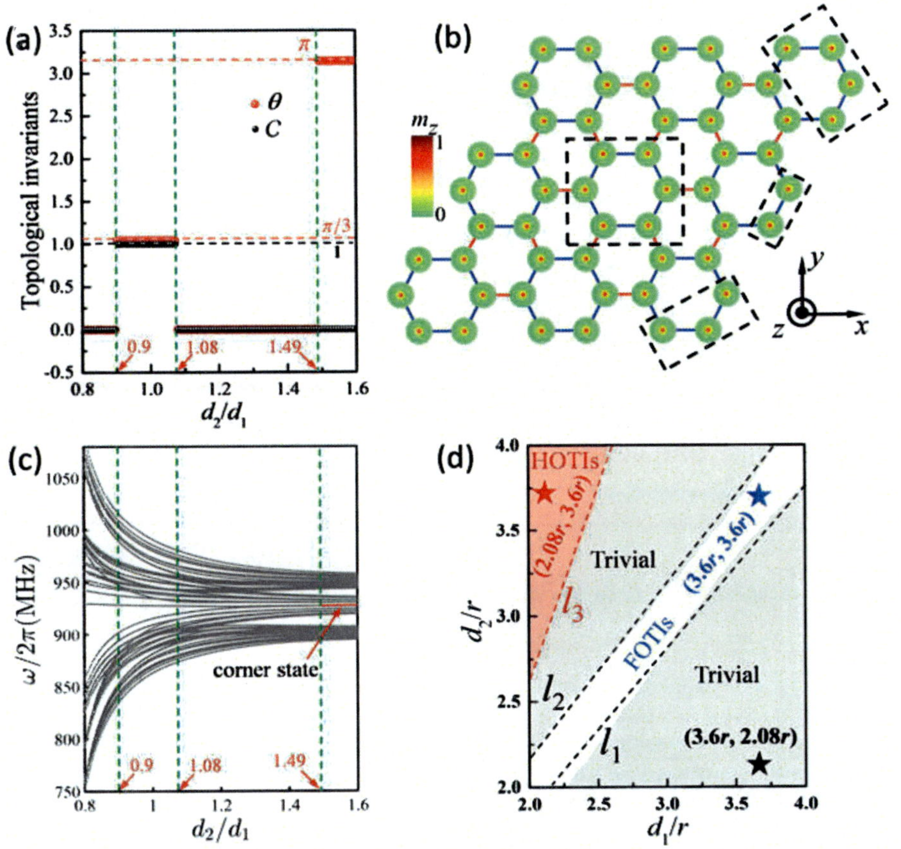

Fig. 15.20 **a** Dependence of the topological invariants Chern number and $\mathbb{Z}_6$ Berry phase on the ratio d_2/d_1 when d_1 is fixed at $2.5r$. **b** Schematic plot of the parallelogram-shaped vortex lattice with armchair edges. **c** Eigenfrequencies of collective vortex gyration under different ratios d_2/d_1 with the red segment denoting the corner state phase. **d** Phase diagram of the system with pentagonal stars of different colors representing three typical parameters of d_1 and d_2 for different phases considered in the subsequent calculations and analyses. Images are taken from [137]

the gapless band structure for perfect graphene nanoribbons, the imaginary second-nearest hopping term opens a gap at the Dirac point and supports a topologically protected first-order chiral edge state [83, 125]. For $d_1 = 2.08r$ and $d_2 = 3.6r$ (red star in Fig. 15.20d), one can clearly see two distinct edge bands, in addition to bulk ones, as shown in Fig. 15.21d. These localized modes are actually not topological because they maintain the bidirectional propagation nature, which is justified by the fact that the wave group-velocity $d\omega/dk_x$ can be either positive or negative at different k_x points. Below, we will show that higher-order topological corner states exactly emerge around these edge bands when the system is decreased to be finite in both dimensions.

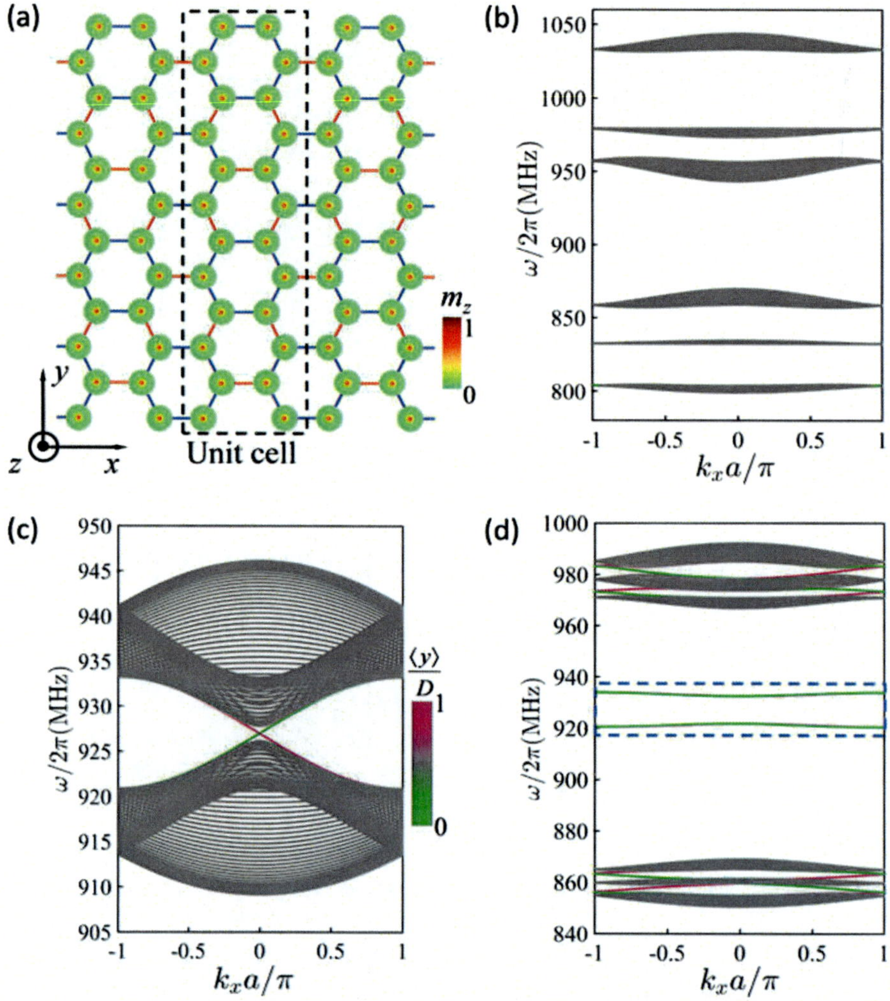

Fig. 15.21 **a** Nanoribbon with armchair edges (closed boundaries in the *x*-direction and open boundaries in the *y*-direction). The dashed black rectangle is the unit cell. Band dispersions with different geometric parameters as denoted in Fig. 15.20d: $d_1 = 3.6r$, $d_2 = 2.08r$ (**b**), $d_1 = d_2 = 3.6r$ (**c**), and $d_1 = 2.08r$, $d_2 = 3.6r$ (**d**). The dashed blue frame in (**d**) indicates the band of non-chiral edge states. *D* is the width of the nanoribbon. Images are taken from [137]

A parallelogram-shaped vortex lattice is considered to visualize the second-order corner states, as shown in Fig. 15.20b, where $d_1 = 2.08r$ and $d_2 = 3.6r$. From the spectrum (see Fig. 15.22a), one can clearly see that there exist a few degenerate modes in the band gap. To distinguish these states, the spatial distribution of vortex gyrations are plotted for each mode in Fig. 15.22b–f. Three types of corner states are confirmed, all of which have oscillations highly localized at obtuse-angled or

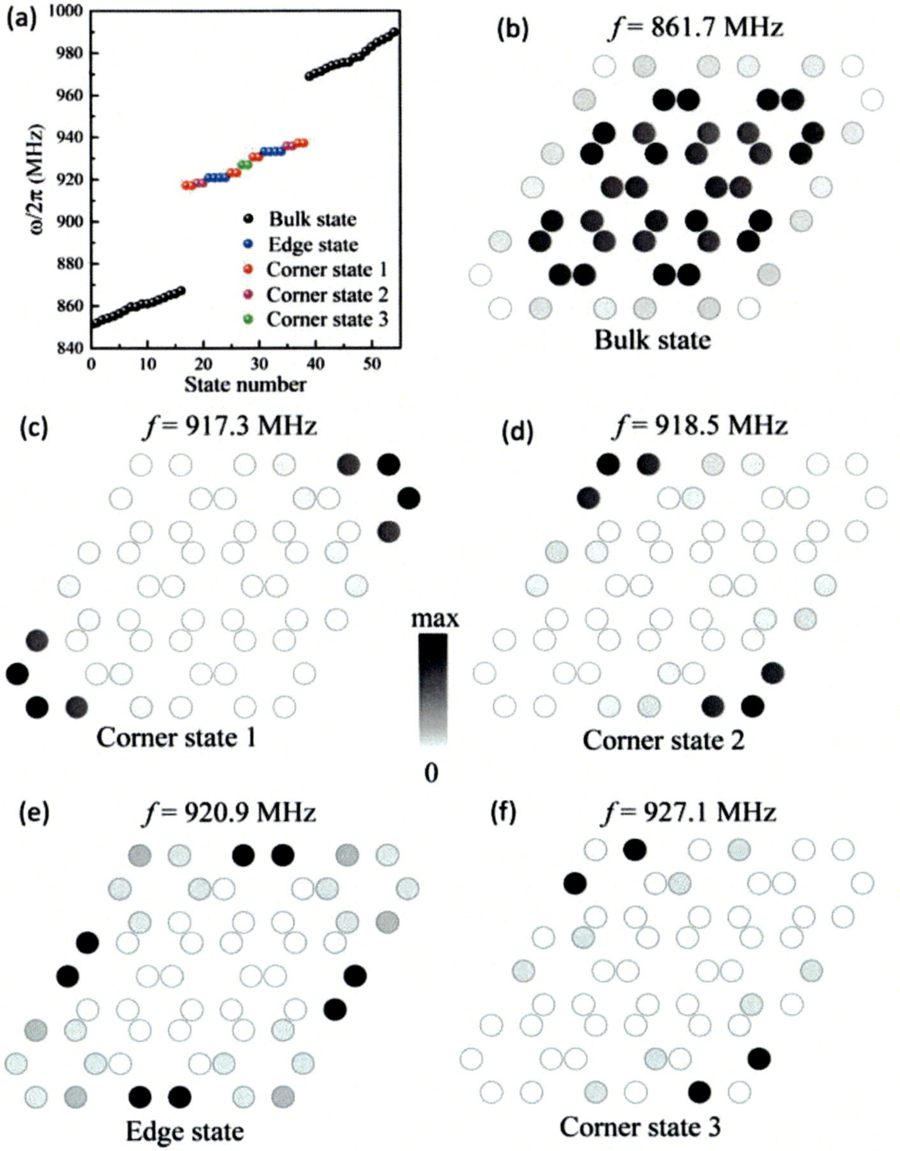

Fig. 15.22 **a** Eigenfrequencies of the finite system with parameters $d_1 = 2.08r$ and $d_2 = 3.6r$ for the parallelogram-shaped structure (see Fig. 15.3b). The spatial distribution of vortex gyrations for the bulk (**b**), corner (**c**, **d** and **f**), and edge (**e**) states of five representative frequencies. Images are taken from [137]

acute-angled corners (see Fig. 15.22c, d, f), where corner states 1 (type I), 2 (type II), and 3 (type III) are denoted by red, magenta, and green balls, respectively. Two degenerate edge modes are denoted by blue balls, in which only two vortices on each edge participate in the oscillation, as shown in Fig. 15.22e. Figure 15.22b shows the bulk state with oscillations spreading over the whole lattice except the boundaries. To judge whether these edge and corner states are topologically protected or not, moderate defects and disorder are introduced into the system and we evaluate the change of the spectrum. It can be found that the eigenfrequency of "zero-energy" corner state 3 at the obtuse-angled corner (see Fig. 15.22f) is well confined around 927 MHz, which means that this corner state is suitably immune from external frustrations. This feature is due to the topological protection from the generalized chiral symmetry. However, the frequencies of other corner modes (Fig. 15.22c, d) have obvious shifts, revealing that these crystalline-symmetry-induced modes are sensitive to disorder. The origin of the edge state (Fig. 15.22e) is again attributed to the so-called Tamm-Shockley mechanism [118, 119].

To figure out why the chiral symmetry-protected (CSP) corner modes emerge at only obtuse-angled corners instead of acute-angled corners, the topological index $\mathcal{N} = |\mathcal{N}_+ - \mathcal{N}_-|$ is introduced, which captures the interplay between the topology of the bulk Hamiltonian and the defect structure [31, 141]. Here, $\mathcal{N}$ counts the number of topologically stable modes bound to corners, and $\mathcal{N}_\pm$ are integers counting the number of eigenstates of the chiral symmetry operator $\hat{\Pi}$ with eigenvalues $+1$ and -1, respectively. In the zero-correlation length limit $d_2 \to \infty$, the breathing honeycomb lattice is then reduced to isolated dimers (see Fig. 15.23b). As long as the gap is not closed, the symmetry and the Berry phase remain, as evidenced by the topological invariant θ. When the system is in the HOTI phase, $\mathcal{N}_+ = \mathcal{N}_- = 1$ in each edge unit cell and $\mathcal{N}_+ = \mathcal{N}_- = 2$ for acute-angled corners, such that $\mathcal{N} = 0$, indicating that there may exist non-CSP modes at acute-angled corners or edges. However, a similar analysis results in totally different outcome for obtuse-angled corners: $\mathcal{N}_+ = 1$ and $\mathcal{N}_- = 2$ or $\mathcal{N}_+ = 2$ and $\mathcal{N}_- = 1$, which leads to $\mathcal{N} = 1$. Thus, CSP or "zero-energy" modes must exist in each obtuse-angled corner. It is worth noting that the "zero-energy" corner state appears at acute-angled corners rather than obtuse-angled corners if the edges of lattice change to zigzag type. This result also can be fully explained in terms of the topological index $\mathcal{N}$ (see Fig. 15.23d, e).

On the other hand, in the limit $d_1 \to \infty$ (see Fig. 15.23c, f), we find that there are no uncoupled vortices. The six corners of the isolated hexagon are equivalent, with no special edge or corner states.

15.5.3 Micromagnetic Simulations

The micromagnetic simulations are implemented to verify theoretical predictions. Here, the parallelogram-shaped breathing honeycomb lattice of magnetic vortices with an armchair edge is considered, as shown in Fig. 15.20b. All material parameters are the same as those for theoretical calculations in Fig. 15.21d. The numerical package MUMAX3 [117] is used to simulate the collective dynamics of vortex lattice.

Figure 15.24a shows the temporal Fourier spectra of the vortex oscillations at different positions, with the black, red, blue, and green curves indicating the positions

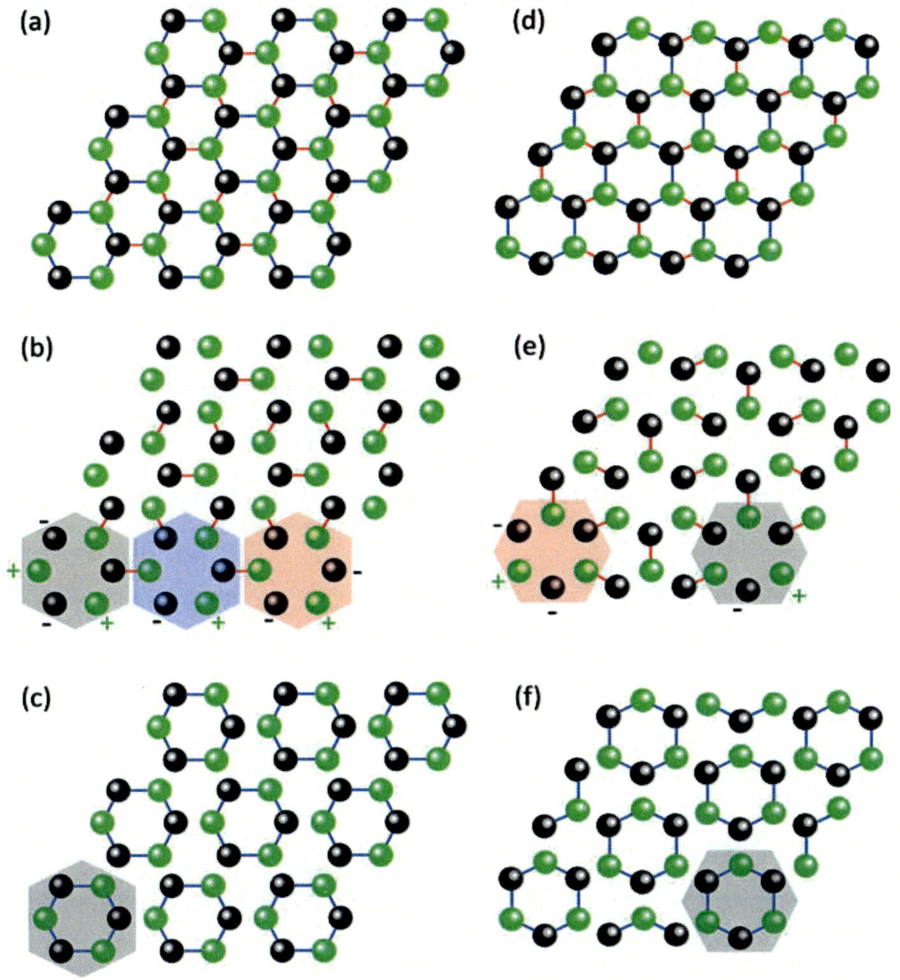

Fig. 15.23 The configuration of breathing honeycomb lattices of magnetic vortices with armchair (**a**) and zigzag edges (**d**). (**b**) and (**e**) [(**c**) and (**f**)] are the corresponding configurations of (**a**) and (**d**) in the zero-correlation length limit $d_2 \to \infty$ ($d_1 \to \infty$), respectively, which consist of isolated dimers (hexamer). Green and black balls indicate eigenvalues of $+1$ and -1 of the chiral-symmetry operator, respectively. Shaded areas represent the unit cell at different positions. Images are taken from [137]

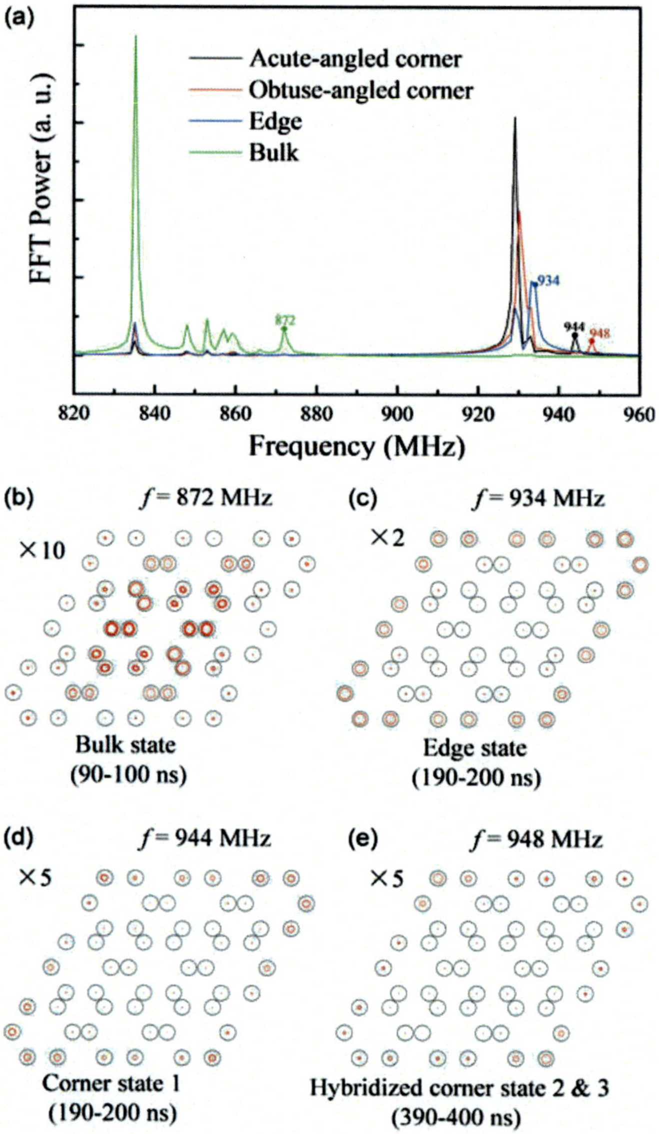

Fig. 15.24 **a** The temporal Fourier spectra of the vortex oscillations at different positions marked by dashed black rectangles in Fig. 15.20b. The gyration path for all vortices under excitation fields with different frequencies, 872 MHz (**b**), 934 MHz (**c**), 944 MHz (**d**), and 948 MHz (**e**). Since the oscillation amplitudes of the vortex centers are too small, we have magnified them by 2, 5 or 10 times, as labeled in each figure. Images are taken from [137]

of acute-angled corner, obtuse-angled corner, edge, and bulk bands, respectively. It can be seen that around the frequency of 944 MHz (948 MHz), the spectra for acute-angled corner (obtuse-angled corner) have an obvious peak, which does not happen for the spectra for edge and bulk bands. Therefore, these two peaks denote two different corner states that are located at acute-angled or obtuse-angled corners. Similarly, the frequency range for bulk and edge states is identified. Further, to visualize the spatial distribution of the vortex oscillations for different modes, four representative frequencies are chosen and are marked by green, blue, black, and red dots: 872 MHz for the bulk state, 934 MHz for the edge state, 944 MHz for the acute-angled corner state, and 948 MHz for the obtuse-angled corner state, respectively. We then stimulate their dynamics by applying a sinusoidal field to the whole system. The 10 ns gyration paths of all vortices are plotted in Fig. 15.24b–e when the excitation field drives a steady-state vortex dynamics. The spatial distribution of vortices motion for the bulk and edge states are shown in Fig. 15.24b, c, respectively. We observe type I corner state with vortex oscillation localized at the acute-angled corner in Fig. 15.24d, which is in good agreement with the theoretical result. Interestingly, we note a strong hybridization between the type II and type III corner states, as shown in Fig. 15.22f, which is because their frequencies are very close to each other and their wavefunctions have a large overlap (see Fig. 15.22).

15.6 Conclusion and Outlook

We have introduced the concept of topological insulator (both first-order and higher-order) based on spin texture metamaterials. The emerging multiband chiral edge modes possessing different handness in skyrmion lattice should be appealing for designing future skyrmionic topological devices. Besides, the predicted second-order insulating phase in vortex lattice can facilitate designing new functional devices based on magnetic solitons. For instance, we can realize a magnetic imaging system by designing one vortex lattice of the desired shape in the HOTI phase surrounded by another vortex lattice in the trivial phase [142, 143]. It is noted that the unique localization property of vortex gyrations strongly depends on the working frequency, which would motivate us to devise magnetic nano-oscillators with high spatiotemporal resolution. Furthermore, the multiband nature of corner modes (spectrum ranging from sub GHz to dozens of GHz) is very useful to design broad-band topological devices. In the present model, we have considered nanodisks with identical radius. When the translational symmetry is broken, for instance, by introducing Kekulé distortions to disk sizes, one may realize novel devices supporting robust Majorana-like zero modes localized in the device's geometric center [144]. From an experimental point of view, the fabrication of artificial vortex or skyrmion lattices is readily within reach of current technology, e.g., electron-beam lithography [78, 79, 145] or X-ray illumination [146]. By tracking the nanometer-scale vortex orbits using the recently developed ultrafast Lorentz microscopy technique in a time-resolved manner [147], one can directly observe the second-order topological corner states.

Acknowledgements This work was supported by the National Natural Science Foundation of China (NSFC) (Grants No. 12074057, 11604041 and 11704060), the National Key Research Development Program under Contract No. 2016YFA0300801, and the National Thousand-Young-Talent Program of China. Z.-X. L. acknowledges the financial support of NSFC under Grant No. 11904048 and the China Postdoctoral Science Foundation under Grant No. 2019M663461. We acknowledged Z.Y.W, C.W. and X.S.W. for helpful discussions.

References

1. M.Z. Hasan, C.L. Kane, Rev. Mod. Phys. **82**, 3045 (2010)
2. X.-L. Qi, S.-C. Zhang, Rev. Mod. Phys. **83**, 1057 (2011)
3. J.E. Moore, Nature **464**, 194–198 (2010)
4. C.L. Kane, E.J. Mele, Phys. Rev. Lett. **95**, 226801 (2005)
5. C.L. Kane, E.J. Mele, Phys. Rev. Lett. **95**, 146802 (2005)
6. L. Lu, J.D. Joannopoulos, M. Soljacic, Nat. Photon. **8**, 821–829 (2014)
7. A.B. Khanikaev, G. Shvets, Nat. Photon. **11**, 763–773 (2017)
8. G. Harari, M. A. Bandres, Y. Lumer et al., Science **359**, eaar4003 (2018)
9. M. A. Bandres, S. Wittek, G. Harari et al., Science **359**, eaar4005 (2018)
10. T. Ozawa, H.M. Price, A. Amo et al., Rev. Mod. Phys. **91**, 015006 (2019)
11. J. Paulose, B.G.G. Chen, V. Vitelli, Nat. Phys. **11**, 153–156 (2015)
12. L.M. Nash, D. Kleckner, A. Read et al., PNAS **112**, 14495 (2015)
13. S.D. Huber, Nat. Phys. **12**, 621–623 (2016)
14. N.P. Mitchell, L.M. Nash, D. Hexner et al., Nat. Phys. **14**, 380–385 (2018)
15. Z. Yang, F. Gao, X. Shi et al., Phys. Rev. Lett. **114**, 114301 (2015)
16. C. He, X. Ni, H. Ge et al., Nat. Phys. **12**, 1124–1129 (2016)
17. R. Fleury, A.B. Khanikaev, A. Alù, Nat. Commun. **7**, 11744 (2016)
18. X. Zhang, M. Xiao, Y. Cheng et al., Commun. Phys. **1**, 97 (2018)
19. L. Zhang, J. Ren, J.-S. Wang et al., Phys. Rev. B **87**, 144101 (2013)
20. A. Mook, J. Henk, I. Mertig, Phys. Rev. B **90**, 024412 (2014)
21. R. Chisnell, J.S. Helton, D.E. Freedman et al., Phys. Rev. Lett. **115**, 147201 (2015)
22. A. Ruckriegel, A. Brataas, R.A. Duine, Phys. Rev. B **97**, 081106(R) (2018)
23. W.A. Benalcazar, B.A. Bernevig, T.L. Hughes, Science **357**, 61–66 (2017)
24. W.A. Benalcazar, B.A. Bernevig, T.L. Hughes, Phys. Rev. B **96**, 245115 (2017)
25. M. Ezawa, Phys. Rev. Lett. **120**, 026801 (2018)
26. Z. Song, Z. Fang, C. Fang, Phys. Rev. Lett. **119**, 246402 (2017)
27. J. Langbehn, Y. Peng, L. Trifunovic et al., Phys. Rev. Lett. **119**, 246401 (2017)
28. F. Schindler, A.M. Cook, M.G. Vergniory et al., Sci. Adv. **4**, eaat0346 (2018)
29. R. Queiro, A. Stern, Phys. Rev. Lett. **123**, 036802 (2019)
30. B.-Y. Xie, H.-F. Wang, H.-X. Wang et al., Phys. Rev. B **98**, 205147 (2018)
31. J. Noh, W.A. Benalcazar, S. Huang et al., Nat. Photon. **12**, 408–415 (2018)
32. A.E. Hassan, F.K. Kunst, A. Moritz et al., Nat. Photon. **13**, 697–700 (2019)
33. S. Mittal, V.V. Orre, G. Zhu et al., Nat. Photon. **13**, 692–696 (2019)
34. X.-D. Chen, W.-M. Deng, F.-L. Shi et al., Phys. Rev. Lett. **122**, 233902 (2019)
35. B.-Y, Xie, G.-X. Su, H.-F. Wang et al., Phys. Rev. Lett. **122**, 233903 (2019)
36. H. Xue, Y. Yang, F. Gao et al., Nat. Mater. **18**, 108–112 (2019)
37. X. Ni, M. Weiner, A. Alu et al., Nat. Mater. **18**, 113–120 (2019)
38. H. Xue, Y. Yang, G. Liu et al., Phys. Rev. Lett. **122**, 244301 (2019)
39. C. He, S.-Y. Yu, H. Wang et al., Phys. Rev. Lett. **123**, 195503 (2019)
40. X. Zhang, H.-X. Wang, Z.-K. Lin et al., Nat. Phys. **15**, 582–588 (2019)
41. S. Imhof, C. Berger, F. Bayer et al., Nat. Phys. **14**, 925–929 (2018)
42. M. Ezawa, Phys. Rev. B **98**, 201402(R) (2018)

43. M. Serra-Garcia, R. Süsstrunk, S.D. Huber, Phys. Rev. B **99**, 020304(R) (2019)
44. J. Bao, D. Zou, W. Zhang et al., Phys. Rev. B **100**, 201406(R) (2019)
45. H. Yang, Z.-X. Li, Y. Liu et al., Phys. Rev. Research **2**, 022028(R) (2020)
46. Y. Xu, R. Xue, S. Wan,. arXiv:1711.09202
47. R.-J. Slager, L. Rademaker, J. Zaanen et al., Phys. Rev. B **92**, 085126 (2015)
48. J. Zak, Phys. Rev. Lett. **62**, 2747 (1989)
49. T. Kariyado, T. Morimoto, Y. Hatsugai, Phys. Rev. Lett. **120**, 247202 (2018)
50. Y. Hatsugai, I. Maruyama, Europhys. Lett. **95**, 20003 (2011)
51. H. Wakao, T. Yoshida, H. Araki et al., Phys. Rev. B **101**, 094107 (2020)
52. H. Araki, T. Mizoguchi, Y. Hatsugai, Phys. Rev. Res. **2**, 012009(R) (2020)
53. A.V. Chumak, V.I. Vasyuchka, A.A. Serga et al., Nat. Phys. **11**, 453–461 (2015)
54. V.V. Kruglyak, S.O. Demokritov, D. Grundler, J. Phys. D **43**, 264001 (2010)
55. R. Shindou, J.-I. Ohe, R. Matsumoto et al., Phys. Rev. B **87**, 174402 (2013)
56. A.L. Chernyshev, P.A. Maksimov, Phys. Rev. Lett. **117**, 187203 (2016)
57. S.K. Kim, H. Ochoa, R. Zarzuela et al., Phys. Rev. Lett. **117**, 227201 (2016)
58. X.S. Wang, Y. Su, X.R. Wang, Phys. Rev. B **95**, 014435 (2017)
59. X.S. Wang, H.W. Zhang, X.R. Wang, Phys. Rev. Appl. **9**, 024029 (2018)
60. L. Chen, J.-H. Chung, B. Gao et al., Phys. Rev. X **8**, 041028 (2018)
61. A. Wachowiak, J. Wiebe, M. Bode et al., Science **298**, 577–580 (2002)
62. B.V. Waeyenberge, A. Puzic, H. Stoll et al., Nature **444**, 461–464 (2006)
63. I. Makhfudz, B. Kruger, O. Tchernyshyov, Phys. Rev. Lett. **109**, 217201 (2012)
64. K.W. Moon, B.S. Chun, W. Kim et al., Phys. Rev. B **89**, 064413 (2014)
65. D. Petit, P.R. Seem, M. Tillette et al., Appl. Phys. Lett. **106**, 022402 (2015)
66. S. Mühlbauer, B. Binz, F. Jonietz et al., Science **323**, 915–919 (2009)
67. W. Jiang, P. Upadhyaya, W. Zhang et al., Science **349**, 283–286 (2015)
68. D.A. Allwood, G. Xiong, C.C. Faulkner et al., Science **309**, 1688–1692 (2005)
69. M. Hayashi, L. Thomas, R. Moriya et al., Science **320**, 209–211 (2008)
70. G. Catalan, J. Seidel, R. Ramesh et al., Rev. Mod. Phys. **84**, 119 (2012)
71. S.S.P. Parkin, M. Hayashi, L. Thomas, Science **320**, 190–194 (2008)
72. S. Parkin, S.-H. Yang, Nat. Nanotech. **10**, 195–198 (2015)
73. F. Jonietz, S. Mühlbauer, C. Pfleiderer et al., Science **330**, 1648–1651 (2010)
74. X.Z. Yu, N. Kanazawa, W.Z. Zhang et al., Nat. Commun. **3**, 988 (2012)
75. V.S. Pribiag, I.N. Krivorotov, G.D. Fuchs et al., Nat. Phys. **3**, 498–503 (2007)
76. G. Hrkac, P.S. Keatley, M.T. Bryan et al., J. Phys. D **48**, 453001 (2015)
77. S. Zhang, J. Wang, Q. Zheng et al., New J. Phys. **17**, 023061 (2015)
78. D.S. Han, A. Vogel, H. Jung et al., Sci. Rep. **3**, 2262 (2013)
79. C. Behncke, M. Hänze, C.F. Adolff et al., Phys. Rev. B **91**, 224417 (2015)
80. M. Hänze, C.F. Adolff, B. Schulte et al., Sci. Rep. **6**, 22402 (2016)
81. J. Kim, J. Yang, Y.-J. Cho et al., Sci. Rep. **7**, 45185 (2017)
82. M. Mruczkiewicz, P. Gruszecki, M. Zelent et al., Phys. Rev. B **93**, 174429 (2016)
83. S.K. Kim, Y. Tserkovnyak, Phys. Rev. Lett. **119**, 077204 (2017)
84. A. Vogel, A. Drews, T. Kamionka et al., Phys. Rev. Lett. **105**, 037201 (2010)
85. H. Jung, K.-S. Lee, D.-E. Jeong et al., Sci. Rep. **1**, 59 (2011)
86. S. Sugimoto, Y. Fukuma, S. Kasai et al., Phys. Rev. Lett. **106**, 197203 (2011)
87. M. Hänze, C.F. Adolff, S. Velten et al., Phys. Rev. B **93**, 054411 (2016)
88. S. Velten, R. Streubel, A. Farhan et al., Appl. Phys. Lett. **110**, 262406 (2017)
89. M.Y. Im, P. Fischer, H.-S. Han et al., NPG Asia Mater. **9**, e348 (2017)
90. A. Fert, V. Cros, J. Sampaio, Nat. Nanotech. **8**, 152–156 (2013)
91. A.A. Thiele, Phys. Rev. Lett. **30**, 230 (1973)
92. F. Büttner, C. Moutafis, M. Schneider et al., Nat. Phys. **11**, 225–228 (2015)
93. X.Z. Yu, Y. Onose, N. Kanazawa et al., Nature **465**, 901–904 (2010)
94. N. Romming, C. Hanneken, M. Menzel et al., Science **341**, 636–639 (2013)
95. X. Zhang, M. Ezawa, D. Xiao et al., Nanotechnology **26**, 225701 (2015)
96. C. Psaroudaki, D. Loss, Phys. Rev. Lett. **120**, 237203 (2018)

97. W. Wang, M. Beg, B. Zhang et al., Phys. Rev. B **92**, 020403(R) (2015)
98. B. Zhang, W. Wang, M. Beg et al., Appl. Phys. Lett. **106**, 102401 (2015)
99. K. Everschor, M. Garst, B. Binz et al., Phys. Rev. B **86**, 054432 (2012)
100. L. Kong, J. Zang, Phys. Rev. Lett. **111**, 067203 (2013)
101. S.-Z. Lin, C.D. Batista, C. Reichhardt et al., Phys. Rev. Lett. **112**, 187203 (2014)
102. W. Yang, H. Yang, Y. Cao et al., Opt. Express **26**, 8778 (2018)
103. Y. Jiang, H.Y. Yuan, Z.-X. Li et al., Phys. Rev. Lett. **124**, 217204 (2020)
104. K. Litzius, I. Lemesh, B. Krüger et al., Nat. Phys. **13**, 170–175 (2017)
105. W. Jiang, X. Zhang, G. Yu et al., Nat. Phys. **13**, 162–169 (2017)
106. X. Zhang, Y. Zhou, M. Ezawa, Nat. Commun. **7**, 10293 (2016)
107. X. Zhang, Y. Zhou, M. Ezawa, Sci. Rep. **6**, 24795 (2016)
108. J. Barker, O.A. Tretiakov, Phys. Rev. Lett. **116**, 147203 (2016)
109. H. Yang, C. Wang, T. Yu et al., Phys. Rev. Lett. **121**, 197201 (2018)
110. X. Wang, X.S. Wang, C. Wang et al., J. Phys. D **52**, 225001 (2019)
111. X. Zhang, M. Ezawa, Y. Zhou, Sci. Rep. **5**, 9400 (2015)
112. W. Kang, Y. Huang, C. Zheng et al., Sci. Rep. **6**, 23164 (2016)
113. S. Luo, M. Song, X. Li et al., Nano Lett. **18**, 1180–1184 (2018)
114. H. Yang, C. Wang, X. Wang et al., Phys. Rev. B **98**, 014433 (2018)
115. R. Tomasello, E. Martinez, R. Zivieri et al., Sci. Rep. **4**, 6784 (2014)
116. S. Seki, X.Z. Yu, S. Ishiwata et al., Science **336**, 198–201 (2012)
117. A. Vansteenkiste, J. Leliaert, M. Dvornik et al., AIP Adv. **4**, 107133 (2014)
118. I. Tamm, Phys. Z. Sowjetunion. **76**, 849 (1932)
119. W. Shockley, Phys. Rev. **56**, 317 (1939)
120. F.G. Mertens, H.J. Schnitzer, A.R. Bishop, Phys. Rev. B **56**, 2510 (1997)
121. B.A. Ivanov, G.G. Avanesyan, A.V. Khvalkovskiy et al., JETP Lett. **91**, 178 (2010)
122. S.S. Cherepov, B.C. Koop, A.Yu. Galkin et al., Phys. Rev. Lett. **109**, 097204 (2012)
123. J. Shibata, K. Shigeto, Y. Otani, Phys. Rev. B **67**, 224404 (2003)
124. J. Shibata, Y. Otani, Phys. Rev. B **70**, 012404 (2004)
125. Z.-X. Li, C. Wang, Y. Cao et al., Phys. Rev. B **98**, 180407(R) (2018)
126. Z.-X. Li, Y. Cao, P. Yan et al., npj Comput. Mater. **5**, 107 (2019)
127. K.S. Lee, H. Jung, D.S. Han et al., J. Appl. Phys. **110**, 113903 (2011)
128. O.V. Sukhostavets, J. González, K.Y. Guslienko, Phys. Rev. B **87**, 094402 (2013)
129. J.P. Sinnecker, H. Vigo-Cotrina, F. Garcia et al., J. Appl. Phys. **115**, 203902 (2014)
130. M.W. Yoo, J. Lee, S.K. Kim, Appl. Phys. Lett. **100**, 172413 (2012)
131. R.D. King-Smith, D. Vanderbilt, Phys. Rev. B **47**, 1651 (1993)
132. D. Vanderbilt, R.D. King-Smith, Phys. Rev. B **48**, 4442 (1993)
133. A.H. Castro Neto, F. Guinea, N.M.R. Peres et al., Rev. Mod. Phys. **81**, 109 (2009)
134. F.D.M. Haldane, Phys. Rev. Lett. **61**, 2015 (1988)
135. Y. Yao, F. Ye, X.-L. Qi et al., Phys. Rev. B **75**, 041401(R) (2007)
136. H. Fan, B. Xia, L. Tong et al., Phys. Rev. Lett. **122**, 204301 (2019)
137. Z.-X. Li, Y. Cao, X. R. Wang et al., Phys. Rev. Appl. **13**, 064058 (2020)
138. J.E. Avron, R. Seiler, B. Simon, Phys. Rev. Lett. **51**, 51 (1983)
139. S.N. Kempkes, M.R. Slot, J.J. van den Broeke et al., Nat. Mater. **18**, 1292–1297 (2019)
140. W.P. Su, J.R. Schrieffer, A.J. Heeger, Phys. Rev. Lett. **42**, 1698 (1979)
141. J.C.Y. Teo, C.L. Kane, Phys. Rev. B **82**, 115120 (2010)
142. Z. Zhang, H. Long, C. Liu et al., Adv. Mater. **31**, 1904682 (2019)
143. Z.-X. Li, Y. Cao, X.R. Wang et al., Phys. Rev. B **101**, 184404 (2020)
144. P. Gao, D. Torrent, F. Cervera et al., Phys. Rev. Lett. **123**, 196601 (2019)
145. L. Sun, R.X. Cao, B.F. Miao et al., Phys. Rev. Lett. **110**, 167201 (2013)
146. Y. Guang, I. Bykova, Y. Liu et al., Nat. Commun. **11**, 949 (2020)
147. M. Möller, J.H. Gaida, S. Schäfer et al., Commun. Phys. **3**, 36 (2020)

Chapter 16
Antiferromagnetic Skyrmions and Bimerons

Laichuan Shen, Xue Liang, Jing Xia, Xichao Zhang, Motohiko Ezawa, Oleg A. Tretiakov, and Yan Zhou

Abstract The topological spin textures, such as magnetic skyrmions and bimerons, are currently a hot topic in condensed matter physics. Magnetic skyrmions are swirling spin textures, which can be stabilized in chiral magnets with perpendicular magnetic anisotropy, while magnetic bimerons can be regarded as the counterpart of skyrmions in magnetic systems with in-plane anisotropy. Both magnetic skyrmions and bimerons have attracted a lot of attentions, because they have small size and low depinning current, and can be used as nonvolatile information carries in future spintronic devices. In this chapter, we mainly present and discuss recent original results about the magnetization dynamics of antiferromagnetic skyrmions and bimerons excited by currents and spatially non-uniform magnetic anisotropy. In addition, we also explore possible applications based on antiferromagnetic skyrmions and bimerons, and review some relevant results.

16.1 Introduction

Antiferromagnets are magnetically ordered spin systems, and antiferromagnetic materials are promising for building advanced spintronic devices [1–4]. Considering an antiferromagnet with two sublattices, its magnetization dynamics is governed by the Landau-Lifshitz-Gilbert (LLG) equations [5], which can also describe the dynamics of a ferromagnet. However, the dynamic equations of antiferromagnets are coupled via the exchange interaction between sublattices, causing that the motion equation of the order parameter is of second order with respect to time [2]. Thus, the

L. Shen · X. Liang · J. Xia · X. Zhang · Y. Zhou (✉)
School of Science and Engineering, The Chinese University of Hong Kong, Shenzhen 518172, Guangdong, China
e-mail: zhouyan@cuhk.edu.cn

M. Ezawa
Department of Applied Physics, The University of Tokyo, 7-3-1 Hongo, Tokyo 113-8656, Japan

O. A. Tretiakov
School of Physics, The University of New South Wales, Sydney 2052, Australia

E. Kamenetskii (ed.), *Chirality, Magnetism and Magnetoelectricity*, Topics in Applied Physics 138,
https://doi.org/10.1007/978-3-030-62844-4_16

dynamic behavior of antiferromagnets should be similar to that described by the classical Newton's kinetic equation, and antiferromagnets may have richer dynamics than ferromagnets. On the other hand, intrinsic resonance frequency of the ferromagnet is typically on the order of gigahertz, while for the antiferromagnet it has high resonance frequency (generally, terahertz) thanks to the presence of strong antiferromagnetic exchange fields [6]. Therefore, the antiferromagnet can be used as an oscillator to generate terahertz oscillation signal [7]. Additionally, the magnetic moments of antiferromagnetic sublattices are compensated, which leads to nearly zero stray fields and insensitivity to the perturbation due to magnetic fields. These advantages make antiferromagnets active in condensed matter physics.

Magnetic textures, such as domain wall [8–10], skyrmion [11–16], bimeron [17–21] and skyrmionium [22–25] have attracted a great deal of attention due to their rich physics and important applications in information storage and computing devices. In addition to ferromagnetic systems, these magnetic textures can exist in antiferromagnetic systems. Figure 16.1 shows the antiferromagnetic domain wall, bimeron, skyrmion and skyrmionium, where antiferromagnetic textures may have some advantages compared to ferromagnetic analogues. For example, the speed of a ferromagnetic domain wall is limited by the Walker breakdown [26]. However, for a domain wall in antiferromagnet, only when its speed is close to the spin wave velocity (it has a very high value, about 10 km/s [9]), the domain wall will undergo the speed limit. On the other hand, skyrmions are topologically protected spin textures that have been experimentally observed in chiral materials, for example, MnSi [27] and Pt/Co/MgO [28]. For a skyrmion in ferromagnet, it has an integer topological charge, so that the

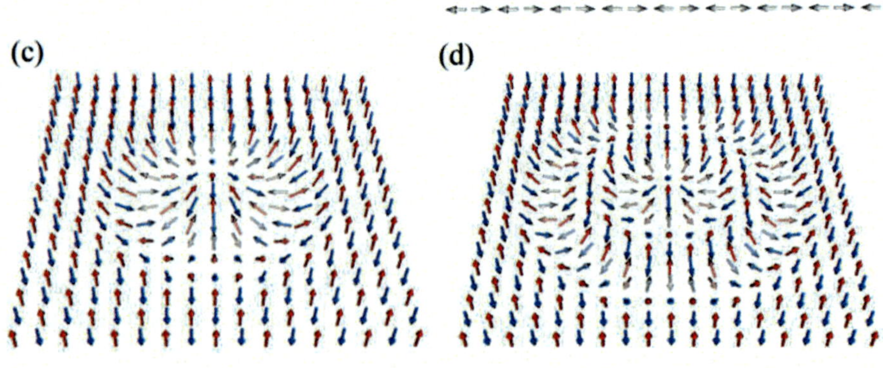

Fig. 16.1. Antiferromagnetic **a** domain wall, **b** bimeron, **c** skyrmion and **d** skyrmionium

Magnus force is always nonzero and leads to the skyrmion Hall effect [29, 30]. Such a skyrmion Hall effect may cause the skyrmion to annihilate at the sample edge, which is detrimental for practical applications. For an antiferromagnetic skyrmion, it consists of two ferromagnetic skyrmions (one on each sublattice) with opposite topological charges for magnetization, so that the net Magnus force is zero and the fast-moving skyrmion will not be destroyed by touching the sample edge.

In order to understand physics of antiferromagnetic textures and explore their applications, it is necessary to discuss their magnetization dynamics. In Sect. 16.2, we discuss the current-induced creation, motion and chaos of antiferromagnetic skyrmions and bimerons. In Sect. 16.3, we review and discuss of the spin torque nano-oscillators based on antiferromagnetic skyrmions. In Sect. 16.4, we focus on the study of synthetic antiferromagnetic skyrmions driven by the spin current. The following two parts, i.e., Sects. 16.5 and 16.6, mainly present and discuss the magnetization dynamics of antiferromagnetic skyrmions in the presence of spatially non-uniform magnetic anisotropy. In Sect. 16.5, we present the calculation results of using a magnetic anisotropy gradient to drive antiferromagnetic skyrmions, where the magnetic anisotropy gradient is applicable in both metals and insulators. Then, in Sect. 16.6, the pinning and depinning processes of antiferromagnetic skyrmions in a racetrack with a defect caused by the local variation of magnetic anisotropy are studied. Finally, in Sect. 16.7, we make a summary of this chapter.

16.2 Current-Driven Creation, Motion, and Chaos of Antiferromagnetic Skyrmions and Bimerons

In this section, we discuss the current-induced creation, motion and chaos of antiferromagnetic skyrmions and bimerons. Magnetic skyrmions are swirling spin textures carrying an integer topological charge, and they can be stabilized in chiral magnets with perpendicular magnetic anisotropy [11, 12]. Magnetic bimeron composed of two merons can be regarded as a counterpart of magnetic skyrmion in systems with in-plane magnetic anisotropy, which shares the merits of skyrmions, such as nanoscale size, nontrivial spin structure and low depinning current [18, 19]. These excellent characteristics make that both skyrmion and bimeron are promising candidates as information carriers in future spintronic devices. In practical applications, the controllable creation and manipulation of skyrmion and bimeron are crucial. In 2016, Zhang et al. [31] reported the creation of an antiferromagnetic skyrmion by applying a vertical spin current. In addition, their calculations show that using a junction geometry, the antiferromagnetic skyrmion is generated from a domain wall pair, as shown in Fig. 16.2. Similar to the skyrmion, an isolated bimeron is created in antiferromagnetic films by applying a vertical spin current to reverse the local magnetic moments, as reported by Shen et al. [19]. It is worth mentioning that in addition to the spin current, the time-dependent magnetic field can also induce the generation of an isolated antiferromagnetic skyrmion [32].

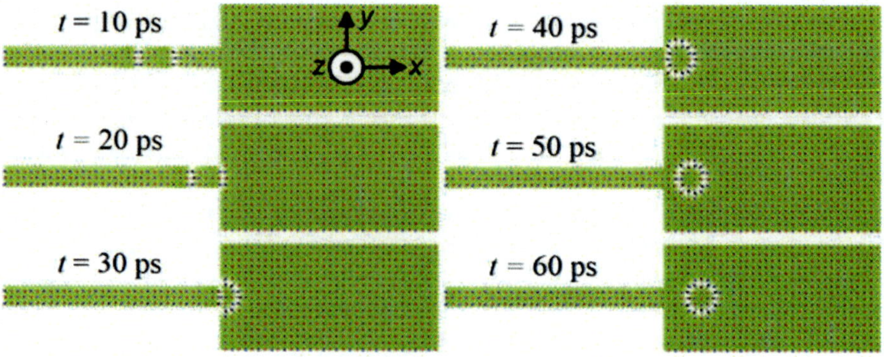

Fig. 16.2. Creation of an isolated antiferromagnetic skyrmion via a domain wall pair driven by a vertical spin current. Reprinted with permission from [31]. CC BY 4.0

On the other hand, manipulating magnetic textures is indispensable in future information storage and computing devices. Various methods have been proposed to drive antiferromagnetic skyrmion and bimeron, such as using spin currents [19, 31, 33, 34], magnetic anisotropy gradients [35] and temperature gradients [32]. In particular, the current-induced spin torque is a common way for manipulating magnetic materials. For current-induced spin-transfer torques including adiabatic and nonadiabatic terms [34, 36], the speed of both antiferromagnetic skyrmion and bimeron is proportional to the nonadiabatic spin-transfer torque parameter, and inversely proportional to the damping constant [33]. For current-induced damping-like spin torque [34], antiferromagnetic skyrmion and bimeron can be effectively driven to move, while for field-like spin torque, their response is weak because perfect antiferromagnetic materials are insensitive to a homogeneous magnetic field. Thanks to the cancellation of the Magnus force, current-induced spin torques can drive the antiferromagnetic skyrmion and bimeron at a speed of a few kilometers per second without showing any transverse drift, so that their motion trajectory is a perfect straight line along the driving force direction and they are ideal information carriers in racetrack-type memory [19, 25, 31, 33]. Note that in the high-speed region, a transverse expansion will be present, as reported in [33, 37, 38]. In 2020, Salimath et al. [37] demonstrated that such an expansion is reminiscent of the well-known Lorentz contraction, which has been identified in antiferromagnetic domain walls [9].

Next, based on the equation of motion, the dynamics of antiferromagnetic bimeron (or skyrmion) induced by the alternating current is discussed. As mentioned earlier, for antiferromagnetic systems, the motion equation of the order parameter is related to the second derivative with respect to time. On the other hand, the motion equation of the systems, such as the Landau-Lifshitz-Gilbert equation [5], is usually nonlinear, resulting in the dynamic behavior being complex or even chaotic. Thus, considering a skyrmion or bimeron in an antiferromagnetic nanodisk, its motion can be described by the Duffing equation, where the nanodisk boundary provides a nonlinear restoring force [19]. Note that for the chaos, the nonlinearity is a necessary condition rather than a sufficient condition, so that not all nonlinear systems will exhibit chaotic behavior. Therefore, under the action of the alternating current, the antiferromagnetic bimeron (or skyrmion) generally does periodic motion. Taking certain parameter values, it shows chaotic behavior. The Lyapunov exponents (LEs) are usually used to judge whether there is chaos [39]. If the largest LE is positive, it means that two close trajectories will be separated. Namely, a small initial error will increase rapidly, resulting in the motion of antiferromagnetic bimeron being sensitive to initial conditions, and its motion behavior cannot be predicted for a long time, i.e., the bimeron does chaotic motion. In 2020, based on the motion equation of an antiferromagnetic bimeron (i.e., the Duffing equation), Shen et al. [19] calculated the bifurcation diagram and LEs, as shown in Fig. 16.3. It can be seen that the periodic and chaotic windows appear at intervals and a small damping constant can lead to the chaotic motion. In addition, the current is also of great importance to induce the occurrence of the chaos, as it can be easily tuned in experiment. Their calculation shows that for small currents, the system exhibits a periodic motion. With increasing currents, the period-doubling phenomenon takes place, and then the system shows chaotic behavior. It is worth mentioning that the chaotic behavior reported by Shen et al. [19] is subject to the boundary-induced restoring force, which depends on both the geometric and magnetic parameters. Since the antiferromagnetic bimerons show chaotic behavior, they can be used as chaotic oscillators, which are promising for various applications, such as detecting weak signals [40], generating random numbers [41] and building chaotic logic gates [42].

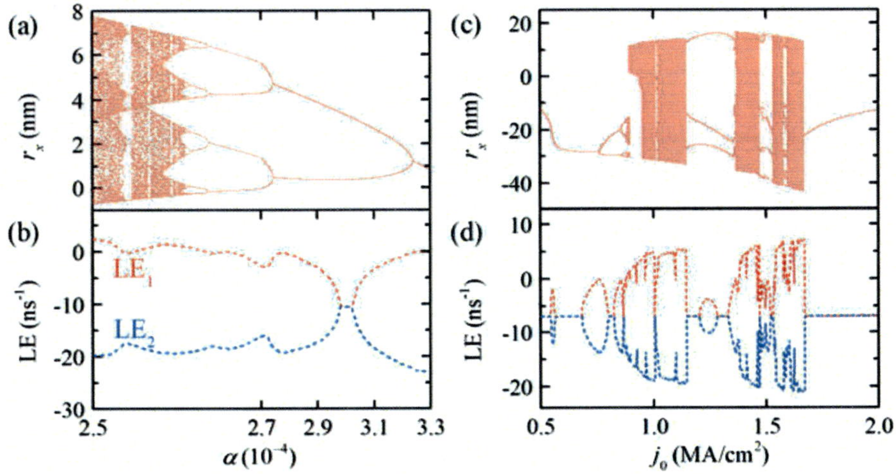

Fig. 16.3. **a** Calculated bifurcation diagram and **b** Lyapunov exponents (LEs) as functions of the damping constant. **c** Calculated bifurcation diagram and **d** LEs as functions of the current. Reprinted with permission from [19]. Copyright © 2020 American Physical Society

16.3 Spin Torque Nano-oscillators Based on Antiferromagnetic Skyrmions

The antiferromagnetic textures can be used not only as chaotic oscillators, but also as microwave signal generators (or spin torque nano-oscillators). In this section, we first review recent advances in skyrmion-based spin torque nano-oscillators, and then focus on the discussion of the nano-oscillators based on antiferromagnetic skyrmions. Spin torque nano-oscillators based on magnetic skyrmions have received great attention, because they can excite microwave signals with small linewidth and are expected to improve the output power [43–46]. In addition, the arrays of such nano-oscillators can be used to perform the neuromorphic computing [47]. In 2015, Zhang et al. [45] firstly demonstrated that applying a uniform current to the nano-contact oscillators, the ferromagnetic skyrmion can be induced to move in a circular motion, and then an oscillating signal is obtained by detecting the skyrmion position and using the magnetoresistance effect. In 2016, Garcia-Sanchez et al. [44] showed an alternative skyrmion-based nano-oscillator, in which a fixed layer with a vortex magnetic configuration is used to generate the spin-polarized current with a vortex-like polarization.

Generally, for the ferromagnetic skyrmion-based nano-oscillators, the oscillation frequencies are low (about 1 gigahertz), because the fast-moving skyrmion results in the presence of a large Magnus force and then the skyrmion will be destroyed at the nanodisk edge. In order to overcome this obstacle, various methods have been proposed. For example, in 2019, Feng et al. [48] showed that the oscillation frequency can be increased by 75%, where the nanodisk edge is enhanced by applying high perpendicular magnetic anisotropy. In addition, modifying the profile of Dzyaloshinskii-Moriya interaction can also lead to the increase in oscillation frequency [49]. Besides, in 2020, Jin et al. [50] showed that creating an annular groove in the surface of the free layer, the frequency tunability of the nano-oscillators reaches to 15.63 GHz. However, the above methods require sophisticated reprocessing and are not favorable from the point of view of device applications. Shen et al. [51] proposed to use the circular motion of an antiferromagnetic skyrmion to create the oscillation signal, where sophisticated reprocessing is not required. In addition to the cancellation of the Magnus force, the antiferromagnetic skyrmion obeys the inertial dynamics, as the motion equation of the antiferromagnetic order parameter is of second order with respect to time (for ferromagnetic dynamic equation it is of first order). Thus, the motion behavior of the antiferromagnetic and ferromagnetic skyrmions in a nanodisk is different, and the oscillation frequency of the antiferromagnetic skyrmion (tens of gigahertz) is higher than that of the ferromagnetic skyrmion [51]. Figure 16.4 shows the comparison of ferromagnetic and antiferromagnetic skyrmion-based nano-oscillators. We can see that the ferromagnetic skyrmion moves toward the nanodisk center when the positive current is applied, while for the negative current, the skyrmion is destroyed at the nanodisk edge. For the antiferromagnetic skyrmion, it moves steadily in the nanodisk under the action of the same currents, and its motion is independent of the sign of the applied current.

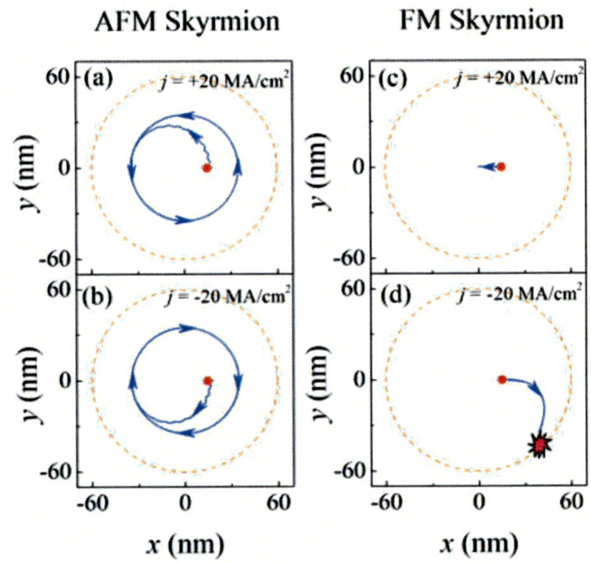

Fig. 16.4. The comparison of ferromagnetic and antiferromagnetic skyrmions in the nanodisk. The trajectory for an antiferromagnetic skyrmion driven by positive (**a**) and negative (**b**) currents. The trajectory for a ferromagnetic skyrmion driven by positive (**c**) and negative (**d**) currents. Reprinted from [51], with the permission of AIP Publishing.

16.4 Synthetic Antiferromagnetic Skyrmions Driven by the Spin Current

In this section, we review and discuss the current-driven dynamics of magnetic skyrmions in the synthetic antiferromagnetic system. Different from the above-mentioned antiferromagnetic system, the synthetic antiferromagnet is made of ferromagnetic thin films but with antiferromagnetic interlayer exchange couplings [52, 53], which can be realized by the Ruderman-Kittel-Kasuya-Yosida (RKKY) interaction mechanism [54]. A basic and simple synthetic antiferromagnetic system can be constructed by a bilayer structure consisting of two antiferromagnetically exchange-coupled ferromagnetic layers with identical material properties and thicknesses. Namely, the total net magnetization of the bilayer system is equal to zero, leading to the zero-stray-field nature of a perfect antiferromagnetic thin film. It should be noted that an ultra-thin metal spacer is usually sandwiched between the two ferromagnetic layers, of which the thickness is adjusted to give rise to the RKKY-type interlayer antiferromagnetic exchange coupling.

The concept of skyrmions in synthetic antiferromagnets with zero net magnetization is originated for the purpose of eliminating the so-called skyrmion Hall effect [29, 30, 55]. The skyrmion Hall effect is a dynamic phenomenon associated with topological number of the skyrmion [29, 30], that is, a current-driven ferromagnetic skyrmion carrying an integer topological charge may experience a Magnus force and move at an angle with respect to the driving current direction. Such a phenomenon could result in the accumulation and/or even the destruction of skyrmions at sample edges, which is detrimental for some skyrmion-based applications such as the skyrmionic racetrack memory [52]. In a synthetic antiferromagnetic bilayer structure, such a

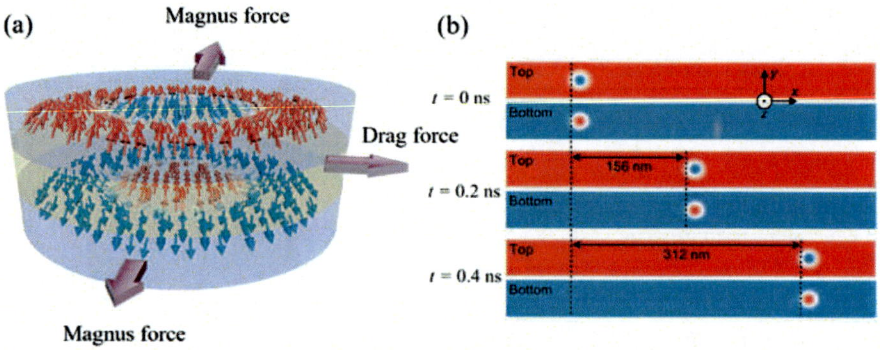

Fig. 16.5. A magnetic skyrmion in the synthetic antiferromagnetic bilayer system. **a** The Magnus forces acted on the top-layer and bottom-layer skyrmions will cancel each other, which results in the elimination of the skyrmion Hall effect. **b** Current-driven motion of a synthetic antiferromagnetic bilayer skyrmion. Reprinted with permission from [52]. CC BY 4.0

skyrmion Hall effect can be effectively eliminated as the skyrmions in the top and bottom layers have opposite skyrmion numbers and therefore, the Magnus forces acted on top and bottom skyrmions could cancel each other since they are identical in magnitude but pointing toward opposite directions, as shown in Fig. 16.5. Consequently, the synthetic antiferromagnetic bilayer skyrmion driven by a current moves in a straight line along the driving force direction, as shown in Fig. 16.5.

The concept of synthetic antiferromagnetic bilayer skyrmion was first purposed in a simulation work by Zhang et al. in 2016 [52]. In the same year, Zhang et al. also studied the skyrmion dynamics in multilayer synthetic antiferromagnetic racetracks [56]. It was found that the current-driven skyrmion shows no transverse motion in multilayer synthetic antiferromagnetic nanotracks packed with even identical ferromagnetic layers. Namely, only when the total skyrmion number equals zero and the system has zero net magnetization, the skyrmion Hall effect is truly eliminated. Zhang et al. [56] also suggest that the synthetic antiferromagnetic skyrmion may have a better thermal stability during its motion in nanotracks due to the absence of skyrmion-edge interaction induced by the skyrmion Hall effect. In 2017, by using micromagnetic simulations Tomasello et al. [57] further studied the performance of racetrack memory based on the synthetic antiferromagnetic skyrmion. They pointed out that the velocity of synthetic antiferromagnetic skyrmions and synthetic antiferromagnetic Néel-type domain walls are of the same order and can reach values larger than 1200 m/s, which is promising for real applications.

It is worth mentioning that the formation and current-induced motion of synthetic antiferromagnetic skyrmion bubbles have been realized in experiments by Dohi et al. in 2019 [58] and the stabilization of antiferromagnetic skyrmions in synthetic antiferromagnets at room temperature has also been experimentally realized by Legrand et al. in 2020 [59]. The two experimental works demonstrated promising features of synthetic antiferromagnetic skyrmions, that is, the thermal stable nanoscale size and negligible skyrmion Hall effect, which highlight the possibility of using synthetic antiferromagnetic skyrmions in future high-performance spintronic devices with higher storage density and operation speed.

16.5 Antiferromagnetic Skyrmions Driven by the Magnetic Anisotropy Gradient

Previous section shows that antiferromagnetic skyrmions can be driven by electric currents. However, using the electric current as a driving source faces the issue of Joule heating, and the dynamics of the insulating antiferromagnet cannot be excited by an electric current. Therefore, alternative methods are crucial and have been explored, for example, using a voltage-controlled magnetic anisotropy gradient. The voltage-controlled magnetic anisotropy gradient is a promising method to drive the magnetic textures, because it has ultralow power consumption and is applicable in both metals and insulators. Recent studies experimentally and theoretically demonstrate that such a voltage-controlled magnetic anisotropy gradient can be used to manipulate the magnetic textures [35, 60–63]. Particularly, in 2019, Ma et al. [60] experimentally presented the electric field-induced creation and motion of domain walls and skyrmion bubbles. In this section, we first describe the calculation results of using a magnetic anisotropy gradient to drive antiferromagnetic skyrmions, and then compare the velocities of antiferromagnetic and ferromagnetic skyrmions.

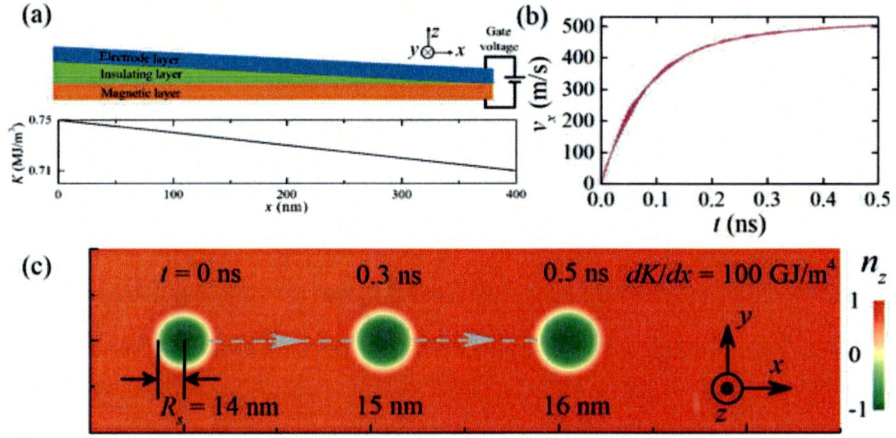

Fig. 16.6. **a** The sketch of the voltage-controlled magnetic anisotropy device, where the magnetic anisotropy K linearly decreases with the increase of the spatial coordinate x. **b** The evolution of the skyrmion speed for an antiferromagnetic skyrmion induced by a magnetic anisotropy gradient $dK/dx = 100$ GJ/m^4. **c** The top view of the antiferromagnetic skyrmion motion, where the color represents the out-of-plane component n_z of the Néel vector. Reprinted with permission from [35]. Copyright © 2018 American Physical Society

As shown in Fig. 16.6a, the magnetic anisotropy gradient ($dK/dx = 100$ GJ/m^4) is assumed to be induced by applying a voltage to the sample with a wedged insulating layer. Taking such an anisotropy gradient, i.e., $dK/dx = 100$ GJ/m^4, the motion of the antiferromagnetic skyrmion is simulated by solving the dynamic equation of antiferromagnet, and the time evolution of the skyrmion speed v_x is shown in Fig. 16.6b. We can see that the antiferromagnetic skyrmion is first accelerated to 450 m/s in 0.2 ns and then its speed increases slowly to 504 m/s by $t = 0.5$ ns [35]. The speed cannot reach a constant value, as the decreasing magnetic anisotropy K gives rise to the change of the skyrmion size. Namely, when the skyrmion moves in the positive x direction, the decreasing K results in the increase of the skyrmion size, so that the skyrmion speed will increase slowly. Fig. 16.6c shows the top view of the skyrmion motion, from which we can see that the radius of the skyrmion at $t =$ 0.5 ns is larger than that of the initial state. Such an effect also exists in the case of ferromagnetic skyrmions driven by an anisotropy gradient, as reported by Tomasello et al. [62].

On the other hand, in antiferromagnetic systems, the magnetic moments of sublattices are compensated, so that the antiferromagnetic skyrmions do not show the skyrmion Hall effect [29, 30] due to zero net Magnus force. However, for ferromagnetic skyrmions, there is a transverse drift, i.e., the skyrmion Hall effect. Therefore, the motion behaviors of antiferromagnetic skyrmions driven by an anisotropy gradient are different from that of ferromagnetic skyrmions. The velocities of antiferromagnetic and ferromagnetic skyrmions are calculated as functions of the magnetic anisotropy gradient dK/dx and damping constant α, as shown in Fig. 16.7. It can be seen that their velocities are proportional to dK/dx and $1/\alpha$, and owing to the cancellation of the Magnus force, the speed of an antiferromagnetic skyrmion is larger than that of a ferromagnetic skyrmion.

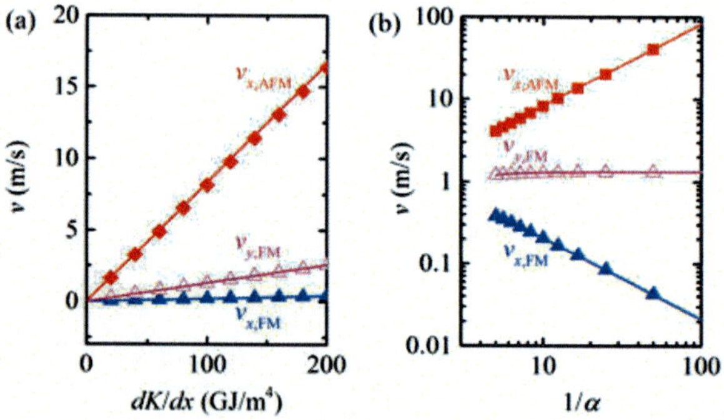

Fig. 16.7. Numerical (symbols) and analytical (lines) velocities of the antiferromagnetic and ferromagnetic skyrmions. **a** The velocities as functions of magnetic anisotropy gradient dK/dx, where the damping constant is set as 0.1. **b** The velocities as functions of the damping constant, where the magnetic anisotropy gradient dK/dx is 100 GJ/m^4. Reprinted with permission from [35]. Copyright © 2018 American Physical Society

As a result, adopting the magnetic anisotropy gradient to drive the skyrmion in pure antiferromagnets, the skyrmion can move at high speeds (up to 500 m/s) without any transverse drift. It is worth mentioning that using the anisotropy gradient as a driving source can be introduced to antiferromagnetically coupled bilayer systems, as reported by Qiu et al. [64]. The above results may open an alternative way for the design of antiferromagnetic skyrmion-based devices, such as the horizontal racetrack-type memory [65]. In addition, when these devices are chained together, it can also be used to build the antiferromagnetic skyrmion-based diode [66].

16.6 Pinning and Depinning of Antiferromagnetic Skyrmions

To understand more antiferromagnetic skyrmion physics and use it in future nanoscale magnetic data storage and logic devices, it is essentially important to know the pinning and depinning of antiferromagnetic skyrmions to defects. In this section, we discuss the effect of defects caused by the local variation of perpendicular magnetic anisotropy on the current-induced dynamics of an antiferromagnetic skyrmion under the framework of micromagnetics.

From [67], the local magnetic anisotropy value in the defect area is specified by $K = K_0\left[1.0 + \lambda/\exp(|\mathbf{r} - \mathbf{r}_d|/R_d)^2\right]$, where K_0 is the magnetic anisotropy constant in the homogeneous area, $|\lambda|$ denotes the amplitude of variation referred as the strength of the defect, R_d represents its characteristic size (i.e., the radius) and $\mathbf{r}_\mathrm{d}$ is the position vector of the defect center. Such an anisotropy profile to describe the inhomogeneity induced by potential impurity sites, might be more realistic than a simple step-like defect or a defect with a linear variation of anisotropy, and is analogous to the model proposed by Kronmüller [68]. In addition, similar distribution of exchange interaction J_ex has also been used to discuss the skyrmion-defect interaction in ferromagnetic film [69, 70]. In all simulations, the coordinate of the antiferromagnetic skyrmion center is represented by (R_x, R_y) with the initial position (0, 0).

First let us consider the effect of the defect on the antiferromagnetic skyrmion without any external driving force. The simulation results show that when the skyrmion is placed at the area where the defect can interact with it, the skyrmion will spontaneously move and eventually stop at the center or off-center of the defect. Moreover, for the same defect, the skyrmion always be pinned at the same position. In particular, the final pinned position depends strongly on the ratio of skyrmion size to the defect size and is not sensitive to the strength of the defect. As shown in Fig. 16.8a, when the radius R_d of defect is smaller than the antiferromagnetic skyrmion radius R_sk, the skyrmion stops at the off-center of the defect, and the distance L gradually decreases as R_d increases. When R_d is close to or exceeds R_sk, L drops sharply and approaches 0, and the skyrmion will stay at the center of the defect. Such dynamic pinning processes are also similar to those of ferromagnetic skyrmions [71].

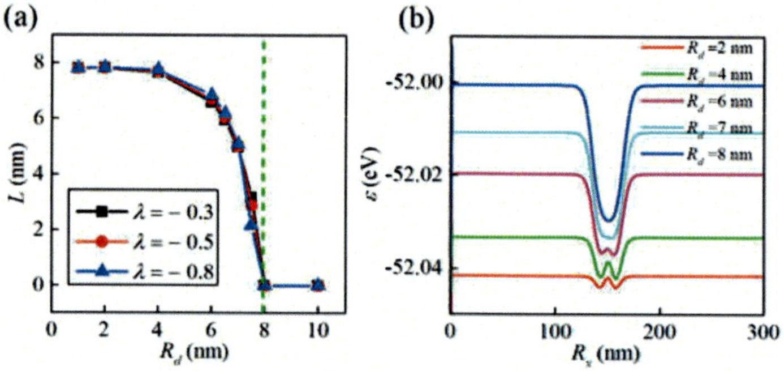

Fig. 16.8. **a** The distance L between the skyrmion final position and the defect center, as a function of the radius of defect for different strengths, where the green dashed line represents the radius of skyrmion; **b** The variation of the total energy with the skyrmion position R_x (with $R_y = 0$ nm) is plotted for different defect sizes, where the defect is initially placed at 150 nm in front of the skyrmion. Reprinted with permission from [67]. Copyright © 2019 American Physical Society.

To understand the physics behind the dynamic pinning process of the antiferromagnetic skyrmion mentioned above, Fig. 16.8b shows the total energy of the system vs the skyrmion position when the strength λ of the defect is fixed at −0.5. It can be found that there are two local minimal values near the defect center when $R_d < R_{sk}$, but only one minimal value at the center of defect when $R_d > R_{sk}$. Note that the skyrmion always prefers to be pinned at the low-energy place which related to the trough in the energy curve. Besides, for the defects with the same size and different strength (when $\lambda < 0$), the corresponding energy curves have the same shapes. Therefore, one can understand why the skyrmion final position is not sensitive to the strength of the defect, but mainly depends on its size.

We further discuss the motion of an antiferromagnetic skyrmion driven by the spin- polarized current in a defective racetrack, and find the transition from a pinned to a depinned state. Considering the force $\mathbf{F}_u$ arising from the defect is obtained by $\mathbf{F}_u = -\nabla\varepsilon$ where ε (R_x, R_y) is the total potential energy of the system, the dynamic behaviors of the antiferromagnetic skyrmion around defect will depend on both the driving current density and the defect. As shown in Fig. 16.9, four motion behaviors of the skyrmion will occur when $\lambda = -0.5$. It also can be found that for different defect sizes, there will always be a critical depinning current density j_c required to drive the skyrmion passing through the defect area successfully along a straight line. Reference [67] has shown that the critical current density not only increases with increasing defect strength, but also is proportional to the radius of the defect. Moreover, it is much larger than that of the homogeneous racetrack where $\lambda = 0$ (~10^6 A m^{-2}), which implies that a proper understanding of how antiferromagnetic skyrmions interact with defects is vital for the development and performance of future spintronic devices.

Generally, the potential defects or impurities in real materials may also contribute to the local increase of the magnetic anisotropy value, or even cause changes in other magnetic parameters, such as the Heisenberg exchange stiffness and the

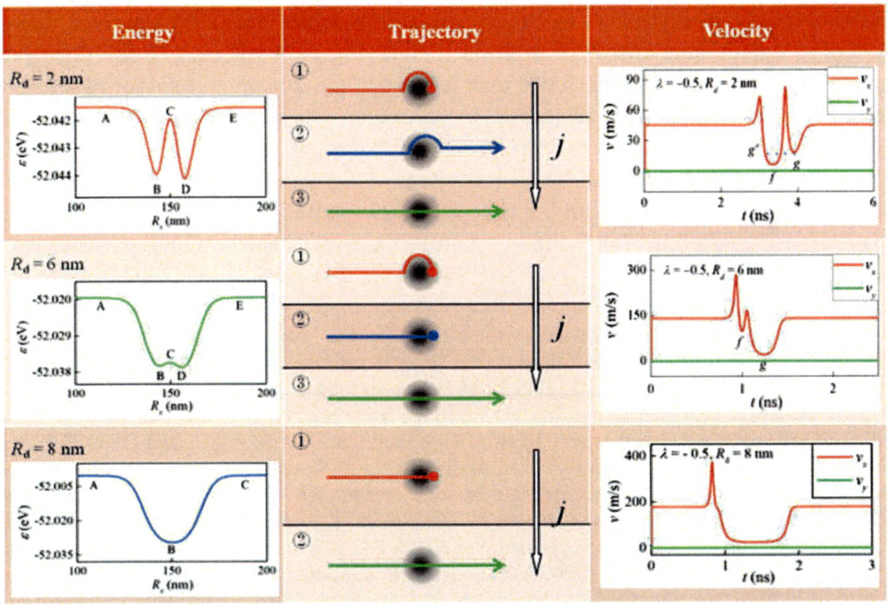

Fig. 16.9. Overview of the different motion behaviors of an antiferromagnetic skyrmion. Note that the black spot denotes the considered defect and the current density increases along the direction of the arrow. Reprinted with permission from [67]. Copyright © 2019 American Physical Society.

Dzyaloshinskii-Moriya interaction constant. Reference [67] has systematically investigated the dynamic pinning and depinning processes of antiferromagnetic skyrmions on a racetrack with different defects.

16.7 Summary

In this chapter, we have introduced and discussed the magnetization dynamics of antiferromagnetic skyrmions and bimerons excited by currents and spatially non-uniform magnetic anisotropy. In addition, we have also proposed possible applications based on antiferromagnetic skyrmions and bimerons, and reviewed some relevant results. We have shown that spin currents can create and drive the antiferromagnetic skyrmions and bimerons. Due to the cancellation of the Magnus force, the antiferromagnetic skyrmion and bimeron will not show the skyrmion Hall effect, so that they are ideal information carriers in racetrack-type memory. In addition, in the synthetic antiferromagnetic bilayer structure, the skyrmion also does not show the skyrmion Hall effect and moves along the driving force direction. Besides, we have demonstrated that the antiferromagnetic skyrmion can be driven by a magnetic anisotropy gradient. Also, we have studied the pinning and depinning processes of antiferromagnetic skyrmions in a racetrack with a defect caused by the local variation of magnetic anisotropy. In addition to studying the dynamics of skyrmions in an antiferromagnetic racetrack, we also discussed the dynamic behavior of antiferromagnetic skyrmions in a nanodisk. In particular, we pointed out that driving an antiferromagnetic skyrmion to move in a circular motion on a nanodisk, one may get an oscillation signal with high frequencies (tens of gigahertz). These results are useful for understanding skyrmion and bimeron physics in antiferromagnetic systems, and may provide guidelines for building spintronic devices based on antiferromagnetic textures.

Acknowledgements X.Z. acknowledges the support by the National Natural Science Foundation of China (Grant No. 12004320), and the Guangdong Basic and Applied Basic Research Foundation (Grant No. 2019A1515110713). M.E. acknowledges the support from the Grants-in-Aid for Scientific Research from JSPS KAKENHI (Grant Nos. JP18H03676, JP17K05490 and JP15H05854) and the support from CREST, JST (Grant Nos. JPMJCR16F1 and JPMJCR1874). O.A.T. acknowledges the support by the Australian Research Council (Grant No. DP200101027), the Cooperative Research Project Program at the Research Institute of Electrical Communication, Tohoku University (Japan), and by the NCMAS grant. Y.Z. acknowledges the support by the Guangdong Special Support Project (Grant No. 2019BT02X030), Shenzhen Peacock Group Plan (Grant No. KQTD20180413181702403), Pearl River Recruitment Program of Talents (Grant No. 2017GC010293), and National Natural Science Foundation of China (Grant Nos. 11974298 and 61961136006).

References

1. T. Jungwirth, X. Marti, P. Wadley, J. Wunderlich, Antiferromagnetic spintronics. Nat. Nanotechnol. **11**, 231 (2016)
2. V. Baltz, A. Manchon, M. Tsoi, T. Moriyama, T. Ono, Y. Tserkovnyak, Antiferromagnetic spintronics. Rev. Mod. Phys. **90**, 015005 (2018)
3. O. Gomonay, V. Baltz, A. Brataas, Y. Tserkovnyak, Antiferromagnetic spin textures and dynamics. Nat. Phys. **14**, 213 (2018)
4. L. Šmejkal, Y. Mokrousov, B. Yan, A.H. MacDonald, Topological antiferromagnetic spintronics. Nat. Phys. **14**, 242 (2018)
5. T.L. Gilbert, A phenomenological theory of damping in ferromagnetic materials. IEEE Trans. Magn. **40**, 3443 (2004)
6. F. Keffer, C. Kittel, Theory of antiferromagnetic resonance. Phys. Rev. **85**, 329 (1952)
7. R. Cheng, D. Xiao, A. Brataas, Terahertz antiferromagnetic spin hall nano-oscillator. Phys. Rev. Lett. **116**, 207603 (2016)
8. O. Gomonay, T. Jungwirth, J. Sinova, High antiferromagnetic domain wall velocity induced by Neel spin-orbit torques. Phys. Rev. Lett. **117**, 017202 (2016)
9. T. Shiino, S.H. Oh, P.M. Haney, S.W. Lee, G. Go, B.G. Park, K.J. Lee, Antiferromagnetic domain wall motion driven by spin-orbit torques. Phys. Rev. Lett. **117**, 087203 (2016)
10. E.G. Tveten, A. Qaiumzadeh, O.A. Tretiakov, A. Brataas, Staggered dynamics in antiferromagnets by collective coordinates. Phys. Rev. Lett. **110**, 127208 (2013)
11. Y. Zhou, Magnetic skyrmions: Intriguing physics and new spintronic device concepts. Natl. Sci. Rev. **6**, 210 (2019)
12. X. Zhang, Y. Zhou, K. Mee Song, T. E. Park, J. Xia, M. Ezawa, X. Liu, W. Zhao, G. Zhao, S. Woo, Skyrmion-electronics: writing, deleting, reading and processing magnetic skyrmions toward spintronic applications, J. Phys.: Condens. Matter **32**, 143001 (2019)

13. U.K. Rossler, A.N. Bogdanov, C. Pfleiderer, Spontaneous skyrmion ground states in magnetic metals. Nature **442**, 797 (2006)
14. N. Nagaosa, Y. Tokura, Topological properties and dynamics of magnetic skyrmions. Nat. Nanotechnol. **8**, 899 (2013)
15. A. Fert, N. Reyren, V. Cros, Magnetic skyrmions: advances in physics and potential applications. Nat. Rev. Mater. **2**, 17031 (2017)
16. K. Everschor-Sitte, J. Masell, R.M. Reeve, M. Kläui, Perspective: Magnetic skyrmions—Overview of recent progress in an active research field. J. Appl. Phys. **124**, 240901 (2018)
17. Y.A. Kharkov, O.P. Sushkov, M. Mostovoy, Bound states of skyrmions and merons near the Lifshitz point. Phys. Rev. Lett. **119**, 207201 (2017)
18. B. Göbel, A. Mook, J. Henk, I. Mertig, O.A. Tretiakov, Magnetic bimerons as skyrmion analogues in in-plane magnets. Phys. Rev. B **99**, 060407(R) (2019)
19. L. Shen, J. Xia, X. Zhang, M. Ezawa, O.A. Tretiakov, X. Liu, G. Zhao, Y. Zhou, Current-induced dynamics and chaos of antiferromagnetic bimerons. Phys. Rev. Lett. **124**, 037202 (2020)
20. X. Zhang, J. Xia, L. Shen, M. Ezawa, O.A. Tretiakov, G. Zhao, X. Liu, Y. Zhou, Static and dynamic properties of bimerons in a frustrated ferromagnetic monolayer. Phys. Rev. B **101**, 144435 (2020)
21. X.Z. Yu, W. Koshibae, Y. Tokunaga, K. Shibata, Y. Taguchi, N. Nagaosa, Y. Tokura, Transformation between meron and skyrmion topological spin textures in a chiral magnet. Nature **564**, 95 (2018)
22. X. Zhang, J. Xia, Y. Zhou, D. Wang, X. Liu, W. Zhao, M. Ezawa, Control and manipulation of a magnetic skyrmionium in nanostructures. Phys. Rev. B **94**, 094420 (2016)
23. F. Zheng, H. Li, S. Wang, D. Song, C. Jin, W. Wei, A. Kovacs, J. Zang, M. Tian, Y. Zhang, H. Du, R.E. Dunin-Borkowski, Direct Imaging of a zero-field target skyrmion and its polarity switch in a chiral magnetic nanodisk. Phys. Rev. Lett. **119**, 197205 (2017)
24. M. Finazzi, M. Savoini, A.R. Khorsand, A. Tsukamoto, A. Itoh, L. Duo, A. Kirilyuk, T. Rasing, M. Ezawa, Laser-induced magnetic nanostructures with tunable topological properties. Phys. Rev. Lett. **110**, 177205 (2013)
25. L. Shen, X. Li, Y. Zhao, J. Xia, G. Zhao, Y. Zhou, Current-induced dynamics of the antiferromagnetic skyrmion and skyrmionium. Phys. Rev. Appl. **12**, 064033 (2019)
26. N.L. Schryer, L.R. Walker, The motion of 180° domain walls in uniform dc magnetic fields. J. Appl. Phys. **45**, 5406 (1974)
27. S. Mühlbauer, B. Binz, F. Jonietz, C. Pfleiderer, A. Rosch, A. Neubauer, R. Georgii, P. Böni, Skyrmion lattice in a chiral magnet. Science **323**, 915 (2009)
28. O. Boulle, J. Vogel, H. Yang, S. Pizzini, D. de Souza Chaves, A. Locatelli, T. O. Mentes, A. Sala, L. D. Buda-Prejbeanu, O. Klein, M. Belmeguenai, Y. Roussigne, A. Stashkevich, S. M. Cherif, L. Aballe, M. Foerster, M. Chshiev, S. Auffret, I. M. Miron, G. Gaudin, Room-temperature chiral magnetic skyrmions in ultrathin magnetic nanostructures, Nat. Nanotechnol. **11**, 449 (2016)
29. W. Jiang, X. Zhang, G. Yu, W. Zhang, X. Wang, M. Benjamin Jungfleisch, John E. Pearson, X. Cheng, O. Heinonen, K. L. Wang, Y. Zhou, A. Hoffmann, Suzanne G. E. te Velthuis, Direct observation of the skyrmion Hall effect, Nat. Phys. **13**, 162 (2017)
30. K. Litzius, I. Lemesh, B. Krüger, P. Bassirian, L. Caretta, K. Richter, F. Büttner, K. Sato, O.A. Tretiakov, J. Förster, R.M. Reeve, M. Weigand, I. Bykova, H. Stoll, G. Schütz, G.S.D. Beach, M. Kläui, Skyrmion Hall effect revealed by direct time-resolved X-ray microscopy. Nat. Phys. **13**, 170 (2017)
31. X. Zhang, Y. Zhou, M. Ezawa, Antiferromagnetic skyrmion: Stability, creation and manipulation. Sci. Rep. **6**, 24795 (2016)
32. R. Khoshlahni, A. Qaiumzadeh, A. Bergman, A. Brataas, Ultrafast generation and dynamics of isolated skyrmions in antiferromagnetic insulators. Phys. Rev. B **99**, 054423 (2019)
33. J. Barker, O.A. Tretiakov, Static and dynamical properties of antiferromagnetic skyrmions in the presence of applied current and temperature. Phys. Rev. Lett. **116**, 147203 (2016)

34. H. Velkov, O. Gomonay, M. Beens, G. Schwiete, A. Brataas, J. Sinova, R.A. Duine, Phenomenology of current-induced skyrmion motion in antiferromagnets. New J. Phys. **18**, 075016 (2016)
35. L. Shen, J. Xia, G. Zhao, X. Zhang, M. Ezawa, O.A. Tretiakov, X. Liu, Y. Zhou, Dynamics of the antiferromagnetic skyrmion induced by a magnetic anisotropy gradient. Phys. Rev. B **98**, 134448 (2018)
36. K.M.D. Hals, Y. Tserkovnyak, A. Brataas, Phenomenology of current-induced dynamics in antiferromagnets. Phys. Rev. Lett. **106**, 107206 (2011)
37. A. Salimath, F. Zhuo, R. Tomasello, G. Finocchio, A. Manchon, Controlling the deformation of antiferromagnetic skyrmions in the high-velocity regime. Phys. Rev. B **101**, 024429 (2020)
38. C. Jin, C. Song, J. Wang, Q. Liu, Dynamics of antiferromagnetic skyrmion driven by the spin Hall effect. Appl. Phys. Lett. **109**, 182404 (2016)
39. Z. Yang, S. Zhang, Y.C. Li, Chaotic dynamics of spin-valve oscillators. Phys. Rev. Lett. **99**, 134101 (2007)
40. G. Wang, D. Chen, J. Lin, X. Chen, The application of chaotic oscillators to weak signal detection. IEEE Trans. Industr. Electron. **46**, 440 (1999)
41. A. Fukushima, T. Seki, K. Yakushiji, H. Kubota, H. Imamura, S. Yuasa, K. Ando, Spin dice: A scalable truly random number generator based on spintronics. Appl. Phys. Express **7**, 083001 (2014)
42. W.L. Ditto, A. Miliotis, K. Murali, S. Sinha, M.L. Spano, Chaogates: morphing logic gates that exploit dynamical patterns. Chaos **20**, 037107 (2010)
43. C. Jin, J. Wang, W. Wang, C. Song, J. Wang, H. Xia, Q. Liu, Array of synchronized nano-oscillators based on repulsion between domain wall and skyrmion. Phys. Rev. Appl. **9**, 044007 (2018)
44. F. Garcia-Sanchez, J. Sampaio, N. Reyren, V. Cros, J.V. Kim, A skyrmion-based spin-torque nano-oscillator. New J. Phys. **18**, 075011 (2016)
45. S. Zhang, J. Wang, Q. Zheng, Q. Zhu, X. Liu, S. Chen, C. Jin, Q. Liu, C. Jia, D. Xue, Current-induced magnetic skyrmions oscillator. New J. Phys. **17**, 023061 (2015)
46. Y. Zhou, E. Iacocca, A.A. Awad, R.K. Dumas, F.C. Zhang, H.B. Braun, J. Akerman, Dynamically stabilized magnetic skyrmions. Nat. Commun. **6**, 8193 (2015)
47. M. Zahedinejad, A.A. Awad, S. Muralidhar, R. Khymyn, H. Fulara, H. Mazraati, M. Dvornik, J. Akerman, Two-dimensional mutually synchronized spin Hall nano-oscillator arrays for neuromorphic computing. Nat. Nanotechnol. **15**, 47 (2020)
48. Y. Feng, J. Xia, L. Qiu, X. Cai, L. Shen, F.J. Morvan, X. Zhang, Y. Zhou, G. Zhao, A skyrmion-based spin-torque nano-oscillator with enhanced edge. J. Magn. Magn. Mater. **491**, 165610 (2019)
49. J.H. Guo, J. Xia, X.C. Zhang, P.W.T. Pong, Y.M. Wu, H. Chen, W.S. Zhao, Y. Zhou, A ferromagnetic skyrmion-based nano-oscillator with modified profile of Dzyaloshinskii-Moriya interaction. J. Magn. Magn. Mater. **496**, 165912 (2020)
50. C. Jin, Y. Ma, C. Song, H. Xia, J. Wang, C. Zhang, Z. Zeng, J. Wang, Q. Liu, High-frequency spin transfer nano-oscillator based on the motion of skyrmions in an annular groove. New J. Phys. **22**, 033001 (2020)
51. L. Shen, J. Xia, G. Zhao, X. Zhang, M. Ezawa, O.A. Tretiakov, X. Liu, Y. Zhou, Spin torque nano-oscillators based on antiferromagnetic skyrmions. Appl. Phys. Lett. **114**, 042402 (2019)
52. X. Zhang, Y. Zhou, M. Ezawa, Magnetic bilayer-skyrmions without skyrmion Hall effect. Nat. Commun. **7**, 10293 (2016)
53. J. Xia, X. Zhang, M. Ezawa, Z. Hou, W. Wang, X. Liu, Y. Zhou, Current-driven dynamics of frustrated skyrmions in a synthetic antiferromagnetic bilayer. Phys. Rev. Appl. **11**, 044046 (2019)
54. S.S. Parkin, R. Bhadra, K.P. Roche, Oscillatory magnetic exchange coupling through thin copper layers. Phys. Rev. Lett. **66**, 2152 (1991)
55. J. Zang, M. Mostovoy, J.H. Han, N. Nagaosa, Dynamics of Skyrmion crystals in metallic thin films. Phys. Rev. Lett. **107**, 136804 (2011)

56. X. Zhang, M. Ezawa, Y. Zhou, Thermally stable magnetic skyrmions in multilayer synthetic antiferromagnetic racetracks. Phys. Rev. B **94**, 064406 (2016)
57. R. Tomasello, V. Puliafito, E. Martinez, A. Manchon, M. Ricci, M. Carpentieri, G. Finocchio, Performance of synthetic antiferromagnetic racetrack memory: domain wall versus skyrmion. J. Phys. D: Appl. Phys. **50**, 325302 (2017)
58. T. Dohi, S. DuttaGupta, S. Fukami, H. Ohno, Formation and current-induced motion of synthetic antiferromagnetic skyrmion bubbles. Nat. Commun. **10**, 5153 (2019)
59. W. Legrand, D. Maccariello, F. Ajejas, S. Collin, A. Vecchiola, K. Bouzehouane, N. Reyren, V. Cros, A. Fert, Room-temperature stabilization of antiferromagnetic skyrmions in synthetic antiferromagnets. Nat. Mater. **19**, 34 (2020)
60. C. Ma, X. Zhang, J. Xia, M. Ezawa, W. Jiang, T. Ono, S.N. Piramanayagam, A. Morisako, Y. Zhou, X. Liu, Electric field-induced creation and directional motion of domain walls and skyrmion bubbles. Nano Lett. **19**, 353 (2019)
61. X. Wang, W.L. Gan, J.C. Martinez, F.N. Tan, M.B.A. Jalil, W.S. Lew, Efficient skyrmion transport mediated by a voltage controlled magnetic anisotropy gradient. Nanoscale **10**, 733 (2018)
62. R. Tomasello, S. Komineas, G. Siracusano, M. Carpentieri, G. Finocchio, Chiral skyrmions in an anisotropy gradient. Phys. Rev. B **98**, 024421 (2018)
63. H. Xia, C. Song, C. Jin, J. Wang, J. Wang, Q. Liu, Skyrmion motion driven by the gradient of voltage-controlled magnetic anisotropy. J. Magn. Magn. Mater. **458**, 57 (2018)
64. L. Qiu, J. Xia, Y. Feng, L. Shen, F.J. Morvan, X. Zhang, X. Liu, L. Xie, Y. Zhou, G. Zhao, Dynamics of antiskyrmions induced by the voltage-controlled magnetic anisotropy gradient. J. Magn. Magn. Mater. **496**, 165922 (2020)
65. S.S. Parkin, M. Hayashi, L. Thomas, Magnetic domain-wall racetrack memory. Science **320**, 190 (2008)
66. J. Wang, J. Xia, X. Zhang, G.P. Zhao, L. Ye, J. Wu, Y. Xu, W. Zhao, Z. Zou, Y. Zhou, Controllable transport of a skyrmion in a ferromagnetic narrow channel with voltage-controlled magnetic anisotropy. J. Phys. D: Appl. Phys. **51**, 205002 (2018)
67. X. Liang, G. Zhao, L. Shen, J. Xia, L. Zhao, X. Zhang, Y. Zhou, Dynamics of an antiferromagnetic skyrmion in a racetrack with a defect. Phys. Rev. B **100**, 144439 (2019)
68. H. Kronmüller, Theory of nucleation fields in inhomogeneous ferromagnets. Phys. Stat. Sol. B **114**, 385 (1987)
69. S.-Z. Lin, C. Reichhardt, C.D. Batista, A. Saxena, Particle model for skyrmions in metallic chiral magnets: Dynamics, pinning, and creep. Phys. Rev. B **87**, 214419 (2013)
70. Y. H. Liu, Y. Q. Li, A mechanism to pin skyrmions in chiral magnets, J. Phys.: Condens. Matter **25**, 076005 (2013)
71. D. Stosic, T.B. Ludermir, M.V. Milošević, Pinning of magnetic skyrmions in a monolayer Co film on Pt(111): Theoretical characterization and exemplified utilization. Phys. Rev. B **96**, 214403 (2017)

Chapter 17
Axion Electrodynamics in Magnetoelectric Media

A. Martín-Ruiz, M. Cambiaso, and L. F. Urrutia

Abstract Topologically ordered media demand a new understanding of the emergent properties of quantum matter. This is a fundamental and technological feat. Topological insulators and Weyl semimetals are materials with topological order. Here we will focus on how these materials interact with sources of the electromagnetic field. We start from the effective field theory of Maxwell's electrodynamics extended by a so-called magnetoelectric term, namely axion electrodynamics and summarize some results we have found exploiting a Green's function approach to solve for the electromagnetic fields. Signals of the magnetolectric effect are minute compared with other electromagnetic responses, therefore precision measurements are required for its detection. Our formulation can be used for topological insulators and Weyl semimetals with planar, cylindrical and spherical geometries interacting with general charges, currents and boundary conditions. Our formulation is exemplified by: (i) the issue of Casimir effect involving a planar topological insulator, (ii) Vavilov-Cherenkov radiation produced in the forward- and backward-direction of a charged particle traversing a planar interface of two magnetoelectric media and (iii) the electromagnetic fields induced by a static electric charge near the surface of a Weyl semimetal. All three applications can yield observable signals that are within experimental sensitivities.

A. Martín-Ruiz · L. F. Urrutia (✉)
Instituto de Ciencias Nucleares, Universidad Nacional Autónoma de México, 04510 Ciudad de México, Mexico
e-mail: urrutia@nucleares.unam.mx

A. Martín-Ruiz
e-mail: alberto.martin@nucleares.unam.mx

M. Cambiaso
Universidad Andres Bello, Departamento de Ciencias Fisicas, Facultad de Ciencias Exactas, Avenida Republica 220, Santiago, Chile
e-mail: mcambiaso@unab.cl

E. Kamenetskii (ed.), *Chirality, Magnetism and Magnetoelectricity*,
Topics in Applied Physics 138,
https://doi.org/10.1007/978-3-030-62844-4_17

17.1 Introduction

Classical and quantum electrodynamics summarize all our understanding of the interaction between matter and radiation. Although these topics have been profusely studied in many different areas since their early discoveries, still today they constitute a fruitful research discipline and an excellent arena with potential for new discoveries. This is specially true when precision measurements are at hand and also when new materials come into play whose novel properties, of ultimate quantum origin, result in new possible forms of interaction between light and such materials. That is the case with topological insulators, as well as other materials with topological order.

Interestingly enough, the interaction between matter characterized by topological order and external electromagnetic fields can be described by an extension of Maxwell's theory. In fact, in electrodynamics there is the possibility of writing two quadratic gauge and Lorentz invariant terms: the first one is the usual electromagnetic density $\mathcal{L}_{\text{EM}} = (\mathbf{E}^2 - \mathbf{B}^2)/8\pi$ which yields Maxwell's equations, and the second one is the magnetoelectric term $\mathcal{L}_\theta = \theta\, \mathbf{E} \cdot \mathbf{B}$, where θ is a coupling field usually termed the axion angle.

Many of the interesting properties of the latter can be recognized from its covariant form $\mathcal{L}_\theta = -(\theta/8)\epsilon^{\mu\nu\rho\lambda} F_{\mu\nu} F_{\rho\lambda}$, where $\epsilon^{\mu\nu\rho\lambda}$ is the Levi-Civita symbol and $F_{\mu\nu} = \partial_\mu A_\nu - \partial_\nu A_\mu$ is the electromagnetic field strength written in terms of the connection A_μ. When θ is globally constant, the θ-term is a total derivative and has no effect on Maxwell's equations. This property, together with its invariance under continuous changes in the connection A_μ, qualify $\mathcal{P} = -(1/8)\epsilon^{\mu\nu\rho\lambda} F_{\mu\nu} F_{\rho\lambda}$ to be a topological invariant. Actually, $\mathcal{P}$ is the simplest example of a Pontryagin density [1], corresponding to the abelian group $U(1)$. This structure, together with its generalization to nonabelian groups, has been relevant in diverse topics in high-energy physics such as anomalies [2], the strong CP problem [3], topological field theories [4] and axions [5], for example.

It is interesting to recall that the coupling to a Pontryagin density has a long story in the development of Yang-Mills theories, with its origin dating back to a problem arising because the axial symmetry of the QCD Lagrangian, in the massless limit, would imply the existence of low mass nucleons with odd parity, which are not found in nature [6]. The alternative realization of this symmetry in the spontaneously broken mode is not possible either because the three lighter mesons in the spectrum (the pions) do not match the required four pseudo Goldstone bosons. The absence of a fourth pseudoscalar meson with a mass similar to that of the pions is what is known as the $U(1)_A$ problem. A solution was proposed in [7, 8] grounded in a more detailed study of the structure of the vacuum in a Yang-Mills theory, which introduces a superposition of topologically nonequivalent vacua as the true ground state of the system: the θ vacuum. This superposition eliminates the $U(1)_A$ problem but requires the CP violating coupling of the θ-vacuum to the non-abelian Pontryagin density in the QCD Lagrangian. Unfortunately, this addition generates a new problem. In particular, the resulting effective pion-nucleon interaction predicts an electric dipole moment for the neutron given by $d_n \sim 3.2 \times 10^{-16}\theta$ e cm. Comparison with the experimental value $(d_n)_{\rm exp} < 3 \times 10^{-26}$ e cm produces the naturalness problem: why should $\theta \sim 10^{-10}$ be so small. This is known as the strong CP problem. A solution was proposed in [9, 10] through the introduction of a new field, the axion $a(x)$, and a new symmetry $U(1)_{\rm PQ}$ realized again in the spontaneously broken mode. The vacuum expectation value of $a(x)$ can be chosen to cancel the previous CP violating term, but a new pseudo Goldstone boson must appear. Estimations of the coupling constant and mass of this new particle reveal extremely low values, thus making its detection very difficult. The abelian sector of the axion coupling in the QCD Lagrangian, $\mathcal{L}_{a\gamma\gamma} = -g_{a\gamma\gamma} a\, \mathbf{E} \cdot \mathbf{B}$, was proposed in [11] as a possibility to detect the axion through its coupling with strong electromagnetic fields available in the laboratory. The addition of this term to the usual electromagnetic Lagrangian, gives rise to what is now known as axion electrodynamics. Recently the axion is considered as a candidate to describe dark matter. Reference [5] contains a recent review of axion physics, including the efforts made towards its detection.

Another subject related to the θ-coupling is the photon sector of the Standard Model Extension (SME), designed to study possible violations of the Lorentz and CPT symmetries [12–14]. There one considers the effective couplings $(k_F)_{\kappa\lambda\mu\nu} F^{\kappa\lambda} F^{\mu\nu}$ and $(k_{AF})^{\kappa} \epsilon_{\kappa\lambda\mu\nu} A^{\lambda} F^{\mu\nu}$ where the constant tensors $(k_F)_{\kappa\lambda\mu\nu}$ and $(k_{AF})^{\kappa}$ define fixed directions in spacetime that break Lorentz symmetry. An alternative way to perform this breaking is by choosing spacetime dependent coupling parameters [15], like $(k_F)_{\kappa\lambda\mu\nu} = \theta(x)\epsilon_{\kappa\lambda\mu\nu}$ and $(k_{AF})_{\kappa} = \partial_{\kappa}\theta(x)$, for example. Both cases yield a version of the axion coupling described previously which can be used as effective theories with real Lorentz symmetry breaking in matter. The point of this comment is to observe that many of the techniques and of the numerous high-precision experiments proposed to test Lorentz symmetry breaking could adapt to the case of matter-photon interaction.

Recently, an important additional application of the Pontryagin extended electrodynamics has been highlighted in condensed matter physics, where a non-dynamical axion angle θ provides an effective field theory describing the electromagnetic response of some materials such as (i) magnetoelectric media [16, 17], (ii) metamaterials when θ is a purely complex function [18], (iii) topological insulators (TIs) when $\theta = (2n + 1)\pi$, with n integer [19–22] and (iv) Weyl semimetals (WSM) when $\theta(\mathbf{x}, t) = 2\mathbf{b} \cdot \mathbf{x} - 2b_0 t$, where $\mathbf{b}$ is the separation in momentum space between the Weyl nodes and b_0 is their separation in energy [23]. Lately, the study of topological insulating and Weyl semimetal phases, either from a theoretical or an experimental perspective has been actively pursued [22–24].

We will devote this chapter to the study of the interaction between magnetoelectric media and the electromagnetic field. To do so we will employ field theoretic techniques predicated on the aforementioned model Lagrangian of Maxwell's theory extended by the magnetoelectric term. The chapter is organized as follows. In Sect. 17.2 we present the general framework of electrodynamics in media characterized by a parameter θ (to be called a θ-medium) and review some of the most important features of magnetoelectric media regarding their electromagnetic properties. Section 17.3 contains a summary of our generalized Green's function method to construct the corresponding electromagnetic fields produced by charges, currents and boundary conditions in systems satisfying the following coordinate conditions: (i) the coordinates can be chosen such that the interface between two media, with different θ, is defined by setting to a constant only one of coordinates and (ii) the Laplacian is separable in such coordinates. At the end of this section the particularly simple case of planar symmetry is discussed. There the reader is referred to the analogous extensions to cylindrical and spherical coordinates. As a specific application of our methods to the case of a planar interface, the Casimir effect between two metallic plates with a topological insulator between them is considered in Sect. 17.4. In Sect. 17.5, we study the problem of Vavilov-Cherenkov radiation produced by an electric charge propagating perpendicularly to the planar interface between two different magnetoelectric media to find that, besides the usual forward-directed radiation, there is a backward-directed emission of radiation, a so-called reversed Vavilov-Cherenkov radiation (RVCR), solely due to the magnetoelectric nature of the media. As yet another application, in Sect. 17.6 we show how a static electric charge induces unexpected electromagnetic fields when placed near the surface of a Weyl semimetal and provide two experimental proposals, feasible for present-day sensitivities, that could be performed to measure the effects associated with the topological character of the WSM. Finally in Sect. 17.7 we present some concluding remarks, provide a summary of important results and highlight the benefits of our approach for solving the electromagnetic response of magnetoelectric media.

Our conventions are taken from [25], where $F_{\mu\nu} = \partial_\mu A_\nu - \partial_\nu A_\mu$, $\tilde{F}^{\mu\nu} = \epsilon^{\mu\nu\alpha\beta} F_{\alpha\beta}/2$, $F^{i0} = E^i$, $F^{ij} = -\epsilon^{ijk} B^k$ and $\tilde{F}^{i0} = B^i$, $\tilde{F}^{ij} = \epsilon^{ijk} E^k$. Also $\mathbf{V} = (V^i) = (V_x, V_y, V_z)$ for any vector $\mathbf{V}$. The metric is $(+, -, -, -)$ and $\epsilon^{0123} = +1 = \epsilon^{123}$.

17.2 Nondynamical Axion Electrodynamics

In this section we discuss the basic features arising from adding to Maxwell's electrodynamics the coupling of the abelian Pontryagin density to a pseudoscalar field $\theta(x)$, leading to a theory that we call θ-electrodynamics (θ-ED), retaining the name of axion-electrodynamics for the case where the axion field θ becomes dynamical. We call the parameter $\theta(x)$ the magnetoelectric polarizability (MEP) of the medium, which we consider in the same footing as its permittivity ε and permeability μ. The nature of the MEP depends on the type of magnetoelectric material under consideration and it is ultimately related to the magnetic symmetries of the substance [26, 27] and/or to the properties of its band structure [19–21]. It can be calculated from a Kubo-type response formula, once a microscopic model Hamiltonian for the material is adopted. The permittivity tensor ϵ is usually understood by the Drude-Lorentz type of single resonance oscillator model [28].

Magnetoelectric media [16, 17] are naturally existing materials like antiferromagnets [30], topological insulators (TIs) [19–22] and Weyl semimetals [23, 24], for example. Leaving aside the remarkable microscopic properties of different magnetoelectric media, we will concentrate on θ-ED as the effective macroscopic theory describing the electromagnetic response in the case of linear, isotropic and homogeneous magnetoelectrics. Let us start from the general formulation of electrodynamics in a material medium according to the Maxwell equations

$$\nabla \cdot \mathbf{D} = 4\pi\rho, \quad \nabla \cdot \mathbf{B} = 0, \quad \nabla \times \mathbf{E} + \frac{1}{c}\frac{\partial \mathbf{B}}{\partial t} = 0, \quad \nabla \times \mathbf{H} - \frac{1}{c}\frac{\partial \mathbf{D}}{\partial t} = \frac{4\pi}{c}\mathbf{J}, \tag{17.1}$$

together with the Lorentz force

$$\mathbf{F}_L = q\left(\mathbf{E} + \frac{\mathbf{v}}{c} \times \mathbf{B}\right). \tag{17.2}$$

The characterization of a specific media is given by the constitutive relations which define the displacement **D** and the magnetic field **H** in terms of the electric field **E** and the magnetic induction **B**, which are the fundamental fields that define the electromagnetic potentials according to the homogeneous equations in (17.1) [25].

The constitutive relations depend on the nature of the material and usually have the form $\mathbf{D} = \mathbf{D}(\mathbf{E}, \mathbf{B})$ and $\mathbf{H} = \mathbf{H}(\mathbf{E}, \mathbf{B})$. For example, in linear nonmagnetoelectric media they are $D_i = \varepsilon_{ij} E_j$ y $B_i = \mu_{ij} H_j$, where ε_{ij} is the permittivity and μ_{ij} is the permeability tensors, respectively, which can depend on the position and time. For isotropic and homogeneous materials $\varepsilon_{ij} = \varepsilon\delta_{ij}$ and $\mu_{ij} = \mu\delta_{ij}$, with ε and μ constants. In the case of magnetoelectrics we will consider media described by the following constitutive relations

$$\mathbf{D} = \varepsilon\mathbf{E} - \frac{\theta\alpha}{\pi}\mathbf{B}, \qquad \mathbf{H} = \frac{1}{\mu}\mathbf{B} + \frac{\theta\alpha}{\pi}\mathbf{E}, \tag{17.3}$$

where $\alpha = e^2/\hbar c \simeq 1/137$ is the fine-structure constant and the MEP θ is an additional parameter of the medium. The extension of these constitutive relations to the anisotropic case, for the optical properties of the material (ε and μ) is direct and for the MEP θ is also possible.

Substituting them in (17.1), we obtain the following modified non-homogeneous Maxwell equations

$$\nabla \cdot (\varepsilon\mathbf{E}) = 4\pi\rho + \frac{\alpha}{\pi}\nabla\theta \cdot \mathbf{B}, \quad \nabla \times (\mathbf{B}/\mu) - \frac{1}{c}\frac{\partial(\varepsilon\mathbf{E})}{\partial t} = \frac{4\pi}{c}\mathbf{J} - \frac{\alpha}{\pi}\nabla\theta \times \mathbf{E} - \frac{1}{c}\frac{\alpha}{\pi}\frac{\partial\theta}{\partial t}\mathbf{B}. \tag{17.4}$$

We observe that the above equations can be derived from the usual Maxwell action extended by the coupling of the abelian Pontryagin density $\mathcal{P}$ to a non-dynamical axion field $\theta(\mathbf{x}, t)$

$$S[\Phi, \mathbf{A}] = \int dt\, d^3\mathbf{x}\left[\frac{1}{8\pi}\left(\varepsilon\mathbf{E}^2 - \frac{1}{\mu}\mathbf{B}^2\right) - \frac{\alpha}{4\pi^2}\theta(\mathbf{x}, t)\,\mathbf{E} \cdot \mathbf{B} - \rho\Phi + \frac{1}{c}\mathbf{J} \cdot \mathbf{A}\right]. \tag{17.5}$$

The electromagnetic fields **E** and **B** are expressed in terms of the electromagnetic potentials Φ and **A** as

$$\mathbf{E} = -\frac{1}{c}\frac{\partial \mathbf{A}}{\partial t} - \nabla\Phi, \qquad \mathbf{B} = \nabla \times \mathbf{A}, \tag{17.6}$$

which solve the homogeneous equations in (17.1). An important consequence of (17.4) is the so-called magnetoelectric effect (MEE), summarized in the appearance of the following effective field dependent charge and current densities

$$\rho_\theta = \frac{\alpha}{4\pi^2}\nabla\theta \cdot \mathbf{B}, \qquad \mathbf{J}_\theta = -\frac{c\alpha}{4\pi^2}\nabla\theta \times \mathbf{E} - \frac{\alpha}{4\pi^2}\frac{\partial\theta}{\partial t}\mathbf{B}, \tag{17.7}$$

evidencing the ability of the magnetic (electric) fields to produce charge (current) densities, respectively. This effect is one of the most remarkable physical consequences of the additional θ coupling. It was predicted in [26] and subsequently observed in [30]. For an updated review of the MEE see, for example, [31]. A universal topological magnetoelectric effect has recently been measured in TIs [32]. Many additional consequences of the MEE have been highlighted using different approaches. For example, electric charges close to the interface between two θ-media induce image magnetic monopoles (and vice versa) [33–36]. Also, the propagation of electromagnetic waves across a θ-boundary have been studied finding that a non trivial Faraday rotation of the polarizations appears [29, 34, 35, 37]. The shifting of the spectral lines in hydrogen-like ions placed in front of a planar TI, as well as the modifications to the Casimir Polder potential in the non-retarded approximation were studied in [38]. The classical dynamics of a Rydberg hydrogen atom near a planar TI has also been investigated [39].

We observe that the dynamical modifications in (17.4) depend on the spatial and temporal gradients of the MEP, as required because the Pontryagin density $\mathcal{P}$ is a total derivative. In this way, the coupling with θ does not affect the equations of motion when the MEP is globally constant. The explicit dependence on θ, instead of $\partial_\mu\theta$, of the constitutive relations may erroneously induce the belief that a globally constant θ could produce dynamical effects. Nevertheless, from this perspective one would identify additional polarization $\mathbf{P}_\theta = -\sigma\mathbf{B}$ and magnetization $\mathbf{M}_\theta = -\sigma\mathbf{E}$ with $\sigma = \theta\alpha/4\pi^2$. When θ is constant, the calculation of the effective sources

$$\rho_{\text{eff}} = \nabla \cdot \mathbf{P}_\theta = \sigma\nabla \cdot \mathbf{B}, \qquad \mathbf{J}_{\text{eff}} = \frac{1}{c}\frac{\partial \mathbf{P}_\theta}{\partial t} + \nabla \times \mathbf{M}_\theta = -\sigma\Big(\frac{1}{c}\frac{\partial \mathbf{B}}{\partial t} + \nabla \times \mathbf{E}\Big), \tag{17.8}$$

yields identically zero due to the homogeneous Maxwell equations.

17.3 The Green Function Approach to the Electromagnetic Response of Linear Isotropic Homogeneous Magnetolectric Media

The knowledge of the Green function (GF) of an electromagnetic system allows the calculation of its response to arbitrary external sources, providing a definite starting point for the algebraic or numerical approximations which are required in most cases. This method supersedes the image approach, frequently used in the literature, which is appealing for interpreting results in terms of the superposition of images of charges and currents. Nevertheless, it requires a good amount of educated guesses which are far from obvious in many interesting cases. We will focus on calculating the GFs for the cases when materials with piecewise constant MEP's exhibit planar, cylindrical and spherical symmetries. Certainly one could solve for the electric and magnetic fields from the modified Maxwell equations together with the boundary conditions, however, just as in ordinary electrodynamics, there might be occasions where information about the sources is unknown and rather we are provided with data of the 4-potential at some given boundaries. In these cases, the GF method provides the general solution to such boundary-value problem (Dirichlet or Neumann) for arbitrary sources.

As an important class of magnetoelectric media with constant MEP we single out TIs, which serve to illustrate some of the phenomena discussed. Three dimensional TIs are a class of topological materials that can host conducting helical surface states each having the dispersion relation of a non-degenerate Dirac cone with a crossing point at/close to the Fermi level. Nevertheless, TIs behave as insulators in the bulk with a finite energy gap. The surface state is further topologically protected by time-reversal symmetry and/or inversion symmetry, coupled with spin-momentum locking properties. The latter means that the spin orientation of the electrons on the surface Dirac cone is always locked perpendicularly to their momentum. A distinguishing feature of 3D TIs among magnetoelectrics, is that the MEP θ is of topological nature and arises from the bulk band structure. It is given by a non-Abelian Berry flux over the Brillouin zone [19]. For 3D time reversal invariant insulators in a manifold without boundaries there are only two possibilities: $\theta = 0$ for normal insulators and $\theta = \pi$ for TIs. In order to continuously connect both classes of insulators, time reversal invariance should be broken at the interface. This can be achieved, for example, by depositing a thin magnetic coating of a few nanometers at the interface. According to the specific nature of the coating, the MEP of the TI will be modified to $\theta = \pi + 2n\pi$, where n is an integer [40].

From a macroscopic perspective we consider TIs as a class of magnetoelectric media described by θ-ED and characterized by the choice of a constant MEP θ. To illustrate the calculation of the GF it will be enough to consider the simplest case where the $(3+1)$ dimensional spacetime $\mathcal{M}$ can be split in such a way that $\mathcal{M} = \mathcal{U} \times \mathbb{R}$, where $\mathcal{U}$ is a three-dimensional manifold and $\mathbb{R}$ is the temporal axis. Moreover, the space $\mathcal{U}$ is partitioned in two region $\mathcal{U}_1$ and $\mathcal{U}_2$, such that $\mathcal{U}_1$ and $\mathcal{U}_2$ have a two-dimensional common interface Σ. In this way $\mathcal{U} = \mathcal{U}_1 \cup \mathcal{U}_2$ and

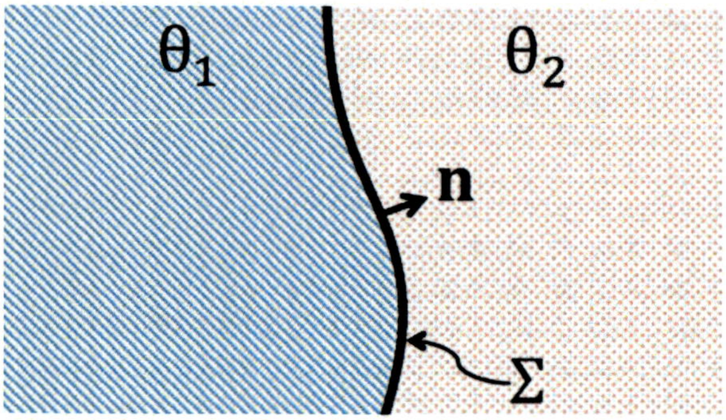

Fig. 17.1 Region over which the electromagnetic field theory is defined, (adapted from [43])

$\Sigma = \mathcal{U}_1 \cap \mathcal{U}_2$, as shown in the Fig. 17.1. Also we assume that the MEP θ is piecewise constant taking the values $\theta = \theta_1$ in the region $\mathcal{U}_1$ and $\theta = \theta_2$ in the region $\mathcal{U}_2$, which is expressed by the characteristic function

$$\theta(\mathbf{x}) = \begin{cases} \theta_1 & , \quad \mathbf{x} \in \mathcal{U}_1 \\ \theta_2 & , \quad \mathbf{x} \in \mathcal{U}_2 \end{cases} . \tag{17.9}$$

The interface Σ is parametrized by a function $F_\Sigma(\mathbf{x}) = 0$, which yields $n_\mu = (0, \mathbf{n}) = \partial_\mu F_\Sigma(\mathbf{x})$, as the normal vector to Σ which is external to the region $\mathcal{U}_1$. In this setup, the action S_θ, corresponding to the second term on the right-hand side of (17.5), is no longer a total derivative and the modified Maxwell equations (17.4) acquire field dependent effective charges and currents with support only at the interface (in the following we set $c = 1$)

$$\nabla \cdot (\epsilon \mathbf{E}) = \tilde{\theta}\delta\left(F_\Sigma(\mathbf{x})\right) \mathbf{B} \cdot \mathbf{n} + 4\pi\rho, \tag{17.10}$$

$$\nabla \times \left(\mathbf{B}/\mu\right) - \frac{\partial(\epsilon \mathbf{E})}{\partial t} = \tilde{\theta}\delta\left(F_\Sigma(\mathbf{x})\right) \mathbf{E} \times \mathbf{n} + 4\pi \mathbf{J}. \tag{17.11}$$

Here $\mathbf{n}$ is the vector normal to Σ external to the region $\mathcal{U}_1$ and $\tilde{\theta} = \alpha\left(\theta_1 - \theta_2\right)/\pi$. Equations (17.10)–(17.11) show that in the bulk regions $\mathcal{U}_{1,2}$ we recover the usual Maxwell equations. The MEE shows up again in (17.10)–(17.11) and the realization of such effect that can provide a way to measure the MEP of a medium, is one of the main goals in the research related to TIs.

In the following we restrict ourselves to contributions of free sources only outside the interface Σ, with no additional boundary conditions (BCs) besides those required at Σ. Assuming that the temporal derivatives of the fields are finite in the neighbourhood of the interface, the field equations (17.10) and (17.11) yield the following boundary conditions

$$\Delta \mathbf{E} \cdot \mathbf{n}\big|_\Sigma = \tilde{\theta}\mathbf{B} \cdot \mathbf{n}\big|_\Sigma, \quad \Delta \mathbf{B} \times \mathbf{n}\big|_\Sigma = -\tilde{\theta}\mathbf{E} \times \mathbf{n}\big|_\Sigma, \tag{17.12}$$

$$\Delta \mathbf{B} \cdot \mathbf{n}\big|_\Sigma = 0, \quad \Delta \mathbf{E} \times \mathbf{n}\big|_\Sigma = 0. \tag{17.13}$$

The notation $\Delta \mathbf{V}_i\big|_\Sigma$ stands for the discontinuity of the i component of the vector $\mathbf{V}$ through the interface Σ, while $\mathbf{V}_j\big|_\Sigma$ indicates the continuous value of the j component evaluated at Σ. The boundary conditions in (17.13) imply that the members of the right hand side in (17.12) are correctly defined, representing field dependent charge densities and surface currents respectively. Again, the magnetoelectric effect is manifest in the boundary conditions (17.12). In order to emphasize the effects of the topological coupling we consider the simplest media having $\theta_1 \neq \theta_2$, but with $\varepsilon = 1$ and $\mu = 1$.

At this stage it is convenient to go back to the four dimensional notation: $A^\mu = (\Phi, \mathbf{A})$, $F_{\mu\nu} = \partial_\mu A_\nu - \partial_\nu A_\mu$, $\tilde{F}^{\mu\nu} = \frac{1}{2}\epsilon^{\mu\nu\alpha\beta}F_{\alpha\beta}$, $j^\mu = (\rho, \mathbf{J})$. In this way, the
inhomogeneous Maxwell equations (17.10) and (17.11) are

$$\partial_\mu F^{\mu\nu} = \tilde{\theta}\delta\left(F_\Sigma(\mathbf{x})\right) n_\mu \tilde{F}^{\mu\nu} + 4\pi j^\nu. \tag{17.14}$$

In the Lorenz gauge $\partial_\mu A^\mu = 0$, the 4-potential satisfies the equation of motion

$$\left[\eta^\mu{}_\nu \partial^2 - \tilde{\theta}\delta\left(F_\Sigma(\mathbf{x})\right) n_\rho \epsilon^{\rho\mu\alpha}{}_\nu \partial_\alpha\right] A^\nu = 4\pi j^\mu, \qquad \partial^2 = \partial_t^2 - \nabla^2 \tag{17.15}$$

together with the boundary conditions

$$\Delta A^\mu\big|_\Sigma = 0, \qquad \Delta\left(\partial_z A^\mu\right)\big|_\Sigma = -\tilde{\theta}\epsilon^{3\mu\alpha}{}_\nu \left(\partial_\alpha A^\nu\right)\big|_\Sigma, \tag{17.16}$$

which reproduce those written in (17.12)–(17.13) for the electric and magnetic fields.

To obtain a general solution for the potential A^μ in the presence of arbitrary external sources $j^\mu(x)$, we introduce the GF $G^\nu{}_\sigma\left(x, x'\right)$ solving (17.15) for a point-like source,

$$\left[\eta^\mu{}_\nu \partial^2 - \tilde{\theta}\delta\left(F_\Sigma(\mathbf{x})\right) n_\rho \epsilon^{\rho\mu\alpha}{}_\nu \partial_\alpha\right] G^\nu{}_\sigma\left(x, x'\right) = 4\pi \eta^\mu_\sigma \delta^4\left(x - x'\right), \tag{17.17}$$

together with the boundary conditions derived from (17.16), in such a way that the solution for the 4-potential in the Lorenz gauge is

$$A^\mu(x) = \int d^4x'\, G^\mu{}_\nu\left(x, x'\right) j^\nu\left(x'\right), \tag{17.18}$$

up to homogeneous contributions.

As we will show in the following, a further simplification arises when the system satisfies the following two coordinate conditions:(i) the coordinate system can be chosen so that the interface Σ is defined by setting constant only one of them and (ii) the Laplacian is separable in these coordinates in such a way that a complete orthonormal set of eigenfunctions can be defined in the subspace orthogonal to the coordinate defining the interface. Three cases show up immediately: (i) a planar interface at fixed z, (ii) a spherical interface at constant r and (iii) a cylindrical interface at constant ρ. In all this cases the characteristic function $\theta(\mathbf{x})$ defined in (17.9) can be written in terms of the Heaviside function H of one coordinate like $H(z-a)$, $H(r-a)$ and $H(\rho-a)$, respectively, with the associated unit vectors $\hat{\mathbf{n}}$ given by $\hat{\mathbf{k}}$, $\hat{\mathbf{r}}$ and $\hat{\boldsymbol{\rho}}$, in each of the adapted coordinate systems. Then (17.17) reduces to

$$\left[\eta^{\mu}{}_{\nu}\partial^2 - \tilde{\theta}\delta(\xi-\xi_0)\epsilon^{\xi\mu\alpha}{}_{\nu}\partial_{\alpha}\right]G^{\nu}{}_{\sigma}(x,x') = 4\pi\eta^{\mu}{}_{\sigma}\delta^4(x-x'), \tag{17.19}$$

where ξ denotes the coordinate defining the interface at $\xi=\xi_0$ and the coupling of the θ-term is given by a one dimensional delta function with support only at ξ_0. Also, the unit vector $\hat{\mathbf{n}}$ will have a component only in the direction ξ.

Let us illustrate the procedure sketched above by taking the static case of a planar interface located at $z=a$, separating two semi-infinite TIs, such that the MEP is $\theta(z)=\theta_1 H(a-z)+\theta_2 H(z-a)$. In the Coulomb gauge the GF $G^{\nu}{}_{\sigma}(\mathbf{x},\mathbf{x}')$ satisfies

$$\left[-\eta^{\mu}{}_{\nu}\nabla^2 - \tilde{\theta}\delta(z-a))\epsilon^{3\mu\alpha}{}_{\nu}\partial_{\alpha}\right]G^{\nu}{}_{\sigma}(\mathbf{x},\mathbf{x}') = 4\pi\eta^{\mu}_{\sigma}\delta^3(\mathbf{x}-\mathbf{x}'), \tag{17.20}$$

together with the boundary conditions (17.16). The coordinates are separated according to $\xi=z$, $\xi_0=a$, plus the two remaining x and y defining the plane parallel to the interface (i.e., perpendicular to the z-axis). The separation of the Laplacian is direct

$$\nabla^2 = \frac{\partial^2}{\partial z^2} + \nabla^2_{\parallel}, \qquad \nabla^2_{\parallel} = \frac{\partial^2}{\partial x^2} + \frac{\partial^2}{\partial y^2}, \tag{17.21}$$

with the operator $\nabla^2_{\parallel}$ having the eigenfunctions

$$\Psi_{\mathbf{p}_{\parallel}}(x,y) = \frac{1}{2\pi}e^{i\mathbf{p}_{\parallel}\cdot\mathbf{x}_{\parallel}}, \tag{17.22}$$

labelled by the momentum $\mathbf{p}_{\parallel}=(p_x,p_y)$ parallel to Σ and where $\mathbf{x}_{\parallel}=(x,y)$. Let us emphasize that the subindex $\parallel$ denotes objects living in the $x-y$ plane, parallel to the interface $z=a$. The eigenfunctions $\Psi_{\mathbf{p}_{\parallel}}(x,y)$ are a complete and orthonormal set in the $x-y$ plane, satisfying [41]

$$\int dxdy\,\Psi^*_{\mathbf{p}_{\parallel}}(x,y)\Psi_{\mathbf{p}'_{\parallel}}(x,y) = \delta^2_{\mathbf{p}_{\parallel},\mathbf{p}'_{\parallel}}, \qquad \int d^2\mathbf{p}_{\parallel}\Psi_{\mathbf{p}_{\parallel}}(x,y)\Psi^*_{\mathbf{p}_{\parallel}}(x,y) = \delta^2(\mathbf{x}_{\parallel}-\mathbf{x}'_{\parallel}). \tag{17.23}$$

We recall that in the full 3D-space we have $d^3\mathbf{x}=d\mathbf{x}_{\parallel}\,dz$ and $\delta^3(\mathbf{x}-\mathbf{x}')=\delta^2(\mathbf{x}_{\parallel}-\mathbf{x}'_{\parallel})\delta(z-z')$. Invariance under translation in the $x-y$ plane together with the the properties in (17.23) allow us to simplify (17.20) introducing the reduced GF $g^{\mu}{}_{\nu}(z,z',\mathbf{p}_{\parallel})$, such that

$$G^{\mu}{}_{\nu}(\mathbf{x},\mathbf{x}') = 4\pi\int\frac{d^2\mathbf{p}_{\parallel}}{(2\pi)^2}e^{i\mathbf{p}_{\parallel}\cdot(\mathbf{x}-\mathbf{x}')_{\parallel}}g^{\mu}{}_{\nu}(z,z',\mathbf{p}_{\parallel}). \tag{17.24}$$

For future use we denote $p^{\alpha} = (0, p_x, p_y, 0) = (0, \mathbf{p}_{\parallel})$. Thus, the final representation of the GF in (17.20) is given in terms of the Fourier transform of the reduced GF in the directions x, y parallel to the plane Σ [43]. Due to the antisymmetry of the Levi-Civita symbol, the partial derivative that appears in the second term of (17.20) for the GF does not introduce derivatives with respect to z, but only in the parallel directions. This allows us to write the equation of the reduced GF as

$$\left[\eta^{\mu}{}_{\nu}\tilde{\nabla}^2 + i\tilde{\theta}\delta\left(z-a\right)\epsilon^{3\mu\alpha}{}_{\nu}p_{\alpha}\right]g^{\nu}{}_{\sigma}\left(z, z', \mathbf{p}_{\parallel},\right) = \eta^{\mu}{}_{\sigma}\delta\left(z-z'\right), \tag{17.25}$$

with $\tilde{\nabla}^2 = \mathbf{p}_{\parallel}^2 - \partial_z^2$, $p^{\alpha}p_{\alpha} = -\mathbf{p}_{\parallel}{}^2$ and $|\mathbf{p}_{\parallel}| = p$. In this way we transform the calculation of the reduced GF into a one-dimensional problem with a delta interaction. Moreover, equation (17.25) can be integrated with the knowledge of an additional reduced GF $\mathfrak{g}^{\mu}{}_{\sigma}\left(z, z', \mathbf{p}_{\parallel}\right) = \eta^{\mu}{}_{\sigma}\mathfrak{g}(z, z', \mathbf{p}_{\parallel})$, satisfying

$$\left(\mathbf{p}_{\parallel}^2 - \frac{\partial^2}{\partial z^2}\right)\mathfrak{g}(z, z', \mathbf{p}_{\parallel}) = \delta\left(z-z'\right), \tag{17.26}$$

together with the required boundary conditions. This auxiliary GF results from the limit $\tilde{\theta} = 0$ in (17.25) and we call it the free reduced GF of the problem, emphasizing that arises when the θ-media are absent. These GFs can be taken directly from the vast literature on electrodynamics and constitute the basis for finding the electromagnetic response of a system with the same symmetries, but now in the presence of a θ-medium whose interface defines the corresponding coordinate system. In the planar case of interest and with the usual BCs at infinity, the option is to take [42]

$$\mathfrak{g}(z, z') = \frac{1}{2p}e^{-p|z-z'|}. \tag{17.27}$$

As a first step in the solution of the (17.25) we obtain the integral equation

$$g^{\mu}{}_{\sigma}\left(z, z'\right) = \eta^{\mu}{}_{\sigma}\mathfrak{g}\left(z, z'\right) - i\tilde{\theta}\epsilon^{3\mu\alpha}{}_{\nu}p_{\alpha}\int dz''\mathfrak{g}\left(z, z''\right)\delta(z''-a)g^{\nu}{}_{\sigma}\left(z'', z'\right), \tag{17.28}$$

where the integration over z'' can be readily performed, reducing the problem to a set of coupled algebraic equations. The solution to (17.28) is obtained as the result of the following steps. First we split the index μ into $\mu = 0$ and $\mu = j = 1, 2, 3$ obtaining

$$g^{0}{}_{\sigma}\left(z, z'\right) = \eta^{0}{}_{\sigma}\mathfrak{g}\left(z, z'\right) - i\tilde{\theta}\epsilon^{30i}{}_{j}p_i\mathfrak{g}\left(z, a\right)g^{j}{}_{\sigma}\left(a, z'\right), \tag{17.29}$$

$$g^{j}{}_{\sigma}\left(z, z'\right) = \eta^{j}{}_{\sigma}\mathfrak{g}\left(z, z'\right) - i\tilde{\theta}\epsilon^{3ji}{}_{0}p_i\mathfrak{g}\left(z, a\right)g^{0}{}_{\sigma}\left(a, z'\right). \tag{17.30}$$

Next we evaluate (17.30) at $z = a$ and substitute $g^{j}{}_{\sigma}\left(a, z'\right)$ in (17.29), yielding

$$g^0{}_\sigma\left(z,z'\right)=\eta^0{}_\sigma\mathfrak{g}\left(z,z'\right)-i\tilde{\theta}\epsilon^{30i}{}_j p_i\eta^j{}_\sigma\mathfrak{g}\left(z,a\right)\mathfrak{g}\left(a,z'\right)-\tilde{\theta}^2p^2\mathfrak{g}\left(z,a\right)\mathfrak{g}\left(a,a\right)g^0{}_\sigma\left(a,z'\right). \tag{17.31}$$

Setting $z=a$ in the above equation, we solve for $g^0{}_\sigma\left(a,z'\right)$ which we substitute back to obtain $g^0{}_\sigma\left(z,z'\right)$ in terms of the free GF $\mathfrak{g}\left(z,z'\right)$. The remaining components are obtained after substituting $g^0{}_\sigma\left(a,z'\right)$ in (17.30). The final result is

$$g^\mu{}_\nu\left(z,z'\right)=\eta^\mu{}_\nu\mathfrak{g}\left(z,z'\right)+A\left(z,z'\right)\left\{\tilde{\theta}\mathfrak{g}\left(a,a\right)\left[p^\mu p_\nu+\left(\eta^\mu{}_\nu+n^\mu n_\nu\right)p^2\right]+i\epsilon^\mu{}_\nu{}^{\alpha 3}p_\alpha\right\}, \tag{17.32}$$

where $n_\mu=(0,0,0,1)$ and

$$A\left(z,z'\right)=-\tilde{\theta}\frac{\mathfrak{g}\left(z,a\right)\mathfrak{g}\left(a,z'\right)}{1+p^2\tilde{\theta}^2\mathfrak{g}^2\left(a,a\right)}. \tag{17.33}$$

Extending the above results to the case where one of the media has arbitrary ε, keeping still nonmagnetic materials, we obtain a new version of the (17.32). Going back to the coordinate representation by calculating the Fourier transform in (17.24) we obtain

$$G^0{}_0(\mathbf{x},\mathbf{x}')=\frac{1}{\varepsilon(z')}\left[\frac{1}{|\mathbf{x}-\mathbf{x}'|}-\frac{\operatorname{sgn}(z')\,2(\varepsilon-1)+\tilde{\theta}^2}{2(\varepsilon+1)+\tilde{\theta}^2}\frac{1}{\sqrt{R^2+Z^2}}\right], \tag{17.34}$$

$$G^i{}_0(\mathbf{x},\mathbf{x}')=G^0{}_i(\mathbf{x},\mathbf{x}')=-\frac{2\tilde{\theta}}{2(\varepsilon+1)+\tilde{\theta}^2}\frac{\epsilon^{0ij3}R_j}{R^2}\left(1-\frac{Z}{\sqrt{R^2+Z^2}}\right), \tag{17.35}$$

$$G^i{}_j(\mathbf{x},\mathbf{x}')=\frac{\eta^i{}_j}{|\mathbf{x}-\mathbf{x}'|}-\frac{\tilde{\theta}^2}{2(\varepsilon+1)+\tilde{\theta}^2}\frac{\Lambda^i{}_j}{\sqrt{R^2+Z^2}}+\frac{\tilde{\theta}^2}{2(\varepsilon+1)+\tilde{\theta}^2}\Lambda^r{}_j\partial_r K^i(\mathbf{x},\mathbf{x}'). \tag{17.36}$$

where $Z=|z-a|+|z'-a|$, $\mathbf{R}=(x-x',y-y')$, $R=|\mathbf{R}|$, $\Lambda^i{}_j=\eta^i{}_j+n^in_j$, and

$$K^i\left(\mathbf{x},\mathbf{x}'\right)=\frac{\sqrt{R^2+Z^2}-Z}{R^2}R^i. \tag{17.37}$$

We observe that the substitution of the free reduced GF (17.27) in the expression (17.33) yields that the θ-dependent contribution to the full reduced GF is a function of Z, instead of $|z-z'|$ as it is in the $\tilde{\theta}=0$ contribution. This property is clearly manifest in the expressions written above for the GF components in coordinate space and will have interesting consequences in the case of radiation to be discussed in Sect. 17.5.

The details of the calculations already summarized are presented in [43, 44]. The results for the static GF with spherical and cylindrical symmetries have been extended to the case of ponderable magnetoelectric media with piecewise constant ε and μ in [45, 46]. The MEE has also been explored in the static case by locating a conducting sphere at constant potential in front of a planar TI [47]. Again, we mention that the extension of the above to the case of anisotropic optical properties is possible, and so is the case for anisotropic MEP. Concretely, the antiferromagnet Cr_2O_3 possess a $\theta_{ij} \sim 1.26 \times 10^{-3}$ (in non-rationalized CGS units), which is however small compared to non magnetoelectric effects. Larger MEE signals can be found in the $TbPO_4$, whose coupling is ~ 0.22 (in non-rationalized CGS units) and the search for giant ME couplings continues mainly in composite multiferroics [48]. An early review of numerous works on the study of infinite linear homogeneous bianisotropic media can be found in [49], which deals mainly with the calculation of the Green functions and the plane-wave propagation in various classes of these media. An alternative way of taking into account magnetoelectric effects in such media would be to introduce interfaces among them. This certainly requires an extension of the methods previously developed. Our GF approach precisely facilitates such extension and this has been one of our motivations.

17.4 The Casimir Effect

The Casimir effect (CE) is one of the most remarkable consequences predicted by quantum field theory as a result of the nonzero energy of the vacuum [50] and has already been experimentally confirmed [51]. For a review of the subject see [52, 53]. Experimental access to probe distances of the order of microns, together with the recent discovery of three-dimensional topological insulators provide additional ground to study the CE [54].

The CE we consider is produced between two perfectly conducting flat surfaces (plates) in vacuum, denoted by P_1 and P_2 and separated by a distance L. Inside this device is placed a planar TI rigidly attached to the surface P_2, as shown in Fig. 17.2. The Σ interface of the TI, located at $z = a < L$, is covered by a very thin magnetized layer. In order to explore the purely topological contribution of the MEP ($\tilde{\theta} \neq 0$) we

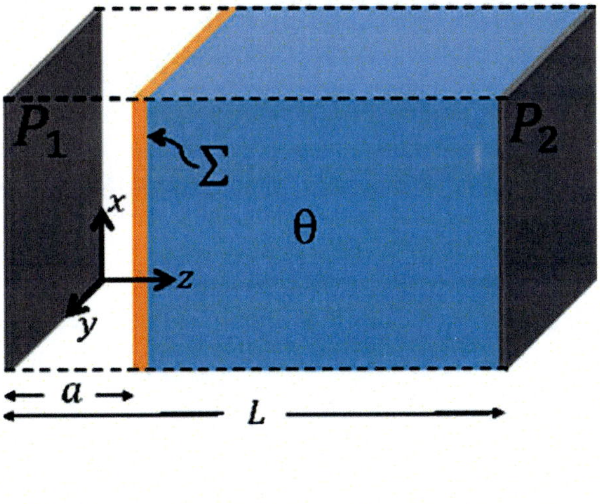

Fig. 17.2 Schematic of the Casimir effect in θ-ED, (adapted from [57])

take both media with $\varepsilon = \mu = 1$, however, non trivial optical properties as well as anisotropy effects can also be considered.

We follow a method similar to that of [42, 55] that begins with the calculation of the corresponding GF, from which the renormalized energy-momentum tensor in the region between the plates is obtained. A subsequent calculation yields the Casimir stress in the interface Σ of the TI. We also consider the limit when the plate P_2 goes back to infinity ($L \to \infty$) to obtain the Casimir stress produced by a single conducting plate in front of a semi-infinite TI. The boundary conditions on the plates P_1 and P_2 are the usual ones for a perfect conductor: $n_\mu \tilde{F}^{\mu\nu}|_{P_{1,2}} = 0$, where $n = (0, 0, 0, 1)$.

This calculation highlights two extensions of our method: (i) the inclusion of time-dependence and (ii) the use of a convenient well-known free reduced GF to obtain the full result. In the absence of free sources at the interface, the GF

$$G^{\mu}{}_{\nu}\left(x, x'\right) = 4\pi \int \frac{d^2\mathbf{p}_{\parallel}}{(2\pi)^2} e^{i\mathbf{p}_{\parallel}\cdot\mathbf{x}_{\parallel}} \int \frac{d\omega}{2\pi} e^{-i\omega(t-t')} g^{\mu}{}_{\nu}\left(z, z'\right), \tag{17.38}$$

satisfies the (17.19), together with the BC of (17.16). From here on we do not write the dependence of the reduced GF $g^{\mu}{}_{\nu}$ on ω and $\mathbf{p}_{\parallel}$. In the Lorenz gauge, the equation for $g^{\mu}{}_{\nu}$ is

$$\left[\eta^{\mu}{}_{\nu}\partial^2 + i\tilde{\theta}\delta\left(z-a\right)\epsilon^{3\mu\alpha}{}_{\nu}p_{\alpha}\right] g^{\nu}{}_{\sigma}\left(z, z'\right) = \eta^{\mu}{}_{\sigma}\delta\left(z - z'\right), \tag{17.39}$$

where now we have $\partial^2 = \mathbf{p}_{\parallel}{}^2 - \omega^2 - \partial_z^2$ and $p^{\alpha} = \left(\omega, \mathbf{p}_{\parallel}, 0\right)$.

The boundary contribution in $z = L$, which is not present in (17.39), is identically zero in the distributional sense due to the BCs on the conducting plate P_2. The next step, which illustrates the flexibility of the method developed in the Sect. 17.3, is to choose a free GF to integrate (17.39). It is clear that we must use the free reduced GF corresponding to two perfectly conducting parallel plates located at $z = 0$ and $z = L$, which is given by [42]

$$\mathfrak{g}_C\left(z, z'\right) = \frac{\sin\left[pz_<\right]\sin\left[p\left(L - z_>\right)\right]}{p\sin\left[pL\right]}, \tag{17.40}$$

where the suffix C is for Casimir, $z_>$ ($z_<$) is the greater (smaller) value between the coordinates z, z', and $p = \sqrt{\omega^2 - \mathbf{p}^2}$. Writting the solution of (17.39) as

$$g^{\mu}{}_{\nu}\left(z, z'\right) = \eta^{\mu}{}_{\nu}\mathfrak{g}_C\left(z, z'\right) + g^{\mu}_{C\nu}\left(z, z'\right). \tag{17.41}$$

we realize that the first term corresponds to the absence of the TI between the plates, while the second term, to be called the reduced θ-GF,

$$g^{\mu}_{C\nu}\left(z,z'\right)=\tilde{\theta}\mathfrak{g}_C\left(a,a\right)\left[p^{\mu}p_{\nu}-\left(\eta^{\mu}{}_{\nu}+n^{\mu}n_{\nu}\right)p^2\right]A_C\left(z,z'\right)+i\,\epsilon^{\mu}{}_{\nu}{}^{\alpha 3}p_{\alpha}A_C\left(z,z'\right),$$

$$A_C\left(z,z'\right)=-\tilde{\theta}\frac{\mathfrak{g}_C\left(z,a\right)\mathfrak{g}_C\left(a,z'\right)}{1-p^2\tilde{\theta}^2\mathfrak{g}_C^2\left(a,a\right)},\tag{17.42}$$

will be responsible for the effect we are looking for. The partition in (17.41) yields the full GF as $G^{\mu}{}_{\nu}\left(x,x'\right)=\eta^{\mu}{}_{\nu}\mathcal{G}\left(x,x'\right)+G^{\mu}_{C\nu}\left(x,x'\right)$, where each term arises from the respective contribution in (17.41). Since the MEP modifies the behavior of the fields only in the interface, the energy-momentum tensor (EMT) $T^{\mu\nu}$ in the bulk retains the original expression of the unmodified Maxwell equations, failing to be conserved only in the interface in the absence of free external sources [45]. Next we determine the vacuum expectation value (VEV) of $T^{\mu\nu}$, to be called the vacuum stress (VE), according to the basic relation $G^{\mu\nu}(x,x)=-i\langle 0|\mathcal{T}A^{\mu}(x)A^{\nu}(x')|0\rangle$, where $\mathcal{T}$ denotes time ordering. Following the standard point-splitting method we obtain

$$\langle T^{\mu\nu}\rangle=\frac{i}{4\pi}\lim_{x\to x'}\Big[-\partial^{\mu}\partial'^{\nu}G^{\lambda}{}_{\lambda}+\partial^{\mu}\partial'_{\lambda}G^{\lambda\nu}+\partial^{\lambda}\partial'^{\nu}G^{\mu}{}_{\lambda}$$
$$-\partial'^{\lambda}\partial_{\lambda}G^{\mu\nu}+\frac{1}{2}\eta^{\mu\nu}\left(\partial^{\alpha}\partial'_{\alpha}G^{\lambda}{}_{\lambda}-\partial^{\alpha}\partial'_{\beta}G^{\beta}{}_{\alpha}\right)\Big].\tag{17.43}$$

Again, we can write $\langle T^{\mu\nu}\rangle=\langle t^{\mu\nu}\rangle+\left\langle T^{\mu\nu}_C\right\rangle$. The first term is the contribution to the VE in the absence of the TI, yielding the well-known result in [55], which is independent of a. After some calculation, the second term $\left\langle T^{\mu\nu}_C\right\rangle$, to be called the θ-vacuum stress (θ-VS), is

$$\left\langle T^{\mu\nu}_C\right\rangle=i\tilde{\theta}\int\frac{d^2\mathbf{p}_{\parallel}}{(2\pi)^2}\int\frac{d\omega}{2\pi}\left(p^{\mu}p^{\nu}+n^{\mu}n^{\nu}p^2\right)\mathfrak{g}_C\left(a,a\right)\lim_{z\to z'}\left(p^2+\partial'_z\partial_z\right)A_C\left(z,z'\right).\tag{17.44}$$

Next we require the renormalized VS: $\langle T^{\mu\nu}\rangle_{\text{ren}}=\langle T^{\mu\nu}\rangle-\langle T_C{}^{\mu\nu}\rangle_{\text{vac}}$, where the first (second) term is the VS in the presence (absence) of conducting plates [55, 56]. We obtain

$$\left\langle T^{\mu\nu}_C\right\rangle_{\text{ren}}=-\frac{\pi^2}{720L^4}\left(\eta^{\mu\nu}+4n^{\mu}n^{\nu}\right)\left[u(\theta,\chi)H\left(a-z\right)+u(\theta,1-\chi)H\left(z-a\right)\right],\tag{17.45}$$

$$u(\theta,\chi)=\frac{120}{\pi^4}\int_0^{\infty}\frac{\tilde{\theta}^2\xi^3\text{sh}\left[\xi\chi\right]\text{sh}^3\left[\xi\left(1-\chi\right)\right]\text{sh}^{-3}\left[\xi\right]}{1+\tilde{\theta}^2\text{sh}^2\left[\xi\chi\right]\text{sh}^2\left[\xi\left(1-\chi\right)\right]\text{sh}^{-2}\left[\xi\right]}d\xi.\tag{17.46}$$

The notation is $\text{sh}(x)=\sinh(x)$ and $\chi=a/L$ with $0<\chi<1$. Our expression (17.45) has the same tensorial structure as that in [55], except for a z dependence because the EMT is not conserved at the interface, implying that the renormalized θ-VS is constant in each bulk but has a discontinuity in $z=a$ consistent with $\partial_z\left\langle T^{00}_C\right\rangle_{\text{ren}}\propto\delta(\Sigma)$.

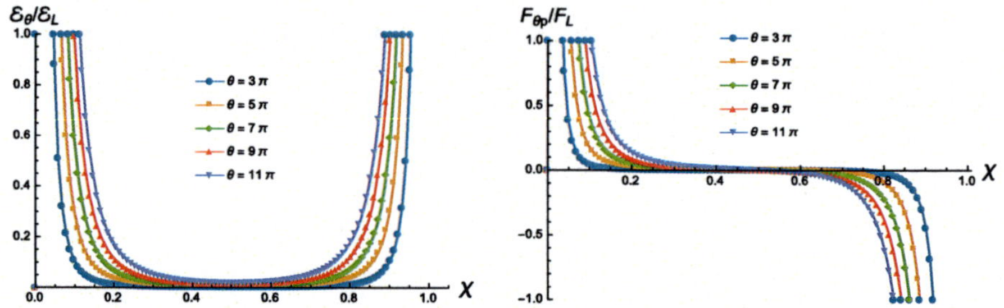

Fig. 17.3 Left panel: the ratio $\mathcal{E}_\theta/\mathcal{E}_L$. Right panel: the Casimir pressure at the interface in units of F_L. Both plots are in terms of the the dimensional parameter $\chi = a/L$ and for different values of θ, (adapted from [57])

The Casimir energy $\mathcal{E} = \mathcal{E}_L + \mathcal{E}_\theta$ is defined as the energy per unit area stored by the electromagnetic field in the region between the conducting plates. Let us recall that $\mathcal{E}_L = -\pi^2/(720L^3)$ is the Casimir energy in the absence of the TI. The relevant contribution is

$$\mathcal{E}_\theta = \int_0^L dz \langle T_C^{00} \rangle_{\text{ren}} = \mathcal{E}_L \left[\chi u(\theta, \chi) + (1-\chi) u(\theta, 1-\chi) \right]. \tag{17.47}$$

The first term is the energy stored between the plate P_1 and the interface, while the second term corresponds to the energy stored in the bulk of the TI. The Casimir pressure $F_{\theta p}$ on the interface is $F_{\theta p} = -d\mathcal{E}_\theta/da$, yielding

$$\frac{F_{\theta p}}{F_L} = -\frac{1}{3}\frac{d}{d\chi} \left[\chi u(\theta, \chi) + (1-\chi) u(\theta, 1-\chi) \right], \qquad F_L = -\pi^2/(240L^4). \tag{17.48}$$

The ratios $\mathcal{E}_\theta/\mathcal{E}_L$ (left panel) and $F_{\theta p}/F_L$ (right panel) as a function of χ, for different values of θ are plotted in Fig. 17.3 [57]. Our setup is a 3D analogue of the Casimir piston [58], and we obtain the similar result that the Casimir pressure tends to pull the interface towards the closest conducting plate, as shown in the right panel of the Fig. 17.3. When the plate P_2 recedes to infinity, $L \to \infty$, our setup describes the Casimir interaction between a perfect conducting plate P_1 at a distance a from a semi infinite TI. The Casimir energy (17.47) is now $\mathcal{E}_\theta^{L\to\infty} = \mathcal{E}_a R(\theta)$, with $\mathcal{E}_a = -\pi^2/720a^3$. The function

$$R(\theta) = \frac{120}{\pi^4} \int_0^\infty \xi^3 \frac{\tilde{\theta}^2}{1 + \tilde{\theta}^2 e^{-2\xi} \sinh^2 \xi} e^{-3\xi} \sinh \xi \, d\xi, \tag{17.49}$$

turns out to be independent of a and it is bounded by its limit $\theta \to \pm\infty$. In this way

Table 17.1 The normalized pressure $f_\theta = F_\theta^{L\to\infty}/F_a = R(\theta)$ for different values of θ

θ	$\pm 7\pi$	$\pm 15\pi$	$\pm 23\pi$	$\pm 31\pi$	$\pm 39\pi$
f_θ	0.0005	0.0025	0.0060	0.0109	0.0172

$$R(\theta) \leq \frac{120}{\pi^4} \int_0^\infty \xi^3 \frac{e^{-\xi}}{\sinh \xi} d\xi = 1. \tag{17.50}$$

Physically, this means that in the limit $\theta \to \infty$, the interface Σ behaves as a conducting plate. This is analogous to the Schwinger prescription that a perfect conductor is obtained in the limit $\varepsilon \to \infty$ of a dielectric material [42]. In the limit $L \to \infty$ the Casimir pressure is $f_\theta = F_\theta^{L\to\infty}/F_a = R(\theta)$, with $F_a = -\pi^2/240a^4$. In Table 17.1 we show some numerical results for f_θ for different values of θ.

The θ dependence of the Casimir pressure could be used to determine the corresponding MEP. This pressure has been measured for separation distances L between the metal plates in the range of 0.5–3.0 µm [51], which will require to prepare TIs with widths smaller than these amounts. In addition, the ratios for f_θ indicated in Table 17.1 allow us to estimate that an increase of several orders of magnitude would be required in the experimental accuracy. The particular values of $\theta = \pm 7\pi, \pm 15\pi$ are suitable for a TI such as $Bi_{1-x}Se_x$ [59]. The CE could be explored in TIs with higher values of $\tilde{\theta} = (2n+1)\pi$, however these material features magnetoelectric couplings that are not considered in our model based on θ-ED [60, 61].

As discussed above, the nondynamical axion electrodynamics can be seen as a particular realization of the photon sector of the Standard Model Extension with the identification $(k_F)_{\kappa\lambda\mu\nu} = \theta(x)\epsilon_{\kappa\lambda\mu\nu}$. The Casimir effect has also been analyzed within the context of the SME, for arbitrary $(k_F)_{\kappa\lambda\mu\nu}$ and $(k_{AF})^\kappa$ Lorentz-violating couplings [62, 63].

17.5 Reversed Vavilov-Cherenkov (VC) Radiation in Naturally Existing Magnetoelectric Media

In this section we summarize the discovery reported in the [64] of reversed VC radiation (RVCR) in naturally occurring magnetoelectric materials. The usual VC radiation is produced when a charge q propagates with velocity $v > c/n$ in a medium with refraction index n, i. e., with a velocity higher than the speed of light in the material [65, 66]. The first theoretical description of such radiation in the framework of Maxwell's theory, developed by Frank and Tamm in [67], revealed its unique polarization and directional properties. In particular, VC radiation is localized in a forward cone with opening angle $\alpha = \arccos(vn/c)$. VC radiation has played a fundamental role in the study of high-energy particle physics, high-power microwave sources and nuclear and cosmic-ray physics [68, 69], both theoretically and experimentally.

In 1968 Veselago theoretically proposed that RVCR could be produced in materials having simultaneously a negative permittivity and permeability, dubbed as left-handed media (LHM). A medium is left-handed when the momentum $\mathbf{k}$ of a propagating electromagnetic wave is antiparallel to its Poynting vector $\mathbf{E} \times \mathbf{B}$, i. e., when the group velocity and the phase velocity of the wave are antiparallel [70]. In this case photons would be emitted in the backward direction with respect to the velocity of the propagating charge. Since these materials are not found in nature, this proposal has generated a great boost to the design and construction of metamaterials, that is, artifacts built in the laboratory that can reproduce these properties in certain frequency ranges. In recent years the study of RVCR has been of considerable interest [71–79]. The realization of a LHM in an interface between an ordinary media was experimentally proved in [80]. It is interesting that Reversed Cherenkov radiation has also been found for sound in topological insulators in [81, 82].

Let us consider two semi infinite magnetoelectric media ($\theta_1 \neq \theta_2$) separated by a planar interface Σ located at $z = 0$. The setup is described by θ-ED according to the conventions in Sect. 17.2 and the modified Maxwell equations are (17.4). This time we take into account the permittivity of the media, but choose $\epsilon_1 = \epsilon_2 = \epsilon$ to avoid the interference with transition radiation. Still we consider non-magnetic media with $\mu_1 = \mu_2 = 1$. These choices provide a first approximation to the problem which aims to single out the effects of the axion coupling, determined by $\tilde{\theta} = \alpha(\theta_2 - \theta_1)/\pi$.

The electric and magnetic fields are obtained from (17.6) after the potential A_μ is expressed in terms of the GF according to (17.18). The GF $G^{\nu}{}_{\sigma}(x, x')$ satisfies the equation

$$\left([\Box^2]^{\mu}{}_{\nu} - \tilde{\theta}\delta(z)\varepsilon^{3\mu\alpha}{}_{\nu}\partial_\alpha\right) G^{\nu}{}_{\sigma}(\mathbf{x}, \mathbf{x}', t - t') = 4\pi\eta^{\mu}{}_{\sigma}\delta^4(x - x') \tag{17.51}$$

and its calculation is analogous to the static case in Sect. 17.3. The main differences with respect to (17.20) arise in the inclusion of the time dependence together with setting $\epsilon \neq 1$ and they are reflected in the changes $-\eta^{\mu}_{\nu}\nabla^2 \to [\Box^2]^{\mu}{}_{\nu} = \left(\epsilon\Box^2, \Box^2\delta^i{}_j\right)$ with $\Box^2 = \epsilon\partial_t^2 - \nabla^2$. An additional Fourier transform involving $\exp i\omega(t - t')$ in (17.24) allows to write the GF in terms of the reduced GF $g^{\nu}{}_{\sigma}(z, z', \mathbf{p}_\parallel, \omega)$ which satisfies an equation similar to (17.25). In this case, the free reduced GF, $\mathcal{F}_0(z, z'; \mathbf{k}_\parallel, \omega)$, satisfies $\left(\mathbf{p}_\parallel^2 - \omega^2\epsilon - \partial_z^2\right)\mathcal{F}_0(z, z'; \mathbf{k}_\parallel, \omega) = \delta(z - z')$, plus the standard BCs at infinity. The result is [87]

$$\mathcal{F}_0(z, z'; \mathbf{p}_\parallel, \omega) = \frac{i e^{ik_z|z-z'|}}{2k_z}, \qquad k_z = \sqrt{\epsilon\omega^2 - \mathbf{p}_\parallel^2}. \tag{17.52}$$

The solution of the subsequent coupled algebraic equations yields a result for $g^{\nu}{}_{\sigma}(z, z', \mathbf{p}_\parallel, \omega)$ in terms of $\mathcal{F}_0(z, z'; \mathbf{p}_\parallel, \omega)$ whose θ-dependent contribution shows a dependence on $|z| + |z'|$, as it was the case in the static situation. The knowledge of the reduced GF yields the coordinate representation $G^{\nu}{}_{\sigma}(\mathbf{x}, \mathbf{x}', \omega)$ in terms of a Fourier transform in the parallel momentum space, as indicated in (17.24), whose components are explicitly given in (43) of [64]. The next step is to evaluate the GF in the far-field approximation corresponding to the coordinate conditions $r \equiv \|\mathbf{x}\| \gg \|\mathbf{x}'\|, \quad R_\parallel \equiv \| \left(\mathbf{x}-\mathbf{x}'\right)_\parallel \| \simeq \|\mathbf{x}_\parallel\|, \quad |z|+|z'| \simeq |z|$. In this approximation the required integrals yielding the Fourier transforms include rapidly oscillating exponential functions whose leading contributions are calculated in the steepest descent method [83–85]. Also we make repeated use of the generic Sommerfeld identity [86]

$$i\int_0^\infty \frac{k_\parallel dk_\parallel}{\sqrt{\tilde{k}_0^2-k_\parallel^2}} J_0(k_\parallel R_\parallel) e^{i\sqrt{\tilde{k}_0^2-k_\parallel^2}|Z|} = \frac{e^{i\tilde{k}_0\mathcal{R}}}{\mathcal{R}}, \qquad \tilde{k}_0 = \sqrt{\epsilon}\omega, \tag{17.53}$$

with $\mathcal{R} = \sqrt{R_\parallel^2 + Z^2}$ and $n = \sqrt{\epsilon}$ being the refraction index. Two cases appear: (i) $Z = |z-z'|$, where $\mathcal{R}$ is denoted by R, corresponding to the usual $\tilde{\theta} = 0$ contribution and (ii) $Z = |z| + |z'|$, where $\mathcal{R}$ is denoted by $\tilde{R}$ and which involves the effects of the axion coupling. It is pertinent to emphasize an important difference in the phase of the exponentials related to the source variables $\mathbf{x}'$ in these two different cases. In the choice (i) we encounter the exponential $e^{i\tilde{k}_0 R}$, which in the coordinate approximation of the far-field zone produces the phase $i\tilde{k}_0(r - \hat{\mathbf{x}}\cdot\mathbf{x}'_\parallel - z'\cos\theta) = i\tilde{k}_0(r - \hat{\mathbf{x}}\cdot\mathbf{x}')$ characteristic of radiation in standard electrodynamics [87]. Here $\hat{\mathbf{x}} = \mathbf{x}/\|\mathbf{x}\|$. On the other hand, the contributions to the GF involving the axion coupling arising from the choice (ii) include the exponential $e^{i\tilde{k}_0\tilde{R}}$ with

$$\begin{aligned}\tilde{R} &= \sqrt{(\mathbf{x}-\mathbf{x}')_\parallel^2 + (|z|+|z'|)^2} = \sqrt{(\mathbf{x}-\mathbf{x}')^2 - (z-z')^2 + (|z|+|z'|)^2} \\ &= \sqrt{(\mathbf{x}-\mathbf{x}')^2 + 2(|zz'| + zz')} \sim r - \hat{\mathbf{x}}\cdot\mathbf{x}'_\parallel + |z'\cos\theta|,\end{aligned} \tag{17.54}$$

in the far-field approximation. From the square root in second line of (17.54) we remark that whenever the sign of $zz' = r\cos\theta z'$ is positive we will have an additional relative phase contributing to the GFs, which will show up in observable quantities as the radiated power for example. The term $(|z|+|z'|)^2$ can ultimately be traced back to the form of the reduced GF, as mentioned before. In the following we will show that RVCR arises precisely due to the contribution $|z'\cos\theta|$ in the phase of the GF. The detailed calculation is presented in the Appendix of [64] and here we only summarize the results emphasizing the phase dependence of the contributions. The full GF $G^\mu{}_\nu$ can be written in terms of the auxiliary function $\bar{G}^\mu{}_\nu = \bar{G}^\mu{}_{ED\,\nu} + \bar{G}^\mu{}_{\tilde{\theta}\,\nu} + \bar{G}^\mu{}_{\tilde{\theta}^2\,\nu}$ such that $G^0{}_0 = \bar{G}^0{}_0/\epsilon$ and $G^\mu{}_\nu = \bar{G}^\mu{}_\nu$ for the remaining cases. In the far field approximation we find

$$\bar{G}^{\mu}_{ED\ \nu}(\mathbf{x},\mathbf{x}';\omega) = \eta^{\mu}{}_{\nu}\frac{e^{i\tilde{k}_0 r}}{r}e^{-i\tilde{k}_0\hat{\mathbf{x}}\cdot\mathbf{x}'},$$

$$\bar{G}^{\mu}_{\tilde{\theta}\ \nu}(\mathbf{x},\mathbf{x}';\omega) = \varepsilon^{\mu}{}_{\nu}{}^{\alpha 3}\frac{2\tilde{\theta}}{4n^2+\tilde{\theta}^2}\frac{s_\alpha}{|\cos\theta|}\frac{e^{i\tilde{k}_0 r}}{r}e^{i\tilde{k}_0\left(-\hat{\mathbf{x}}\cdot\mathbf{x}'_{\parallel}+|z'\cos\theta|\right)},$$

$$\bar{G}^{\mu}_{\tilde{\theta}^2\ \nu}(\mathbf{x},\mathbf{x}';\omega) = \frac{\tilde{\theta}^2}{4n^2+\tilde{\theta}^2}\frac{e^{i\tilde{k}_0 r}}{r\cos^2\theta}C^{\mu}{}_{\nu}(\theta,\phi,n)\times e^{i\tilde{k}_0\left(-\hat{\mathbf{x}}\cdot\mathbf{x}'_{\parallel}+|z'\cos\theta|\right)}, \tag{17.55}$$

Here $s_\alpha = (1/n, \hat{\mathbf{x}})$ and θ, ϕ are the standard spherical angles labelling the unit vector $\hat{\mathbf{x}}$.

Now we choose our source by considering a point charge q moving with a constant velocity $v\hat{\mathbf{u}}$ in direction perpendicular to the interface Σ which is described by the charge and current densities $\varrho(\mathbf{x}';\omega) = \frac{q}{v}\delta(x')\delta(y')e^{i\omega\frac{z'}{v}}$ and $\mathbf{j}(\mathbf{x}';\omega) = \varrho v\hat{\mathbf{u}}$, respectively. Here $\hat{\mathbf{u}}$ is the unit vector perpendicular to the the interface and directed from the region $z < 0$ to the region $z > 0$. In the following we assume $v > 0$ and consider the motion in the interval $z \in (-\zeta, \zeta)$, with $\zeta \gg v/\omega$, to then take the limit $\zeta \to \infty$. After a long calculation in the far field approximation, we get the electric field starting from the potential A_μ obtained from the convolution of the given source with the GF resulting from (17.55). In this way we are in position to determine the spectral distribution (SD) of the radiation given by

$$d^2E/d\omega d\Omega = (\mathbf{E}^* \cdot \mathbf{E})\, nr^2/4\pi^2 \tag{17.56}$$

in the limit $r \to \infty$. The main point to stress is that the resulting $\mathbf{E}$ turns out to be linear in the following integrals

$$\mathcal{I}_1(\omega,\theta) = \int_{-\zeta}^{\zeta} dz' e^{i\frac{\omega z'}{v}(1-vn\cos\theta)} = \frac{2\sin(\zeta\,\Xi_-)}{\Xi_-}, \tag{17.57}$$

$$\mathcal{I}_2(\omega,\theta) = \int_{-\zeta}^{\zeta} dz' e^{in\omega|z'\cos\theta|+i\omega\frac{z'}{v}} = \frac{\sin(\zeta\,\tilde{\Xi}_-)}{\tilde{\Xi}_-} + \frac{\sin(\zeta\,\tilde{\Xi}_+)}{\tilde{\Xi}_+} + 2i\frac{\sin^2(\frac{\zeta}{2}\tilde{\Xi}_-)}{\tilde{\Xi}_-} - 2i\frac{\sin^2(\frac{\zeta}{2}\tilde{\Xi}_+)}{\tilde{\Xi}_+}. \tag{17.58}$$

The notation is $\Xi_- = \frac{\omega}{v}(1 - vn\cos\theta)$ and $\tilde{\Xi}_\pm = \frac{\omega}{v}(1 \pm vn|\cos\theta|)$. Moreover, in the limit $\zeta \to \infty$, the right hand side in (17.57) and (17.58) yields contributions of the type $\sin(\zeta\,\rho N)/(\rho N)$ which behave like $\pi\delta(\rho N)$. This implies that the non-zero contributions to the electric field arise from those terms whose arguments Ξ_- and $\tilde{\Xi}_\pm$ can take the value zero. With our previous conventions ($vn > 0$) this condition eliminates the dependence on $\tilde{\Xi}_+$ as a possible candidate. The remaining possibilities are (i) $1 - vn\cos\theta = 0$ and/or (ii) $1 - vn|\cos\theta| = 0$. The first case contributes to the standard VC radiation yielding $\cos\theta_C = 1/(nv)$. Case (ii) opens up the possibility that $\cos\theta < 0$, yielding a radiation cone with angle $\theta_R = \pi - \theta_C$ which signals the reversed VC radiation. In other words, the term containing $\tilde{\Xi}_-$ makes possible the production of radiation in the backward direction with respect to the moving charge. The total SD of the radiation is the sum of the following three contributions

$$\frac{d^2E_1}{d\omega d\Omega} = \frac{n\omega^2 q^2}{\pi^2}\left(1 - \frac{1}{v^2n^2}\right)\frac{\sin^2(\zeta\,\Xi_-)}{\Xi_-^2}, \tag{17.59}$$

$$\frac{d^2E_{12}}{d\omega d\Omega} = -\frac{n\omega^2}{\pi^2}\frac{\tilde{\theta}^2 q^2}{4n^2+\tilde{\theta}^2}\left(1 - \frac{1}{v^2n^2}\right)\frac{\sin(\zeta\,\Xi_-)\sin\left(\zeta\,\tilde{\Xi}_-\right)}{\Xi_-\tilde{\Xi}_-}, \tag{17.60}$$

$$\frac{d^2E_2}{d\omega d\Omega} = \frac{n\omega^2}{4\pi^2}\frac{\tilde{\theta}^2 q^2}{4n^2+\tilde{\theta}^2}\left(1 - \frac{1}{v^2n^2}\right)\left[\frac{\sin^2\left(\zeta\,\tilde{\Xi}_-\right)}{\tilde{\Xi}_-^2} + \frac{\sin^4\left(\frac{\zeta}{2}\tilde{\Xi}_-\right)}{\frac{1}{4}\tilde{\Xi}_-^2}\right]. \tag{17.61}$$

Equation (17.59) gives the SD of standard VC radiation ($\cos\theta_C = 1/(nv) > 0$), which acquires $\tilde{\theta}$-dependent corrections from the remaining contributions. Let us observe that (17.60) contributes only to the forward radiation. The reversed VC radiation, which is strictly zero for normal insulators, arises only from the term in (17.61) which nevertheless also contributes to the forward output.

The total energy per unit frequency and per unit length radiated by the charge on its path from $-\zeta$ to $+\zeta$ is calculated along the steps in [88] integrating the solid angle and taking into account that the delta-like behavior of the integrands shows that the radiation is sharply localized in a main lobe around the angles θ_C of the forward/backward cone. Such lobes produce the final cones of radiation when $\zeta \to \infty$. The results for the forward and reversed VC radiation are:

$$\frac{d^2E_{\mathrm{FVCR}}}{d\omega\, dL} = q^2\omega\left(1 - \frac{1}{v^2n^2}\right)\left[1 - \frac{1}{2}\frac{\tilde{\theta}^2}{4n^2+\tilde{\theta}^2}\right], \tag{17.62}$$

$$\frac{d^2E_{\mathrm{RVCR}}}{d\omega\, dL} = q^2\omega\left(1 - \frac{1}{v^2n^2}\right)\left[\frac{1}{2}\frac{\tilde{\theta}^2}{4n^2+\tilde{\theta}^2}\right], \tag{17.63}$$

respectively. We have introduced the total length $L = 2\zeta$ travelled by the particle.

To illustrate our results, we consider medium 2 as the topological insulator $\mathrm{TlBiSe_2}$, with $n_2 = 2$, together with a normal insulator as medium 1, characterized by $n_1 = n_2$. We consider the radiation emitted at a frequency of $\omega = 2.5$ eV (500 nm), which constitutes an average in the Cherenkov radiation spectrum. The angular distribution of the total radiation is shown in Fig. 17.4. In the left panel we plot an enlargement of the angular distribution in the backward direction, which shows the appearance of the reversed VC radiation and makes its suppression evident with respect to the forward output, which is given by the ratio $\tilde{\theta}^2/8n^2$ between the results (17.63) and (17.62). A comparison with measurements of reversed VC radiation in metamaterials can be established by interpreting this suppression as due to the detection of radiation at an effective frequency $\omega_{\mathrm{eff}} = \omega\tilde{\theta}^2/8n^2$. Taking an average of 500 nm (2.5 eV) in the Cherenkov spectrum, we would expect

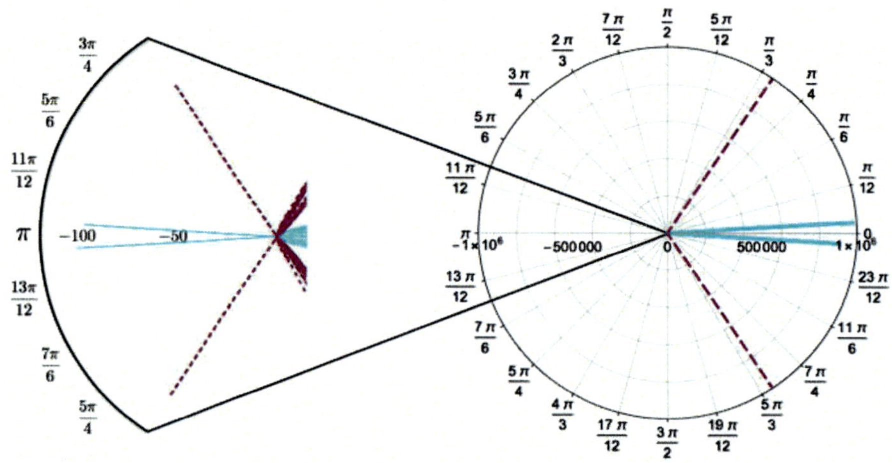

Fig. 17.4 Angular distribution for the total radiated energy per unit frequency for the full VC radiation for the choices for $n = 2$, $\omega = 2.48$ eV and $\tilde{\theta} = 11\alpha$. The dashed purple line corresponds to $v = 0.9$ and $\zeta = 343\ \mathrm{eV}^{-1}$, and the solid cyan line to $v = 0.5009$ and $\zeta = 4830\ \mathrm{eV}^{-1}$. The scale of the polar plot is in arbitrary dimensions and runs from 0 to 10^6. In the left side of the figure we plot an enlargement in the backward direction showing the onset of the reversed VC radiation. Here the radial scale goes from zero to 10^2. The charge moves from left to right, (adapted from [64])

the production of reversed VC radiation at effective frequencies ω_{eff} in the range from 4×10^{-3} meV for $\tilde{\theta} = \alpha$ to 0.5 meV for $\tilde{\theta} = 11\alpha$ respectively, using $\mathrm{TlBiSe_2}$ as a θ-medium. Recent measurements of reversed VC radiation in metamaterials show that this estimations are within the experimental capabilities. Reversed VC radiation has been measured at a frequency of 2.85 GHz (1.2×10^{-2} meV), in an all-metal metamaterial consisting of a square waveguide loaded with complementary electric split ring resonators [78]. Likewise, reversed VC radiation in the range $(3.4 - 3.9) \times 10^{-2}$ meV has also been experimentally verified in a phased electromagnetic dipole array used to model a moving charged particle [76]. We have also estimated that $d^2E_{\mathrm{RVCR}}/dtd\omega = vd^2E_{\mathrm{RVCR}}/d\omega\, dL$ is within the range $10^{-3} - 10^{-2}$ μW/eV for the frequency interval $\omega = 2 - 8$ eV. Such values are smaller by a factor of $10^{-4} - 10^{-3}$ than the maximum output of $\sim 10\ \mu$W/eV theoretically predicted to occur in the narrow interval of $5.7 - 6.5$ eV in a metal-insulator-metal waveguide [79]. In such a waveguide with a core thickness of $a = 20$ nm, surface plasmon polaritons excited by an electron moving at $v = 0.8$ produce reversed VC radiation.

The main features of the RVCR we have found are: (i) the threshold condition $v > 1/n$ must be satisfied as in the usual case, (ii) RVCR occurs for all frequencies in the Cherenkov spectrum and is always accompanied by forward radiation with the same frequency, (iii) the angular distribution of the RVCR is suppressed with respect to the forward emission by a factor $\tilde{\theta}^2/8n^2$. The details of this development are reported in [64].

17.6 Electromagnetic Response of Weyl Semimetals

In addition to the topological insulators, which are characterized by a gapped bulk and protected boundary modes that are robust against disorder [19–22], we have recently learned that gapless semimetallic states may be topologically nontrivial in the same sense as gapped insulators. Weyl semimetals are topological states of matter in which the 3D bulk contains Weyl points (band crossing points) protected by crystalline symmetries and whose low-energy quasiparticles are linearly dispersing massless Weyl fermions. Weyl semimetals possess protected gapless surface states on disconnected Fermi arcs with end points at the projection of the bulk nodes onto the surface Brillouin zone [24].

Besides their spectroscopic distinguishing features, WSMs also exhibit unusual electromagnetic responses that are a direct macroscopic manifestation of the non-trivial topology of their band structure. Mathematically, the anomalous Hall effect and related effects to the chiral anomaly may be compactly expressed as an induced θ term in the action of the electromagnetic field, when chiral fermions are integrated out [89–91]:

$$S_\theta = \frac{\alpha}{4\pi^2} \int \theta(\mathbf{r}, t)\, \mathbf{E} \cdot \mathbf{B}\, dt\, d^3\mathbf{r}, \tag{17.64}$$

where $\alpha = e^2/\hbar c$ is the fine-structure constant and $\theta(\mathbf{r}, t)$ is the axion field, with the following form

$$\theta(\mathbf{r}, t) = 2\mathbf{b} \cdot \mathbf{r} - 2b_0 t, \tag{17.65}$$

where $2\mathbf{b}$ is the separation, in momentum space, between Weyl nodes and $2b_0$ their separation in energy. Topological response of WSMs is thus described by an action similar to that of axion-electrodynamics [36]. The relevant equations of motion are obtained by varying the full action (S_θ plus the nontopological Maxwell action):

$$\nabla \cdot \mathbf{D} = 4\pi \left(\rho - \frac{\alpha}{2\pi^2} \mathbf{b} \cdot \mathbf{B} \right), \tag{17.66}$$

and

$$\nabla \times \mathbf{H} - \frac{1}{c} \frac{\partial \mathbf{D}}{\partial t} = \frac{4\pi}{c} \left(\mathbf{J} + \frac{\alpha}{2\pi^2} c\mathbf{b} \times \mathbf{E} - \frac{\alpha}{2\pi^2} b_0 \mathbf{B} \right), \tag{17.67}$$

where $\mathbf{D} = \varepsilon\,\mathbf{E}$ and $\mathbf{B} = \tilde{\mu}\,\mathbf{H}$. Faraday's law, $\nabla \times \mathbf{E} = -c^{-1}\partial\mathbf{B}/\partial t$, and the equation stating the absence of magnetic monopoles, $\nabla \cdot \mathbf{B} = 0$, remain unaltered. Here, as in ordinary metals, $\varepsilon = \epsilon + i\sigma_{xx}(\omega)/\omega$ and $\tilde{\mu} = 1 + \chi_m$, where ϵ is the static permittivity, $\sigma_{xx}(\omega)$ is the longitudinal conductivity and χ_m is the magnetic susceptibility that we assume is negligible for the WSM. Again, we will treat the WSM material as linear and isotropic, but the extension to the linear and anisotropic case is possible.

There are other distinguishing effects of Weyl semimetals not fully captured by axion electrodynamics, such as the chiral magnetic effect and the chiral separation effect. These can be derived by using, for example, the semiclassical Boltzmann transport theory. In short, if we have chiral fermions in a magnetic field with chemical potentials μ_{L} and μ_{R} for left- and right-handed fermions, respectively, there are two additional $\mathbf{B}$-dependent current terms, namely,

$$\mathbf{J}^{(\mathrm{B})} = \frac{\alpha}{2\pi^2}\mu_5\mathbf{B} \quad , \quad \mathbf{J}_5^{(\mathrm{B})} = \frac{\alpha}{2\pi^2}\mu\mathbf{B}, \tag{17.68}$$

where $\mu_5 = (\mu_{\mathrm{L}} - \mu_{\mathrm{R}})/2$ and $\mu = (\mu_{\mathrm{L}} + \mu_{\mathrm{R}})/2$ are the chiral and the electric chemical potentials, respectively.

The anomalous Hall effect, which is expected to occur in a Weyl semimetal with broken time-reversal (TR) symmetry, is described by the $\mathbf{b}$-dependent terms in (17.66) and (17.67). In addition, the b_0-dependent term that arises in Weyl semimetals with broken inversion symmetry, describes only one part of the celebrated chiral magnetic effect, namely, the generation of an electric current driven by an applied magnetic field. The second part is given by $\mathbf{J}^{(\mathrm{B})}$ in (17.68), which arises from an imbalance between chemical potentials of right- and left-handed fermions. The total contribution to the chiral magnetic effect is thus given by $\mathbf{J}_{\mathrm{CME}} = \frac{\alpha}{2\pi^2}\,(\mu_5 - b_0)\,\mathbf{B}$, that vanishes for $b_0 = \mu_5$ in which case the WSM is said to be at the equilibrium state [92]. The electric current $\mathbf{J}_5^{B}$ in (17.68) is identified as the chiral separation effect, which vanishes for $\mu = 0$, condition that defines the neutrality point [92].

17.6.1 Electric Charge Near a Weyl Semimetal

Recent publications have tackled a number of physical effects on the basis of the above theory. Among them we find the magneto-optical Faraday and Kerr rotations [93] and the Casimir effect [94], as well as the appearance of plasmon polaritons [95] and helicons [96] at the sample's surface. Here, inspired by the image magnetic monopole effect of topological insulators [33, 45, 97], we shall briefly discuss the problem of a pointlike electric charge near the surface of a topological WSM in the equilibrium state and at the neutrality point. This means that the material's response will be a direct consequence of the anomalous Hall effect in the bulk. Charge neutrality is not an unrealistic assumption for WSMs, since it can be attained under specific circumstances, as shown both theoretically and experimentally in [98].

Let us consider a topological Weyl semimetal with a pair of nodes separated along the k_z-direction in the bulk BZ occupying the half-space $z < 0$. The region $z > 0$ is occupied by a dielectric fluid. An electric charge is brought near the surface that does not support Fermi-arc electronic states, in this case the xy-plane for $\mathbf{b} = b\hat{\mathbf{e}}_z$. Neglecting all frequency dependence on the conductivities (since we deal with a static problem), the electromagnetic response of the WSM is fully captured by (17.66) and (17.67), with $b_0 = \mu_5$ and $\mu = 0$. Since $\theta(z = 0) = 0$, there are no surface currents, and the resulting material is just a bulk Hall medium with current responses given by the transverse Hall conductivity

$$\sigma_{xy} = \frac{e^2 b}{2\pi^2 \hbar}. \tag{17.69}$$

Due to the gauge invariance of the theory, we can introduce the electromagnetic potentials as usual: $\mathbf{E} = -\nabla\Phi$ and $\mathbf{B} = \nabla \times \mathbf{A}$. In the Coulomb gauge $\nabla \cdot \mathbf{A} = 0$, for a pointlike electric charge of strength q at $\mathbf{r}' = z'\hat{\mathbf{e}}_z$ with $z' > 0$ (that is the charge lies in the dielectric fluid), the electromagnetic potentials satisfy the coupled equations of motion

$$-\nabla \cdot [\epsilon(z)\nabla\Phi] + \frac{4\pi}{c}\sigma_{xy}(z)\,\hat{\mathbf{e}}_z \cdot \nabla \times \mathbf{A} = 4\pi q\delta(z - z'), \tag{17.70}$$

$$-\nabla^2\mathbf{A} + \frac{4\pi}{c}\sigma_{xy}(z)\,\hat{\mathbf{e}}_z \times \nabla\Phi = 0, \tag{17.71}$$

where $\sigma_{xy}(z) = \sigma_{xy} H(-z)$ is the bulk Hall conductivity and $\epsilon(z) = \epsilon_1 H(-z) + \epsilon_2 H(z)$ is the static permittivity of the system. The differential equations (17.70) and (17.71), along with the appropriate boundary conditions at the interface $z = 0$ and at the singular point $z = z'$, constitute a complete boundary value problem, which can be solved with standard techniques of electromagnetism [42]. The solution is simple, but not straightforward [99]. On the one hand, the final result for the electrostatic potential beneath the surface is

$$\Phi_{z<0} = 2q \int_0^\infty \frac{(\lambda_+ + k)\cos(\lambda_- z) + \lambda_- \sin(\lambda_- z)}{\epsilon_1\left(\lambda_+^2 + \lambda_-^2\right) + \epsilon_2 k^2 + k\lambda_+(\epsilon_1 + \epsilon_2)} k J_0(k\rho) e^{\lambda_+ z - kz'} dk, \tag{17.72}$$

and, above the surface, we find

$$\begin{aligned}\Phi_{z>0} = {} & \frac{q}{\epsilon_2}\frac{1}{\sqrt{\rho^2 + (z - z')^2}} + \frac{q}{\epsilon_2}\frac{\epsilon_2 - \epsilon_1}{\epsilon_2 + \epsilon_1}\frac{1}{\sqrt{\rho^2 + (z + z')^2}} \\ & - \frac{2q\epsilon_1}{\epsilon_1 + \epsilon_2}\int_0^\infty \frac{\lambda_+^2 + \lambda_-^2 - k^2}{\epsilon_1\left(\lambda_+^2 + \lambda_-^2\right) + \epsilon_2 k^2 + k\lambda_+(\epsilon_1 + \epsilon_2)} J_0(k\rho) e^{-k(z+z')} dk.\end{aligned} \tag{17.73}$$

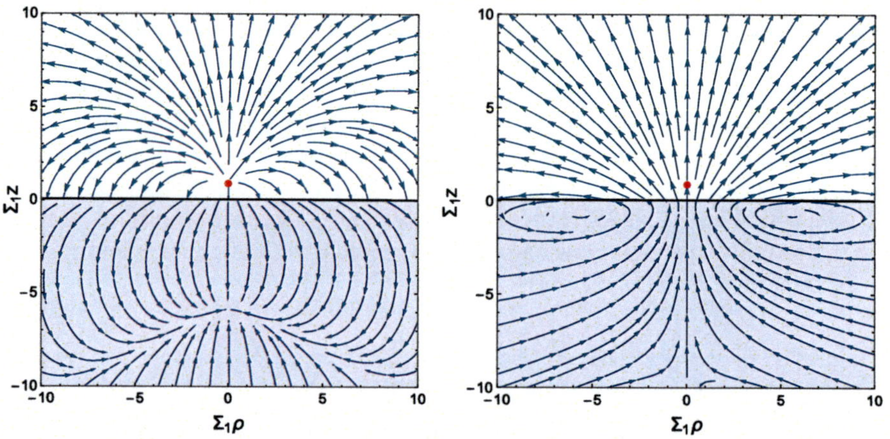

Fig. 17.5 Illustration of the electric (left) and magnetic (right) fields (in units of $q\Sigma^2$), as a function of Σz and $\Sigma\rho$, induced by an electric charge of strength q at $z' = 1/\Sigma$ (marked with the red sphere) above the WSM TaAs (marked with the blue shaded region), (adapted from [99])

In these expressions $\lambda_\pm^2 = (k/2)\left[\sqrt{k^2+\Sigma^2} \pm k\right]$, $\Sigma = 4\pi\sigma_{xy}/(c\sqrt{\epsilon_1})$, $\rho^2 = x^2 + y^2$ and J_n is the nth order Bessel function of the first kind. We observe that in the dielectric fluid, the electric potential can be interpreted as due to the original electric charge of strength q at z', an image electric charge of strength $q(\epsilon_2 - \epsilon_1)/(\epsilon_2 + \epsilon_1)$ at $-z'$, and an additional term arising from the nontrivial topology of the WSM. Inside the material, the electric potential has no simple interpretation. However, as evidenced by the exponential term in (17.72), it is attenuated inside the WSM due to the metallic character of the material. This can also be seen in the limit $\epsilon_1 \to \infty$, for which $\Phi_{z<0} = 0$, as in a perfect conductor.

The electric field is obtained from the above potentials as $\mathbf{E} = -\nabla\Phi$. The left panel of Fig. 17.5 illustrates the electric field $\mathbf{E}$ (in units of $q\Sigma^2$) generated by an electric charge in vacuum ($\epsilon_2 = 1$) at $z' = 1/\Sigma$ (red sphere) close to the WSM TaAs [100] as a function of the dimensionless coordinates $\Sigma\rho$ and Σz. We observe that the electric field outside the WSM is similar to that generated by the original electric charge, with deviations close to the interface due to the screening of the field inside the material. This behaviour is similar to that produced by an electric charge close to an ordinary metal or a dielectric, but the electric field beneath the surface is more complicated than in these nontopological cases. For example, the electric field within a uniform and isotropic dielectric is a radially directed field (with the charge outside the material as its source); while the field inside an ordinary metal is zero. In the present case, as shown in Fig. 17.5, the electric field is remarkably different as evidenced by the curved field lines inside.

On the other hand, the resulting vector potential beneath the surface is

$$\mathbf{A}_{z<0} = 2q\epsilon_1 \hat{\mathbf{e}}_\varphi \int_0^\infty \frac{\lambda_- \cos(\lambda_- z) - (\lambda_+ + k)\sin(\lambda_- z)}{\epsilon_1 \left(\lambda_-^2 + \lambda_+^2\right) + \epsilon_2 k^2 + k\lambda_+ (\epsilon_1 + \epsilon_2)} k J_1(k\rho) e^{\lambda_+ z - kz'} dk, \tag{17.74}$$

and, above the surface, we obtain

$$\mathbf{A}_{z>0} = 2q\epsilon_1 \hat{\mathbf{e}}_\varphi \int_0^\infty \frac{\lambda_-}{\epsilon_1 \left(\lambda_-^2 + \lambda_+^2\right) + \epsilon_2 k^2 + k\lambda_+ (\epsilon_1 + \epsilon_2)} k J_1(k\rho) e^{-k(z+z')} dk. \tag{17.75}$$

The magnetic field is obtained as $\mathbf{B} = \nabla \times \mathbf{A}$. The right panel of Fig. 17.5 shows the magnetic field $\mathbf{B}$ (in units of $q\Sigma^2$) induced by an electric charge in vacuum at $z' = 1/\Sigma$ close to the WSM TaAs as a function of the dimensionless coordinates $\Sigma\rho$ and Σz. We observe that in the present case, the behaviour of the field lines is significantly different from the radially directed field lines appearing when the charge is close to a TI. We will discuss this point later.

The physical origin of the magnetic field is the bulk Hall current circulating around the symmetry axis, $\mathbf{J}_{\mathrm{Hall}} = \sigma_{xy} \left(\hat{\mathbf{e}}_\rho \cdot \mathbf{E}_{z<0}\right) \hat{\mathbf{e}}_\varphi$, which is induced by the in-plane component of the electric field produced by the charge. In this way, each cross section (perpendicular to the symmetry axis) of the bulk Hall current resembles the surface Hall current induced by an electric charge near to a topological insulator. This suggests that a 3D Weyl semimetallic phase can be understood as an infinite number of 2+1 Dirac subsystems (one for each value of z in the bulk) supporting a surface Hall current [100].

Now we go back to the discussion of the behaviour of the magnetic field lines in Fig. 17.5. A close inspection reveals that below the surface of the WSM, centered at the position of the image charge, the $\mathbf{B}$-field lines wind in an axisymmetric way as if about a loop of current, similar to those of a physical magnetic dipole. This suggests that we consider a multipole expansion of the magnetic field and determine the dominant contribution. Still, we recall that the source of the magnetic field is not localized (since $\mathbf{J}_{\mathrm{Hall}}$ is defined in the whole bulk), and hence the standard multipole expansion for localized sources does not necessarily applies. In the far zone, the integral in (17.75) can be approximated in powers of $1/r$. The dominant contributions are

$$\mathbf{A}^{(1)} \approx \frac{3}{32}\frac{q}{\Sigma}\sqrt{\frac{\pi}{2}} \frac{\tan(\theta/2)(13 + 3\cos\theta)}{r^{3/2}} \hat{\mathbf{e}}_\varphi, \qquad \mathbf{A}^{(2)} \approx -\frac{q}{\Sigma}\left(1 + \frac{\epsilon_2}{\epsilon_1}\right) \frac{\sin\theta}{r^2} \hat{\mathbf{e}}_\varphi, \tag{17.76}$$

where θ is the angle from the z-axis to the observation point, i.e. $r\cos\theta = \hat{\mathbf{e}}_z \cdot \mathbf{r}$ and $r = \sqrt{\rho^2 + (z + z')^2}$. We observe that the leading term is a fractional multipole, with no analogue in standard electromagnetism. The measurement of this contribution would be a unique signature of the anomalous Hall effect in the bulk of 3D WSMs. The second term can be successfully compared with the vector potential produced by a magnetic dipole of strength $m = -(q/\Sigma)(1 + \epsilon_2/\epsilon_1)$ located at the image point $-z'$, thus confirming the qualitative expectation that a magnetic dipole is induced.

17.6.2 Experimental Proposals

Now we discuss two specific fingerprints of the induced magnetic field which could, in principle, be measured.

Angle-resolved measurement

In the static case considered above, the WSM attracts the charge toward the surface in the direction perpendicular to it with a force $\mathbf{F}_e = q\mathbf{E}_{z>0}(\mathbf{r}')$, where $\mathbf{r}' = z'\hat{\mathbf{e}}_z$ is the position of the charge. However, interesting phenomena appear when we consider the dynamics of the external charge. Let us consider, for instance, a steady electron beam drifting at a distance z' above the surface of the WSM. To make our solution still valid, we have to consider the electrons slow enough with respect to the Fermi velocity in the solid, in such a way that the induced polarization and magnetization of the material rearranges infinitely fast. So, besides the electrostatic force $\mathbf{F}_e$, if the charge q moves with a uniform velocity $\mathbf{v}$ above the surface of the WSM, a magnetic force of the form $\mathbf{F}_m = q\frac{\mathbf{v}}{c} \times \mathbf{B}_{z>0}(\mathbf{r}')$ will also act on the charge due to the induced magnetic field. For an electron beam moving along the x-direction (with velocity $\mathbf{v} = v_x\hat{\mathbf{e}}_x$) we find

$$\mathbf{F}_m = -2q^2\epsilon_1(v_x/c)\,\hat{\mathbf{e}}_y \int_0^\infty \frac{k^2\lambda_-\, e^{-2kz'}}{\epsilon_1\left(\lambda_+^2 + \lambda_-^2\right) + \epsilon_2\, k^2 + k\lambda_+\,(\epsilon_1 + \epsilon_2)} dk. \tag{17.77}$$

Remarkably, this anomalous force is orthogonal to the electrons' motion as well as to the electric contribution $\mathbf{F}_e$, and hence these effects can be distinguished from each other. In fact, the effect of the anomalous force (17.77) on the electrons' motion is a deflection in the y-direction which could, in principle, be measurable. For a rough estimate, we take $v_x \sim 10^7$ cm/s (which is appropriate for the steady electron beam emitted from a low-energy electron gun) [101], $z' \sim 1\,\mu$m and $L \sim 1$ cm for the sample's size. So, for a beam of electrons drifting above the WSM TaAs (for which $\epsilon_1 \sim 6$ and $b \sim 10^9\,\mathrm{m}^{-1}$ [100]), the resulting transverse drift would be 3.2 μm. This deflection can be easily traced by angle-resolved measurement [101]. If this experiment were carried out with a Dirac semimetal by applying an external magnetic field, instead of a genuine WSM such as the TaAs, the induced magnetic field will be overwhelmed by the external one, and so would its contribution to the Lorentz force on a moving charge. For example, by considering the Dirac semimetal Cd_3As_2 in the presence of a magnetic field of 1T (for which $b = 5 \times 10^8\mathrm{m}^{-1}$ and $\epsilon_1 = 12$), the resulting transverse drift is 10^7 larger than the purely topological contribution, thus making its detection challenging. We then conclude that an angle-resolved measurement is appropriate for experimental realization only if it were possible to consider a genuine WSM, for which no external magnetic field is needed.

Scanning SQUID magnetometry

Scanning SQUID (Superconducting Quantum Interference Device) magnetometry provides an alternative for measuring the induced magnetic field. SQUIDs are very sensitive magnetometers based on superconducting loops containing Josephson junctions, and they are used to measure extremely low magnetic fields (of the order of 5×10^{-18}T [102]). In general terms, these devices measure the magnetic flux through a loop (parallel to the surface) placed at a fixed distance above the material, i.e. $\Phi_{\mathbf{B}} = \int_S \mathbf{B} \cdot d\mathbf{S}$, where S is the surface of the loop.

In the present case, a simple calculation produces $\Phi_{\mathbf{B}}(R, z) = 2\pi R\, \hat{\mathbf{e}}_{\varphi} \cdot \mathbf{A}_{z>0}(R, z)$, where R is the radius of the loop and $\mathbf{A}_{z>0}$ is given by (17.75). Of course, $\Phi_{\mathbf{B}} = 0$ at $R = 0$. Furthermore, since the magnetic field lines start at the WSM surface and go back again to the surface (see the right panel of Fig. 17.5), we then have $\Phi_{\mathbf{B}} \to 0$ as $R \to \infty$. This behaviour implies the existence of a maximum flux at a critical radius R_c, which can be determined in the usual manner (i.e. by solving $\partial_R \Phi_{\mathbf{B}}|_{R=R_c} = 0$ for R_c). For a numerical estimate of the magnetic flux we consider a charge $q = n_e|e|$ placed at a distance $z' = 1\mu$m above the surface of the WSM TaAs and a SQUID of radius $R = 10\mu$m located at $z = 10\mu$m. We find $\Phi_{\mathbf{B}} \approx 7n_e \times 10^{-14}\text{T} \cdot \text{cm}^2$, which is measurable with present day attainable sensitivities of SQUIDs [102]. One of the key challenges for the experimental detection of this flux profile would be to find a way to fix and localize the charge above the surface.

If this experiment were carried out with a Dirac semimetal instead of a genuine WSM, the required external magnetic field will overwhelm the topological contribution to the total magnetic flux, as before. Nevertheless, in this case it is still possible to disentangle these effects by using the fact that the contribution to the flux produced by the external magnetic field, say $\Phi_{\mathbf{B}}^{\text{ext}}$, is constant in space and time, but not $\Phi_{\mathbf{B}}$. A sensitive magnetometer as the SQUID will be capable to measure small variations of the flux which amounts to eventually measuring the induced electromotive force $\mathcal{E}$ in the loop. Therefore, this allows for isolating the topological contribution, for example, by producing a controlled displacement of the SQUID along the z-axis at speed v_z, namely: $\mathcal{E} = -\frac{d}{dt}\left(\Phi_{\mathbf{B}}^{\text{ext}} + \Phi_{\mathbf{B}}\right) = -v_z \frac{d\Phi_{\mathbf{B}}}{dz}$, where the z-dependence is read-off from (17.75).

17.7 Conclusions

In this chapter we have given an overview as to how the properties of magnetoelectric media associated with topological order, that arise from subtleties in their electronic structure, can be understood as macroscopic optical properties in regards to their electromagnetic response. We have presented the reader with the theoretical framework that allows to describe the electromagnetic response of magnetoelectric media by means of axion-like extended electrodynamics with a non-dynamical axion field θ. Among the benefits of such effective field theory approach is that we can exploit a Green's function formulation to solve for the electromagnetic fields with rather arbitrary configuration of charges, currents, boundary conditions and/or combination of these. Our tools and techniques have been applied to linear isotropic and homogeneous magnetoelectric media with mild coordinate restrictions that are, however, plausible from an experimental point of view. Other benefit of the GF formulation is that the method can be extended to the case of anisotropic media. To illustrate our tools and techniques, we summarize some previously reported results that include: (i) a detailed assessment of the boundary value problem of axion electrodynamics and the modified boundary conditions for the electromagnetic fields across the magnetoelectric interface; (ii) the application to the case of a topological insulator with planar geometry located between two parallel conducting plates to compute the ensuing modification to the Casimir effect; (iii) as an example on how to extend the formulation beyond the static cases, we review some results showing that this technique can be applied to the case of an electric charge traversing from one magnetoelectric medium to another at a constant speed and perpendicular to the interface between them to reveal that, besides the usual forward-directed Vavilov-Cherenkov emission of radiation, a novel feature due entirely to the magnetoelectric effect is an additional backward-directed radiation, termed reversed Vavilov-Cherenkov radiation. Last (iv) we employ the same approach to study the induced electromagnetic fields due to a static electric charge near the surface of a Weyl semimetal in the equilibrium state and at neutrality point to find that inside the WSM medium the induced electric field is nothing like inside a conductor nor insulator and the induced magnetic field outside the WSM has among its multipole contributions one term that corresponds to a magnetic dipole field, as if below the surface of the WSM the electric charge induced a stack of alternating axisymmetric circulating currents centered at the projection of the electric charge. Two different experimental setups are proposed to measure distinctive characteristics of the induced magnetic field as smoking-gun signals of the magnetoelectric effect in WSM and we argue that these observable signals could be measured given the present-day experimental sensitivities available.

As mentioned in the beginning, the study of topological phases of matter is both a fundamental and technological challenge. We think that with our exposition we contributed in both fronts. Our approach allows to shed lights on the understanding of the interaction of these new states of matter with electromagnetic radiation. At the same time, and as a corollary of our approach, we have provided new means to eventually measure observable signals of the magnetoelectric effect using topological insulators or Weyl semimetals. Though we have provided but a few examples, the generality of the Green's function method paves the way for new proposals to similar ends.

Acknowledgements A.M.R. acknowledges support from DGAPA-UNAM project IA101320. M.C. has been partially supported by UNAB DGID under Grant No. DI-33-17/RG. L.F.U. acknowledges support from DGAPA-UNAM Project No. IN103319. We thank to F.M.D. Pellegrino and A. Santonocito for useful remarks.

References

1. C. Nash, S. Sen, *Topology and Geometry for Physicists* (Academic Press Inc, London, 1983)
2. K. Fujikawa, H. Suzuki, *Path Integrals and Quantum Anomalies* (Clarendon Press, Oxford, 2004)
3. M. Dine, TASI Lectures on the Strong CP Problem. arXiv:0011376 [hep-ph] (2000)
4. D. Birmingham, M. Blau, M. Radowski, G. Thompson, Topological field theory. Phys. Rep. **209**, 129 (1991)
5. M. Kuster, G. Raffelt, B. Beltrán (eds), *Axions: Theory, Cosmology, and Experimental Searches* (*Lecture Notes in Physics* vol 741) (Springer, Berlin, 2008)
6. S. Weinberg, The U(1) problem. Phys. Rev. D **11**, 3583 (1975)
7. G. t'Hooft: Symmetry breaking through Bell-Jackiw anomalies. Phys. Rev. Lett. **37**, 8 (1976)
8. G. t'Hooft, Computation of the quantum effects due to a four-dimensional pseudoparticle. Phys. Rev. D **14**, 3432 (1976)
9. R. Peccei, H. Quinn, CP conservation in the presence of pseudoparticles. Phys. Rev. Lett. **38**, 1440 (1977)
10. R. Peccei, H. Quinn, Constraints imposed by CP conservation in the presence of pseudoparticles. Phys. Rev. D **16**, 1791 (1977)
11. P. Sikivie, Experimental tests of the "Invisible" axion. Phys. Rev. Lett. **51**, 1415 (1983)
12. D. Colladay, V.A. Kostelecký, CPT violation and the standard model. Phys. Rev. D **55**, 6760 (1997)
13. D. Colladay, V.A. Kostelecký, Lorentz-violating extension of the standard model. Phys. Rev. D **58**, 116002 (1998)
14. V.A. Kostelecký, M. Mewes, Signals for Lorentz violation in electrodynamics. Phys. Rev. D **66**, 056005 (2002)
15. V.A. Kostelecký, R. Lehnert, M.J. Perry, Spacetime-varying couplings and Lorentz violation. Phys. Rev. D **68**, 123511 (2003)
16. T.H. O'Dell, *The Electrodynamics of Magneto-Electric Media* (North-Holland, Amsterdam, 1970)
17. L.D. Landau, E.M. Lifshitz, L.P. Pitaevskii, *Electrodynamics of Continuous Media* (*Course of Theoretical Physics* vol 8) (Oxford: Pergamon Press, 1984)
18. E. Plum, J. Zhou, J. Dong, V.A. Fedotov, T. Koschny, C.M. Soukoulis, N.I. Zheludev, Metamaterial with negative index due to chirality. Phys. Rev. B **79**, 035407 (2009)
19. X.-L. Qi, T.L. Hughes, S.-C. Zhang, Topological field theory of time-reversal invariant insulators. Phys. Rev. B **78**, 195424 (2008)
20. X.-L. Qi, S.-C. Zhang, Topological insulators and superconductors. Rev. Mod. Phys. **83**, 1057 (2011)
21. M.Z. Hasan, C.L. Kane, Colloquium: Topological insulators. Rev. Mod. Phys. **82**, 3045 (2010)
22. X.-L. Qi, Field-theory foundations of topological insulators, in *Topological Insulators (Contemporary Concepts of Condensed Matter Science)*, Vol. 6, eds. by M. Franz, L. Molenkamp (Elsevier, Amsterdam, 2013)
23. M.M. Vazifeh, M. Franz, Electromagnetic response of weyl semimetals. Phys. Rev. Lett. **111**, 027201 (2013)
24. N.P. Armitage, E.J. Mele, A. Vishwanath, Weyl and Dirac semimetals in three-dimensional solids. Rev. Mod. Phys. **90**, 015001 (2018)
25. J.D. Jackson, *Classical Electrodynamics* (Wiley, Hoboken NJ, 1999)
26. I.E. Dzyaloshinskii, On the magneto-electrical effect in antiferromagnets. JETP **37**, 881 (1959)
27. J.P. Rivera, A short review of the magnetoelectric effect and related experimental techniques on single phase (multi-) ferroics. Eur. Phys. J. B **71**, 299 (2009)
28. G.L. Klimchtskaya, U. Mohideen, V.M. Mostepanenko, The Casimir force between real materials: experiment and theory. Rev. Mod. Phys. **81**, 1827 (2009)
29. Y.N. Obukhov, F.W. Hehl, Measuring a piecewise constant axion field in classical electrodynamics. Phys. Lett. A **341**, 357 (2005)

30. D.N. Astrov, The magneto-electrical effect in antiferromagnets. JETP **38**, 984 (1960)
31. M. Fiebig, Revival of the magnetoelectric effect. J. Phys. D: Appl. Phys. **38**, R123 (2005)
32. V. Dziom, A. Shuvaev, A. Pimenov, G.V. Astakhov, C. Ames, K. Bendias, J. Böttcher, G. Tkachov, E.M. Hankiewicz, C. Brüne, H. Buhmann, L.W. Molenkamp, Observation of the universal magnetoelectric effect in a 3D topological insulator. Nat. Commun. **8**, 15297 (2017)
33. X.-L. Qi, R. Li, J. Zang, S.-C. Zhang, Inducing a magnetic monopole with topological surface States. Science **323**, 1184 (2009)
34. C. Kim, E. Koh, K. Lee, Janus and multifaced supersymmetric theories. J. High Energy Phys. **0806**, 040 (2008)
35. C. Kim, E. Koh, K. Lee, Janus and multifaced supersymmetric theories II. Phys. Rev. D **79**, 126013 (2009)
36. F. Wilczek, Two applications of axion electrodynamics. Phys. Rev. Lett. **58**, 1799 (1987)
37. L. Huerta, J. Zanelli, Optical properties of a θ vacuum. Phys. Rev. D **85**, 085024 (2012)
38. A. Martín-Ruiz, L.F. Urrutia, Interaction of a hydrogenlike ion with a planar topological insulator. Phys. Rev. A **97**, 022502 (2018)
39. A. Martín-Ruiz, E. Chan-López, Dynamics of a Rydberg hydrogen atom near a topologically insulating surface. Eur. Phys. Lett. **119**, 53001 (2017)
40. A.M. Essin, J.E. Moore, D. Vanderbilt, Magnetoelectric polarizability and axion electrodynamics in crystalline insulators. Phys. Rev. Lett. **102**, 146805 (2009)
41. A. Martín-Ruiz, M. Cambiaso, L.F. Urrutia, The magnetoelectric coupling in electrodynamics. Int. J. Mod. Phys. A **34**, 1941002 (2019)
42. J. Schwinger, L. DeRaad, K. Milton, W. Tsai, *Classical Electrodynamics* (Perseus Books, Advanced Book Program, 1998)
43. A. Martín-Ruiz, M. Cambiaso, L.F. Urrutia, Green's function approach to Chern-Simons extended electrodynamics: An effective theory describing topological insulators. Phys. Rev. D **92**, 125015 (2015)
44. A. Martín-Ruiz, M. Cambiaso, L.F. Urrutia, Electro- and magnetostatics of topological insulators as modeled by planar, spherical, and cylindrical θ boundaries: Green's function approach. Phys. Rev. D **93**, 045022 (2016)
45. A. Martín-Ruiz, M. Cambiaso, L.F. Urrutia, Electromagnetic description of three-dimensional time-reversal invariant ponderable topological insulators. Phys. Rev. D **94**, 085019 (2016)
46. A. Martín-Ruiz, Magnetoelectric effect in cylindrical topological insulators. Phys. Rev. D **98**, 056012 (2018)
47. A. Martín-Ruiz, O. Rodríguez-Tzompantzi, J.R. Maze, L.F. Urrutia, Magnetoelectric effect of a conducting sphere near a planar topological insulator. Phys. Rev. A **100**, 042124 (2019)
48. U. Özgür, Y. Alivov, H. Morkoç, Microwave ferrites, part 2: Passive components and electrical tuning. J. Mater. Sci.: Mater. Electron. **20**, 911 (2009)
49. F. Olyslager, I.V. Lindell, Electromagnetics and exotic media: A quest for the Holy Grail. IEEE Antenna's Propagat. Mag. **44**, 48 (2002)
50. H.B.G. Casimir, On the attraction between two perfectly conducting plates. Proc. K. Ned. Akad. Wet. **51**, 793 (1948)
51. G. Bressi, G. Carugno, R. Onofrio, G. Ruoso, Measurement of the Casimir Force between parallel metallic surfaces. Phys. Rev. Lett. **88**, 041804 (2002)
52. K.A. Milton, *The Casimir Effect: Physical Manifestation of Zero-Point Energy* (World Scientific, Singapore, 2001)
53. M. Bordag, G.L. Klimchitskaya, U. Mohideen, V.M. Mostepanenko, *Advances in Casimir Effect* (Oxford University Press, Great Britain, 2009)
54. L. Fu, C.L. Kane, E.J. Mele, Topological insulators in three dimensions. Phys. Rev. Lett. **98**, 106803 (2007); D. Hsieh, D. Qian, L. Wray, Y. Xia, Y.S. Hor, R.J. Cava, M.Z. Hasan, A topological Dirac insulator in a quantum spin Hall phase. Nature **452**, 970 (2008)
55. L.S. Brown, G.J. Maclay, Vacuum stress between conducting plates: an image solution. Phys. Rev. **184**, 1272 (1969)
56. D. Deutsch, P. Candelas, Boundary effects in quantum field theory. Phys. Rev. D **20**, 3063 (1979)

57. A. Martín-Ruiz, M. Cambiaso, L.F. Urrutia, A Green's function approach to the Casimir effect on topological insulators with planar symmetry. Eur. Phys. Lett. **113**, 60005 (2016)
58. R.M. Cavalcanti, Casimir force on a piston. Phys. Rev. D **69**, 065015 (2004)
59. X. Zhou, J. Zhang, X. Ling, S. Chen, H. Luo, S. Wen, Photonic spin Hall effect in topological insulators. Phys. Rev. A **88**, 053840 (2013)
60. A.G. Grushin, A. Cortijo, Tunable Casimir Repulsion with three-dimensional topological insulators. Phys. Rev. Lett. **106**, 020403 (2011)
61. A.G. Grushin, P. Rodriguez-Lopez, A. Cortijo, Effect of finite temperature and uniaxial anisotropy on the Casimir effect with three-dimensional topological insulators. Phys. Rev. B **84**, 045119 (2011)
62. A. Martín-Ruiz, C.A. Escobar, Casimir effect between ponderable media as modeled by the standard model extension. Phys. Rev. D **94**, 076010 (2016)
63. A. Martín-Ruiz, C.A. Escobar, Local effects of the quantum vacuum in Lorentz-violating electrodynamics. Phys. Rev. D **95**, 036011 (2017)
64. O.J. Franca, L.F. Urrutia, O. Rodríguez-Tzompantzi, Reversed electromagnetic erenkov radiation in naturally existing magnetoelectric media. Phys. Rev. D **99**, 116020 (2019)
65. P.A. Čerenkov: Visible luminescence of pure liquids under the influence of γ-radiation. Dokl. Akad. Nauk SSSR **2**, 451 (1934)
66. S.I. Vavilov, On the possible causes of blue γ-glow of liquids. Dokl. Akad. Nauk SSSR **2**, 457 (1934)
67. I.M. Frank, I.E. Tamm: Coherent visible radiation of fast electrons passing through matter. Dokl. Akad. Nauk. **14**, 107 (1937) [Compt. Rend. (Dokl) **14**, 109 (1937)]
68. V.P. Jelley, Cherenkov radiation and its applications. Br. J. Appl. Phys. **6**, 227 (1955)
69. V.P. Jelley, *Cherenkov Radiation and its Applications* (Pergamon, Oxford, 1958)
70. V.G. Veselago, The electrodynamics of substances with simultaneously negative values of ϵ and μ. Soviet Physics Uspekhi **10**, 509 (1968)
71. J.B. Pendry, A.J. Holden, W.J. Stewart, I. Youngs, Extremely low frequency plasmons in metallic mesostructures. Phys. Rev. Lett. **76**, 4773 (1996)
72. J.B. Pendry, A.J. Holden, D.J. Robbins, W.J. Stewart, Magnetism from conductors and enhanced nonlinear phenomena. IEEE Trans. Microwave Theory Tech. **47**, 2075 (1999)
73. J. Lu, T.M. Grzegorczyk, Y. Zhang, J. Pacheco Jr., B.-I. Wu, J.A. Kong, M. Chen, Čerenkov radiation in materials with negative permittivity and permeability. Opt.Exp. **11**, 723 (2003)
74. C. Luo, M. Ibanescu, S.G. Johnson, J.D. Joannopoulos, Cerenkov radiation in photonic crystals. Science **229**, 368 (2003)
75. Z.Y. Duan, B.-I. Wu, S. Xi, H.S. Chen, M. Chen, Research progress in reversed Cherenkov radiations in double-negative metamaterials. Prog. Electromagn. Res. **90**, 75 (2009)
76. S. Xi, H. Chen, T. Jiang, L. Ran, J. Huangfu, B.-I. Wu, J.A. Kong, M. Chen, Experimental verification of reversed Cherenkov radiation in left-handed metamaterial. Phys. Rev. Lett. **103**, 194801 (2009)
77. H. Chen, M. Chen, Flipping photons backward: reversed Cherenkov radiation. Materials Today **14**, 34 (2011)
78. Z. Duan, X. Tang, Z. Wang, Y. Zhang, X. Chen, M. Chen, Y. Gong, Observation of the reversed Cherenkov radiation. Nat. Commun. **8**, 14901 (2017)
79. J. Tao, Q.J. Wang, J. Zhang, Y. Luo, Reverse surface-polariton Cherenkov radiation. Sci. Rep. **6**, 30704 (2016)
80. R.A. Shelby, D.R. Smith, S. Schultz, Experimental verification of a negative index of refraction. Science **292**, 77 (2001)
81. S. Smirnov, Cherenkov sound on a surface of a topological insulator. Phys. Rev. B **88**, 205301 (2013)
82. S. Smirnov, Asymmetric Cherenkov acoustic reverse in topological insulators. Phys. Rev. B **90**, 125305 (2014)
83. W.C. Chew, *Waves and Fields in Inhomogenous Media* (IEEE, New York, 1990)
84. W.C. Chew, A quick way to approximate a Sommerfeld-Weyl-type integral (antenna far-field radiation). IEEE Trans. Antenas Propag. **36**, 1654 (1988)

85. L. Mandel, E. Wolf, *Optical Coherence and Quantum Optics* (Cambridge University Press, Cambridge, England, 1995)
86. A. Sommerfeld, *Partial Differential Equations in Physics* (Academic Press, New York, 1964)
87. J. Schwinger, L. DeRaad and K. Milton: Casimir effect in dielectrics. Ann. Phys. (N.Y.) **115**, 1 (1978)
88. W.K.H. Panofsky, M. Phillips, *Classical Electricity and Magnetism*, 2nd edn. (Addison-Wesley, Reading, MA, 1962)
89. A.A. Zyuzin, A.A. Burkov, Topological response in Weyl semimetals and the chiral anomaly. Phys. Rev. B **86**, 115133 (2012)
90. A.A. Zyuzin, S. Wu, A.A. Burkov, Weyl semimetal with broken time reversal and inversion symmetries. Phys. Rev. B **85**, 165110 (2012)
91. P. Goswami, S. Tewari, Axionic field theory of (3+ 1)-dimensional Weyl semimetals. Phys. Rev. B **88**, 245107 (2013)
92. K. Landsteiner, Notes on anomaly induced transport. Acta Phys. Pol. B **47**, 2617 (2016)
93. M. Kargarian, M. Randeria, N. Trivedi, Theory of Kerr and Faraday rotations and linear dichroism in Topological Weyl Semimetals. Sci. Rep. **5**, 12683 (2015)
94. J.H. Wilson, A.A. Allocca, V. Galitski, Repulsive Casimir force between Weyl semimetals. Phys. Rev. B **91**, 235115 (2015)
95. J. Hofmann, S.D. Sarma, Surface plasmon polaritons in topological Weyl semimetals Phys. Rev. B **93**, 241402(R) (2016)
96. F.M.D. Pellegrino, M.I. Katsnelson, M. Polini, Helicons in Weyl semimetals. Phys. Rev. B **92**, 201407(R) (2015)
97. A. Karch, Electric-magnetic duality and topological insulators. Phys. Rev. Lett. **103**, 171601 (2009)
98. J.-R. Soh, F. de Juan, M.G. Vergniory, N.B.M. Schröter, M.C. Rahn, D.Y. Yan, J. Jiang, M. Bristow, P. Reiss, J.N. Blandy, Y.F. Guo, Y.G. Shi, T.K. Kim, A. McCollam, S.H. Simon, Y. Chen, A.I. Coldea, A.T. Boothroyd, Ideal Weyl semimetal induced by magnetic exchange. Phys. Rev. B **100**, 201102(R) (2019)
99. A. Martín-Ruiz, M. Cambiaso, L.F. Urrutia, Electromagnetic fields induced by an electric charge near a Weyl semimetal. Phys. Rev. B **99**, 155142 (2019)
100. S.-M. Huang, S.-Y. Xu, I. Belopolski, C.-C. Lee, G. Chang, B. Wang, N. Alidoust, G. Bian, M. Neupane, C. Zhang, S. Jia, A. Bansil, H. Lin, M.Z. Hasan, A Weyl Fermion semimetal with surface Fermi arcs in the transition metal monopnictide TaAs class. Nat. Commun. **6**, 7373 (2015)
101. J. Zang, N. Nagaosa, Monopole current and unconventional Hall response on a topological insulator. Phys. Rev. B **81**, 245125 (2010)
102. J.C. Allred, R.N. Lyman, T.W. Kornack, M.V. Romalis, High-sensitivity atomic magnetometer unaffected by spin-exchange relaxation. Phys. Rev. Lett. **89**, 130801 (2002)

Chapter 18
Purcell Effect in PT-Symmetric Waveguides

Alina Karabchevsky, Andrey Novitsky, and Fyodor Morozko

Abstract This chapter overviews the principles of the spontaneous emission rate increase, that is the Purcell effect, in relation to the photonic parity-time (PT) symmetry. Being focused on the system of coupled PT-symmetric optical waveguides, we consider behaviors of the Purcell factor in PT-symmetric and broken-PT-symmetric regimes. Surprisingly, exceptional points in a coupled waveguide do not influence on the Purcell factor.

18.1 Introduction

By exploring the interplay between loss and gain as well as the coupling mechanism in waveguide-emitter systems, one can generate and control light on a chip. This chapter introduces the underlying physics of Purcell effect for emitters in PT-symmetric waveguides. In general, physical world exhibits symmetries lying behind the conservation laws of physics. They help to control the structure of matter and define interactions. The laws of physics are required to be invariant under changes of redundant degrees of freedom dictated by the symmetries. There are several fundamental symmetries including the charge conjunction or *C symmetry* for a particle and its anti-particle, parity or *P symmetry* for a system and its mirror image and time reversal or *T symmetry* for the time running forward and backward. Despite the fact that the laws of physics are dictated by symmetries, it is the symmetry breaking that

A. Karabchevsky (✉) · F. Morozko
School of Electrical and Computer Engineering, Ben-Gurion University of the Negev, 8410501 Beer-Sheva, Israel
e-mail: alinak@bgu.ac.il

F. Morozko
e-mail: fyodormorozko95@gmail.com

A. Novitsky · F. Morozko
Department of Theoretical Physics and Astrophysics, Belarusian State University, Nezavisimosti Avenue 4, 220030 Minsk, Belarus
e-mail: andreyvnovitsky@gmail.com

E. Kamenetskii (ed.), *Chirality, Magnetism and Magnetoelectricity*, Topics in Applied Physics 138,
https://doi.org/10.1007/978-3-030-62844-4_18

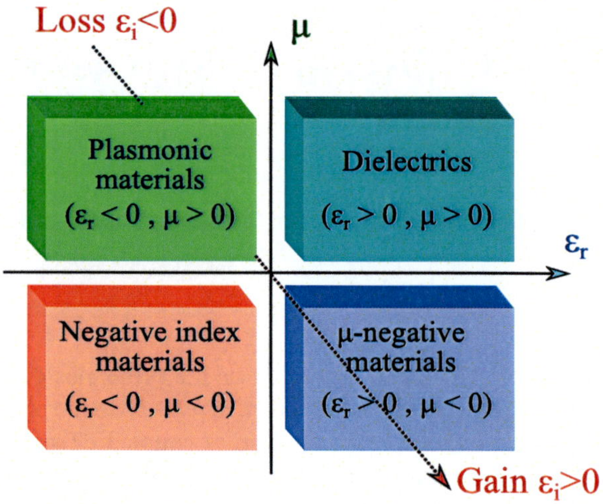

Fig. 18.1 Schematics of materials characterisation in terms of loss and gain

creates nontrivial physics by lifting the degeneracies. A number of intriguing properties in photonics are related to the PT-symmetry usually described by non-Hermitian systems. Non-Hermitian Hamiltonians possessing parity-time (PT) symmetry that is the symmetry with respect to the simultaneous coordinate and time reversal [1]. There is still a debate whether PT-symmetry is a fundamental feature or shares common properties with naturally occurring symmetries. Also questionable is the phenomenon of phase transition and it is important to understand the spectral degeneracies induced by PT-symmetry named exceptional points (EP) which is a point in parameter space at which phase transition occurs.

Controlling the magnetic permeability μ and the real part of the dielectric permittivity ε_r has enabled novel functionalities. PT symmetry and non-Hermitian photonics open new possibilities by controlling the imaginary part of the dielectric permittivity (ε_i), and by considering gain and loss. Figure 18.1 schematically shows this interplay when characterising materials in terms of the gain and loss.

The chapter is organised in the following way: In Sect. 18.2, we introduce the principles of PT-symmetry. PT-symmetric photonic devices are presented in Sect. 18.3 such as those based on coupled-mode theory, two-dimensional photonic waveguide lattices, multilayer structures, and microresonators. Purcell effect in PT-symmetric waveguides is described in Sect. 18.4. Eventually, the Sect. 18.5 summarises the chapter and gives an outlook to future research.

18.2 Principles of PT Symmetry

In 1998 Bender and Bötcher [2] have shown that quantum systems with non-Hermitian Hamiltonians can have entirely real spectra. Such Hamiltonians are referred to as pseudo-Hermitian. The first known class of pseudo-Hermitian Hamiltonians are PT-symmetric ones i.e. those commuting with $\hat{P}\hat{T}$ operator where $\hat{P}$ and $\hat{T}$ are correspondingly coordinate and time reversal operators

$$\hat{P}\hat{T}\hat{H} = \hat{H}\hat{P}\hat{T}. \tag{18.1}$$

Operator $\hat{P}$ changes sign of coordinates whereas $\hat{T}$ operator changes sign of time and performs complex conjugation [1]. This means that PT-invariance of the Hamiltonian can be stated in the form

$$\hat{H}(\hat{\mathbf{p}}, \hat{\mathbf{r}}, t) = \hat{H}^*(\hat{\mathbf{p}}, -\hat{\mathbf{r}}, -t). \tag{18.2}$$

For Hamiltonians of the form

$$\hat{H} = \frac{\hat{\mathbf{p}}^2}{2m} + V(\mathbf{r}), \tag{18.3}$$

where $\hat{\mathbf{p}}$ is the momentum operator, m is mass, and V is complex potential, action of $\hat{P}\hat{T}$ operator results in the Hamiltonian

$$\hat{P}\hat{T}\hat{H} = \hat{H}\hat{P}\hat{T} = \frac{\hat{\mathbf{p}}^2}{2m} + V^*(-\mathbf{r}). \tag{18.4}$$

Therefore, for the Hamiltonian (18.3) to be PT-invariant it is needed that potential energy $V(\mathbf{r})$ satisfies the condition

$$V(\mathbf{r}) = V^*(-\mathbf{r}). \tag{18.5}$$

In other words, real part of the potential energy must be even function of coordinates whereas imaginary part must be odd function.

It can be shown that if the eigenfunctions $|\psi_n\rangle$ of the PT-symmetric Hamiltonian $\hat{H}$,

$$\hat{H}|\psi_n\rangle = E_n|\psi_n\rangle, \tag{18.6}$$

with the corresponding eigenvalues E_n are also eigenfunctions of the $\hat{P}\hat{T}$ operator

$$\hat{P}\hat{T}|\psi_n\rangle \equiv \sigma_n|\psi_n\rangle \tag{18.7}$$

with some (complex) eigenvalues σ_n, the eigenvalues E_n of the Hamiltonian are real. Condition (18.7) is necessary and sufficient for eigenvalues of the Hamiltonian [1] to be real. Hence, if the eigenvalues E_n are real, the eigenfunctions are PT-symmetric and the system is considered to be in PT-symmetric regime (phase). Contrarily, if the eigenvalues are complex, the eigenfunctions are essentially not PT-symmetric and the system is in PT-symmetry-broken regime.

In contrast to Hermitian case, eigenfunctions of pseudo-Hermitian Hamiltonians are not orthogonal under conventional inner product. Instead, they obey more general biorthogonality relations. Orthogonality can be re-established by modifying the inner product. Discussion of quantum mechanics based on biorthogonal states is given in [3–6].

18.2.1 Phase Transition in PT-Symmetric Systems

One of the most intriguing features of PT-symmetric systems is a phase transition from the PT-symmetric to PT-symmetry-broken phase. If the Hamiltonian of the system $\hat{H}(\hat{\mathbf{p}}, \hat{\mathbf{r}}, t, p)$ depends on some parameter p, the Hamiltonian can have real as well as complex eigenvalues being, either in PT-symmetric or non-PT-symmetric states, respectively. When due to variation of the parameter p the system's spectrum changes from real to complex and vice-versa, one can study the phase transition related to the spontaneous breaking of PT symmetry. The point in parameter space $p = p_c$ at which phase transition occurs is named as an exceptional point (EP). At the EP, both eigenvalues and eigenfunctions coalesce.

18.2.2 PT-Symmetry in Optics

Quantum-mechanical concept of the PT symmetry can be realised in optics due to the fact that Maxwell's equations in case of two- and one-dimensional photonic structures can be reformulated into an equation formally coinciding with the Schrödinger equation.

With light propagation along these structures it is convenient to introduce the so called slowly varying envelope (SVE) field, where most of the electromagnetic field variation is extracted by defining a suitably selected reference propagation constant [7]. Specifically, if the structure is invariant in z, SVE $\mathbf{e}$ for electric field is defined as

$$\mathbf{E}(x, y, z) = \mathbf{e}(x, y, z)\mathrm{e}^{-ik_0 n_0 z} \tag{18.8}$$

where $k_0 = \omega/c$ is the vacuum wavenumber and n_0 is the reference (background) refractive index. SVE $\mathbf{h}$ for magnetic field is defined analogously as

$$\mathbf{H}(x, y, z) = \mathbf{h}(x, y, z)\mathrm{e}^{-ik_0 n_0 z}. \tag{18.9}$$

Within the slowly varying envelope approximation (SVEA) it is assumed that

$$\frac{\partial^2}{\partial z^2}\begin{pmatrix}\mathbf{e}\\ \mathbf{h}\end{pmatrix} \ll 2k_0 n_0 \frac{\partial}{\partial z}\begin{pmatrix}\mathbf{e}\\ \mathbf{h}\end{pmatrix} \tag{18.10}$$

and the second-order z-derivative terms are neglected. Due to z-invariance of the structure transverse and longitudinal components of $\mathbf{e}$ and $\mathbf{h}$ decouple. The transverse components of SVE fields $\mathbf{e}_t$ and $\mathbf{h}_t$, hence, satisfy first-order equations with respect to the z derivative. The above can be summarised in a Schrödinger-like equation

$$\mathrm{i}\frac{\partial}{\partial z}|\psi\rangle = \hat{H}|\psi\rangle, \tag{18.11}$$

for an optical state-vector $|\psi\rangle$ defined as

$$|\psi\rangle = \begin{pmatrix} \mathbf{e}_t \\ \mathbf{h}_t \end{pmatrix}. \tag{18.12}$$

$\hat{H}$ is an optical Hamiltonian governing the z-evolution of SVE fields. Generally, $\hat{H}$ is represented by a 4×4 matrix joining operators describing evolution of $\mathbf{e}_t$ and $\mathbf{h}_t$. Explicit form of these operators found in [7]. For waveguide structures with very small index contrast in both transverse directions equation (18.11) can be reduced to a scalar equation. Within the scalar approximation the Hamiltonian $\hat{H}$ takes the form

$$\hat{H} = \frac{1}{2k_0 n_0}\left(\frac{\partial^2}{\partial x^2} + \frac{\partial^2}{\partial y^2} + V(x, y)\right). \tag{18.13}$$

Quantity

$$V(x, y) = k_0^2\left(\varepsilon(x, y) - n_0^2\right) \tag{18.14}$$

can be associated with a potential of the Schödinger equation. From the condition of the PT symmetry in quantum mechanics $V(x, y) = V^*(-x, -y)$ we arrive at the similar condition in optics $\varepsilon(x, y) = \varepsilon^*(-x, -y)$. Therefore, in optical systems the PT symmetry can be established by judiciously incorporating gain and loss. Thus, the refractive index profile now plays the role of the complex potential.

18.2.3 Inner Product for PT-Symmetric Optical Systems

We define the inner product as a cross product of the bra-electric and ket-magnetic fields integrated over the cross-section $z = \text{const}$:

$$\langle\phi_1|\phi_2\rangle \equiv \int (\mathbf{E}_1 \times \mathbf{H}_2) \cdot \hat{\mathbf{z}} dx dy \tag{18.15}$$

Such a definition is justified by the non-Hermitian nature of PT-symmetric systems. In the above and following relations we can drop t subscripts because z component of the vector products depends only on transverse components and $\left(\mathbf{E}_{t,1} \times \mathbf{H}_{t,2}\right) \cdot \hat{\mathbf{z}} = (\mathbf{E}_1 \times \mathbf{H}_2) \cdot \hat{\mathbf{z}}$.

It is well known that the modes of Hermitian systems are orthogonal in the sense

$$\langle i|j^*\rangle = \int \left(\mathbf{e}_i \times \mathbf{h}_j^*\right) \cdot \hat{\mathbf{z}} \mathrm{d}x\mathrm{d}y \sim \delta_{ij}, \tag{18.16}$$

where δ_{ij} is the Kronecker delta. Here and below $|\psi^*\rangle = (\mathbf{e}_t^*, \mathbf{h}_t^*)^T$. Relationship (18.16) is often referred to as power orthogonality, because $\frac{1}{2}\,\mathrm{Re}\,\langle i|i^*\rangle$ is the power carried by the mode $|i\rangle$. However, the loss and gain in the non-Hermitian systems break power orthogonality. In this case, one should use a non-conjugate inner product [8–10] bringing us to the orthogonality relationship

$$\langle i|j\rangle = \int \left(\mathbf{e}_i \times \mathbf{h}_j\right) \cdot \hat{\mathbf{z}} \mathrm{d}x\mathrm{d}y = 2N_i\delta_{ij}, \tag{18.17}$$

where N_i is a normalisation parameter. We want to stress that orthogonality relation (18.17) is valid not only for PT-symmetric but for arbitrary non-Hermitian systems.

Forward and backward transverse modal fields $\mathbf{e}_{t,i}$ and $\mathbf{h}_{t,i}$ satisfy the symmetry relations

$$\mathbf{e}_{t,-i} = \mathbf{e}_{t,i}, \qquad \mathbf{h}_{t,-i} = -\mathbf{h}_{t,i} \tag{18.18}$$

both in the case of Hermitian and non-Hermitian systems.

This means that the inner product of the modes also meets the symmetry relations for its bra- and ket-parts:

$$\langle -i|j\rangle = \langle i|j\rangle, \tag{18.19}$$

$$\langle i|-j\rangle = -\langle i|j\rangle. \tag{18.20}$$

18.2.4 Petermann Factor

It is common to express non-orthogonality of the modes quantitatively in terms of Petermann factor [11–15]. Petermann factor is defined as the squared ratio between Hermitian and non-Hermitian norms. In our notation Petermann factor K_i of the mode $|i\rangle$ reads as

$$K_i = \frac{|\langle i|i^*\rangle|^2}{|\langle i|i\rangle|^2}. \tag{18.21}$$

Petermann factor obviously equals to unity in Hermitian case since in this case transverse modal fields always can be rescaled to be real. Hence, non-Hermitian norm is equal to the Hermitian norm and to the power carried by the mode.

18.2.5 Eigenmodes of PT-Symmetric Optical Systems

To get some insight on the eigenstates of photonic PT-symmetric systems, let us analyse the system of two coupled waveguides using the coupled mode theory. Coupled waveguides are the simplest systems proposed at the beginning of the era of optical PT symmetry [16]. As schematically shown in Fig. 18.5a, they consist of gain and lossy waveguides having identical geometrical parameters at a distance g one from another. The waveguide can be either slab, rectangular, circular or gradient one, yet the physics behind the coupling mechanism is the same.

We express the total field in the coupled system in terms of the modes $|g\rangle = (\mathbf{e}_{g,t}, \mathbf{h}_{g,t})^T$ and $|l\rangle = (\mathbf{e}_{l,t}, \mathbf{h}_{l,t})^T$ of isolated gain and loss waveguides with corresponding z-dependent amplitudes g and l as

$$|\psi\rangle = g(z)|g\rangle + l(z)|l\rangle. \tag{18.22}$$

We assume that the overlap between the modes of isolated waveguides is negligible (weak coupling condition), therefore, the modes are orthogonal and normalised as follows

$$\langle g|l\rangle = \langle g|l^*\rangle = 0, \tag{18.23}$$

$$\langle g|g\rangle = \langle l|l\rangle = 1. \tag{18.24}$$

$\hat{P}\hat{T}$ operator converts the mode of isolated lossy waveguide to the mode of the isolated gain waveguide and vice versa, and so

$$\hat{P}\hat{T}|g\rangle = |l\rangle, \tag{18.25a}$$

$$\hat{P}\hat{T}|l\rangle = |g\rangle. \tag{18.25b}$$

Coupled mode theory for optical PT-symmetric systems can be formulated on the basis of Lagrangian formalism [17] or by using Lorentz reciprocity theorem [18].

Spatial evolution of amplitudes is governed by the system of coupled equations

$$\mathrm{i}\frac{\mathrm{d}}{\mathrm{d}z}\begin{pmatrix} g \\ l \end{pmatrix} = \begin{pmatrix} \mathrm{Re}\,(\beta+\delta) - \mathrm{i}\alpha/2 & \kappa \\ \kappa & \mathrm{Re}\,(\beta+\delta) + \mathrm{i}\alpha/2 \end{pmatrix}\begin{pmatrix} g \\ l \end{pmatrix} \tag{18.26}$$

where β is a propagation constant, κ is a coupling coefficient, δ is a correction to the propagation constant, α is an effective gain (or loss). It can be shown that due to the weak coupling and relations (18.25) the coupling constant κ is real [17, 18].

Matrix in the right hand side of (18.26) is the matrix of the system's Hamiltonian in the basis $|g\rangle$, $|l\rangle$.

The eigenvalues of this Hamiltonian are the propagation constants of the sytem's eigenmodes. They read as

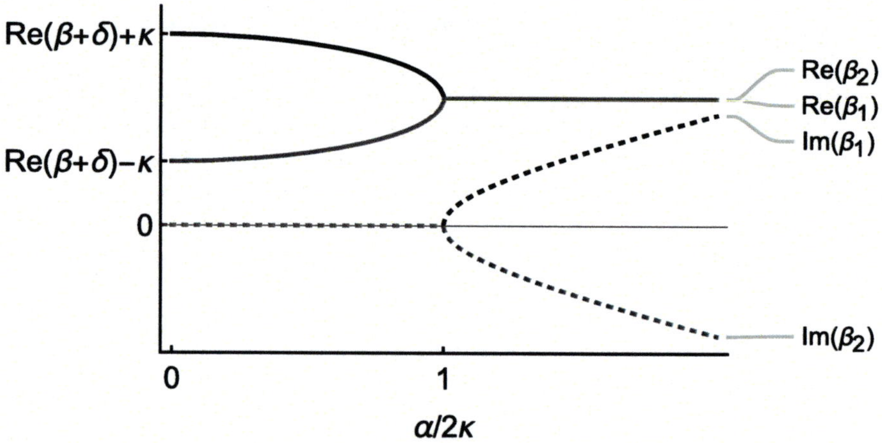

Fig. 18.2 Eigenvalues of the coupled waveguide system's Hamiltonian versus non-Hermiticity parameter $\alpha/2\kappa$. Black curves correspond to the first supermode, grey curves correspond to the second supermode. Solid curves correspond to real parts, dotted curves correspond to imaginary parts

$$\beta_{1,2} = \mathrm{Re}\,(\beta + \delta) \pm \sqrt{\kappa^2 - \alpha^2/4}. \tag{18.27}$$

Clearly, the system behaves differently depending on whether $\alpha/2$ is less or greater than κ. When $\alpha/2$ is less than κ both propagation constants are real. When $\alpha/2$ is greater than κ the eigenvalues constitute complex-conjugate pair and one mode experiences gain whereas the other one experiences loss. When $\alpha = \alpha_c = 2\kappa$ modes degenerate. Therefore the point $\alpha = \alpha_c$ corresponds to exceptional point (EP). The situation when α passes through α_c is called the phase transition. Phase diagram of a PT-symmetric coupled waveguide system in Fig. 18.2 shows distribution of real and imaginary parts of system's eigenvalues. It illustrates a typical picture of the phase transition in a PT-symmetric system.

In PT-symmetric regime, the eigenvalues can be written as

$$\beta_{1,2} = \mathrm{Re}\,(\beta + \delta) \pm \kappa \cos\theta, \tag{18.28}$$

where $\sin\theta = \alpha/2\kappa$. With this parametrization supermodes take the form

$$|1, 2\rangle = |g\rangle \pm \mathrm{e}^{\pm i\theta}|l\rangle. \tag{18.29}$$

It can be seen from (18.25) that the states (18.29) are indeed the eigenstates of the $\hat{P}\hat{T}$ operator

$$\hat{P}\hat{T}|1\rangle = |l\rangle + \mathrm{e}^{-i\theta}|g\rangle = \mathrm{e}^{-i\theta}|1\rangle, \tag{18.30}$$
$$\hat{P}\hat{T}|2\rangle = |l\rangle - \mathrm{e}^{+i\theta}|g\rangle = \mathrm{e}^{+i\theta}|2\rangle. \tag{18.31}$$

In the PT-broken regime, eigenvalues can be written as

$$\beta_{1,2} = \mathrm{Re}\,(\beta + \delta) \pm \mathrm{i}\kappa \sinh\theta, \tag{18.32}$$

where $\cosh\theta = \alpha/2\kappa$. Supermodes then read as

$$|1, 2\rangle = |g\rangle + \mathrm{i}e^{\mp\theta}|l\rangle. \tag{18.33}$$

The eigenmodes in PT-broken regime are not longer the eigenstates of the $\hat{P}\hat{T}$ operator. Instead, in this regime $\hat{P}\hat{T}$ operator relates $|1\rangle$ and $|2\rangle$ as follows

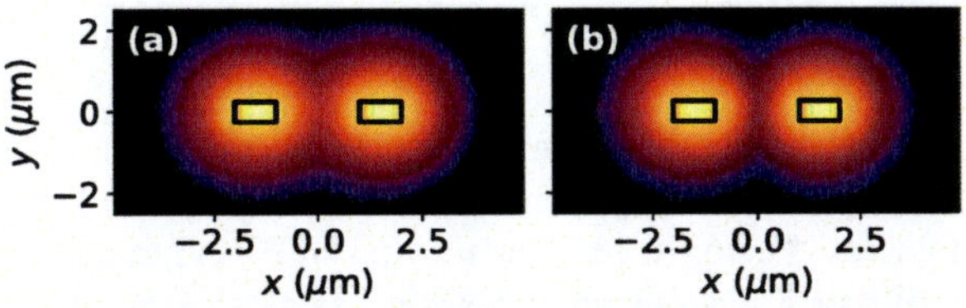

Fig. 18.3 Mode profiles in PT-symmetric regime: **a** mode $|1\rangle$ and **b** mode $|2\rangle$

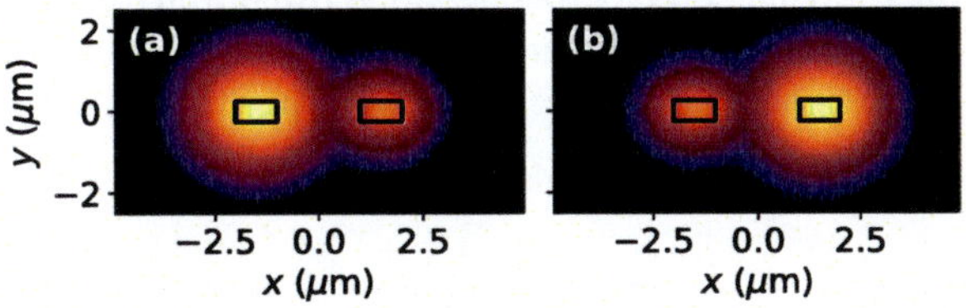

Fig. 18.4 Mode profiles in PT-symmetry-broken regime: **a** mode $|1\rangle$ and **b** mode $|2\rangle$

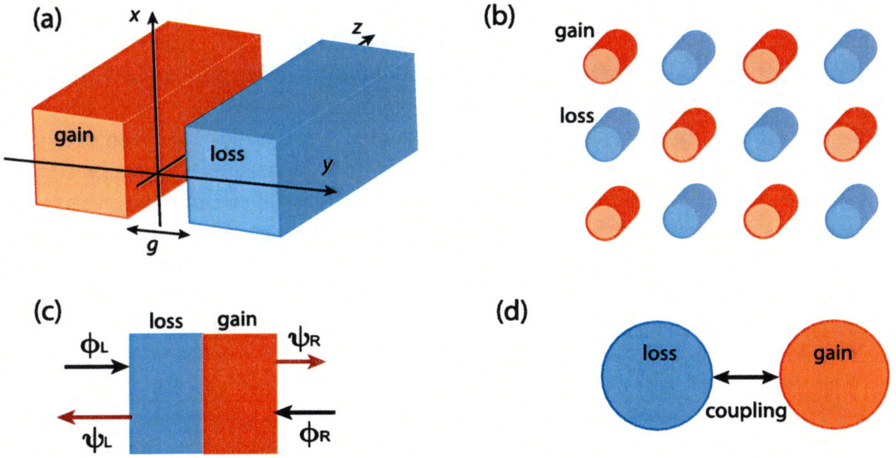

Fig. 18.5 Schematics of state-of-art PT-symmetric structures: **a** coupled waveguides, **b** two-dimensional photonic lattices, **c** multilayer systems and **d** microresonators

$$\hat{P}\hat{T}|1\rangle = |l\rangle - \mathrm{i}e^{-\theta}|g\rangle = \mathrm{i}e^{-\theta}|2\rangle, \tag{18.34}$$

$$\hat{P}\hat{T}|2\rangle = |g\rangle - \mathrm{i}e^{\theta}|g\rangle = -\mathrm{i}e^{\theta}|1\rangle. \tag{18.35}$$

Typical mode profiles for the coupled waveguide system (see Fig. 18.5a) in PT-symmetric and PT-broken regimes are shown in Fig. 18.3 and in Fig. 18.4.

When the parameter α approaches the value $\alpha_c = 2\kappa$ corresponding to the EP, eigenmodes $|1, 2\rangle$ merge to become $|g\rangle + \mathrm{i}|l\rangle$. Interestingly, that the modes become self-orthogonal as $\langle 1|1\rangle = \langle 2|2\rangle = 0$ at the EP. Self-orthogonality is responsible for singularity of Petermann factor due to zero in denominator in (18.21).

18.3 PT-Symmetric Photonic Devices

Photonics is an excellent platform for experimental verification of the fundamental concept of the parity-time symmetry discussed earlier. Novel photonic devices can be fabricated using several basic types of PT-symmetric structures, such as waveguides, multilayer systems and photonic crystals. A number of remarkable applications of the PT symmetry have been proposed and well studied including unidirectional invisibility, lasing, sensing and coherent perfect absorption. In this section, we overview the recent PT-symmetric photonic devices with the application perspective.

18.3.1 Coupled Waveguide Systems

Coupled waveguides were the first candidates for observation of the parity-time symmetry. In [19] it was fabricated a gradient-index system with balanced loss and gain, the gain being guaranteed by the photorefractive nonlinearity of Fe-doped $LiNbO_3$. The detailed description of the physics of coupled waveguides has been provided in the previous section. Here we discuss a gain-free technique for observation of the PT symmetry demonstrated in practice in [20]. The idea behind the passive PT symmetry is to carry out transformation of the fields in a purely lossy system with the aim of reducing the governing equation to that describing PT-symmetric systems. In fact, using the gauge transformation $|\psi\rangle = \exp(-\gamma z)|\tilde{\psi}\rangle$, we can rewrite (18.11) of the passive system as

$$\mathrm{i}\frac{\partial}{\partial z}|\tilde{\psi}\rangle = \hat{\tilde{H}}|\tilde{\psi}\rangle \tag{18.36}$$

Now we claim that (18.36) describes a PT-symmetric system with the effective Hamiltonian $\hat{\tilde{H}}$ which corresponds to the system with the effective permittivity

$$\tilde{\varepsilon}(x, y) = \varepsilon(x, y) + 2\mathrm{i}\frac{n_0\gamma}{k_0}. \tag{18.37}$$

$\tilde{\varepsilon}$ satisfies condition $\tilde{\varepsilon}(x, y) = \tilde{\varepsilon}^*(-x, -y)$ and the permittivity of the passive system meets

$$\varepsilon(x, y) = \varepsilon^*(-x, -y) - 4\mathrm{i}\frac{n_0\gamma}{k_0}. \tag{18.38}$$

Assuming that one of the waveguides is transparent (lossless) with $\varepsilon_g = \varepsilon_g^*$, one can easily determine the permittivity of the lossy waveguide as $\varepsilon_l = \varepsilon_g - 4\mathrm{i}n_0\gamma/k_0$. In spite of the gain-free system does not have a true PT symmetry, it still possesses some features inherent in PT-symmetric systems as PT-symmetry breaking observed in [20]. Passive PT symmetry is a smart technique to ease validation of PT-symmetry effects and its applicability.

A number of surprising effects arise in guiding systems under gain-loss modulation in a PT-symmetric manner. The modulation shifts positions of exceptional points resulting in the directional amplification, when the phase transition is made with a threshold in one direction, being thresholdless in the opposite direction [21]. Similarly, on the boundary between the metallic substrate and PT-symmetrically modulated dielectric slab, there is a unidirectional excitation of the surface plasmon-polaritons [22]. In a waveguide as an isotropic slab sandwiched between oppositely biased gyrotropic media, topologically protected guided modes arise. PT symmetry in such a system introduces exceptional points, where electromagnetic modes are slow-light and linearly-growing [23]. Slow light phenomenon is known to be associated with degeneracy of the modes (matching of their propagation constants). In PT-symmetric systems, the degeneration is realised at exceptional points of mode coalescence [24].

PT-symmetric laser waveguide was fabricated in [25]. Gain and loss are electrically controlled to achieve a lasing threshold in the range of PT-symmetry violation. By interplay of two guided modes there were distinguished several phases, the lasing within which being confirmed experimentally.

Two coupled waveguides experience optical forces originating from excited eigenmodes. The forces qualitatively change at exceptional points and may result in pushing and pulling of one waveguide to another. When the forces deflect the waveguides, they may induce the phase transition through changing a gap between them [26]. An unusual power flow in the PT-symmetric coupled waveguide results in an off-diagonal stress tensor components causing the shear along the mode propagation direction [27].

18.3.2 Two-Dimensional Photonic Waveguide Lattices

An array of parallel waveguides can be arranged in nodes of a lattice as demonstrated in Fig. 18.5b. Such a two-dimensional photonic crystal is a natural generalization of a pair of coupled waveguides. To engage a PT symmetry in the lattice, the gain and lossy waveguides should be disposed periodically. As other PT-symmetric systems, the lattice has exceptional points bordering phases of broken and unbroken PT-symmetric states. At the same time, light beams propagating in lattices demonstrate beam splitting, power oscillations, nonreciprocity and secondary emissions [28]. These diffraction properties are explained by nonorthogonality of the Floquet-Bloch modes of the periodic structure.

In the system of PT-symmetric periodically arranged cylinders situated at the interface between two semi-infinite media, unidirectional transmission without reflection can be achieved. It was investigated in [29] using the perturbation and scattering matrix theories. Photonic graphene lattice of waveguides described using the coupled-mode techniques can be represented as the Dirac equation. The PT symmetry in such a system requires the corresponding Hamiltonian is non-Hermitian exhibiting unbroken and broken PT-symmetry phases. These theoretical findings are well confirmed in experiments [30]. In [31] photonic lattices were used for the proof of existence of topological interface states on a defect waveguide between two PT-symmetric media. The results propose a way of excitation of topologically protected localised states. A PT-symmetric photonic crystal can be also designed as a group of gain cylinders paired with a group of lossy cylinders. Then surface electromagnetic waves emerge at the gain-loss interface, while exceptional points can be tuned to coalesce forming higher-order exceptional points [32].

18.3.3 Multilayer Structures

To illustrate the basic principles of the PT symmetry, a simple multilayer system may be harnessed. The PT symmetry in multilayer structures is introduced in direction of the wave propagation, and multilayers as open systems can be described by a scattering matrix. The scattering matrix connects the input and output channels. For the multilayer system there are two input channels and two output channels as demonstrated in Fig. 18.5c. The output fields $\psi_R = t\phi_L + r_R\phi_R$ and $\psi_L = t\phi_R + r_L\phi_L$ can be arranged as

$$\begin{pmatrix} \psi_R \\ \psi_L \end{pmatrix} = \begin{pmatrix} r_R & t \\ t & r_L \end{pmatrix} \begin{pmatrix} \phi_R \\ \phi_L \end{pmatrix} \quad \text{or} \quad \begin{pmatrix} \psi_L \\ \psi_R \end{pmatrix} = \begin{pmatrix} t & r_L \\ r_R & t \end{pmatrix} \begin{pmatrix} \phi_R \\ \phi_L \end{pmatrix}, \qquad (18.39)$$

where $\phi_{L,R}$ are the input fields, t is the transmission coefficent and r_L and r_R are the reflectrion coefficients to the left and to the right sides, respectively. Transmission coefficient t does not depend on the direction of wave incidence owing to reciprocity of the system.

Equation (18.39) shows that the scattering matrix as a matrix between input and output fields can be defined in two different ways by means of the permutation of the output channels. Such a nominal designation is expected to be unimportant. However, since the scattering matrix eigenvalues are different for two matrices in (18.39), but, as generally accepted, should predict exceptional points, a dilemma arises, which scattering matrix is appropriate [33, 34]. The problem of uniqueness of the scattering matrix of the PT-symmetric system has been successfully solved in [35] using the direct connection of the scattering matrix $\hat{S}$ with the PT-symmetric Hamiltonian $\hat{H}$ of the one-dimensional multilayer system as $\hat{S} = \exp(\mathrm{i}\hat{H}(t - t_0)/\hbar)$. Correct positions of the exceptional points then read as $s_{1,2} = t \pm \sqrt{r_L r_R}$, where the scattering matrix defined by the right-hand equation in (18.39) is employed. Exceptional points of the scattering matrix with permuted channels given by the another scattering matrix approximate the lasing onset.

PT-symmetric multilayer systems are widely studied with the aim of enhancement of physical phenomena near exceptional points. A PT-symmetric bilayer exhibits a giant Goos-Hänhen shift at specific angles of incidence [36]. The enhancement is explained by excitation of surface modes at the interface between the gain and lossy slabs facilitating the quasi-BIC (quasi-bound state in the continuum) states. An ordinarily weak spin-orbit interaction of light can be also significantly intensified in vicinity of exceptional points of the PT-symmetric bilayer. Interaction of light spin and orbital momenta is coined as the spin Hall or Imbert-Fedorov effect and results in a lateral shift of a light beam. Near exceptional points of the PT-symmetric bilayer the reflection coefficient experiences negligible values and abrupt phase shift enhancing the lateral beam displacement, though it takes zero value at the exceptional point [37]. Graphene sheets in PT-symmetric multilayer systems can be used for modulation of an exceptional point position via tuning their surface conductivity [38]. Light also makes a bilayer to move. Optical forces exerting on PT-symmetric multilayer structures can be both pushing and pulling depending on the direction of light and realization of the broken or unbroken PT-symmetric phase [39].

A PT-symmetric multilayer structure can be used as a laser. More precisely, the laser must be simultaneously a coherent perfect absorber [40, 41]. In the broken PT-symmetry phase, an illumination from one side is normally amplified, while a coherent illumination from both sides of the multilayer system is absorbed owing to interference. Laser-absorber modes arise, when a pole and a zero of a scattering matrix approach each other on the real axis in the plane of complex frequency. Finite-difference-time-domain (FDTD) calculations basically confirm predictions of the transfer-matrix and scattering-matrix approaches and show an enormous increase in the output intensity, when the laser threshold is achieved [35]. If the pole is not on the real axis, the lasing onset is still feasible at a greater threshold. In realistic PT-symmetric systems, realization of the gain occurs in quantum systems and requires accounting for the saturation effect. This means that the PT-symmetric system has to be nonlinear. In [42] it was considered a two-level resonant medium described by the Maxwell-Bloch equations. Due to the saturation, the condition for PT symmetry is approximately valid and the system can be named as a non-Hermitian one [43]. Saturation may result in novel effects, e.g., locking of the light propagation by the PT-symmetry breaking [42]. Steady-state solutions for PT-symmetric multilayer structures with nonlinear refractive indices of gain and loss media are investigated in [44] using a modified transfer-matrix method. The bistable behavior of the transmitted and reflected intensities was studied together with unidirectional invisibility and coherent perfect absorption versus the input and saturation intensities.

18.3.4 Microresonators

PT symmetry can be realised on a resonator platform promising for interesting applications on a chip. In Fig. 18.5d we show a typical system comprising two coupled gain and loss cavities. It is shown in [45] that a non-Hermitian optical microring resonator coupled to a waveguide can be used as an asymmetric absorber, if a mirror is placed on one side of the waveguide. Robustness of the asymmetric absorption is explained by the emergence of the chiral exceptional surface, which can be also exploited for directional absorption control. Unidirectional lasing and coherent perfect absorption can be achieved using unidirectional destructive interferences being realised with an asymmetrically coupled passive resonator chain [46]. Asymmetry in coupling breaks the reciprocity in transmission due to the destructive interference. A PT-symmetric side-coupled resonator can be realised using unidirectional lasing together with perfectly absorbing resonators and can result in simultaneous unidirectional lasing and perfect absorption effects. In a similar fashion, the spectral singularities of scattering matrix can be investigated in a PT-symmetric two-arm Aharonov-Bohm interferometer [46]. Such spectral singularities can be tailored to realise symmetric, asymmetric and unidirectional lasing onsets.

Non-Hermitian resonators are widely used as extremely sensible sensors at exceptional points. The sensitivity can be even more increased for higher-order exceptional points, at which more than two eigenvalues of a non-Hermitian Hamiltonian coincide. In this case, the frequency splitting stems from a perturbation $\varepsilon \ll 1$ follows the law $\varepsilon^{1/N}$, where N is the order of the exceptional point [47]. Since the susceptibility $d(\varepsilon^{1/N})/d\varepsilon$ diverges at $\varepsilon = 0$, the sensitivity can be arbitrarily high. This idea was experimentally validated in a PT-symmetric ternary (loss-neutral-gain) micro-ring system [48] and micro-toroid cavity [49], the non-Hermiticity of the latter being introduced by a scatterer resulting in coupling eigenmodes of the cavity. Sensitivity of parameter estimation can be analysed using the formalism of quantum Fisher information without referring to a specific measurement scheme. The average of all merging eigenstates cancels out the divergence at the singularity resulting to a finite value at the exceptional point [38].

PT-symmetric [50] and anti-PT-symmetric [51] optical gyroscopes were proposed on the basis of ring resonators coupled to a waveguide. Frequency splitting and, hence, sensitivity in gyroscopes are normally limited by the ring dimensions. In PT-symmetric gyroscopes, the frequency splitting is independent of the ring radius and, therefore, the phase shift of interference fringes is enhanced in vicinity of exceptional points. A unique "superluminal" lasing may be used as a sensor and can be obtained in a broken PT-symmetry phase of the white-light cavity that consists of gain and lossy coupled micro-resonators [52].

Finally, PT-symmetric microcavities possess indispensable nanophotonic properties for suppression of spontaneous relaxation rate [53]. In the next section we will discuss this subject in detail.

18.4 Purcell Effect in PT-Symmetric Waveguides

In 1946, E. M. Purcell predicted that the spontaneous emission rate of a light source is not solely an intrinsic property of the source but is affected by the optical environment [54]. This effect is now referred to as Purcell effect. The Purcell factor is defined as

$$F_p = \frac{P_{\text{system}}}{P_0}, \tag{18.40}$$

where P_{system} denotes the power of an emitter radiated into a particular optical system and P_0 is the power of the same emitter radiated into vacuum or free space. Purcell factor is a common figure of merit to describe the emission enhancement induced by feedback of the source with a particular optical system. Alternatively, Purcell factor can be defined in terms of spontaneous emission rate

$$F_p = \frac{\tau_0}{\tau_{\text{system}}}, \tag{18.41}$$

where τ_0 is the spontaneous emission lifetime in vacuum and τ_{system} is the lifetime of the emitter in the particular system of interest.

The interaction between the emitter and its environment is formally described by Fermi's golden rule which states that the probability for spontaneous emission is proportional to the (photonic) local density of states (LDOS). LDOS, in turn, is proportional to the imaginary part of Green's dyadic $\hat{G}$ at the position of the emitter [55]

$$\rho_p(\mathbf{r}_0, \omega) = \frac{6\omega}{\pi c^2} \left[\hat{\mathbf{p}} \, \text{Im} \, \hat{G}(\mathbf{r}_0, \mathbf{r}_0; \omega) \hat{\mathbf{p}} \right], \tag{18.42}$$

where $\mathbf{r}_0$ is emitter position and $\hat{\mathbf{p}}$ denotes unit vector of the dipole orientation.

It is well known since the Purcell's work [54] that the strong Purcell enhancement occurs in resonant systems where the light is confined to small volumes. More recent work suggests that giant enhancements can occur via the less familiar Petermann effect [11, 12, 14]. The Petermann enhancement factor is a measure of non-orthogonality of the modes in non-Hermitian systems and it appears to diverge when two modes coalesce at an exceptional point (EP). In the work of Pick et al. [15] authors develop a general theory of the spontaneous emission at exceptional points. They show that traditional theories of spontaneous emission fail in case of degenerate resonances occurring at EPs and lead to infinite Purcell factors. Approach presented in [15] is based on the perturbation theory which properly accounts for degeneracies at EPs by using Jordan vectors. Within this approach authors prove that actual enhancement factors is finite, but can still be significant (about hundreds) in gain-aided and higher-order EP systems.

Interestingly, that not only enhancement but rather suppression of spontaneous decay rate can occur in PT-symmetric systems. Akbarzadeh et al. in [53] show that a PT-symmetric planar cavity is able to suppress the spontaneous relaxation rate of a two-level atom below the vacuum level. Recent work of Khanbekyan and Wiersig reports on decay suppression of spontaneous emission of a single emitter in a high-Q cavity at exceptional points [56].

The Purcell factor can be calculated separately for each of the discrete scattering channels. For instance, just a couple of years ago, the Purcell effect in the mode of the basic element of PIC planar waveguide was introduced within the scattering matrix formalism [57].

It has recently been shown in the context of single molecule detections that the power emitted from a molecule into a single mode fiber can be elegantly calculated using the reciprocity theorem of electromagnetic theory [58]. In the work [59] authors propose a reciprocity approach to calculate the emission enhancement for emitters coupled to arbitrary resonant or non-resonant open optical systems. They calculate the *modal Purcell factor*—the quantity which measures the power emitted by an emitter situated in the vicinity of a device into a particular propagating mode normalised by the power radiated by the same emitter into the free space.

18.4.1 Reciprocity Approach

In this section, we generalise the reciprocity approach formulated in [59] to the case when the propagating modes are not orthogonal. We probe the method by calculation of the modal Purcell factor in PT-symmetric coupled waveguide system. In the following section we will obtain closed-form expressions for modal Purcell factor describing the system in terms of coupled modes.

We consider a current source (current density distribution $\mathbf{J}_1$) situated in the vicinity of some optical system with two exit ports at z_1 and z_n [59]. For brevity, we use optical state-vector notation for 4-component vector joining transverse electric and magnetic fields introduced in (18.12). In this way we can describe the fields of guiding (and leaking) modes. For the ith mode we write

$$|M_i(z)\rangle = \begin{pmatrix} \mathbf{E}_{t,i}(x, y, z) \\ \mathbf{H}_{t,i}(x, y, z) \end{pmatrix} = |i\rangle \mathrm{e}^{-\mathrm{i}\beta_i z}, \tag{18.43}$$

where

$$|i\rangle = \begin{pmatrix} \mathbf{e}_{t,i}(x, y) \\ \mathbf{h}_{t,i}(x, y) \end{pmatrix} \tag{18.44}$$

and

$$\begin{pmatrix} \mathbf{E}_{t,i}(x, y, z) \\ \mathbf{H}_{t,i}(x, y, z) \end{pmatrix} = \begin{pmatrix} \mathbf{e}_{t,i}(x, y) \\ \mathbf{h}_{t,i}(x, y) \end{pmatrix} \mathrm{e}^{-\mathrm{i}\beta_i z}. \tag{18.45}$$

The fields excited by the current source $\mathbf{J}_1$ at the cross-section of exit ports can be expanded into a set of modes as follows

$$|\psi_1(z_1)\rangle = \sum_i A_{i,z_1}|i, z_1\rangle,$$
$$|\psi_1(z_n)\rangle = \sum_i A_{-i,z_n}|-i, z_n\rangle. \tag{18.46}$$

Here A_{i,z_1} and A_{-i,z_n} are the amplitudes of the modes propagating forward to port z_1 and backward to port z_n, respectively, $|i, z_1\rangle$, $|-i, z_n\rangle$ are respectively eigenmodes of ports z_1 and z_n propagating from the cavity.

In our notations the Lorentz reciprocity theorem

$$\int_{\delta V} (\mathbf{E}_1 \times \mathbf{H}_2 - \mathbf{E}_2 \times \mathbf{H}_1) \cdot \hat{\mathbf{z}} dx dy = \int_V (\mathbf{E}_2 \cdot \mathbf{J}_1 - \mathbf{E}_1 \cdot \mathbf{J}_2)\, dV. \tag{18.47}$$

should be rewritten as

$$\langle\psi_1(z_1)|\psi_2(z_1)\rangle - \langle\psi_2(z_1)|\psi_1(z_1)\rangle - \langle\psi_1(z_n)|\psi_2(z_n)\rangle + \langle\psi_2(z_n)|\psi_1(z_n)\rangle$$
$$= \int_V (\mathbf{E}_2 \cdot \mathbf{J}_1 - \mathbf{E}_1 \cdot \mathbf{J}_2)\, dV, \tag{18.48}$$

where δV is the surface enclosing the cavity volume V between two planes $z = z_1$ and $z = z_n$. In (18.48), $\mathbf{J}_1$ and $|\psi_1\rangle$ are defined above, while the source $\mathbf{J}_2$ and the fields $|\psi_2\rangle$ produced by it can be chosen as we need. Let the source current $\mathbf{J}_2$, being outside the volume V ($\mathbf{J}_2 = 0$), excite a single mode $|-k, z_1\rangle$. In general, this mode is scattered by the cavity V and creates the set of transmitted and reflected modes as discussed in [59]:

$$|\psi_2(z_1)\rangle = B_{-k,z_1}|-k, z_1\rangle + \sum_i B_{i,z_1}|i, z_1\rangle, \tag{18.49}$$

$$|\psi_2(z_n)\rangle = \sum_i B_{-i,z_n}|-i, z_n\rangle. \tag{18.50}$$

Using the orthogonality condition (18.17) and the symmetry relations (18.20) we obtain the inner products of the fields

$$\langle\psi_1(z_1)|\psi_2(z_1)\rangle = \sum_i A_{i,z_1} B_{-k,z_1}\langle i, z_1|-k, z_1\rangle + \sum_{i,j} A_{i,z_1} B_{j,z_1}\langle i, z_1|j, z_1\rangle$$
$$= -2A_{k,z_1} B_{-k,z_1} N_k + 2\sum_i A_{i,z_1} B_{i,z_1} N_{i,z_1}. \tag{18.51a}$$

$$\langle\psi_2(z_1)|\psi_1(z_1)\rangle = \sum_i A_{i,z_1} B_{-k,z_1} \langle -k, z_1 | i, z_1\rangle + \sum_{i,j} A_{i,z_1} B_{j,z_1} \langle i, z_1 | j, z_1\rangle$$
$$= 2A_{k,z_1} B_{-k,z_1} N_{k,z_1} + 2\sum_i A_{i,z_1} B_{i,z_1} N_{i,z_1}. \quad (18.51b)$$

$$\langle\psi_1(z_n)|\psi_2(z_n)\rangle = \langle\psi_2(z_n)|\psi_1(z_n)\rangle = \sum_{i,j} A_{-i,z_n} B_{-j,z_n} \langle -i, z_n | -j, z_n\rangle$$
$$= 2\sum_i A_{-i,z_n} B_{-i,z_n} N_{i,z_n}, \quad (18.51c)$$

where $N_{i,z_{1(n)}}$ the norm of the mode $|i, z_{1(n)}\rangle$ as defined in (18.17).

By substituting these equations into (18.48), we arrive at the amplitude A_{k,z_1} of the mode excited by the source current $\mathbf{J}_1$

$$A_{k,z_1} = -\frac{1}{4B_{-k,z_1} N_{k,z_1}} \int_V \mathbf{E}_{2,-k} \cdot \mathbf{J}_1 \mathrm{d}V, \quad (18.52)$$

where $\mathbf{E}_{2,-k} = B_{-k,z_1} \mathbf{e}_{-k}(x, y) \mathrm{e}^{\mathrm{i}\beta_k(z-z_1)}$ is the electric field created by the excitation of the system with reciprocal mode $| -k, z_1\rangle$ at the port z_1.

As an emitter we consider a point dipole oscillating at the circular frequency ω and having the current density distribution

$$\mathbf{J}_1(\mathbf{r}) = \mathrm{i}\omega\mathbf{p}\delta(\mathbf{r} - \mathbf{r}_0), \quad (18.53)$$

where $\mathbf{p}$ is the dipole moment of the emitter and $\mathbf{r}_0$ is its position. Then we are able to carry out the integration in (18.52) and obtain

$$A_{k,z_1} = -\frac{\mathrm{i}\omega}{4B_{-k,z_1} N_{k,z_1}} \mathbf{E}_{2,-k}(\mathbf{r}_0) \cdot \mathbf{p}. \quad (18.54)$$

Here we observe a difference compared to the Hermitian case considered in [59]. This difference appears due to the fact that now the expansion coefficients A_{k,z_1} are not directly related to the powers carried by the modes. Finding a power carried by a specific mode is a challenge. To circumvent this challenge, we propose a calculation of the total power carried by the set of modes as we describe below.

The power emitted by the current source $\mathbf{J}_1$ into the port z_1 can be written as

$$P = \frac{1}{2}\mathrm{Re} \int_{z=z_1} (\mathbf{E}_1 \times \mathbf{H}_1^*) \cdot \hat{\mathbf{z}} \mathrm{d}x\mathrm{d}y = \frac{1}{2}\mathrm{Re}\,\langle\psi_1(z_1)|\psi_1^*(z_1)\rangle. \quad (18.55)$$

Expanding the electromagnetic fields $|\psi_1(z_1)\rangle$ according to (18.46) we represent the power transmitted through the port (18.55) as follows

$$P = \mathrm{Re} \sum_{k,l} A_{k,z_1} A^*_{l,z_1} P_{kl}, \tag{18.56}$$

where P_{kl} is the so called cross-power equal to the Hermitian inner product of the modal fields

$$P_{kl,z_1} = \frac{1}{2}\langle k, z_1 | l^*, z_1 \rangle = \frac{1}{2} \int\limits_{z=z1} \left(\mathbf{e}_{k,z_1} \times \mathbf{h}^*_{l,z_1}\right) \cdot \hat{\mathbf{z}} \mathrm{d}x\mathrm{d}y. \tag{18.57}$$

For $k = l$ the cross-power reduces to the mode power $P_k = \mathrm{Re}\, P_{kk}$. By considering the expansion coefficients (18.54) we rewrite the power (18.56) in terms of the reciprocal fields $\mathbf{E}_{2,-k}$ as

$$\begin{aligned} P &= \frac{\omega^2}{16} \mathrm{Re} \sum_{k,l} \frac{(\mathbf{E}_{2,-k}(\mathbf{r}_0) \cdot \mathbf{p})(\mathbf{E}^*_{2,-l}(\mathbf{r}_0) \cdot \mathbf{p}^*)}{B_{-k} B^*_{-l} N_k N^*_l} P_{kl} \\ &= \frac{\omega^2}{16} \mathrm{Re} \sum_{k,l} \frac{(\mathbf{e}_{-k}(x_0, y_0) \cdot \mathbf{p})(\mathbf{e}^*_{-l}(x_0, y_0) \cdot \mathbf{p}^*)}{N_k N^*_l} P_{kl}. \end{aligned} \tag{18.58}$$

The last equality is the consequence of the substitution of $\mathbf{E}_{2,-k}$ at the emitter position $\mathbf{r}_0 = (x_0, y_0, z_0)$ considering the negligible dimensions of the cavity $z_1 \approx z_n \approx z_0$. Note that here we dropped z_1 subscripts.

To find the Purcell factor we divide (18.58) by the power emitted by the same dipole into the free space

$$P_0 = \frac{\mu_0}{12\pi c} \omega^4 |p|^2, \tag{18.59}$$

where μ_0 is the vacuum permeability and c is the speed of light in vacuum. The dipole moment, located in the xy plane, can be presented using the unit vector $\hat{\mathbf{p}}$ as follows

$$\mathbf{p} = p\hat{\mathbf{p}}, \tag{18.60}$$

therefore,

$$\mathbf{E}_{2,-k}(\mathbf{r}_0) \cdot \mathbf{p} = \mathbf{E}_{2,-k}(\mathbf{r}_0) \cdot \hat{\mathbf{p}} p = E_{p,k}(\mathbf{r}_0) p. \tag{18.61}$$

Here $E_{p,k}$ denotes projection of the vector $\mathbf{E}_{2,-k}$ onto the dipole orientation vector $\hat{\mathbf{p}}$

$$E_{p,k} = \mathbf{E}_{2,-k} \cdot \hat{\mathbf{p}}. \tag{18.62}$$

Then the Purcell factor reads

$$F_p = \frac{P}{P_0} = \frac{3\pi c}{4\omega^2 \mu_0} \mathrm{Re} \sum_{k,l} \frac{e_{p,k}(x_0, y_0)\, e^*_{p,l}(x_0, y_0)}{N_k N^*_l} P_{kl}. \tag{18.63}$$

It is convenient to rewrite (18.63) through the normalised fields as

$$F_p = \frac{3\pi c}{4\omega^2 \mu_0} \mathrm{Re} \sum_{kl} \hat{e}_{p,k} \hat{e}^*_{p,l} K_{kl} \hat{P}_{kl}, \tag{18.64}$$

where we have introduced normalised modal electric fields

$$\hat{\mathbf{e}}_{2,i} = \frac{\mathbf{e}_{2,i}}{\sqrt{N_i^h}} \tag{18.65}$$

and normalised cross-power coefficients

$$\hat{P}_{kl} = \frac{1}{\sqrt{N_k^h N_l^h}} P_{kl} \tag{18.66}$$

Where $N_i^h = \frac{1}{2}\langle i|i^*\rangle$ is the Hermitian norm of the mode $|i\rangle$ defined by the Hermitian inner product (18.16). Here we generalise the Petermann factor defined in Sect. 18.2.4

$$K_i = K_{ii} \tag{18.67}$$

defining cross-mode Petermann factor

$$K_{kl} = \frac{\langle k|k^*\rangle}{\langle k|k\rangle} \frac{\langle l|l^*\rangle^*}{\langle l|l\rangle^*}. \tag{18.68}$$

The modal Purcell factor can be naturally divided into two parts, the first of which is the sum of all diagonal ($k = l$) terms, while the second part is the sum of off-diagonal ($k \neq l$) terms:

$$F_p = F_{p,\mathrm{diag}} + F_{p,\mathrm{off-diag}} = \sum_k F_{p,k} + \sum_{k \neq l} F_{p,kl}, \tag{18.69}$$

where

$$F_{p,i} = \frac{3\pi c}{4\omega^2 \mu_0} |\hat{e}_{p,k}|^2 K_i, \tag{18.70}$$

$$F_{p,kl} = \frac{3\pi c}{4\omega^2 \mu_0} \hat{e}_{p,k} \hat{e}^*_{p,l} K_{kl} \hat{P}_{kl}. \tag{18.71}$$

In the Hermitian case, the off-diagonal terms (18.71) reduce to zero due to the regular orthogonality of the modes expressed by $\hat{P}_{kl} = \delta_{kl}$. That is why the Purcell factor (18.64) applied to Hermitian systems coincides with the expression in [59].

18.4.2 Modal Purcell Factor Within the Coupled Mode Theory

PT-symmetric regime

To find the modal Purcell factor for the coupled waveguide system in PT-symmetric regime we substitute the modes in the form (18.29) into expression (18.64).

One more assumption is introduced for the sake of simplicity:

$$\langle g|g^*\rangle = \langle l|l^*\rangle = 1. \tag{18.72}$$

It implies that the Hermitian norms of the isolated modes are equal to the non-Hermitian norms or, in other words, the Petermann factors for the modes equal unity.

Then the quantities K_{kl} and $\hat{P}_{kl}$ can be written in the closed form as

$$K_1 = \frac{|\langle 1|1^*\rangle|^2}{|\langle 1|1\rangle|^2} = \frac{2}{1+\cos 2\theta}, \tag{18.73a}$$

$$K_2 = \frac{|\langle 2|2^*\rangle|^2}{|\langle 2|2\rangle|^2} = \frac{2}{1+\cos 2\theta}, \tag{18.73b}$$

$$K_{12} = \frac{\langle 1|1^*\rangle}{\langle 1|1\rangle}\frac{\langle 2|2^*\rangle^*}{\langle 2|2\rangle^*} = \frac{2(1+\mathrm{e}^{-\mathrm{i}2\theta})^2}{(1+\cos 2\theta)^2}, \tag{18.73c}$$

$$K_{21} = \frac{\langle 2|2^*\rangle}{\langle 2|2\rangle}\frac{\langle 1|1^*\rangle^*}{\langle 1|1\rangle^*} = \frac{2(1+\mathrm{e}^{+\mathrm{i}2\theta})^2}{(1+\cos 2\theta)^2}, \tag{18.73d}$$

$$\hat{P}_{12} = \frac{\langle 1|2^*\rangle}{\sqrt{\langle 1|1^*\rangle\langle 2|2^*\rangle}} = \frac{1}{2}(1-\mathrm{e}^{\mathrm{i}2\theta}), \tag{18.74a}$$

$$\hat{P}_{21} = \frac{\langle 2|1^*\rangle}{\sqrt{\langle 1|1^*\rangle\langle 2|2^*\rangle}} = \frac{1}{2}(1-\mathrm{e}^{-\mathrm{i}2\theta}). \tag{18.74b}$$

Normalised field projections $\hat{e}_{p,k}$ in the basis of isolated modes read

$$\hat{e}_{p,1} = \frac{1}{\sqrt{\frac{1}{2}\langle 1|1^*\rangle}}(\hat{e}_{p,g} + \mathrm{e}^{\mathrm{i}\theta}\hat{e}_{p,l}) = \hat{e}_{p,g} + \mathrm{e}^{\mathrm{i}\theta}\hat{e}_{p,l}, \tag{18.75a}$$

$$\hat{e}_{p,2} = \frac{1}{\sqrt{\frac{1}{2}\langle 2|2^*\rangle}}(\hat{e}_{p,g} - \mathrm{e}^{-\mathrm{i}\theta}\hat{e}_{p,l}) = \hat{e}_{p,g} - \mathrm{e}^{-\mathrm{i}\theta}\hat{e}_{p,l}. \tag{18.75b}$$

In above expressions $\hat{e}_{p,g}$ and $\hat{e}_{p,l}$ denote projections of the fields of backward-propagating isolated modes onto dipole orientation. If the emitter dipole moment is perpendicular to $\hat{z}$, projections of backward-propagating modal fields are equal to the projections of forward-propagating ones.

Performing calculation of the modal Purcell factor (18.64) using relations (18.73)–(18.74) we obtain

$$F_p = F_{p,\mathrm{diag}} + F_{p,\mathrm{off-diag}} = \frac{6\pi c}{\omega^2 \mu_0}(|\hat{e}_{p,g}|^2 + |\hat{e}_{p,l}|^2). \tag{18.76}$$

Diagonal and off-diagonal terms separately take the form

$$F_{p,\mathrm{diag}} = \frac{3\pi c}{4\omega^2 \mu_0}\frac{4}{1+\cos 2\theta}(|\hat{e}_{p,g}|^2 + |\hat{e}_{p,l}|^2), \tag{18.77a}$$

$$F_{p,\mathrm{off-diag}} = -\frac{3\pi c}{4\omega^2 \mu_0}\frac{2(1-\cos 2\theta)}{1+\cos 2\theta}(|\hat{e}_{p,g}|^2 + |\hat{e}_{p,l}|^2). \tag{18.77b}$$

It is curious that although both diagonal and off-diagonal terms (18.77) are singular at the EP corresponding to $\theta_{EP} = \pi/2$ and $\cos 2\theta_{EP} = -1$, the singularities cancel each other making the modal Purcell factor finite and independent of θ. The modal Purcell factor (18.76) depends solely on the mode profiles of the isolated modes in PT-symmetric regime.

PT-symmetry-broken regime

To obtain the modal Purcell factor in PT-symmetry-broken regime we substitute the modes in the form (18.33) into expression (18.64).

Calculating the Petermann factors

$$K_1 = \coth^2\theta, \tag{18.78a}$$

$$K_2 = \coth^2\theta, \tag{18.78b}$$

$$K_{12} = -\coth^2\theta, \tag{18.78c}$$

$$K_{21} = -\coth^2\theta, \tag{18.78d}$$

normalised cross-powers

$$\hat{P}_{12} = \frac{1}{\cosh\theta}, \tag{18.79a}$$

$$\hat{P}_{21} = \frac{1}{\cosh\theta}, \tag{18.79b}$$

and reciprocal modal field projections

$$\hat{e}_{p,1} = \frac{1}{\sqrt{\frac{1}{2}(1+\mathrm{e}^{-2\theta})}}(\hat{e}_{p,g} + \mathrm{i}\mathrm{e}^{-\theta}\hat{e}_{p,l}), \tag{18.80a}$$

$$\hat{e}_{p,2} = \frac{1}{\sqrt{\frac{1}{2}(1+\mathrm{e}^{2\theta})}}(\hat{e}_{p,g} + \mathrm{i}\mathrm{e}^{\theta}\hat{e}_{p,l}) \tag{18.80b}$$

we straightforwardly derive the diagonal and off-diagonal terms

$$F_{p,\text{diag}} = \frac{3\pi c}{4\omega^2 \mu_0} \frac{2\cosh\theta}{\sinh^2\theta} \left((|\hat{e}_{p,g}|^2 + |\hat{e}_{p,l}|^2)\cosh\theta - 2\,\text{Im}\,(\hat{e}^*_{p,g}\hat{e}_{p,l}) \right), \quad (18.81a)$$

$$F_{p,\text{off-diag}} = -\frac{3\pi c}{4\omega^2 \mu_0} \frac{2}{\sinh^2\theta} \left(|\hat{e}_{p,g}|^2 + |\hat{e}_{p,l}|^2 - 2\cosh\theta\,\text{Im}\,(\hat{e}^*_{p,g}\hat{e}_{p,l}) \right) \quad (18.81b)$$

as well as the modal Purcell factor

$$F_p = F_{p,\text{diag}} + F_{p,\text{off-diag}} = \frac{6\pi c}{\omega^2 \mu_0} (|\hat{e}_{p,g}|^2 + |\hat{e}_{p,l}|^2). \quad (18.82)$$

The main result of this section is that although diagonal and off-diagonal terms of the modal Purcell factor diverge at the EP, the modal Purcell factor itself does not exhibit a singular behavior when approaching to the EP either from the left or right side.

Though we do not carry out a rigorous analysis of the behavior at the EP accounting for the degeneracy of the modes as it was done in [15], the developed approach leads to the well-defined expressions (18.76) and (18.82) for F_p at the exceptional point.

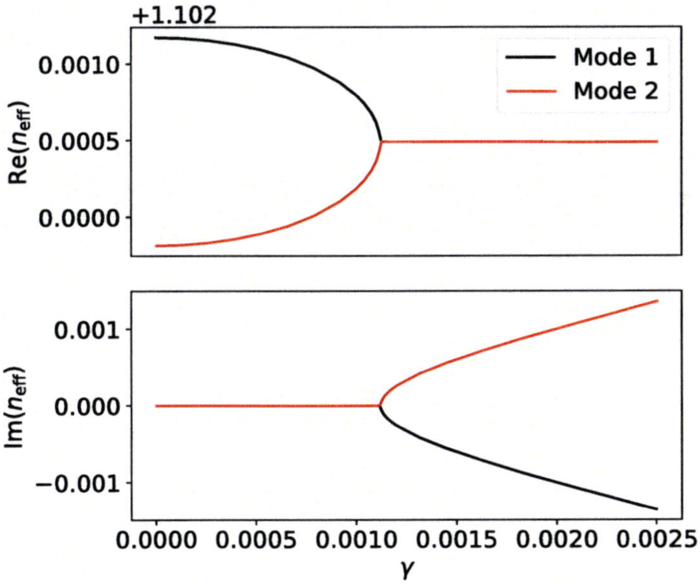

Fig. 18.6 Effective mode indices versus the non-Hermiticity parameter γ. Black curves correspond to the mode $|1\rangle$. Red curves correspond to the mode $|2\rangle$

18.4.3 Numerical Example: PT-Symmetric Coupler

This section presents an example of utilising the theory developed in the previous section. Here, we analyse an optical system consisting of two coupled rectangular waveguides with width w and height h separated by the distance g as schematically shown in Fig. 18.5a. We assume that the complex refractive indices of the left (Gain) and right (Loss) waveguides are $n_l = n_{co} + i\gamma$ and $n_r = n_{co} - i\gamma$ respectively. n_{co} is the real part of the refractive index and $\gamma > 0$ is the gain/loss (non-Hermiticity) parameter. Thus, the system of the coupled waveguides satisfies PT-symmetry condition $n(x, y) = n^*(-x, -y)$. The refractive index of the background is assumed to be unity.

We take parameters of the waveguide coupler as $g = 2$ μm, $w = 1$ μm, $h = 0.5$ μm, and $n_{co} = 1.44$. The coupler has two quasi-TE supermodes at this wavelength. We calculated the field distribution of the guided modes of these waveguides, shown in Figs. 18.3 and 18.4.

By increasing the gain/loss parameter γ the system passes through the regime of propagation (PT-symmetric state) for two non-decaying supermodes to the regime of decay/amplification (PT-symmetry-broken state). This behavior, shown by the curves in Fig. 18.6.

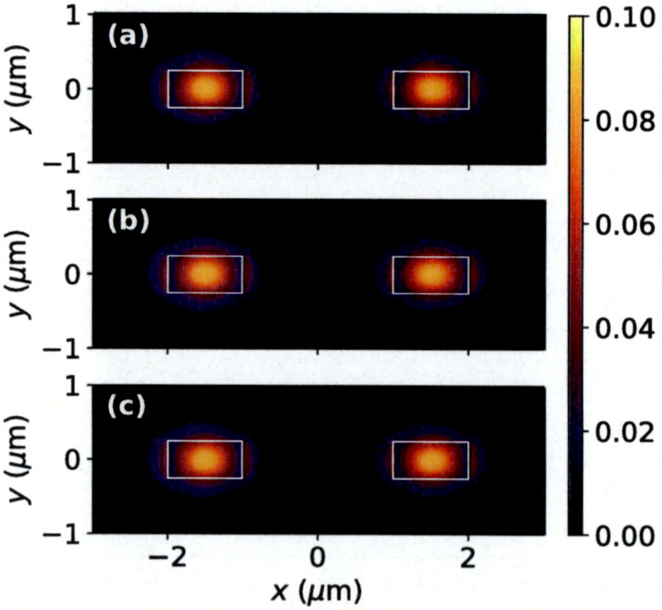

Fig. 18.7 Purcell factor distribution in the plane (x, y) **a** for the Hermitian system characterised by $\gamma = 0$, **b** in the PT-symmetric phase ($\gamma = 1.02 \times 10^{-3}$), **c** in the broken-PT-symmetric state ($\gamma = 2.5 \times 10^{-3}$). Parameters of the waveguide coupler: $g = 2.0$ μm, $w = 1$ μm, $h = 0.5$ μm, and $n_{co} = 1.44$

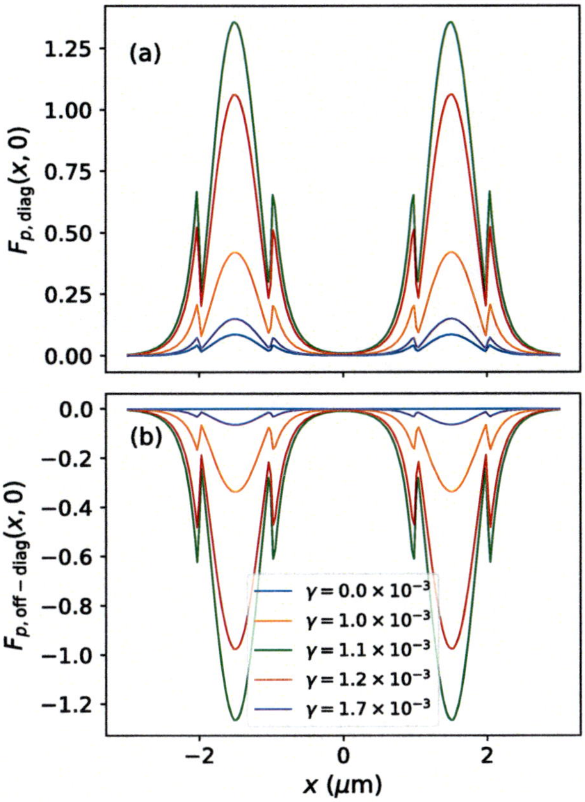

Fig. 18.8 Distribution of the Purcell factor **a** diagonal and **b** off-diagonal terms depending on the emitter position x_0 at $y_0 = 0$ for different values of γ. Parameters of the coupled waveguide are given in the caption of Fig. 18.7

For the studied system the value of γ corresponding to EP is $\gamma_{\mathrm{EP}} = 1.12 \times 10^{-3}$.

Next, we explore the modal Purcell factor F_p for the pair of quasi-TE modes. According to (18.63), the Purcell factor is defined by the fields of the reciprocal modes at the dipole position (x_0, y_0, $z_0 \approx z_1 \approx z_n$). In Fig. 18.7, we demonstrate the Purcell factor for an x-oriented dipoles as a function of x_0 and y_0 for different values of non-Hermiticity parameter γ.

From Fig. 18.7 we conclude that the modal Purcell factor is symmetric in (a) Hermitian regime as well as in (b) PT-symmetric and (c) PT-symmetry broken regimes. In all three cases, the modal Purcell factor F_p distribution is the same and finite (taking maximum value of approximately 0.085 in the middle of the waveguides) despite the fact that both diagonal and off-diagonal terms experience enhancement as shown in Fig. 18.8. According to the Eqs. (18.77) and (18.81) this enhancement is direct consequence of non-orthogonality. Opposite signs and close absolute values of diagonal and off-diagonal terms observed in Fig. 18.8 result in cancellation of

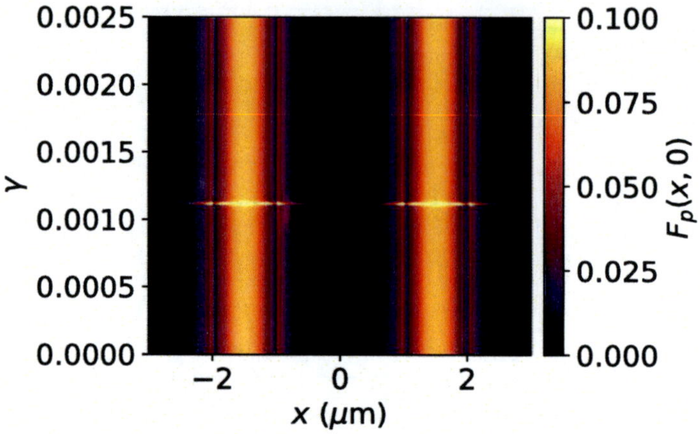

Fig. 18.9 Distribution of the Purcell factor at the line $y = 0$ as function of the emitter position x and non-Hermiticity parameter γ. Parameters of the coupled waveguide are given in the caption of Fig. 18.7

divergent terms in modal Purcell factor. This explains small values of the modal Purcell factor and its independence on the non-Hermiticity parameter γ demonstrated in Fig. 18.9. Independence on the non-Hermiticity parameter also confirms the analytical predictions given by (18.76) and (18.82). Note: a tiny spike observed near the EP is a numerical artefact. It arises due to amplification of terms $F_{p,\text{diag}}$ and $F_{p,\text{off-diag}}$.

Such a behavior well agrees with the result obtained in Sect. 18.4.2 utilising the coupled-mode theory, namely, the numerically observed distribution of the modal Purcell factor is similar in Hermitian, PT-symmetric, and PT-symmetry broken regimes.

18.5 Summary and Outlook

To summarise, one of the challenges in integrated photonics, is to develop on-chip optical devices for efficient light manipulation finding its use in emerging applications such as data processing, quantum technologies, healthcare, security and sensing. Purcell effects in PT-symmetry can be utilised in variety of applications on a chip for instance for lasing. Lasing like behaviour can be realised based on multilayer system releasing the pumped energy in the form of powerful pulses [60]. Similar approach studied in [60] can be implemented on a chip. Figure 18.10 shows the concept of Transmission and Reflection through the multilayered waveguide core, composed from Loss and Gain media.

Although efficient on-chip light manipulation can be achieved by engineering artificial materials (metamaterials) with unique optical permittivities and permeabilities, PT-symmetric photonics allows to tune the complex refractive index and control the

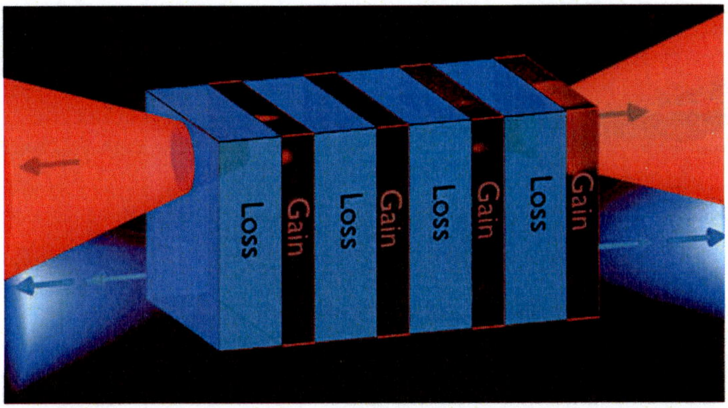

Fig. 18.10 Schematics of the proposed concept with N-periods multilayer waveguide core of alternating Loss and Gain media for on-chip lasing based on PT-symmetry effect

interplay between the phase (real part of complex refractive index) and attenuation or loss (in case the imaginary part of complex refractive index is negative); or amplification/gain in case the imaginary part of complex refractive index is positive. The major advantage of PT-symmetric systems is to confine and guide light in coupled passive waveguides as was first shown in [20]. Then, the active fully PT-symmetric system with gain and loss was demonstrated using two coupled waveguides fabricated from Fe-doped $LiNbO_3$ [19] in such a way that the transmission always appeared at the output of the active waveguide regardless of the input waveguide. This effect is named non-reciprocal meaning that power oscillations between the coupled waveguides are asymmetric. The degree of non-reciprocity in such nonlinear devices depends on the intensity of the signal. However in [19], Lorentz reciprocity still holds as long as no nonlinearity builds up.

PT-optomechanics is another interesting way to go and explore the interaction between the optical fields and mechanical option in PT-symmetric systems in presence of a quantum emitter. In coupled mechanical resonators with optically induced loss and gain, a combination of nonlinear saturation and noise leads to preserved or weakly broken PT-symmetry, and a transition occurs from a thermal to a lasing state with small amplitude [16, 61].

Systems with exceptional points, particularly, PT-symmetric systems are known to be able to enhance [15] and suppress [53, 56] the spontaneous emission rate in optical systems when operating near exceptional point. Analysis of the spontaneous emission enhancement and coupling to the guided modes of the PT-symmetric coupled waveguide system shown that, interestingly, for this class of systems the modal enhancement factor (modal Purcell factor) does not depend on the non-Hermiticity even at the EP.

In conclusion, although the PT symmetry and non-Hermiticity in integrated photonics research has already established novel ways of utilising gain, loss and their coupling to control light transport, there is still a room for new direction to go, when considering a Purcell effect in PT-symmetric waveguides.

Acknowledgements AK acknowledges the support of Israel Science Foundation (ISF) Grant no. 2598/20.

References

1. A. Zyablovsky, A.P. Vinogradov, A.A. Pukhov, A.V. Dorofeenko, A.A. Lisyansky, Physics-Uspekhi **57**(11), 1063 (2014)
2. C.M. Bender, S. Boettcher, Phys. Rev. Lett. **80**(24), 5243 (1998). https://doi.org/10.1103/PhysRevLett.80.5243
3. S. Weigert, Phys. Rev. A **68**(6), 062111 (2003). https://doi.org/10.1103/PhysRevA.68.062111
4. A. Mostafazadeh, Int. J. Geometric Methods Modern Phys. **07**(07), 1191 (2010). https://doi.org/10.1142/S0219887810004816
5. N. Moiseyev, *Non-Hermitian Quantum Mechanics* (Cambridge University Press, Cambridge; New York, 2011)
6. D.C. Brody, J. Phys. A: Math. Theor. **49**(10), 10LT03 (2016). 10.1088/1751-8113/49/10/10LT03
7. G.L. Pedrola, *Beam Propagation Method for Design of Optical Waveguide Devices*, 1st edn. (Wiley, New York, 2015)
8. A.W. Snyder, J.D. Love, *Optical Waveguide Theory* (Springer, US, Boston, MA, 1984)
9. G.K. Svendsen, M.W. Haakestad, J. Skaar, Physical Review A **87**, 013838 (2013). https://doi.org/10.1103/PhysRevA.87.013838
10. B. Wu, Z. Wang, W. Chen, Z. Xiong, J. Xu, Y. Chen, Opt. Exp. **27**(13), 17648 (2019). https://doi.org/10.1364/OE.27.017648
11. K. Petermann, IEEE J. Quantum Electron. **15**(7), 566 (1979). https://doi.org/10.1109/JQE.1979.1070064
12. A.E. Siegman, Phys. Rev. A **39**, 1253 (1989). https://doi.org/10.1103/PhysRevA.39.1253
13. M.V. Berry, J. Modern Opt. **50**(1), 63 (2003). https://doi.org/10.1080/09500340308234532
14. G. Yoo, H.S. Sim, H. Schomerus, Phys. Rev. A **84**(6), 063833 (2011). https://doi.org/10.1103/PhysRevA.84.063833
15. A. Pick, B. Zhen, O.D. Miller, C.W. Hsu, F. Hernandez, A.W. Rodriguez, M. Soljačić, S.G. Johnson, Opt. Exp. **25**(11), 12325 (2017). https://doi.org/10.1364/OE.25.012325
16. Ş. Özdemir, S. Rotter, F. Nori, L. Yang, Nat. Mater.**18**(8), 783 (2019)
17. R. El-Ganainy, K.G. Makris, D.N. Christodoulides, Z.H. Musslimani, Opt. Lett. **32**(17), 2632 (2007). https://doi.org/10.1364/OL.32.002632
18. Shun-Lien Chuang, J. Lightwave Technol. **5**(1), 5 (1987)
19. C.E. Rüter, K.G. Makris, R. El-Ganainy, D.N. Christodoulides, M. Segev, D. Kip, Nat. Phys.**6**(3), 192 (2010). https://doi.org/10.1038/nphys1515
20. A. Guo, G.J. Salamo, D. Duchesne, R. Morandotti, M. Volatier-Ravat, V. Aimez, G.A. Siviloglou, D.N. Christodoulides, Phys. Rev. Lett. **103**, 093902 (2009). https://doi.org/10.1103/PhysRevLett.103.093902
21. A.Y. Song, Y. Shi, Q. Lin, S. Fan, Phys. Rev. A **99**, 013824 (2019). https://doi.org/10.1103/PhysRevA.99.013824
22. W. Wang, L.Q. Wang, R.D. Xue, H.L. Chen, R.P. Guo, Y. Liu, J. Chen, Phys. Rev. Lett. **119**, 7 (2017). https://doi.org/10.1103/PhysRevLett.119.077401

23. S.A. Hassani Gangaraj, F. Monticone, Phys. Rev. Lett. **121**, 093901 (2018). https://doi.org/10.1103/PhysRevLett.121.093901
24. T. Goldzak, A.A. Mailybaev, N. Moiseyev, Phys. Rev. Lett. **120**, 013901 (2018). https://doi.org/10.1103/PhysRevLett.120.013901
25. R. Yao, C.S. Lee, V. Podolskiy, W. Guo, Laser Photonics Rev. **13**(1), 1800154 (2019). https://doi.org/10.1002/lpor.201800154
26. X. Xu, L. Shi, L. Ren, X. Zhang, Opt. Exp. **26**(8), 10220 (2018)
27. M.A. Miri, M. Cotrufo, A. Alù, Opt. Lett. **44**(14), 3558 (2019). https://doi.org/10.1364/OL.44.003558
28. K.G. Makris, R. El-Ganainy, D.N. Christodoulides, Z.H. Musslimani, Phys. Rev. Lett. **100**(10), 103904 (2008). https://doi.org/10.1103/PhysRevLett.100.103904
29. L. Yuan, Y.Y. Lu, Phys. Rev. A **100**, 053805 (2019). https://doi.org/10.1103/PhysRevA.100.053805
30. M. Kremer, T. Biesenthal, L.J. Maczewsky, M. Heinrich, R. Thomale, A. Szameit, Nature Communications **10**(1), 435 (2019). https://doi.org/10.1038/s41467-018-08104-x
31. S. Weimann, M. Kremer, Y. Plotnik, Y. Lumer, S. Nolte, K.G. Makris, M. Segev, M.C. Rechtsman, A. Szameit, Nat. Mater. **16**(4), 433 (2017). https://doi.org/10.1038/nmat4811
32. X. Cui, K. Ding, J.W. Dong, C.T. Chan, Phys. Rev. B **100**(11), 115412 (2019). https://doi.org/10.1103/PhysRevB.100.115412
33. S. Longhi, Phys. Rev. A **82**, 031801 (2010). https://doi.org/10.1103/PhysRevA.82.031801
34. L. Ge, Y.D. Chong, A.D. Stone, Phys. Rev. A **85**, 2 (2012). https://doi.org/10.1103/PhysRevA.85.023802
35. A. Novitsky, D. Lyakhov, D. Michels, A.A. Pavlov, A.S. Shalin, D.V. Novitsky, Phys. Rev. **101**(4), 043834 (2020)
36. Y. Cao, Y. Fu, Q. Zhou, Y. Xu, L. Gao, H. Chen, Opt. Exp. **27**(6), 7857 (2019)
37. X. Zhou, X. Lin, Z. Xiao, T. Low, A. Alù, B. Zhang, H. Sun, Phys. Rev. B **100**(11), 115429 (2019)
38. C. Chen, L. Jin, R.B. Liu, New J. Phys. **21**(8), 083002 (2019)
39. R. Alaee, J. Christensen, M. Kadic, Phys. Rev. Appl. **9**(1), 014007 (2018)
40. Y. Chong, L. Ge, A.D. Stone, Phys. Rev. Lett. **106**(9), 093902 (2011)
41. Z.J. Wong, Y.L. Xu, J. Kim, K. O'Brien, Y. Wang, L. Feng, X. Zhang, Nat. Photonics **10**(12), 796 (2016). https://doi.org/10.1038/nphoton.2016.216
42. D.V. Novitsky, A. Karabchevsky, A.V. Lavrinenko, A.S. Shalin, A.V. Novitsky, in *Journal of Physics: Conference Series*, vol. 1092 (IOP Publishing, 2018), p. 012100
43. D.R. Barton III, H. Alaeian, M. Lawrence, J. Dionne, Phys. Rev. B **97**(4), 045432 (2018)
44. P. Witoński, A. Mossakowska-Wyszyńska, P. Szczepański, IEEE J. Quantum Electron. **53**(6), 1 (2017)
45. Q. Zhong, S. Nelson, Ş. Özdemir, R. El-Ganainy, Opt. Lett. **44**(21), 5242 (2019)
46. L. Jin, Phys. Rev. A **97**, 033840 (2018). https://doi.org/10.1103/PhysRevA.97.033840
47. J. Wiersig, Phys. Rev. Lett. **112**(20), 203901 (2014)
48. H. Hodaei, A.U. Hassan, S. Wittek, H. Garcia-Gracia, R. El-Ganainy, D.N. Christodoulides, M. Khajavikhan, Nature **548**(7666), 187 (2017). https://doi.org/10.1038/nature23280
49. W. Chen, Ş. KayaÖzdemir, G. Zhao, J. Wiersig, L. Yang, Nature **548**(7666), 192 (2017). https://doi.org/10.1038/nature23281
50. J. Ren, H. Hodaei, G. Harari, A.U. Hassan, W. Chow, M. Soltani, D. Christodoulides, M. Khajavikhan, Opt. Lett. **42**(8), 1556 (2017)
51. M. De Carlo, F. De Leonardis, L. Lamberti, V.M. Passaro, Opt. Lett. **44**(16), 3956 (2019)
52. J. Scheuer, Opt. Exp. **26**(24), 32091 (2018). https://doi.org/10.1364/OE.26.032091
53. A. Akbarzadeh, M. Kafesaki, E.N. Economou, C.M. Soukoulis, J.A. Crosse, Phys. Rev. A **99**, 3 (2019). https://doi.org/10.1103/PhysRevA.99.033853
54. E. Purcell, Phys. Rev. **69**(11–12), 674 (1946). https://doi.org/10.1103/PhysRev.69.674.2
55. L. Novotny, B. Hecht, *Principles of Nano-Optics*, 2nd edn. (Cambridge University Press, 2012)
56. M. Khanbekyan, J. Wiersig, Phys. Rev. Res. **2**, 023375 (2020). https://doi.org/10.1103/PhysRevResearch.2.023375

57. K.A. Ivanov, A.R. Gubaidullin, K.M. Morozov, M.E. Sasin, M.A. Kaliteevskii, Opt. Spectroscopy **122**(5), 835 (2017). https://doi.org/10.1134/S0030400X17050095
58. P. Then, G. Razinskas, T. Feichtner, P. Haas, A. Wild, N. Bellini, R. Osellame, G. Cerullo, B. Hecht, Phys. Rev. A **89**(5), 053801 (2014). https://doi.org/10.1103/PhysRevA.89.053801
59. K.M. Schulz, D. Jalas, A.Y. Petrov, M. Eich, Opt. Exp. **26**(15), 19247 (2018). https://doi.org/10.1364/OE.26.019247
60. D.V. Novitsky, A. Karabchevsky, A.V. Lavrinenko, A.S. Shalin, A.V. Novitsky, Phys. Rev. B **98**(12), 125102 (2018)
61. K.V. Kepesidis, T.J. Milburn, J. Huber, K.G. Makris, S. Rotter, P. Rabl, New J. Phys. **18**(9), 095003 (2016)

Chapter 19
Magnetoelectric Near Fields

Eugene Kamenetskii

Abstract Similar to electromagnetic (EM) phenomena, described by Maxwell equations, physics of magnetoelectric (ME) phenomena deals with the fundamental problems of the relationship between electric and magnetic fields. The different nature of these two notions is especially evident in dynamic regimes. Analyzing the EM phenomena inside the ME material, the question arises: What kind of the near fields, originated from a sample of such a material, can be measured? Observation of the ME states requires an experimental technique characterized by a violation of spatial and temporal inversion symmetries in a subwavelength region. This presumes the existence of specific near fields. Recently, such field structures, called ME fields, were found as the near fields of a quasi-2D subwavelength-size ferrite disk with magnetic-dipolar-mode (MDM) oscillations. The key physical characteristics that determine the configurations of the ME near fields are the spin and orbital angular momenta of the quantum states of the MDM spectra. This leads to the appearance of subwavelength power-flow vortices. By virtue of unique topology, the ME quantum fluctuations in vacuum are different from virtual EM photons. While preserving the ME properties, one observes strong enhancing the near-field intensity. The main purpose of this chapter is to review and analyze the studies of the ME fields. We consider the near-field topological singularities originated from the MDM ferrite-disk particle. These topological features can be transmitted to various types of nonmagnetic material structures.

19.1 Introduction

Symmetry principles play an important role in the laws of nature. Maxwell added an electric displacement current to put into a symmetrical form the equations which couple together the electric and magnetic fields. The dual symmetry between electric

E. Kamenetskii (✉)
Microwave Magnetic Laboratoty, Ben Gurion University of the Negev, Beersheba, Israel
e-mail: kmntsk@bgu.ac.il

E. Kamenetskii (ed.), *Chirality, Magnetism and Magnetoelectricity*,
Topics in Applied Physics 138,
https://doi.org/10.1007/978-3-030-62844-4_19

and magnetic fields underlies the conservation of energy and momentum for electromagnetic fields [1]. Recently, it was shown that this dual symmetry determines the conservation of optical (electromagnetic) chirality [2, 3]. Based on an analysis of the interaction of chiral light and chiral specimens, new mechanisms of enantiomer discrimination and separation in optics have been proposed [2–8]. Since chiroptical effects are usually hampered by weak chiral light-matter interaction, it is argued that to enhance the chiral effects it is necessary that the near field remains chiral in the process. Different plasmonic and dielectric nanostructures have recently been proposed as a viable route for near-field enhancement of chiral light-matter interactions [9–11]. In a more general sense, one can say that this is an attempt to enhance the near-field intensity while preserving the ME properties. However, the following fundamental questions arise: Can one really observe effects of the near-field magnetoelectricity in dynamic regimes. What are the symmetry properties of such dynamic ME fields?

The question on relationships between magnetoelectricity and electromagnetism is a subject of a strong interest and numerous discussions in microwave and optical wave physics and material sciences. The problem of the near-field magnetoelectricity in electromagnetism is a topical problem. Evanescent fields are oscillating fields whose energy is spatially concentrated in the vicinity of the oscillating currents. In classical electrodynamics we know only two types of local (subwavelength) electric currents: linear and circular. These types of currents determine elementary electric and magnetic dipole oscillations in matter. The electric polarization is parity odd and time-reversal even. At the same time, the magnetization is parity even and time-reversal odd [1]. These symmetry relation cast doubt on the idea of a local (subwavelength) coupling of two, electric and magnetic, small dipoles. When the violation of the invariances under space reflection parity and time inversion are necessary conditions for the emergence of the ME effect, the same symmetry properties should be observed for the near fields—the ME near fields.

The uniqueness of the proposed ME near fields can be shown by analyzing vacuum near fields originated from a scatterer made of a ME structure. In this connection, it is worth noting that in a case of usual (non-ME) material structures one can distinguish two kinds of the EM near fields: (a) near fields originated from EM wave resonances and (b) near fields originated from dipole-carrying resonances. The former fields, abbreviated as EM NFs, are obtained based on the full-Maxwell-equation solutions with use of Mie theory [12]. The latter fields, abbreviated as DC NFs, are observed when the electric or magnetic dipole-carrying oscillations (such, for example, as surface plasmons [13–15] and magnons [16–18]) take place. Notably, in accordance with Mie theory one can observe EM NFs with magnetic responses originated from small nonmagnetic dielectric resonators, both in microwaves [19] and optics [20–22]. In a case of DC NFs, strong coupling of EM waves with electric or magnetic dipole-carrying excitations, called polaritons, occur [16, 23]. Importantly, the spatial scale of the DC NFs is much smaller than the spatial scale of the EM NFs in the same frequency range. Due to strong coupling of EM waves with dipole-carrying excitations and temporal dispersion of the material, polaritons display enhanced field localization to surfaces and edges. Properties of vacuum near-fields originated from a small non-ME (dielectric or magnetic) sample become evident when this sample has sizes significantly smaller than the EM wavelength *in all three spatial dimensions*. The matter of fact is that near such a scatter we can only measure the electric $\vec{E}$ or the magnetic field $\vec{H}$ with accuracy. As volumes smaller than the wavelength are probed, measurements of EM energy become uncertain, highlighting the difficulty with performing measurements in this regime. There is Heisenberg's uncertainty principle binding $\vec{E}$ and $\vec{H}$ fields of the EM wave [13, 24].

Taking all this into account, let us consider now a *subwavelength ME sample*. The near-field structure of such a point scatterer is dominated by two types of the fields: the electric and magnetic fields, which are *mutually coupled* due to the intrinsic properties of a ME material. This fact gives us much greater uncertainty in probing of the fields. In total, such fields can be represented as the structures of cross $\vec{E} \times \vec{H}$ or dot $\vec{E} \cdot \vec{H}$ products in a subwavelength region. Due to $\mathcal{PT}$ symmetry of ME structure, the ME near fields of a subwavelength sample should be characterized by a certain pseudoscar parameter. Moreover, supposing that in a subwavelength region *both* structures of cross $\vec{E} \times \vec{H}$ and dot $\vec{E} \cdot \vec{H}$ products exist, one should assume the presence of helicity properties of the fields. It is evident that such a near-field structure—the ME-field structure—is beyond the frames of the Maxwell theory description [1].

When we are talking on ME dynamics, we have to refer also to an analysis of artificial structures—bianisotropic metamaterials. The notion "bianisotropic media" had been introduced to generalize different effects of coupling between magnetic and electric properties [25]. The local bianisotropic media is supposed as the media composed by structural subwavelength elements with "glued" pairs of electric and magnetic dipoles. The consideration of high-order quadrupole and multipole transitions is actually an account of spatial dispersion [26, 27]. It is assumed that bianisotropy (chirality) in metamaterials arises from a "local ME effect" [28–31]. Such a "first-principle", "microscopic-scale" ME effect of a structure composed by "glued" pairs of electric and magnetic dipoles raises a basic question on the ways of probing the dynamic parameters, since the near field structure of such a probe should violate both the spatial and temporal inversion symmetries. However, in metamaterial bianisotropic (chiral) structures, the known experimental retrieval of the cross-polarization parameters is via *far-field measurement* of the scattering-matrix characteristics [32–34]. Far-field retrieved permittivity and permeability frequently retain non-physical values, especially in the regions of the metamaterial resonances where most interesting features are expected. Far-field retrieved cross-polarization parameters of "bianisotropic particles" retain much greater non-physical value. The observed far-field phenomena of bianisotropy (chirality) can be very weakly related to the near-field manipulation effects. We can say that the cross-polarization properties of small "glued-pair" bianisotropic particles are incompatible with the effects of Rayleigh scattering.

In this chapter, we consider near fields originated from subwavelength resonators, that are the systems with quantum-confinement effects of dipolar-mode quasistatic oscillations. We analyze the possibilities of these resonances to exhibit near-field ME properties. An analysis of such dipole-carrying excitations allows finding a proper way in realizing polariton structures with properties of strong ME interactions.

19.2 Subwavelength Resonators with Dipole-Carrying Excitations

An interaction between the photon and medium dipole-carrying excitation becomes strong enough near the resonance between the light mode and the mode of the medium excitation. At the resonance region, the dispersion curves of these modes transform into two split polaritonic branches showing anticrossing behavior. The examples are exciton polaritons, surface-plasmon polaritons, and magnon polaritons. Semiclassically, polaritons are described using Maxwell equations and constitutive relations that include the frequency dependent response functions. Quantum mechanically, polaritons are described as hybrid collective excitations that are linear superpositions of matter collective excitations and photons. There are the effects of *interaction between real and virtual photons*. When dipole-carrying excitations are observed in a high-quality confined structure, the coupling modes can appear as composite bosons. Strong long-range dipole-dipole interactions significantly modify the mean-field predictions of the quantum phases of microscopic short-range excitations by stabilizing the condensate phase. It can persist up to densities high enough to support quantum liquidity with very long lifetimes. In *exciton-polariton condensates*, in particular, this effect leads to sustained trapping of the emitted photon [35–39].

Excitons in semiconductor resonators are dipole-carrying oscillations. Plasmons and magnons in confined structures are also dipole-carrying oscillations. Plasmons are optical responses of metal structures arising from collective oscillations of their conduction electrons. The microwave responses of ferrite samples—magnons—arise from collective oscillations of their precessing electrons. Both plasmons and magnons are bosons. In increasing the capabilities of the optical and microwave techniques further into the subwavelength regime, small plasmon and magnon resonant structures have attracted considerable interest. These oscillations in subwavelength resonators, however, are not *composite bosons*, as in the case of exciton resonances. No dipole-dipole plasmon condensate and dipole-dipole magnon condensate in confined resonant structures are observed, to the best to our knowledge. The problem of creating a condensate with linked (electric *and* magnetic) dipole-carrying excitations confined in a high-quality resonant structure appears as a lot more exotic. Is it even possible to observe tightly bound ME excitations, which turns into a composite boson (or fermion) and behaves as quasiparticle? Can we, in general, solve the problem of creation of the ME-polariton condensate? To answer these questions, we should analyze the possibility of finding ME properties in subwavelength resonators with quasistatic (dipolar-mode) oscillations.

A. On the possibility to observe the quantum confinement effects of electrostatic and magnetostatic oscillations

Electromagnetic (EM) responses of plasmon oscillations in optics and magnon oscillations in microwaves give rise to a strong enhancement of local fields near the surfaces of subwavelength resonators. We can classify these oscillations as electrostatic (ES) and magnetostatic (MS) resonances, respectively [18, 40]. In ES resonances in small metallic samples, one neglects a time variation of magnetic energy in comparison with a time variation of electric energy. It means that one neglects a magnetic displacement current and an electric field is expressed via an ES potential, $\vec{E} = -\vec{\nabla}\phi$ [40]. However, the Ampere–Maxwell law gives the presence of a curl magnetic field. In like manner, in the case of MS resonances in ferrite samples, one neglects a time variation of electric energy in comparison with a time variation of magnetic energy. It means that the MS-resonance problem is considered as zero-order approximation of Maxwell's equations when one neglects the electric displacement current and expresses a magnetic field via a MS potential, $\vec{H} = -\vec{\nabla}\psi$ [18]. While Faraday's law gives the presence of a curl electric field. Importantly, from a classical electrodynamics point of view, one does not have a physical mechanism describing the reverse effect of transformation of a curl magnetic field to a potential electric field in the case of ES resonances. Also, one does not have a physical mechanism describing the reverse effect of transformation of a curl electric field to a potential magnetic field a case of the MS resonance [1, 41]. It means that, fundamentally, subwavelength sizes of the particles should eliminate any EM retardation effects. We can say that for an EM wavelength λ and particle of a characteristic size a, the quasistatic approximation $2\pi a/\lambda \ll 1$ implies the transition to a small EM phase.

What kind of the time-varying field structure one can expect to see when an electric or magnetic displacements currents are neglected and so the electromagnetic-field symmetry (dual symmetry) of Maxwell equations is broken? When one neglects a displacement current (magnetic or electric) and considers the scalar-function [$\phi(\vec{r}, t)$ or $\psi(\vec{r}, t)$] solutions, as the wave-propagation solutions, one has to accept the possibility to observe the quantum confinement effects of electrostatic and magnetostatic oscillations. Such an analysis of quasistatic resonances is based on postulates about a physical meaning of scalar function as a complex scalar wavefunction, which presumes a long-range phase coherence in dipole–dipole interactions. These solutions should be based on the Schrödinger-like equation.

For quasi-ES resonances in subwavelength metal structures characterised by non-homogeneous scalar permittivity, we have Poison's equation [42–45]

$$\nabla^2\phi + \vec{\nabla} \cdot \left((\varepsilon(\vec{r}) - 1)\vec{\nabla}\phi\right) = 0. \tag{19.1}$$

At the same time, for quasi-MS resonances in subwavelength microwave ferrite structures with tensor permeability $\overleftrightarrow{\mu}$, there is Walker's equation [18, 46]:

$$\nabla^2\psi + \vec{\nabla}\cdot\left(\left(\frac{\overleftrightarrow{\mu}}{\mu_0} - \overleftrightarrow{I}\right)\cdot\vec{\nabla}\psi\right) = 0. \qquad (19.2)$$

Solutions of both these equations are harmonic functions. Nevertheless, it appears that in spite of a certain similarity between (19.1) and (19.2), the physical properties of the ES and MS oscillation spectra are fundamentally different in many aspects. The most important factor distinguishing the MS resonance from the ES resonance is the tensorial form of permeability and the presence off-diagonal gyrotropic elements in this tensor.

In [47] it was discussed that in a case of the surface plasmon resonances in subwavelength optical metallic structures no retardation processes characterized by the electric dipole-dipole interaction and described exclusively by electrostatic wave function $\phi(\vec{r}, t)$ take place. There is no possibility to describe these resonances by the Schrödinger-equation energy eigenstate problem. Nevertheless, for MS resonances in ferrite specimens we have bulk wave process, which are determined by a scalar wave function $\psi(\vec{r}, t)$. Due to the retardation processes caused by the magnetic dipole-dipole interaction in a subwavelength ferrite particle, we have a possibility to formulate the energy eigenstate boundary problem based on the Schrödinger-like equation for scalar-wave eigenfunctions $\psi(\vec{r}, t)$. Such a behavior can be obtained in a ferrite particle in a form of a quasi-2D disk. The oscillations in a quasi-2D ferrite disk, analyzed as spectral solutions for the MS-potential scalar wave function $\psi(\vec{r}, t)$, have evident quantum-like attributes. Quantized forms of such oscillations we call the MS magnons or the magnetic-dipolar-mode (MDM) magnons. The macroscopic nature of MDMs, involving the collective motion of a many-body system of precessing electrons, does not destroy a quantum behavior. The long-range dipole-dipole correlation in positions of electron spins can be treated in terms of collective excitations of a system as a whole.

Analyzing the confinement effects of electrostatic and magnetostatic oscillations in subwavelength resonators, it is also worth making another important remark. Considering light interaction with photonic and plasmonic resonances, the authors of review in [48] noted that as the optical mode becomes deeply subwavelength in all three dimensions, independent of its shape, the Q-factor of the resonances is limited to about 10 or less. As they argue, the reason is that in such small volumes, self-sustaining oscillations are no longer possible between the electric-field and magnetic-field energies and, at the same time, no effects of the electric dipole-dipole oscillations can be assumed. At the same time, in the case of a microwave MDMs in a deeply subwavelength ferrite disk resonator, we have the Q-factor about several thousand [49–51]. For such MDM resonances, subwavelength sizes of the ferrite particle allow eliminate any electromagnetic retardation effects and consider only the magnetic dipole-dipole interaction effects.

To make the MDM spectral problem analytically integrable, two approaches were suggested. These approaches, distinguished by differential operators and boundary conditions used for solving the spectral problem, give two types of MDM oscillation spectra in a quasi-2D ferrite disk. These two approaches are conditionally called as the G and L modes in the magnetic dipolar spectra [52–57]. The MS-potential wave function $\psi(\vec{r}, t)$ manifests itself in different manners for each of these types of spectra. In the case of the G-mode spectrum, where the physically observable quantities are energy eigenstates, the MS-potential wave function appears as a Hilbert-space scalar wave function. In the case of the L modes, the MS-potential wave function is considered as a generating function for the vector harmonics of the magnetic and electric fields.

B. Spectral problems for MDM magnetostatic oscillations: *G* modes

The MDM-resonance spectral solutions obtained from the second-order differential equation—the Walker's equation [18, 46]—are constructed in accordance with basic symmetry considerations for the sample geometry. For an open quasi-2D ferrite disk normally magnetized along the z axis, we can use separation of variables. In cylindrical coordinate system (z, r, θ), the solutions are represented as [52–57]

$$\psi_{p,\nu,q} = A_{p,\nu,q}\xi_{p,\nu,q}(z)\tilde{\eta}_{\nu,q}(r, \theta), \tag{19.3}$$

where $A_{p,\nu,q}$ is a dimensional amplitude coefficient, $\xi_{p,\nu,q}(z)$ is a dimensionless function of the MS-potential distribution along z axis, and $\tilde{\eta}_{\nu,q}(r, \theta)$ is dimensionless membrane function. The membrane function $\tilde{\eta}$ is defined by a Bessel-function order ν and a number of zeros of the Bessel function corresponding to a radial variations q. The dimensionless "thickness-mode" function $\xi(z)$ is determined by the axial-variation number p.

In a quasi-2D ferrite disk, one can formulate the energy eigenstate boundary problem based on the Schrödinger-like equation for scalar-wave eigenfunctions $\psi(\vec{r}, t)$ with using the Dirichlet-Neumann (ND) boundary conditions. The energy eigenvalue problem for MDMs is defined by differential equation

$$\hat{G}_{\perp}\tilde{\eta}_n = E_n\tilde{\eta}_n, \tag{19.4}$$

where $\hat{G}_{\perp}$ is a two-dimensional (on the r, θ disk plane) differential operator. The quantity E_n is interpreted as density of accumulated magnetic energy of mode n. This is the average (on the RF period) energy accumulated in the ferrite-disk region of unit in-plane cross-section and unit length along z axis [52–57]. The operator $\hat{G}_{\perp}$ and quantity E_n are defined as

$$\frac{g\mu_0}{4}\mu_n\nabla_{\perp}^2\tilde{\eta}_n = E_n\tilde{\eta}_n, \tag{19.5}$$

where

$$E_n = \frac{g\mu_0}{4}(\beta_n)^2. \quad (19.6)$$

Here $\nabla_\perp^2$ is the two-dimensional (on the circular cross section of a ferrite-disk region) Laplace operator, g is a dimensional normalization coefficient (with the unit of dimension ψ^2) for mode n and β_n is the propagation constant of mode n along the disk axis z. The parameter μ_n (which is a diagonal component of the permeability tensor [18]) should be considered as an eigenvalue. Outside a ferrite $\mu_n = 1$. The operator $\hat{G}_\perp$ is a self-adjoint operator only for negative quantities μ_n in a ferrite. For self-adjointness of operator $\hat{G}_\perp$, the membrane function $\tilde{\eta}_n(r,\theta)$ must be continuous and differentiable with respect to the normal to lateral surface of a ferrite disk. The homogeneous boundary conditions—the ND boundary conditions—for the membrane function are:

$$(\tilde{\eta}_n)_{r=\mathcal{R}^-} - (\tilde{\eta}_n)_{r=\mathcal{R}^+} = 0 \quad (19.7)$$

and

$$\mu\left(\frac{\partial\tilde{\eta}_n}{\partial r}\right)_{r=\mathcal{R}^-} - \left(\frac{\partial\tilde{\eta}_n}{\partial r}\right)_{r=\mathcal{R}^+} = 0, \quad (19.8)$$

where $\mathcal{R}$ is the disk radius. MDM oscillations in a ferrite disk are described by real eigenfunctions: $\tilde{\eta}_{-n} = \tilde{\eta}_n^*$. For modes n and n', the orthogonality conditions are expressed as

$$\int_{S_c} \tilde{\eta}_n \tilde{\eta}_{n'}^* dS = \delta_{nn'}, \quad (19.9)$$

where S_c is a square of a circular cross section of a ferrite-disk region and $\delta_{nn'}$ is the Kronecker delta. The spectral problem gives the energy orthogonality relation for MDMs:

$$(E_n - E_{n'}) \int_{S_c} \tilde{\eta}_n \tilde{\eta}_{n'}^* dS = 0. \quad (19.10)$$

Since the space of square integrable functions is a Hilbert space with a well-defined scalar product, we can introduce a basis set. A dimensional amplitude coefficient we write as $A_n = c' a_n$, where c' is a dimensional unit coefficient and a_n is a normalized dimensionless amplitude. The normalized scalar-wave membrane function $\tilde{\eta}$ can be represented as $\tilde{\eta} = \sum_n a_n \tilde{\eta}_n$. The amplitude is defined as $|a_n|^2 = \left|\int_{S_c} \tilde{\eta}\tilde{\eta}_n^* dS\right|^2$. The mode amplitude can be interpreted as the probability to find a system in a certain state n. Normalization of membrane function is expressed as $\sum_n |a_n|^2 = 1$ [52–57].

The analysis of discrete-energy eigenstates of the MDM oscillations, resulting from structural confinement in a normally magnetized ferrite disk, is based on a continuum model. Using the principle of wave-particle duality, one can describe this oscillating system as a collective motion of quasiparticles. There are "flat-mode" quasiparticles at a reflexively-translational motion behavior between the lower and upper planes of a quasi-2D disk. Such quasiparticles are called "light" magnons. In our study we consider MS magnons in ferromagnet as quanta of collective MS spin waves that involves the precession of many spins on the long-range dipole-dipole interactions. It is different from the short-range magnons for exchange-interaction spin waves with a quadratic character of dispersion. The meaning of the term "light", used for the condensed MDM magnons, arises from the fact that effective masses of these quasiparticles are much less, than effective masses of "real" magnons—the quasiparticles describing small-scale exchange-interaction effects in magnetic structures. The effective mass of the "light" magnon for a monochromatic MDM is defined as [53]:

$$\left(m_{lm}^{(eff)}\right)_n = \frac{\hbar}{2}\frac{\beta_n^2}{\omega}. \tag{19.11}$$

In solving boundary value problems for MS resonances, one encounters some questions when using boundary conditions. As is known, in solving a boundary value problem that involves the eigenfunctions of a differential operator, the boundary conditions must be in a definite correlation with the type of this differential operator [58, 59]. In an analysis of MDM resonances in a ferrite disk, we used the homogeneous ND boundary conditions, which mean continuity of the MS wave functions together with continuity of their first derivatives on the sample boundaries. Only in this case the functions form a complete set of orthogonal basis functions and thus the field expansion in terms of orthogonal MS-potential functions can be employed.

However, the ND boundary condition (19.7), (19.8) are not the EM boundary conditions. While the considered above ND boundary conditions are the so-called essential boundary conditions, the EM boundary conditions are the natural boundary conditions [58]. For the EM boundary conditions, on a lateral surface of a ferrite disk we have to have continuity of membrane function $\tilde{\eta}$ and a radial component of the magnetic flux density $B_r = \mu_0\left(\mu\frac{\partial\tilde{\eta}_n}{\partial r} + \mu_a\frac{1}{r}\frac{\partial\tilde{\eta}_n}{\partial\theta}\right)$. Here μ and μ_a are diagonal component and off-diagonal components of the permeability tensor [18]. With such EM boundary conditions, it becomes evident that the membrane function $\tilde{\eta}$ must not only be continuous and differentiable with respect to a normal to the lateral surface of a disk, but (because of the presence of a gyrotropy term μ_a) be also differentiable with respect to a tangent to this surface. There is evidence of the presence of an azimuthal magnetic field on the border circle with clockwise and counterclockwise rotation asymmetry. In this case, the membrane functions $\tilde{\eta}$ cannot be considered as single-valued functions, and the question arises of the validity of the energy orthogonality relation for MS-wave modes.

To restore the ND boundary conditions and thus the completeness of eigenfunctions $\tilde{\eta}$, we need introducing a certain surface magnetic current $\left(j_s^{(m)}\right)_{top}$ circulating on a lateral surface of the disk. This is a topological current, which compensates the term $\left(i\mu_a \frac{1}{r}\frac{\partial\tilde{\eta}}{\partial\theta}\right)_{r=\mathcal{R}^-}$ in the EM boundary conditions [55, 56, 60]. Evidently, for a given direction of a bias magnetic field (that is, for a given sign of μ_a), there can be two, clockwise and counterclockwise, quantities of the circulating magnetic current. The topological current $\left(j_s^{(m)}\right)_{top}$ is defined by the velocity of an irrotational border flow. This flow is observable via the circulation integral of the gradient $\vec{\nabla}_\theta \delta = \frac{1}{\mathcal{R}}\left(\frac{\partial\delta_\pm}{\partial\theta}\right)_{r=\mathcal{R}} \vec{e}_\theta$, where $\delta_\pm$ is a double-valued edge wave function on contour $\mathcal{L} = 2\pi\mathcal{R}$. On a lateral surface of a quasi-2D ferrite disk, one can distinguish two different functions $\delta_\pm$, which are the counterclockwise and clockwise rotating-wave edge functions with respect to a membrane function $\tilde{\eta}$. The spin-half wave-function $\delta_\pm$ changes its sign when the regular-coordinate angle θ is rotated by 2π. As a result, one has the eigenstate spectrum of MDM oscillations with topological phases accumulated by the edge wave function δ. A circulation of gradient $\vec{\nabla}_\theta \delta$ along contour $\mathcal{L} = 2\pi\mathcal{R}$ gives a non-zero quantity when an azimuth number is a quantity divisible by $\frac{1}{2}$. A line integral around a singular contour $\mathcal{L}$: $\frac{1}{\Re}\oint\limits_{\mathcal{L}}\left(i\frac{\partial\delta_\pm}{\partial\theta}\right)(\delta_\pm)^* \, d\mathcal{L} = \int\limits_0^{2\pi}\left[\left(i\frac{\partial\delta_\pm}{\partial\theta}\right)(\delta_\pm)^*\right]_{r=\Re} d\theta$ is an observable quantity. Because of the existing the geometrical phase factor on a lateral boundary of a ferrite disk, MDM oscillations are characterized by a pseudo-electric field (the gauge field) $\vec{\epsilon}$. The pseudo-electric field $\vec{\epsilon}$ can be found as $\vec{\epsilon}_\pm = -\vec{\nabla}\times \left.\vec{\Lambda}_\epsilon^{(m)}\right)_\pm$. The field $\vec{\epsilon}$ is the Berry curvature. The corresponding flux of the gauge field $\vec{\epsilon}$ through a circle of radius $\Re$ is obtained as: $K\int\limits_S\left(\vec{\epsilon}\right)_\pm \cdot d\vec{S} = K\oint\limits_{\mathcal{L}}\left(\vec{\Lambda}_\epsilon^{(m)}\right)_\pm \cdot d\vec{\mathcal{L}} = K\left(\Xi^{(e)}\right)_\pm = 2\pi q_\pm$, where $\left(\Xi^{(e)}\right)_\pm$ are quantized fluxes of pseudo-electric fields, K is the normalization coefficient. Each MDM is quantized to a quantum of an emergent electric flux. There are the positive and negative eigenfluxes. These different-sign fluxes should be nonequivalent to avoid the cancellation. It is evident that while integration of the Berry curvature over the regular-coordinate angle θ is quantized in units of 2π, integration over the spin-coordinate angle θ' $\left(\theta' = \frac{1}{2}\theta\right)$ is quantized in units of π. The physical meaning of coefficient K concerns the property of a flux of a pseudo-electric field. The Berry mechanism provides a microscopic basis for the surface magnetic current at the interface between gyrotropic and nongyrotropic media. Following the spectrum analysis of MDMs in a quasi-2D ferrite disk one obtains pseudo-scalar axion-like fields and edge chiral magnetic currents. The anapole moment for every mode n is calculated as [55, 56, 60].

$$\left(a_{\pm}^{(e)}\right)_n \propto \mathcal{R} \int_0^d \oint_{\mathcal{L}} \left[\left(\vec{j}_s^{(m)}(z)\right)_{top}\right]_n \cdot d\vec{l} dz, \tag{19.12}$$

where d is the disk thickness. The edge magnetic current $\left(j_s^{(m)}\right)_{top}$ is a persistent current appearing due to the mesoscopic effect: the magnitude of such a resonant current becomes appreciable when the parameters of the ferrite disk are reduced to the scale of the dipole-dipole quantum phase coherence length of precessing electrons. At the MDM resonances, one has the "spin-orbit" interaction between precessing magnetic dipoles and a persistent orbital magnetic current. When the frequency of the orbital rotation of MDM resonances in a ferrite disk is close to the ferromagnetic resonance frequency, the precessing magnetic dipoles become strongly correlated and one observers fermionization of the system composed of bosons. There are the macroscopic quantum phenomena related to the collective motion of magnetic dipoles coalescing into the same quantum state, described by a single coherent wave-function of the condensate. This is a fundamentally distinctive feature from a boson condensate created by small-scale exchange-interaction magnons [61].

It is worth noting that along with the circulation of the surface magnetic current $\left(j_s^{(m)}\right)_{top}$ caused by the edge wave function, there is also the quadratic-form circulation due to this function. For the double-valued edge wave functions $\delta_{\pm}$, we have the following orthonormality condition on contour $\mathcal{L} = 2\pi\mathcal{R}$ [55]:

$$\begin{gathered}\mathcal{R} \int_0^{2\pi} \left[\left(i\mu_a \frac{\partial(\delta_{\pm})_n}{\partial\theta}\right)\left(\delta_{\pm}^*\right)_{n'} - (\delta_{\pm})_n \left(i\mu_a \frac{\partial(\delta_{\pm})_{n'}}{\partial\theta}\right)^*\right]_{r=\mathcal{R}} d\theta \\ = \mathcal{R}\mu_a\left[(q_{\pm})_n - (q_{\pm})_{n'}\right] \int_0^{2\pi} \left[(\delta_{\pm})_n \left(\delta_{\pm}^*\right)_{n'}\right]_{r=\mathcal{R}} d\theta = 0\end{gathered} \quad . \tag{19.13}$$

For mode n, there are the normalisation relations for the edge functions

$$\int_0^{2\pi} \left[(\delta_+)_n \left(\delta_+^*\right)_n\right]_{r=\mathcal{R}} d\theta = (N_+)_n \tag{19.14}$$

and

$$\int_0^{2\pi} \left[(\delta_-)_n \left(\delta_-^*\right)_n\right]_{r=\mathcal{R}} d\theta = (N_-)_n \tag{19.15}$$

where $(N_+)_n$ and $(N_-)_n$ are real quantities, which we characterize as surface power flow density of the MDM mode. For a given direction of a bias magnetic field, the wave described by functions $(\delta_+)_n$ propagates only in one direction along the edge. Also, the wave described by function $(\delta_-)_n$—propagates in one direction (opposite to the former case) along the edge. It is evident that a complex conjugate 'particle' configuration is an 'antiparticle' configuration and vice versa. This fact is related to the existence of only a single edgemode excitation for each wavenumber q. In other words, the 'particle' configurations are their own 'antiparticle' configurations. This resembles the well-known properties of the Majorana fermions [62]. So, we have a "chiral Majorana fermion field" [63, 64] on the lateral wall of the MDM ferrite disk.

Due to the presence of surface power flow density, membrane eigenfunction $\tilde{\eta}$ of every MDM rotates around the disk axis. When for every MDM we introduce the notion of an effective mass $\left(m_{lm}^{(eff)}\right)_n$, expressed by (19.11), we can assume that for every MDM there exists also an effective moment of inertia $\left(I_z^{(eff)}\right)_n$. With this assumption, an orbital angular momentum a mode is expressed as $(L_z)_n = \left(I_z^{(eff)}\right)_n \omega$. At the first approximation, let us suppose that the membrane eigenfunction $\tilde{\eta}_n$ is viewed as an infinitely thin homogenous disk of radius $\mathcal{R}$. In other words, we assume that for every MDM, the radial and azimuth variation of the MS-potential function, are averaged. In such a case, we can write

$$\left(I_z^{(eff)}\right)_n = \frac{1}{2}\left(m_{lm}^{(eff)}\right)_n \mathcal{R}^2 d. \tag{19.16}$$

and

$$\left(L_z^{(eff)}\right)_n = \left(I_z^{(eff)}\right)_n \omega = \frac{\hbar}{4}\beta_n^2 \mathcal{R}^2 d. \tag{19.17}$$

C. Spectral problems for MDM magnetostatic oscillations: *L* modes

In the above spectral analysis of the *G* modes, we used the ND boundary conditions. To bridge this spectral problem with the EM boundary conditions, we introduced the contour integrals determining surface magnetic current $\left(j_s^{(m)}\right)_{top}$ and surface power flow density of the MDM modes. However, the MDM spectral can be solved directly based on the EM boundary conditions. This approach is called the *L*-mode spectral analysis. The solution for MS-potential wave function of a *L*-mode is written as

$$\psi_{p,\nu,q} = C_{p,\nu,q}\xi_{p,\nu,q}(z)\tilde{\varphi}_{\nu,q}(r,\theta), \tag{19.18}$$

where $C_{p,\nu,q}$ is a dimensional amplitude coefficient and $\tilde{\varphi}$ is a membrane function. For solutions in a cylindrical coordinate system, one uses the following the boundary condition on a lateral surface of a ferrite disk [52, 53, 55, 56]:

$$\mu\left(\frac{\partial\tilde{\varphi}}{\partial r}\right)_{r=\mathcal{R}^-} - \left(\frac{\partial\tilde{\varphi}}{\partial r}\right)_{r=\mathcal{R}^+} + i\frac{\mu_a}{\mathcal{R}}\left(\frac{\partial\tilde{\varphi}}{\partial\theta}\right)_{r=\mathcal{R}^-} = 0, \tag{19.19}$$

which are different from the boundary conditions (19.8). A function $\tilde{\varphi}$ is not a single-valued function. It changes a sign when angle θ is turned on 2π. For any mode n, the function $\tilde{\varphi}_n$ is a two-component sprinor pictorially denoted by two arrows:

$$\tilde{\varphi}_n^{\uparrow\downarrow}(\vec{r},\theta) = \begin{bmatrix} \tilde{\varphi}_n^{\uparrow} \\ \tilde{\varphi}_n^{\downarrow} \end{bmatrix} = \tilde{\eta}_n(\vec{r},\theta) \begin{bmatrix} e^{-\frac{1}{2}i\theta} \\ e^{+\frac{1}{2}i\theta} \end{bmatrix} \tag{19.20}$$

For MS waves in a ferrite medium, described by a L-mode scalar wave function $\psi(\vec{r},t)$, we define a magnetic flux density: $\vec{B} = \overleftrightarrow{\mu} \cdot \vec{H} = -\overleftrightarrow{\mu} \cdot \vec{\nabla}\psi$. In this case, the power flow density can be viewed as a current density, is expressed as [55]:

$$\vec{\mathcal{J}} = \frac{i\omega}{4}\left(\psi \vec{B}^* - \psi^* \vec{B}\right). \tag{19.21}$$

Such a power flow can appear because of dipole-dipole interaction of magnetic dipoles. With use of separation of variables and taking into account a form of tensor $\overleftrightarrow{\mu}$ [18], we decompose a magnetic flux density by two components:

$$\vec{B} = \vec{B}_\perp + \vec{B}_{||}. \tag{19.22}$$

The component $\vec{B}_\perp$ are given as

$$\vec{B}_\perp = -\overleftrightarrow{\mu}_\perp \cdot \vec{\nabla}_\perp \psi = -C_n \xi_n(z)\left[\overleftrightarrow{\mu}_\perp \cdot \vec{\nabla}_\perp \tilde{\varphi}_n(r,\theta)\right]\vec{e}_\perp, \tag{19.23}$$

where $\vec{e}_\perp$ is a unit vector laying in the r,θ plane, and

$$\overleftrightarrow{\mu}_\perp = \mu_0 \begin{bmatrix} \mu & i\mu_a \\ -i\mu_a & \mu \end{bmatrix}. \tag{19.24}$$

For the component $\vec{B}_{||}$ we have

$$\vec{B}_{||} = -\mu_0 \cdot \vec{\nabla}_{||}\psi = -\mu_0 C_n \frac{\partial \xi_n(z)}{\partial z} \tilde{\varphi}_n(r,\theta)\vec{e}_z, \tag{19.25}$$

where $\vec{e}_z$ is a unit vector directed along the z axis.

The above representations allow considering two components of the power flow density (current density). For mode n, we can write (19.21) as

$$\vec{\mathcal{J}}_n = \left(\vec{\mathcal{J}}_\perp\right)_n + \left(\vec{\mathcal{J}}_{||}\right)_n, \tag{19.26}$$

where $\left(\vec{\mathcal{J}}_{\perp}\right)_n = \frac{i\omega}{4}\left[\psi_n\left(\vec{B}_{\perp}^*\right)_n - \psi_n^*\left(\vec{B}_{\perp}\right)_n\right]$ and $\left(\vec{\mathcal{J}}_{\parallel}\right)_n = \frac{i\omega}{4}\left[\psi_n\left(\vec{B}_{\parallel}^*\right)_n - \psi_n^*\left(\vec{B}_{\parallel}\right)_n\right]$. Along every of the coordinates $\vec{r}, \vec{\theta}$, and $\vec{z}$, we have the power flows (currents):

$$\left(\vec{\mathcal{J}}_r\right)_n = -\frac{i\omega}{4}|C_n|^2|\xi_n|^2\mu_0\left\{\tilde{\varphi}_n\left(\mu\frac{\partial\tilde{\varphi}_n}{\partial r} + i\mu_a\frac{1}{r}\frac{\partial\tilde{\varphi}_n}{\partial\theta}\right)^* - \tilde{\varphi}_n^*\left(\mu\frac{\partial\tilde{\varphi}_n}{\partial r} + i\mu_a\frac{1}{r}\frac{\partial\tilde{\varphi}_n}{\partial\theta}\right)\right\}\vec{e}_r, \tag{19.27}$$

$$\left(\vec{\mathcal{J}}_\theta\right)_n = -\frac{i\omega}{4}|C_n|^2|\xi_n|^2\mu_0\left\{\tilde{\varphi}_n\left(-i\mu_a\frac{\partial\tilde{\varphi}_n}{\partial r} + \mu\frac{1}{r}\frac{\partial\tilde{\varphi}_n}{\partial\theta}\right)^* - \tilde{\varphi}_n^*\left(-i\mu_a\frac{\partial\tilde{\varphi}_n}{\partial r} + \mu\frac{1}{r}\frac{\partial\tilde{\varphi}_n}{\partial\theta}\right)\right\}\vec{e}_\theta, \tag{19.28}$$

$$\left(\vec{\mathcal{J}}_z\right)_n = -\frac{i\omega}{4}\mu_0|C_n|^2|\tilde{\varphi}|^2\left[\xi_n\left(\frac{\partial\xi_n}{\partial z}\right)^* - \xi_n^*\frac{\partial\xi_n}{\partial z}\right]\vec{e}_z, \tag{19.29}$$

where $\vec{e}_r$, $\vec{e}_\theta$, and $\vec{e}_z$ are the unit vectors. We can see that for membrane function $\tilde{\varphi}$, defined by (19.20), there is a non-zero real azimuth component of the power-flow density. So, there is a non-zero quantity of the power flow circulation (clockwise or counterclockwise) around a circle $L = 2\pi r$, where $0 < r \leq \mathcal{R}$. At the same time, homogeneous EM boundary conditions imposed on a ferrite disk on the r and z axes give standing waves without real power flows.

19.3 Near Fields of MDM Oscillations—the ME Near Fields

The L-mode wave function $\psi(\vec{r}, t)$ can define a magnetic flux density in a ferrite disk, as shown above. This scalar wave function is considered as a generating function for other types of the fields both inside and outside a ferrite disk. It allows analyzing complex topological properties of vectorial fields, associated with orbital angular momentum properties of MDM resonances.

When the spectral problem for the MS-potential scalar wave function $\psi(\vec{r}, t)$, expressed by (19.18), is solved, distribution of magnetization in a ferrite disk is found as $\vec{m} = -\overleftrightarrow{\chi} \cdot \vec{\nabla}\psi$, where $\overleftrightarrow{\chi}$ is the susceptibility tensor of a ferrite [18]. Based on the known magnetization $\vec{m}$, one can find the magnetic field distribution at any point outside a ferrite disk [1, 65]:

$$\vec{H}(\vec{r}) = \frac{1}{4\pi}\left(\int\limits_V \frac{\left(\vec{\nabla}' \cdot \vec{m}(r')\right)(\vec{r} - \vec{r}')}{|\vec{r} - \vec{r}'|^3} dV' - \int\limits_S \frac{(\vec{n}' \cdot \vec{m}(r'))(\vec{r} - \vec{r}')}{|\vec{r} - \vec{r}'|^3} dS'\right). \tag{19.30}$$

Also, the electric field in any point outside a ferrite disk is defined as [56, 65]

$$\vec{E}(\vec{r}) = -\frac{1}{4\pi}\int\limits_V \frac{\vec{j}^{(m)}(\vec{r}') \times (\vec{r} - \vec{r}')}{|\vec{r} - \vec{r}'|^3} dV', \tag{19.31}$$

where $\vec{j}^{(m)} = i\omega\mu_0\vec{m}$ is the density of a bulk magnetic current and frequency ω is the MDM resonance frequency. In (19.30) and (19.31), V and S are a volume and a surface of a ferrite sample, respectively. Vector $\vec{n}'$ is the outwardly directed normal to surface S.

Depending on a direction of a bias magnetic field, we can distinguish the clockwise and counterclockwise topological-phase rotation of the fields. At the MDM resonances, for the magnetic and electric fields defined by (19.30) and (19.31) one can compose a vector

$$\left\langle \vec{S}_{MDM} \right\rangle^{\uparrow\downarrow} \equiv \frac{1}{2}\mathrm{Re}\left(\vec{E} \times \vec{H}^*\right). \tag{19.32}$$

The vector $\left\langle \vec{S}_{MDM} \right\rangle$ can be considered as a power flow density vector. Really, based on the vector relation $\vec{\nabla} \cdot \left(\vec{E} \times \vec{H}^*\right) = \vec{H}^* \cdot \vec{\nabla} \times \vec{E}^* - \vec{E} \cdot \vec{\nabla} \times \vec{H}^*$ with taking into account equations $\vec{\nabla} \times \vec{E} = -i\omega\vec{B}$, $\vec{H} = -\vec{\nabla}\psi$ and $\nabla \cdot \vec{B} = 0$, one has as a result $\vec{\nabla} \cdot \left(\vec{E} \times \vec{H}^*\right) = i\omega\vec{\nabla} \cdot \left(\psi^*\vec{B}\right)$. The right-hand side of this equation is a divergence of the power flow density of monochromatic MS waves [55]. So, vector $\left\langle \vec{S}_{MDM} \right\rangle$ can be interpreted as the power flow density as well. Nevertheless, this is not the "EM Poynting vector". Compare to the case of EM wave propagation (with both curl electric and curl magnetic fields), we have here the modes with *curl* electric and *potential* magnetic fields. As we noted above, that there are no EM laws describing transformation of the curl electric field to the potential magnetic field.

In the MDM resonance, the orbital angular-momentum of the power flow density is expressed as

$$\vec{\mathcal{L}}_z = \frac{1}{2}\mathrm{Re}\left[\vec{r} \times \left(\vec{E} \times \vec{H}^*\right)\right]. \tag{19.33}$$

Depending on a direction of a bias magnetic field, we can distinguish the clockwise and counterclockwise topological-phase rotation of the fields outside the ferrite disk. The direction of an orbital angular-momentum $\vec{\mathcal{L}}_z$ is correlated with the direction of a bias magnetic field $\vec{H}_0$ (along $+z$ axis or $-z$ axis). The active power flow of the field

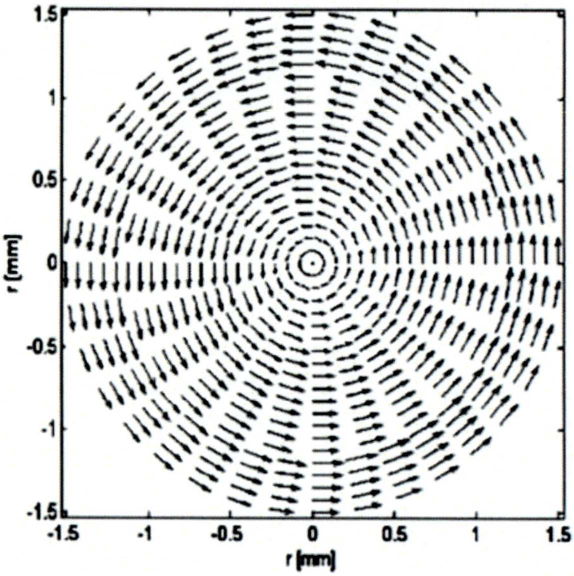

Fig. 19.1 Analytically derived power-flow-density distribution for the main MDM inside a ferrite disk of dimeter 3 mm (arbitrary units)

both inside and outside a subwavelength ferrite disk has the vortex topology. In [65–67] it was shown that for every MDM mode, the power flow circulation calculated by (19.28) have the same distributions on the r, θ plane, as the circulation of the power flow vector $\left\langle \vec{S}_{MDM} \right\rangle$. Such the analytically derived distributions coincide with the numerical patterns of the power flows. The analysis was made for a YIG ferrite disk of a 3 mm diameter and thickness of 0.050 mm at the frequency region 8–9 GHz. Figure 19.1 gives an example of the analytically derived power-flow-density distribution. Figure 19.2 shows some numerical patterns of the power flows. One can see a strong confinement of the fields arising from the vortices of the MDM resonances. In Fig. 19.3, we give a schematic representation of the circulation of the power flow, depicted on the surface of the vacuum sphere and on the surface of the solid angle. Direction of an orbital angular-momentum of a ferrite disk is correlated with the direction of a bias magnetic field.

A persistent edge magnetic current circulating along the contour $\mathcal{L} = 2\pi\mathcal{R}$ on a lateral surface of ferrite disk determines an angular momentum—the anapole moment. Another type of an angular momentum is associated with the power-flow circulation. In a lossless ferrite disk, circulation of the power flow density can be considered as a persistent current as well. The divergentless power-flow-density persistent current, circulating on the r, θ plane, is an intrinsic property of the fields at the MDM resonances unrelated to the rigid-body rotation of a ferrite-disk. In the Introduction, we asked a question about the possibility of observing a dot product

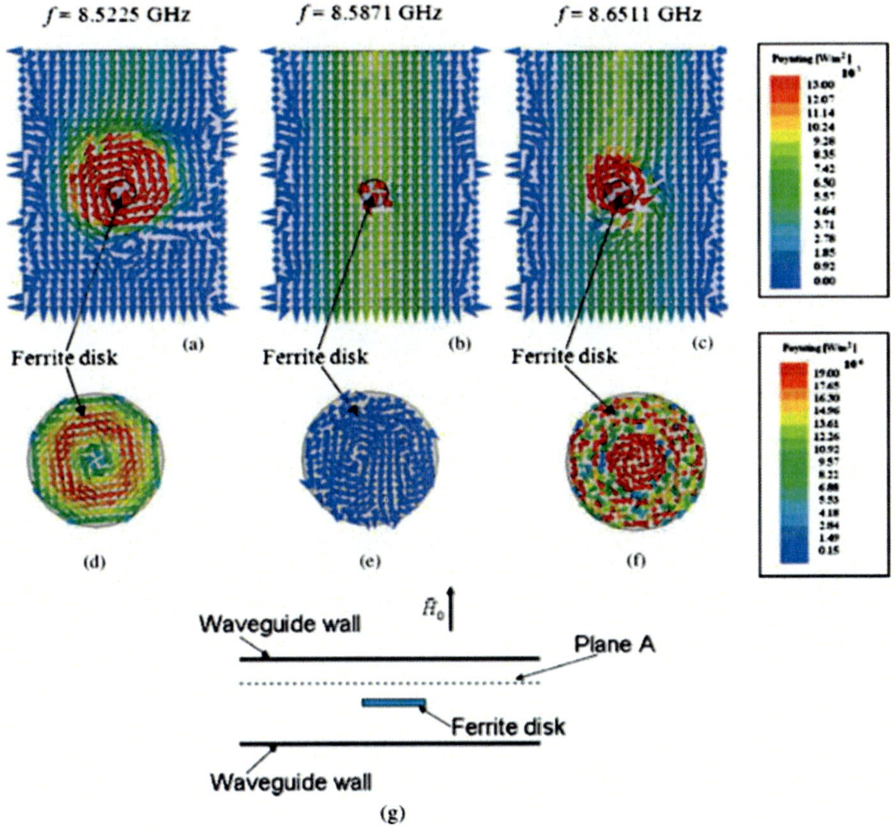

Fig. 19.2 Field confinement originating from the MDM vortices in a ferrite disk. **a** The Poynting vector distributions for the field on plane A at the frequency (f = 8.5225 GHz) of the first resonance. **b** The same at the frequency (f = 8.5871 GHz) between the resonances. **c** The same at the frequency (f = 8.6511 GHz) of the second resonance. **d** The Poynting vector distributions inside a ferrite disk at the frequency of the first resonance. **e** The same at the frequency between resonances. **f** The same at the frequency of the second resonance. **g** The plane A is a vacuum plane inside a waveguide situated above an upper plane of a MDM ferrite disk [67]

$\vec{E} \cdot \vec{H}$ together with a cross product $\vec{E} \times \vec{H}$ in the near-field region of a subwavelength sample. This question concerned the samples with ME properties. When, for MDM oscillations in a subwavelength ferrite disk, we observe the cross-product of the fields, can we classify this field structure as the ME fields, which are also characterized by the properties of $\mathcal{PT}$-symmetry and dot-product $\vec{E} \cdot \vec{H}$ of the fields? In [56] it was shown that in the near-field region adjacent to the MDM ferrite disk, there exists also another quadratic parameter determined by the scalar product between the electric and magnetic field components:

$$F = \frac{\varepsilon_0}{4} \mathrm{Im}\left[\vec{E} \cdot \left(\nabla \times \vec{E}\right)^*\right] = \frac{\omega \varepsilon_0 \mu_0}{4} \mathrm{Re}\left(\vec{E} \cdot \vec{H}^*\right) \neq 0. \tag{19.34}$$

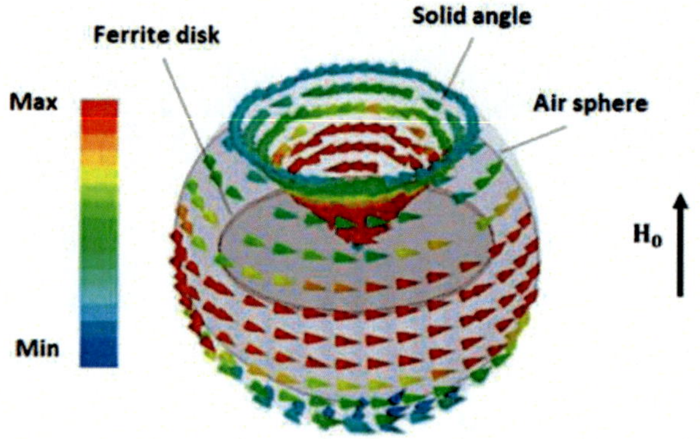

Fig. 19.3 Schematic representation of the circulation of the power flow, depicted on the surface of the vacuum sphere and on the surface of the solid angle. Direction of an orbital angular-momentum of a ferrite disk is correlated with the direction of a bias magnetic field

This effect is due to the presence of both the curl and potential electric fields in the subwavelength region of the MDM ferrite disk. At the same time, the magnetic near field is pure potential. Parameter F is the ME-field helicity density. It appears only at the MDM resonances. A sign of the helicity parameter depends on a direction of a bias magnetic field. Because of time-reversal symmetry breaking, all the regions with positive helicity become the regions with negative helicity (and vice versa), when one changes a direction of a bias magnetic field:

$$F^{\vec{H}_0\uparrow} = -F^{\vec{H}_0\downarrow}. \tag{19.35}$$

An integral of the ME-field helicity over an entire near-field vacuum region should be equal to zero [68, 69]. This "helicity neutrality" can be considered as a specific conservation law of helicity. The helicity parameter F is a pseudoscalar: to come back to the initial stage, one has to combine a reflection in a ferrite-disk plane and an opposite (time-reversal) rotation about an axis perpendicular to that plane. The helicity-density distribution is related to the angle between the spinning electric and magnetic fields. Figures 19.4 and 19.5 show the magnetic and electric field distributions on the upper plane of a ferrite disk for the first MDM resonance at different time phases. For such a field structure one can observe both the cross $\vec{E} \times \vec{H}$ and dot $\vec{E} \cdot \vec{H}$ products in the near-field region. The dot-product distributions (the helicity density distributions) are showed in Fig. 19.6. When one moves from the ferrite surfaces, above or below a ferrite disk, one observes reduction of the field amplitudes and also variation of the angle between spinning electric and magnetic fields. This angle varies from 0° or 180° (near the disk surfaces) to 90° (sufficiently far from a ferrite disk). The "source" of the helicity factor is the pseudoscalar quantity of the magnetization distribution in a ferrite disk at the MDM resonances [68]

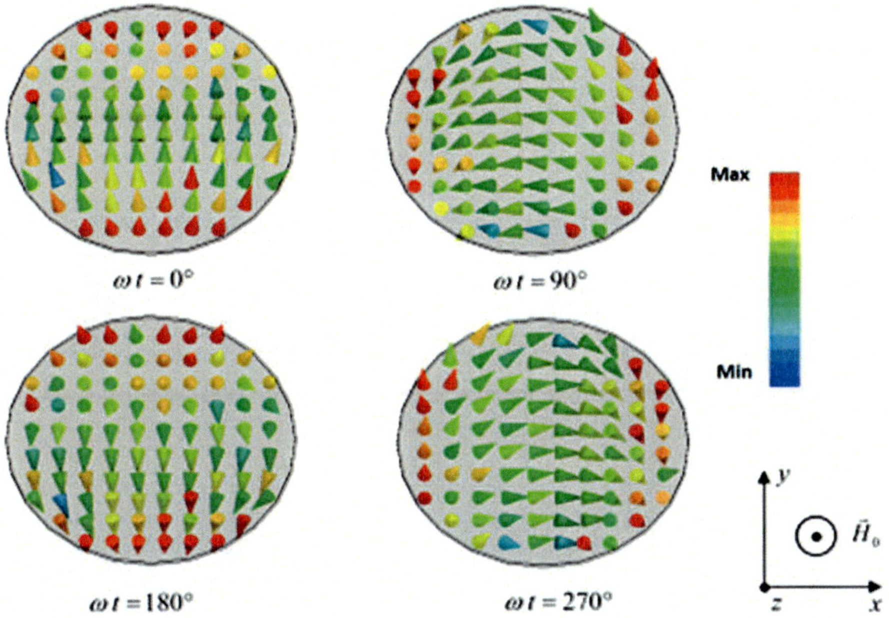

Fig. 19.4 Magnetic field distributions on the upper plane of a ferrite disk for the first MDM resonance at different time phases

$$\mathrm{Im}\int_{V^{(\pm)}}\left[\vec{m}\cdot\left(\vec{\nabla}\times\vec{m}\right)^{*}\right]dV\neq 0, \tag{19.36}$$

where $V^{(\pm)}$ are volumes of the upper and lower halves of the ferrite disk. These magnetization parameters are distributed asymmetrically with respect to the z-axis (see Fig. 19.7). Thus, the distribution of the helicity factor is also asymmetric. The regions with nonzero helicity factors we can characterize as the regions with nonzero ME energies. The area with positive helicity factor $F^{(+)}$ is the area with positive ME energy, $W_{ME}^{(+)}$. The area with negative helicity factor $F^{(-)}$ is the area with negative ME energy, $W_{ME}^{(-)}$. The total "ME potential energy" is related to the "ME kinetic energy" of the power-flow rotation. It a symmetrical structure, we have "magnetoelectrically neutral" condensate.

At the MDM resonances, both the power-flow vortices and the helicity states of ME fields are topologically protected quantumlike states. In [69], it was shown that the power-flow density and the helicity are the complex quantities. In the absence of losses and sources, there exist also the vector $\mathrm{Im}\vec{E}\times\vec{H}^{*}$. This vector can be classified as the reactive power flow density. Figure 19.8 illustrates the active and reactive power flows distributions at the MDM resonance above and below a ferrite disk. We can see that while the active power flow is characterized by the vortex topology, the reactive power flow has a source which is originated from a ferrite

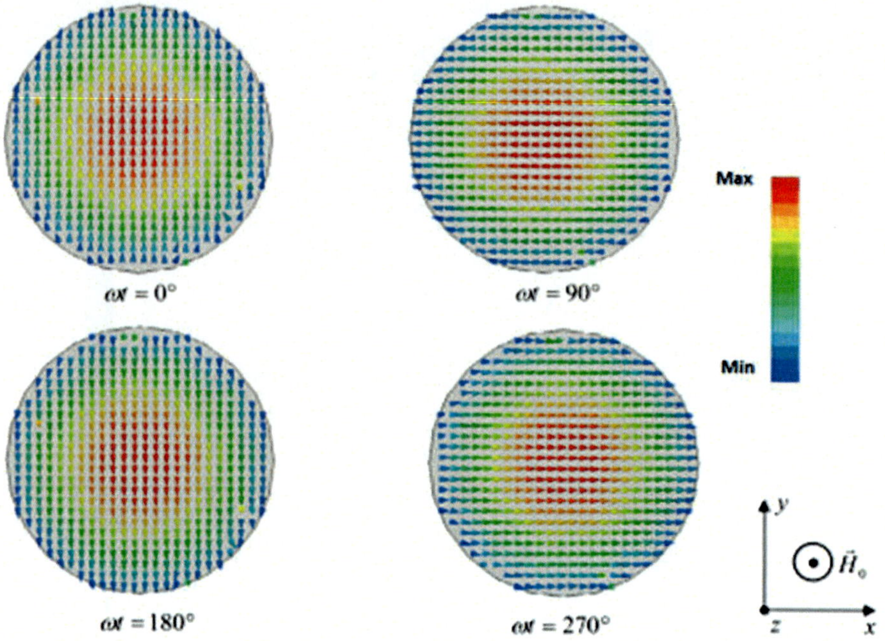

Fig. 19.5 Electric field distributions on the upper plane of a ferrite disc for the first MDM resonance at different time phases

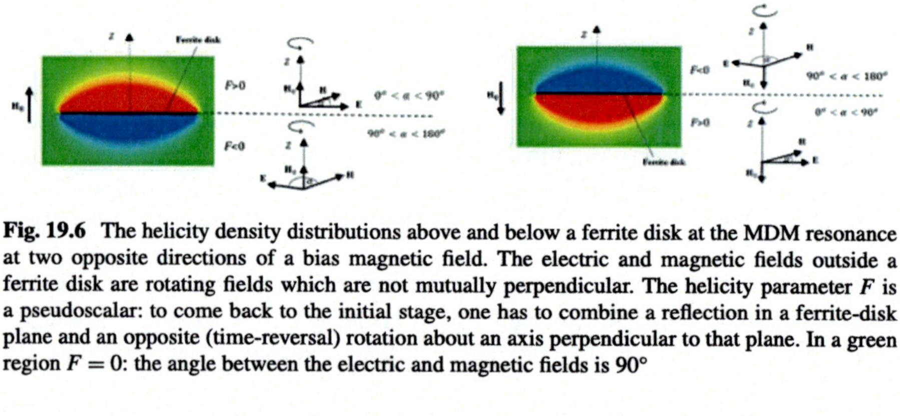

Fig. 19.6 The helicity density distributions above and below a ferrite disk at the MDM resonance at two opposite directions of a bias magnetic field. The electric and magnetic fields outside a ferrite disk are rotating fields which are not mutually perpendicular. The helicity parameter F is a pseudoscalar: to come back to the initial stage, one has to combine a reflection in a ferrite-disk plane and an opposite (time-reversal) rotation about an axis perpendicular to that plane. In a green region $F = 0$: the angle between the electric and magnetic fields is 90°

disk. The regions of localization of the active and reactive power flows are different. While the active power flow is localized at the disk periphery, the reactive power flow is localized at a central part of the disk. It was shown [69] that above and below a quasi-2D ferrite disk, the real part of the helicity density (defined by (19.34)) is related to an imaginary part of the complex power-flow density:

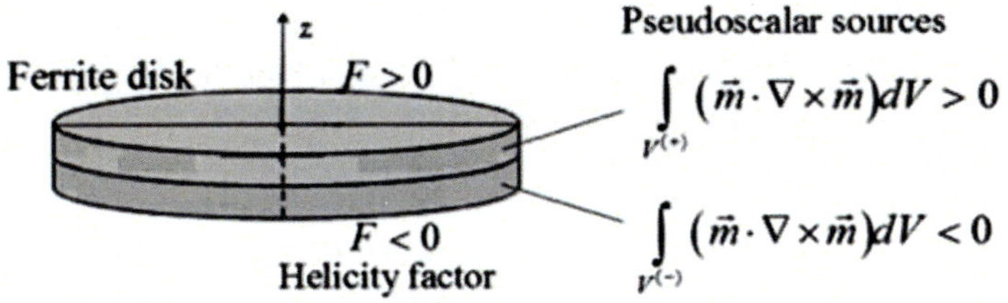

Fig. 19.7 Pseudoscalar quantity of the magnetization in a ferrite disk as a "source" of the helicity factor at the MDM resonance

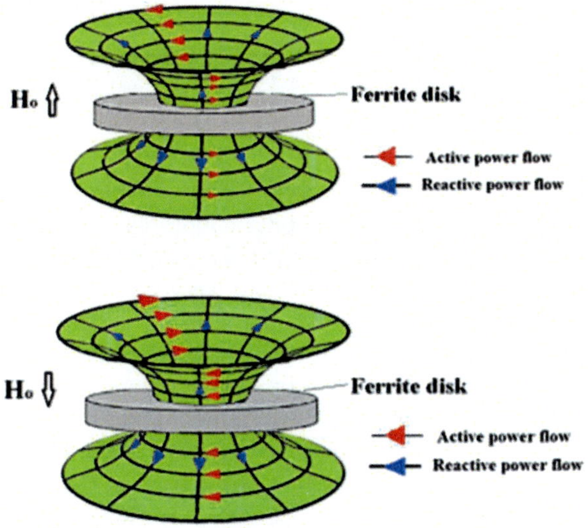

Fig. 19.8 The active and reactive power flows of the ME field at the MDM resonance. **a** An upward directed bias magnetic field; **b** a downward directed bias magnetic field. The active and reactive power flows are mutually perpendicular. These flows constitute surfaces, which can be considered as deformed versions of the complex planes, i.e., as Riemann surfaces. When one changes a direction of a bias field, the active power flow changes its direction as well. At the same time, the reactive power flow does not change its direction when the direction of a bias field is changed

$$\frac{1}{2}\mathrm{Re}\left|\vec{E}\cdot\vec{H}^*\right| = \frac{1}{2}\mathrm{Im}\left|\left[\vec{E}\times\vec{H}^*\right]_z\right| \tag{19.37}$$

The numerical results in [69] clearly show that in a vacuum region where the helicity-density factor exist, the reactive power flow is observed as well. So, in a region near a ferrite disk, the reactive power flow is accompanied by the helicity factor or, in other words, by the ME-energy density.

The pseudoscalar parameter (19.36) and the helicity factor F, arise due to spin-orbit interaction. Such PT-symmetric parameters, mixing electric and magnetic fields, are associated with the axion-electrodynamic term, leading to modification of inhomogeneous Maxwell equations [57, 70, 71]. It means that the ME fields appear as the fields of axion electrodynamics. With such a unique topological structure of ME near fields, two types of polaritons should be observed: right-handed ME polaritons and left-handed ME polaritons. When an external microwave structure is geometrically symmetrical, the two types of ME polaritons are indistinguishable. Otherwise, different microwave responses could be observed depending on the direction of the bias magnetic field. When an external microwave structure contains any elements with geometrical chirality, the right-hand and left-hand ME polaritons becomes nondegenerate, and microwave responses depend on the direction of the bias magnetic field. This fact was confirmed both numerically and experimentally [56].

19.4 MDM Particles Inside Waveguides and Cavities

In microwaves, we are witnesses that long-standing research in coupling between electrodynamics and magnetization dynamics noticeably reappear in recent studies of strong magnon-photon interaction [72–76]. In a small ferromagnetic particle, the exchange interaction can lead to the fact that a very large number of spins to lock together into one macrospin with a corresponding increase in oscillator strength. This results in strong enhancement of spin-photon coupling. In a structure of a microwave cavity with a yttrium iron garnet (YIG) sphere inside, the avoided crossing in the microwave reflection spectra verifies the strong coupling between the microwave photon and the macrospin magnon. In these studies, the Zeeman energy is defined by a coherent state of the macrospin-photon system when a magnetic dipole is in its antiparallel orientation to the cavity magnetic field. Together with an analysis of the strong coupling of the electromagnetic modes of a cavity with the fundamental Kittel modes, coupling with non-uniform modes—the Walker modes—in a YIG sphere was considered. In the microwave experiments, identification of the Walker modes in the sphere was made based an effect of overlapping between the cavity and spin waves due to relative symmetries of the fields [77, 78]. Nevertheless, the experimentally observed effects of strong magnon-photon interaction, cannot be described properly in terms of a single magnon-photon coupling process. In a view of these aspects, the theory based on solving coupled Maxwell and Landau-Lifshitz-Gilbert equations without making the conventional magnetostatic approximation have been suggested [79, 80]. Currently, the studies of strong magnon-photon interaction are integrated in a new field of research called cavity spintronics (or spin cavitronics) [81].

The coupling strength in the magnon-photon system is proportional to the probability of conversion of a photon to a magnon and vice versa. An effective way for strong coupling is to confine both magnons and photons to a small (subwavelength) region. Long-range spin transport in magnetic insulators demonstrates that the dipolar interactions alone generate coherent spin waves on the scales that are much larger than the exchange-interaction scales and, at the same time, much smaller than the electromagnetic-wave scales. Because of symmetry breakings, the MDM ferrite disk,

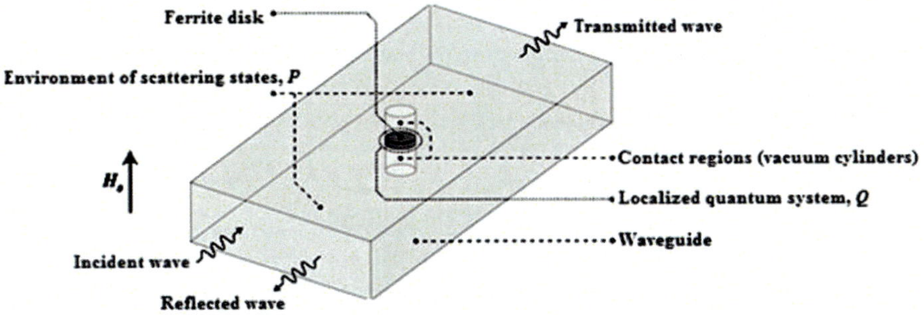

Fig. 19.9 An interaction of a MDM ferrite disk with a microwave waveguide. The structure is viewed as the $P + Q$ space. It consists of a localized quantum system (the MDM ferrite disk), denoted as the region Q, which is embedded within an environment of scattering states (the microwave waveguide), denoted as the regions P. The coupling between the regions Q and P is regulated by means of the two "contact regions" in the waveguide space

being a very small particle compared to the free-space electromagnetic wavelength, is a singular point for electromagnetic fields in a waveguide or cavity. When we consider a ferrite disk in vacuum environment, the unidirectional power-flow circulation might seem to violate the law of conservation of an angular momentum in a mechanically stationary system. In a microwave structure with an embedded ferrite disk, an orbital angular momentum, related to the power-flow circulation, must be conserved in the process. It can be conserved if topological properties of electromagnetic fields in the entire microwave structure are taken into account. Thus, if power-flow circulation is pushed in one direction in a ferrite disk, then the power-flow circulation on metal walls of the waveguide or cavity to be pushed in the opposite direction at the same time. It means that, in a general consideration, the model of MDM-vortex polaritons appears as an integrodifferential problem. Figure 19.9 presents a schematic picture of an interaction of a MDM ferrite disk with an external microwave structure. In [68] it was shown that due to the topological action of the azimuthally unidirectional transport of energy in a MDM-resonance ferrite sample there exists the opposite topological reaction (opposite azimuthally unidirectional transport of energy) on a metal screen placed near this sample. It is obvious that the question of the interaction of a MDM ferrite disk with an external microwave structure is far from trivial. To illustrate this nontriviality in more details, we adduce here some topological problems related to our studies.

A. On Rayleigh scattering by a thin ferrite rod

In the above studies of the MDM oscillation spectra in an open quasi-2D ferrite disk, the separation of variables in a cylindrical coordinate system was used. Analytically, we cannot apply a 2D model to consider scattering of EM waves by a subwavelength ferrite disk. Nevertheless, based on a simple qualitative analysis of a 2D structure, we can illustrate the role of topology in the EM-wave scattering by a ferrite sample. For this purpose, we will view some properties of the EM-wave scattering by a thin

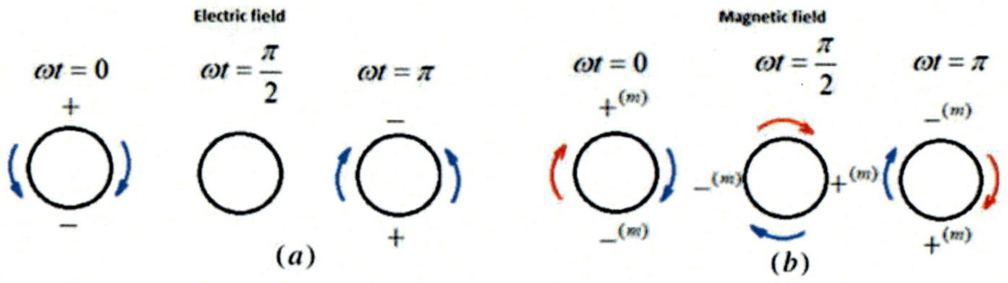

Fig. 19.10 Schematic illustration of charges and currents on the cross-section of the rods at the dipole-like scattering. **a** Electric charges and currents on a surface of a thin metal rod induced by RF electric field in a r, θ plane. **b** Magnetic charges and currents on a surface of a thin ferrite rod induced by RF magnetic field in a r, θ plane. Magnetic current of a polaritonic structure is conventionally represented as a circle composed by the blue and red arrows. In this case, the entire cycle of rotation corresponds to the π-shift of a dynamic phase

endless ferrite rod in comparison with the EM-wave scattering by a thin endless metal rod.

In Fig. 19.10, we give a schematic illustration of charges and currents on the cross-section of the rods at the dipole-like scattering. Let us consider, initially, Rayleigh scattering by a thin endless cylindrical rod made from a perfect electric conductor (PEC). A rod oriented along the z axis is acted upon by an external alternating electric field in the plane r, θ of a plane electromagnetic wave. Assuming that the rod diameter is much less than the EM wavelength, the analysis can be viewed as a quasi-electrostatic problem. The electric field of the EM wave induces positive and negative electric charges on diametrically opposite points of the r, θ plane, which cause two, clockwise (CW) and counter clockwise (CCW), azimuthal electric currents on the rod surface. Creating an azimuthally symmetric structure, each of these surface currents passes over a regular-coordinate angle π. In such a structure we have both the azimuthal and time symmetries. In Fig. 19.10a, the +, − surface electric charges correspond to the maximum, minimum of the charge distributions in the azimuth coordinates. One can adduce other examples of the azimuthal and time symmetries (the *PT* symmetry) at the dipole-like scattering from subwavelength structures. This includes also the electric-dipole eigenmodes of the surface plasmon resonances [82–84].

We consider now a thin endless cylindrical rod made from a magnetic insulator, YIG. A plane EM wave propagates along the rod axis. The rod diameter is much less than the EM wavelength and the analysis is considered as a quasi-magnetostatic problem. The rod is axially magnetized up to saturation by a bias magnetic field directed along the z axis. Due to the anisotropy (gyrotropy) induced by bias magnetic field, the RF magnetic field of the EM wave, which lies in the r, θ plane, causes the precessional motion of the alternating magnetization vector about the z axis. In this structure, magnetic charges at diametrically opposite points of a ferrite rod can appear due to the divergence of magnetization. It is known that at the ferromagnetic resonance frequency in an infinite medium, no divergence of the magnetization exists. Also, it is known that a divergence of the DC magnetization exists in a ferrite ellipsoid (in an endless cylindrical ferrite rod, in particular) with the homogeneous-precession mode (Kittel's mode) [18]. The divergence of both the DC and RF magnetizations may occur in a ferrite sample in a case of nonhomogeneous-precession modes. These magnetic-dipole modes—Walker's modes—in a thin endless cylindrical ferrite rod were studied in [85]. For these eigenmodes, the RF magnetic field of the incident EM wave induces a magnetic dipoles, which lies in the r, θ plane of a ferrite rod. In Fig. 19.10b, the $+^{(m)}$, $-^{(m)}$ surface magnetic charges correspond to the maximum, minimum of the charge distributions in the azimuth coordinates. Due to the time-reversal symmetry breaking, these surface magnetic charges cannot cause two, clockwise (CW) *and* counter clockwise (CCW), azimuthal magnetic currents. For the given direction of bias magnetic field, we may have only CW *or* CCW induced magnetic current, which passes over a regular-coordinate angle π at the time phase of π. In Fig. 19.10b, this is shown as the CW blue-arrow current. We can suppose, however, that there exists a polaritonic structure with an additional (non-electromagnetic) phase shift, when the gradient of twisting angle plays the role of the phase gradient. A global phase texture with coflowing an EM-wave induced magnetic current and topological magnetic current will provide us with the possibility to have rotational symmetry by a turn over a regular-coordinate angle 2π at the time phase of π. This situation is shown in Fig. 19.10b, where the CW magnetic current of a polaritonic structure is conventionally represented as a circle composed by the blue and red arrows. It is worth noting, however, that one can view such a phenomenon not in a ferrite rod, but in the r, θ plane of a ferrite disk with MDM oscillations, where the non-electromagnetic torque is caused by the topological-phase effects. In the MDM ferrite-disk resonator, the non-zero circulation of such a magnetic current, observed at the time phase shift of π, results in appearance of a constant angular momentum directed along the z axis. This is possible due to an additional phase shift of the magnetic current along the z axis. The magnetic currents have a helical structure. When such helical currents (and so helical waves) cannot be observed in a smooth ferrite rod, they can be seen in a MDM ferrite disk [86].

B. Testing the topological properties of the ME field with small metal rods and rings in a microwave waveguide

In a structure of a MDM particle embedded in a microwave waveguide, photons interact strongly and coherently with magnetic excitations. The creation of certain non-classical states in such a macroscopic system can be observed with help of small metallic elements placed inside a microwave waveguide near the ferrite disk. Here we show some of the topological properties of the ME field using small metallic rods and rings.

A structure of microwave waveguide with a ferrite disk and small metallic rod, shown in Fig. 19.11a, was studied experimentally in [87] and numerically in [88]. The rod is oriented along an electric field of a rectangular waveguide. Its diameter is a very small compared to the disk diameter and to the free-space electromagnetic wavelength. On the basis of a comparative analysis of experimental oscillation spectra, it was argued in [87] that the fact that an additional small capacitive coupling

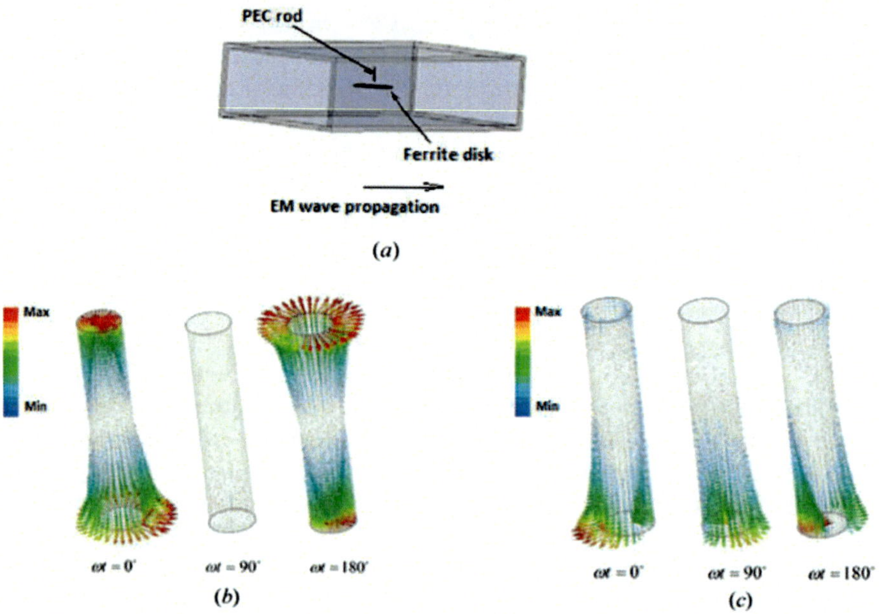

Fig. 19.11 **a** A structure of microwave waveguide with a ferrite disk and small metallic rod. **b** Electric field on a small PEC rod for the frequency far from the MDM resonance at different time phases. There is a trivial picture of the fields of a small electric dipole inside a waveguide. **c** Electric field on a small PEC rod in the MDM resonance at different time phases. A PEC rod behaves as a small line defect on which rotational symmetry is violated. The observed evolution of the radial part of polarization gives evidence for the presence of a geometrical phase in the vacuum-region field of the MDM-vortex polariton

(due to a piece of a nonmagnetic wire) strongly affects magnetic oscillation proves the presence of the electric-dipole moments (anapole moments) of the MDMs in a quasi-2D ferrite disk. In numerical studies [88], a metal rod is made of a PEC. At frequencies far from the MDM resonances, the field structure of an entire waveguide is not noticeably disturbed. The electric field on the rod demonstrates a trivial picture of the field induced on a small electric dipole inside a waveguide (Fig. 19.11b). At the same time, in the case of the MDM resonance, there is a strong reflection of electromagnetic waves in a waveguide. The PEC rod behaves as a small line defect on which rotational symmetry is violated. The observed evolution of the radial part of the electric polarization, giving, as a result, a circulating electric current, indicates the presence of a geometrical phase in the vacuum-region field of the MDM-vortex polariton (Fig. 19.11c).

Let us bend the metallic rod into a ring and rigidly connect the ends. At the MDM resonance, rotating electric charges and circulating electric currents arise on a ring placed above the ferrite disk. [89]. Figure 19.12 shows circulation of a surface electric current along a PEC ring. The ferrite disk has a diameter of 3 mm. A metallic ring made from a wire of a diameter of 0,05 mm, has a diameter of 1.5 mm. The ring is

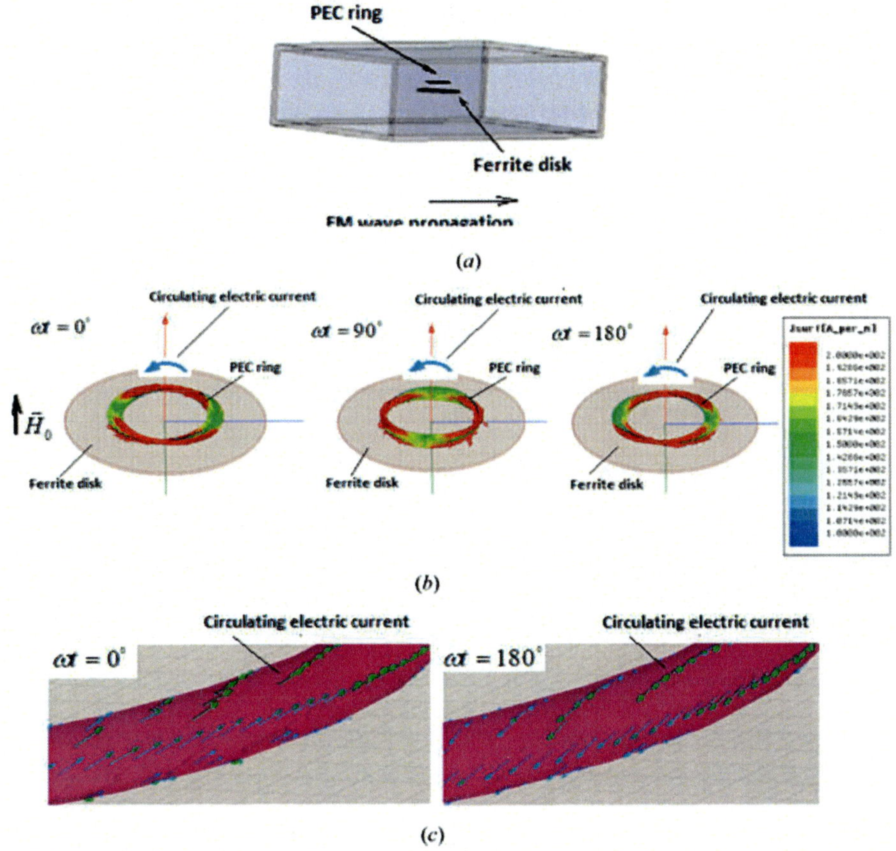

Fig. 19.12 **a** A structure of microwave waveguide with a ferrite disk and small metallic ring. **b** Circulating surface electric current on a metallic ring. **c** The electric current on the surface of a metallic ring has the spin degree of freedom

located above a ferrite disk at a distance of 0.05 mm. The circulating current will give an angular-momentum flux. The intensity of the flux is proportional to the gradient of twisting angle, which plays the role of the phase gradient. A critical phase gradient is required to enable the process. This occurs only at the MDM resonance. In this case, the persistent charge current in the ring is correlated to the persistent magnetic current in a ferrite disk. The electric current on the surface of the metallic ring has the spin degree of freedom (see Fig. 19.12c). When using 2D models in our main studies of MDM oscillations, we can conclude now that, generally, an analysis of ME fields should be made based on the 3D model. At the MDM resonance, the current induces on a test metal ring is a topological soliton structure which is quantized simultaneously in poloidal and toroidal directions. This 3D continuous vector-field structure—a hopfion (or Hopf soliton)—cannot be unknotted without cutting [90, 91]. It is also worth noting that the helicity properties of the 3D structure of the ME field in vacuum reflect its own topologically nontrivial structure at *each mode* of the MDM-oscillation spectrum.

It is worth noting that in a remarkable paper [92], the authors had measured the low-temperature magnetization response of an isolated mescoscopic copper ring to a slowly varying magnetic flux. They showed that the total magnetization response oscillates as a function of the enclosed magnetic flux on the scale of *half a flux quantum*. In our study, a numerical analysis made in [89] shows that the currents induced on a metal ring at the MDM resonances, strongly perturb the electric, but not the magnetic, field in a vacuum region above the ferrite disk. This means that the ring is threaded mainly by an electric flux. Taking into account the above-analyzed properties of the anapole, we have evidence of the presence of the enclosed *electric* flux on the scale of *half a flux quantum*.

C. MDM cavity electrodynamics

For the case of MDM resonances in a small ferrite disk, characterizing by non-uniform magnetization dynamics, the above-mentioned model of coherent states of the macrospin/photon system in a ferrite sphere [72–76, 81], is not applicable. In [57, 93], it was shown that multiresonance microwave oscillations observed in experiments [49–51, 93], are related to the fact that magnetization dynamics of MDM oscillations in a quasi-2D ferrite disk have a strong impact on the phenomena associated with the quantized energy fluctuation of microwave photons in a cavity. Figure 19.13 gives a sketch showing the relationship between quantized states of microwave energy in a cavity and magnetic energy in a MDM ferrite disk. The

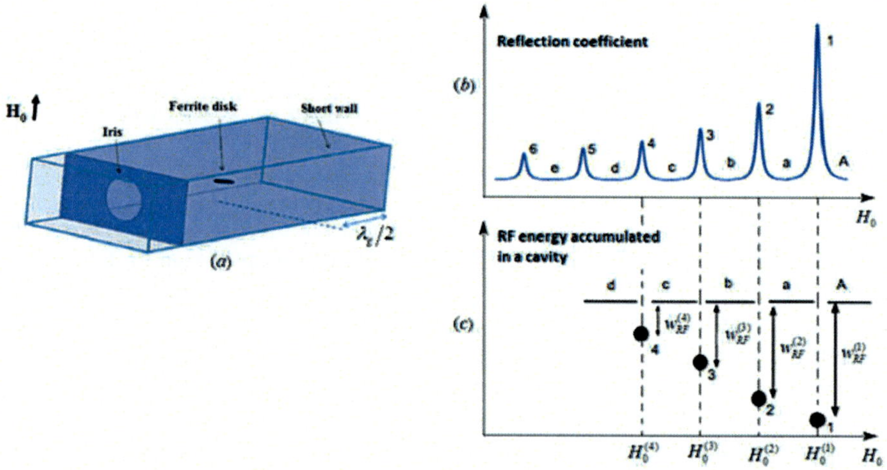

Fig. 19.13 A sketch showing the relationship between quantized states of microwave energy in a cavity and magnetic energy in a ferrite disk. **a** A structure of a rectangular waveguide cavity with a normally magnetized ferrite-disk sample. **b** A typical multiresonance spectrum of modulus of the reflection coefficient. **c** Microwave energy accumulated in a cavity; $w_{RF}^{(n)}$ are jumps of electromagnetic energy at MDM resonances

microwave structure is a rectangular waveguide cavity with a normally magnetized ferrite-disk sample. The operating frequency, which is a resonant frequency of the cavity, is constant. The only external parameter, which varies in the experiment, is a bias magnetic field. The observed discrete variation of the cavity impedances is related to discrete states of the cavity fields. Since the effect was obtained at a given resonant frequency, the shown resonances are *not conventional cavity modes* related to the frequency-dependent quantization of the photon wave vector. These resonances are caused by the quantized variation of energy of a ferrite disk, which appear due to variation of energy of an external source—the bias magnetic field.

At the regions of a bias magnetic field, designated in Fig. 19.13 as *A, a, b, c, d,* ..., we do not have MDM resonances. In these regions, a ferrite disk is "seen" by electromagnetic waves, as a very small obstacle which, practically, does not perturb a microwave cavity. In this case, the cavity (with an embedded ferrite disk) has good impedance matching with an external waveguiding structure and a microwave energy accumulated in a cavity is at a certain maximal level. At the MDM resonances, the reflection coefficient sharply increases (the states designated in Fig. 19.13 by numbers 1, 2, 3, ...). The input impedances are real, but the cavity is strongly mismatched with an external waveguiding structure. It means that at the MDM-resonance peaks, the cavity receives less energy from an external microwave source. In these states of a bias magnetic field, the microwave energy stored in a cavity sharply decreases, compared to its maximal level in the *A, a, b, c, d,* ... Since the only external parameter, which varies in this experiment, is a bias magnetic field, such a sharp release of the microwave energy accumulated in a cavity to an external waveguiding structure should be related to the emission of discrete portions of energy from a ferrite disk. This means that at the MDM resonances, a strong and sharp decrease in the magnetic energy of the ferrite sample should be observed.

When speaking about the eigenstates of the microwave-cavity spectrum observed at the bias-field variation and constant frequency, we should answer the question about the eigenfunctions of this spectrum. In general, microwave resonators with the time-reversal symmetry breakings give an example of a nonintegrable, i.e., path dependent, system. The time-reversal symmetry breaking effect leads to creation of the Poynting-vector vortices in a vacuum region the microwave resonators with enclosed lossless ferrite samples [94–97]. In an analysis of the cavity eigenfunctions, it makes no sense to consider the reflection of electromagnetic waves from magnetized ferrites from the standpoint of energy flow and ray propagation [98]. One cannot use an interpretation which allows viewing the modes as pairs of two bouncing electromagnetic plane waves. This interpretation clearly shows that for a structure with an enclosed magnetized ferrite sample given, for example, in Fig. 19.14, there can be no identity between the rays $1 \rightarrow F \rightarrow 1'$ and $1' \rightarrow F \rightarrow 1$ in the sense that these rays can acquire different phases when are reflected by the ferrite.

At the same time, it is argued [99] that in quantum mechanics the distinction between integrable and nonintegrable systems does not work any longer. The initial conditions are defined only within the limits of the uncertainty relation $\Delta x \Delta p \geq \frac{1}{2}\hbar$. Since the Schrödinger equation is linear, a quantum mechanical wave packet can be constructed from the eigenfunctions by the superposition principle. What do we have

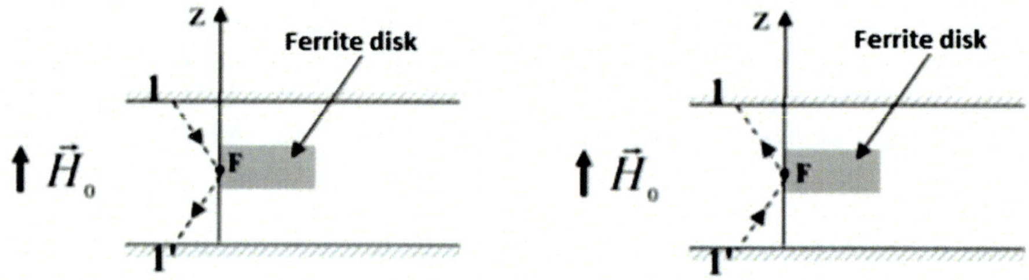

Fig. 19.14 The rays $1 \to F \to 1'$ and $1' \to F \to 1$ acquire different phases when are reflected by the ferrite

in our structure of a MDM ferrite disk in a cavity? We use the Walker equation for a MS scalar wave function. It also allows to construct a wave packet from the eigenfunctions by the superposition principle. We have to use a description of the spectral response functions of the system with respect to two external parameters—a bias magnetic field H_0 and a signal frequency ω— and analyze the correlations between the spectral response functions at different values of these external parameters. It means that, in neglect of losses, there should exist a certain *uncertainty limit* stating that

$$\Delta f \Delta H_0 \geq uncertainty\, limit \tag{19.38}$$

This uncertainty limit is a constant which depends on the disk size parameters and the ferrite material property (such as saturation magnetization) [57]. Beyond the frames of the uncertainty limit (38) one has continuum of energy. The fact that there are different mechanisms of quantization allows to conclude that for MDM oscillations in a quasi-2D ferrite disk both discrete energy eigenstate and a continuum of energy can exist. In quantum mechanics, the uncertainty principle says that the values of a pair of canonically conjugate observables cannot both be precisely determined in any quantum state. In a formal harmonic analysis in classical physics, the uncertainty principle can be summed up as follows: A nonzero function and its Fourier transform cannot be sharply localized. This principle states also that there exist limitations in performing measurements on a system without disturbing it. Basically, formulation of the main statement of the MDM-oscillation theory is impossible without using a classical microwave structure. If a MDM particle is under interaction with a "classical electrodynamics" object, the states of this classical object change. The character and value of these changes depend on the MDM quantized states and so can serve as its qualitative characteristics. The microwave measurement reflects interaction between a microwave cavity and a MDM particle. It is worth noting that for different types of subwavelength particles (ferrite disks), the uncertainty limits may be different.

The fact of the existence of the uncertainty limit (38) is indirectly confirmed by the experimental results presented in [100]. In the microwave structure shown in Fig. 19.15a, a ferrite disk is placed in a cavity with a very low Q factor. The wide bandwidth is due to losses caused by the test samples embedded in the cavity.

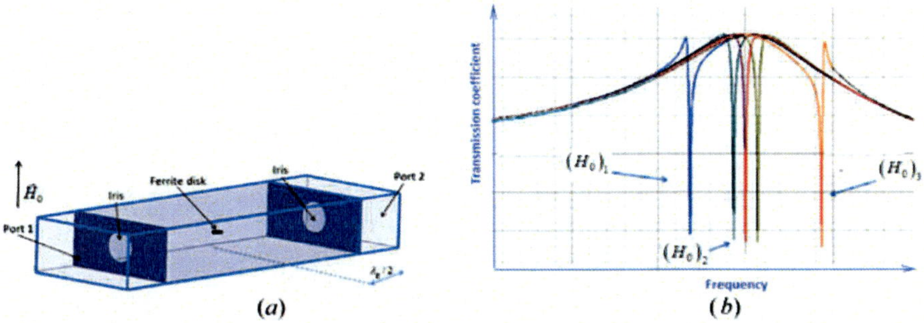

Fig. 19.15 **a** A structure of a rectangular waveguide cavity. **b** Modification of the Fano-resonance shape. At variation of a bias magnetic field, $(H_0)_1 < (H_0)_2 < (H_0)_3$, the Fano line shape of a MDM resonance can be completely damped. The scattering cross section of a single Lorentzian peak corresponds to a pure dark mode

Figure 19.15b shows how a bias magnetic field tunes the shape of the MDM resonance. It can be seen that as one approaches the top of the cavity resonance curve, the effect of Fano resonance collapses, the Fano line shape is completely decays, and a single Lorentz peak is observed. The Lorentzian response is a narrow, highly symmetric peak. The scattering cross section corresponds to a pure dark mode. All this means that, within uncertainty limit (19.38), it is possible to carry out observations for a very wide linewidth of the cavity mode (Δf is very big) and an extremely narrow linewidth of the MDM resonance peak (ΔH_0 is very small).

In the above studies, we considered the G modes (with a scalar MS membrane function $\tilde{\eta}$ and the ND BC) and the L modes (with vector MS membrane function $\tilde{V} \equiv \begin{pmatrix} \vec{\tilde{B}} \\ \tilde{\varphi} \end{pmatrix}$ and the EM BC). The G-mode spectral analysis is more appropriate to use at the regime of a constant frequency and the bias magnetic-field variation, while the L-mode analysis—at a constant bias magnetic field and the frequency variation. These two spectral problems are bridged within uncertainty limit (19.38).

19.5 Transfer of Angular Momentum to Dielectric Materials, Metals and Biological Structures from MDM Resonators

Due to unique structures of twisted ME near fields, one can observe angular momentums (spin and orbital) transfer to electric polarization in a dielectric sample (Fig. 19.16). Experiments [101, 102] show explicit shifts of the MDM resonance peaks due to the dielectric loading of the ferrite disk. This effect was explained in [56, 57, 102]. The mechanical torque exerted on a given electric dipole in a dielectric sample is defined as a cross product of the MDM electric field and the electric

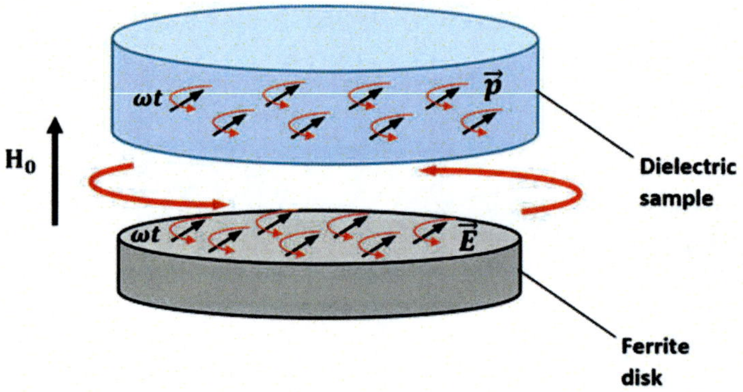

Fig. 19.16 Angular momentums (spin and orbital) transfer to electric polarization in a dielectric sample

moment of the dipole. The torque exerting on the electric polarization in a dielectric sample due to the MDM electric field should be equal to reaction torque exerting on the magnetization in a ferrite disk. Because of this reaction torque, the precessing magnetic moment density of the ferromagnet will be under additional mechanical rotation at a certain frequency Ω. The frequency Ω is defined based on both, spin and orbital, momentums of the fields of MDM oscillations. It was shown experimentally that the chiral structure of near-field ME provides the potential for microwave chirality discrimination in chemical and biological objects [103].

Because of a chiral topology of near fields originated from MDM oscillations in a ferrite disk, one has helical electric currents induced on a surface of a metal wire electrode placed on a surface of a ferrite disk. On a butt end of a wire probe one can observe twisted near fields (Fig. 19.17). The handedness of these fields depends on a direction of a bias magnetic field applied to a MDM ferrite resonator [102]. Using helical electric currents induced on a metal wire electrode, one can obtain the angular momentum transfer to localized regions in dielectric samples (Fig. 19.18).

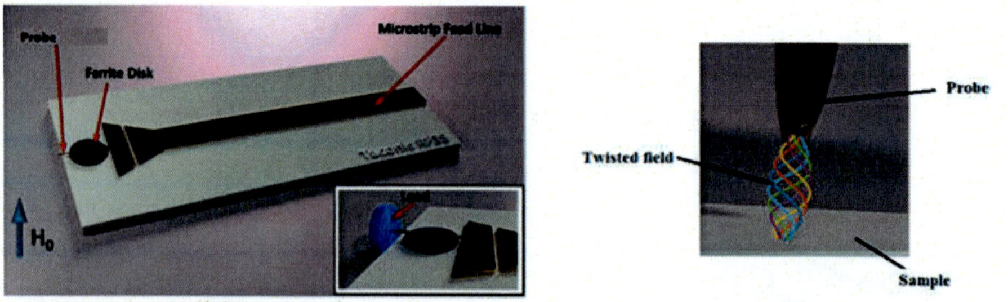

(*a*) (*b*)

Fig. 19.17 **a** Microwave probing structure with a MDM ferrite-disk resonator and a wire electrode. **b** Schematic illustration of twisted near fields

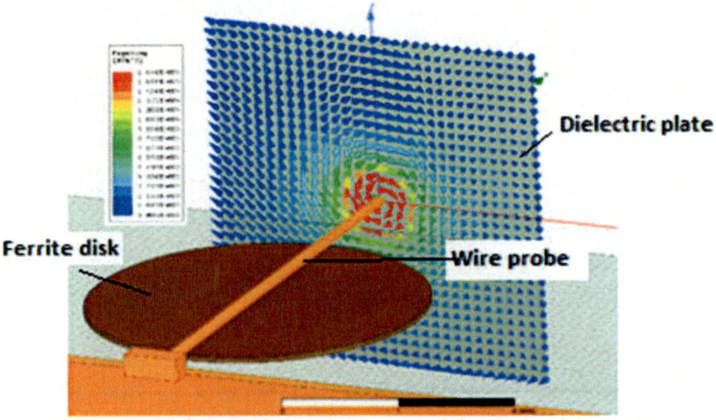

Fig. 19.18 Angular momentum transfer to localized regions in dielectric samples

Due to strong reflection and absorption of electromagnetic waves in conductive layers and biological tissue, standard microwave techniques cannot be used for testing such structures. Twisted microwave near fields with strong energy concentration, originated from MDM ferrite disk with a metal wire electrode, allow probing effectively high absorption conductive layers. This effect can be explained by a simple physical model. When the electromagnetic wave incidents on a conductive material, the induced electric current is almost parallel to the electric field (Ohm's law). Joule losses in conductive materials are defined by a scalar product of an induced electric current and an electric field. When, however, a conductive material is placed in a twisted microwave near field, the RF electric current and an RF electric field become mutually nonparallel. It means that for Joule losses, one has $\vec{J} \cdot \vec{E} = J \cdot E \cos\delta$ with $\cos\delta \ll 1$. Extremely small Joule losses result in strong enlargement of a penetration length—the skin depth—in a sample. Figure 19.19 presents numerical results illustrating the effect of penetration of the twisted-field microwave power through a thin metal screen [104].

19.6 Conclusion

ME fields are subwavelength-domain fields with specific properties of violation of spatial and temporal inversion symmetry. When searching for such fields, we consider near fields originated from subwavelength resonators, that are the systems with quantum-confinement effects of dipolar-mode quasistatic oscillations. We show that the near fields of a quasi-2D subwavelength-size ferrite disk with magnetic-dipolar-mode (MDM) oscillations have the properties of ME fields. The ME fields, being originated from magnetization dynamics at MDM resonances, appear as the pseudoscalar axionlike fields. Whenever the pseudoscalar axionlike fields, is introduced

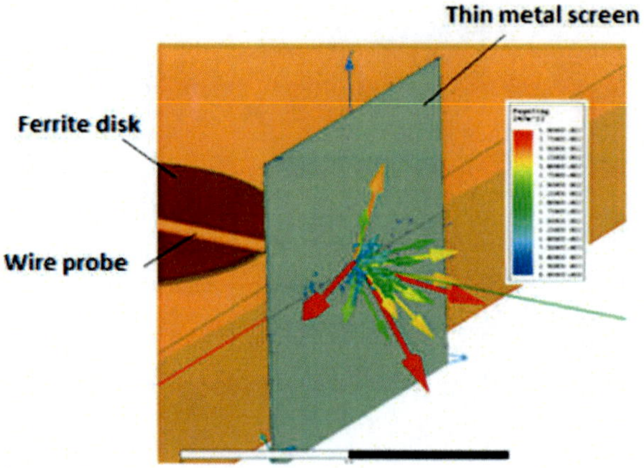

Fig. 19.19 Numerical results showing the twisted-field effect of penetration microwave power through a thin metal screen

in the electromagnetic theory, the dual symmetry is spontaneously and explicitly broken. This results in non-trivial coupling between pseudoscalar quasistatic ME fields and the EM fields in microwave structures with an embedded MDM ferrite disk.

Long range magnetic dipole-dipole correlation can be treated in terms of collective magnetostatic excitations of the system. In small ferromagnetic-resonance ferrite disk, macroscopic quantum coherence can be observed. In a case of a quasi-2D ferrite disk, the quantized forms of these collective matter oscillations—the MDM magnons—were found to be quasiparticles with both wave-like and particle-like behaviors, as expected for quantum excitations. With use of MS-potential scalar wave function ψ we formulate properly the energy eigenstate problem based on the Schrödinger-like equation. We obtain currents (fluxes) for MS modes. We show that in a subwavelength ferrite-disk particle one can observe an angular momentum due to the power-flow circulation of double-valued edge MS-wave functions. For incident electromagnetic wave, this magnon subwavelength particle emerges as a singular point carrying quanta of angular momenta. In a ferrite-disk sample, the magnetization has both the spin and orbital rotations. There is the spin-orbit interaction between these angular momenta. The MDMs are characterized by the pseudoscalar magnetization helicity parameter, which can be considered as a certain source of the helicity properties of ME fields.

Quantized ME fields arising from nonhomogeneous ferromagnetic resonances with spin-orbit effect, suggest a conceptually new microwave functionality for material characterization. Due to unique structures of twisted ME near fields, one can observe angular momentums (spin and orbital) transfer to electric polarization in a dielectric sample. The chiral structure of near-field ME provides the potential for microwave chirality discrimination in chemical and biological objects. Twisted

ME fields allow deep penetration of the microwave power into materials with high conductivity.

References

1. J.D. Jackson, *Classical Electrodynamics*, 2nd edn. (Wiley, New York, 1975)
2. Y. Tang, A.E. Cohen, Optical chirality and its interaction with matter. Phys. Rev. Lett. **104**, 163901 (2010)
3. K.Y. Bliokh, A.Y. Bekshaev, F. Nori, Dual electromagnetism: Helicity, spin, momentum and angular momentum. New J. Phys. **15**, 033026 (2013)
4. E. Hendry, T. Carpy, J. Johnston et al., Ultrasensitive detection and characterization of biomolecules using superchiral fields. Nat. Nanotechnol. **5**, 783 (2010)
5. M.M. Coles, D.L. Andrews, Chirality and angular momentum in optical radiation. Phys. Rev. A **85**, 063810 (2012)
6. R.P. Cameron, S.M. Barnett, A.M. Yao, Discriminatory optical force for chiral molecules. New J. Phys. **16**, 013020 (2014)
7. A. Canaguier-Durand, J.A. Hutchison, C. Genet, T.W. Ebbesen, Mechanical separation of chiral dipoles by chiral light. New J. Phys. **15**, 123037 (2013)
8. A. Canaguier-Durand, C. Genet, Chiral route to pulling optical forces and left-handed optical torques. Phys. Rev. A **92**, 043823 (2015)
9. E. Mohammadi, K.L. Tsakmakidis, A.N. Askarpour, P. Dehkhoda, A. Tavakoli, H. Altug, Nanophotonic platforms for enhanced chiral sensing. ACS Photonics **5**, 2669 (2018)
10. S. Droulias, L. Bougas, Surface plasmon platform for angle-resolved chiral sensing. ACS Photon. **6**, 1485 (2019)
11. T.V. Raziman, R.H. Godiksen, M.A. Müller, A.G. Curto, Conditions for enhancing chiral nanophotonics near achiral nanoparticles. ACS Photon. **6**, 2583 (2019)
12. C.F. Bohren, D.R. Huffman, *Absorption and Scattering of Light by Small Particles* (John Wiley & Sons, New York, 1983)
13. W.L. Barnes, A. Dereux, T.W. Ebbesen, Surface plasmon subwavelength optics. Nature **424**, 824 (2003)
14. E. Ozbay, Plasmonics: Merging photonics and electronics at nanoscale dimensions. Science **311**, 189 (2006)
15. S.A. Maier, *Plasmonics: Fundamentals and Applications* (Springer, New York, 2007)
16. M.I. Kaganov, N.B. Pustyl'nik, T.N. Shalaeva, Magnons, magnetic polaritons, magnetostatic waves. Phys. Usp. **40**, 181 (1997)
17. V.V. Kruglyak, S.O. Demokritov, D. Grundler, Magnonics. J. Phys. D Appl. Phys. **43**, 264001 (2010)
18. A.G. Gurevich, G.A. Melkov, *Magnetic Oscillations and Waves* (CRC Press, New York, 1996)
19. S.B. Cohn, Microwave bandpass filters containing high-Q dielectric resonators. IEEE Trans. Microw. Theor. Techn. **MTT-16**, 218 (1968)
20. J.C. Ginn, I. Brener, D.W. Peters, J.R. Wendt, J.O. Stevens, P.F. Hines, L.I. Basilio, L.K. Warne, J.F. Ihlefeld, P.G. Clem, M.B. Sinclair, Realizing optical magnetism from dielectric metamaterials. Phys. Rev. Lett. **108**, 097402 (2012)
21. A.I. Kuznetsov, A.E. Miroshnichenko, Y.H. Fu, J.B. Zhang, B. Luk'yanchuk, Magnetic light. Sci. Rep. **2**, 492 (2012)
22. G.T. Papadakis, D. Fleischman, A. Davoyan, P. Yeh, H.A. Atwater, Optical magnetism in planar metamaterial heterostructures. Nat. Commun. **9**, 296 (2018)
23. J.M. Pitarke, V.M. Silkin, E.V. Chulkov, P.M. Echenique, Theory of surface plasmons and surface-plasmon polaritons. Rep. Prog. Phys. **70**, 1 (2007)
24. B. Bêchea, E. Gaviot, About the Heisenberg's uncertainty principle and the determination of effective optical indices in integrated photonics at high sub-wavelength regime. Optik **127**, 3643 (2016)

25. J.A. Kong, Theorems of bianisotropic media. Proc. IEEE **60**, 1036 (1972)
26. C.E. Kriegler, M.S. Rill, S. Linden, M. Wegener, Bianisotropic photonic metamaterials. IEEE J. Select. Top. Quant. Electron. **16**, 367 (2010)
27. N. Guth, B. Gallas, J. Rivory, J. Grand, A. Ourir, G. Guida, R. Abdeddaim, C. Jouvaud, J. de Rosny, Optical properties of metamaterials: Influence of electric multipoles, magnetoelectric coupling, and spatial dispersion. Phys. Rev. B **85**, 115138 (2012)
28. S.A. Tretyakov, C.R. Simovski, M. Hudlička, Bianisotropic route to the realization and matching of backward-wave metamaterial slabs. Phys. Rev. B **75**, 153104 (2007)
29. A. Kwadrin, A.F. Koenderink, Probing the electrodynamic local density of states with magnetoelectric point scatterers. Phys. Rev. B **87**, 125123 (2013)
30. J. Proust, N. Bonod, J. Grand, B. Gallas, Optical monitoring of the magnetoelectric coupling in individual plasmonic scatterers. ACS Photonics **3**, 1581 (2016)
31. C. Caloz, A. Sihvola, Electromagnetic chirality, arXiv:1903.09087
32. X. Chen, B.-I. Wu, J.A. Kong, T.M. Grzegorczyk, Retrieval of the effective constitutive parameters of bianisotropic metamaterials. Phys. Rev. E **71**, 046610 (2005)
33. Z.F. Li, K. Aydin, E. Ozbay, Determination of the effective constitutive parameters of bianisotropic metamaterials from reflection and transmission coefficients. Phys. Rev. E **79**, 026610 (2009)
34. Y. Moritake, T. Tanaka, Bi-anisotropic Fano resonance in three-dimensional metamaterials. Scient. Rep. **8**, 9012 (2018)
35. C. Weisbuch, M. Nishioka, A. Ishikawa, Y. Arakawa, Observation of the coupled exciton-photon mode splitting in a semiconductor quantum microcavity. Phys. Rev. Lett. **69**, 3314 (1992)
36. J. Kasprzak, M. Richard, S. Kundermann, A. Baas, P. Jeambrun, J. M. J. Keeling, F. M. Marchetti, M. H. Szymańska, R. André, J. L. Staehli, V. Savona, P. B. Littlewood, B. Deveaud, L. S. Dang, Bose–Einstein condensation of exciton polaritons, Nature **443**, (2006)
37. H. Deng, H. Haug, Y. Yamamoto, Exciton-polariton Bose-Einstein condensation. Rev. Mod. Phys. **82**, 1489 (2010)
38. D. Snoke, P. Littlewood, Polariton condensates. Phys. Today **63**, 42 (2010)
39. T. Byrnes, N.Y. Kim, Y. Yamamoto, Exciton–polariton condensates. Nature Phys. **10**, 803 (2014)
40. T.J. Davis, D.E. Gómez, Colloquium: An algebraic model of localized surface plasmons and their interactions. Rev. Mod. Phys. **89**, 011003 (2017)
41. L.D. Landau, E.M. Lifshitz, *Electrodynamics of Continuous Media* (Pergamon Press, Oxford, 1960)
42. D.J. Bergman, The dielectric constant of a composite material—A problem in classical physics. Phys. Rep. **43**, 377 (1978)
43. M.I. Stockman, S.V. Faleev, D.J. Bergman, Localization versus delocalization of surface plasmons in nanosystems: Can one state have both characteristics? Phys. Rev. Lett. **87**, 167401 (2001)
44. V. Klimov, G.-Y. Guo, M. Pikhota, Plasmon resonances in metal nanoparticles with sharp edges and vertices: a material independent approach. J. Phys. Chem. C **118**, 13052 (2014)
45. R. Yu, L.M. Liz-Marzán, F.J. Garcia de Abajo, Universal analytical modeling of plasmonic nanoparticles. Chem. Soc. Rev. **46**, 6710 (2017)
46. L.R. Walker, Magnetostatic modes in ferromagnetic resonance. Phys. Rev. **105**, 390 (1957)
47. E.O. Kamenetskii,Quasistatic oscillations in subwavelength particles: Can one observe energy eigenstates?, Ann. Phys. (Berlin) **531**, 1800496 (2019)
48. P. Lalanne, W. Yan, K. Vynck, C. Sauvan, J.-P. Hugonin, Light interaction with photonic and plasmonic resonances. Laser Photon. Rev. **12**, 1700113 (2018)
49. J.F. Dillon Jr., Magnetostatic modes in disks and rods. J. Appl. Phys. **31**, 1605 (1960)
50. T. Yukawa, K. Abe, FMR spectrum of magnetostatic waves in a normally magnetized YIG disk. J. Appl. Phys. **45**, 3146 (1974)
51. E.O. Kamenetskii, A.K. Saha, I. Awai, Interaction of magnetic-dipolar modes with microwave-cavity electromagnetic fields. Phys. Lett. A **332**, 303 (2004)

52. E.O. Kamenetskii, Energy eigenstates of magnetostatic waves and oscillations. Phys. Rev. E **63**, 066612 (2001)
53. E.O. Kamenetskii, R. Shavit, M. Sigalov, Quantum wells based on magnetic-dipolar-mode oscillations in disk ferromagnetic particles. Europhys. Lett. **64**, 730 (2003)
54. E.O. Kamenetskii, M. Sigalov, R. Shavit,Quantum confinement of magnetic-dipolar oscillations in ferrite discs, J. Phys.: Condens. Matter **17**, 2211 (2005)
55. E.O. Kamenetskii, Vortices and chirality of magnetostatic modes in quasi-2D ferrite disc particles. J. Phys. A: Math. Theor. **40**, 6539 (2007)
56. E.O. Kamenetskii, R. Joffe, R. Shavit, Microwave magnetoelectric fields and their role in the matter-field interaction. Phys. Rev. E **87**, 023201 (2013)
57. E.O. Kamenetskii, Quantization of magnetoelectric fields. J. Modern Opt. **66**, 909 (2019)
58. S.G. Mikhlin, *Variational Methods in Mathematical Physics* (McMillan, New York, 1964)
59. M.A. Naimark, *Linear Differential Operators* (Frederick Ungar Publishing, New York, 1967)
60. E.O. Kamenetskii, The anapole moments in disk-form MS-wave ferrite particles. Europhys. Lett. **65**, 269 (2004)
61. S.O. Demokritov, V.E. Demidov, O. Dzyapko, G.A. Melkov, A.A. Serga, B. Hillebrands, A.N. Slavin, Bose-Einstein condensation of quasi-equilibrium magnons at room temperature under pumping. Nature **443**, 430 (2006)
62. A.S. Davydov, *Quantum Mechanics*, 2nd edn. (Pergamon, Oxford, 1976)
63. D.A. Ivanov, Non-Abelian statistics of half-quantum vortices in p-wave superconductors. Phys. Rev. Lett. **86**, 268 (2001)
64. N. Read, D. Green, Paired states of fermions in two dimensions with breaking of parity and time-reversal symmetries and the fractional quantum Hall effect. Phys. Rev. B **61**, 10267 (2000)
65. R. Joffe, E.O. Kamenetskii, R. Shavit, Microwave magnetoelectric fields: An analytical study of topological characteristics. J. Magn. Magn. Mater. **392**, 6 (2015)
66. M. Sigalov, E.O. Kamenetskii, R. Shavit, Magnetic-dipolar and electromagnetic vortices in quasi-2D ferrite discs, J. Phys.: Condens. Matter **21**, 016003 (2009)
67. E.O. Kamenetskii, M. Sigalov, R. Shavit, Manipulating microwaves with magnetic-dipolar-mode vortices. Phys. Rev. A **81**, 053823 (2010)
68. R. Joffe, E.O. Kamenetskii, R. Shavit, Azimuthally unidirectional transport of energy in magnetoelectric fields: Topological Lenz's effect. J. Mod. Opt. **64**, 2316 (2017)
69. E.O. Kamenetskii, M. Berezin, R. Shavit, Microwave magnetoelectric fields: Helicities and reactive power flows. Appl. Phys. B: Lasers Opt. **121**, 31 (2015)
70. F. Wilczek, Two applications of axion electrodynamics. Phys. Rev. Lett. **58**, 1799 (1987)
71. L. Visinelli, Dual axion electrodynamic, arXiv 1111.2268 (2011)
72. Ö.O. Soukal, M.E. Flatté, Strong field interactions between a nanomagnet and a photonic cavity, Phys. Rev. Lett. 104, 077202 (2010); Size dependence of strong coupling between nanomagnets and photonic cavities, Phys. Rev. B 82, 104413 (2010)
73. X. Zhang, C.L. Zou, L. Jiang, H.X. Tang, Strongly coupled magnons and cavity microwave photons. Phys. Rev. Lett. **113**, 156401 (2014)
74. Y. Tabuchi, S. Ishino, T. Ishikawa, R. Yamazaki, K. Usami, Y. Nakamura, Hybridizing ferromagnetic magnons and microwave photons in the quantum limit. Phys. Rev. Lett. **113**, 083603 (2014)
75. M. Goryachev, W.G. Farr, D.L. Creedon, Y. Fan, M. Kostylev, M.E. Tobar, High-cooperativity cavity QED with magnons at microwave frequencies. Phys. Rev. Appl. **2**, 054002 (2014)
76. J.A. Haigh, N.J. Lambert, A.C. Doherty, A.J. Ferguson, Dispersive readout of ferromagnetic resonance for strongly coupled magnons and microwave photons. Phys. Rev. B **91**, 104410 (2015)
77. N.J. Lambert, J.A. Haigh, A.J. Ferguson, Identification of spin wave modes in yttrium iron garnet strongly coupled to a co-axial cavity. J. Appl. Phys. **117**, 053910 (2015)
78. A. Leo, A. Grazia Monteduro, S. Rizzato, L. Martina, G. Maruccio, Identification and time-resolved study of ferrimagnetic spin-wave modes in a microwave cavity in the strong-coupling regime, Phys. Rev. B **101**, 014439 (2020)

79. Y. Cao, P. Yan, H. Huebl, S.T.B. Goennenwein, G.E.W. Bauer, Exchange magnon-polaritons in microwave cavities. Phys. Rev. B **91**, 094423 (2015)
80. B. Zare Rameshti, Y, Cao, G.E.W. Bauer, Magnetic spheres in microwave cavities, Phys. Rev. B **91**, 214430 (2015)
81. M. Harder, C.-M. Hu, Cavity spintronics: An early review of recent progress in the study of magnon–photon level repulsion. Solid State Phys. **69**, 47 (2018)
82. D. Tzarouchis, A. Sihvola, Light scattering by a dielectric sphere: Perspectives on the Mie resonances. Appl. Sci. **8**, 184 (2018)
83. I.D. Mayergoyz, D.R. Fredkin, Z. Zhang, Electrostatic (plasmon) resonances in nanoparticles. Phys. Rev. B **72**, 155412 (2005)
84. B.S. Luk'yanchuk, V. Ternovsky, Light scattering by a thin wire with a surface-plasmon resonance: Bifurcations of the Poynting vector field, Phys. Rev. B **73**, 235432 (2006)
85. R.I. Joseph, E. Schlomann, Theory of magnetostatic modes in long, axially magnetized cylinders. J. App. Phys. **32**, 1001 (1961)
86. E.O. Kamenetskii,Helical-mode magnetostatic resonances in small ferrite particles and singular metamaterials, J. Phys.: Condens. Matter **22**, 486005 (2010)
87. M. Sigalov, E.O. Kamenetskii, R. Shavit, Eigen electric moments and magnetic–dipolar vortices in quasi-2D ferrite disks. Appl. Phys. B **93**, 339 (2008)
88. E.O. Kamenetskii, R. Joffe, R. Shavit, Coupled states of electromagnetic fields with magnetic-dipolar-mode vortices: Magnetic-dipolar-mode vortex polaritons. Phys. Rev. A **84**, 023836 (2011)
89. E.O. Kamenetskii, Y. Peretz, Topological properties of magnetoelectric fields, Unpublished paper (2015)
90. Y.M. Bidasyuk, A.V. Chumachenko, O.O. Prikhodko, S.I. Vilchinskii, M. Weyrauch, A.I. Yakimenko, Stable Hopf solitons in rotating Bose-Einstein condensates. Phys. Rev. A **92**, 053603 (2015)
91. B.A. Malomed, Vortex solitons: Old results and new perspectives, Physica D: Nonlinear Phenom. **399**, 108 (20019)
92. L.P. Lévy, G. Dolan, J. Dunsmuir, H. Bouchiat, "Magnetization of mesoscopic copper rings: Evidence for persistent currents, Phys. Rev. Lett. **64**, 2074 (1990)
93. E.O. Kamenetskii, G. Vaisman, R. Shavit, Fano resonances in microwave structures with embedded magneto-dipolar quantum dots. J. Appl. Phys. **114**, 173902 (2013)
94. P. So, S.M. Anlage, E. Ott, R.N. Oerter, Wave chaos experiments with and without time reversal symmetry: GUE and GOE statistics. Phys. Rev. Lett. **74**, 2662 (1995)
95. M. Vraničar, M. Barth, G. Veble, M. Robnik, H.-J. Stöckmann, Persistent currents' and eigenfunctions in microwave resonators with broken time-reversal symmetry. J. Phys. A: Math. Gen. **35**, 4929 (2002)
96. H. Schanze, H.-J. Stöckmann, M. Martinez-Mares, C.H. Lewenkopf, Universal transport properties of open microwave cavities with and without time-reversal symmetry. Phys. Rev. E **71**, 016223 (2005)
97. E.O. Kamenetskii, M. Sigalov, R. Shavit, Microwave whirlpools in a rectangular waveguide cavity with a thin ferrite disk. Phys. Rev. E **74**, 036620 (2006)
98. S.S. Gupta, N.C. Srivastava, Physics of microwave reflection at a dielectric-ferrite interface. Phys. Rev. B **19**, 5403 (1979)
99. H.-J. Stöckmann, Microwave billiards and quantum chaos. Scholarpedia **5**, 10243 (2010)
100. G. Vaisman, E.O. Kamenetskii, R. Shavit, Magnetic-dipolar-mode Fano resonances for microwave spectroscopy of high absorption matter. J. Phys. D Appl. Phys. **48**, 115003 (2015)
101. M. Sigalov, E.O. Kamenetskii, R. Shavit, Electric self-inductance of quasi-two-dimensional magnetic-dipolar-mode ferrite disks. J. Appl. Phys. **104**, 053901 (2008)
102. R. Joffe, E.O. Kamenetskii, R. Shavit, Novel microwave near-field sensors for material characterization, biology, and nanotechnology. J. Appl. Phys. **113**, 063912 (2013)

103. E. Hollander, E.O. Kamenetskii, R. Shavit, Microwave chirality discrimination in enantiomeric liquids. J. Appl. Phys. **122**, 034901 (2017)
104. E.O. Kamenetskii, A. Davidov, Twisted microwave near fields for probing high absorption conductive layers and biological tissue, Unpublished paper (2019)

Subject Index

E. Kamenetskii (ed.), *Chirality, Magnetism and Magnetoelectricity*,
Topics in Applied Physics 138,
https://doi.org/10.1007/978-3-030-62844-4

T

U

V

W